HILL GROUP

Bill Groom

# Malaria Parasites

## Genomes and Molecular Biology

Edited by

## Andrew P. Waters

## Chris J. Janse

*Leiden University Medical Centre,*
*Leiden, The Netherlands*

Copyright © 2004
Caister Academic Press
32 Hewitts Lane
Wymondham
Norfolk NR18 0JA
England

**www.caister.com**

**British Library Cataloguing-in-Publication Data**

A catalogue record for this book is available from the British
Library

ISBN: 0-9542464-6-2

*Printed and bound in Great Britain*

# Contents

# Books of Related Interest

# Contributors

* indicates coresponding author

**Y.I.H.Alavi**
Dept Biol Sci
Imperial College of Sci Technol Med
South Kensington
London SW7 2AZ

**Casilda G. Black**
Dept Microbiol &
Victorian Bioinformatics Consortium
Building 53
Monash University 3800
Victoria
Australia

**G.A.Butcher**
Dept Biol Sciences
Imperial College of Sci Technol Med
South Kensington
London SW7 2AZ

**\* Jane Carlton**
The Institute for Genomic Research
9712 Medical Center Drive
Rockville
MD 20850 USA
carlton@tigr.org

**Daniel J. Carucci**
NMRC Malaria Program (IDD)
503 Robert Grant Avenue
Silver Spring, MD 20910-7500
USA
caruccid@nmrc.navy.mil

**Debopam Chakrabarti**
Dept. of Molecular Biology and
Microbiology
University of Central Florida
12722 Research Parkway
Orlando FL32826, USA
dchak@pegasus.cc.ucf.edu

**\* Alister G. Craig**
Liverpool School of Tropical Medicine
Pembroke Place
Liverpool L3 5QA
UK
agcraig@liverpool.ac.uk

**\* Ross L. Coppel**
Dept Microbiol &
Victorian Bioinformatics Consortium
Building 53
Monash University 3800
Victoria
AUSTRALIA
Ross.Coppel@med.monash.edu.au

**\* Kirk W. Deitsch**
Dept Microbiol Immunol
Weill Medical College of Cornell University
1300 York Avenue
W-704 Box 62
New York, NY 10021, USA
kwd2001@med.cornell.edu

**J.T.Dessens**
Dept Biol Sciences
Imperial College of Sci Technol Med
South Kensington
London SW7 2AZ

**\* Christian Doerig**
Inserm Team
Wellcome Centre for Molecular Parasitology
University of Glasgow
56 Dumbarton Road
Glasgow G11 6NU, Scotland
cdoer001@udcf.gla.ac.uk

**Teresa Gil Carvalho**
Unité de Biologie et Génétique du Paludisme
Institut Pasteur
25 Rue du Dr Roux
75724 Paris Cedex 15
France

**Jun Fang**
Laboratory of Malaria Research
National Inst Allergy and Infec Dis
National Institutes of Health
Bethesda
Maryland 20892
USA

* **David A. Fidock**
Dept Microbiol Immunol
Albert Einstein College of Medicine
Forchheimer 403,
1300 Morris Park Avenue
Bronx, NY 10461, USA
dfidock@aecom.yu.edu

**Luisa M. Figueiredo**
Lab of Molecular Parasitology
The Rockefeller University
1230 York Avenue
New York
NY-10021, USA
Luisa.Figueiredo@mail.rockefeller.edu

**Lúcio H. Freitas-Junior**
UNIFESP-EPM
Parasitology
862, Rua Botucata 6th floor
04023-062 São Paulo
Brazil

*Patricia M. Goodwin,
The Wellcome Trust,
183 Euston Road
London NW1 2BE
UK
p.goodwin@wellcome.ac.uk

**Michael Gottlieb,**
Parasitol and International Programs Branch
Natl Inst Allergy and Infect Dis
National Institutes of Health
Dept Health and Human Services
6610 Rockledge Drive/Room 5099
Bethesda, MD 20892-6604
USA
mg35s@nih.gov

* **Neil Hall**
The Wellcome Trust Sanger Institute
Hinxton
Cambridge CB10 1SA, UK
nh1@sanger.ac.uk

**Malcolm J. Gardner**
Parasite Genomics Group
The Institute for Genomic Research
9712 Medical Center Drive
Rockville, MD 20850
USA

* **Chris Janse**
Laboratory of Parasitology
Centre for Infectious Diseases
Leiden University Medical Centre
Albinusdreef 2
2333 ZA Leiden
The Netherlands
C.J.Janse@lumc.nl

* **Deirdre Joy**
Laboratory of Malaria and Vector Res
National Institutes of Health
4 Center Drive, MSC 0425
Bethesda, MD 20892-0425
USA
djoy@niaid.nih.gov

**Jessica C. Kissinger**
Ctr Tropical and Emerging Global Dis
and Department of Genetics,
University of Georgia
Athens
Georgia 30602, USA

**Sylvia Lee**
Dept Microbiol Immunol
Albert Einstein College of Medicine
1300 Morris Park Avenue
Bronx, NY 10461
USA

* **Thomas F. McCutchan**
Laboratory of Malaria Research
Nat Inst Allergy and Infect Dis
National Institutes of Health
Bethesda,
Maryland 20892, USA
TMCCUTCHAN@niaid.nih.gov

* **Geoff I. McFadden**
Plant Cell Biology Research Centre
University of Melbourne
3010 Australia
g.mcfadden@botany.unimelb.edu.au

**Victoria Mcgovern**
Burroughs Wellcome Fund
Post Office Box 13901
Research Triangle Park
NC 27709- 3901, USA
vmcgovern@bwfund.org

**Markus Meissner**
Dept Biol Sci
Imperial College of Science
Technology and Medicine
Sir Alexander Fleming Building
Imperial College Road
London SW7 2AZ, UK
m.meissner@imperial.ac.uk

* **Robert Ménard**
Unité de Biologie et Génétique du Paludisme
Institut Pasteur
25 Rue du Dr Roux
75724 Paris Cedex 15
France
rmenard@pasteur.fr

**Jianbing Mu**
Laboratory of Malaria and Vector Res
National Institutes of Health
4 Center Drive, MSC 0425
Bethesda, MD 20892-0425
USA

**J.D.Raine**
Dept Biol Sciences
Imperial College of Sci Technol Med
South Kensington
London SW7 2AZ

* **David S. Roos**
Department of Biology
University of Pennsylvania
415 South University Avenue
Philadelphia
PA 19104-6018, USA
droos@sas.upenn.edu

* **Artur Scherf**
Unité de Biol Interactions Hôte-Parasite
CNRS URA 1960
Institut Pasteur
25 rue du Dr. Roux
75724 Paris Cedex 15
France
ascherf@pasteur.fr

**Joana Silva**
The Institute for Genomic Research
9712 Medical Center Drive
Rockville
MD 20850 USA

* **R.E. Sinden**
Dept Biol Sci
Sir Alexander Fleming Building
Imperial College of Sci Technol Med
Imperial College Road
London SW7 2AZ, UK
r.sinden@ic.ac.uk

**Joseph D. Smith**
Seattle Biomedical Institute
Four Nickerson Street
Suite 200
Seattle, WA 98109-1651
USA

* **Dominique Soldati**
Dept Biol Sciences
Imperial College of Sci Technol Med
Sir Alexander Fleming Building
Imperial College Road
London SW7 2AZ, UK
d.soldati@imperial.ac.uk

**Xin-zhuan Su**
Laboratory of Malaria and Vector Res
National Institutes of Health
4 Center Drive, MSC 0425
Bethesda, MD 20892-0425
USA

**Agnieszka E. Topolska**
Dept Microbiol &
Victorian Bioinformatics Consortium
Building 53
Monash University 3800
Victoria
AUSTRALIA
Agnieszka.Topolaska@med.monash.edu.au

**H.E. Trueman**
Dept Biol Sci
Imperial College of Sci Technol Med
South Kensington
London SW7 2AZ

**Rosalinda van Spaendonk**
Laboratory of Parasitology
Centre for Infectious Diseases
Leiden University Medical Centre
Albinusdreef 2
2333 ZA Leiden
The Netherlands

**Karena L. Waller**
Department of Cardiology
Albert Einstein College of Medicine
1300 Morris Park Avenue
Bronx, NY 10461
USA
kwaller@aecom.yu.edu

**Ross F. Waller**
Botany Department
University of British Columbia
3529-6270 University Boulevard
Vancouver, BC
V6T 1Z4 Canada
rfwaller@interchange.ubc.ca

**Lina Wang**
Dept Microbiol &
Victorian Bioinformatics Consortium
Building 53
Monash University 3800
Victoria
AUSTRALIA

\* **Andrew P. Waters**
Laboratory of Parasitology
Centre for Infectious Diseases
Leiden University Medical Centre
Albinusdreef 2
2333 ZA Leiden
The Netherlands
A.P.Waters@lumc.nl

# Preface

The landmark that many malaria researchers had been eagerly anticipating was finally officially reached in October 2002 with the publication in Nature of the complete sequence for the entire genome of a laboratory strain of the pernicious human malaria parasite, *Plasmodium falciparum*, the first parasite genome to be sequenced. The immediate bounty of such work is quite correctly publically proclaimed as enabling searches for new drug targets and vaccine candidates but, the laudability of this possibility notwithstanding, the goals of the funding agencies (explained in the Introduction) were also focussed on the longer term and we hope this is reflected in this book. It is the firm believe of many fundamental biologists (many of whom who contributed chapters to this book) that it is only through a more complete understanding of the biology of *Plasmodium* and its active relationship with both the mammalian host and the Anopheline vector that we can hope to design effective interdiction strategies. Indeed these strategies might ultimately turn out not to be based solely on drugs and vaccines. Therefore, this book seeks to emphasise not the vaccine and drug seeking potential of the genome but rather the boon that it represents to furthering our understanding of the living, dynamic parasite. Thus three chapters are devoted to a description of the *P. falciparum* genome, the additional benefits that comparitive *Plasmodium* genomics might bestow and how the data boom must be managed, organised and made accessible to the scientist, now and in the future as post-genome analyses provide increasingly complex datasets.

In subsequent chapters, a flavour of the applications to which the genome resources can be put is given. This is not intended to be exhaustive and readers may miss chapters on proteomics and transcription profiling but these are such fast evolving fields that it is only now that the best approaches are becoming clear. However vital post genome technologies are described such as reverse genetics in both *Plasmodium* as well as in another apicomplexan (*Toxoplasma gondii*) that besides its intrinsic medical and veterinary interest serves as a useful model for *Plasmodium*. Chapters 6-10 concentrate on aspects of molecular biology from the powerful applications of microsatellite analysis to the fundamental analysis of chromosome structure, cell cycle to RNA polymerase I and II mediated gene expression. Cell biology will also undergo a revolution. This is most apparent in the apicoplast where the contribution of the nuclear genome to the biology of this fascinating organelle was truly emphasised by genome analysis. However this will also be true of the surface and cell biology of the stage specific forms of the parasite as it progresses through its life cycle and several examples are given in chapters 12-15. Lastly although we have not included a chapter on the search for new drugs, the apicoplast as a target-rich source notwithstanding, the genome, genome technologies and reverse genetics can also shed a light on the possible result of using drugs, namely the development of drug resistance. We learn how insights into drug resistance might not only allow us to understand the specificities of the phenomeon (in the given example of chloroquine) but give hope that such understanding might help alternative strategies/compounds to be developed.

Whilst it would have been gratifying to attempt to produce a book that dealt with every aspect of malaria research and therapy that the availability of the *P. falciparum* genome sequence will affect, we hope that the examples given here provide some flavour of the excitement and hope engendered that the authors of these chapters clearly feel (to each of them infinite thanks for their contributions) about the advent of the post genome era of *Plasmodium* research. The hoped for drugs, vaccines and as yet undreamt of therapies may not be with us yet but their development is immensely empowered by the landmark that is the *P. falciparum* genome.

Andy Waters
Chris Janse

*November 2003*

From: Malaria Parasites: Genomes and Molecular Biology
Edited by: A.P. Waters and C.J. Janse

# Introduction

## The *Plasmodium falciparum* Genome Project

## 1. The Birth of the Project

Malaria is responsible for more than a million deaths a year, and up to 500 million people suffer from this disease. In response to this growing global crisis a group of 30 international researchers and funders gathered in May 1996 to discuss an exciting but daunting new project – obtaining the complete genome sequence of a malaria pathogen. At that time only small viral genomes and the bacterium *Haemophilus influenzae* (1.83Mb) had been completely sequenced. Although there were on-going projects to sequence some larger eukaryotic genomes of model organisms, for example the 12 Mb genome of the yeast *Saccharomyces cerevisiae* and the 97Mb genome of the nematode worm, *Caenorhabditis elegans*, their completion required a very large international effort; and in 1996 it was inconceivable that a draft human genome sequence would be available by 2000. Sequencing the 14 chromosomes (30Mb of DNA) of *Plasmodium falciparum*, the most lethal of the malaria parasites infecting humans, was then a considerable technical challenge because of the AT-rich nature of the DNA, but the potential value of genomic approaches for drug and vaccine development were strong motivators for the multiyear, multimillion-dollar investment. Initially the Burroughs Wellcome Fund (BWF), The National Institute of Allergy and Infectious Diseases (NIAID), the U.S. Department of Defense (DoD) and the Wellcome Trust (WT) agreed to work collaboratively and complementarily on funding the project (Hoffman *et al.,* 1997). The World Health Organization (WHO) joined the project in 1998.

After some debate it was agreed that the organism of choice for the sequencing project should be a clone of *P. falciparum* known as 3D7 that is culturable through all life stages and that has been used for genetic crosses. Initial efforts began in late 1996 at three centres - the Sanger Centre, Stanford University and The Institute for Genomic Research (TIGR), which worked in collaboration with the Naval Medical Research Center (NMRC). Whole genome shotgun sequencing of large eukaryotes had not yet been attempted so the strategy adopted was to purify the chromosomes using pulse field gels and then sequence them individually. With funding secured, the chromosomes were divided among the centers, with Stanford sequencing chromosome number 12, TIGR/NMRC sequencing chromosomes 2,10,11,and 14, and Sanger sequencing chromosomes 1,3,4,5,6,7,8,9,and 13. Chromosomes 6-8, which had become known as 'the blob' because of their appearance on pulse field gels, do not separate well and were handled as a group.

Critical to the success of the project were the regularly scheduled meetings of individuals from the funding agencies, the sequencing centers and the malaria and broader scientific communities. These meetings, held approximately every six months, provided opportunities for scientific exchange on technical advances among the sequencing centers and other interested scientists. The meetings also promoted discussions on matters of data release and use, the establishment of a common database and, latterly, publication of the annotated sequence.

## 2. Problems and Solutions

The physical and genetic properties of *P. falciparum's* AT-rich DNA presented substantial challenges for the teams. Many bacterial cloning strains do not tolerate such AT-rich DNA and this often leads to deletions and rearrangements in the resultant clones. By using optimized methods of library construction, including careful avoidance of exposure of the DNA to damaging ultraviolet light, representative sequencing libraries were constructed successfully. Substantial effort was required to optimize the sequencing conditions to minimize artifacts associated with the highly AT- rich DNA and to close the hundreds of "gaps" remaining in the sequence.

The difficult task of assembly - the process of putting the fragmentary sequence data into order – was aided by the construction of maps to provide context and landmarks for assembly. An international consortium funded by the Wellcome Trust constructed a physical map (Dame *et al.*, 1996); Xin-Zhuan Su generated a genetic map (Su *et al.*, 1999), with phenotypically scorable markers mapped to different chromosomes; and David Schwartz's group built optical restriction maps (Jing *et al.*, 1999; Lai *et al.*, 1999) which are used much like physical maps. Later the application of 'Happy Mapping', developed by Paul Dear (Dear *et al.*, 1998), to 'the blob' facilitated progress on the three chromosomes which are difficult to separate.

Chromosome 2 was completed in 1998 (Gardner *et al.*, 1998) and chromosome 3 in 1999 (Bowman *et al.*, 1999). Finally, in 2002 the sequences of the remaining chromosomes, together with an analysis of the genome were published in Nature (Gardner *et al.*, 2002a, 2002b; Hall *et al.*, 2002).

# 3. Data Access

An important feature of the project has been the immediate and continual release of data from the three sequencing centers, which has enabled the entire community to access the latest, albeit imperfect, sequence data. However, by 1999 it was recognised that there was an urgent need for a common database accessible via the Internet, and the Burroughs Wellcome Fund provided a $1M grant funding for this to be developed at the University of Pennsylvania for the use of the international malaria community by David Roos' and the late Chris Overton's groups, with important input coming from Ross Coppel's group at Monash University, which has long aimed to develop an "encyclopedia" of malaria. PlasmoDB is appropriately viewed as an element of this emerging encyclopedia. PlasmoDB, launched in June 2000, provides centralized access to analysis tools and data (see Roos, pp 65-100). Gene searching, graphical maps and intuitive "click through" web navigation enhance site use. The database developers are eager to help researchers become comfortable with using the sequence.

# 4. Malaria Research and Reference Reagent Resource

As the sequencing centers were beginning the sequencing effort, a large contingent of malaria researchers and health science administrators from various funding agencies were meeting in Dakar, Senegal in January 1997. The primary goal of the meeting was to identify gaps in knowledge and scientific opportunities that would accelerate progress towards the development of new approaches to lessen the burden of malaria disease in Africa. In addition, the meeting sought to promote collaboration among scientists and clinicians from the Northern Industrialized countries and the malaria endemic areas. One of the major recommendations arising from the Dakar conference was the need for a resource for well-standardized *Plasmodium* and related reagents, including those of mosquito vectors. To meet the challenge of this recommendation, the NIAID has established the Malaria Research and Reference Reagent Resource (http://www.malaria.mr4.org/). The MR4 provides registered investigators with access to parasite isolates and nucleic acid libraries and clones directly relevant to the sequencing project as well as many other valuable reagents. In addition to supplying reagents, the MR4 supports and conducts workshops, for example on bioinformatics related to the genome project. This has enabled many more investigators, especially those from the endemic areas, to gain the knowledge and expertise needed to derive benefit from the sequence data. The MR4 also supplies reagents and protocols to identify polymorphisms in the parasite genome relevant to drug resistance and vaccine development. It is anticipated that the MR4 will provide many additional resources as the functional analysis of the sequence data proceeds, as exemplified by providing microarray slides to researchers.

# 5. Next Steps for the Project

Once the sequencing of strain 3D7 was well underway it became clear that it would be very valuable to sequence some other Plasmodium species, particularly those which are used experimentally as models for human malaria. To this end, the US Department of Defense sponsored the comparative sequencing of *P. yoelii* and has initiated the sequencing of *P. vivax* at TIGR. The Wellcome Trust funded 3-fold coverage of the genomes of *P. berghei*, *P chabaudi*, *P. knowlesi* and *P. gallinaceum*, and of a field isolate of *P. falciparum*, together with sequencing of expressed sequence tags from *P. falciparum* and *P. knowlesi* asynchronous blood stage parasites, from *P. berghei* ookinetes and from *P. chabaudi* late trophozoites. Comparative genomics of Plasmodium species is discussed by Carlton, Silva and Hall (pp 33-64). Other reviews in this issue demonstrate how knowledge of the genome sequence is having a major impact on our understanding of the genetics and cell biology of the malaria parasite and of its interactions with both the human and mosquito hosts. This, in turn, will facilitate the development of new drugs and vaccines.

# 6. Concluding Remarks

We, as funders, would like to take this opportunity of thanking all the members of the genome centers, academic and government laboratories who have worked so hard on the various aspects of this difficult sequencing project. We have enjoyed seeing this project progress (with the inevitable ups and downs) to fruition and look forward to seeing the outcomes in our increased understanding of the biology of malaria, as well as in the development of new drugs and vaccines.

The malaria genome sequencing project succeeded because of substantial international collaboration between researchers and between funders. As this project was underway, a separate multi-partner effort yielded the genomic sequence and map of the malaria vector mosquito *Anopheles gambiae*.These projects have both been remarkable steps forward for science aimed at understanding malaria. With the human genome sequence in hand as well, it is time to work toward understanding the coordinated host, vector, and parasite genetics, biochemistry and physiology that lead to the disease. Continued and enhanced coordination between researchers, between funders, and between fields are critical.

CAPT Daniel Carucci, MC, USN
*US Department of Defense*

Pat Goodwin
*The Wellcome Trust*

Michael Gottlieb
*NIAID, National Institutes of Health*

Victoria McGovern
*Burroughs Wellcome Fund*

The opinions expressed are those of the authors and do not reflect the official policy of the Department of the Navy, Department of Defense, the U.S. government or their funding agencies.

# 7. References

Bowman, S., Lawson, D., Basham, D., Brown, D., Chillingworth, T., Churcher, C.M., Craig, A., Davies, R.M., Devlin, K., Feltwell, T., Gentles, S., Gwilliam, R., Hamlin, N., Harris, D., Holroyd, S., Hornsby, T., Horrocks, P., Jagels, K., Jassal, B., Kyes, S., McLean, J., Moule, S., Mungall, K., Murphy, L., and Barrell, B.G. 1999. The complete nucleotide sequence of chromosome 3 of *Plasmodium falciparum*. Nature. 400 (6744): 532-538.

Dame, J.B., Arnot, D.E., Bourke, P.F., Chakrabarti, D., Christodoulou, Z., Coppel, R.L. Cowman, A.F., Craig, A.G., Fischer, K., Foster, J., Goodman, N., Hinterberg, K., Holder, A.A., Holt, D.C., Kemp, D.J., Lanzer, M., Lim, A., Newbold, C.I., Ravetch, J.V., Reddy, G.R., Rubio, J., Schuster, S.M., Su, X.Z., Thompson, J.K., and Werner, E.B. 1996. Current status of the *Plasmodium falciparum* genome project. Molec. Biochem. Parasitol. 79(1):1-12.

Dear, P.H., Bankier, A.T., and Piper, M.B. 1998. A high-resolution metric HAPPY map of human chromosome 14. Genomics. 48(2): 232-241.

Gardner, M.J., Tettelin, H., Carucci, D.J., Cummings, L.M., Aravind, L., Koonin, E.V., Shallom, S., Mason, T., Yu, K., Fujii, C., Pederson, J., Shen, K., Jing, J., Aston, C., Lai, Z., Schwartz, D.C., Pertea, M., Salzberg, S., Zhou, L., Sutton, G.G., Clayton, R., White, O., Smith, HO., Fraser, C.M., and Hoffman, S.L. 1998. Chromosome 2 sequence of the human malaria parasite *Plasmodium falciparum*. Science. 282(5391): 1126-1132.

Gardner, M.J., Hall, N., Fung, E., White, O., Berriman, M., Hyman, R.W., Carlton, J.M., Pain, A., Nelson, K.E., Bowman, S., Paulsen, I.T., James, K., Eisen, J.A., Rutherford, K., Salzberg, S.L., Craig, A., Kyes, S., Chan, M.S., Nene, V., Shallom, S.J., Suh, B., Peterson, J., Angiuoli, S., Pertea, M., Allen, J., Selengut, J., Haft, D., Mather, M.W., Vaidya, A.B., Martin, D.M., Fairlamb, A.H. Fraunholz, M.J., Roos, D.S., Ralph, S.A., McFadden, G.I., Cummings, L.M., Subramanian, G.M., Mungall, C., Venter, J.C., Carucci, D.J., Hoffman, S.L. Newbold, C., Davis,

R.W., Fraser, C.M., and Barrell, B. 2002a. Genome sequence of the human malaria parasite *Plasmodium falciparum*. Nature. 419(6906): 498-511.

Gardner, M.J., Shallom, S.J., Carlton, J.M., Salzberg, S.L., Nene, V., Shoaibi, A., Ciecko, A., Lynn, J., Rizzo, M., Weaver, B., Jarrahi, B., Brenner, M., Parvizi, B., Tallon, L., Moazzez, A., Granger, D., Fujii, C., Hansen, C., Pederson, J., Feldblyum, T., Peterson, J., Suh, B., Angiuoli, S., Pertea, M., Allen, J., Selengut, J., White, O., Cummings, L.M., Smith, H.O., Adams, M.D., Venter, J.C., Carucci, D.J., Hoffman, S.L., and Fraser, C.M. 2002b. Sequence of *Plasmodium falciparum* chromosomes 2, 10, 11 and 14. Nature. 419(6906): 531-534.

Hall, N., Pain, A., Berriman, M., Churcher, C., Harris, B., Harris, D., Mungall, K., Bowman, S., Atkin, R., Baker, S., Barron, A., Brooks, K., Buckee, C.O., Burrows, C., Cherevach, I., Chillingworth, C., Chillingworth, T., Christodoulou, Z., Clark, L., Clark, R., Corton, C., Cronin, A., Davies, R., Davis, P., Dear, P., Dearden, F., Doggett, J., Feltwell, T., Goble, A., Goodhead, I., Gwilliam, R., Hamlin, N., Hance, Z., Harper, D., Hauser, H., Hornsby, T., Holroyd, S., Horrocks, P., Humphray, S., Jagels, K., James, K.D., Johnson, D., Kerhornou, A., Knights, A., Konfortov, B., Kyes, S., Larke, N., Lawson, D., Lennard, N., Line, A., Maddison, M., McLean, J., Mooney, P., Moule, S., Murphy, L., Oliver, K., Ormond, D., Price, C., Quail, M.A., Rabbinowitsch, E., Rajandream, M.A., Rutter, S., Rutherford, K.M., Sanders, M., Simmonds, M., Seeger, K., Sharp, S., Smith, R., Squares, R., Squares, S., Stevens, K., Taylor, K., Tivey, A., Unwin, L., Whitehead, S., Woodward, J., Sulston, J.E., Craig, A., Newbold, C., and Barrell, B.G. 2002. Sequence of *Plasmodium falciparum* chromosomes 1, 3-9 and 13. Nature. 419(6906): 527-531.

Hoffman, S.L., Bancroft, W.H., Gottlieb, M., James, S.L., Burroughs, E.C., Stephenson, J.R., and Morgan, M.J. 1997. Funding for malaria genome sequencing. Nature. 387(6634): 647.

Jing, J., Lai, Z,. Aston, C., Lin, J., Carucci, D.J., Gardner, M.J., Mishra, B., Anantharaman, T.S., Tettelin, H., Cummings, L.M., Hoffman, S.L., Venter, J.C., and Schwartz, D.C. 1999. Optical mapping of *Plasmodium falciparum* chromosome 2. Genome Research. 9(2): 175-181.

Lai, Z., Jing, J., Aston, C., Clarke, V., Apodaca, J., Dimalanta, E.T., Carucci, D.J., Gardner, M.J., Mishra, B., Anantharaman, TS., Paxia, S., Hoffman, S.L., Venter, J.C, Huff, E.J., and Schwartz, D.C. 1999. A shotgun optical map of the entire *Plasmodium falciparum* genome. Nature Genetics. 23(3): 309-313.

Su, X., Ferdig, M.T., Huang, Y., Huynh, C.Q., Liu, A., You, J., Wootton, J.C., and Wellems, T.E. 1999. A genetic map and recombination parameters of the human malaria parasite *Plasmodium falciparum*. Science. 286(5443): 1351-1353.

From: Malaria Parasites: Genomes and Molecular Biology
Edited by: A.P. Waters and C.J. Janse

# Chapter 1

## The Genome of
## *Plasmodium falciparum*

Neil Hall and Malcolm J. Gardner

## Abstract

The *Plasmodium falciparum* genome project, initiated in 1996, was the first attempt to sequence a genome of a eukaryotic pathogen. Not only has the genome told us a lot about the biology of the parasite but since the sequence data started to be released onto public web sites, it has changed the way many *Plasmodium* laboratories have approached their research. Here we outline how the sequencing project was approached, and describe what the major findings have been from being able to look at the unusual genome of *P. falciparum*. We also attempt to foresee what impact genomics and comparative genomics will have on the field of malaria research.

## 1. Initiation and Management of the
## *P. falciparum* Genome Sequencing Project

The effort to sequence the genome of the human malaria parasite *P. falciparum* began in 1996, soon after the publication of the first complete microbial genome sequences (Bult *et al.*, 1996; Fleischmann *et al.*, 1995; Fraser *et al.*, 1995; Kaneko *et al.*, 1996). These publications generated tremendous excitement, not just because of the magnitude of the technical accomplishments in high-throughput sequencing and genome assembly they represented, but primarily because of the wealth of information these projects generated and the new biological insights they provided (Bloom,

1995). Subsequently, investigators in the U.K. and the U.S. began to explore the possibility of sequencing the *P. falciparum* genome. This parasite infects hundreds of millions of people every year and kills up to 3 million people annually, primarily children in sub-Saharan Africa (Breman, 2001). Drug resistance in the parasite and mosquito resistance to insecticides had become major problems, the malaria situation was worsening, and few new antimalarial drugs were available or under development. Development of effective, practical malaria vaccines seemed to be years away. Furthermore, research into the basic biology and biochemistry of the malaria parasite was hampered by its complex life cycle and intracellular nature. Determination of the genome sequence offered a means to facilitate and accelerate research efforts that might lead to the development of novel drugs and vaccines.

Relatively little was known about the *P. falciparum* genome in 1996. The number of chromosomes in malaria parasites, 14, had been determined by the counting of kinetochores in electron micrographs of serial sections of parasites (Prensier and Slomianny, 1986). This number was subsequently confirmed by pulsed field gel electrophoresis (Triglia *et al.*, 1992), which also provided an estimate of the genome size, ~30 Mb, similar to what had been measured previously by buoyant density centrifugation (Pollack *et al.*, 1982). Hybridization of cloned genes to Southern blots of *P. falciparum* genomic DNA on pulsed field gels, and the Wellcome Trust funded genome mapping project (Foster and Thompson, 1995), provided rough physical maps of the genome. In addition, two organellar genomes had been identified and sequenced – a 6 kb linear mitochondrial genome (Suplick *et al.*, 1988; Vaidya *et al.*, 1989) and a 35 kb circular plastid genome-like molecule (Gardner *et al.*, 1991; Wilson *et al.*, 1996) that was subsequently localized in a novel organelle called the apicoplast (Kohler *et al.*, 1997).

A worrying potential obstacle to the determination of the genome sequence was the AT-richness (~80%) of *P. falciparum* nuclear DNA (Pollack *et al.*, 1982). The high AT-content was apparently responsible for the inability to clone and manipulate many *P. falciparum* genes in *E. coli*, despite the use of different vectors and recombination-deficient *E. coli* host strains. Cloning of fragments > 5 kb was particularly difficult such that construction of representative genomic libraries was impossible. Many experienced malariologists, including some involved in the sequencing effort, were concerned that the high AT content would prevent the successful completion of the genome sequence. In view of the potential benefits that might accrue if the project were successful, however, the sequencing centers and funding agencies agreed to initiate pilot projects to evaluate the strategies and methods to be used for determination of the genome sequence (Hoffman *et al.*, 1997). Once the pilot projects appeared to be succeeding the project was enlarged to sequence the entire *P. falciparum* genome.

## 2. Sequencing Strategy and Status of the Genome Sequence

At the time the *Plasmodium falciparum* sequencing project was initiated it was unlikely that a single institute or funding body would be able to take on the determination of entire genome sequence independently. When the project commenced in mid-1996, only 4 microbial genomes had been sequenced (*Haemophilus influenzae, Mycoplasma genitalium, Methanococcus jannaschii,* and *Synechocystis* sp.) and the largest of these genomes was 3.5 Mb (Bult *et al.*, 1996; Fleischmann *et al.*, 1995; Fraser *et al.*, 1995; Kaneko *et al., 1996*). Based on the estimated genome size of 30 Mb, the proposed *P. falciparum* project was 3-fold larger than any other completed genome that had been sequenced via a whole-genome shotgun approach. Because of the high estimated cost of the project and technical limitations such as the lack of genome assembly software able to assemble the entire genome, the project was subdivided such that each sequencing centre (and funding agency) were assigned different chromosomes: The Institute of Genome Research (TIGR) and the Naval Medical Research Center (NMRC) sequenced chromosomes 2, 10, 11 and 14, The Sanger Institute sequenced 1, 3, 9, and 13 and Stanford Genome Technology Center sequenced chromosome 12.

Each centre used the chromosome shotgun method to sequence the genome, which involved taking purified chromosomal DNA that had been separated by Pulsed Field Gel Electrophoresis (PFGE), shattering it by mechanical shearing or sonication into 1-2 kb fragments, and cloning of the fragments to make small insert libraries. The clones were then sequenced from either end and the resulting short sequences (~400-700bp) assembled using specialised computer software to generate long contiguous sequences (Bowman *et al.*, 1999; Gardner *et al.*, 2002b; 1998; Hall *et al.*, 2002; Hyman *et al.*, 2002). Where there is a region of DNA that is not covered by a sequencing read this is known as a sequence gap. Where there is a region that is not covered by a clone this is known as a physical gap. The principle of shotgun sequencing is shown in Figure 1. The genome was sequenced to average depth (or coverage) of 14.5 fold; this means that every base in the genome is represented in the shotgun, on average, 14.5 times (Gardner *et al.*, 2002a).

However, because the genome of *P. falciparum* is so AT rich (80.6%), many regions are very difficult to clone, and hence will not be represented in the shotgun. This problem is made worse because A+T rich DNA is difficult to sequence hence each read is shorter than would be the case for most other genome projects. As *P. falciparum* DNA can't be stably cloned into bacterial large insert vectors, such as bacterial artificial chromosomes (BACs), gaps could not be bridged by these means, which would otherwise allow contigs to be ordered and orientated on either side of the gap.

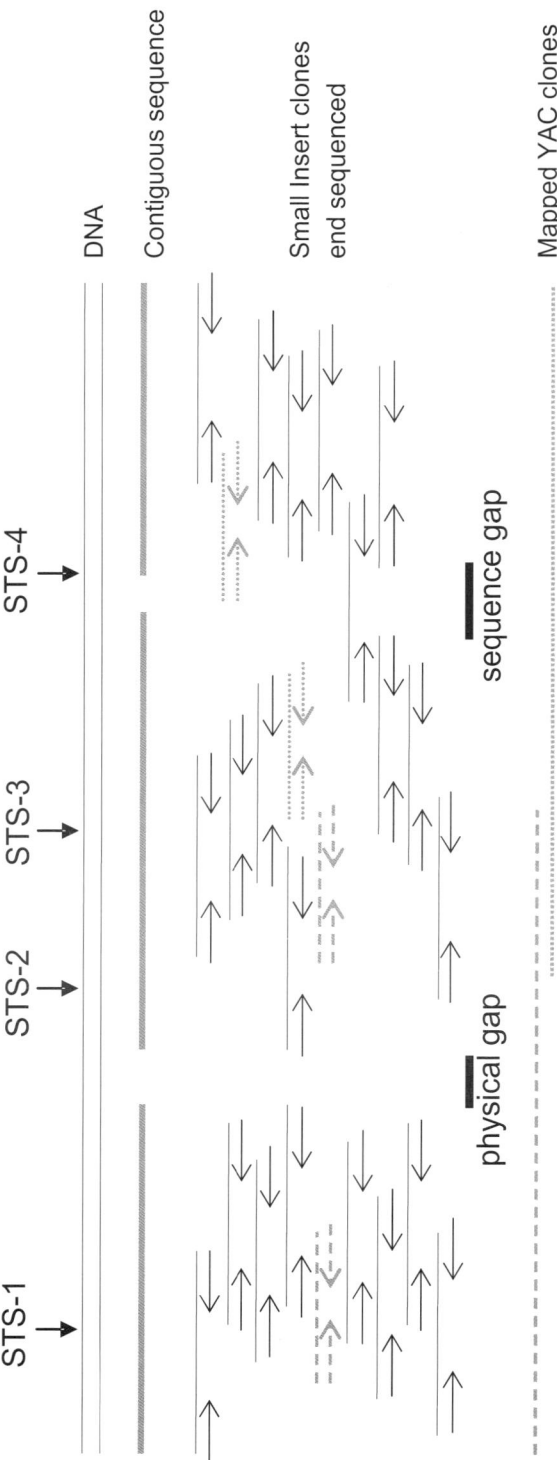

**Figure 1.** Shotgun Sequencing. Chromosomal DNA is represented by parallel lines at top with positions of mapped STS markers 1-4) represented by arrows. The DNA is shattered (shotgunned) into 2-4kb fragments which were cloned into plasmid vectors and sequenced from both ends (represented by arrows). The DNA sequences are assembled to form contiguous sequences (contigs) represented by bold black lines. Mapped YAC clones, represented by the dashed and dotted lines, were also shotgun sequenced generating reads (dotted arrows) that assemble into the genome derived contigs allowing these contigs to be aligned to the YAC map. The contigs can be positioned further using the STS markers or read pairs. A sequence gap is where a pair or reads span a gap so a clone covers the gap. A physical gap is where a no clones span a gap.

Because of the difficulties of working with *Plasmodium* DNA, mapping data was used to assist in the ordering of contigs. In total, four different maps, generated by different techniques, were employed to map the genome: genetically mapped microsatellite markers, optical maps, yeast artificial chromosome (YAC) tiling paths and a HAPPY map.

## 2.1. The Optical Map

The optical maps were generated by immobilising large DNA molecules on a surface and digesting them with restriction enzymes. For the *P. falciparum* project, surface mounted chromosomes were digested using *Bam*HI and *Nhe*I and the resulting restriction fragments were measured and ordered optically using digital fluorescence microscopy. The molecules were then assembled into an ordered set of restriction fragments that allowed the contig assemblies to be verified and to help order and orientate the larger contigs (Jing *et al.*, 1999; Lai *et al.*, 1999).

## 2.2. Genetic Map

The genetic map was generated from a cross between the Dd2 and HB3 strains of *P. falciparum*. Microsatellite markers were screened in the progeny of this cross and recombination frequencies and linkage groups were calculated. The resulting map contains 901 markers with an average spacing of 25.5 kb (Su *et al.*, 1999). While such maps do not give clear indications of physical distance, they do help to order and orientate contigs during the closure process. However this map has not only been useful for the genome project but has also been a vital tool for forward genetic research, particularly in the field of drug resistance (see Ferdig and Su, 2000, for review).

## 2.3. YAC Map

YACs were mapped onto the chromosomes using either sequence tagged site (STS) markers, or the optical map. Tiling paths could then be constructed using restriction mapping or hybridisation (Rubio *et al.*, 1995; Thompson *et al.*, 1999; Thompson and Cowman, 1997). The Sanger Institute and Stanford used these YACs by shotgun sequencing them to 2x coverage. These reads were then assembled in with the chromosome shotgun reads thus allowing the contigs containing YAC reads to be grouped together. At the outset of the project, the YAC map covered around 60-70% of the genome, but as the genome project progressed the sequence generated was used to identify more YACs to add to the map.

## 2.4. Happy Map

As chromosomes 6,7 and 8 could not be separated by PFGE, they proved to be particularly difficult to assemble. Thus, a new mapping technique was employed to group contigs from particular chromosomes. HAPPY maps are generated by taking genomic DNA which is mechanically sheared and aliquoted into 96 samples, each containing less than one genome equivalent of DNA. Linkers are ligated onto the DNA fragments and the the fragments are then amplified and replicated to create a mapping panel This panel can then be screened for STS markers that were generated from the shotgun sequencing reads using PCR amplification. Those markers that are located close to each other in the genome will tend to be segregated together in the panel (Dear *et al.*, 1998; Konfortov *et al.*, 2000).The linkage groups generated from HAPPY mapping of chromosomes 6, 7, and 8 were combined with other mapping data, which allowed previously unlocalized contigs to be anchored to specific chromosomes, which greatly simplified the assembly and closure of each chromosome. reducing the problem to three small chromosome assemblies as opposed to on large assembly.

Although the genome of *P. falciparum* has been published it is not yet complete. At the time of writing the genome contains 21 gaps. Most, but not all, of these gaps have been positioned with respect to the contigs. In most cases these gaps will be small and will contain only repetitive tracts of AT rich DNA; others will be larger and probably contain genes, hence we are not yet able to determine with confidence whether any particular gene is not present in the *Plasmodium falciparum* genome. Table 1 gives an overview of the state of the genome and its content. The sequencing centres are continuing work to complete the genome and update its annotation.

# 3. The Genome Composition and Structure

The nuclear genome of *P. falciparum* is comprised of 14 chromosomes, is 22.8 Mb in size, and encodes about 5,300 genes (Gardner *et al.*, 2002a) Fifty-four percent of the genes that were identified contained introns. In comparison to two well-studied eukaroytes whose genomes have been sequenced, *S. cerevisiae* and *S. pombe*, the *P. falciparum* genome is about twice as large but contains roughly similar numbers of genes, giving *P. falciparum* a lower gene density than either *S. pombe* or *S. cerevisiae*. On average, the coding regions of *P. falciparum* genes were longer than in other sequenced eukaryotes such as *S. pombe* (2.3 kb vs 1.4 kb). The reason for this increased gene length is not understood but may be due to the regions of low amino acid complexity found in many *P. falciparum* proteins. These low complexity regions are much more numerous in *P. falciparum* than in other sequenced eukaryotes. Fully 60% of the encoded proteins have little or no similarity to proteins in other organisms and are of unknown function. The

**Table 1.** Status and content of the *P. falciparum* genome

| | Sequencing Centre | Size | Status | Number of tRNAs* | Number of rRNAs*§ | Number of Protein Coding Genes* |
|---|---|---|---|---|---|---|
| 1 | Sanger Institute | 643,292 | Finished | 0 | 1 | 143 |
| 2 | TIGR | 947,102 | Finished | 1 | 0 | 233 |
| 3 | Sanger Institute | 1,060,087 | Finished | 2 | 0 | 239 |
| 4 | Sanger Institute | 1,204,112 | Finished | 5 | 0 | 237 |
| 5 | Sanger Institute | 1,343,552 | Finished | 5 | 1 | 312 |
| 6 | Sanger Institute | 1418244 | Finished | 3 | 0 | 312 |
| 7 | Sanger Institute | 1391523 | 6 gaps | 7 | 1 | 277 |
| 8 | Sanger Institute | 1235471 | 6 gaps | 0 | 2 | 295 |
| 9 | Sanger Institute | 1,541,723 | Finished | 0 | 0 | 265 |
| 10 | TIGR | 1,687,655 | Finished | 0 | 0 | 403 |
| 11 | TIGR | 2,038,337 | Finished | 2 | 1 | 492 |
| 12 | Stanford | | Finished | 3 | 0 | 526 |
| 13 | Sanger Institute | 2,988,383 | 18 gaps | 5 | 1 | 672 |
| 14 | TIGR | 3,291,871 | Finished | 2 | 3 (5S) | 769 |

*figures are taken from the published annotation which has not been updated for the most recent assemblies.
§total numbers of operons and 5S rRNA genes are given.

proportion of genes encoding these so-called hypothetical proteins is higher in *P. falciparum* than in other organisms, which is probably a reflection of the greater evolutionary distance between *Plasmodium* and other sequenced eukaryotes. A more practical implication is that having spent 20+ years studying the molecular biology of this organism, determining the genome sequence, and cataloging most of the parasite's genes and predicted gene products, we still have a very limited understanding of parasite biology. Thus there is a great deal yet to learn, but we are much better equipped to make progress than we were just a few years ago.

## 3.1. Centromeres

The centromeres of different organisms vary in structure and compositon, although they are generally A+T rich. For example, *Saccharomyces cerevisiae* has small (A+T) rich centromeres that can be as short as 125 bp in length. Most other higher eukaryotes contain large centromeres containing complex repeats over regions that may be as large as 4 Mb in humans (for review see Sullivan *et al.*, 2001). Centromeres have not been experimentally characterised in *Plasmodium*, but a putative centromere has been identified by comparison of chromosomes 2 and 3. This structure has since been identified on all complete chromosomes. The region is characterised by a ~3 kb region of extreme A+T content (>97%) and occurs within non-coding regions of up to 11 kb. Importantly, there is never more than one such region on any chromosome. A similar region can be seen in the other apicomplexan parasites *Theileria annulata* and *Theileria parva* (unpublished data). If these regions are indeed acting as centromeres in *P. falciparum*, they resemble most closely the centromeres of *S. cerevisiae*.

## 3.2. Telomeres

The telomeres of *P. falciparum* are much more than simply the ends of chromosomes. As well as the telomeric hexamer repeat, the sub-telomeric region contains many other complex repeat units and many members of the gene families involved in antigenic variation and immune avoidance. It is believed that many of the genes encoding proteins that interact with the host reside in the subtelomere because these regions recombine more readily than other parts of the genome, thus generating variation in gene products that play a role in avoidance of the host immune system (reviewed in Scherf *et al.*, 2001).

By comparing chromosomes one can subdivide the telomeres into specific subtelomeric blocks (SBs) defined by conserved regions. SB1 to SB3 contain the previously described subtelomeric repeats TARE1-6 (Telomere Associated Repetitive Elements) which are highly conserved between chromosomes (Figueiredo *et al.*, 2000). SB1 and TARE1 are equivalents containing the *P. falciparum* telomeric repeat containing the G rich consensus sequence of GGGTT (T/C)A this sequence is unusual in being a non-perfect repeat as in most other eukaryotes. SB3 is made up of a 21 base pair repeat known as rep20 (TARE 6). Plasmids containing rep20 have been shown to be more stably maintained in episomal transformations of merazoites than plasmids that do not contain the repeat, suggesting a role for rep20 in chromosome interaction at mitosis (O'Donnell *et al.*, 2002), however rep20 is not found in any other *Plasmodium* species (Figueiredo *et al.*, 2000) and deletion of rep20 can readily occur in *P. falciparum* (Patarapotikul and Langsley, 1988). Unlike SB1-3, SB-4 and SB-5 contain protein coding genes. SB-4 contains the telomeric var genes and several repeat elements.

SB5 does not contain repeats but it is conserved in half of the chromosomes within the genome (Figure 2). It is quite possible that these conserved regions are required to promote recombination at the telomeres and hence diversity in the surface antigen genes. Recently, the formation of a chimeric *var* gene generated during meiotic recombination has been observed (Freitas-Junior *et al.*, 2000). Telomeres of yeast have been observed to anchor to the nuclear periphery and form a distinct nuclear compartment (Cockell and Gasser, 1999). This is believed to be important for regulation of telomeric genes and DNA recombination. Nuclease digestion of *Plasmodium* genomic DNA suggests that the telomere has a different nucleosome structure than the rest of the genome (Cary *et al.*, 1994).

While the sequence of *P. falciparum* has massively increased our knowledge of the telomeric structure of *Plasmodium*, changes are known to have occurred in the telomeres of the 3D7 sequencing strain in culture and we don't know how much of the telomeric structures of 3D7 are reflected in field isolates. A sequencing project of a field isolate of *P. falciparum* is planned for the future.

## 4. tRNA and rRNA Genes

In most organisms sequenced to date the copy number of specific tRNA's reflects the codon usage of the organism (Duret, 2000; Moriyama and Powell, 1997), *i.e* common codons will have abundant anti-codons. It has been demonstrated in many unicellular organisms such as *E. coli* and yeast that the cellular level of tRNAs is related to copy number (Ikemura, 1981; Ikemura, 1982) and that protein coding genes can be regulated at translation by adapting there codon usage to abundant anticodons for highly expressed genes and rare anticodons for less highly expressed genes (Gouy and Gautier, 1982; Grantham *et al.*, 1981; Ikemura and Wada, 1991).

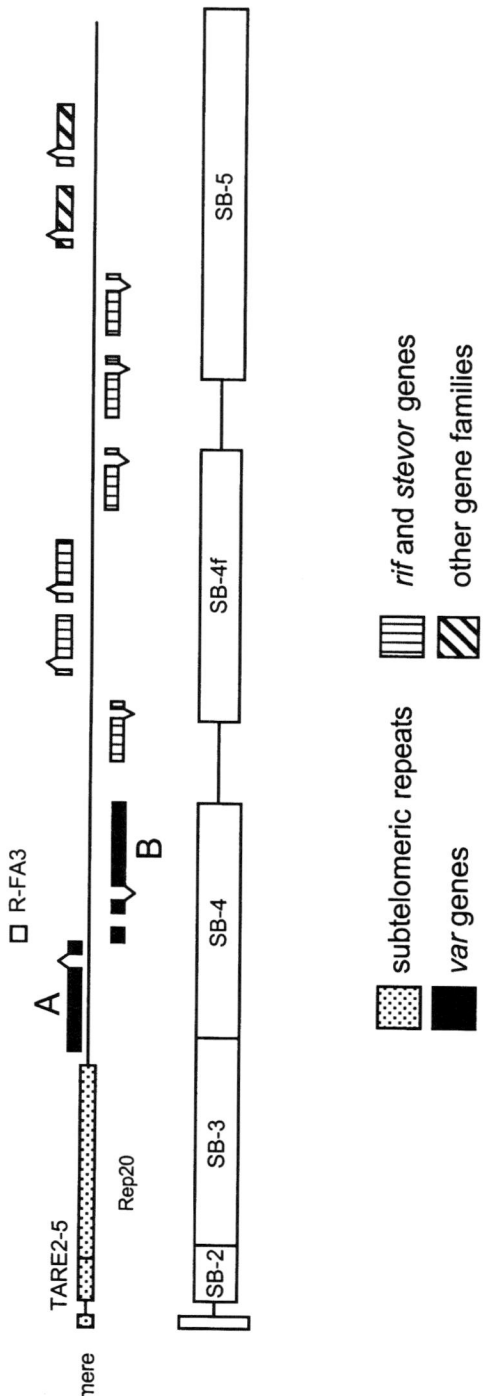

**Figure 2.** Gene order and repeat structures at the telomeres of *Plasmodium falciparum* chromosomes. The top line shows the gene order with respect to the main subtelomeric repeats. Adjacent to the telomere is TAREs 2-5 which includes Rep20. The most telomeric gene is VAR in all chromosomes and in many cases there are two genes in positions A and B as shown.. The RIF and STEVOR genes are found in arrays internal to the VAR genes and in many cases other gene families are found within 150kb of the telomere. The line below shows the approximate position of the subtelomeric blocks (SBs). These blocks are conserved between chromosomes.

Since the complete genome of *P. falciparum* has been sequenced we have been able to identify the entire tRNA compliment. Surprisingly it contains a minimal set of genes encoding tRNAs; very few tRNA genes are present in more than one copy. In many cases they appear to be organised into small clusters, but they are not tandemly arrayed. In fact, for most codons there is only the minimal complement of isoaccepters; so that all codons ending in C and T appear to be read by single tRNAs with a G in the first position, which is likely to read both codons by T:U wobble. Although the tRNA for the translation of TGT and TGC (coding for Cys) was not identified we have found the tRNA ligase genes for all tRNAs; hence it is likely that this missing gene resides in one of the remaining gaps in the genome.

As there is only a minimal set of tRNA genes in the genome we must also assume that some of the tRNAs are transported into the mitochondria to allow translation of the 3 mitochondrial genes. As, unlike mitochondria from other species, the *Plasmodium* mitochondria contains no tRNA genes. The mechanism for this process is not yet known but similar systems have been observed in other organisms (Tan *et al.*, 2002; Tarassov and Martin, 1996). It would seem therefore that in *Plasmodium* each tRNA is highly expressed to a level where tRNA abundance may not be a limiting step in translation and codon usage is unlikely to play a role in regulation of expression.

*Plasmodium* spp. contain different isoforms of large subunit rRNA which are expressed at different times throughout their life cycle (reviewed in (Li *et al.*, 1994; Waters, 1994; Waters *et al.*, 1989)). This presumably acts as a global regulator for translation of different gene sets required at different stages. In *Plasmodium falciparum* two isotypes have been characterised the A type expressed during asexual development and the S type expressed primarily in the mosquito vector. In *Plasmodium vivax* a third type of rRNA has been identified, the O-type which is expressed in the ookinete (Li *et al.*, 1997).

The genome sequence revealed 7 rRNA loci encoding genes for 18S, 5.8S, and 28S rRNAs, and three genes encoding 5S rRNAs. The organisation of the rRNA genes in *P. falciparum* is different to that seen in other eukaryotes as they are dispersed throughout the genome in single copies as opposed to being in long tandem arrays. The A type rRNA genes reside on chromosomes 11 and 13 and the S type are on chromosomes 5 and 7. Nothing resembling O-type rRNA has been identified, however an unusual 28s rRNA gene was identified in the unit on chromosome 1 and two unusual units reside on chromosome 8 that contain no 28s rRNA. The function, if any, of these divergent units is unknown.

# 5. The Proteome of *Plasmodium falciparum*

The genome sequence was annotated by the use of several gene finding programs to predict gene structures, followed by the manual curation of gene models and manual assignment of potential functions to the predicted proteins (Gardner *et al.*, 2002b; Hall *et al.*, 2002; Hyman *et al.*, 2002). Of the 5,268 predicted proteins, about 60% had little or no similarity to proteins in other organisms and were of unknown function (the so-called "hypothetical proteins"). The proportion of the proteome consisting of hypothetical proteins was higher than in many other sequenced eukaryotes; this may be due to the great evolutionary distance between *P. falciparum* and other sequenced eukaryotes, or to the biased amino acid composition of *P. falciparum* proteins resulting from the high (A+T) content of the genome. As described below, about 4% of *P. falciparum* genes encoded proteins involved in antigenic variation.

Analysis of the predicted proteome provided an overview of metabolism and transport in malaria parasites. However, some features of parasite metabolism remained unclear due to the absence of some enzymes or enzyme subunits, and the predicted subcellular localization of some enzymes differed from the known localization of the enzymes in other organisms, making it difficult to reconstruct the metabolism of the parasite with certainty. Nevertheless, analyis of the genome sequence provided valuable insights into the biochemistry of the parasite and pinpointed areas that require further investigation in the laboratory.

*P. falciparum* appears to have much reduced capacities for metabolism and for the transport of organic nutrients and ions than free-living microbes. Sequence similarity searches with sequences of known enzymes revealed that only 14% of the predicted proteins encoded enzymes, a smaller proportion than has been observed in other sequenced eukaryotes. The *P. falciparum* proteome also contained a much smaller repertoire of membrane transporters than other eukaryotic microbes such as *S. cerevisiae* and *S. pombe*. The *P. falciparum* genome encoded enzymes for the complete glycolytic pathway from glucose-6-phosphate to pyruvate, and for the conversion of pyruvate to lactate, confirming biochemical studies suggesting that erythrocytic stage parasites rely primarily on glycolysis for ATP production. All of the enzymes of the tricarboxylic acid (TCA) cycle were found, but the potential function of the TCA cycle was unclear. Pyruvate dehydrogenase, which in other organisms is usually localized in the mitochondrion and which converts pyruvate to acetyl CoA required for the first step of the TCA cycle, was predicted to be located within the apicoplast, a relict plastid, in *P. falciparum*. Moreover, malate dehydrogenase was predicted to have a cytoplasmic rather than the typical mitochondrial subcellular localization and was proposed to be replaced in the *Plasmodium* TCA cycle by mitochondrial malate-quinone oxidoreductase. These unusual properties of the TCA cycle suggested that, at least in erythrocytic stages, it may be used to supply intermediates for biosynthetic pathways such as heme biosynthesis rather than for the complete oxidation of the products of glycolysis. The role of the TCA cycle in the exoerythrocytic or mosquito stages of the life cycle is unknown. However, proteomic studies (Florens *et al.*, 2002; Lasonder *et al.*, 2002) suggested that some TCA cycle enzymes are more abundant in gametocytes than in erythrocytic parasites, implying that the TCA cycle may be more important in the non-erythrocytic stages. Another unusual finding was the apparent absence of the $F_0$ a and b subunits of ATP synthase. Their absence implied that the mitochondrial ATP synthase might be non-functional in *Plasmodium*, although the genes encoding these proteins are short and may be located within the currently unsequenced regions of the genome. Alternatively, the amino acid sequences of the *Plasmodium* $F_0$ a and b subunits may be quite different from those in other organisms and have been undetectable by the sequence similarity approaches that were used for the annotation of the genome sequence.

Other unusual features of *P. falciparum* metabolism deduced from analysis of the genome sequence were the absence of gluconeogenesis and enzymes for the biosynthesis of amino acids, except for several enzymes that perform interconversions of amino acids. The lack of amino acid biosynthesis, and the absence of homologs of known amino acid transporters, emphasized the parasite's dependence on the host for amino acids, at least in the erythrocytic stages in which amino acids are obtained by the digestion of hemoglobin in the food vacuole. How the parasite obtains amino acids during the mosquito stages of the life cycle is not known, but some as yet unknown mechanism must be used to obtain amino acids from the extracellular environment. Biosynthesis of fatty acids and isoprenoids occurs in the apicoplast, a relict plastid that is essential to parasite survival. Both of these pathways resemble the pathways used by plants and bacteria rather than animals and offer several potential targets for novel antimalarials (Jomaa *et al.*, 1999; Surolia *et al.*, 2002; Surolia and Surolia, 2001). Almost 10% of nuclear genes encode proteins that are destined for import in the apicoplast (Foth *et al.*, 2003), suggesting that additional as yet unknown metabolic functions may be associated with this organelle. Overall, the metabolic and transport capabilities of *Plasmodium* are much reduced compared to free-living organisms, which may be a reflection of its parasitic life style. This reduction in overall metabolic and transport potential was also observed in the intracellular microsporidian parasite *Encephalitozoon cuniculi* (Katinka *et al.*, 2001). The genome of this organism is only 2.9 Mb and encodes just 1,997 genes. The reduction of metabolic capacity is more extreme than in *P. falciparum*, with even the TCA cycle, fatty acid biosynthesis, and the $F_0$-$F_1$ ATP synthase (and the mitochondrion) being absent. Reduced metabolic potential is probably a feature characteristic of other intracellular parasitic protozoa.

# 6. Antigens and Antigenic Variation

After invading an erythrocyte, the *Plasmodium* cell grows to maturity within the red cell which coincides with the appearance of parasite-derived variant proteins on the surface of the infected cell. These proteins elicit a protective immune response from the host. To evade the immune response the parasite undergoes antigenic variation by successively changing these proteins and thus remaining one step ahead of the host immune system. The most well characterised parasite protein that is presented on the erythrocyte surface is *P. falciparum* erythrocyte membrane protein 1 (PfEMP1). These proteins also cause mature red blood cells to bind specific host receptors on vascular endothelial cells that cause the red blood cells to be sequestered away from the peripheral circulation (for review see (Kyes *et al.*, 2001)). It is this property of PfEMP1 proteins that makes *P. falciparum* infections so life threatening as the accumulation of large numbers of infected red blood cells in vital organs can be fatal.

The completed genome sequence will allow high throughput molecular biological techniques to be deployed to identify new antigen targets. However, the sequence has already told us a lot about the sequence diversity of antigens that are already known. We now have a good idea of the number and organisation of *var* genes (which code for PfEMP-1 protein) in *P. falciparum* strain 3D7. We have also characterised their genomic context and diversity. There are 59 intact *var* genes in *P. falciparum* clone 3D7; every telomere has either a *var* gene, a *var* gene fragment or a *var* psedogene proximal to the rep20 repeat. There is often a second *var* gene internal to the telomeric *var* in the reverse orientation. Interestingly the most telomeric *var* genes (position A in Figure 2) appear to have a very conserved domain structure compared to the *var* genes that are found adjacent to them (Position 2). Also most of the *var* genes that had been previously characterised due to there specific adhesive phenotypes come from the pool of genes found in position 2 (Smith *et al.*, 2000).

The telomeres also contain arrays of *rif* genes encoding Repeat Interspersed Family proteins (rifins) which are also expressed on the surface of the erythrocyte and known to undergo antigenic variation (Fernandez *et al.*, 1999; Kyes *et al.*, 1999) although unlike the *var* genes, they do not yet have another characterised function. Alignment of all r*ifin* peptides shows that they can be categorised into two types depending on their length, however we don't know if the two types have different functions. Stevors (Sub-TElomeric Variable Open Reading frame) are a family of proteins related to the rifins but are less variable in amino acid sequence (Cheng *et al.*, 1998); the function of stevors is not known although they appear to localise in Maurers clefts (Kaviratne *et al.*, 2002). One hundred and forty nine rifin genes and 28 stevor genes have been identified in the genome. Figure 2 shows the usual telomeric organisation. of *var*, *rif* and *stevor* genes. There are also 4 clusters of *var* genes in internal regions chromosomes 4,7,8 and 12 and a single internal *var* gene on chromosome 6.

# 7. Gene Regulation

The current annotation of the *P. falciparum* genome suggests that it contains very few transcriptional regulators compared to other eukaryotic genomes such as *S. cerevisae* (10 compared to 270). This anomaly may have two possible explanations: 1.) *P falciparum* does not rely on regulation of gene expression at the level of transcription. 2.) The transcriptional regulators in *P. falciparum* are so divergent from characterised regulators that they have not been identified. It would seem on the face of it that hypothesis 1 is unlikely as *P. falciparum* has to adapt to life in a number of different environments during its life cycle and proteomic analysis suggests that the proteins expressed in these environments are very different (Hayward *et al.*, 2000; Kappe *et al.*, 2001). Comparison of microarray data to proteomic data suggests that regulation is occurring at the level of transcript accumulation (Bozdech *et al.*, 2003; Le Roch *et al.*, 2003). Hence it is more likely that *P. falciparum* does have other transcriptional regulators that do not conform to the typical domain structure seen in the model organisms. It is also likely that because *P. falciparum* exists in highly controlled environments throughout its life cycle that it has limited need for many pathway specific regulatory networks and much of its gene regulation could occur using a few global regulators. Recent microarry data demonstrates that there is highly co-ordinated regulation of large numbers of genes during intra-erythocytic development (Bozdech *et al.*, 2003).

There is some evidence that in *P. falciparum* epigenic mechanisms, such as chromatin structure, play an important role in transcriptional control. It has been demonstrated that developmentally regulated gene expression of *gbp110* was lost when expressed on an episomal plasmid that does not preserve the genomic nucleosome organisation (Horrocks *et al.*, 2002). Proteomic analysis suggests that chromosomal location could be important for co-regulation of genes, as genes that share particular expression patterns are seen to cluster together (Florens *et al.*, 2002); although there is not yet enough data to make this statistically conclusive. While analysis of the genetic content of individual chromosomes has shown that genes associated with similar functions cluster on particular chromosomes (Hall *et al.*, 2002) this may be due to expansion of parologous gene families rather than actual functional clustering. Although microarray analysis has also demonstrated that co regulated genes are rarely tightly clustered (Bozdech *et al.*, 2003), one study has suggested that genes at telomeres are more likely to be involved in erythrocyte remodelling while internal genes are more likely to be involved in cell growth or maintenance (Le Roch *et al.*, 2003).

# 8. Comparative Genomics

*P. yoelii yoelii* is a rodent malaria parasite has been used extensively to study the mechanisms of protective immunity to malaria, to identify and study antigens that are the targets of immunity, and for the testing of vaccines against these antigens. Pulsed field gels and hybridization to chromosome-specific probes have been used to examine the conservation of gene synteny in different species of rodent, primate, and human malaria parasites (Carlton *et al.*, 1999; 1998; Janse *et al.*, 1994). These studies identified regions of gene synteny between species, but suggested that the extent of conservation of gene synteny declined as the evolutionary distance between the species increased. Thus the level of synteny between *P. falciparum* and rodent malaria parasites was less than that between more closely related rodent parasites.

As a result of the rapid decline in sequencing costs in recent years, TIGR and the Naval Medical Research Center were able to use funds originally awarded for the *P. falciparum* genome project for the sequencing of the *P. yoelii* genome to 5X coverage. The 23 Mb *P. yoelii* genome was found to encode approximately 5,900 genes, about 300 more genes than were identified in *P. falciparum*, probably due to 838 small genes (*yir*) in *P. yoelii* that encode variant antigens. A comparative analysis of the *P. yoelii* and *P. falciparum* genomes was performed (Carlton *et al.*, 2002). About half of the predicted *P. yoelii* proteins had orthologs in *P. falciparum*. Almost all of the putative orthologs were located in the central portions of the chromosomes, whereas many non-orthologous genes such as the *P. yoelii yir* genes and the *rif*, *stevor*, and *var* genes in *P. falciparum*, were located in the subtelomeric regions of chromosomes.

Determination of the *P. falciparum* and *P. yoelii yoelii* genome sequences enabled studies of gene synteny to be performed using bioinformatic techniques across the entirety of both genomes. The MUMer program (Delcher *et al.*, 2002) was used to translate each genome sequence in all 6 reading frames and identify all unique exact matches longer than 5 amino acids between the species. Over 70% of the *P. yoelii* genome could be tiled against the *P. falciparum* genome; most of the conserved *P. yoelii* contigs mapped to the central regions of the *P. falciparum* chromosomes. The *P. yoelii* contigs that were tiled along the *P. falciparum* genome were subsequently linked into 457 groups of contigs by the use of mate-pair information, PCR amplification of products from the gaps between contigs, and identification of genes that spanned the gaps. These groups of contigs represented regions of conserved gene synteny and were up to 800 kb in length. Mapping of the *P. yoelii* contigs to *P. yoelii* chromosomes with chromosome specific markers then provided a map of conserved gene synteny and identified potential syntenic breakpoints. About 70% of *P. y. yoelii* genes occurred in the same order and orientation as the putative orthologous genes in *P. falciparum*. Syntenic breakpoints in *P. yoelii* were often associated with loci encoding rRNA genes, suggesting that chromosome breakage and recombination at the rRNA loci may have been responsible for the disruption of gene synteny in malaria parasites. Analysis of the structures of orthologous genes in syntenic regions shared by the two species revealed remarkable conservation of exon structures in genes. Thus, as was noted previously, alignments of orthologous genes from different *Plasmodium* species may be very helpful in the elucidation of gene structures (del Portillo *et al.*, 2001; van Lin *et al.*, 2001).

Comparison of the *P. falciparum* and *P. yoelii* genomes has revealed a great deal about genome structure and organization in malaria parasites, and has highlighted differences between specieis in terms of the gene products involved in the evasion of the host immune response. Similar analyses of other malaria parasites and related apicomplexans will shed further light on these subjects. The Sanger Institute is sequencing the genomes of the rodent malaria parasites, *P. berghei* and *P. chabaudi*, are currently being sequenced to 3X coverage, as well as the genome of the primate malaria parasite *P. knowlesi*. TIGR and the NMRC are also sequencing the genome of the second most important human malaria parasite *P. vivax* (Carlton, 2003). In addition, genome sequencing projects for *Toxoplasma gondii*, *Theileria parva* (Nene *et al.*, 2000), *T. annulata*, and *Crytosporidium parvum* are well underway or nearing completion (see www.ncbi.nlm.nih.gov/genomes/static/EG_T.html for a comprehensive listing of links to the web pages for eukaryote genome projects).

# 9. Future Directions

While the genome sequence has heralded a quantum leap in our understanding of the parasites biology, it has also highlighted how little we know about the parasite, with 60% of the predicted genes having unknown functions. The annotation of the genome could be enhanced by improving gene finding tools and sequencing more full length cDNAs to improve existing gene models. One of the major challenges facing the scientific community will be to begin to understand what these genes do and how they are regulated. While high throughput technologies such as microarrays and proteomics will be important tools, comparative genomics will also have a role. By identifying what genes that are rapidly evolving between closely related *Plasmodium* species and strains we may identify antigenic proteins. Conserved sequences may highlight unknown genes and regulatory sequences. Comparative genomics, microarrays and proteomic analysis will only generate information to direct experimentation and there is still no single technology that will elucidate the function of all orphan genes. Scientists will be unravelling the information contained within the genome sequence for decades to come.

As the genome of *P. falciparum* and other malaria parasites is completed and more functional information is generated, it is important that the genome annotation is kept up to date and made available to all. Gene function and expression analyses need to be integrated with the genome data so that researches have access to a complete picture of the organisms' biology.

The PlasmoDB database has been instrumental in allowing the community to easily queryable data from the sequencing centres and other labs. However if this is to remain a useful resource the genome must be curated and updated so that the annotation associated with it remains relevant as the knowledge base changes.

## 9.1. Impact of Genomics on Malaria Research

The decision, in 1997, to form a consortium to sequence the genome of *P. falciparum* was a controversial one within the malaria community. Many investigators felt that completion of the genome sequence would be difficult or impossible due to the extreme AT-richness of the genome and the effort would fail. Others thought that the estimated price tag of $15 million was too high and that funding of the genome project would divert funds from basic research. Another argument against the sequencing effort was that it would not rapidly lead to new drugs or vaccines, and that the money would be far better spent on existing control measures. Despite these reservations, the sequencing consortium went ahead, with support from public and private funding agencies, in the belief that the genome sequence would provide valuable information about this parasite that would support research efforts by scientists. In addition, after much discussion between the sequencing centers, malaria researchers, and funding agencies a data release policy was adopted that facilitated the release of preliminary sequence information to the community in advance of the publication of the genome sequence. The community also requested, and the Burroughs Wellcome Fund supported, the development of a database and user-friendly web site (PlasmoDB) that provided access to all of the sequence data and annotation produced by the three sequencing centers (Bahl *et al.*, 2003). The sequencing of the genome, the release of preliminary data, and the development of PlasmoDB have had a profound stimulatory effect on malaria research. The reports on the sequences of chromosomes 2 and 3 have been among the most highly cited papers in the malaria field. Dozens of publications have also acknowledged the use of the preliminary sequence data that was released years before the publication of the genome sequence itself; many of these provide functional information on novel genes and gene products that would have not been discovered as rapidly without the availability of the genome sequence. Many of these publications are by scientists who have not previously published in the malaria field, suggesting that availability of the genome sequence may have attracted new investigators to work on malaria.

Now that the complete genome sequence is in hand, additional hypothesis-driven research utilizing the genome "parts list" is required to generate a better understanding of the parasite and its interactions with the human and mosquito hosts. This knowledge must subsequently be applied to the development of new drugs and vaccines to combat malaria. Despite some highly encouraging initial successes, such as the identification of novel drug targets associated with the apicoplast (Ridley, 1998; 1999), this process has hardly begun. More than half of the predicted proteins encoded by the *P. falciparum* genome are of unknown function. Elucidation of their function will be a painstaking process potentially lasting many years, but this effort must be undertaken and supported financially if the dividends from the investment in the genome sequence are to be realized.

## 10. Acknowledgements

The authors thank their many colleagues at The Sanger Institute, TIGR, the Naval Medical Research Insitute, and the Stanford Genome Technology Center for their contributions to the *P. falciparum* genome project. The *P. falciparum* genome sequencing project was supported by the Wellcome Trust, the National Institute for Allergy and Infectious Diseases, the Burroughs Wellcome Fund, the Naval Medical Research Center, and the U.S. Army Medical Research and Materiel Command.

## 11. References

Bahl, A., Brunk, B., Crabtree, J., Fraunholz, M.J., Gajria, B., Grant, G.R., Ginsburg, H., Gupta, D., Kissinger, J.C., Labo, P., Li, L., Mailman, M.D., Milgram, A.J., Pearson, D.S., Roos, D.S., Schug, J., Stoeckert, C.J., Jr., and Whetzel, P. 2003. PlasmoDB: the *Plasmodium* genome resource. A database integrating experimental and computational data. Nucleic Acids Res. 31: 212-215.

Bloom, B.R. 1995. A microbial minimalist. Nature. 378: 236.

Bowman, S., Lawson, D., Basham, D., Brown, D., Chillingworth, T., Churcher, C.M., Craig, A., Davies, R.M., Devlin, K., Feltwell, T., Gentles, S., Gwilliam, R., Hamlin, N., Harris, D., Holroyd, S., Hornsby, T., Horrocks, P., Jagels, K., Jassal, B., Kyes, S., McLean, J., Moule, S., Mungall, K., Murphy, L., Barrell, B.G. 1999. The complete nucleotide sequence of chromosome 3 of *Plasmodium falciparum*. Nature. 400: 532-538.

Bozdech, Z., Llinas, M., Pulliam, B.L., Wong, E.D., Zhu, J., and DeRisi, J.L. 2003. The transcriptome of the intraerythrocytic developmental cycle of *Plasmodium falciparum*. PLoS Biol. 1: 5.

Breman, J.G. 2001. The ears of the hippopotamus: manifestations, determinants, and estimates of the malaria burden. Am. J. Trop. Med Hyg. 64: 1-11.

Bult, C.J., White, O., Olsen, G.J., Zhou, L., Fleischmann, R.D., Sutton, G.G., Blake, J.A., FitzGerald, L.M., Clayton, R.A., Gocayne, J.D., Kerlavage, A.R., Dougherty, B.A., Tomb, J.F., Adams, M.D., Reich, C.I., Overbeek, R., Kirkness,

E.F., Weinstock, K.G., Merrick, J.M., Glodek, A., Scott, J.L., Geoghagen, N.S.M., and Venter, J.C. 1996. Complete genome sequence of the methanogenic archaeon, *Methanococcus jannaschii*. Science. 273: 1058-1073.

Carlton, J. 2003. The *Plasmodium vivax* genome sequencing project. Trends Parasitol. 19: 227-231.

Carlton, J.M., Angiuoli, S.V., Suh, B.B., Kooij, T.W., Pertea, M., Silva, J.C., Ermolaeva, M.D., Allen, J.E., Selengut, J.D., Koo, H.L., Peterson, J.D., Pop, M., Kosack, D.S., Shumway, M.F., Bidwell, S.L., Shallom, S.J., van Aken, S.E., Riedmuller, S.B., Feldblyum, T.V., Cho, J.K., Quackenbush, J., Sedegah, M., Shoaibi, A., Cummings, L.M., Florens, L., Yates, J.R., Raine, J.D., Sinden, R.E., Harris, M.A., Cunningham, D.A., Preiser, P.R., Bergman, L.W., Vaidya, A.B., van Lin, L.H., Janse, C.J., Waters, A.P., Smith, H.O., White, O.R., Salzberg, S.L., Venter, J.C., Fraser, C.M., Hoffman, S.L., Gardner, M.J., and Carucci, D.J. 2002. Genome sequence and comparative analysis of the model rodent malaria parasite *Plasmodium yoelii yoelii*. Nature. 419: 512-519.

Carlton, J.M., Galinski, M.R., Barnwell, J.W., and Dame, J.B. 1999. Karyotype and synteny among the chromosomes of all four species of human malaria parasite. Mol. Biochem. Parasitol. 101: 23-32.

Carlton, J.M.R., Vinkenoog, R., Waters, A.P., and Walliker, D. 1998. Gene synteny in species of *Plasmodium*. Mol. Biochem. Parasitol. 93: 285-294.

Cary, C., Lamont, D., Dalton, J.P., and Doerig, C. 1994. *Plasmodium falciparum* chromatin: nucleosomal organisation and histone-like proteins. Parasitol. Res. 80: 255-258.

Cheng, Q., Cloonan, N., Fischer, K., Thompson, J., Waine, G., Lanzer, M., and Saul, A. 1998. *stevor* and *rif* are *Plasmodium falciparum* multicopy gene families which potentially encode variant antigens. Mol. Biochem. Parasitol. 97: 161-176.

Cockell, M., and Gasser, S.M. 1999. Nuclear compartments and gene regulation. Curr. Opin. Genet. Dev. 9: 199-205.

Dear, P.H., Bankier, A.T., and Piper, M.B. 1998. A high-resolution metric HAPPY map of human chromosome 14. Genomics. 48: 232-241.

del Portillo, H.A., Fernandez-Becerra, C., Bowman, S., Oliver, K., Preuss, M., Sanchez, C.P., Schneider, N.K., Villalobos, J.M., Rajandream, M.A., Harris, D., Pereira da Silva, L.H., Barrell, B., and Lanzer, M. 2001. A superfamily of variant genes encoded in the subtelomeric region of *Plasmodium vivax*. Nature. 410: 839-842.

Delcher, A.L., Phillippy, A., Carlton, J., and Salzberg, S.L. 2002. Fast algorithms for large-scale genome alignment and comparison. Nucleic Acids Res. 30: 2478-2483.

Duret, L. 2000. tRNA gene number and codon usage in the *C. elegans* genome are co-adapted for optimal translation of highly expressed genes. Trends Genet. 16: 287-289.

Ferdig, M.T., and Su, X.Z. 2000. Microsatellite markers and genetic mapping in *Plasmodium falciparum*. Parasitol. Today. 16: 307-312.

Fernandez, V., Hommel, M., Chen, Q., Hagblom, P., and Wahlgren, M. 1999. Small, clonally variant antigens expressed on the surface of the *Plasmodium falciparum*-infected erythrocyte are encoded by the *rif* gene family and are the target of human immune responses. J. Exp. Med. 190: 1393-1404.

Figueiredo, L.M., Pirrit, L.A., and Scherf, A. 2000. Genomic organisation and chromatin structure of *Plasmodium falciparum* chromosome ends. Mol. Biochem. Parasitol. 106: 169-174.

Fleischmann, R.D., Adams, M.D., White, O., Clayton, R.A., Kirkness, E.F., Kerlavage, A.R., Bult, C.J., Tomb, J.F., Dougherty, B.A., Merrick, J.M., McKenney, K., Sutton, G., *et al.* 1995. Whole-genome random sequencing and assembly of *Haemophilus influenzae* Rd. Science. 269: 496-512.

Florens, L., Washburn, M.P., Raine, J.D., Anthony, R.M., Grainger, M., Haynes, J.D., Moch, J.K., Muster, N., Sacci, J.B., Tabb, D.L., Witney, A.A., Wolters, D., Wu, Y., Gardner, M.J., Holder, A.A., Sinden, R.E., Yates, J.R., 3rd, and Carucci, D.J. 2002. A proteomic view of the *Plasmodium falciparum* life cycle. Nature. 419: 520-526.

Foster, J., and Thompson, J. 1995. The *Plasmodium falciparum* genome project: a resource for researchers. Parasitol. Today. 11: 1-4.

Foth, B.J., Ralph, S.A., Tonkin, C.J., Struck, N.S., Fraunholz, M., Roos, D.S., Cowman, A.F., and McFadden, G.I. 2003. Dissecting apicoplast targeting in the malaria parasite *Plasmodium falciparum*. Science. 299: 705-708.

Fraser, C., Gocayne, J.D., White, O., Adams, M.D., Clayton, R.A., Fleischmann, R.D., Bult, C.J., Kerlavage, A.R., Sutton, G., Kelley, J.M., Fritchman, J.L., Weidman, J.F., Small, K.V., Sandusky, M., Fuhrmann, J., Nguyen, D., Utterback, T.R., Saudek, D.M., Phillips, C.A., Merrick, J.M., Tomb, J.F., Dougherty, B.A., Bott, K.F., Hu, P.C., Lucier, T.S., Peterson, S.N., Smith, H.O., Hutchinson, C.A., and Venter, J.C. 1995. The minimal gene complement of *Mycoplasma genitalium*. Science. 270: 397-403.

Freitas-Junior, L.H., Bottius, E., Pirrit, L.A., Deitsch, K.W., Scheidig, C., Guinet, F., Nehrbass, U., Wellems, T.E., and Scherf, A. 2000. Frequent ectopic recombination of virulence factor genes in telomeric chromosome clusters of *P. falciparum*. Nature. 407: 1018-1022.

Gardner, M.J., Hall, N., Fung, E., White, O., Berriman, M., Hyman, R.W., Carlton, J.M., Pain, A., Nelson, K.E., Bowman, S., Paulsen, I.T., James, K., Eisen, J.A., Rutherford, K., Salzberg, S.L., Craig, A., Kyes, S., Chan, M.-S., Nene, V., Shallom, S.J., Suh, B., Peterson, J., Angiuoli, S., Pertea, M., Allen, J., Selengut, J., Haft, D., Mather, M.W., Vaidya, A.B., Martin, D., Fairlamb, A.H., Fraunholz, M.J., Roos, D.S., Ralph, S., McFadden, G.I., Cummings, L.M., Subramanian, G.M., Mungall, C., Venter, J.C., Carucci, D.J., Hoffman, S.L., Newbold, C., Davis, R.W., Fraser, C.M., and Barrell, B. 2002a. Genome sequence of the human malaria parasite *Plasmodium falciparum*. Nature. 419: 498-511.

Gardner, M.J., Shallom, S.J., Carlton, J.M., Salzberg, S.L., Nene, V., Shoaibi, A., Ciecko, A., Lynn, J., Rizzo, M., Weaver, B., Jarrahi, B., Brenner, M., Parvizi, B., Tallon, L., Moazzez, A., Granger, D., Fujii, C., Hansen, C., Pederson, J., Feldblyum, T., Peterson, J., Suh, B., Angiuoli, S., Pertea, M., Allen, J., Selengut, J., White, O., Cummings, L.M., Smith, H.O., Adams, M.D., Venter, J.C., Carucci, D.J., Hoffman, S.L., and Fraser, C.M. 2002b. Sequence of *Plasmodium falciparum* chromosomes 2, 10, 11 and 14. Nature. 419: 531-534.

Gardner, M.J., Tettelin, H., Carucci, D.J., Cummings, L.M., Aravind, L., Koonin, E.V., Shallom, S., Mason, T., Yu, K., Fujii, C., Pederson, J., Shen, K., Jing, J., Aston, C., Lai, Z., Schwartz, D.C., Pertea, M., Salzberg, S., Zhou, L., Sutton, G.G., Clayton, R., White, O., Smith, H.O., Fraser, C.M., Hoffman, S.L., and *et al.* 1998. Chromosome 2 sequence of the human malaria parasite *Plasmodium falciparum*. Science. 282: 1126-1132.

Gardner, M.J., Williamson, D.H., and Wilson, R.J.M. 1991. A circular DNA in malaria parasites encodes an RNA polymerase like that of prokaryotes and chloroplasts. Mol. Biochem. Parasitol. 44: 115-124.

Gouy, M., and Gautier, C. 1982. Codon usage in bacteria: correlation with gene expressivity. Nucleic Acids Res. 10: 7055-7074.

Grantham, R., Gautier, C., Gouy, M., Jacobzone, M., and Mercier, R. 1981. Codon catalog usage is a genome strategy modulated for gene expressivity. Nucleic Acids Res. 9: r43-74.

Hall, N., Pain, A., Berriman, M., Churcher, C., Harris, B., Harris, D., Mungall, K., Bowman, S., Atkin, R., Baker, S., Barron, A., Brooks, K., Buckee, C.O., Burrows, C., Cherevach, I., Chillingworth, C., Chillingworth, T., Christodoulou, Z., Clark, L., Clark, R., Corton, C., Cronin, A., Davies, R., Davis, P., Dear, P., Dearden, F., Doggett, J., Feltwell, T., Goble, A., Goodhead, I., Gwilliam, R., Hamlin, N., Hance, Z., Harper, D., Hauser, H., Hornsby, T., Holroyd, S., Horrocks, P., Humphray, S., Jagels, K., James, K.D., Johnson, D., Kerhornou, A., Knights, A., Konfortov, B., Kyes, S., Larke, N., Lawson, D., Lennard, N., Line, A., Maddison, M., McLean, J., Mooney, P., Moule, S., Murphy, L., Oliver, K., Ormond, D., Price, C., Quail, M.A., Rabbinowitsch, E., Rajandream, M.A., Rutter, S., Rutherford, K.M., Sanders, M., Simmonds, M., Seeger, K., Sharp, S., Smith, R., Squares, R., Squares, S., Stevens, K., Taylor, K., Tivey, A., Unwin, L., Whitehead, S., Woodward, J., Sulston, J.E., Craig, A., Newbold, C., and Barrell, B.G. 2002. Sequence of *Plasmodium falciparum* chromosomes 1, 3-9 and 13. Nature. 419: 527-531.

Hayward, R.E., Derisi, J.L., Alfadhli, S., Kaslow, D.C., Brown, P.O., and Rathod, P.K. 2000. Shotgun DNA microarrays and stage-specific gene expression in *Plasmodium falciparum* malaria. Mol. Microbiol. 35: 6-14.

Hoffman, S.L., Bancroft, W.H., Gottlieb, M., James, S.L., Bond, E.C., Stephenson, J.R., and Morgan, M.J. 1997. Funding for malaria genome sequencing. Nature. 387: 647.

Horrocks, P., Pinches, R., Kriek, N., and Newbold, C. 2002. Stage-specific promoter activity from stably maintained episomes in *Plasmodium falciparum*. Int. J. Parasitol. 32: 1203-1206.

Hyman, R.W., Fung, E., Conway, A., Kurdi, O., Mao, J., Miranda, M., Nakao, B., Rowley, D., Tamaki, T., Wang, F., and Davis, R.W. 2002. Sequence of *Plasmodium falciparum* chromosome 12. Nature. 419: 534-537.

Ikemura, T. 1981. Correlation between the abundance of *Escherichia coli* transfer RNAs and the occurrence of the respective codons in coli transfer RNAs and the occurrence of the respective codons in optimal for the *E. coli* translational system. J. Mol. Biol. 151: 389-409.

Ikemura, T. 1982. Correlation between the abundance of yeast transfer RNAs and the occurrence of the respective codons in protein genes: differences in synonymous codon choice patterns of yeast and *Escherichia coli* with reference to the abundance of isoaccepting transfer RNAs. J. Mol. Biol. 158: 573–597.

Ikemura, T., and Wada, K. 1991. Evident diversity of codon usage patterns of human genes with respect to chromosome banding patterns and chromosome numbers; relation between nucleotide sequence data and cytogenetic data. Nucleic Acids Res. 19: 4333-4339.

Janse, C.J., Carlton, J.M.-R., Walliker, D., and Waters, A.P. 1994. Conserved location of genes on polymorphic chromosomes of four species of malaria parasites. Mol. Biochem. Parasitol. 68: 285-296.

Jing, J., Lai, Z., Aston, C., Lin, J., Carucci, D.J., Gardner, M.J., Mishra, B., Anantharaman, T.S., Tettelin, H., Cummings, L.M., Hoffman, S.L., Venter, J.C., and Schwartz, D.C. 1999. Optical mapping of *Plasmodium falciparum* chromosome 2. Genome Res. 9: 175-181.

Jomaa, H., Wiesner, J., Sanderbrand, S., Altincicek, B., Weidemeyer, C., Hintz, M., I, T.r., Eberl, M., Zeidler, J., Lichtenthaler, H.K., Soldati, D., and Beck, E.

1999. Inhibitors of the nonmevalonate pathway of isoprenoid biosynthesis as antimalarial drugs. Science. 285: 1573-1576.

Kaneko, T., Sato, S., Kotani , H., Tanaka, A., Asamizu, E., Nakamura, Y., Miyajima, N., Hirosawa, M., Sugiura, M., Sasamoto, S., Kimura, T., Hosouchi, T., Matsuno, A., Muraki, A., Nakazaki, N., Naruo, K., Okumura, S., Shimpo, S., Takeuchi, C., Wada, T., Watanabe, A., Yamada, M., Yasuda, M., and Tabata, S. 1996. Sequence analysis of the genome of the unicellular cyanobacterium *Synechocystis* sp. Strain PCC6803. II. Sequence determination of the entire genome and assignment of potential protein-coding regions. DNA Research. 3: 109-136.

Kappe, S.H., Gardner, M.J., Brown, S.M., Ross, J., Matuschewski, K., Ribeiro, J.M., Adams, J.H., Quackenbush, J., Cho, J., Carucci, D.J., Hoffman, S.L., and Nussenzweig, V. 2001. Exploring the transcriptome of the malaria sporozoite stage. Proc. Natl. Acad. Sci. USA. 98: 9895-9900.

Katinka, M.D., Duprat, S., Cornillot, E., Metenier, G., Thomarat, F., Prensier, G., Barbe, V., Peyretaillade, E., Brottier, P., Wincker, P., Delbac, F., El Alaoui, H., Peyret, P., Saurin, W., Gouy, M., Weissenbach, J., and Vivares, C.P. 2001. Genome sequence and gene compaction of the eukaryote parasite *Encephalitozoon cuniculi*. Nature. 414: 450-453.

Kaviratne, M., Khan, S.M., Jarra, W., and Preiser, P.R. 2002. Small Variant STEVOR Antigen Is Uniquely Located within Maurer's Clefts in *Plasmodium falciparum*-Infected Red Blood Cells. Eukaryot. Cell. 1: 926-935.

Kohler, S., Delwiche, C.F., Denny, P.W., Tilney, L.G., Webster, P., Wilson, R.J.M., Palmer, J.D., and Roos, D.S. 1997. A plastid of probable green algal origin in apicomplexan parasites. Science. 275: 1485-1489.

Konfortov, B.A., Cohen, H.M , Bankier, A.T., and Dear, P.H. 2000. A high-resolution HAPPY map of *Dictyostelium discoideum* chromosome 6. Genome Res. 10: 1737-1742.

Kyes, S., Horrocks, P., and Newbold, C. 2001. Antigenic variation at the infected red cell surface in malaria. Annu. Rev. Microbiol. 55: 673-707.

Kyes, S.A., Rowe, J.A., Kriek, N., and Newbold, C.I. 1999. Rifins: a second family of clonally variant proteins expressed on the surface of red cells infected with *Plasmodium falciparum*. Proc. Natl. Acad. Sci. USA. 96: 9333-9338.

Lai, Z., Jing, J., Aston, C., Clarke, V., Apodaca, J., Dimalanta, E.T., Carucci, D.J., Gardner, M.J., Mishra, B., Anantharaman, T.S., Paxia, S., Hoffman, S.L., Craig Venter, J., Huff, E.J., and Schwartz, D.C. 1999. A shotgun optical map of the entire *Plasmodium falciparum* genome. Nat. Genet. 23: 309-313.

Lasonder, E., Ishihama, Y., Andersen, J.S., Vermunt, A.M., Pain, A., Sauerwein, R.W., Eling, W.M., Hall, N., Waters, A.P., Stunnenberg, H.G., and Mann, M. 2002. Analysis of the *Plasmodium falciparum* proteome by high-accuracy mass spectrometry. Nature. 419: 537-542.

Le Roch, K.G., Zhou, Y., Blair, P.L., Grainger, M., Moch, J.K., Haynes, J.D., De La Vega, P., Holder, A.A., Batalov, S., Carucci, D.J., and Winzeler, E.A. 2003. Discovery of gene function by expression profiling of the malaria parasite life cycle. Science. 301: 1503-1508.

Li, J., Gutell, R.R., Damberger, S.H., Wirtz, R.A., Kissinger, J.C., Rogers, M.J., Sattabongkot, J., and McCutchan, T.F. 1997. Regulation and trafficking of three distinct 18 S ribosomal RNAs during development of the malaria parasite. J. Mol. Biol. 269: 203-213.

Li, J., McConkey, G.A., Rogers, M.J., Waters, A.P., and McCutchan, T.R. 1994. *Plasmodium*: the developmentally regulated ribosome. Exp. Parasitol. 78: 437-441.

Moriyama, E.N., and Powell, J.R. 1997. Codon usage bias and tRNA abundance in Drosophila. J. Mol. Evol. 45: 514-523.

Nene, V., Bishop, R., Morzaria, S., Gardner, M.J., Sugimoto, C., ole-MoiYoi, O.K., Fraser, C.M., and Irvin, A. 2000. *Theileria parva* genomics reveals an atypical apicomplexan genome. Int. J. Parasitol. 30: 465-474.

O'Donnell, R.A., Freitas-Junior, L.H., Preiser, P.R., Williamson, D.H., Duraisingh, M., McElwain, T.F., Scherf, A., Cowman, A.F., and Crabb, B.S. 2002. A genetic screen for improved plasmid segregation reveals a role for Rep20 in the interaction of *Plasmodium falciparum* chromosomes. EMBO J. 21: 1231-1239.

Patarapotikul, J., and Langsley, G. 1988. Chromosome size polymorphism in *Plasmodium falciparum* can involve deletions of the subtelomeric pPFrep20 sequence. Nucleic Acids Res. 16: 4331-4340.

Pollack, Y., Katzen, A.L., Spira, D.T., and Golenser, J. 1982. The genome of *Plasmodium falciparum*. I: DNA base composition. Nucleic Acids Res. 10: 539-546.

Prensier, G., and Slomianny, C. 1986. The karyotype of *Plasmodium falciparum* determined by ultrastructural serial sectioning and 3D reconstruction. J. Parasitol. 72: 731-736.

Ridley, R.G. 1998. Planting new targets for antiparasitic drugs. Nat. Med. 4: 894-995.

Ridley, R.G. 1999. Planting the seeds of new antimalarial drugs. Science. 285: 1502-1503.

Rubio, J.P., Triglia, T., Kemp, D.J., de Bruin, D., Ravetch, J.V., and Cowman, A.F. 1995. A YAC contig map of *Plasmodium falciparum* chromosome 4: characterization of a DNA amplification between two recently separated isolates. Genomics. 26: 192-198.

Scherf, A., Figueiredo, L.M., and Freitas-Junior, L.H. 2001. *Plasmodium* telomeres: a pathogen's perspective. Curr. Opin. Microbiol. 4: 409-414.

Smith, J.D., Subramanian, G., Gamain, B., Baruch, D.I., and Miller, L.H. 2000. Classification of adhesive domains in the *Plasmodium falciparum* erythrocyte membrane protein 1 family. Mol. Biochem. Parasitol. 110: 293-310.

Su, X., Ferdig, M.T., Huang, Y., Huynh, C.Q., Liu, A., You, J., Wootton, J.C., and Wellems, T.E. 1999. A genetic map and recombination parameters of the human malaria parasite *Plasmodium falciparum*. Science. 286: 1351-1353.

Sullivan, B.A., Blower, M.D., and Karpen, G.H. 2001. Determining centromere identity: cyclical stories and forking paths. Nat. Rev. Genet. 2: 584-596.

Suplick, K., Akella, R., Saul, A., and Vaidya, A.B. 1988. Molecular cloning and partial sequence of a 5.8 kilobase pair repetitive DNA from *Plasmodium falciparum*. Mol. Biochem. Parasitol. 30: 289-290.

Surolia, N., RamachandraRao, S.P., and Surolia, A. 2002. Paradigm shifts in malaria parasite biochemistry and anti-malarial chemotherapy. Bioessays. 24: 192-196.

Surolia, N., and Surolia, A. 2001. Triclosan offers protection against blood stages of malaria by inhibiting enoyl-ACP reductase of *Plasmodium falciparum*. Nat. Med. 7: 167-173.

Tan, T.H., Pach, R., Crausaz, A., Ivens, A., and Schneider, A. 2002. tRNAs in *Trypanosoma brucei*: genomic organization, expression, and mitochondrial import. Mol. Cell. Biol. 22: 3707-3717.

Tarassov, I.A., and Martin, R.P. 1996. Mechanisms of tRNA import into yeast mitochondria: an overview. Biochimie. 78: 502-510.

Thompson, J.K., Caruana, S.R., and Cowman, A.F. 1999. YAC contigs and restriction maps of chromosomes 4 and 5 from the cloned line 3D7 of *Plasmodium falciparum*. Mol. Biochem. Parasitol. 102: 197-204.

Thompson, J.K., and Cowman, A.F. 1997. A YAC contig and high resolution restriction map of chromosome 3 from *Plasmodium falciparum*. Mol. Biochem. Parasitol. 90: 537-542.

Triglia, T., Wellems, T.E., and Kemp, D.J. 1992. Towards a high-resolution map of the *Plasmodium falciparum* genome. Parasitol. Today. 8: 225-229.

Vaidya, A.B., Akella, R., and Suplick, K. 1989. Sequences similar to genes for two mitochondrial proteins and portions of ribosomal RNA in tandemly arrayed 6-kilobase-pair DNA of a malaria parasite. Mol. Biochem. Parasitol. 35: 97-107.

van Lin, L.H., Pace, T., Janse, C.J., Birago, C., Ramesar, J., Picci, L., Ponzi, M., and Waters, A.P. 2001. Interspecies conservation of gene order and intron-exon structure in a genomic locus of high gene density and complexity in *Plasmodium*. Nucleic Acids Res. 29: 2059-2068.

Waters, A.P. 1994. The ribosomal RNA genes of Plasmodium. Adv. Parasitol. 34: 33-79.

Waters, A.P., Syin, C., and McCutchan, T.F. 1989. Developmental regulation of stage-specific ribosome populations in *Plasmodium*. Nature. 342: 438-440.

Wilson, R.J.M., Denny, P.W., Preiser, P.R., Rangachari, K., Roberts, K., Roy, A., Whyte, A., Strath, M., Moore, D.J., Moore, P.W., and Williamson, D.H. 1996. Complete gene map of the plastid-like DNA of the malaria parasite *Plasmodium falciparum*. J. Mol. Biol. 261: 155-172.

From: Malaria Parasites: Genomes and Molecular Biology
Edited by: A.P. Waters and C.J. Janse

# Chapter 2

# The Genome of Model Malaria Parasites, and Comparative Genomics

Jane Carlton, Joana Silva
and Neil Hall

## Abstract

The field of comparative genomics of malaria parasites has recently come of age with the completion of the whole genome sequences of the human malaria parasite *Plasmodium falciparum* and a rodent malaria model, *Plasmodium yoelii yoelii*. With several other genome sequencing projects of different model and human malaria parasite species underway, comparing genomes from multiple species has necessitated the development of improved informatics tools and analyses. Results from initial comparative analyses reveal striking conservation of gene synteny between malaria species within conserved chromosome cores, in contrast to reduced homology within subtelomeric regions, in line with previous findings on a smaller scale. Genes that elicit a host immune response are frequently found to be species-specific, although a large variant multigene family is common to many rodent malaria species and *Plasmodium vivax*. Sequence alignment of syntenic regions from multiple species has revealed the similarity between species in coding regions to be high relative to non-coding regions, and phylogenetic footprinting studies promise to reveal conserved motifs in the latter. Comparison of non-synonymous substitution rates between orthologous genes is proving a powerful technique for identifying genes under selection pressure, and may be useful for vaccine design. This is a stimulating time for comparative genomics of model and human malaria parasites, which promises to produce useful results for the development of antimalarial drugs and vaccines.

# 1. Introduction

Model malaria parasites have proven invaluable in the study of the human form of the disease, where host specificity represents a major constraint for laboratory-based experimentation. Before the development of *in vitro* cultivation of *Plasmodium falciparum,* animal models of malaria were widely used and provided researchers with a means to develop a better understanding of the biology of the parasite and its interactions with the mammalian host and vector (Waters, 2002). Their role in providing biological insight continues today, since certain aspects of malaria pathology and biology, for example invasion of hepatocytes by sporozoites (Mota *et al.*, 2001), cannot be studied without the use of an animal model system. In many instances they also provide the only source of biological material for several life-cycle stages, such as ookinetes and zygotes (Janse *et al.*, 1995). Moreover, their usefulness in functional characterization of genes through gene knock-out and modification studies is well established (de Koning-Ward *et al.*, 2000). Three groups of model systems can be identified: (1) simian malaria species that naturally parasitize non-human primates, for example the *Plasmodium knowlesi*/macaque monkey model system; (2) species of bird malaria that infect domestic fowl, for example the *Plasmodium gallinaceum*/domestic chicken model system; and (3) species of African thicket rat parasite that have been adapted for growth in laboratory rodents. The latter group, consisting of four species *Plasmodium berghei*, *Plasmodium chabaudi*, *Plasmodium vinckei* and *Plasmodium yoelii*, have been the most widely used as models for the study of *P. falciparum* malaria, primarily due to the ease of handling and maintaining rats and mice in the laboratory. In terms of evolutionary relatedness, studies involving the comparison of homologous genes from different *Plasmodium* species have shown that the genus is comprised of several deep branches. The four human malaria species *P. falciparum*, *Plasmodium vivax*, *Plasmodium malariae* and *Plasmodium ovale* form separate clades, with *P. vivax* showing distinct clustering with monkey malaria parasites such as *P. knowlesi*, and *P. falciparum* more closely associated with avian malaria species (Escalante *et al.*, 1994; Waters *et al.*, 1991). The rodent malaria species also form a distinct clade.

The notion of a universal model for the study of all human malaria species has been shown to be untenable, and instead, a view of model species selection based upon the complement of genes within the model that best fit the phenotypic trait under study, is more appropriate. With the completion of the *P. falciparum* genome sequencing project, undertaken by an international consortium of sequencing centers and malaria researchers, additional genome sequencing projects have started to generate substantial information from other model and human *Plasmodium* species, enabling the full gene complement to be identified within each species. Thus, comparative analysis of genome data from multiple malaria species is now a tangible prospect.

A current list of malaria parasite genome initiatives is given in Table 1. The genomes of two species have been sequenced and published to date, the complete finished sequence of *P. falciparum* (Gardner *et al.*, 2002), and the partial sequence of one of the four species of rodent malaria parasites, *Plasmodium yoelii yoelii* (Carlton *et al.*, 2002). Other current sequencing projects include partial shotgun coverage of the monkey malaria parasite *P. knowlesi* and two more rodent malaria parasite species *P. berghei* and *P. chabaudi chabaudi*, and the finished genome sequence of a second human malaria parasite, *Plasmodium vivax* (Carlton, 2003), with publications describing the annotation and comparative analysis of the genomes expected before the close of 2004 (definitions of genomic terms used throughout this chapter can be found in Box 1). All sequence data are being released by the sequencing centers to researchers in advance of final publication so that biological experimentation can be 'jump-started'. This has proven highly successful in the case of prior release of the *P. falciparum* genome sequence data, resulting in identification of parasite-specific pathways which may represent unique targets for intervention strategies (see for example Jomaa *et al.*, 1999), while acknowledging the perogative of the sequencing centers to publish a whole genome analysis of the final data.

Comparative analyses of genome data, or 'comparative genomics', encompasses several areas of research. Prior to the production of large-scale genome sequencing data, comparative gene mapping studies showed that relative gene location and order can be conserved over large regions of chromosomes of different species (Graves, 1998). This area of research established criteria for defining homologies between genes of different species, which are still adhered to today (Box 1). With the advent of computational biology, algorithms such as the BLAST series for pairwise sequence alignment (Altschul *et al.*, 1990) and the development of the International Nucleotide Sequence Database, comprising DDBJ, EMBL and GenBank, tools were available for comparative analysis of nucleotide and protein sequence data from different species *in silico*. With the arrival of the genomics and bioinformatics revolution, comparative genomics has scaled up and whole genome comparisons are now used to describe relative genome composition, genome organization, identify orthologous and paralogous genes, classify species-specific genes, and chart the evolution of the organisms being compared, in all three domains of life: bacterial (Fraser *et al.*, 2000), archaeal (Nelson *et al.*, 2000) and eukaryotic (Rubin *et al.*, 2000).

Comparative genomics of malaria parasite genomes is still a science in its infancy. This chapter will focus primarily on the rodent models of malaria and comparative genomic studies with the human malaria species

**Table 1.** Malaria parasite genome, transcriptome and proteome projects

| Species and strain, line or clone | Genome project goal (current status) | Sequencing Center | [a]ESTs in GenBank dbEST | [b]GSSs in GenBank dbGSS | Microarray projects | Proteomic projects |
|---|---|---|---|---|---|---|
| *P. falciparum* 3D7 clone | Finished genome (complete and published) | TIGR, Sanger Institute, Stanford University | 17,939 | 1,782 | cDNA arrays gDNA arrays oligo arrays SAGE | asexual stages food vacuole sporozoite gametocyte gamete |
| *P. falciparum* patient isolate | Partial to 3X (underway) | Sanger Institute | | | | - |
| *P. vivax* Salvador I strain | Full shotgun and closure to 8X (gap closure) | TIGR | - | 10,682 | - | - |
| *P. yoelii yoelii* 17XNL clone | Partial to 5X (complete and published) | TIGR | 15,562 | - | cDNA arrays oligo arrays | asexual stages sporozoite gametocyte ookinete |

| Species | | | | | | |
|---|---|---|---|---|---|---|
| *P. chabaudi chabaudi* AS clone | Partial to 3X (complete) | Sanger Institute | - | - | - | - |
| *P. berghei* ANKA clone | Partial to 3X (complete) | Sanger Institute | 5,329 | 5,476 | gDNA arrays | asexual stages sporozoite gametocyte ookinete oocyst |
| *P. knowlesi* H | Partial to 5X (complete) | Sanger Institute | - | - | - | - |
| *P. reichenowi* Oscar | Partial to 3X (1X) | Sanger Institute | - | - | - | - |
| *P. gallinaceum* A | Partial to 3X (underway) | Sanger Institute | - | - | - | - |

[a] ESTs: expressed sequence tags from cDNA libraries
[b] GSSs: genome survey sequences from mung bean nuclease-digested gDNA libraries
Data may be accessed through PlasmoDB at http://plasmodb.org (Bahl *et al.*, 2003)

## Box 1. Glossary of Genome Sequencing and Comparative Genomics Terms

### Genome Sequencing Terms

| | |
|---|---|
| Raw sequence: | Unassembled sequence reads produced from sequencing of inserts from individual recombinant clones of a genomic DNA library. |
| Finished sequence: | Complete sequence of a genome with no gaps and an accuracy of > 99.9%. |
| Genome coverage: | Average number of times a nucleotide is represented by a high-quality base in random raw sequence. |
| Full shotgun coverage: | Genome coverage in random raw sequence required to produce finished sequence, usually 8-10 fold ('8-10X'). |
| Partial shotgun coverage: | Typically 3-6X random coverage of a genome which produces sequence data of sufficient quality to enable gene identification but which is not sufficient to produce a finished genome sequence |
| Paired reads: | Sequence reads determined from both ends of a cloned insert in a recombinant clone. |
| Contig: | Contiguous DNA sequence produced from joining overlapping raw sequence reads. |
| Singleton: | Single sequence read that cannot be joined ('assembled') into a contig. |
| Scaffold: | A group of ordered and orientated contigs known to be physically linked to each other by paired read information. |
| EST: | Expressed sequence tag generated by sequencing one end of a recombinant clone from a cDNA library. |
| GSS: | Genome survey sequence generated by sequencing one end of a recombinant clone from a genomic DNA library. |
| SNP: | Single nucleotide polymorphism, i.e a single nucleotide position for which two or more alternative alleles are present at a certain frequency. |
| ORF: | Open reading frame, stretches of codons in the same reading frame uninterrupted by STOP codons and calculated from a six-frame translation of DNA sequence. |

### Comparative Genomics Terms

| | |
|---|---|
| Homologs: | Genes related to each other by descent from a common ancestral DNA sequence. |
| Orthologs: | Homologous genes generated by speciation, i.e related to each other by vertical descent. |
| Paralogs: | Homologous genes generated by duplication, i.e related to each other by horizontal descent. |
| Conserved synteny: | Three or more genes located on the same chromosome in different species regardless of gene order. |
| Conserved linkage: | A group of genes conserved in synteny and order between species. |

*P. falciparum*, since these are the most advanced. A brief, general background concerning the *Plasmodium* genome and a description of published studies in comparative genomics are given, but since much of this has been recently reviewed (van Lin *et al.*, 2000; Waters, 2002), a greater emphasis will be placed on more recent developments and future directions for research.

## 2. The Nuclear Genome and Gene Complement of Malaria Parasites

What does the nuclear genome of a typical malaria parasite look like? By taking data from a number of genome projects both partial and finished, it is now possible to create a generalized view (Table 2A). The haploid genome has a standard size of approximately 22-26 Mb (Carlton *et al.*, 2002; Gardner *et al.*, 2002), distributed among 14 linear chromosomes that range in size from 500 kb to over 3 Mb (Carlton *et al.*, 1999; Janse *et al.*, 1994; Kemp *et al.*, 1987). Note: Karyotype data is not available for all *Plasmodium* species, however it is unlikely that any species deviates significantly from this number. Genome composition varies from species to species, and is not host lineage-specific. For example, the (A+T) genome composition of *P. falciparum* is 81% (Gardner *et al.*, 2002) compared to 62% in *P. vivax* (Carlton, 2003). The rodent malaria species have similarly high (A+T)-rich genomes compared with *P. falciparum*, whereas *P. knowlesi* and *P. vivax* are less biased. The genomes of some species have an additional higher order structuring, in that sections of the genome are compartmentalized into discrete regions or 'isochores' of differing (A+T) content (McCutchan *et al.*, 1984). An example is the simian malaria parasite *Plasmodium cynomolgi* in which protein coding genes have been localized to (G+C)-rich isochores, whereas chromosome ends containing the telomeres appear to be located in (A+T)-rich isochores (McCutchan *et al.*, 1988). Evidence from the complete sequence of two *P. vivax* YACs, one containing a telomeric chromosome segment (del Portillo *et al.*, 2001), and the other a more central chromosome region (Tchavtchitch *et al.*, 2001), supports a similar organization of the *P. vivax* genome. In contrast, *P. falciparum* has a uniform genome composition, with the exception of short regions of >97% (A+T) on each chromosome which most likely contain the centromeres (Hall *et al.*, 2002), and the bias exhibited between coding and non-coding regions (described below). It is tempting to speculate that isochores may encode genes that mediate phenomena specific to the pathophysiology of the species that harbour them (McCutchan *et al.*, 1984), but evidence for this has yet to emerge.

Each *Plasmodium* species appears to have 5,000-6,000 predicted genes per genome (Buckee, 2002; Carlton *et al.*, 2002; Gardner *et al.*, 2002). Of these, 60% represent orthologous genes between the species, as determined by reciprocal best-match *BLAST* analysis (Buckee, 2002; Carlton *et al.*, 2002). Many of the genes unique to each species are located within

**Table 2.** *Plasmodium* genome characteristics

A. Comparison of general genome characteristics from six *Plasmodium* genome datasets.

| | *P. falciparum* | *P. knowlesi* | *P. y. yoelii* | *P. c. chabaudi* | *P. berghei* | *P. vivax* |
|---|---|---|---|---|---|---|
| Genome size (Mb) | 22.9 | 23.0 | 23.1 | ~23-24[b] | ~23-24[b] | 25.8 |
| No. chromosomes | 14 | 14 | 14 | 14 | 14 | 14 |
| % (A+T) | 80.6 | 61.9 | 77.4 | 75.6 | 76.1 | 62.4 |
| Isochore structure | Absent | Absent | Absent | Absent | Absent | Present |
| No. genes | 5,268 | 5,281 | 5,878 [a] | ND | ND | ND |
| Copy no. of largest gene families | 149 *rif* <br> 59 *var* <br> 28 *stevor* | 194 SICA*var*[a] <br> 36 hypothetical1[a] <br> 33 *kir*[a] | ~800 *yir*[a] <br> ~168 *pyst-a*[a] <br> ~57 *pyst-b*[a] | ND | ND | 600-1000 *vir* |
| Centromeres | Functionally uncharacterized | ND | ND | ND | ND | ND |
| Telomeric repeat | AACCCTA | AACCCTA | AACCCTG | AACCCT(G/A) | AACCCT(G/A) | AACCCT(G/A) |
| Complex subtelomeric repeats | Identified | Identified | Limited | Identified | Identified | Absent |

[a] Likely to be an over-estimate due to inclusion of partial genes;
[b] Determined from karyotype data; ND: not determined.

**B.** Comparison of *Plasmodium* coding and non-coding regions.

| Characteristic | *P. falciparum* | *P. y. yoelii* |
|---|---|---|
| No. predicted genes | 5,268 | 5,878 |
| Mean gene length (bp) | 2,283 | 1,298 |
| Gene density (bp per gene) | 4,338 | 2,566 |
| Percent coding | 52.6 | 50.6 |
| Genes with introns (%) | 53.9 | 54.2 |
| Mean no. exons per gene | 2.4 | 2.0 |
| Exon (G+C) content (%) | 23.7 | 24.8 |
| Exon mean length (bp) | 949 | 641 |
| Intron (G+C) content (%) | 13.5 | 21.1 |
| Intron mean length (bp) | 179 | 209 |
| Intergenic (G+C) content (%) | 13.6 | 20.7 |
| Intergenic mean length (bp) | 1,694 | 859 |

subtelomeric regions, and many are known to code for immunodominant antigens. The difference in gene number between species is due to (a) gene expansion/contraction in different lineages, for example the *pyst-a* gene family which has more than 150 members in *P. y. yoelii* but only one copy in *P. falciparum* (Table 2A); and (b) the presence of a large variant gene family in some *Plasmodium* species, predicted to be involved in antigenic variation. The family was first described in *P. vivax* [the *vir* family (del Portillo *et al.*, 2001)], and latterly in *P. yoelii* [the *yir* family (Carlton *et al.*, 2002)], *P. berghei* (the *bir* family) and *P. chabaudi* [the *cir* family (Janssen *et al.*, 2002)], and *P. knowlesi* [the *kir* family (Buckee, 2002)]. True homologs of this family so far have not been identified in *P. falciparum*, which contains other gene families involved in antigenic variation and evasion of immune responses [the *var*, *rif*, *clag* and *stevor* gene families (Craig *et al.*, 2001)]. In *P. knowlesi*, the *SICAvar* gene family has also been described (al-Khedery *et al.*, 1999) which is expressed on the surface of infected erythrocytes and is implicated in antigenic variation in this species. No significant homology exists between the *var* and *SICAvar* genes. As a cautionary note however, discrepancies in the number of predicted genes between species may also reflect the incomplete nature of partial genome data, which can exacerbate the problems associated with accurate gene prediction.

A comparison of the *P. falciparum* and *P. y. yoelii* coding and non-coding regions (Table 2B), suggests that different *Plasmodium* species exhibit similar characteristics for these regions (Carlton *et al.*, 2002). For example, coding regions of the genome have a lower (A+T) content (76%) than non-coding regions (80-87%), and a similar percentage of genes contain introns (54%). The main exception appears to be the mean length of genes, which in *P. falciparum* is almost twice the size of the gene length in *P. y. yoelii*, and also larger than the mean length of genes in the budding yeast *Saccharomyces cerevisiae* and the fission yeast *Schizosaccharomyces pombe* (Gardner *et al.*, 2002), both lower eukaryotes. The explanation for increased gene length in *P. falciparum* is at present not known.

Besides gene families involved in antigenic variation, comparative analysis of several other nuclear gene families in different *Plasmodium* species is ongoing. For example, members of the *P48/45* gene super family have been identified in *P. falciparum*, *P. berghei*, *P. vivax* and *P. yoelii* (Thompson *et al.*, 2001). This is a large conserved family of proteins expressed during the sexual stages of which there are ten members in *P. falciparum* (J. Thompson, pers. comm), and it is likely that a similar number will be found in the other species. Given the stage-specific expression and role in the development of transmission blocking vaccines of *P48/45*, rodent model orthologs of the family are proving to be immensely valuable in functional analyses of the genes, for example by gene knock-out (van Dijk *et al.*, 2001). Members of the *Py235* multi-gene family, first identified in *P. yoelii* as exhibiting a novel form of clonal antigenic variation whereby each merozoite from a parental schizont has the propensity to express a different Py235 protein (Preiser *et al.*, 2002), have been identified in *P. falciparum* and *P. vivax* (Khan *et al.*, 2001). Examination of their role in merozoite attachment and invasion of specific erythrocytes is proving to be of value for the determination of the mechanism of erythrocyte invasion in different species, such as *P. vivax*, which is restricted to growth in reticulocytes positive for the Duffy blood group antigen complex. Another group of genes involved in *Plasmodium* merozoite invasion and specific recognition of host cell receptors is the *ebl* gene family, which contains six members in *P. falciparum*: *baebl*, *eba-175*, *ebl-1*, *jesebl*, *maebl* and *pebl*, and the *P. vivax* and *P. knowlesi* Duffy-binding proteins (Adams *et al.*, 2001). The *Plasmodium ebl* genes are single copy, have a multi-exon structure encoding distinct functional domains, and conserved exon–intron splice junctions. Gene duplication has been found to be a common characteristic of the family, providing the molecular basis for the development of alternative invasion pathways. Cross-species analysis of the conserved cysteine-rich domains in members of the gene family has identified certain of the genes as having ancient origins which predate the speciation of *Plasmodium* (Michon *et al.*, 2002).

# 3. Comparative Gene Expression Studies

Table 1 lists gene and protein expression data being generated for different life-stages of various *Plasmodium* species. Large-scale sequencing projects have generated a substantial number of ESTs and full-length cDNA sequences from *P. falciparum* (Chakrabarti *et al.*, 1994; Watanabe *et al.*, 2002), *P. berghei* (Carlton *et al.*, 2001b; Matuschewski *et al.*, 2002) and *P. y. yoelii* (Kappe *et al.*, 2001), as well as several thousand mung bean nuclease GSSs from *P. vivax*, *P. falciparum* and *P. berghei* (Carlton *et al.*, 2002; Reddy *et al.*, 1993), enabling some preliminary comparative analyses of the transcriptome and proteome of malaria parasites. In one study, clustering algorithms were used to assemble the data and to create several thousand concensus sequences which were compared between *P. falciparum*, *P. berghei* and *P. vivax* (Carlton *et al.*, 2001b). This comparison of partial data identified many protein motifs and signatures as being conserved between the species. Comparison of the Gene Ontology terms [GO terms represent a vocabulary designed to describe all known genes (Ashburner *et al.*, 2000)] assigned to proteins of each species showed similar numbers of proteins in each class for each species, with the exception of the Cell Process and Defense and Immunity classes. This finding was later confirmed by whole proteome comparative analysis of *P. falciparum* and *P. y. yoelii* (Carlton *et al.*, 2002), and reflects the non-homologous Nature of the proteins involved in antigenic variation and evasion of immune responses in *Plasmodium* species. In another study, comparative analysis of genes expressed in salivary gland sporzoites versus those expressed in oocyst sporozoites identified genes that were upregulated in the former, signifying possible developmental changes in the infectious transmission stage of *Plasmodium* (Matuschewski *et al.*, 2002).

Several microarray analyses of gene expression of whole *P. falciparum* chromosomes (Le Roch *et al.*, 2002) and the complete genome at different developmental stages (Ben Mamoun *et al.*, 2001; Hayward *et al.*, 2000) have been published, as well as serial analysis of gene expression (SAGE) studies (Patankar *et al.*, 2001). The latter study purported to find a significant number of antisense messages in asexual and sexual stages, the first time this has been reported in species of *Plasmodium*. Microarrays of other rodent model species are also in progress (M. Karras and A. Waters, unpublished), with the specific aim of comparing results to the *P. falciparum* expression studies, and enabling a transcriptional profile of orthologous *Plasmodium* genes to be created. Microarays of the mosquito vector have been constructed too, and pilot studies undertaken to determine mosquito genes induced through infection with *P. berghei* (Dimopoulos *et al.*, 2002).

Two large-scale, high-throughput mass spectrometric analyses of *P. falciparum* proteins from sporozoite, merozoite, trophozoite, gametocyte and gamete stages were recently published (Florens *et al.*, 2002; Lasonder *et al.*, 2002), and a smaller analysis of the proteins in sporozoite and gametocyte stages of *P. y. yoelii* (Carlton *et al.*, 2002). These datasets provide validation of gene predictions in both species (52% of predicted *P. falciparum* genes were confirmed by proteomic data). A comparative analysis between these and other ongoing *Plasmodium* proteome projects is underway (D. Raine, L. Florens, R. Sinden and J. Yates, unpublished). Finally, data from a number of transcriptome and proteome projects, and mapping of the expression data to the genome sequence, will facilitate a thorough investigation of the phenomenon of "coordinated gene expression clustering", as shown to exist in certain eukaryotes (Blumenthal *et al.*, 2002; Caron *et al.*, 2001; Cohen *et al.*, 2000). Gene clustering can be defined in a number of different ways (Carlton, 1999), depending upon whether the genes under study are functionally related, polycistronically transcribed, expressed in the same pathway, or paralogous gene copies generated by gene duplication events. Preliminary evidence exists for some undefined level of synchronized gene expression (Florens *et al.*, 2002), but to what extent and what consequence remains to be determined.

# 4. Chromosome Structure, Comparative Mapping and Gene Synteny Studies

Several features of chromosome structure appear to be well conserved in all *Plasmodium* species. All possess telomeres consisting of degenerate, canonical, tandem repeats, the most common motif being AACCCT(A/G) (Scherf *et al.*, 2001). The mean length of the telomeric array (~800 to ~6700 bp) varies from species to species, although it remains remarkably constant within species (Figueiredo *et al.*, 2002). Subtelomeric regions of *Plasmodium* chromosomes consist of a variable number of species-specific repeats that extend 10-40 kb towards the internal part of chromosomes, and which have extensive large-scale similarity between chromosomes, indicative of intra-chromosomal exchange (Carlton *et al.*, 2002; Gardner *et al.*, 2002). Low restriction maps and high-resolution YAC contig maps, in conjunction with the *P. falciparum* and *P. y. yoelii* finished sequence data, have now established that species-specific gene families coding for immunodominant antigens and proteins known to be involved in antigenic variation are predominantly found within these regions, whereas conserved 'housekeeping' genes are located within central chromosome regions. Thus *Plasmodium* chromosomes consist of a central conserved core flanked at each end by less conserved regions containing antigen genes. This chromosomal organization has been confirmed at the genomic level by construction of a SNP map of *P. falciparum* chromosome 2 from several parasite isolates using an oligonucleotide array (Volkman *et al.*, 2002). Recently, *P. falciparum* chromosome ends were shown to cluster at the periphery of the nucleus, facilitating ectopic recombination among heterologous subtelomeric chromosome regions and thus providing a mechanism for the generation of different repertoires of antigen genes (Freitas-Junior *et al.*, 2000). Whether this represents a common mechanism shared by other *Plasmodium* species remains to be seen, but it is interesting to note the shared features of chromosome organization between species which would facilitate this.

The chromosomes of *P. falciparum* (Kemp *et al.*, 1985; van der Ploeg *et al.*, 1985), *P. vivax* (Langsley *et al.*, 1988) and rodent malaria species (Janse, 1993) are known to vary considerably in length. Such 'chromosome size polymorphisms' are found to occur in field isolates, most likely as a result of unequal recombination between homologous chromosomes of different parasite clones during meiosis, but also by gene amplification, and deletion and insertion of repeat sequences. *P. falciparum* chromosomes are also found to vary in size during *in vitro* culture, due to chromosome breakage followed by healing of the blunt end by the addition of telomeric repeats (Bottius *et al.*, 1998). Most of these large-scale chromosomal rearrangements affect non-coding repeat sequences in the subtelomeric regions, since the conserved core of the chromosome appears less prone to rearrangement. An exception are the genome rearrangements that occur in parasites under selective pressure, which have caused changes in ploidy as well as 'amplicons' containing copies of the same gene in tandem (Carlton *et al.*, 2001a). Thus, chromosomal rearrangements in *Plasmodium* are important for the evolution of the genome, although to what extent this occurs in natural populations of the parasite remains to be determined.

Given the range and diversity of karyotypes seen among different species, a surprising result of chromosome mapping experiments has been the high degree of conservation of gene synteny between *Plasmodium* species. Initial studies involving mapping of conserved genes to separations of *Plasmodium* chromosomes showed that gene location (conserved synteny) and gene order (conserved linkage) are preserved over large regions between all four species of rodent malaria (Janse *et al.*, 1994), between species of rodent malaria and *P. falciparum* (Carlton *et al.*, 1998), and between all four human malaria species (Carlton *et al.*, 1999). These studies have now been extended and show that even exon/intron boundaries and the fine-scale organization of genes can be conserved between species (Tchavtchitch *et al.*, 2001; van Lin *et al.*, 2001; Vinkenoog *et al.*, 1995). The degree of conservation of synteny is greatest when comparing genomes of more closely related species. The rodent malaria parasites, for example, show conservation of whole chromosome synteny (Janse *et al.*, 1994), whereas synteny is reduced to the level of conservation of large chromosomal blocks between *P. falciparum* and the rodent malaria species (Carlton *et al.*, 1998).

The initial comparative mapping studies of *Plasmodium* species described above involved hybridization of a limited number of conserved genes to chromosome separations, and the construction of partial genome synteny maps. With the advent of genome technology and bioinformatics, and the availability of large *Plasmodium* genome datasets, it is now possible to use computational methods for whole genome comparative analyses, as described below.

# 5. Computational Algorithms for Cross-species Comparisons

To some extent, the availability of sequence data from a number of species has outpaced the computational and experimental methods used to compare and decode the information within the data. Whole genome shotgun sequencing has progressed so much as to be a high-throughput science, however, the computational and analytical software to analyze the data coming from the pipeline has not developed in a similar fashion. Comparative genomics tools are being designed to specifically address this problem (Frazer *et al.*, 2003).

The first step in a comparison of two or more sequences from evolutionarily-related genomes is to align the sequences in order to identify conserved regions. Two types of alignment programs exist, 'local' and 'global'. Local alignment tools produce optimal similarity scores between subregions of sequences, for example in cases where sequences exhibit conservation of gene synteny but jumbled order. These algorithms find short common segments between sequences first, and then extend the match as far as possible. Examples of local alignment tools are *BLASTZ* (Schwartz *et al.*, 2000) and *MUMmer2* (Delcher *et al.*, 2002; used to generate the alignment depicted in Figure 1). Global alignment tools produce optimal similarity scores over the entire length of the sequences being compared, for example in cases where the sequences are expected to share similarity over

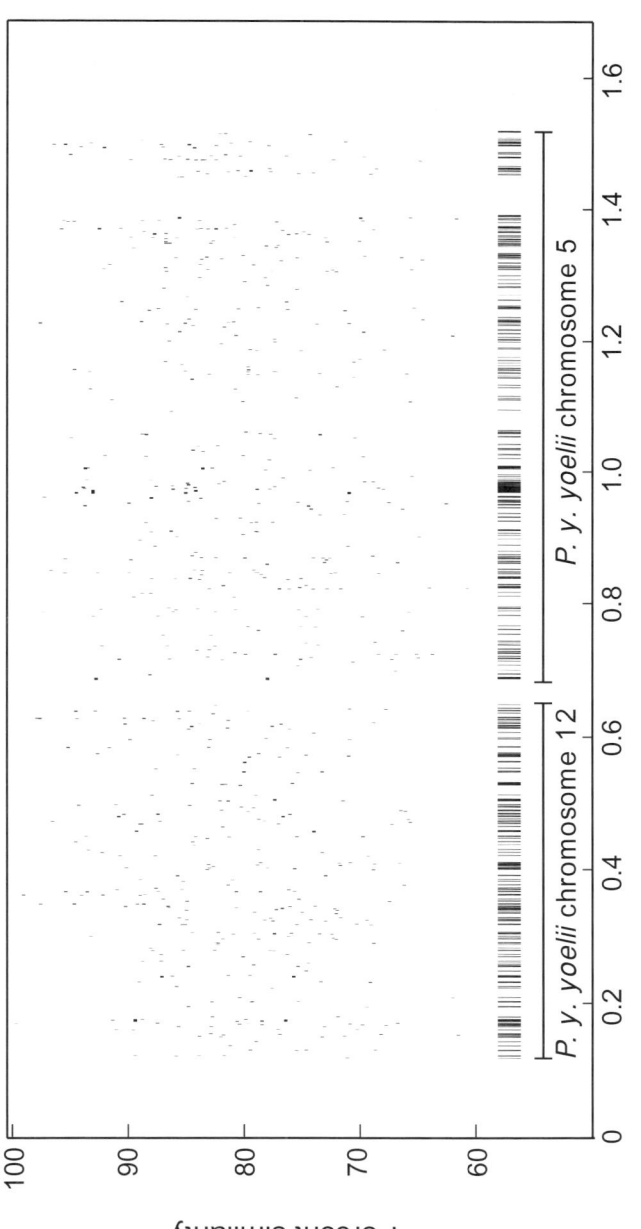

**Figure 1.** Schematic showing the tiling path of *P. y. yoelii* contigs along chromosome 10 of *P. falciparum* determined using *MUMmer*. The x-axis represents chromosome 10 (1.7 Mb), with vertical bars representing each *P. y. yoelii* contig that matches the *P.falciparum* chromosome. The percent similarity at the amino-acid level is given for each contig on the y-axis. The majority of contigs could be linked by PCR into two syntenic groups, and physical map data identified these as being located on *P. y. yoelii* chromosomes 12 and 5. Note the paucity of *P. y. yoelii* contigs with matches to the telomeric/subtelomeric ends of the *P. falciparum* chromosome, indicative of the species-specific immunodominant antigen genes located there.

their full length. These methods attempt to find an all-inclusive map between sequences, but can be memory intensive and time consuming. Examples of global alignment tools are *AVID* (Bray *et al.*, 2003), *GLASS* (Batzoglou *et al.*, 2000) and *OWEN* (Ogurtsov *et al.*, 2002); used to generate the global alignment depicted in Figure 4). Several visualization software tools are available for the production of graphical views of alignments, using either local (eg *PipMaker* (Schwartz *et al.*, 2000); *ACT* http://www.sanger.ac.uk/ Software/ACT/, used to visualize the alignment in Figure 3) or global [eg *VISTA* (Mayor *et al.*, 2000)] alignment software. Both local and global approaches to aligning sequences are informative; however, a comparison of alignment programs and servers is outside the scope of this essay. *MUMmer2*, *OWEN* and the alignment visualization tool *ACT* have all been used for comparative analysis of *Plasmodium* sequences by the authors and examples of these are given below.

# 6. Whole Genome Synteny Maps of *Plasmodium*

Using a mixture of computational algorithms and laboratory-based methods, a whole genome synteny map of the complete sequence of *P. falciparum* and the partial sequence of *P. y. yoelii* (Carlton *et al.*, 2002) has been created. *MUMmer2* was used to identify local matches of at least five amino acids long from six-frame translations of both sequences; these seed matches were extended to create a tiling path of *P. y. yoelii* contigs against the *P. falciparum* chromosomes. The contigs were linked where possible by means of 'paired reads' and PCR amplification of the intervening sequence between contigs. The syntenic groups were assigned to a *P. y. yoelii* chromosome through the use of physical map data. An example of the tiling path of *P. y. yoelii* contigs against one *P. falciparum* chromosome is shown in Figure 1. From a total of 4,787 *P. y. yoelii* genes in the tiling path, 3,525 (74%) were found to be conserved in order between the two species using *Position Effect* (Carlton *et al.*, 2002) software. This compares with 41/48 (85%) of genes found to be conserved in order in a 200 kb region syntenic between *P. falciparum* and *P. vivax* (Tchavtchitch *et al.*, 2001).

The *P. y. yoelii*/*P. falciparum* syntenic map has identified long contiguous sections of the *P. y. yoelii* genome, and by extension, of the other rodent malaria parasite genomes, and their syntenic regions in *P. falciparum*. Studies are underway to complete and extend the map, which can be seen in its current form as an 'Oxford Grid' (a conventional method for displaying synteny between two species) in Figure 2. The construction of synteny maps between other *Plasmodium* species is also ongoing, although this is limited by the nature of the genome data. For example, creation of a map using partial genome data requires that one of the genomes be finished or at least in megabase 'scaffolds' and preferably with some karyotype and chromosome mapping data available. A synteny map of two human malaria species,

## Plasmodium falciparum

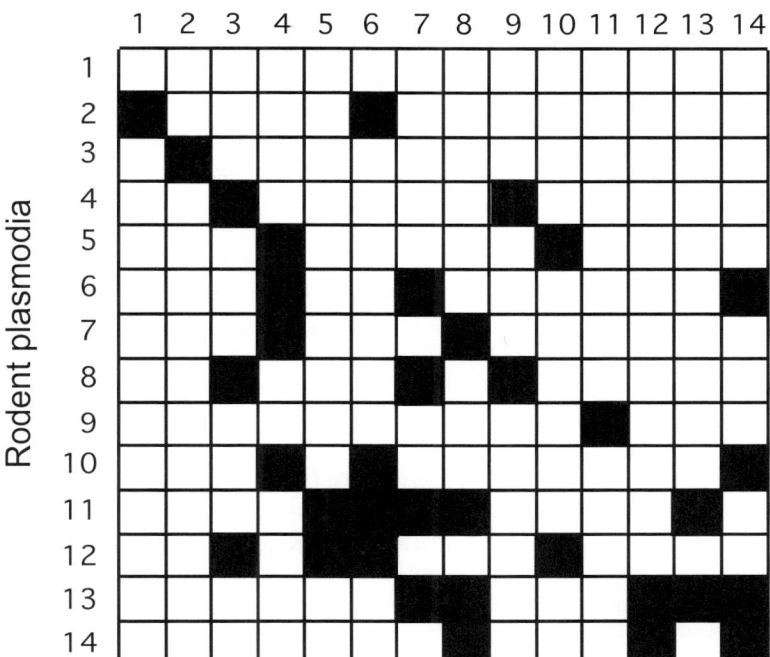

**Figure 2**. Genome-wide synteny map of *Plasmodium* plotted as an Oxford Grid. Syntenic regions conserved between *P. falciparum* and the rodent malaria species are shaded. For example, chromosome 8 in the rodent malaria parasites is syntenic to blocks of *P. falciparum* chromosomes 3, 7 and 9. The grid is incomplete as not all syntenic regions between the two species have been assigned to a rodent malaria chromosome. (Additional chromosome mapping data provided by T. Kooij and A. Waters, unpublished.)

*P. falciparum* and *P. vivax*. is planned, with preliminary tiling paths already suggesting a high degree of conservation of synteny between the two (Carlton, 2003). Synteny maps between *Plasmodium* species are particularly valuable for a number of studies: (1) as a means to chart the evolution of the genus, since syntenic break-points represent ancient evolutionary events that most likely occurred prior to speciation of the organisms being compared; (2) as a method of identifying true orthologs between species, for the comparison of molecular mechanisms underlying shared phenotypes; (3) for refinement of gene predictions through simultaneous annotation of multiple *Plasmodium* genomes; (4) for comparative analysis of gene expression, for example through identification of conserved non-coding regions of the *Plasmodium* genome ("phylogenetic footprinting"), and the evaluation of coordinated gene expression; and (5) as a means for the classification of genes under different evolutionary pressures. Examples of some of these studies are given below, many of which are works in progress due to the preliminary nature of *Plasmodium* comparative genomics.

# 7. Comparative Genomic Studies of *Plasmodium* Species

## 7.1. Molecular Evolution Studies of the *Plasmodium* Genus

Comparison of syntenic regions between *Plasmodium* species can aid in the creation of an evolutionary map of the genus. For example, DNA alterations leading to the generation of paralogous gene families, or to the loss or gain of genes in certain lineages, can be identified. Figure 3 shows an analysis of three genomes using *ACT*, a tool which reads annotated DNA sequences and *BLAST* analyses of the sequences and generates a visual map of syntenic regions. Two genes identified as coding for reticulocyte binding protein-2 (*RBP-2*) proteins are present in the *P. falciparum* genomic segment but absent in the *P. y. yoelii* and *P. knowlesi* contigs. These represent a gene family that appears to have been gained in *P. falciparum* or lost from the other species. In close proximity, the tandemly arrayed *MSP7* gene family appears to have undergone different degrees of gene expansion in the three species.

Generation of synteny maps between species can also help identify chromosomal rearrangement events that may have led to speciation. Several of the breaks in synteny between the *P. falciparum* and *P. y. yoelii* genomes were found to be located within areas containing the rRNA18S-5.8S-28S gene units, of which there are seven in *P. falciparum* (Gardner *et al.*, 2002), as well as in regions of the *P. falciparum* genome containing internal *var* and *rif* genes. Thus preliminary evidence exists for one possible mechanism underlying evolution of the *Plasmodium* genus, that of chromosome breakage and recombination at sites of rRNA genes (Carlton *et al.*, 2002).

Finally, the evolution of *P. falciparum* has been a matter of much debate (Hartl *et al.*, 2002), with one camp firmly of the view that *P. falciparum* is an ancient species, and the other that the species is of recent origin having emerged through one or several genetic bottlenecks. A genomics approach to tackling the question was undertaken recently with the creation of a SNP map of *P. falciparum* chromosome 3 from five parasite clones, which gave further credence to the view that the parasite is a genetically diverse and ancient species (Mu *et al.*, 2002). Comparative SNP studies of the syntenic region in *P. vivax* are underway and have provided evidence that *P. vivax* too has a highly diverse genome with an evolutionary history possibly parallel to that of *P. falciparum* (Feng *et al.*, 2003).

## 7.2. Comparative Studies of Molecular Mechanisms Underlying Shared Phenotypes

Identification in one species of the ortholog of a candidate gene from a second species is important for cross-species comparison of gene function, and evaluation of molecular mechanisms associated with a shared phenotype. As an example, identification of the ortholog of the *P. falciparum* chloroquine

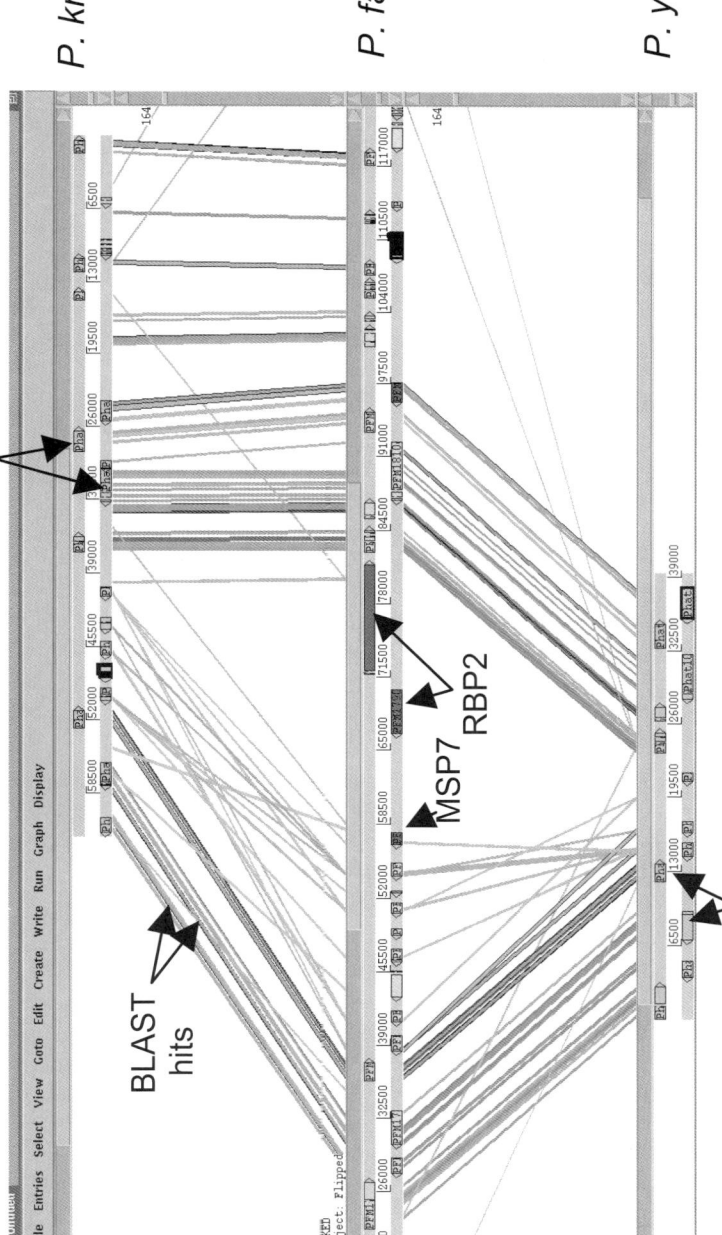

**Figure 3.** Graphical display generated by ACT of an alignment of a section of *P. falciparum* chromosome 13 compared to the syntenic regions in *P. y. yoelii* and *P. knowlesi*. Shaded, directional boxes indicate predicted genes on either DNA strand; *P. falciparum* gene predictions were manually curated. Vertical lines signify BLAST hits between the genomes. Two genes identified as coding for reticulocyte binding protein 2 (RBP2) proteins are present in the *P. falciparum* chromosome segment but absent in the *P. y. yoelii* and *P. falciparum* contigs. These represent a gene family that appears to have been inserted in *P. falciparum* or lost from the other species. The tandemly arranged *MSP7* gene family, located next to the *RBP2* genes, show various levels of gene expansion in the three species (present as three copies in *P. y. yoelii*, four copies in *P. knowlesi* and five copies in *P. falciparum*).

resistance gene, *pfcrt*, in the *P. vivax* genome enabled comparison of the molecular mechanism of resistance to chloroquine in both species (Nomura *et al.*, 2001). A 350 kb YAC containing the *P. vivax* ortholog *pvcg10* was partially sequenced, and orthologs of genes in the same order and orientation as those flanking the *pfcrt* gene in the *P. falciparum* genome were identified, distinguishing *pvcg10* gene as the true ortholog of *pfcrt*. However, mutations in the *pvcg10* gene did not correlate with chloroquine resistance in *P. vivax* isolates, demonstrating that all pleiotropic functions are not necessarily shared between orthologs. Orthologs of the *cg10* gene from *P. knowlesi* and *P. berghei* were also sequenced and used to infer the ancestral haplotype of *pfcrt*. Since chloroquine-resistant *P. falciparum* isolates contain *pfcrt* alleles that deviate significantly from this haplotype, construction of the canonical sensitive allele through analysis of the gene in other model malaria species, enabled identification of the gene as being under strong selective pressure in *P. falciparum*.

Although in this instance the molecular mechanism underlying chloroquine resistance in two human malaria species was found to be different, rodent malaria models in particular have been used widely to study drug resistance in *P. falciparum* (Carlton *et al.*, 2001a). While the mechanism of resistance in some instances has been found to be remarkably similar between the species (such as the molecular basis for pyrimethamine resistance, which in many malaria species involves a single point mutation in the drug target dihyrofolate-reductase), the fact that the molecular mechanism can vary among different species does not negate the value of investigation into the phenotype in *Plasmodium* models. Such exploration provides an additional level of insight into the biology of the organism which may be valuable in other areas of *Plasmodium* research.

### 7.3. Gene Prediction and Annotation Refinement

Comparative genomics lends itself readily to the simultaneous annotation of syntenic regions in multiple species. Both gene models and accurate exon/ intron boundaries can be difficult to predict in cases where little experimental evidence exists for verification, and where genome bias confounds the issue, as has been the case for gene prediction in *P. falciparum* (Gardner *et al.*, 2002; Hall *et al.*, 2002; Hyman *et al.*, 2002). Access to gene models from two or more species provides a way to check and improve on existing models, as shown in Figure 4. A global alignment of a 40 kb syntenic region from *P. falciparum*, *P. vivax* and *P. y. yoelii* shows that the structure and length of the gene models predicted in the three species using various gene prediction algorithms are in good agreement with each other. Four gene models in *P. falciparum* were altered to match those in *P. vivax* and *P. y. yoelii*; in all cases, the alternative *P. falciparum* model corresponded to an initial prediction made by one of the algorithms and subsequently discarded as a candidate for the final model. One gene model in *P. falciparum* (between genes 8 and 9) was excluded since

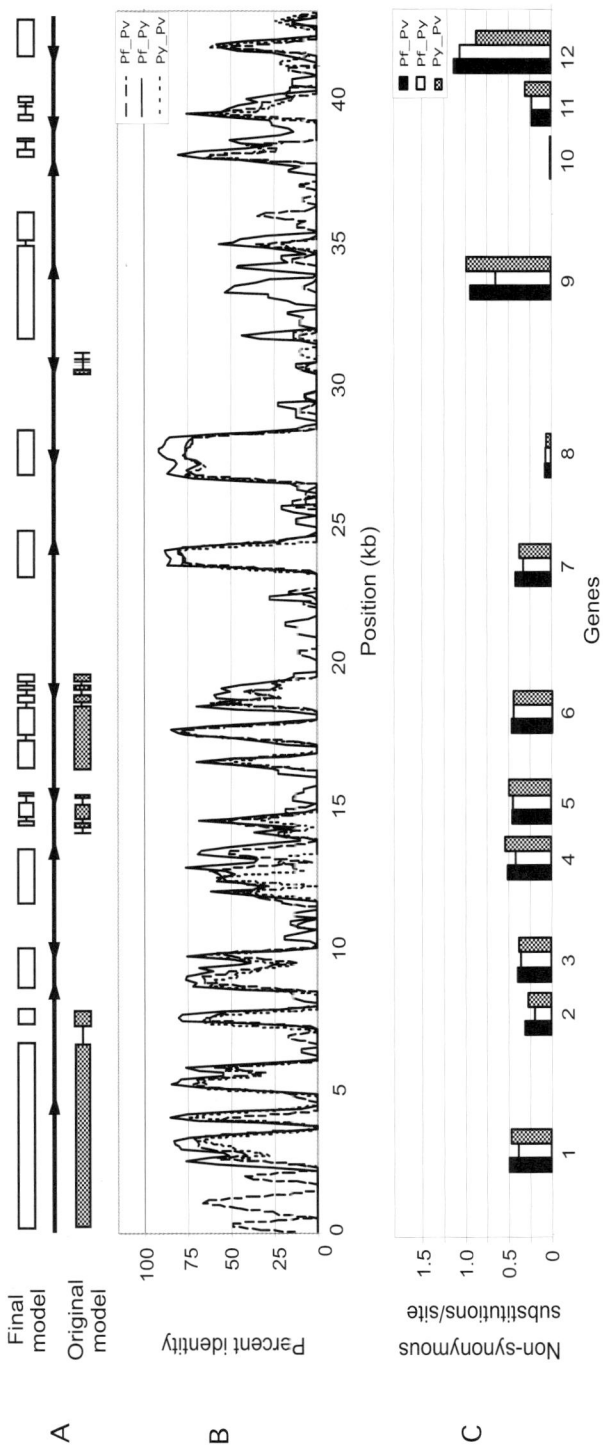

**Figure 4.** Global alignment of a 40 kb syntenic segment between three species of malaria parasite, *P. falciparum* (chromosome 3, at coordinates 178 kb to 220 kb), *P. vivax* (YAC1H14, at coordinates 95 kb to 135 kb) and *P. y. yoelii* (contigs MALPY2504, MALPY141, MALPY1025), encompassing twelve orthologous genes and 13 intergenic regions. (A) Structure of gene models used to estimate divergence are shown above the DNA strand (horizontal black line), and *P. falciparum* gene models refined through comparison with orthologous *P. vivax* and *P. y. yoelii* gene models are shown below the DNA strand. Gene orientation is represented by arrows. (B) Percent identity of the three pairwise nucleotide alignments, constructed using OWEN (Ogurtsov *et al.*, 2002) and computed using a sliding window of 250 bases with an overlap of 60%. Note the regions of conserved nucleotides in intergenic regions which may represent conserved non-coding regulatory regions. (C) Number of non-synonymous mutations per non-synonymous site plotted for all pairwise comparisons of the three species (see Carlton *et al.*, 2002 for methodology). Synonymous sites are saturated in all pairwise comparisons that include *P. vivax* (data not shown).

it was not detected in either of the other species by any of the gene prediction algorithms. Thus, annotation of multiple *Plasmodium* genomes can aid in the verification and perfection of gene models in syntenic regions.

## 7.4. Phylogenetic Footprinting

Figure 4 also shows the power of global alignments for identification of conserved intergenic motifs (phylogenetic footprints) that may be involved in gene regulation. Little is known concerning DNA elements that direct the transciption of *Plasmodium* genes (Horrocks *et al.*, 1998; van Lin *et al.*, 2000). However, promoter elements from one species can function in other species (Crabb *et al.*, 1996), which indicates a significant functional conservation of elements between different *Plasmodium* species. As outlined above, alignment at the DNA level shows coding regions to be highly conserved between *Plasmodium* species, as shown by overlapping peaks and troughs of the pairwise comparisons that coincide with exons in the gene models in Figure 4B. (An exception in the example shown is gene 9, annotated as a hypothetical gene, for which very little similarity is found at the nucleotide level between *P. vivax* and the other two species. This difference is due at least in part to a marked shift in amino acid composition in this protein, with the (A+T)-rich codons coding for amino acids isoleucine, tyrosine, asparagine and lysine making up 50% of the protein in *P. falciparum* but only 20% in *P. vivax*, which exhibits a more balanced amino acid composition.) However, the pattern of conservation within the coding regions differs markedly between genes; while some are conserved in their entirety (e.g., gene 10), others demonstrate fluctuation of conservation along the length of the gene (e.g., gene 1). In contrast, the similarity between species in intergenic regions is almost negligible, a situation mirrored in syntenic comparisons of mouse and human (Jareborg *et al.*, 1999). Since non-coding and silent positions in intergenic regions are mostly saturated (Carlton *et al.*, 2002), sequence similarity in these positions must be restricted to regions under selection. Prime phylogenetic footprint candidates are motifs conserved across all three species, some examples of which can be seen in Figure 4B. Phylogenetic footprinting has already been used successfully to detect conserved motifs in several eukaryotic lineages (Bergman *et al.*, 2001; Wasserman *et al.*, 2000; Webb *et al.*, 2002). Studies in *Plasmodium* will continue and expand to encompass alignment of genes known to be expressed at certain stages of the life-cycle (J. Silva and J. Carlton, unpublished).

## 7.5. Identification of Genes Under Selection Pressure

Multiple alignments of syntenic regions can be used in conjunction with simple molecular evolution methods to group *Plasmodium* genes according to the degree of selective pressure acting upon them. Similar methodology has been used on other organisms (Endo *et al.*, 1996), and in a few single gene studies in species of *Plasmodium* (Black *et al.*, 1999; Escalante *et al.*, 1998). With the release of large *Plasmodium* genome datasets, however, this can now be achieved on an automated whole-genome scale. As a detailed example, Figure 4C shows the number of non-synonymous substitutions (those that give rise to a change in amino acid) per non-synonymous site ($d_N$) for each of twelve orthologs in *P. vivax*, *P. falciparum* and *P. y. yoelii*. The degree of similarity in non-synonymous sites is roughly the same in the three pairwise comparisons for each gene, which suggests that these three species are approximately equidistant in evolutionary terms. However, the genes exhibit a wide spectrum of evolutionary rates, with some genes evolving under very strong 'stabilizing selection' (e.g., gene 10; $d_N = 0.01$) while others seem to be evolving under 'diversifying selection' (e.g., gene 12; $d_N > d_S > 1.0$). Differences in evolutionary rate among genes can be attributed to differences in the nature and degree of the selective constraints acting upon each gene. Comparison of $d_N$ rates with gene function for genes 10 and 12 reveals that the highly conserved gene 10 codes for the 60S ribosomal protein L44, a member of a highly conserved protein family found in widely divergent taxa such as mammals, protozoa and Archaea. In contrast, the highly divergent gene 12 codes for the circumsporozoite surface (CS) protein, a molecule found on the surface of *Plasmodium* sporozoites and known to interact directly with the host immune system. This class of gene is expected to differ greatly between species since its evolution is fast and dependent on interactions between each *Plasmodium* species and its host.

Since proteins of genes evolving under strong diversifying selection are likely to be in contact with the host immune system or to be targets of drug therapy, they represent good candidates for further study. Studies are underway to use this method to identify additional genes in this class (J. Silva and J. Carlton, unpublished). Furthermore, this analysis should identify species-specific genes that appear to be under diversifying selection in one species but not in others. In addition, extending this evolutionary analysis to encompass the whole genome will allow us to determine whether a non-synonymous divergence rate of 30% to 50% between the oldest branches of the malaria tree is indeed the norm.

## 8. The Future of *Plasmodium* Comparative Genomics

Comparative studies of model malaria parasites with the human malaria species they exemplify provide an invaluable additional level of insight into the biology of the organism and its interaction with host and vector. There is no doubt that model malaria species provide important knowledge through analogy or contrast with what is known concerning human malaria species. This interaction is set to be transfomed over the next few years as genome-wide comparisons of malaria species become possible on a scale not previously seen. Through the construction of genome-wide synteny maps, it will be possible to identify orthologs of human and model malaria parasites even in cases where sequence similarity is low in less well conserved genes, as is the case for many genes that encode surface-expressed proteins. Gene expression data from different transcriptome and proteome studies will enable the expression profile of a gene to be catalogued and compared in a variety of different species. However, further development of genetic manipulation technologies for use in *Plasmodium* will become increasingly necessary as a means to determine gene function and phenotype. High-throughput methods in particular, such as those developed for gene deletion-mutants in yeast (Giaever *et al.*, 2002) and RNAi in *Caenorhabditis elegans* (Kamath *et al.*, 2003), will be of immense value if they are transferable for use in *Plasmodium*.

# 9. References

\*     indicates a paper of special interest in the field of comparative genomics
\*\*   indicates a paper of outstanding interest in the field of comparative genomics

Adams, J.H., Blair, P.L., Kaneko, O., and Peterson, D.S. 2001. An expanding ebl family of *Plasmodium falciparum*. Trends Parasitol. 17: 297-299.

al-Khedery, B., Barnwell, J.W., and Galinski, M.R. 1999. Antigenic variation in malaria: a 3' genomic alteration associated with the expression of a *P. knowlesi* variant antigen. Mol. Cell 3: 131-141.

Altschul, S.F., Gish, W., Miller, W., Myers, E.W., and Lipman, D.J. 1990. Basic local alignment search tool. J. Mol. Biol. 215: 403-410.

Ashburner, M., Ball, C.A., Blake, J.A., Botstein, D., Butler, H., Cherry, J.M., Davis, A.P., Dolinski, K., Dwight, S.S., Eppig, J.T., *et al.*, 2000. Gene ontology: tool for the unification of biology. The Gene Ontology Consortium. Nat. Genet. 25: 25-29.

\* Bahl, A., Brunk, B., Crabtree, J., Fraunholz, M.J., Gajria, B., Grant, G.R., Ginsburg, H., Gupta, D., Kissinger, J.C., Labo, P., *et al.*, 2003. PlasmoDB: the Plasmodium genome resource. A database integrating experimental and computational data. Nucleic Acids Res. 31: 212-215.
Description of the official Plasmodium genome database which can be used for comparative analyses of deposited data.

Batzoglou, S., Pachter, L., Mesirov, J.P., Berger, B., and Lander, E.S. 2000. Human and mouse gene structure: comparative analysis and application to exon prediction. Genome Res. 10: 950-958.

Ben Mamoun, C., Gluzman, I.Y., Hott, C., MacMillan, S.K., Amarakone, A.S., Anderson, D.L., Carlton, J.M., Dame, J.B., Chakrabarti, D., Martin, R.K., *et al.*, 2001. Co-ordinated programme of gene expression during asexual intraerythrocytic development of the human malaria parasite *Plasmodium falciparum* revealed by microarray analysis. Mol. Microiol. 39: 26-36.

Bergman, C.M., and Kreitman, M. 2001. Analysis of conserved noncoding DNA in *Drosophila* reveals similar constraints in intergenic and intronic sequences. Genome Res. 11: 1335-1345.

Black, C.G., Wang, L., Hibbs, A.R., Werner, E., and Coppel, R.L. 1999. Identification of the *Plasmodium chabaudi* homologue of merozoite surface proteins 4 and 5 of *Plasmodium falciparum*. Infect. Immun. 67: 2075-2081.

Blumenthal, T., Evans, D., Link, C.D., Guffanti, A., Lawson, D., Thierry-Mieg, J., Thierry-Mieg, D., Chiu, W.L., Duke, K., Kiraly, M., and Kim, S.K. 2002. A global analysis of *Caenorhabditis elegans* operons. Nature. 417: 851-854.

Bottius, E., Bakhsis, N., and Scherf, A. 1998. *Plasmodium falciparum* telomerase: *de novo* telomere addition to telomeric and nontelomeric sequences and role in chromosome healing. Mol. Cell Biol. 18: 919-925.

Bray, N., Dubchak, I., and Pachter, L. 2003. AVID: A Global Alignment Program. Genome Res. 13: 97-102.

Buckee, C. 2002. *Plasmodium knowlesi* versus *Plasmodium falciparum*: A Comparative Analysis, MSc Thesis, University of York, York.

Carlton, J. 2003. The *Plasmodium vivax* genome sequencing project. Trends Parasitol. 19: 227-231.

**Carlton, J.M., Angiuoli, S.V., Suh, B.B., Kooij, T.W., Pertea, M., Silva, J.C., Ermolaeva, M.D., Allen, J.E., Selengut, J.D., Koo, H.L., *et al.*, 2002. Genome sequence and comparative analysis of the model rodent malaria parasite *Plasmodium yoelii yoelii*. Nature. 419: 512-519.
The first published whole genome comparison between a human malaria species and a rodent malaria parasite used as a model. The paper describes relative genome composition, genome organization, identifies orthologous and paralogous genes, and describes a syntenic map between the two species.

* Carlton, J.M., Hayton, K., Cravo, P.V., and Walliker, D. 2001a. Of mice and malaria mutants: unravelling the genetics of drug resistance using rodent malaria models. Trends Parasitol. 17: 236-242.
Review outlining comparative aspects of drug resistance genotypes and phenotypes in rodent models of malaria and *P. falciparum*.

* Carlton, J.M., Muller, R., Yowell, C.A., Fluegge, M.R., Sturrock, K.A., Pritt, J.R., Vargas-Serrato, E., Galinski, M.R., Barnwell, J.W., Mulder, N., *et al.* 2001b. Profiling the malaria genome: a gene survey of three species of malaria parasite with comparison to other apicomplexan species. Mol. Biochem. Parasitol. 118: 201-210.
An initial endeavor at a comparative analysis of partial expression data in three species of *Plasmodium*.

Carlton, J.M.R. 1999. Gene synteny across *Plasmodium* spp: could 'operon-like' structures exist? Parasito. Today 15: 178-179.

* Carlton, J.M.R., Galinski, M.R., Barnwell, J.W., and Dame, J.B. 1999. Karyotype and synteny among the chromosomes of all four species of human malaria parasite. Mol. Biochem. Parasitol. 101: 23-32.
The first description of the degree of conservation of gene synteny between the genomes of all four human malaria species.

* Carlton, J.M.R., Vinkenoog, R., Waters, A.P., and Walliker, D. 1998. Gene synteny in species of *Plasmodium*. Mol. Biochm. Parasitol. 93: 285-294.
A first attempt at describing the degree of conservation of gene synteny between *P. falciparum* and the rodent models of malaria, which revealed conservation of large chromosome blocks between the species.

Caron, H., van Schaik, B., van der Mee, M., Baas, F., Riggins, G., van Sluis, P., Hermus, M.C., van Asperen, R., Boon, K., Voute, P.A., *et al.* 2001. The human transcriptome map: clustering of highly expressed genes in chromosomal domains. Science. 291: 1289-1292.

Chakrabarti, D., Reddy, G.R., Dame, J.B., Almira, E.C., Laipis, P.J., Ferl, R.J., Yang, T.P., Rowe, T.C., and Schuster, S.M. 1994. Analysis of expressed sequence tags from *Plasmodium falciparum*. Mol. Biochem. Parasitol. 66: 97-104.

Cohen, B.A., Mitra, R.D., Hughes, J.D., and Church, G.M. 2000. A computational analysis of whole-genome expression data reveals chromosomal domains of gene expression. Nat. Genet. 26: 183-186.

Crabb, B.S., and Cowman, A.F. 1996. Characterization of promoters and stable transfection by homologous and nonhomologous recombination in *Plasmodium falciparum*. Proc. Natl. Acad. Sci. USA. 93: 7289-7294.

Craig, A., and Scherf, A. 2001. Molecules on the surface of the *Plasmodium falciparum* infected erythrocyte and their role in malaria pathogenesis and immune evasion. Mol. Biochem. Parasitol. 115: 129-143.

de Koning-Ward, T.F., Janse, C.J., and Waters, A.P. 2000. The development of genetic tools for dissecting the biology of malaria parasites. Annu. Rev. Microbiol. 54: 157-185.

* del Portillo, H.A., Fernandez-Becerra, C., Bowman, S., Oliver, K., Preuss, M., Sanchez, C.P., Schneider, N.K., Villalobos, J.M., Rajandream, M.A., Harris, D., *et al.*, 2001. A superfamily of variant genes encoded in the subtelomeric region of *Plasmodium vivax*. Nature. 410: 839-842.

  The first description of a 150 kb subtelomeric region of the *P. vivax* genome, which identified the *vir* super-family of genes, homologs of which have subsequently been found in many *Plasmodium* species.

Delcher, A.L., Phillippy, A., Carlton, J., and Salzberg, S.L. 2002. Fast algorithms for large-scale genome alignment and comparison. Nucleic Acids Res. 30: 2478-2483.

Dimopoulos, G., Christophides, G.K., Meister, S., Schultz, J., White, K.P., Barillas-Mury, C., and Kafatos, F.C. 2002. Genome expression analysis of *Anopheles gambiae*: responses to injury, bacterial challenge, and malaria infection. Proc. Natl. Acad. Sci. USA. 99: 8814-8819.

Endo, T., Ikeo, K., and Gojobori, T. 1996. Large-scale search for genes on which positive selection may operate. Mol. Biol. Evol 13: 685-690.

Escalante, A.A., and Ayala, F.J. 1994. Phylogeny of the malarial genus *Plasmodium*, derived from rRNA gene sequences. Proc. Natl. Acad. Sci. USA. 91: 11373-11377.

Escalante, A.A., Lal, A.A., and Ayala, F.J. 1998. Genetic polymorphism and natural selection in the malaria parasite *Plasmodium falciparum*. Genetics 149: 189-202.

Figueiredo, L.M., Freitas-Junior, L.H., Bottius, E., Olivo-Marin, J.C., and Scherf, A. 2002. A central role for *Plasmodium falciparum* subtelomeric regions in spatial positioning and telomere length regulation. EMBO J. 21: 815-824.

Feng, X., Carlton, J.M., Joy, D.A., Mu, J., Furuya, T., Suh, B., Wang, Y., Barnwell, J.W., and Su, X, 2003. Single nucleotide polymorphisms and genome diversity in *Plasmodium vivax*. Proc. Natl. Acad. Sci. USA. 100: 8502-8507.

* Florens, L., Washburn , M.P., Raine, D.J., Anthony, R.M., Grainger, M., Yhaynes, J.D., Moch, J.K., Muster, N., Sacci, J.B., Tabb, D.L., *et al.*, 2002. A proteomic view of the *Plasmodium falciparum* life cycle. Nature. 419:520-526.

  One of two mass spectrometric analyses of the *P. falciparum* proteome, which used protein extracts of the sporozoite, merozoite, trophozoite, and gametocyte stages of the parasite. Data from this study also suggest that some gene clustering of co-expressed genes occurs in the *P. falciparum* genome, although these remain uncharacterized.

Fraser, C.M., Eisen, J., Fleischmann, R.D., Ketchum, K.A., and Peterson, S. 2000. Comparative genomics and understanding of microbial biology. Emerg. Infect. Dis. 6: 505-512.

* Frazer, K.A., Elnitski, L., Church, D.M., Dubchak, I., and Hardison, R.C. 2003. Cross-species sequence comparisons: a review of methods and available resources. Genome Res. 13: 1-12.

Interesting review that outlines the strategy for choosing DNA sequences from different species for comparative analyses, and describes publicly available resources and methods available for these studies.

Freitas-Junior, L.H., Bottius, E., Pirrit, L.A., Deitsch, K.W., Scheidig, C., Guinet, F., Nehrbass, U., Wellems, T.E., and Scherf, A. 2000. Frequent ectopic recombination of virulence factor genes in telomeric chromosome clusters of *P. falciparum*. Nature. 407: 1018-1022.

** Gardner, M.J., Hall, N., Fung, E., White, O., Berriman, M., Hyman, R.W., Carlton, J.M., Pain, A., Nelson, K.E., Bowman, S., *et al.*, 2002. Genome sequence of the human malaria parasite *Plasmodium falciparum*. Nature. 419: 498-511.

Description of the complete genome sequence of *P. falciparum* including analysis of chromosome structure, metabolism, proteins involved in transport, immune evasion and the apicoplast.

* Gardner, M.J., Shallom, S.J., Carlton, J.M., Salzberg, S.L., Nene, V., Shoaibi, A., Ciecko, A., Lynn, J., Rizzo, M., Weaver, B., et al., 2002. Sequence of *Plasmodium falciparum* chromosomes 2, 10, 11 and 14. Nature. 419: 531-534.

In depth details concerning the strategy and methodology used for sequencing four of the 14 *P. falciparum* chromosomes.

Giaever, G., Chu, A.M., Ni, L., Connelly, C., Riles, L., Veronneau, S., Dow, S., Lucau-Danila, A., Anderson, K., Andre, B., *et al.*, 2002. Functional profiling of the *Saccharomyces cerevisiae* genome. Nature. 418: 387-391.

* Graves, J.A. 1998. Background and Overview of Comparative Genomics. Ilar J. 39: 48-65.

Excellent review concerning primarily comparative mamalian genomics but also alluding to vertebrate evolution and phylogeny.

* Hall, N., Pain, A., Berriman, M., Churcher, C., Harris, B., Harris, D., Mungall, K., Bowman, S., Atkin, R., Baker, S., *et al.*, 2002. Sequence of *Plasmodium falciparum* chromosomes 1, 3-9 and 13. Nature. 419: 527-531.

Description of sequence methodologies and analysis of eight of the 14 chromosomes of *P. falciparum*. Centromere structures have been identified on these and all other chromosomes, consisting of a region of very high (A+T) content and a core region of slightly higher (G+C) content, within non-coding regions of 8-11 kb.

Hartl, D.L., Volkman, S.K., Nielsen, K.M., Barry, A.E., Day, K.P., Wirth, D.F., and Winzeler, E.A. 2002. The paradoxical population genetics of *Plasmodium falciparum*. Trends Parasitol. 18: 266-272.

Hayward, R.E., Derisi, J.L., Alfadhli, S., Kaslow, D.C., Brown, P.O., and Rathod, P.K. 2000. Shotgun DNA microarrays and stage-specific gene expression in *Plasmodium falciparum* malaria. Mol. Microiol. 35: 6-14.

* Hyman, R.W., Fung, E., Conway, A., Kurdi, O., Mao, J., Miranda, M., Nakao, B., Rowley, D., Tamaki, T., Wang, F., and Davis, R.W. 2002. Sequence of *Plasmodium falciparum* chromosome 12. Nature. 419: 534-537.

Description of the methodology behind sequencing the third largest *P. falciparum* chromosome.

Horrocks, P., Dechering, K., and Lanzer, M. 1998. Control of gene expression in *Plasmodium falciparum*. Mol. Biochem. Parasitol. 95: 171-181.

Janse, C.J. 1993. Chromosome size polymorphism and DNA rearrangements in *Plasmodium*. Parasitol. Today. 9: 19-22.

* Janse, C.J., Carlton, J.M.R., Walliker, D., and Waters, A.P. 1994. Conserved location of genes on polymorphic chromosomes of four species of malaria parasites. Mol. Biochem. Parasitol. 68: 285-296.

The first description of the degree of synteny between the genomes of all four rodent malaria species. From a total of 50 genes only three were found not to be conserved in chromosome location, an example of conservation of whole chromosome synteny.

Janse, C.J., and Waters, A.P. 1995. *Plasmodium berghei*: the application of cultivation and purification techniques to molecular studies of malaria parasites. Parasitol. Today 11: 138-143.

Janssen, C.S., Barrett, M.P., Turner, C.M., and Phillips, R.S. 2002. A large gene family for putative variant antigens shared by human and rodent malaria parasites. Proc. R. Soc. Lond. B. Biol. *Sci.* 269: 431-436.

Jareborg, N., Birney, E., and Durbin, R. 1999. Comparative analysis of noncoding regions of 77 orthologous mouse and human gene pairs. Genome Res. 9: 815-824.

Jomaa, H., Wiesner, J., Sanderbrand, S., Altincicek, B., Weidemeyer, C., Hintz, M., Turbachova, I., Eberl, M., Zeidler, J., Lichtenthaler, H.K., et al., 1999. Inhibitors of the nonmevalonate pathway of isoprenoid biosynthesis as antimalarial drugs. Science. 285: 1573-1576.

Kamath, R.S., Fraser, A.G., Dong, Y., Poulin, G., Durbin, R., Gotta, M., Kanapin, A., Le Bot, N., Moreno, S., Sohrmann, M., et al., 2003. Systematic functional analysis of the *Caenorhabditis elegans* genome using RNAi. Nature. 421: 231-237.

Kappe, S.H., Gardner, M.J., Brown, S.M., Ross, J., Matuschewski, K., Ribeiro, J.M., Adams, J.H., Quackenbush, J., Cho, J., Carucci, D.J., et al., 2001. Exploring the transcriptome of the malaria sporozoite stage. Proc. Natl. Acad. Sci. USA. 98: 9895-9900.

Kemp, D.J., Corcoran, L.M., Coppel, R.L., Stahl, H.D., Bianco, A.E., Brown, G.V., and Anders, R.F. 1985. Size variation in chromosomes from independent cultured isolates of *Plasmodium falcipaum*. Nature. 315: 347-350.

Kemp, D.J., Thompson, D.J., Walliker, D., and Corcoran, L.M. 1987. Molecular karyotype of *Plasmodium falciparum*: conserved linkage groups and expendable histidine-rich proteins. Proc. Natl. Acad. Sci. USA. 84: 7672-7676.

Khan, S.M., Jarra, W., and Preiser, P.R. 2001. The 235 kDa rhoptry protein of *Plasmodium (yoelii) yoelii*: function at the junction. Mol. Biochem. Parasitol. 117: 1-10.

Langsley, G., Patarapotikul, J., Handunnetti, S., Khoury, E., Mendis, K.N., and David, P.H. 1988. *Plasmodium vivax*: karyotype polymorphism of field isolates. Exp. Parasitol. 67: 301-306.

* Lasonder, E., Ishihama, Y., Andersen, J.S., Vermunt, A.M., Pain, A., Sauerwein, R.W., Eling, W.M., Hall, N., Waters, A.P., Stunnenberg, H.G., and Mann, M. 2002. Analysis of the *Plasmodium falciparum* proteome by high-accuracy mass spectrometry. Nature. 419: 537-542.

One of two mass spectrometric analyses of the *P. falciparum* proteome, this analysis used protein extracts of the asexual, gametocyte and gamete stages of the parasite.

* Le Roch, K.G., Zhou, Y., Batalov, S., and Winzeler, E.A. 2002. Monitoring the chromosome 2 intraerythrocytic transcriptome of *Plasmodium falciparum* using oligonucleotide arrays. Am J. Trop Med Hyg 67: 233-243.

First chromosome-wide analysis of gene expression in *P. falciparum* during the asexual stages.

* Matuschewski, K., Ross, J., Brown, S.M., Kaiser, K., Nussenzweig, V., and Kappe, S.H. 2002. Infectivity-asscciated changes in the transcriptional repertoire of the malaria parasite sporozoite stage. J. Biol. Chem. 277: 41948-41953.

Interesting article that uses expression data obtained from one rodent species to scan genome data from another and identify potential genes that are upregulated during development of the sporozoite.

Mayor, C., Brudno, M., Schwartz, J.R., Poliakov, A., Rubin, E.M., Frazer, K.A., Pachter, L.S., and Dubchax, I. 2000. VISTA : visualizing global DNA sequence alignments of arbitrary length. Bioinformatics. 16: 1046-1047.

McCutchan, T.F., Dame, J.B., Gwadz, R.W., and Vernick, K.D. 1988. The genome of *Plasmodium cynomolgi* is partitioned into separable domains which appear to differ in sequence stability. Nucleic Acids Res. 16: 4499-4510.

McCutchan, T.F., Dame, J.B., Miller, L.H., and Barnwell, J. 1984. Evolutionary relatedness of *Plasmodium* species as determined by the structure of DNA. Science. 225: 808-811.

Michon, P., Stevens, J.R., Kaneko, O., and Adams, J.H. 2002. Evolutionary relationships of conserved cysteine-rich motifs in adhesive molecules of malaria parasites. Mol. Biol. Evol 19: 1128-1142.

Mota, M.M., Pradel, G., Vanderberg, J.P., Hafalla, J.C., Frevert, U., Nussenzweig, R.S., Nussenzweig, V., and Rodriguez, A. 2001. Migration of *Plasmodium* sporozoites through cells before infection. Science. 291: 141-144.

Mu, J., Duan, J., Makova, K.D., Joy, D.A., Huynh, C.Q., Branch, O.H., Li, W.H., and Su, X.Z. 2002. Chromosome-wide SNPs reveal an ancient origin for *Plasmodium falciparum*. Nature. 418: 323-326.

Nelson, K.E., Paulsen, I.T., Heidelberg, J.F., and Fraser, C.M. 2000. Status of genome projects for nonpathogenic bacteria and archaea. Nat. Biotechnol 18: 1049-1054.

* Nomura, T., Carlton, J.M.R., Baird, J.K., Del Portillo, H.A., Fryauff, D.J., Rathore, D., Fidock, D.A., Su, X., Collins, W.E., McCutchan, T.F., *et al.*, 2001. Evidence for different mechanisms of chloroquine resistance in two *Plasmodium* species that cause human malaria. J. Infect. Dis. 183: 1653-1661.

Elegant example of comparative genomics in action, where identification of the genes and mutations associated with a phenotype in one species can be used as a probe for the molecular investigation of a similar phenotype in other species.

Ogurtsov, A.Y., Roytberg, M.A , Shabalina, S.A., and Kondrashov, A.S. 2002. OWEN: aligning long collinear regions of genomes. Bioinformatics. 18: 1703-1704.

Patankar, S., Munasinghe, A., Shoaibi, A., Cummings, L.M., and Wirth, D.F. 2001. Serial analysis of gene expression in *Plasmodium falciparum* reveals the global expression profile of erythocytic stages and the presence of anti-sense transcripts in the malarial parasite. Mol. Biol. Cell 12: 3114-3125.

Preiser, P.R., Khan, S., Costa, F.T., Jarra, W., Belnoue, E., Ogun, S., Holder, A.A., Voza, T., Landau, I., Snounou, G., and Renia, L. 2002. Stage-specific transcription of distinct repertoires of a multigene family during *Plasmodium* life cycle. Science. 295: 342-345.

Reddy, G.R., Chakrabarti, D., Schuster, S.M., Ferl, R.J., Almira, E.C., and Dame, J.B. 1993. Gene sequence tags from *Plasmodium falciparum* genomic DNA fragments prepared by the "genease" activity of mung bean nuclease. Proc. Natl. Acad. Sci. USA. 90: 9867-9871.

Rubin, G.M., Yandell, M.D., Wortman, J.R., Gabor Miklos, G.L., Nelson, C.R., Hariharan, I.K., Fortini, M.E., Li, P.W., Apweiler, R., Fleischmann, W., *et al.*, 2000. Comparative genomics of the eukaryotes. Science. 287: 2204-2215.

Scherf, A., Figueiredo, L.M., and Freitas-Junior, L.H. 2001. *Plasmodium* telomeres: a pathogen's perspective. Curr. Opin. Microbiol. 4: 409-414.

Schwartz, S., Zhang, Z., Frazer, K.A., Smit, A., Riemer, C., Bouck, J., Gibbs, R., Hardison, R., and Miller, W. 2000. PipMaker--a web server for aligning two genomic DNA sequences. Genome Res. 10: 577-586.

** Tchavtchitch, M., Fischer, K., Huestis, R., and Saul, A. 2001. The sequence of a 200 kb portion of a *Plasmodium vivax* chromosome reveals a high degree of conservation with *Plasmodium falciparum* chromosome 3. Mol. Biochem. Parasitol. 118: 211-222.

The first description of a 200 kb region of the *P. vivax* genome syntenic to a section of *P. falciparum* chromosome 3 containing the gene coding for the CS protein. Of the 48 genes identified in the region, 41 were conserved in order, structure and orientation between the two species.

Thompson, J., Janse, C.J., and Waters, A.P. 2001. Comparative genomics in Plasmodium: a tool for the identification of genes and functional analysis. Mol. Biochem. Parasitol. 118: 147-154.

van der Ploeg, L.H.T., Smits, M., Ponnudurai, T., Vermulen, A., Meuwissen, J.H.E.T., and Langsley, G. 1985. Chromosome-sized DNA molecules of *Plasmodium falciparum*. Science. 229: 658-661.

van Dijk, M.R., Janse, C.J., Thompson, J., Waters, A.P., Braks, J.A., Dodemont, H.J., Stunnenberg, H.G., van Gemert, G., Sauerwein, R.W., and Eling, W. 2001. A central role for P48/45 in malaria parasite male gamete fertility. Cell. 104: 153-164.

* van Lin, L.H., Janse, C.J., and Waters, A.P. 2000. The conserved genome organisation of non-falciparum malaria species: the need to know more. Int. J. Parasitol. 30: 357-370.

In-depth review of the genome organization of non-falciparum malaria parasites.

* van Lin, L.H., Pace, T., Janse, C.J., Birago, C., Ramesar, J., Picci, L., Ponzi, M., and Waters, A.P. 2001. Interspecies conservation of gene order and intron-exon structure in a genomic locus of high gene density and complexity in *Plasmodium*. Nucleic Acids Res. 29: 2059-2068.

One of the initial publications outlining the usefulness of comparative studies for genome annotation of multiple malaria species.

Vinkenoog, R., Veldhuisen, B., Speranca, M.A., del Portillo, H.A., Janse, C., and Waters, A.P. 1995. Comparison of introns in a *cdc2*-homologous gene within a number of *Plasmodium* species. Mol. Biochem. Parasitol. 71: 233-241.

Volkman, S.K., Hartl, D.L., Wirth, D.F., Nielsen, K.M., Choi, M., Batalov, S., Zhou, Y., Plouffe, D., Le Roch, K.G., Abagyan, R., and Winzeler, E.A. 2002. Excess polymorphisms in genes for membrane proteins in *Plasmodium falciparum*. Science. 298: 216-218.

Wasserman, W.W., Palumbo, M., Thompson, W., Fickett, J.W., and Lawrence, C.E. 2000. Human-mouse genome comparisons to locate regulatory sites. Nat. Genet. 26: 225-228.

Wantanabe, J., Sasaki, M., Suzuki, Y., and Sugano, S. 2002. Analysis of transcriptomes of human malaria parasite *Plasmodium falciparum* using full-length enriched library: identification of novel genes and diverse transcription start sites of messenger RNAs. Gene 291: 105-113.

* Waters, A.P. 2002. Orthology between the genomes of *Plasmodium falciparum* and rodent malaria parasites: possible practical applications. Philos. Trans. R. Soc. Lond. B. Biol. Sci 357: 55-63.

  Excellent review of the advantages of studying rodent malaria species in parallel with studies on *P. falciparum*.

Waters, A.P., Higgins, D.G., and McCutchan, T.F. 1991. *Plasmodium falciparum* appears to have arisen as a result of lateral transfer between avian and human hosts. Proc. Natl. Acad. Sci. USA. 88: 3140-3144.

Webb, C.T., Shabalina, S.A., Ogurtsov, A.Y., and Kondrashov, A.S. 2002. Analysis of similarity within 142 pairs of orthologous intergenic regions of *Caenorhabditis elegans* and *Caenorhabditis briggsae*. Nucleic Acids Res. 30: 1233-1239.

From: Malaria Parasites: Genomes and Molecular Biology
Edited by: A.P. Waters and C.J. Janse

# Chapter 3

## Getting the Most Out of Bioinformatics Resources

Jessica C. Kissinger
and David S. Roos

## Abstract

The recent publication of a complete reference sequence for the *Plasmodium falciparum* genome is a momentous event for malaria researchers. In addition, genomic and functional genomics data is now available for six further *Plasmodium* species and eight non-*Plasmodium* species of apicomplexan parasites. These datasets can greatly expedite the identification of candidate targets for drug, vaccine and diagnostic development, in addition to enhancing our basic understanding of malaria parasites. But how can researchers most effectively access and exploit genomic-scale data, integrating this information with the results from other experiments? Bioinformatics research is fundamentally no different from 'wet lab' experiments conducted at the bench, requiring an understanding of the starting reagents (databases), the strengths and weaknesses of experimental (computational) methods, and a critical analysis of the results obtained. This chapter discusses the nature and organization of data resources, strategies for data mining, and the interpretation of computational results.

# 1. Introduction

Now is an exciting time to be engaged in malaria research. Significant technological advances, research effort, and financial investment have produced a complete reference genome for *Plasmodium falciparum* (Gardner *et al.*, 2002b), effectively complete genome sequences for *P. yoelii, knowlesi, vivax,* and several other apicomplexan parasite species (see below), and complete genome sequences for both the human (and mouse) host and the *Anopheles gambiae* vector (Holt *et al.*, 2002). Advances in genomics and bioinformatics affect all malaria researchers, from the molecular biologist interested in rifin gene organization, to the developmental biologist studying stage-specific gene expression, to the cell biologist investigating protein trafficking to Maurer's clefts, to the evolutionary biologist exploring the origins of cytoadherence ligands, to the immunologist seeking a target for vaccine development, to the population geneticist studying allelic variation for evidence of positive or negative selection. How can we as parasitologists access, manage, and utilize the surfeit of emerging data, integrating 'dry-lab' computational research with 'wet-lab' studies at the laboratory bench?

The search for – and functional analysis of – genes is increasingly moving towards high-throughput studies of the whole genome in parallel, generating huge data sets related to transcript expression and transcriptional regulation, protein translation and steady-state levels, protein-protein interactions, polymorphic diversity, etc. All of these data need to be stored, analyzed, and made widely accessible. Many useful datasets and analysis tools are now accessible on-line, offering great potential for expediting malaria research and stimulating new lines of investigation, but these resources are scattered throughout the internet. One purpose of this chapter is to provide a compendium of on-line resources for the malaria researcher.

Of course, the development of new kinds of data also brings the need for new tools and skills to navigate the continually changing information landscape, whether to find the 5′ end of your favorite gene, determine the hypothetical function for cDNAs on a microarray, or identify a potential target for drug, vaccine, or diagnostic development. Indeed, it is increasingly possible – and at times even essential – to conduct malaria research *in silico*. While this does not obviate the need for wet-lab experimentation, computational approaches often provide a useful complement, and can be much faster than conventional benchwork; the effective integration of wet-lab and computational approaches can produce dramatic research advances. A second purpose of this chapter is to provide a bioinformatics primer for malaria researchers.

Bioinformatics for the average bench scientist and expert alike has evolved rapidly over the last decade, and need not be a "black box". The inner workings of the many common tools are well described in various textbooks (Baxevanis and Ouellette, 2001; Gibson and Muse, 2001; Mount, 2001; Baxevanis *et al.*, 2003) and will not be discussed here in any detail. The field has also accumulated experience with the efficient application of bioinformatics tools and approaches. Bioinformatics techniques, like laboratory techniques, can generate misleading results, and it is therefore critical to understand the strengths and weaknesses of the methods employed. A properly designed bioinformatics experiment includes controls and safeguards designed to detect potential artifacts.

Other contributions to this book describe laboratory approaches and techniques now being applied to *Plasmodium*, many of which employ and/or generate genomic-scale data. In this chapter, we take a 'behind the scenes' look at how genomic data are generated, stored and analyzed, in hopes that this will enable researchers to design experiments that use these data effectively. We begin with a brief introduction to the types of data that one can expect to find, including some relevant information on how these data are processed. The following section describes currently available *Plasmodium* resources, with notes on data organization. We then discuss the power of database queries, illustrated by a series of queries intended to elucidate candidate vaccine antigens (with suggestions for effective data-mining and notes on common pitfalls). Finally, we provide a few notes on newly emerging data types, and opportunities for the future.

## 2. The Datasets

Genomic-scale datasets are currently available for several *Plasmodium* species, including *P. berghei, chabaudi, falciparum, knowlesi, reichenowi, vivax, yoelii*. Table 1 indicates which data types are publicly available for each species as of October 2003. A large amount of data is also available for other related apicomplexan species, including *Babesia bovis, Theileria annulata, Theileria parva, Toxoplasma gondii, Eimeria tenella, Neospora caninum, Sarcocystis neurona* and *Cryptosporidium parvum* (Table 2), and for the human and mouse hosts and the insect vector *Anopheles gambiae* (Table 3).

Data types available for *Plasmodium falciparum* include information on:

Nucleotide Sequences
- Genomic sequence (del Portillo *et al.*, 2001; Tchavtchitch *et al.*, 2001; Carlton *et al.*, 2002; Gardner *et al.*, 2002a) and Genome Survey Sequences (GSS) (Carlton and Dame, 2000; Janssen *et al.*, 2001).
- Expressed Sequence Tags (EST) (Watanabe *et al.*, 2001; Li *et al.*, 2003).

RNA Expression
- ESTs and Serial Analysis of Gene Expression (SAGE) tags (Munasinghe *et al.*, 2001).

**Table 1.** Data sources for *Plasmodium* and other Apicomplexan parasite species

| Species | Data type(s) | URL |
|---|---|---|
| All *Plasmodium* species | multiple | http: //PlasmoDB.org |
| *P. berghei* | genomic | http: //www.sanger.ac.uk/Projects/P_berghei/<br>http: //parasite.vetmed.ufl.edu/<br>http: //www.tigr.org/tdb/tgi/pbgi/<br>http: //www.GeneDB.org |
| *P. chabaudi* | genomic | http: //www.sanger.ac.uk/Projects/P_chabaudi/<br>http: //www.GeneDB.org |
| *P. falciparum* | genomic | http: //sequence-www.stanford.edu/group/malaria/<br>http: //www.tigr.org/tdb/e2k1/pfa1/<br>http: //www.sanger.ac.uk/Projects/P_falciparum/<br>http: //www.GeneDB.org |
|  | EST | http: //fullmal.ims.u-tokyo.ac.jp/<br>http: //parasite.vetmed.ufl.edu/falc.htm<br>http: //www.cbil.upenn.edu/paradbs-servlet/index.html |
|  | GSS | http: //parasite.vetmed.ufl.edu/ |
|  | μsatellite map | http: //www.ncbi.nih.gov/projects/Malaria/Mapsmarkers/PfGMap/pfgmap.html |
|  | optical map | http: //www.lmcg.wisc.edu/research/research.html#plasmodium |
|  | oligonucl. array | http: //malaria.ucsf.edu/ |
|  | Affymetrix array | http: //www.scripps.edu/cb/winzeler/malariatext.html |
| *P. knowlesi* | genomic | http: //www.sanger.ac.uk/Projects/P_knowlesi/ |
| *P. reichenowi* | genomic | http: //www.sanger.ac.uk/Projects/P_reichenowi/ |
| *P. vivax* | genomic | http: //www.tigr.org/tdb/e2k1/pva1/intro.shtml |
|  | GSS | http: //parasite.vetmed.ufl.edu/viva.htm |
|  | YAC | http: //www.sanger.ac.uk/Projects/P_vivax/ |
| *P. yoelii* | genomic | http: //www.tigr.org/tdb/e2k1/pya1/ |
|  | EST | http: //www.tigr.org/tdb/tgi/pygi/ |
| *Babesia bovis* | EST | http: //www.sanger.ac.uk/Projects/B_bovis/ |
| *Cryptosporidium parvum* | multiple | http: //CryptoDB.org |
|  | genomic | http: //www.cbc.umn.edu/ResearchProjects/AGAC/Cp/index.htm<br>http: //www.parvum.mic.vcu.edu/ |
|  | GSS | http: //medsfgh.ucsf.edu/id/CpDemoProj/ |
|  | EST | http: //www.ncbi.nlm.nih.gov |
| *Eimeria tenella* | genomic | http: //www.sanger.ac.uk/Projects/E_tenella/ |
|  | EST | http: //www.cbil.upenn.edu/paradbs-servlet/index.html<br>http: //www.genome.wustl.edu/est/index.php?eimeria=1 |
| *Neospora caninum* | EST | http: //www.cbil.upenn.edu/paradbs-servlet/index.html<br>http: //www.genome.wustl.edu/est/index.php?neospora=1 |
| *Sarcocystis neurona* | EST | http: //www.cbil.upenn.edu/paradbs-servlet/index.html<br>http: //www.genome.wustl.edu/est/index.php?sarcocystis=1 |
| *Theileria annulata* | genomic | http: //www.sanger.ac.uk/Projects/T_annulata/ |
| *Theileria parva* | genomic | http: //www.tigr.org/tdb/e2k1/tpa1/ |
| *Toxoplasma gondii* | multiple | http: //ToxoDB.org |
|  | genomic | http: //www.tigr.org/tdb/t_gondii/ |
|  | EST | http: //www.cbil.upenn.edu/paradbs-servlet/index.html<br>http: //www.genome.wustl.edu/est/index.php?toxoplasma=1 |
|  | BAC-end | http: //www.sanger.ac.uk/Projects/T_gondii/ |

**Table 2.** Bioinformatics resources for *Plasmodium* and other Apicomplexan parasite species

| Source | Database | Tools & Features | URL |
|---|---|---|---|
| PlasmoDB | relational & flat file | query, BLAST, pattern finding, browse | http: //PlasmoDB.org |
| GeneDB | relational & flat file | query, BLAST, links to literature & domain databases, browse | http: //www.GeneDB.org |
| ToxoDB | flat file | BLAST, pattern finding | http: //ToxoDB.org |
| CryptoDB | flat file | BLAST, pattern finding, browse | http: //CryptoDB.org |
| Sanger | flat file | BLAST, software | http: //www.sanger.ac.uk |
| Stanford | flat file | BLAST, browse | http: //sequence-www.stanford.edu/group/malaria/ |
| TIGR | flat file | BLAST | http: //www.tigr.org/ |
| EMBL-EBI | flat file | query, BLAST, pattern finding, classifications | http: //www.ebi.ac.uk/parasites/parasite-genome.html http: //www.ebi.ac.uk/parasites/PlasGN/Proteome/ proteome.html |
| TIGR Gene Indices | flat file & relational | query, BLAST | http: //www.tigr.org/tdb/tgi/protist.shtml |
| NCBI | flat file | BLAST, physical maps, browse | http: //www.ncbi.nlm.nih.gov/projects/Malaria/ |
| Genome Atlases | flat file | browse | http: //www.cbs.dtu.dk/services/GenomeAtlas/ |
| Metabolic pathways | flat file | browse | http: //sites.huji.ac.il/malaria/ |
| Protein structure | relational | query, browse, view 3-D protein models | http: //bioinfo.icgeb.res.in/codes/model.html |
| SANBI | flat file | BLAST | http: //www.sanbi.ac.za/malaria-genesearch/ |
| WEHI | relational & flat file | browse | http: //www.wehi.edu.au/MalDB-www/who.html |
| MR4 | relational & flat file | query, browse | http: //www.malaria.mr4.org/mr4pages/index.html |
| PlasmoCyc | relational | enzymatic reactions, structures, list of drug targets (w/refs) | http: //plasmocyc.stanford.edu http: //www.smi.stanford.edu/projects/helix/malaria.html |
| UCSF | relational | query, browse transcriptome | http: //malaria.ucsf.edu |

**Table 3.** Selected vector and host resources

| Resource or Species | Contents | URL |
|---|---|---|
| AGRIP | database of *Anopheles* resources including mutants, methods, culture information | http: //konops.imbb.forth.gr/AnoDB/Mirror/Mbenedic/ |
| AnoBase (formerly AnoDB) | genome database and reference center | http: //skonops.imbb.forth.gr/AnoBase/ |
| Ensembl (Metazoan Genome Database) | genome database containing human, mouse and *Anopheles* data (among others) | http: //www.ensembl.org/ |
| Mouse - MGI | comprehensive mouse genome and resource database (Jackson Laboratories) | http: //www.informatics.jax.org/ |
| Mouse | multiple resources including genome, SNP, clone, gene expression,HomoloGene, LocusLink | http: //www.ncbi.nlm.nih.gov/genome/guide/mouse/ |
| Mouse | genetic maps, physical maps, genome | http: //www.mgc.har.mrc.ac.uk/ |
| Human | multiple resources including genome, unigene EST assemblies, physical map viewer, locus link, OMIM, SNP databases | http: //www.ncbi.nlm.nih.gov/genome/guide/human/ |
| Human | general genome project information, Gene Gateway | http: //www.ornl.gov/TechResources/Human_Genome/home.html |
| Human | UK human genome mapping project resource center, with broad collection of analysis tools | http: //www.hgmp.mrc.ac.uk/ |
| Human (Golden Path Database) | queries, BLAT, annotation for human and other genomes | http: //genome.ucsc.edu/ |

- Microarrays based on DNA or cDNA clones (Hayward *et al.*, 2000; Mamoun *et al.*, 2001), or oligonucleotides in either glass slide (Bozdech *et al.*, 2003a; 2003b) or photolithographic (Affymetrix) format (Le Roch *et al.*, 2002; 2003).

Protein Expression
- Tandem mass-spectrometry (Florens *et al.*, 2002; Lasonder *et al.*, 2002).

Genetic Organization and Population Structure
- Optical maps (Lai *et al.*, 1999).
- Microsatellites (Su *et al.*, 1999).
- Single-Nucleotide Polymorphisms (SNPs) (Mu *et al.*, 2002).

While it is beyond the scope of this chapter to provide a detailed description of each data type, it is worth a short digression to introduce some of the relevant technology and terms. Familiarity with the processes involved in data generation facilitates recognition of potential artifacts and sources of error, minimizing the chance of error propagation *in silico* – a risk that is inherent in computational bioinformatics.


## 2.1. Genome Sequence and Assembly

Two strategies are commonly employed for genome sequencing: a hierarchical approach in which the genome is broken down into smaller mapped fragments for sequencing, and a shotgun approach in which the whole-genome is subjected to random sequencing and assembly *en masse*. The former may consume considerable resources in mapping and/or fractionation of pure genomic fragments, while the latter poses greater computational problems in assembling the resulting sequence data.

The *P. falciparum* genome was sequenced using a hierarchical approach, in which chromosomes were separated in pulse field gels (and some chromosomes were further sub-cloned into YACs) prior to the production of random clone libraries for sequencing (Gardner *et al.*, 1998; Bowman *et al.*, 1999; Gardner *et al.*, 2002a; 2002b; Hall *et al.*, 2002; Hyman *et al.*, 2002). The *P. yoelii* genome was sequenced using a whole-genome shotgun approach (Carlton *et al.*, 2002).

In either approach, sequences generated by random sequencing (of either the entire genome, or individual chromosomes or smaller fragments) must be reassembled into larger pieces of contiguous DNA, or "contigs", and several software packages are available for contiguating sequence reads. Contigs are then organized into larger "scaffolds" containing gaps between the individual contigs, based on information from mapping data, end sequences from large-insert clones, etc. Overall. genome assembly is a tricky business; common problems and sequence artifacts include:

- Repetitive regions of the genome can cause mis-assembly errors, i.e. sequences that are not adjacent to one another in the genome can become artificially merged if they contain identical stretches of sequence (repeats). While the *P. falciparum* genome is relatively small (by eukaryotic standards), and 'satellite' DNA and other repetitive sequences are not particularly abundant, the extremely high A+T nucleotide content raises similar problems: unique sequences are hard to find in a two letter alphabet!

- When hierarchical sequencing approaches are used, DNA sequenced from one fraction (e.g. one chromosome) may be contaminated with DNA from another fraction (chromosome). Thus, genes may initially appear to be located on the wrong chromosome, or on multiple different chromosomes. Such sequences usually do not contiguate with the majority of the sequences and appear as "singlets" or orphan sequences until sequences for the entire genome are pooled and compared for final assembly. Early stages of assembly of the *P. falciparum* genome contained many overlapping fragments of HRP2, a known single-copy gene. Once the entire genome is assembled, most the remaining orphan sequences are likely to be attributable to contaminating DNA, mis-assembled sequence reads, and other artifacts ... but these sequences will undoubtedly include some valid sequences as well, including RNA- and protein-coding genes.

- Shotgun sequencing provides an extremely cost-efficient means to identify most sequences, but the laws of probability and combinatorics ensure that some sequences will be missed. Moreover, because genomic sequences differ in their clonability, not all will be represented in the library (a problem that is particularly acute for A+T-rich genomes), shotgun sequencing rarely achieved the theoretical level of sequence coverage. Thus, while the 5X random shotgun sequence available for *P. yoelii* means that 5 genome equivalents of DNA have been sequenced (>100 Mb), the assembled sequence still contains many gaps (>5000). Closing such gaps is a laborious and expensive process, and as of this date the *P. falciparum* genome still contains a few physical gaps and a few "unmapped" regions of sequence that need to be correctly placed in the genome.

- If the sequence reassembles into multiple pieces, how should these pieces be ordered and oriented? In the case of the *P. falciparum* genome, two sets of physical anchors or genome landmarks were available to help order the fragments along the chromosomes into scaffolds: a microsatellite map (Su *et al.*, 1999) and an optical map (Lai *et al.*, 1999). *P. falciparum* chromosomes 6-8 (affectionately known as the "BLOB") could not be resolved on a pulse-field gel, and posed a particular challenge. Additional "Happy maps" were therefore constructed to facilitate the ordering and assembly of these chromosomes (Hall *et al.*, 2002).

- Contaminating sequences from cloning vectors (plasmids, transposons), cloning hosts (*E. coli*, yeast), human DNA and other organisms being sequenced may enter into raw sequence output. Users of sequence data should investigate any suspected cases of horizontal gene transfer very carefully, especially at the DNA nucleotide level and via genomic Southern blots, to guard against such sequencing artifacts.

Each of the above difficulties can be resolved, but users examining pre-publication data are cautioned to be aware of potential sequence artifacts. *P. falciparum* sequences have been cleaned of most artifacts, but sequences for other species still have problems. For example, the *P. reichenowi* sequence is known to be heavily contaminated with monkey DNA.

## 2.2. Maps

Genetic and physical maps consist of markers at specific genetic or physical locations within the genome. Classically, these have been constructed based on cytogenetic banding patterns, or by using genetic crosses to map loci responsible for various phenotypes. Unfortunately, because the nuclear envelope does not break down during mitosis in *Plasmodium* (as in most protozoa), it is not possible to isolate condensed chromosomes for the analysis of banding patterns. Genetic mapping studies based on classical genetic crosses are feasible, however, and 1 cM in *P. falciparum* has been measured as ~17 kb (Su *et al.*, 1999). (Centimorgans provide a standard measure of recombination frequency; 1 cM represents an average distance of 10 Mb in humans). Due to the difficulty of conducting classical genetic crosses and mapping of phenotypes in *Plasmodium,* however, alternative approaches have also been developed. Both microsatellite and optical maps of the *Plasmodium* genome have facilitated ordering of the hundreds of genomic contigs onto chromosome scaffolds used for genome closure.

Optical maps rely on novel imaging technology to construct a chromosome-scale restriction enzyme map. Large fragments of chromosomal DNA are attached to a solid surface under gentle fluid flow. After adhesion, a restriction enzyme is added to cleave the DNA *in situ,* leaving an ordered line of fragments, whose size can be assessed microscopically based on labeling with an intercalating DNA dye (providing a quantitative measure proportional to DNA content). Optical maps have been created for *Plasmodium falciparum* strain 3D7 using two different restriction enzymes (Lai *et al.*, 1999).

Microsatellite maps are based on the use of PCR primer pairs that amplify regions differing in length between the two parental genomes, converting a sequence (length) polymorphism into a genetic marker (Su *et al.*, 1999). By examining the lengths of the PCR products in the progeny of a genetic cross, it can be determined which regions of the genome came from which parent. Restriction polymorphisms can also be employed for genetic mapping studies, although these have proved more cumbersome in *Plasmodium,* in part because of the high A+T content of the genome. Careful association of microsatellite patterns with phenotypic or genetic markers permits the construction of an integrated genetic and physical map, linking individual microsatellites to particular regions of specific chromosomes. In addition to their utility for genetic mapping, including the analysis of Quantitative Trail Loci (QTL) for specific phenotypes (Ferdig and Su, 2000; Wootton *et al.*, 2002), microsatellite markers are also extremely useful tools for population surveys.

## 2.3. Expressed Sequence Tags

Expressed Sequence Tags (ESTs) are sequences obtained from reverse-transcribed mRNAs (cDNAs). As such, they can be used to determine gene structure (exons and introns), and identify open reading frames that are likely to encode protein sequences. By focusing on transcribed sequences and minimizing the problems associated with splice-site prediction, EST projects are extremely cost-efficient, delivering a large number of protein predictions for relatively small cost. Moreover, the abundance of EST sequences obtained for individual genes provides a crude indication of transcript abundance in the original library. EST sequences are currently available from several *Plasmodium* cDNA libraries, representing various life cycle stages and species. While the collection and analysis of such data has been discussed elsewhere (Ajioka *et al.*, 1998; Li *et al.*, 2003), several issues likely to impact on bioinformatics experiments are worth considering here.

- Because EST sequences are derived from specific libraries, they represent only the individual strains and life-cycle stages from which these libraries were generated, and the transcripts produced by those parasites. Thus, while random sequencing of genomic DNA can in theory approach complete representation of the parasite genome, no EST project is likely to provide a complete catalog of all genes. Representation is sometimes enhanced, however, by using normalized libraries in which highly abundant sequences have been suppressed (using a variety of strategies).

- EST abundance may be able to provide a crude estimate of relative transcript abundance, but such estimates are likely to be biased by library amplification, and completely invalid in normalized libraries (depending on the method employed). In general, hybridization with RNA, RT-PCR, SAGE, and microarray analysis (see below) provide more suitable methods for transcript profiling.

- EST sequences are often incomplete. To facilitate gene discovery, most EST projects use libraries of directionally-cloned cDNAs (although note that up to 30% of inserts may be cloned in the inverted orientation), and produce only a single sequence read from the presumed 5' end of the cDNA. Because many cDNA clones do not represent full-length mRNAs, however, the start of the sequence may not provide the transcript initiation site, especially for long mRNAs. A single EST sequencing reaction typically yields ~350 nt, and is therefore very unlikely to provide the complete cDNA sequence. Repeated sampling of the library often yields multiple overlapping ESTs derived from the same gene, and clustering of these may yield a longer consensus sequence (see below), but most of these sequences still remain incomplete.

- EST sequences are likely to contain a higher error rate than genomic sequences, for a variety of reasons. For example, because EST projects focus on gene discovery rather than high-fidelity genome assembly, individual cDNAs are generally sequenced only once, although transcripts derived from the same gene may be sequenced multiple times in a given library, as noted above. Clustering of ESTs can help to extend sequence length and reduce error, but inclusion of data from multiple isolates may yield an inaccurate consensus whenever allelic polymorphisms are present. (Indeed, correlation of multiple sequence alignments with strain information provides an excellent source of microsatellite and SNP markers for genetic analysis.) It is therefore important to distinguish between individual ESTs and consensus sequences, and to recognize that even sequences derived from the consensus of many ESTs may be incorrect.

- Differentially-spliced genes – while far less common than in metazoan species – are nevertheless well known in *Plasmodium,* and it is important to recognize when differentially-spliced transcripts derive from the same gene, as opposed to paralogous genes or strain-specific allelic variants. Note, however, that incomplete intron excision is quite common, producing many cDNAs that are unlikely to be fully translated.

- As with genomic sequences, EST libraries may contain a low frequency of contaminating sequences. The lack of redundant sequencing makes it difficult to distinguish rare transcripts from contaminating DNA, however; putative transcripts should always be validated by comparison with genome sequence (when available) and hybridization with genomic DNA. cDNA libraries may also be contaminated with incompletely processed sequences, and with genomic DNA, yielding apparent transcripts that are unlikely to be translated and may not even be transcribed. This is particularly problematic for *Plasmodium,* where the high A+T content may lead to false priming by oligo-dT. In addition, an early Genome Survey Sequence (GSS) project using mung-bean nuclease libraries to provide a 'genes-first' approach to sequencing of *P. falciparum* DNA produced sequences that were initially mislabeled as ESTs.

- Because EST sequencing is often a continuing project, the identifiers associated with assembled sequences for an individual gene may change frequently, producing considerable confusion. In practical terms, it is often most convenient to find the new consensus sequence (and name) via a BLAST query with the old sequence. It is also critical to note the data release date and database version on which any analysis is based.

When ESTs are analyzed in bioinformatics experiments, it is often necessary to understand how the data were generated in order to interpret the results correctly. What species, strain, stage? Is the library directional?

amplified? normalized? How were contaminating sequences removed? How were individual ESTs assembled? How many sequences are represented in an individual cDNA assembly, how well are they aligned, and how deep is the alignment as a function of position?

## 2.4. SAGE Tags

SAGE (Serial Analysis of Gene Expression) was developed to provide a "snapshot" of mRNA abundance at a given time in a given cell or tissue type (Velculescu *et al.*, 1995). Rather than conducting a full-scale EST project, sequencing cDNAs in their entirety, each EST is reduced to a short oligonucleotide sequence tag, normally near the 3' end of the gene. These tags are then ligated into large concatemers, so that an individual sequencing reaction can identify tags derived from dozens of individual cDNAs, rather than the single cDNA represented by an individual EST sequence. Computational analysis is used to determine which gene/mRNA matches which specific SAGE tag. Once again, the high A+T content of the *Plasmodium* genome poses a problem. For example, the GC-rich 10mer SAGE tag GGTTCAGGGT is predicted to occur 0.59 times by chance in the *P. falciparum* genome (based on the observed 80% frequency of A+T), while the AT-rich tag ATCATATAAG is predicted to occur 150 times by chance alone. Thus, the mapping of SAGE tags and other short oligonucleotides to the *P. falciparum* genome may be a "one-to-one" or a "one to many" relationship, but when SAGE tags are combined with other data (gene predictions, EST sequences, BLAST similarities, etc) it is often possible to determine the true sites of expression (Munasinghe *et al.*, 2001; Pleasance *et al.*, 2003) .

## 2.5. Transcript Expression Profiling

Several types of microarrays may be employed to examine the expression of many individual genes in parallel. All of these methods involve immobilized gene-specific nucleic acids: genomic DNA clones, cDNA clones (ESTs), or synthetic oligonucleotides. Plasmid clones can easily be isolated from genomic libraries, and libraries constructed using mung-bean nuclease (which cleaves preferentially in extremely AT-rich DNA) may favor clones containing individual genes or gene exons. cDNA clones can be isolated from plasmid libraries as well, and offer the advantage of being unequivocally derived from individual mRNAs (subject to the quality of the library), although they are unlikely to represent the genome as a whole – with highly-expressed genes represented many times, and other genes not represented at all. All clone-based reagent sets are problematic from the standpoint of quality control: reliably propagating, quantitating, and tracking thousands of individual plasmids is a daunting task. As a result, most array projects have now moved to oligonucleotide-based microarrays, provided that effectively complete genome sequence is available – as is indeed the case for *P. falciparum* (and many related species).

Two alternative formats for oligonucleotide-based microarrays are in common use. In the first, oligonucleotides are synthesized in 96- or 384-well format, and robotically spotted onto glass slides. This format offers several advantages, including the ability to design custom microarrays tailored to individual experimental needs, and the ability to design new probes to take advantage of improved genome annotation and new experimental approaches. Disadvantages include the cost of oligonucleotide synthesis (or purchase), the potential for error in reagent generation/storage/tracking, and difficulties in maintaining the spotting robots for reproducible array production (although many large research centers now support microarray facilities).

Alternatively, oligonucleotides can be synthesized directly on the microarray, using proprietary photolithographic methodology. This format offers the ability to print features at higher densities (typically 500,000 features/chip, vs ~15,000 for glass slide arrays), with greater reproducibility. Disadvantages include the proprietary nature of the technology involved, the cost and inflexibility of photolithographic array design, and the high cost of arrays (obtainable only through the Affymetrix Corporation), and the need for sufficiently large orders to justify printing. Facilities for reading microarrays in both formats are generally available at most large research centers.

For *P. falciparum,* glass slide microarrays have been produced containing cDNA sequences (Mamoun *et al.*, 2001), spotted mungbean nuclease fragments (Hayward *et al.*, 2000), and 70mer oligonucleotides representing the vast majority of predicted genes in the genome (Bozdech *et al.*, 2003a; 2003b). Oligonucleotide probe sets for *P. falciparum* are now available commercially (www.qiagen.com/arrays/oligosets_malaria.php). An Affymetrix chip containing shorter oligonucleotides for nearly every predicted exon in the *P. falciparum* genome, as well and non-coding and opposite strand regions (at a lower frequency) has also been designed (Le Roch *et al.*, 2002; 2003). Many of the resulting expression data sets have been deposited in the *Plasmodium* Genome Database (http://PlasmoDB.org), and the sequences used to create these arrays or oligos have been mapped to the *P. falciparum* genome. For example, the expression profile for the major merozoite surface protein (MSP1) of *P. falciparum* is shown in Figure 1.

An in depth discussion of all bioinformatics aspects of microarrays is beyond the scope of this chapter, but as with all studies producing genomics-scale datasets, it is critical that experiments be well-controlled, reproducible, and understood by any user hoping to make sense of this data, particularly as data from different types of experiments are often stored and accessed in different ways. For example, in comparing two whole-genome expression

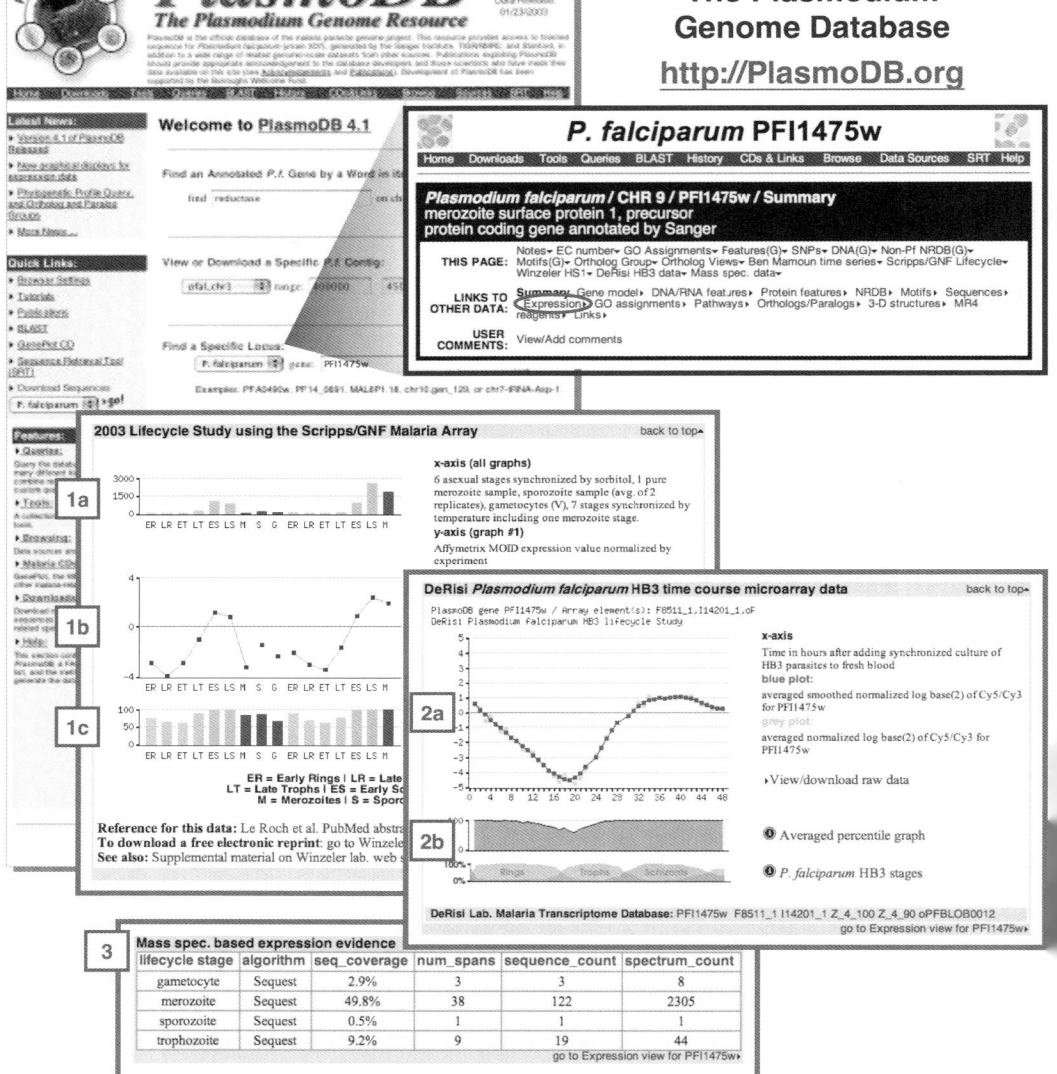

**Figure 1.** Browsing *P. falciparum* genes in PlasmoDB 4.1. Multiple alternative views are available for individual genes. The "expression" view of MSP1 (merozoite surface protein 1; PFI1475w) presents information on RNA and protein expression, including data from both Affymetrix and glass-slide microarrays, and proteomics analysis by MS/MS. See text for further details.

profiling datasets for *P. falciparum* that are accessible via PlasmoDB, the "Scripps/GNF" Affymetrix array (Le Roch *et al.*, 2002; 2003) provides absolute expression values (Figure 1, graph 1a) for seven time points spanning the intraerythrocytic life cycle (using two independent synchronization methods), in addition to data on expression in sporozoites and gametocytes. The spotted glass slide arrays reported in (Bozdech *et al.*, 2003a; 2003b) use a different parasite strain (HB3 vs 3D7), and provide higher time resolution: 48 hourly time points across the intraerythrocytic life cycle. Because experimental variability is high for glass slide arrays, expression values are normalized to a common pooled control, yielding a graph of expression induction rather than absolute values (Figure 1, graph 2a). In order to enable direct comparison between these two microarray platforms, the Scripps/GNF dataset is also presented in the form of of induction ratios (Figure 1, graph 1b), and absolute expression levels are presented as a percentile of all genes in each experiment (Figure 1, graphs 1c and 2b). Both experiments indicate high abundance and strong up-regulation of steady-state MSP1 transcript levels in late schizonts. Raw data can also be downloaded, and links are provided to the home sites for all relevant data sources. All probes are mapped to the parasite genome, enabling convenient comparison.

## 2.6. Proteomics

The production of large-scale proteomic datasets has been made possible by technological advances in mass spectrometry, combined with the availability of complete genome sequences. Analysis of complex protein mixtures (as opposed to purified proteins) and the determination of putative peptide sequences (as opposed to the masses of proteins or peptide fragments), permits comparison with predicted sequences emerging from genome sequencing projects (although the scale of whole genome computational analysis can be problematic). To date, two major genomic-scale proteomic analyses have been published for *P. falciparum* (Florens *et al.*, 2002; Lasonder *et al.*, 2002). Such studies provide a snapshot of the protein repertoire at a given time, and have permitted recognition of >40% of all annotated proteins in the parasite genome.

As with the analysis of microarray data, it is imperative to understand the nature of the data obtained, and limitations of the available results. Because peptide recognition depends on gene predictions, protein sequences associated with incorrectly assigned gene models will not be recognized. Searches of all open reading frames in the genome may lead to the discovery of a gene that was expressed but not predicted in the genome sequence. Even when gene models are accurate, many factors may influence the ability to detect peptide sequences, including protein abundance; post-translational modification; efficiency of solubilization; proteolytic digestion, and ionization; etc. Thus, positive data is likely to indicate peptide presence, but negative data is far less informative. Note also that analyses conducted to date provide no reliable quantitation of abundance, although the number of peptides recognized may provide a crude indication: MSP1 was identified in all samples, but was represented by far more peptides, covering far more of the gene, in merozoites (Figure 1, graph 3).

# 3. Data Repositories and Organization

The *Plasmodium* Genome Database, PlasmoDB (http://PlasmoDB.org), provides the largest and most comprehensive single collection of *Plasmodium*-related data (Bahl *et al.*, 2002; Kissinger *et al.*, 2002). This community resource currently houses genome sequence for several *Plasmodium* species; multiple alternative gene predictions; automated and curated annotation, including controlled vocabulary Gene Ontologies; GSS and EST sequences; SAGE data; mungbean, cDNA and oligonucleotide microarray data based on both glass slide and Affymetrix platforms; MS/MS proteomic data; microsatellite and physical mapping data; and comparative genomic analyses. Much of the data available in PlasmoDB is also available on CD-ROM. Depending on the application, the reader will also benefit from various other *Plasmodium* and apicomplexan parasite resources, as discussed below. Table 1 summarizes sites where data are stored and often available for download. Table 2 lists sites supporting bioinformatics analysis and data queries.

Each of the sequencing centers involved in the generation of the *P. falciparum* genome (the Sanger Institute, Stanford University, and The Institute for Genome Research; TIGR) maintains a BLAST searchable website and an FTP download site where sequences generated by that center may be obtained. Gene predictions and features may be queried at the Sanger Institute via GeneDB. TIGR maintains an EST-based gene index (Quackenbush *et al.*, 2001) for *P. falciparum* and *P. yoelii* (as well as several other apicomplexan parasites), offering a non-redundant view of transcripts analyzed computationally to provide information on potential cellular roles and function. In order to illustrate the logical sequence of events required for developing a bioinformatics resource, Box 1 summarizes the strategy for Gene Index production.

Several databases are dedicated to metabolic pathways and drug discovery. The "Malaria Parasite Metabolic Pathways" site (Table 2) provides curated graphical snapshots of *Plasmodium* metabolic processes organized by pathway. Direct links are provided from each enzyme E.C. number to Expasy-NiceZyme views, Brenda (Schomberg *et al.*, 2002) and PlasmoDB databases. PlasmoCyc contains graphical and searchable representations of *P. falciparum* metabolic pathways, and a whole-cell overview of metabolic pathways along with tools for between-species comparisons. The resource

---

**Box 1. The TIGR Gene Index Protocol for Assembly of ESTs and Transcripts**
**(http: //www.tigr.org/tdb/tgi/definitions.html)**

---

**Preparation of EST data**

- Extract sequences from dbEST and subject to quality control screening (vector, *E. coli*, polyA, T, or CT removal, minimum length = 100 bp, < 3% N).

**Preparation of transcript (ET) database**

- Extract all sequences from the appropriate division of GenBank.
- Discard non-coding sequences.
- Save cDNAs and coding sequences from genomic entries.
- Store sequences and related information in Expressed Gene Anatomy Database (EGAD).
- Make curated ET data set available as a multiple FastA format file (see EGAD main page).

**Assembly**

- Combine cleaned EST sequences and non-redundant transcript (ET) sequences.
- Assemble sequences into contigs using Paracel Transcript Assembler Program. TCs are consensus sequences based on two or more ESTs (and possibly an ET) that overlap ≥40 bases with ≥94% sequence identity (strict criteria help minimize creation of chimeric contigs).
- Assign contigs a TC (Tentative Consensus) number. TCs may comprise ESTs derived from different tissues.
- Assign best hits for TCs by searching against a non-redundant amino acid database (nrAA) using BLAT.
- Select and display top five hits (based on score) for each TC.

**Caveats**

- TCs are only as good as the underlying ESTs; unspliced or chimeric ESTs will produce aberrant TCs.
- The TC set contains some redundancy because sequences will not be combined unless they exhibit a high % identity and match end-to-end.
- TS directionality should not be assumed.
- Not all TCs contain protein-coding regions.

---

can also display individual enzymatic reactions with substrate and reactant structures, cellular localization, information regarding the association of protein subunits into complexes and a list of predicted drug targets (with links to the papers describing them).

Protein annotations have been used to search the protein structure database (PDB), and several *P. falciparum* protein structures have been modeled and are viewable. Microarray data are available from several sources (Table 1) and the UCSF site provides extensive viewing and analysis capabilities (Table 2). DNA structural analyses (repeat content, DNA "bendabilty", etc.) have been calculated for *P. falciparum* and can be viewed using the genome atlases maintained at the Center for Biological Sequence Analysis (CBS). The WHO/TDR Malaria database contains searchable genome annotation and an electronic repository of *Plasmodium* strain information, antigen and other multiple sequence alignments, and an extensive malaria antigen literature database; this database is also available on CD-ROM.

The Malaria Research and Reference Reagent Resource Center, MR4 (Adams *et al.*, 2000) is both an electronic and physical repository for quality controlled malaria-related reagents and information. Registered users can obtain parasites, mosquito vectors, antibodies, antigens, clones and gene libraries. MR4 resources are searchable via the web and many reagents and/ or genes have been linked to PlasmoDB (and vice-versa).

Dozens of other databases are extremely useful for *Plasmodium* research. While it is impossible to describe all of these sites here, Tables 1-3 provide a compendium of several such resources, and most are described in detail in the annual database issue of Nucleic Acids Research (Baxevanis, 2003). Useful databases include (but are not limited to): the NCBI GenBank and EMBL databases, containing large sequence repositories and a variety of tools for accessing these data; *Anopheles,* mouse and human genome databases (Table 3); the SMART (Simple Modular Architecture Research Tool) database for examining protein domain architectures; the BIND (Biomolecular Interaction Network Database) of molecular interactions culled from the literature and high-throughput analyses; the PFAM (Protein Families) database, containing a collection of Hidden Markov Models (HMMs) used to screen protein sequences and identify protein family members based on conserved patterns; and the ProDom and InterPro protein domain databases for searching and identifying protein domains.

The data described above is stored in a variety of formats. "Flat file" text documents can be opened in a word processor or spreadsheet program, and are therefore easy to share. Search functions are generally limited to "Find" commands to locate key words, and/or "Sort" commands to arrange and/or manipulate data. Even very large datasets, such as BLAST databases are typically stored in flat-file format. Other database types – relational and object-oriented – permit more sophisticated functions, and are usually managed by a database management system (DBMS) such as Oracle, DB2 to keep track of data access, deposition, security etc. Such management systems keep track of data records and requests for their access, preventing (for example) simultaneous withdrawal of checking account funds in excess of the amount on deposit.

## 3.1. Relational Databases and Queries

Relational databases store data in tables that are designed to accommodate specific data types, as shown in Figure 2. For example, one table might contain the names of all students in a school, another table might contain the names of all professors, a third the names of all the classes offered, and a fourth, the list of rooms in which classes are taught. Each of these tables

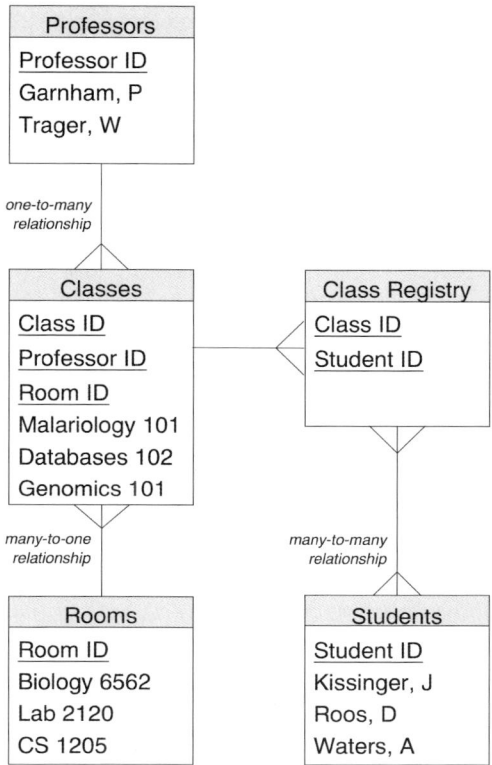

**Figure 2.** A relational database schema. Each data type (Professors, Classes, etc) is stored in a distinct table. Conceptually, tables contain both columns and rows. Tables can hold multiple entries for each data type: a unique identifier ID, the data itself (e.g. names of professors), and identifiers linking data in one table to an ID in another. Tables are associated with each other according to the type of relationship (one-to-one, one-to-many, many-to-one, many-to-many). In this example, each class is taught by a single professor, but each professor may teach multiple classes. Classes have multiple students, and students have multiple classes. An additional Class Registry table links each student to each class. Each class is associated with a single classroom, but an individual classroom may be used for multiple classes. Professors are not directly linked to rooms or students, but a relationship is defined via the classes that they teach.

can accept certain values (e.g. names consisting of alphabetical characters up to 50 characters in length, or alpha-numeric codes like Biology 6562 for classroom location). Relational databases, in addition to storing the data in defined tables, also relate the data contained in the tables to one another. For example, in the example shown, professors teach specific classes, classes have students enrolled in them, and classes are taught in specific classrooms. These relationships are not random, but clearly specified: each class has only one classroom, only one instructor, but multiple students. Students enroll in classes, and classes are taught in classrooms, but there is no direct link between students and classrooms; these two tables are only related via the classes offered. If the appropriate relationships are specified in the design of the database, then users of the database can ask questions (called queries) that require data from multiple tables. For example, one might want to know all classes taught by professor P.C.C. Garnham, all classes taught in the Biology 6562 classroom, or the names of all students enrolled in Malariology 101.

Extending this analogy to consider the very diverse kinds of biological data relevant to *Plasmodium* parasites, database tables can be generated to hold genomic sequences, EST sequences, SAGE tags, translated protein sequences, protein domains, peptides identified by MS/MS, and microarray oligonucleotides, etc. Relationships can then be defined between these tables: protein domains and peptides determined by MS/MS may relate to particular EST sequences or gene models; oligonucleotides from expression studies can be related to predicted or annotated genes; these genes can be related to proteins and these proteins can be related to function via Gene Ontology classifications; etc. The PlasmoDB database is built on a relational schema (Genomics Unified Schema; GUS) that currently contains more than 200 tables (Davidson *et al.*, 2001).

### 3.2. Controlled Vocabularies

Meaningful comments or data analysis requires controlled vocabularies – a standard set of terms that are applied to equivalent genes or processes. For example: it is often necessary to search for a gene by name, but what is that name? MSP-1, Merozoite Surface Protein-1 and PFI1475w are synonymous. In order to create automated systems for comparison it is necessary to agree upon a common vocabulary that is used for all organisms, such as the enzyme commission (E.C.) classification system.

Gene Ontology (GO) terms provide another example of controlled vocabularies. GO terms are created and maintained by the "Gene Ontology" Consortium (http://www.geneontology.org), as hierarchies of increasingly generalized terms around the concepts of "Molecular Function", "Biological Process", and "Cellular Component" (Ashburner *et al.*, 2000). By design, these definitions are sufficiently flexible that they can evolve as new information becomes available.

For example, GO terms for MSP-1 include:
- Biological process:
  - GO:007154, cell communication
  - GO:0030260, cell invasion
- Cellular component:
  - GO:0005623, cell
  - GO:0016020, membrane
- Functional assignment: Not yet defined

Each of these assignments was made by an annotator and comes with an evidence code describing the basis used when assigning the term (in this case all are labeled "TAS", or traceable author statement). Other GO evidence codes include: IC, inferred by curator; IDA, inferred from direct assay; IEA, inferred from electronic annotation; IEP, inferred from expression pattern; IGI, Inferred from genetic interaction; IMP, inferred from mutant phenotype; IPI, inferred from physical interaction; ISS, inferred from sequence or structural similarity; NAS, non-traceable author statement; ND, no biological data available; and TAS, traceable author statement. See http://www.geneontology.org/doc/GO.evidence.html for full explanation.

Once GO terms are applied to gene products, many of the problems related to data searching, integration and comparisons become much simpler. Searches can be performed using GO terms, and genes can easily be related – even across species boundaries –using GO identifiers. Evidence codes provide users with a clear statement as to the origin and confidence associated with each assignment. The combined information provided by GO term classifications and their evidence codes informs database users of what is known about any given assignment.

## 3.3. Data Integration

Data integration is the process of relating one type of data to another. As a simple example, gene annotations are related to a particular genome sequence, and protein features to protein sequences. Of course, integration can also involve more diverse data types: sequences may be related to physical maps, microarray oligonucleotide probes, or proteomic fragments. Defining data integration linkages is a laborious task, requiring extreme attention to detail. One common type of data integration performed by researchers on a regular basis is the association of gene names (and hence putative function) with a given sequence. Such relationships are often inferred on the basis of similarity to other sequences or the presence of particular motifs, in which case, a similarity search (e.g. BLAST) may provide the information necessary to link two diverse pieces of data. Sequence similarity searches are also commonly used to relate EST sequences to genomic sequences.

Applying such processes to genomic-scale datasets can yield very large networks of integrated data. Since nearly all data types can be related to the genome sequence either directly or indirectly, the genome sequence becomes a "bridge" that allows diverse data types to be integrated. Recall the database example provided in Figure 2. While professors are not directly related to classrooms, these distinct data types are related and integrated via the classes offered. Applying the same reasoning to genomic data, it is possible to link a proteomic mass profile to microarray expression levels, along the following path: collision-induced ionization of peptide fragments produces masses, the difference between these masses corresponds to a particular amino acid sequence, this sequence can be found in the data set of predicted proteins, predicted proteins are related to open reading frames, mRNAs are related via gene predictions to regions of genomic sequence, portions of the genomic sequence may correspond to cDNAs or oligonucleotides on a microarray, and elements on this microarray are linked to transcript expression levels, completing the path.

Controlled vocabularies have been developed for a variety of data types, and greatly facilitate the establishment of relationships and integration of diverse data types. With the advent of GO terms and motif identifier numbers and names, these data can be used to quickly relate homologous genes across species, and to identify all predicted proteins with a particular domain arrangement. Further integrating ProDom motif terms with GO identifiers may be able to ease the laborious task of assigning GO terms (Schug *et al.*, 2002): if protein motifs in a predicted sequence can be associated with ProDom terms, and a particular motif order and/or combination can be associated with a GO terms, then automated integration could facilitate annotation. Other data types require unique solutions, such as establishing the locations of all potential SAGE tag origination sites, or the location of enzyme restriction sites that would give rise to the fragments (± error) observed in the optical mapping experiments. Making the data from such analyses accessible through a relational database allows the full power of the integration to be realized.

Great care is essential in defining database schema architecture, and loading data into the database, however, as errors in data integration (such as applying gene prediction coordinates from one genome assembly to another) can be disastrous. Such challenges are a particular concern in a highly networked database architecture where the underlying data is constantly changing. When the reference genome sequence changes, for example, all data must be re-integrated. To minimize potential problems, it is important to record version numbers for each and every data source utilized. This subject is discussed more fully below.

## 3.4. Working in a Mixed Database World

The term "database" can be applied to many forms of collected data that can be downloaded, browsed, analyzed and/or queried. Tables 1-3 list many of the web-accessible data sources and analysis tools available for Apicomplexa and some of their host and vector species. In general, web sites incorporating search tools where user-specified text (or items from pull-down menus) can be used as search terms are likely to be based on a relational architecture. Such databases may indicate they were built using Oracle, Sybase, mySQL or PostgreSQL. Any database that indicates queries can be constructed using a Structured Query Language (SQL) is relational.

It is often impossible to identify the functionality and/or data types available at any given site without exploring the site and reading the introductory and/or tutorial pages, as database "look and feel" is more an indication of artistic style than functionality. The appearance of a web-based "front-end" may be held constant, even when the underlying database architecture is changed. For example, many aspects of PlasmoDB initially handled by smaller flat-file databases have become incorporated into the GUS relational schema. Conversely, changes in the appearance of a web-based "front-end" need not reflect any change in the underlying architecture. The appearance of the NCBI GenBank and PlasmoDB have both changed over the years, while maintaining similar core services and taking on additional functionalities; a little exploration reveals how new tools have been implemented. Similarly, quite different interfaces may be used to access similar databases. For example, the GUS architecture employed for PlasmoDB has recently been adopted by the Sanger Institute to drive GeneDB, providing access to the various organisms sequenced and annotated by Sanger's Pathogen Sequencing Unit. This development should greatly facilitate the development and exchange of software for browsing, visualizing, analyzing, mining and querying data.

Because genomic datasets change frequently, databases – like the web pages used to access them – typically display version numbers or release dates, which should be noted in any publications that depend on these resources, and any communications aimed at identifying problems in data access, analysis, or integration. It is not uncommon to discover that different identifiers are used for identical data stored in different databases, or that the same identifiers may be used for different data in different databases, or different releases of the same database. With the ever-expanding availability of internet resources, the possibilities for confusion are endless! It is therefore critical that bioinformatics researchers keep track of information on data release dates, database versions, etc, and make this information available in any publications (print or electronic) that may result. It is equally important that database developers provide resources that allow published data to be tracked and updated (enabling the correlation of new and old EST assemblies, for example), or at least maintain the ability to access old release data.

Particular attention should be paid to the specific data sets and analysis tools provided in the various available databases. For example, NCBI BLAST and Washington University BLAST (WU-BLAST) are different implementations of the same local sequence alignment algorithm; both work well, but they employ different default settings, arguments and DNA scoring matrices, and will therefore yield slightly different results. Most large genome databases provide a combination of human curated and computationally generated automated analyses. The availability of such diverse data types is what makes bioinformatics analysis possible, but it is important to know exactly how the data have been curated and/or analyzed (see Box 1 for an example).

# 4. Queries: Powerful Tools for Developing Bioinformatics Prowess

Although most databases permit the data that they house to be examined (for example, scrolling through all genes on *P. falciparum* chromosome 1), and many provide tools for data analysis (e.g. BLAST searches against *P. vivax* sequences, or listing all abundant transcripts in gametocytes in a particular experiment), the real power of a relational database lies in the ability to form integrated queries that depend on the relationships between multiple data types, as defined in the database schema. A wide range of queries are available in PlasmoDB, including queries related to

- curated or automated annotations (including GO function, process and component)
- chromosomal location
- results from BLAST searches against GenBank and/or other *Plasmodium* species
- DNA sequence features: low complexity sequence, AT content, coding potential, etc
- gene structure (intron-exon architecture)
- protein sequence features: secretion and organellar targeting signals, transmembrane domains, Pfam/ProDom/other motifs, secondary structure, predicted CD8 epitopes, etc
- the presence of strain-specific nucleotide and/or amino acid sequence polymorphisms
- expression data: EST or SAGE abundance, RNA expression levels/ induction/timing (on several platforms), evidence for protein expression (from MS/MS analysis)
- phylogenetic cross-comparisons (in comparison with other *Plasmodium* species, other Apicomplexa, other eukaryotic species, etc)
- availability of a predicted protein structural model
- involvement in specific metabolic pathways
- availability of reagents

In developing a successful query, it is crucial to translate biological knowledge into computationally-accessible terms. For example, a query for "drug targets" is not particularly well-defined, but a query for enzymes for which a structural model is available and that are known to be expressed in the erythrocytic stages at both the RNA and protein level, can take advantage of GO terminology, the *Plasmodium* protein structural model database, and both microarray and proteomics datasets (Kissinger *et al.*, 2002).

To pursue a vaccine-related example, one might wish to look for surface antigens based on the presence of a predicted secretory signal sequence (and/or one or more transmembrane domains), as shown in Figure 3. In addition, one could look for antigens shared between *P. falciparum* and

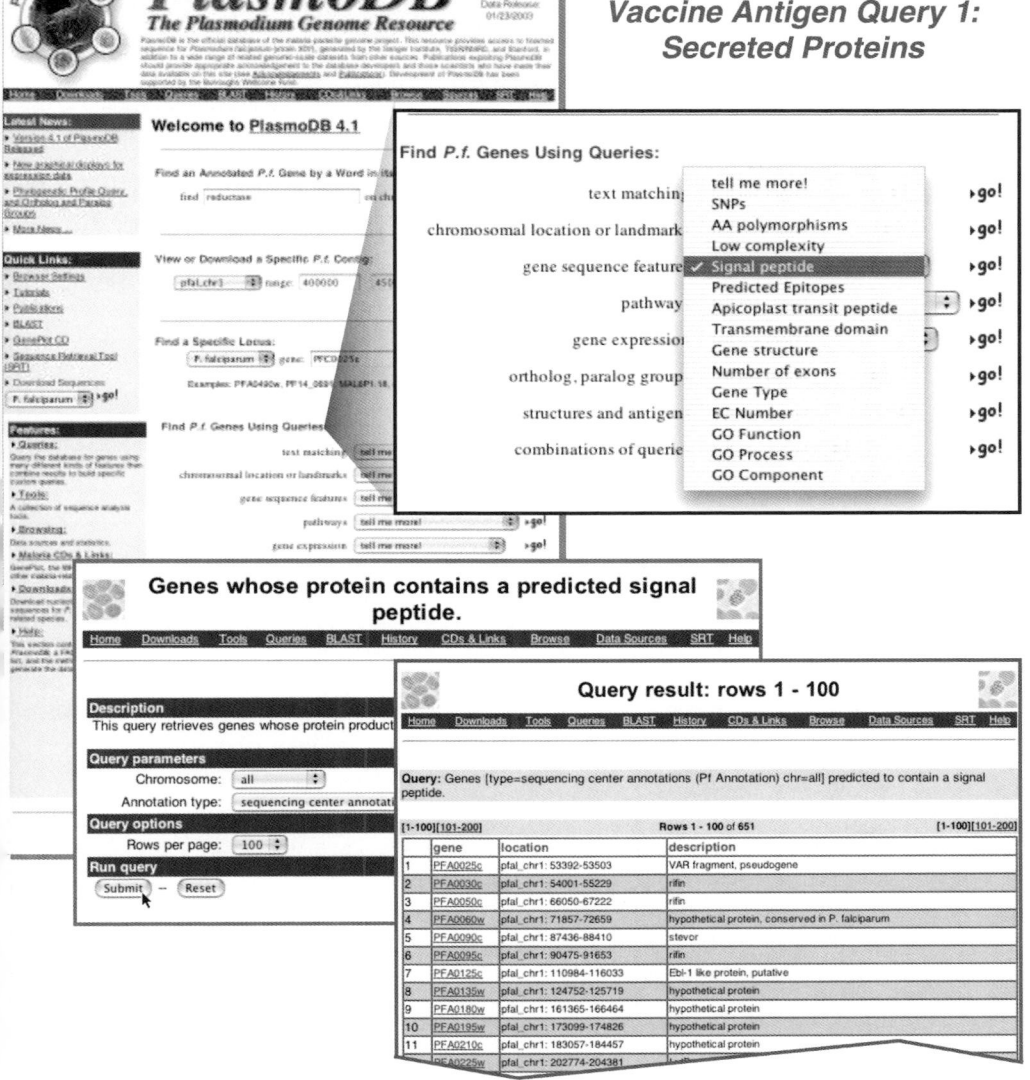

**Figure 3.** Querying the curated annotation of *P. falciparum* genes in PlasmoDB 4.1 for signal peptides (identified using the neural net work program SignalP) yields 651 proteins that are predicted to be secreted. Because accurate prediction of secretory signal sequences requires accurate assignment of the translational initiation, it is likely that this query misses many secreted proteins. Further refinements might include searching alternative gene models, or including proteins with predicted transmembrane domains.

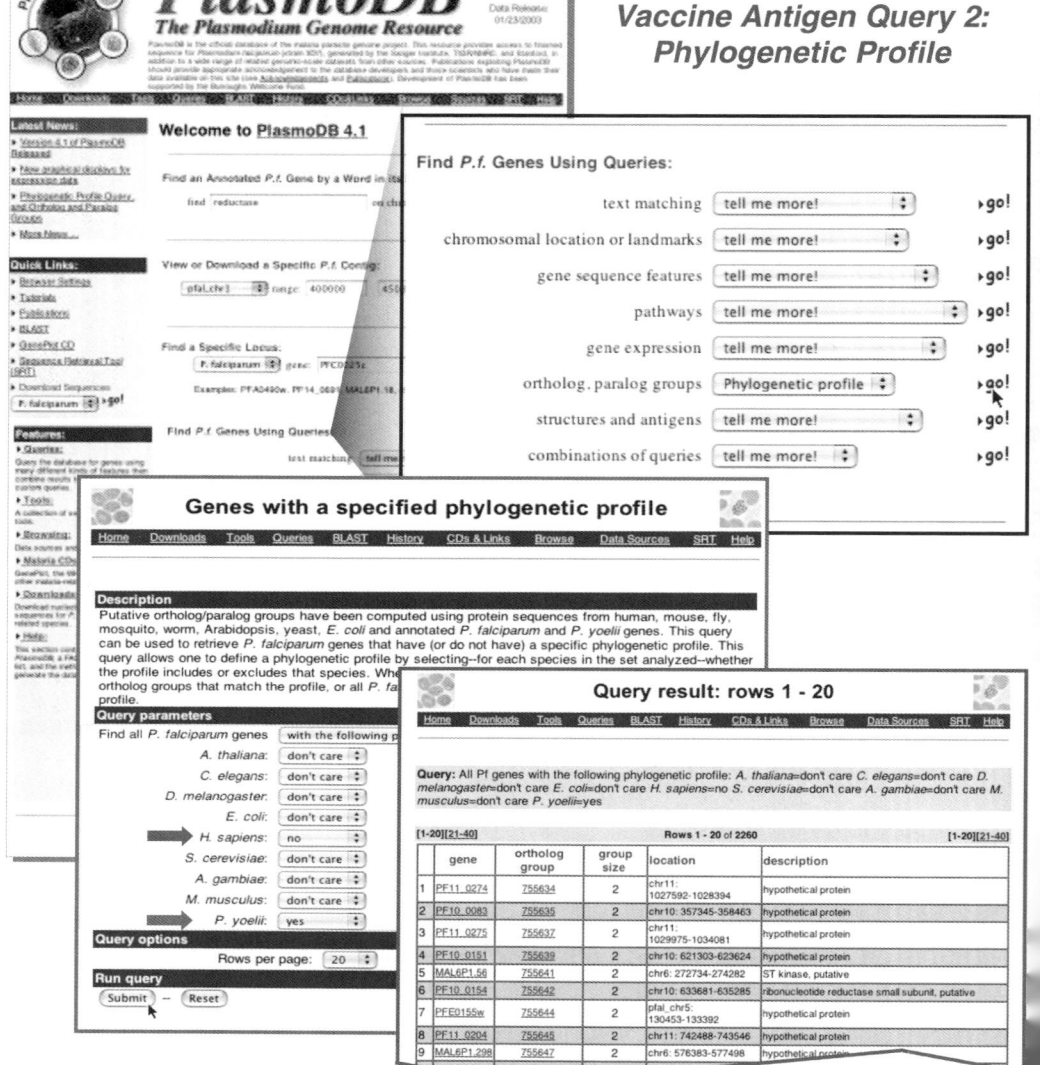

**Figure 4.** Phylogenomic cross-comparisons with other genome sequence data can identify putative orthologous genes (Li *et al*., 2003b), and genes that are phylogenetically-restricted in their distribution (Ajioka *et al*., 1998). In seeking candidate vaccine targets, one might wish to identify antigens that are highly conserved between *P. falciparum* and *P. yoelii*, but not shared with the human host. This query yields 2260 hits (>40% of the parasite genome).

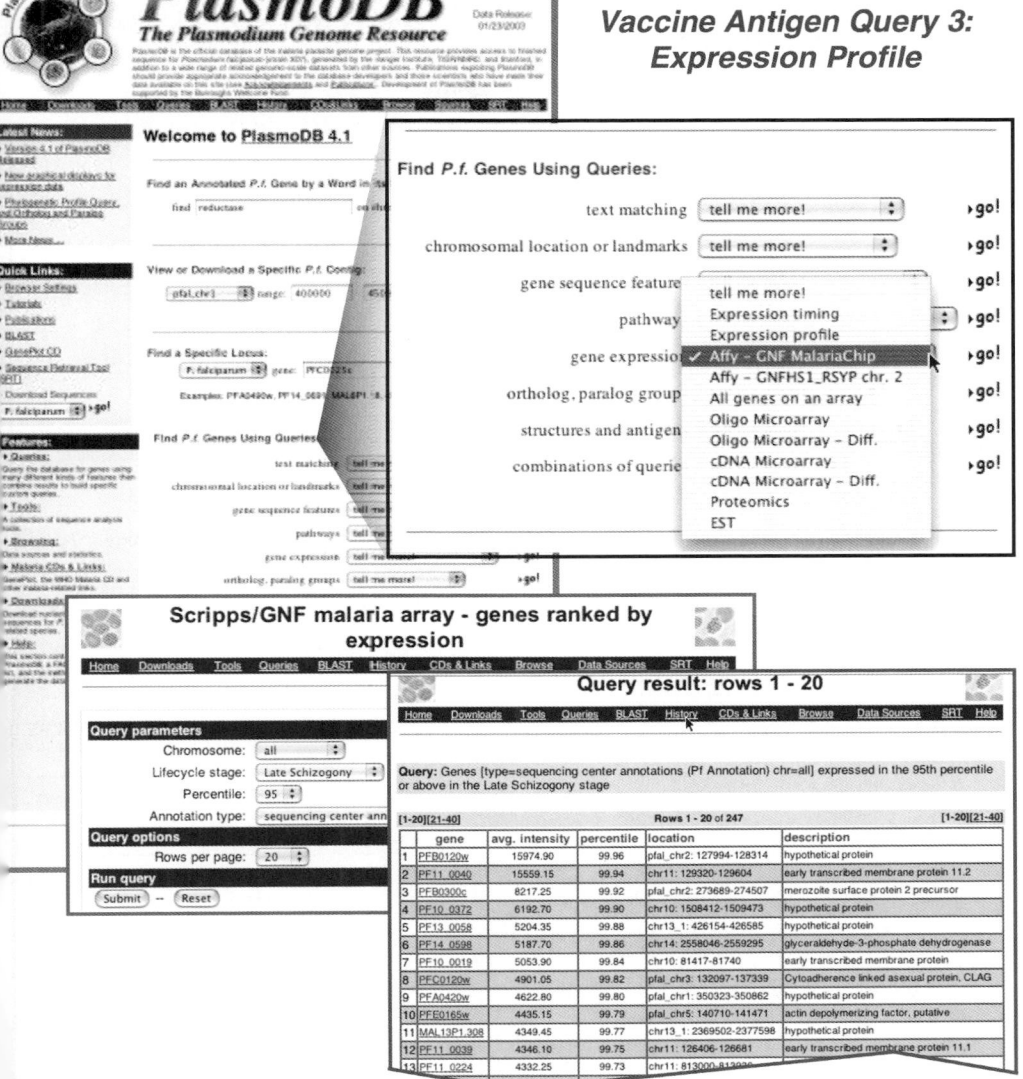

**Figure 5.** In seeking a blood-stage vaccine, one might wish to prioritize antigens that are abundantly expressed in the extra-erythrocytic merozoite stage. This query focuses on experiments conducted by Le Roch *et al.* (2003b) using an Affymetrix microarray to examine expression across the erythrocytic life cycle, and seeks genes that are among the top 5% in steady-state transcript abundance during late schizogony (many other expression datasets and query strategies can also be envisioned).

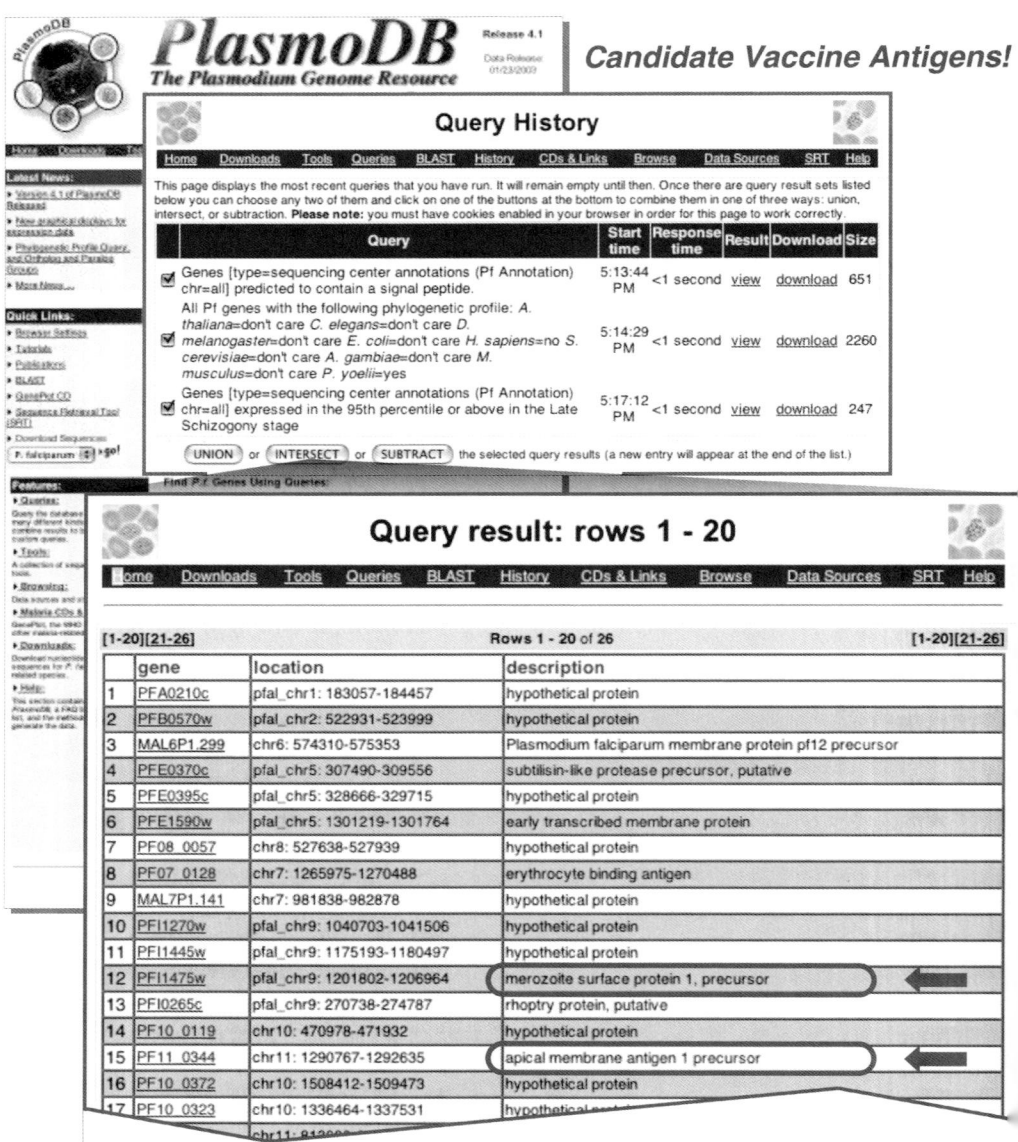

**Figure 6.** Using the 'Query History' feature of PlasmoDB to combine the queries illustrated in Figures 1-3 identifies only 26 genes exhibiting all three desired characteristics: antigens that are secreted, restricted to *Plasmodium* species, and abundantly transcribed just before merozoite emergence. Among these genes are both of the leading erythrocytic vaccine candidates: MSP1 and AMA1.

*P. yoelii,* but absent from the human genome, as shown in Figure 4. One might further wish to restrict consideration to abundant transcripts expressed in late schizonts, based on the GNF photolithographic array, as shown in Figure 5. Each of these queries yields a list of several hundred (or thousand) genes, but exploiting the "History" function of PlasmoDB permits taking the intersection of these queries, yielding 26 hits (Figure 6) ... including two of the leading vaccine antigens now undergoing trials (MSP1 and AMA1). The remaining proteins (mostly annotated as hypothetical proteins) would be interesting to explore as candidate vaccine antigens.

Of course, there are many other ways to configure this search, including refining the desired expression pattern, considering chromosomal location, evaluating potential efficacy, or seeking for evidence for positive selection from population genetic studies, etc (although the data is not yet in place for all of these queries). The point of this exercise is not that analysis *in silico* is ever likely to take the place of laboratory analysis (particularly in the case of vaccine antigen discovery!) Rather, the point is that computational tools can rapidly filter available options, providing each gene in the dataset with a set of credentials that can be assessed for potential vaccine efficacy. Overall, the goal is to let computers do what computers do well (integrating and analyzing large-scale datasets), and let people do what people do well (experimental validation at the laboratory bench).

## 4.1. Future Directions

*Plasmodium* bioinformatics resources are growing daily, and we can anticipate the incorporation of new data at an ever-accelerating rate. The year following completion of reference sequences for *P. falciparum* and *A. gambiae* saw the release of effectively complete genome sequence for *P. yoelii* (and other apicomplexan parasite species), extensive sequence information for several other *Plasmodium* species, whole-genome RNA and protein expression data (on several platforms; Florens *et al.*, 2003; Le Roch *et al.*, 2002; 2003; Bozdech *et al.*, 2003a; 2003b), and new algorithms for cross-genome comparisons (Li *et al.*, 2003b).

The coming year is likely to bring further sequence data for a field isolate of *P. falciparum* and additional *Plasmodium* species (and other apicomplexan parasites); revised and updated annotation for *P. falciparum* and *P. yoelii;* next-generation computational analyses of these genomes, incorporating new algorithms for orthologous group identification; syntenic analysis and other comparisons across species boundaries; genome-wide SNP markers for genetic studies; incorporation of greatly explanded EST datasets, representing several life-cycle stages; additional transcript profiling data, including studies on additional strains, life cycle stages, and treatments; additional proteomics data, including quantitative data from various life cycle stages, and preliminary analysis of protein modifications.

In future, we can also anticipate the availability of additional data types, from population genetic data, to clinical records, to structural genomics results, to publications records. The computational challenge will be to integrate these emerging data types with existing database resources, and develop analysis tools for effective database mining. The biological challenge is to consider how to effectively translate biological questions into computationally accessible terms. What questions do *you* want to ask?

- What features characterize potential drug targets, vaccine antigens, diagnostics?
- How to get a handle on understudied life cycle stages?
- How to map virulence genes and other loci of interest?
- What features define parasite proteins likely to interact with the red cell, liver, and host endothelium?
- What genes to target for genetic knock-outs, knock-downs, etc?
- How best to explore parasite population biology?
- How best to compare *P. falciparum* with other *Plasmodium*, apicomplexan, and other eukaryotic pathogen species: gene families, taxonomically- or functionally-restricted genes?
- What information can we usefully extract from the *P. vivax* genome?
- How to exploit genomic information for host and vector species (*Plasmodium* vs. human, mouse, *Anopheles*)?
- What information is of greatest interest for studying eukaryotic biology and evolution?
- How best to link genomics data to publication records?
- How to integrate clinical data?
- What new 'omics'-scale datasets would be useful?

# 5. Concluding Comments: Bioinformatics Research, and Where to Look for Further Assistance

Now is an exciting time to be engaged in malaria research. The availability of genome sequences for the parasite host and vector, along with emerging expression analyses and anticipated population data, are providing unprecedented insight into the biology of *Plasmodium* and its interaction with its hosts. Utilization of this data requires proper storage, retrieval, analysis and integration of these and new data types.

Databases offer a tremendous asset for biomedical research, but they do not obviate the need for critical thinking; the same analytical approach is required for bioinformatics experiments conducted *in silico* as for experimental work conducted at the laboratory bench. Problems are likely to arise whenever the exact nature of the data type or bioinformatics analysis tool is not understood. Database users should therefore endeavor to fully explore database resources, and should not be reluctant to contact the database developers whenever questions or problems arise that are not clearly explained in the documentation provided.

Bioinformatics assistance is readily available in multiple forms. Many of the major databases provide tutorial, "How To" and "Frequently Asked Questions" help pages. A few minutes spent reading this material can save hours of frustration or misuse/misinterpretation of the data contained in the database. If questions or doubts still remain, contact the database directly via e-mail. Clearly state your question(s), referring to the exact pages or tools, your computer platform (Mac, Windows, Unix) and browser type and version. Many "bugs" are platform- or browser-specific. Explanations of specialty databases can be found in the annual Nucleic Acids Research database issue published each January. Tutorials and in-depth explanations of analysis tools can usually be found on the tool's web site or in bioinformatics books. Classes pertaining to the use of malaria related resources are routinely offered by the Malaria Research and Reference Reagent Resource Center (MR4 - http://www.malaria.mr4.org) and the WHO/TDR (http://www.who.int/tdr/). Workshops on how to use malaria-related databases are offered at several international meetings (Molecular Parasitology, Tropical Medicine and others), participants are encouraged to check meeting agendas and contact meeting organizers.

Finally, it is important to keep in mind that computational analysis is accessible to everyone … you can do it! The advantage of bioinformatics research relative to most bench work is that computational experiments can often be run quickly, at negligible cost, and with no risk of damaging the starting material or wasting reagents. It should also be noted that there are usually many, many routes to an answer: many approaches to predicting genes, many methods for defining protein features, many expression datasets, many methods for analyzing expression data, etc. As long as the raw data is available, new analyses and re-analysis can be and should be performed as new techniques and experimental strategies develop.

# 6. Acknowledgements

The authors would like to thank the numerous researchers and students who have generated genomic-scale datasets and made these resources publicly available. We would particularly like to thank the many malaria researchers whose questions – whether submitted electronically, during workshops and training sessions, or in the laboratory – have contributed to the success of PlasmoDB and other databases. We also thank members of our laboratories, Bindu Gajria, Philip Labo, and Boris Striepen for useful comments on this manuscript.

# 7. References

Adams, J. H., Wu, Y., and Fairfield, A. 2000. Malaria Research and Reference Reagent Resource Center. Parasitol. Today. 16: 89.

Ajioka, J. W., Boothroyd, J. C., Brunk, B. P., Hehl, A., Hillier, L., Manger, I. D., Marra, M., Overton, G. C., Roos, D. S., Wan, K. L., Waterston, R., and Sibley, L. D. 1998. Gene discovery by EST sequencing in *Toxoplasma gondii* reveals sequences restricted to the Apicomplexa. Genome Res. 8: 18-28.

Ashburner, M., Ball, C. A., Blake, J. A., Botstein, D., Butler, H., Cherry, J. M., Davis, A. P., Dolinski, K., Dwight, S. S., Eppig, J. T., Harris, M. A., Hill, D. P., Issel-Tarver, L., Kasarskis, A., Lewis, S., Matese, J. C., Richardson, J. E., Ringwald, M., Rubin, G. M., and Sherlock, G. 2000. Gene ontology: tool for the unification of biology. The Gene Ontology Consortium. Nature Genet. 25: 25-29.

Bahl, A., Brunk, B., Coppel, R. L., Crabtree, J., Diskin, S. J., Fraunholz, M. J., Grant, G. R., Gupta, D., Huestis, R. L., Kissinger, J. C., Labo, P., Li, L., McWeeney, S. K., Milgram, A. J., Roos, D. S., Schug, J., and Stoeckert, C. J., Jr. 2002. PlasmoDB: the *Plasmodium* genome resource. An integrated database providing tools for accessing, analyzing and mapping expression and sequence data (both finished and unfinished). Nucl. Acids Res. 30: 87-90.

Baxevanis, A. D. 2003. The Molecular Biology Database Collection: 2003 update. Nucl. Acids Res. 31: 1-12.

Baxevanis, A. D., Davison, D. B., Page, R. D. M., Petsko, G., Stein, L., and Stormo, G. D. 2003. Current Protocols in Bioinformatics. Rockville.

Baxevanis, A. D., and Ouellette, B. F. 2001. Bioinformatics: A Paractical Guide to the Analysis of Genes & Proteins. Wiley-Interscience, New York.

Bowman, S., Lawson, D., Basham, D., Brown, D., Chillingworth, T., Churcher, C. M., Craig, A., Davies, R. M., Devlin, K., Feltwell, T., Gentles, S., Gwilliam, R., Hamlin, N., Harris, D., Holroyd, S., Hornsby, T., Horrocks, P., Jagels, K., Jassal, B., Kyes, S., McLean, J., Moule, S., Mungall, K., Murphy, L., Barrell, B. G., and *et al.* 1999. The complete nucleotide sequence of chromosome 3 of *Plasmodium falciparum*. Nature. 400: 532-538.

Bozdech, Z., Zhu, J., Joachimiak, M. P., Cohen, F. E., Pulliam, B., and DeRisi, J. L. 2003a. Expression profiling of the schizont and trophozoite stages of *Plasmodium falciparum* with a long-oligonucleotide microarray. Genome Biol. 4: R9.

Bozdech, Z., Llinas, M., Pulliam, B.L., Wong, E.D., Zhu, J., and DeRisi, J.L. 2003b. The transcriptome of the intraerythrocytic developmental cycle of *Plasmodium falciparum*. PLoS Biol. 1: 85: 100.

Carlton, J., and Dame, J. 2000. The *Plasmodium vivax* and *P. berghei* gene sequence tag projects. Parasitol. Today 16: 409.

Carlton, J. M., Angiuoli, S. V., Suh, B. B., Kooij, T. W., Pertea, M., Silva, J. C., Ermolaeva, M. D., Allen, J. E., Selengut, J. D., Koo, H. L., Peterson, J. D., Pop, M., Kosack, D. S., Shumway, M. F., Bidwell, S. L., Shallom, S. J., van Aken, S. E., Riedmuller, S. B., Feldblyum, T. V., Cho, J. K., Quackenbush, J., Sedegah, M., Shoaibi, A., Cummings, L. M., Florens, L., Yates, J. R., Raine, J. D., Sinden, R. E., Harris, M. A., Cunningham, D. A., Preiser, P. R., Bergman, L. W., Vaidya, A. B., van Lin, L. H., Janse, C. J., Waters, A. P., Smith, H. O., White, O. R., Salzberg, S. L., Venter, J. C., Fraser, C. M., Hoffman, S. L., Gardner, M. J., and Carucci, D. J. 2002. Genome sequence and comparative analysis of the model rodent malaria parasite *Plasmodium yoelii yoelii*. Nature. 419: 512-519.

Davidson, S. B., Crabtree, J., Brunk, B., Schug, J., Tannen, V., Overton, G. C., and Stoeckert, C.J. 2001. K2/Kleisli and GUS: Experiments in integrated access to genomic data sources. IBM Syst. J. 40: 512-531 del Portillo, H. A., Fernandez-Becerra, C., Bowman, S., Oliver, K., Preuss, M., Sanchez, C. P., Schneider, N. K., Villalobos, J. M., Rajandream, M. A., Harris, D., Pereira da Silva, L. H., Barrell, B., and Lanzer, M. 2001. A superfamily of variant genes encoded in the subtelomeric region of *Plasmodium vivax*. Nature. 410: 839-842.

Ferdig, M. T., and Su, X. Z. 2000. Microsatellite markers and genetic mapping in *Plasmodium falciparum*. Parasitol. Today 16: 307-312.

Florens, L., Washburn, M. P., Raine, J. D., Anthony, R. M., Grainger, M., Haynes, J. D., Moch, J. K., Muster, N., Sacci, J. B., Tabb, D. L., Witney, A. A., Wolters, D., Wu, Y., Gardner, M. J., Holder, A. A., Sinden, R. E., Yates, J. R., and Carucci, D. J. 2002. A proteomic view of the *Plasmodium falciparum* life cycle. Nature. 419: 520-526.

Gardner, M. J., Hall, N., Fung, E., White, O., Berriman, M., Hyman, R. W., Carlton, J. M., Pain, A., Nelson, K. E., Bowman, S., Paulsen, I. T., James, K., Eisen, J. A., Rutherford, K., Salzberg, S. L., Craig, A., Kyes, S., Chan, M. S., Nene, V., Shallom, S. J., Suh, B., Peterson, J., Angiuoli, S., Pertea, M., Allen, J., Selengut, J., Haft, D., Mather, M. W., Vaidya, A. B., Martin, D. M., Fairlamb, A. H., Fraunholz, M. J., Roos, D. S., Ralph, S. A., McFadden, G. I., Cummings, L. M., Subramanian, G. M., Mungall, C., Venter, J. C., Carucci, D. J., Hoffman, S. L., Newbold, C., Davis, R. W., Fraser, C. M., and Barrell, B. 2002a. Genome sequence of the human malaria parasite *Plasmodium falciparum*. Nature. 419: 498-511.

Gardner, M. J., Shallom, S. J., Carlton, J. M., Salzberg, S. L., Nene, V., Shoaibi, A., Ciecko, A., Lynn, J., Rizzo, M., Weaver, B., Jarrahi, B., Brenner, M., Parvizi, B., Tallon, L., Moazzez, A., Granger, D., Fujii, C., Hansen, C., Pederson, J., Feldblyum, T., Peterson, J., Suh, B., Angiuoli, S., Pertea, M., Allen, J., Selengut, J., White, O., Cummings, L. M., Smith, H. O., Adams, M. D., Venter, J. C., Carucci, D. J., Hoffman, S. L., and Fraser, C. M. 2002b. Sequence of *Plasmodium falciparum* chromosomes 2, 10, 11 and 14. Nature. 419: 531-4.

Gardner, M. J., Tettelin, H., Carucci, D. J., Cummings, L. M., Aravind, L., Koonin, E. V., Shallom, S., Mason, T., Yu, K., Fujii, C., Pederson, J., Shen, K., Jing, J., Aston, C., Lai, Z., Schwartz, D. C., Pertea, M., Salzberg, S., Zhou, L., Sutton, G. G., Clayton, R., White, O., Smith, H. O., Fraser, C. M., Hoffman, S. L., *et al*. 1998. Chromosome 2 sequence of the human malaria parasite *Plasmodium falciparum*. Science. 282: 1126-1132.

Gibson, G., and Muse, S. 2001. A Primer of Genome Science. Sinauer, Sunderland.

Hall, N., Pain, A., Berriman, M., Churcher, C., Harris, B., Harris, D., Mungall, K., Bowman, S., Atkin, R., Baker, S., Barron, A., Brooks, K., Buckee, C. O., Burrows, C., Cherevach, I., Chillingworth, C., Chillingworth, T., Christodoulou, Z., Clark, L., Clark, R., Corton, C., Cronin, A., Davies, R., Davis, P., Dear, P., Dearden, F., Doggett, J., Feltwell, T., Goble, A., Goodhead, I., Gwilliam, R., Hamlin, N., Hance, Z., Harper, D., Hauser, H., Hornsby, T., Holroyd, S., Horrocks, P., Humphray, S., Jagels, K., James, K. D., Johnson, D., Kerhornou, A., Knights, A., Konfortov, B., Kyes, S., Larke, N., Lawson, D., Lennard, N., Line, A., Maddison, M., McLean, J., Mooney, P., Moule, S., Murphy, L., Oliver, K., Ormond, D., Price, C., Quail, M. A., Rabbinowitsch, E., Rajandream, M. A., Rutter, S., Rutherford, K. M., Sanders, M., Simmonds, M., Seeger, K., Sharp, S., Smith, R., Squares, R., Squares, S., Stevens, K., Taylor, K., Tivey, A., Unwin, L., Whitehead, S., Woodward, J., Sulston, J. E., Craig, A., Newbold, C., and Barrell, B. G. 2002. Sequence of *Plasmodium falciparum* chromosomes 1, 3-9 and 13. Nature. 419: 527-531.

Hayward, R. E., Derisi, J. L., Alfadhli, S., Kaslow, D. C., Brown, P. O., and Rathod, P. K. 2000. Shotgun DNA microarrays and stage-specific gene expression in *Plasmodium falciparum* malaria. Mol. Microbiol 35: 6-14.

Holt, R. A., Subramanian, G. M., Halpern, A., Sutton, G. G., Charlab, R., Nusskern, D. R., Wincker, P., Clark, A. G., Ribeiro, J. M., Wides, R., Salzberg, S. L., Loftus, B., Yandell, M., Majoros, W. H., Rusch, D. B., Lai, Z., Kraft, C. L., Abril, J. F., Anthouard, V., Arensburger, P., Atkinson, P. W., Baden, H., de Berardinis, V., Baldwin, D., Benes, V., Biedler, J., Blass, C., Bolanos, R., Boscus, D., Barnstead, M., Cai, S., Center, A., Chatuverdi, K., Christophides, G. K., Chrystal, M. A., Clamp, M., Cravchik, A., Curwen, V., Dana, A., Delcher, A., Dew, I., Evans, C. A., Flanigan, M., Grundschober-Freimoser, A., Friedli, L., Gu, Z., Guan, P., Guigo, R., Hillenmeyer, M. E., Hladun, S. L., Hogan, J. R., Hong, Y. S., Hoover, J., Jaillon, O., Ke, Z., Kodira, C., Kokoza, E., Koutsos, A., Letunic, I., Levitsky, A., Liang, Y., Lin, J. J., Lobo, N. F., Lopez, J. R., Malek, J. A., McIntosh, T. C., Meister, S., Miller, J., Mobarry, C., Mongin, E., Murphy, S. D., O'Brochta, D. A., Pfannkoch, C., Qi, R., Regier, M. A., Remington, K., Shao, H., Sharakhova, M. V., Sitter, C. D., Shetty, J., Smith, T. J., Strong, R., Sun, J., Thomasova, D., Ton, L. Q., Topalis, P., Tu, Z., Unger, M. F., Walenz, B., Wang, A., Wang, J., Wang, M., Wang, X., Woodford, K. J., Wortman, J. R., Wu, M., Yao, A., Zdobnov, E. M., Zhang, H., Zhao, Q., *et al.* 2002. The genome sequence of the malaria mosquito *Anopheles gambiae*. Science. 298: 129-149.

Hyman, R. W., Fung, E., Conway, A., Kurdi, O., Mao, J., Miranda, M., Nakao, B., Rowley, D., Tamaki, T., Wang, F., and Davis, R. W. 2002. Sequence of *Plasmodium falciparum* chromosome 12. Nature. 419: 534-537.

Janssen, C. S., Barrett, M. P., Lawson, D., Quail, M. A., Harris, D., Bowman, S., Phillips, R. S., and Turner, C. M. R. 2001. Gene discovery in *Plasmodium chabaudi* by genome survey sequencing. Mol. Biochem. Parasitol. 113: 251-260.

Kissinger, J. C., Brunk, B. P., Crabtree, J., Fraunholz, M. J., Gajria, B., Milgram, A. J., Pearson, D. S., Schug, J., Bahl, A., Diskin, S. J., Ginsburg, H., Grant, G. R., Gupta, D., Labo, P., Li, L., Mailman, M. D., McWeeney, S. K., Whetzel, P., Stoeckert, C. J., and Roos, D. S. 2002. The *Plasmodium* genome database. Nature 419: 490-492.

Lai, Z., Jing, J., Aston, C., Clarke, V., Apodaca, J., Dimalanta, E. T., Carucci, D. J., Gardner, M. J., Mishra, B., Anantharaman, T. S., Paxia, S., Hoffman, S. L., Craig Venter, J., Huff, E. J., and Schwartz, D. C. 1999. A shotgun optical map of the entire *Plasmodium falciparum* genome. Nature Genet. 23: 309-313.

Lasonder, E., Ishihama, Y., Andersen, J. S., Vermunt, A. M., Pain, A., Sauerwein, R. W., Eling, W. M., Hall, N., Waters, A. P., Stunnenberg, H. G., and Mann, M. 2002. Analysis of the *Plasmodium falciparum* proteome by high-accuracy mass spectrometry. Nature. 419: 537-542.

Le Roch, K. G., Zhou, Y., Batalov, S., and Winzeler, E. A. 2002. Monitoring the chromosome 2 intraerythrocytic transcriptome of *Plasmodium falciparum* using oligonucleotide arrays. Am. J. Trop. Med. Hyg. 67: 233-243.

Le Roch, K.G., Zhou, Y., Blair, P.L., Grainger, M., Moch, J.K., Haynes, J.D., De La Vega, P., Holder, A.A., Batalov, S., Carucci, D.J., and Winzeler, E.A. 2003. Discovery of gene function by expression profiling of the malaria parasite life cycle. Science. 301: 1503-1508.

Li, L., Brunk, B. P., Kissinger, J. C., Pape, D., Martin, J., Wylie, T., Dante, M., Tang, K., Cole, R., Fogarty, S. J., Howe, D. K., Liberator, P. A., Diaz, C., White, M.,

Jerome, M. E., Johnson, E. A., Radke, J. A., Waterston, R., Clifton, S., Roos, D. S., and Sibley, L. D. 2003a. Gene Discovery in the Apicomplexa as Revealed by EST Sequencing and Assembly of a Comparative Gene Database. Genome Res. 13: 443-454.

Li, L., Stoeckert, C. J., and Roos, D. S. 2003b. OrthoMCL: Identification of ortholog groups for eukaryotic genomes. Genome Res. 13: 2178-2190.

Mamoun, C. B., Gluzman, I. Y., Hott, C., MacMillan, S. K., Amarakone, A. S., Anderson, D. L., Carlton, J. M.-R., Dame, J. B., Chakrabarti, D., Martin, R. K., Brownstein, B. H., and Goldberg, D. E. 2001. Co-ordinated programme of gene expression during asexual intraerythrocytic development of the human malaria parasite *Plasmodium falciparum* revealed by microarray analysis. Mol. Microbiol. 39: 26-36.

Mount, D. 2001. Bioinformatics: Sequence and Genome Analysis. Cold Spring Harbor Laboratory Press, Cold Spring Harbor, New York.

Mu, J., Duan, J., Makova, K. D., Joy, D. A., Huynh, C. Q., Branch, O. H., Li, W. H., and Su, X. Z. 2002. Chromosome-wide SNPs reveal an ancient origin for *Plasmodium falciparum*. Nature. 418: 323-326.

Munasinghe, A., Patankar, S., Cook, B. P., Madden, S. L., Martin, R. K., Kyle, D. E., Shoaibi, A., Cummings, L. M., and Wirth, D. F. 2001. Serial analysis of gene expression (SAGE) in *Plasmodium falciparum*: application of the technique to A-T rich genomes. Mol. Biochem. Parasitol. 113: 23-34.

Pleasance, E. D., Marra, M., and Jones, S. J. M. 2003. Assessment of SAGE in Transcript Identification. Genome Res. 13: 1203-1215.

Quackenbush, J., Cho, J., Lee, D., Liang, F., Holt, I., Karamycheva, S., Parvizi, B., Pertea, G., Sultana, R., and White, J. P. 2001. The TIGR Gene Indices: analysis of gene transcript sequences in highly sampled eukaryotic species. Nucl. Acids Res. 29: 159-164.

Schomburg, I., Chang, A., and Schomburg, D. 2002. BRENDA, enzyme data and metabolic information. Nucl. Acids Res. 30: 47-49.

Schug, J., Diskin, S., Mazzarelli, J., Brunk, B. P., and Stoeckert, C. J., Jr. 2002. Predicting gene ontology functions from ProDom and CDD protein domains. Genome Res. 12: 648-655.

Su, X., Ferdig, M. T., Huang, Y., Huynh, C. Q., Liu, A., You, J., Wootton, J. C., and Wellems, T. E. 1999. A genetic map and recombination parameters of the human malaria parasite *Plasmodium falciparum*. Science. 286: 1351-1353.

Tchavtchitch, M., Fischer, K., Huestis, R. L., and Saul, A. 2001. The sequence of a 200 kb portion of a *Plasmodium vivax* chromosome reaveals a high degree of conservation with *Plasmodium falciparum* chromosome 3. Mol. Biochem. Parasitol. 118: 211-222.

Velculescu, V. E., Zhang, L., Vogelstein, B., and Kinzler, K. W. 1995. Serial analysis of gene expression. Science 270: 484-487.

Watanabe, J., Sasaki, M., Suzuki, Y., and Sugano, S. 2001. FULL-malaria: a database for a full-length enriched cDNA library from human malaria parasite, *Plasmodium falciparum*. Nucl. Acids Res. 29: 70-71.

Wootton, J. C., Feng, X., Ferdig, M. T., Cooper, R. A., Mu, J., Baruch, D. I., Magill, A. J., and Su, X. Z. 2002. Genetic diversity and chloroquine selective sweeps in *Plasmodium falciparum*. Nature. 418: 320-323.

From: Malaria Parasites: Genomes and Molecular Biology
Edited by: A.P. Waters and C.J. Janse

# Chapter 4

## Manipulating the *Plasmodium* Genome

Teresa Gil Carvalho
and Robert Ménard

## Abstract

Genome manipulation, the primary tool for assigning function to sequence, will be essential for understanding *Plasmodium* biology and malaria pathogenesis in molecular terms. The first success in transfecting *Plasmodium* was reported almost ten years ago. Gene-targeting studies have since flourished, as *Plasmodium* is haploid and integrates DNA only by homologous recombination. These studies have shed new light on the function of many proteins, including vaccine candidates and drug resistance factors. However, many essential proteins, including those involved in parasite invasion of erythrocytes, cannot be characterized in the absence of conditional mutagenesis. Proteins also cannot be identified on a functional basis as random DNA integration has not been achieved. We overview here the ways in which the *Plasmodium* genome can be manipulated. We also point to the tools that should be established if our goal is to address parasite infectivity in a systematic way and to conduct refined structure-function analysis of selected products.

# 1. Introduction

It is safe to predict that the wealth of information revealed by the sequence of the *Plasmodium falciparum* genome will benefit many areas of malaria research. New drug targets will be identified by capitalizing on the comprehensive view of parasite metabolism, as was already done to demonstrate the anti-malarial activities of fosmidomycin and triclosan (Jomaa *et al.*, 1999; Surolia and Surolia, 2001). Another much anticipated impact of the genome sequence is on vaccine development, via the formulation of new 'vaccinomic' approaches (Hoffman *et al.*, 1998; 2002). Comparative genomics will soon be possible as the genome sequence of more *Plasmodium* species and other Apicomplexa is completed, and will provide insights into the evolution of these protozoan parasites and adaptation to their hosts.

To what extent will the sequence help us to understand *Plasmodium* biology? Encompassing 14 chromosomes, the ~25-megabase *Plasmodium* genome is predicted to encode ~5,000 genes. Apicomplexa are part of one of the most ancient eukaryotic lineages, phylogenetically distant from the model organisms already sequenced. They have unique structural features and have evolved distinct solutions to basic problems; for example they divide by multiple fission, locomote by gliding and induce the formation of new membrane compartments in the host cell. Not surprisingly, the proportion of *Plasmodium* products that have homologs in other organisms is the lowest among sequenced genomes. Annotation of *P. falciparum* chromosome 2 (Gardner *et al.*, 1998) and 3 (Bowman *et al.*, 1999) left about two-thirds of the predicted genes without function, either having no detectable homolog or a Plasmodium/Apicomplexa-specific homolog for which we have no functional information. Function was tentatively assigned to only a third of the predicted genes, but most of these significant matches remain only partially informative. They may reveal the biochemical activity of the product, inherent to the protein and irrespective of cellular context, for example a kinase or a phosphatase activity. They may also indicate the presence of a domain of known function, but in an otherwise unique molecular context. Obviously, homology searches lead to physiological function only for proteins that are involved in one of the core biological processes common to all eukaryotes.

In studying *Plasmodium* biology, the major questions concern the molecular basis of the features that define Apicomplexa protozoa, the traits that are specific to *Plasmodium*, and the parameters that influence disease such as transmission and virulence. Thus the central challenge is to be able to identify the parasite products that are critical to biological processes of interest. For this, we need molecular genetic tools for manipulating and questioning the genome in a variety of ways.

# 2. *Plasmodium* Transfection: A Brief Account of the First Milestones

*Plasmodium* was the last protozoan of medical importance to become amenable to molecular genetics. Transfecting *Plasmodium* was not an easy task, as the parasite spends most of its life located intracellularly within a vacuole, its nucleus being separated from the environment by four membranes. Also, *Plasmodium* DNA is particularly A/T-rich and unstable in *Escherichia coli*, which complicates preparation of transforming constructs. The first success in transfecting *Plasmodium* was reported in 1993, when D. Wirth and collaborators obtained transient gene expression after electroporation of extracellular gametes and zygotes in *P. gallinaceum*, an avian *Plasmodium* species (Goonewardene *et al.*, 1993). The decisive breakthroughs came in 1995, when the groups of T. Wellems working on *P. falciparum* and of C. Janse and A. Waters working on *P. berghei*, a species that infects rodents, could transfect erythrocytic stages of the parasite. Transfection was transient in *P. falciparum* (Wu *et al.*, 1995), and stable in *P. berghei* by means of a pyrimethamine-resistance gene (van Dijk *et al*, 1995). Using a similar selection system, three studies published the next year described integrative transfection, in *P. falciparum* (Wu *et al.*, 1996; Crabb and Cowman, 1996) and *P. berghei* (van Dijk *et al.*, 1996), which indicated a large if not complete dominance of homologous integration in both species. These seminal studies were then rapidly followed by reports on the inactivation of genes of interest (Ménard *et al.*, 1997; Crabb *et al.*, 1997a; Sultan *et al.*, 1997).

Although transfection has since been described in other *Plasmodium* species, *P. falciparum* and *P. berghei* have been the subject of all functional studies. A variety of molecular genetic approaches can now be taken in the two species. Theoretically, episomal or integrative transfection can each be used for either characterizing or identifying genes, as outlined in Figure 1. To propose a complete view of the *Plasmodium* genetic toolbox, we will consider these four situations successively.

# 3. Gene Characterization Using Episomal Transfection

## 3.1. Transient Transfection

Since its first use in 1993, transient transfection has been largely used to study gene expression in *Plasmodium*. Transient transfection plasmids only need to contain a reporter gene flanked by the sequences under study (Figure 1), and reporter genes encoding chloramphenicol acetyltransferase, firefly luciferase or green fluorescent protein (GFP) have been used to analyze the untranslated regions of many genes. The goal of most of these studies was to define by deletion mapping the minimal 5' and 3' regions that

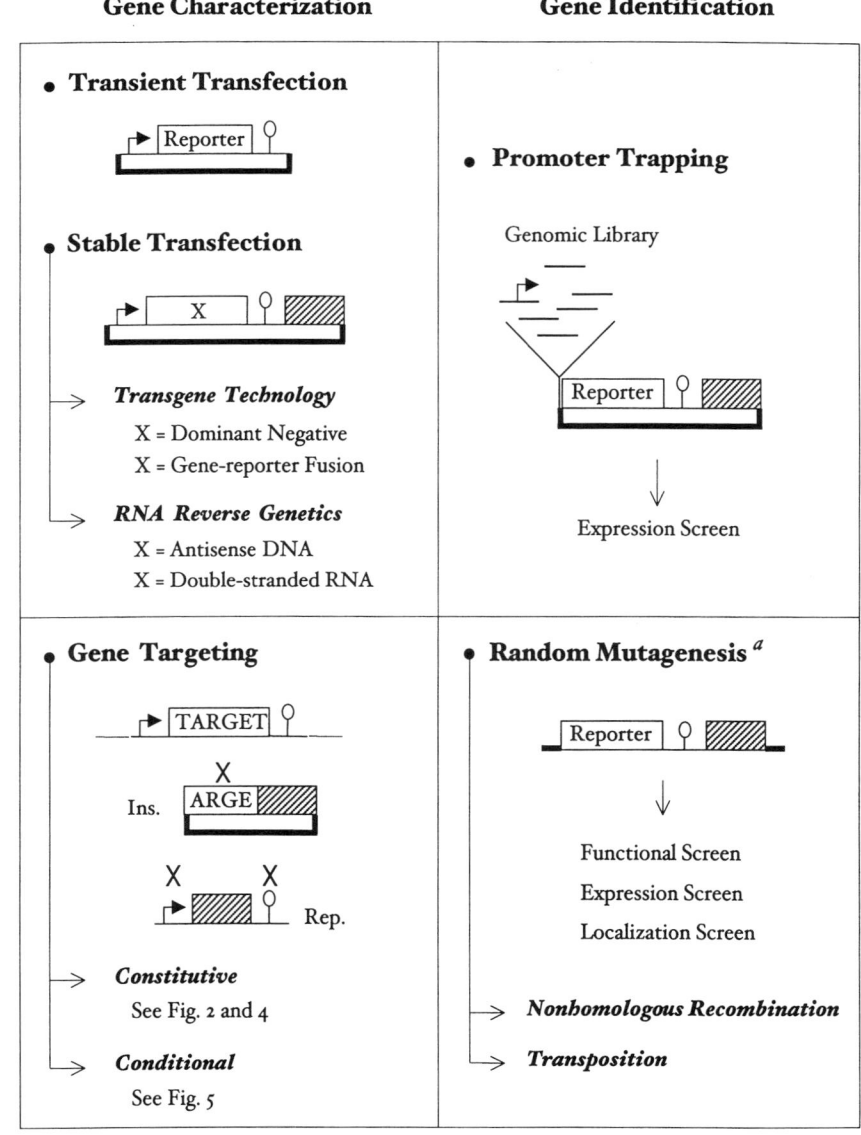

**Figure 1.** A theoretical view of molecular genetic techniques. The selectable marker and its expression sequences are symbolized by a grey box, the bacterial plasmid by thick lines, the gene promoter by an arrow, and the 3' untranslated sequences necessary for gene expression by an open circle. Ins, insertion plasmid; Rep, replacement plasmid. See text for details.

[a]Genes of interest can be identified by random insertional mutagenesis using a mutagenizing DNA (plasmid or transposon) that contains a reporter gene at its 5' end, as shown. Screens may then identify gene function (via gene inactivation), gene expression (using promoter-trap constructs, with a reporter lacking a promoter), or localization of the product (using gene-trap constructs, with a reporter lacking a start codon).

retained the capacity to efficiently drive gene expression (Wu *et al.*, 1995; Crabb and Cowman, 1996; Crabb *et al.*, 1997b; Horrocks and Kilbey, 1996; Dechering *et al.*, 1997). This information has been particularly valuable for constructing expression or resistance cassettes of minimum size. More recent transient transfection studies have initiated characterization of the DNA elements involved in gene expression (Horrocks and Lanzer, 1999) and stage-specific expression (Dechering *et al.*, 1999), or of the mechanisms that ensure expression of a single member of the *var* gene family (Deitsch *et al.*, 1999; 2001). Another study used transient transfection to demonstrate that readthrough of an internal stop codon was occurring in a *Pf60* gene (Bischoff *et al.*, 2000).

Still, little is known of how gene transcription is controlled in *Plasmodium*. *Plasmodium* promoters superficially resemble classical eukaryotic promoters transcribed by RNA polymerase II, consisting of a core promoter region controlled by upstream enhancer elements. However, they are functionally distinct from other eukaryotic promoters, as they do not function in mammalian COS cells and their sequences share no homology with any known transcription factor-binding site of eukaryotes (Horrocks *et al.*, 1998). In addition, promoters of SV40 or other viruses, which are ubiquitously active in higher eukaryotes, fail to drive reporter expression in *P. falciparum* (Horrocks *et al.*, 1998). What adds stage-specificity to gene expression in *Plasmodium* is also mysterious, although upstream elements may be involved (Dechering *et al.*, 1999). Unraveling the transcriptional machinery in *Plasmodium* will be important, as it may reveal new schemes of gene expression and lead to the development of new tools for timely expression of transgenes or mutations.

## 3.2. Stable Transfection

Multiple selectable markers are now available for stable episomal transfection in both *P. falciparum* and *P. berghei*. The most commonly used markers remain the original *Plasmodium* or *Toxoplasma DHFR-TS* variants that confer resistance to pyrimethamine, present in a variety of resistance cassettes (Wu *et al.*, 1996; Waters *et al.*, 1997; Crabb *et al.*, 1997b). In both *Plasmodium* species, transfectants can also be selected via resistance to the antifolate drug WR99210 encoded by a human *DHFR* gene (Fidock and Wellems, 1997; de Koning-Ward *et al.*, 2000a). Derivative markers now exist that confer both drug resistance and fluorescence via a GFP fusion (Sultan *et al.*, 1999b; Kadekoppala *et al.*, 2000). Other selectable markers that act independently from the folate pathway have been developed for stable episomal transfection in *P. falciparum* (Ben Mamoun *et al.*, 1999; de Koning-Ward *et al.*, 2001).

The fate of stably maintained plasmids is different in *P. berghei* and *P. falciparum*. In *P. berghei*, plasmids replicate as unrearranged monomeric units, with an average copy number of 15 per nucleus (van Dijk *et al.*, 1997). These plasmids appear to be fairly stable, even in the absence of drug pressure. In *P. falciparum*, however, the situation is more complex. Plasmids rapidly form large concatemers (Kadekoppala *et al.*, 2001), which probably emerge from inter plasmid homologous recombination. The structure and properties of these concatemers also appear to change with time (O'Donnell *et al.*, 2001). Initially small and unevenly segregated to the daughter merozoites, they become larger structures that after a few months are stably replicated even in the absence of selective pressure. Recently, a 1.4-kb sequence composed of 21-bp degenerate repeats, Rep20, has been shown to improve plasmid maintenance and to allow efficient segregation of plasmids in *P. falciparum* (O'Donnell *et al.*, 2002).

### 3.2.1. Transgene Expression

Gene function can be approached using transgene expression in several ways (Figure 1). (*i*) Over-expressing dominant-negative forms of a protein can generate a defective phenotype and thus inform on protein function. This strategy are is far limited to a few well-known protein families, and has the drawback of possible unspecific effects. (*ii*) GFP fusions of a protein can be produced to analyze its secretory pathway, as in the case of the insightful studies on the apicoplast-targeted and KAHRP proteins (Waller *et al.*, 2000; Wickham *et al.*, 2001). (*iii*) Modified versions of a gene can be expressed in a corresponding null mutant, although a serious limitation of episomes for addressing subtle structure-function relationships is the gene dosage effect due to the high number of replicating units.

Whenever possible, the transgene should be expressed from the natural expression regions of the target gene to minimize artifacts due to temporal misexpression of the product (Kocken *et al.*, 1998). In the future, tools for controlled gene expression should greatly help to refine episomal approaches, particularly for studies on the erythrocytic stages of the parasite. Transgene technology may nonetheless be limited by episome instability when studying mosquito stages of the parasite, on which drug pressure cannot easily be applied.

### 3.2.2. RNA Reverse Genetics

Another tool for probing gene function using episomes is antisense technology. Target gene expression can be suppressed by the annealing of antisense molecules to complementary transcripts, by a poorly understood mechanism that may affect transcript stability, processing, transport and/or translation. Both approaches of electroporating single-stranded antisense oligodeoxynucleotides (Barker *et al.*, 1996; 1998) and stably over-expressing antisense transcripts (Gardiner *et al.*, 2000) have been used with success in *Plasmodium* to disrupt endogenous mRNA function. Stage-specific or inducible expression of antisense RNAs may thus represent an alternative to gene manipulation for investigating protein function. It remains that the inhibitory activity of a given antisense molecule is difficult to predict, and that antisense approaches face the possible limitations of questionable specificity and incomplete efficacy.

RNA interference (RNAi) has emerged as a powerful alternative to antisense technology for specific degradation of target mRNA. RNAi appears to follow the processing of long, double-stranded RNA into 'short interfering' RNAs (21-23 nucleotide fragments), which guide the cleavage of homologous mRNA by the silencing complex RISC (Hammond *et al.*, 2001; Sharp, 2001). This evolutionarily conserved pathway, which may be part of a basic surveillance system that degrades transposon or viral messages, has already been harnessed as a reverse genetics tool. Injecting or expressing double-stranded RNAs causes the silencing of the corresponding gene in many systems tested, from protozoa to multicellular organisms. Degradation of target mRNA is specific and efficient, even with low concentrations of double-stranded RNA and regardless of the sequence chosen in the target gene. The silencing effect is particularly stable, which obviates the need for the extensive chemical modifications that are necessary for enhancing the half-life of antisense oligodeoxynucleotides.

The single RNAi study undertaken in *Plasmodium* is encouraging, showing a partial but apparently specific reduction of target mRNA levels (McRobert and McConkey, 2002). One limiting factor might be the low transfection frequencies in *Plasmodium*. Selection of interfered parasites would necessitate expressing double-stranded RNA from a selectable episome. A variety of constructs have already been devised for inducing stable interference in other systems. For example transcription can occur through inverted DNA repeats, giving rise to hairpin single-stranded RNA mimicking double-stranded RNA (Tavernarakis *et al.*, 2000; Shi *et al.*, 2000; Chuang and Meyerowitz, 2000), or from two opposing promoters, each giving rise to one strand of the double-stranded RNA (Wang *et al.*, 2000). Controllable and stage-specific expression of interfering constructs may thus become a handy tool for probing gene function in *Plasmodium*.

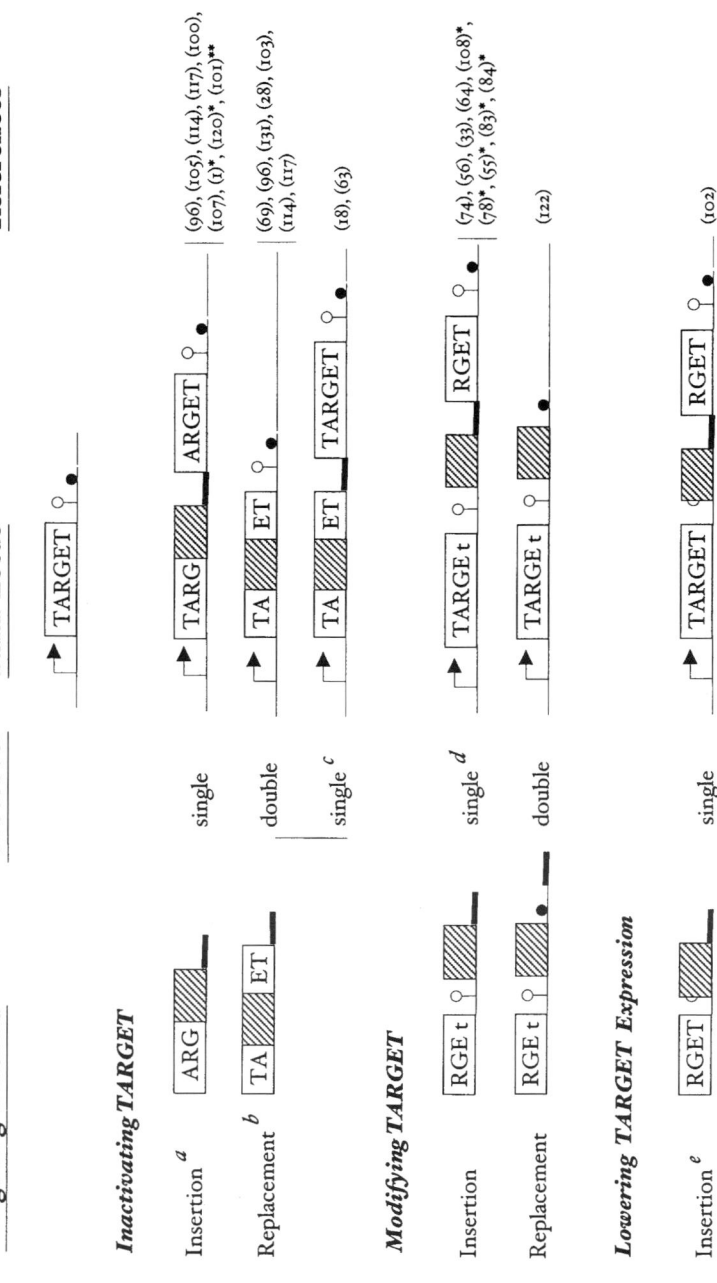

**Targeting Plasmid**     **Crossover**     **Final Locus**     **References**

*Inactivating TARGET*

Insertion [a]

single — (96), (105), (114), (117), (100), (107), (1)*, (120)*, (101)**

Replacement [b]

double — (69), (96), (131), (28), (103), (114), (117)

single [c] — (18), (63)

*Modifying TARGET*

Insertion

single [d] — (74), (56), (33), (64), (108)*, (78)*, (55)*, (83)*, (84)*

Replacement

double — (122)

*Lowering TARGET Expression*

Insertion [e]

single — (102)

**Figure 2.** Gene-targeting strategies used in *Plasmodium*. Gene targeting relies on homologous recombination between a genomic sequence (TARGET gene) and its homolog in the targeting construct (shown on the left). The latter can carry a single region of homology (insertion type) or two regions flanking the marker (replacement type). Insertion plasmids insert via a single crossover (SCO) between the pair of homologs (plasmid retained), while replacement fragments replace the target sequence via a double crossover (DCO) between pairs of homologs (plasmid lost). In *P. berghei*, these events are favored by linearizing the transformed DNA. In *P. falciparum*, linear DNA does not promote recombination, and circular replacement plasmids insert preferentially via a SCO between one pair of homologous sequences. Symbols are as in Figure 1; downstream 3' untranslated sequences are symbolized by a closed circle.

*a*The insertion plasmid must contain an internal fragment of the gene to generate two truncated gene duplicates in the final locus. References *, the first truncated copy is expressed because of the presence of a downstream 3' UTR sequence; **, one duplicate is the full-length gene.

*b*The replacement plasmid should be designed to delete part or all of the target coding sequence.

*c*The SCO shown involves the 5' regions of homology (TA). In this or the reverse case (a SCO between the 3' regions of homology, ET), a full-length target gene copy is created.

*d*For the gene modification (t) to be recovered in the first, expressed and full-length duplicate, the SCO must occur upstream from the modification. Reference *, the 3' UTR is not the natural 3' UTR of the gene.

*e*The 3' sequence necessary for gene expression is truncated.

# 4. Gene Characterization Using Integrative Transfection

## 4.1. Gene Targeting: Current Status

Gene targeting by homologous recombination is arguably the most informative approach to protein *in vivo* function. It is more reliable and predictable than antisense or dominant-negative approaches, and permits a detailed analysis of protein structure-function relationships. The *Plasmodium* genome is haploid, contains mostly single-copy genes and integrates exogenous DNA by ~100% homologous recombination. Thus, for most genes a single recombination event is sufficient for generating a modified parasite clone. In fact, despite the relative youth of transfection technology in *Plasmodium*, and its relative inefficiency (stable transfection frequencies have been evaluated at ~$10^{-6}$ in both *P. falciparum* and *P. berghei*), numerous gene-targeting studies have already been performed. They have revealed important insights into such diverse processes as drug resistance, cell invasion by the various invasive stages of the parasite, sexual differentiation, or cytoadherence of infected erythrocytes. Although usually genes have been inactivated, several genes have been modified and in one case expression levels have been diminished. Figure 2 illustrates the strategies that have been used and Figure 3 shows the *Plasmodium* loci that have been targeted, as of february 2002. Previous reviews have described construct design and selection protocols (Waters *et al.*, 1997; Ménard and Janse, 1997; Tomas *et al.*, 1998; de Koning-Ward *et al.*, 2000b; Ménard and Nussenzweig, 2000).

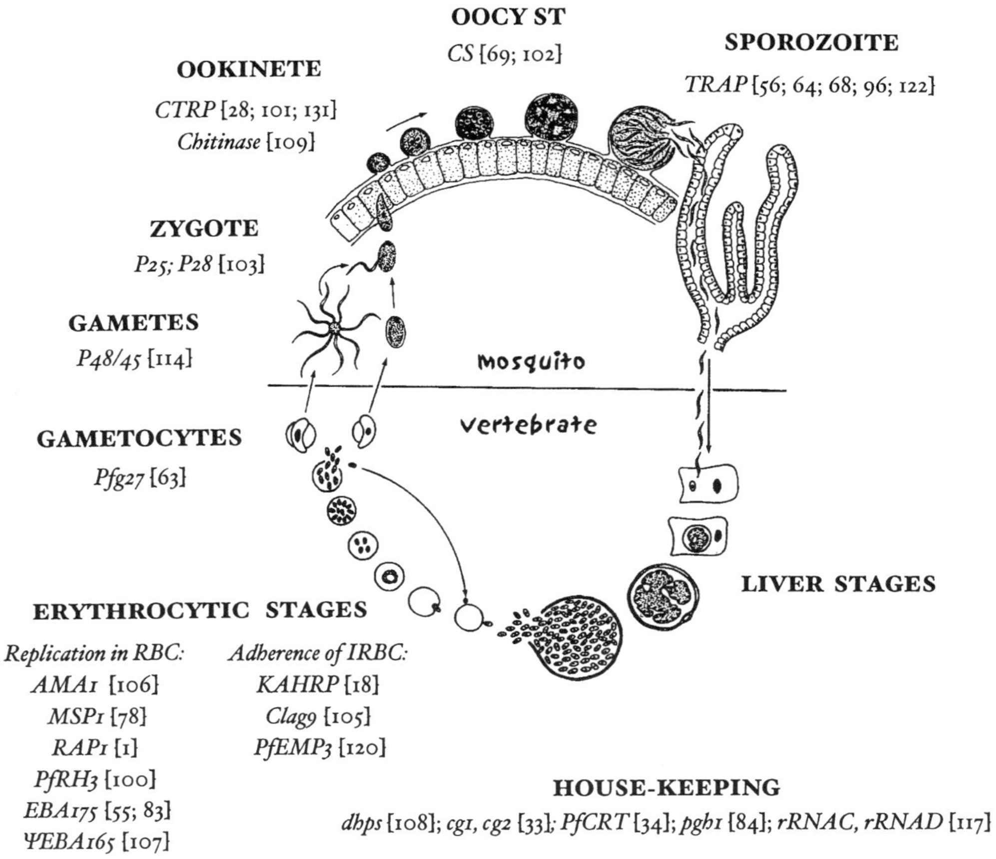

**OOCY ST**
*CS* [69; 102]

**OOKINETE**
*CTRP* [28; 101; 131]
*Chitinase* [109]

**SPOROZOITE**
*TRAP* [56; 64; 68; 96; 122]

**ZYGOTE**
*P25; P28* [103]

**GAMETES**
*P48/45* [114]

**GAMETOCYTES**
*Pfg27* [63]

**ERYTHROCYTIC STAGES**

*Replication in RBC:*
*AMA1* [106]
*MSP1* [78]
*RAP1* [1]
*PfRH3* [100]
*EBA175* [55; 83]
*ΨEBA165* [107]

*Adherence of IRBC:*
*KAHRP* [18]
*Clag9* [105]
*PfEMP3* [120]

**LIVER STAGES**

**HOUSE-KEEPING**
*dhps* [108]; *cg1, cg2* [33]; *PfCRT* [34]; *pgh1* [84]; *rRNAC, rRNAD* [117]

**Figure 3.** Targeted *Plasmodium* genes. The *Plasmodium* life cycle in the two hosts, a vertebrate (lower part) and a mosquito (upper part), is shown. The parasite genes that have been manipulated by gene targeting are indicated below the respective parasite stage, with references in parentheses.

Homologous recombination provides a versatile system for manipulating the *Plasmodium* genome. On the one hand, double crossovers can span and delete tens of kilobases, and be used to introduce large deletions at chromosome ends (Pace *et al.*, 2000). On the other hand, as few as ~300 bp of homology (and possibly less) are sufficient for crossover formation and plasmid insertion is associated with short gene conversion tracts (Nunes *et al.*, 1999). Therefore, point mutations can be introduced using small insertion plasmids. Also, strict homology between the targeting and target sequences is not required for productive recombination. For example, transfection in *P. berghei* with targeting vectors containing sequences from the C-rRNA gene resulted in disruption of the C- as well as the D-rRNA gene, which differ in ~5% of their sequence (van Spaendonk *et al.*, 2001). This implies that a targeting vector may occasionally integrate by homologous recombination elsewhere than at the expected locus, especially when recombination occurs between highly A/T-biased 3' or 5' untranslated sequences. Finally, the multiplicity of selectable markers permits complementation experiments, which provide definitive proof for the involvement of a protein in a defective phenotype (Sultan et al., 1999a; 2001; Thathy *et al.*, 2002). The tools are thus available to perform DNA reverse genetics in *Plasmodium* according to the molecular Koch postulates of S. Falkow (1988).

There are important differences between gene-targeting procedures and their outcomes in *P. falciparum* and *P. berghei*. The targets of electroporation are intraerythrocytic forms of the parasite (rings to schizonts) or extracellular merozoites, and selection occurs *in vitro* or in rodents, respectively. Crucially, linear DNA is the preferred substrate for homologous recombination in *P. berghei*, but not in *P. falciparum*. Linear DNA is presumably degraded in *P. falciparum* when it crosses the four membranes to the parasite nucleus, and this has two important consequences. One is that the time required for selecting integrants is longer in *P. falciparum* (3-4 months versus 2 weeks in *P. berghei*), because circular plasmids preferentially replicate episomally than integrate into the genome. Another consequence is that double crossover events can hardly be selected in *P. falciparum*, because circular replacement plasmids preferentially integrate via single crossovers (see Figure 2).

One way to recover the rare double crossovers that may occur when transfecting circular replacement plasmids is to use a negative marker to counterselect the other transfection products (Figure 4). Two negative selectable markers, cytosine deaminase and thymidine kinase, have already been developed in *P. falciparum* (Duraisingh *et al.*, 2002). They will also serve for conducting more reliable protein structure-function analysis in both *Plasmodium* species. Indeed, so far all subtle gene modifications have been introduced in the presence of a selectable marker, which as a new transcription unit may affect gene expression in the targeted or unlinked loci in unpredictable ways. Figure 4 shows the classical 'hit and run' procedure employed to circumvent this drawback and to introduce mutations in a final locus devoid of exogenous sequence, based on the sequential use of positive and negative selection.

## 4.2. Limitations of Gene Targeting in *Plasmodium*

Despite these exciting achievements, there are still numerous genes whose function cannot be properly investigated. This is the case of genes involved in parasite replication in erythrocytes (on which selection is based), including those important for merozoite invasion of erythrocytes, the most scrutinized step of the parasite life cycle and a primary vaccine target. Loss-of-function mutants in these genes die or are overgrown by non-targeted parasites. Therefore, with currently available tools, the best possible evidence that a gene is important for invasion of erythrocytes is when it can be targeted with

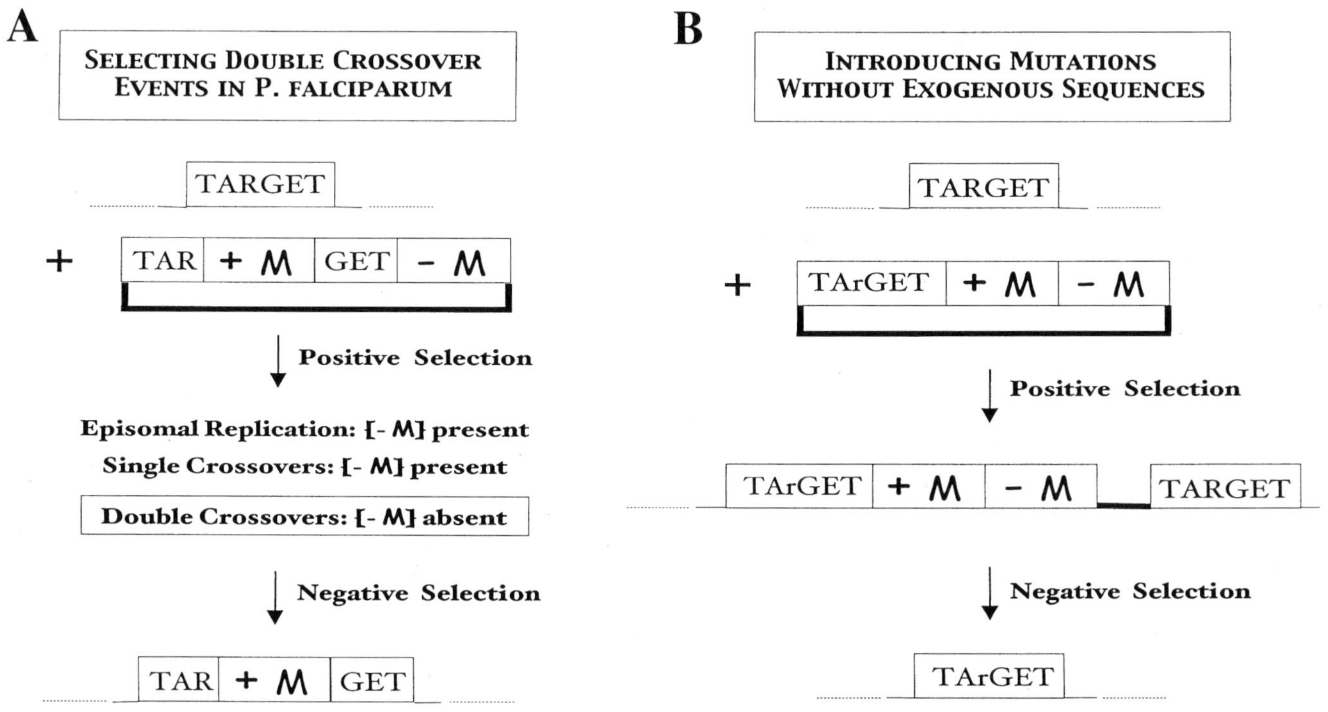

**A** SELECTING DOUBLE CROSSOVER EVENTS IN P. FALCIPARUM

TARGET

+

| TAR | + M | GET | – M |

↓ Positive Selection

Episomal Replication: [– M] present
Single Crossovers: [– M] present

Double Crossovers: [– M] absent

↓ Negative Selection

| TAR | + M | GET |

**B** INTRODUCING MUTATIONS WITHOUT EXOGENOUS SEQUENCES

TARGET

+

| TArGET | + M | – M |

↓ Positive Selection

| TArGET | + M | – M | TARGET

↓ Negative Selection

| TArGET |

**Figure 4.** Using negative selectable markers in *Plasmodium* gene targeting. A negative marker allows for selecting parasites that do not express the marker, generally by conferring susceptibility to a drug. The negative marker is symbolized –M, the positive marker by +M, and the bacterial plasmid by a thick line. A) Upon positive selection, circular replacement plasmids will preferentially replicate episomally, rarely integrate via a single crossover (TAR or GET), and should also integrate via a double crossover (TAR and GET). For recovering the latter, a negative marker can be placed in the construct as shown, and negative pressure applied to counterselect episomes and single crossover integrations that all maintain the negative marker. B) Shown here is the modification of a target gene (R to r) via a hit and run procedure. Positive selection recovers integration of the insertion plasmid (hit), which introduces the modification (r, shown here ending in the first gene duplicate after SCO between TA regions). Negative selection on such an integrant clone will recover parasites that undergo intrachromosomal recombination and plasmid excision (run), while leaving the modifcation (after SCO between GET regions). The reverse situation, i.e., a hit step via SCO between GET regions and a run step via SCO between TA regions also leads to a modified gene.

a nondisruptive construct but not with a disruptive construct (Cowman *et al.*, 2000). This was reported for *MSP-1* and *AMA-1*, along with direct evidence for their role in merozoite invasion via gain-of-function mutants created by trans-species exchange between human and rodent homologs (O'Donnell *et al.*, 2000; Triglia *et al.*, 2000). Nonetheless, failure to select loss-of-function mutants may be a misleading criterium for identifying important genes among uncharacterized sequences, given the poor targeting frequencies in *Plasmodium*. In addition, since impaired mutants cannot be selected, the defective phenotype and actual protein function cannot be studied.

Other genes that cannot be at present fully characterized are those encoding multifunctional proteins. Knocking-out these genes only reveals the earliest non-redundant role of their product. One example is CS, known to be involved in sporozoite adhesion to the mosquito salivary glands and to mammalian hepatocytes, but which is first essential for sporozoite formation in the oocyst (Ménard *et al.*, 1997; Thathy *et al.*, 2002). Thus the role of sporozoite surface-associated CS, the leading vaccine candidate against pre-erythrocytic stages of the parasite, cannot be dissected genetically. It is clear that unrestricted functional analysis of the genome requires the tools for activating or silencing genes at will.

## 4.3. Missing Tools: A Brief Overview of Conditional Mutagenesis

Two types of tools have been widely used for conditional gene expression in eukaryotes: transcriptional regulators and site-specific recombinases. Their basic mechanisms of action and some of the possible applications for studying *Plasmodium* essential genes are illustrated in Figure 5.

The most popular transcriptional regulatory systems are derived from the tetracycline resistance operon of bacterial Tn*10* (Gossen and Bujard, 1992; Baron *et al.*, 1999; Urlinger *et al.*, 2000). They have been developed

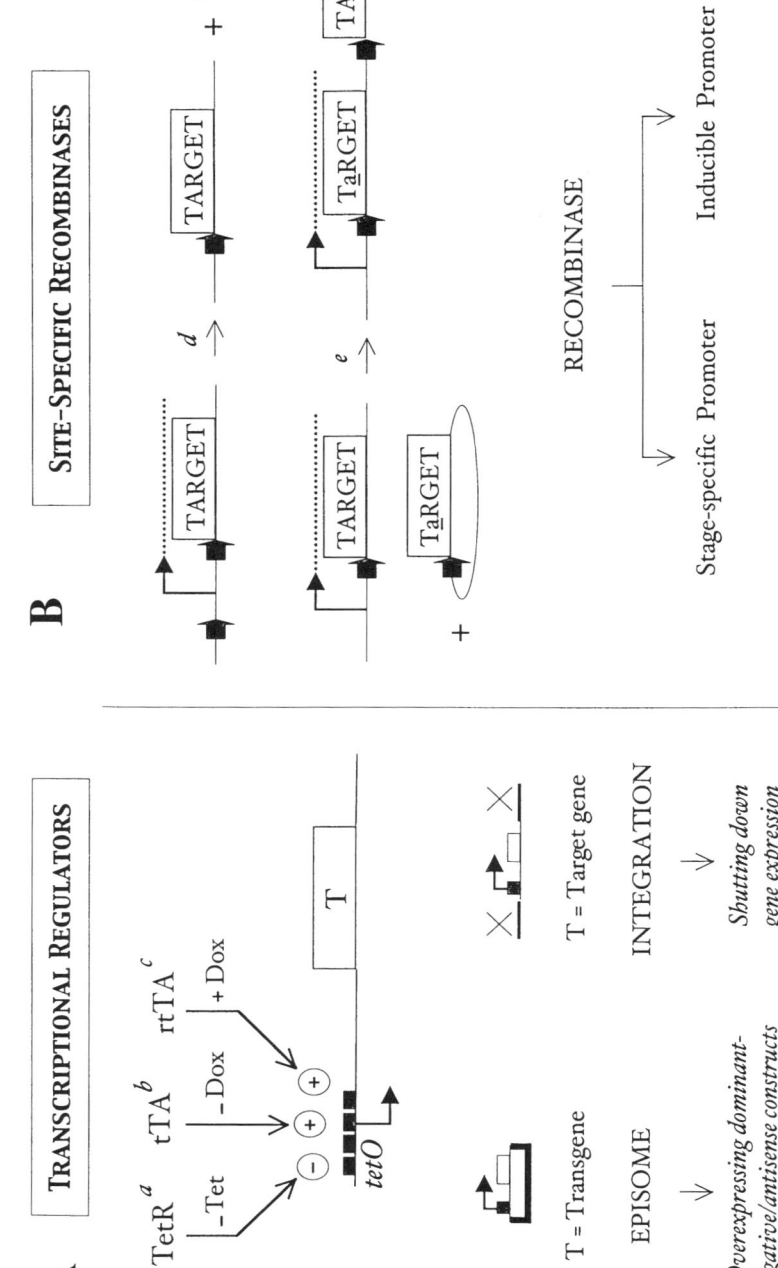

**A**  TRANSCRIPTIONAL REGULATORS

TetR[a]   tTA[b]   rtTA[c]

−Tet   −Dox   +Dox

T = Transgene        T = Target gene

EPISOME              INTEGRATION

*Overexpressing dominant-*        *Shutting down*
*negative/antisense constructs*    *gene expression*

**B**  SITE-SPECIFIC RECOMBINASES

RECOMBINASE

Stage-specific Promoter        Inducible Promoter

**Figure 5.** Tools for conditional gene expression. A) A promoter can be made drug-responsive by inserting 2 to 7 copies of 19-bp *tetO* (tetracycline operator) sequences around the transcriptional initiation site of the promoter, which are recognized by a regulator: TetR, tTA or rtTA. Left, a drug-responsive transgene can be borne by an episome and used to express dominant-negative constructs. Right, a controllable copy of a target gene can also be used to replace its chromosomal copy by double crossover. Ideally, the desired state (overexpression or tight repression) should be obtained by adding rather than removing the effector, because the former situation is associated with more rapid kinetics of expression switch.
[a]TetR (tetracycline repressor) is limited by a narrow range of control. TetR dissociates from tetO upon tetracycline binding, leading to gene transcription.
[b]tTA (tetracycline-controlled transactivator) is best suited for rapid repression of gene expression and knock-out approaches. In the absence of doxycycline, the gene is expressed; upon doxycycline addition, the gene is rapidly repressed.
[c]rtTA (reverse tetracycline-controlled transactivator) is best suited for rapid expression of a transgene and dominant-negative approaches. In the absence of doxycycline, rtTA does not bind to *tetO* and the transgene is not expressed; upon doxycycline addition, it is rapidly expressed.
B) Cre (Flp) catalyses a recombination reaction between two identical 34-bp recognition sites called *lox*P (*FRT*). When the two sites are located on the same molecule (chromosome), recombination will excise (invert) intervening DNA if the sites are in the same (opposite) orientation. Shown here is the deletion of the target gene promoter. When one site is on the linear chromosome and the other on a circular plasmid, recombinase inserts the plasmid at the chromosomal site. Shown here is plasmid integration leading to gene modification. Timely expression of the recombinase may rely on stage-specific or inducible promoters, or other approaches.
[d]intrachromosomal deletion/excision is reversible but is energetically favoured over intermolecular integration.
[e]the inherently unstable insertion product can be obtained by limited expression of the recombinase or by using mutant sites that are refractory to further excision.

into increasingly efficient tools for controlling gene expression in model organisms from yeast to rodents, and have been used with success in various protozoan parasites (Wirtz and Clayton, 1995; Hamann *et al.*, 1997; Wirtz *et al.*, 1999; Meissner *et al.*, 2001; Yan *et al.*, 2002). They consist of (i) a regulator (repressor or activator), (ii) operator sequences, which must be positioned around the transcriptional start site, and (iii) an effector (tetracycline or derivative) for modulating the regulator-operator interaction and turning 'on' or 'off' gene expression. When a tetracycline-responsive copy of a gene is borne by an episome, it can only be used for timely over-expression of dominant-negative, antisense or interfering constructs. A more direct approach is to insert the controllable copy in place of the endogenous gene by homologous recombination. This enables to shut down gene expression in all recombinants at a chosen time, and to directly assess the consequences of the gradual loss of the product. Inducible promoters come with the clear advantages of reversibility, in allowing to generate truly conditional 'on' and 'off' states in one clone, and flexibility, in enabling to create intermediary or temporary states that may be as informative on protein function as a constitutively 'off' state. In some situations, however, their efficiency will ultimately depend on whether complete repression can be obtained, as well as on the kinetics of repression after effector addition/ removal. Studies on essential products of parasite erythrocytic stages would greatly benefit from these tools, particularly the transactivators suited to conditional gene silencing (see Figure 5 legend). The situation seems ideal for *P. falciparum*, which replicates in erythrocytes *in vitro* where effector levels can be more easily controlled.

A second way to inactivate a gene in a temporally restricted manner is offered by site-specific recombinases. Two site-specific recombinases of the λ integrase family have been used for this purpose in a variety of eukaryotes: Cre of bacteriophage P1 and Flp of yeast (Sauer, 1998; Porter, 1998). These enzymes catalyse a reciprocal conservative recombination between two identical 34-bp target sequences and, depending on their position and orientation, recombination will insert, invert, or delete DNA (see the Figure 5 legend). Therefore, these systems offer the primary advantage of enabling not only to inactivate but also to modify or swap genes, and thus to investigate protein structure-function relationships. Recombination occurs regardless of DNA topology and host environment, and the Cre/*lox*P system has been shown to function in the apicomplexan *Toxoplasma gondii* (Brecht *et al.*, 1999). Although it should be easy to design modifications and insert the *LoxP* or *FRT* site(s) into the *Plasmodium* genome by homologous recombination, the challenge is to express the recombinase conditionally. A first possibility would be to use natural stage-specific promoters. If these prove not to be leaky before being activated, they would then be useful for truly conditionally inducing gene rearrangements at a defined stage of the parasite life. They would allow studies on essential genes not only in that particular stage, but also in erythrocytic stages after complete cycling of the parasite. A gene important for merozoite invasion could for example be deleted in a mosquito stage of the parasite, and after transmission to the mammalian host its function could be assessed in merozoite formation in the liver and in subsequent merozoite invasion of erythrocytes. Directly applying recombinase systems to parasite erythrocytic stages would require expressing the recombinase from an inducible promoter or using one of the recombinase variants that can be activated by an exogenous factor (Metzger and Chambon, 2001). It may become possible to use, as was recently performed with mammalian cells (Jo *et al.*, 2001), a cell-permeable recombinase that could be directly added to cells/parasites bearing a manipulated ('floxed' or 'flrted') gene. As increasingly sophisticated site-specific recombination systems are being developed, their usefulness in *Plasmodium* should be evaluated as they would offer virtually unlimited ways of analyzing the function of *Plasmodium* essential genes.

# 5. Gene Identification

*Plasmodium* has been transfected almost exclusively for testing promoter activity or gene function, and reverse genetic techniques are now well established. On the other hand, attempts to identify genes and develop forward genetic screens have been scarce.

## 5.1. Using Episomal Transfection (Promoter Trapping)

In bacteria, episomal transfection has been widely used to identify genes that are induced in response to defined conditions. The basic method consists in fusing a genomic library to a promoterless reporter gene whose product confers a selectable or easily screenable phenotype, for example antibiotic resistance (Figure 1). A simple variant, called differential fluorescence induction, uses green fluorescent protein as the reporter and relies on fluorescence-activated cell sorting (FACS) to isolate bacteria with active transcriptional fusions (Valdivia and Falkow, 1997). In *Plasmodium*, a similar promoter-trap strategy could in theory be used for isolating promoters that are active during any step of the parasite life. A *Plasmodium* genomic library of 1-kb average insert size could be scanned in a few minutes by FACS, and active promoters rescued from fluorescent parasites. However, the tens of thousands of clones necessary to cover the genome still represent many individual transfections, given the low transfection frequencies. Also, although it is clear that stage-specific promoters can be active when carried by episomes (Sultan *et al.*, 1999a; 2001), little is known of their regulation throughout the cycle. In *P. falciparum*, episomes apparently do not properly assemble chromatin (Horrocks *et al.*, 1998), a requirement for the correct developmental expression of many eukaryotic genes. Another problem is that distinct transfected plasmids may assemble into concatemers (Kadekoppala *et al.*, 2001). Therefore episomal transfection is not presently a suitable approach for identifying *Plasmodium* genes based on their expression profile.

## 5.2. Using Integrative Transfection (Insertional Mutagenesis)

A powerful way to identify genes that mediate biological processes, particularly in haploid organisms, is based on random mutagenesis and screening the resulting mutants for a defect in a phenotype of interest. Mutagenesis is typically induced by nonhomologous integration of a plasmid or insertion of a transposon, two methods that tag the mutated site and facilitate its recovery. In theory, saturation mutagenesis permits identification of the function of any gene whose inactivation is not immediately lethal, and for which an appropriate selection or screen is available (for examples in bacterial pathogenesis studies, see Strauss and Falkow, 1997; Chiang *et al.*, 1999; Wren, 2000). In protozoa, efficient random mutagenesis has been reported only in *Toxoplasma* and *Leishmania*, allowing in both cases to select for gene fusions and trap new genes. In *Toxoplasma gondii*, nonhomologous recombination can be obtained by incorporating discontinuous genomic DNA in transfection constructs, and current screens are targeting parasite genes induced by the transition from the tachyzoite to the bradyzoite stage (Roos *et al.*, 1997). In *Leishmania major*, the *Mos1* element of the *mariner*/Tc1 family of transposons, which are ubiquitous elements of eukaryotic genomes, transposes efficiently ($\sim10^{-4}$; Gueiros-Filho and Beverley, 1997).

Unfortunately, such methods are not in sight in *Plasmodium*. Nonhomologous recombination does not occur using currently used vectors, or with frequencies incompatible with gene discovery. Transposition was reported only once, using the *mariner* element, but with apparently low efficiency (Ben Mamoun *et al.*, 2000). Even if tools for random DNA insertion into the *Plasmodium* genome can be established, their utility for gene discovery would also necessitate increasing frequencies of transfection.

### 5.3. What are the Prospects for Functional Genomics in *Plasmodium*?

In the absence of appropriate molecular genetic tools, genomic techniques will be crucial for classifying genes according to their pattern of expression (Figure 6). High-redundancy methods can be useful for providing transcriptome snapshots, such as massive cDNA sequencing projects (Carlton *et al.*, 2001) and serial analysis of gene expression (Munasinghe *et al.*, 2000; Patankar *et al.*, 2001). Several genome-wide techniques that compare relative levels of mRNAs in two conditions are also being applied to *Plasmodium*, including DNA microarray hybridization (Hayward *et al.*, 2000; Ben Mamoun *et al.*, 2001), subtractive suppressive hybridization (Dessens *et al.*, 2000), and differential display (Lau *et al.*, 2000; Cui *et al.*, 2001). So far these techniques have been used mainly for analyzing expression profiles in erythrocytic stages of the parasite, the only stages that yield the necessary amounts of mRNA. To facilitate similar studies with mosquito or liver stages, tools are being developed that should help to purify the small available numbers of parasites by FACS (Natarajan *et al.*, 2001) or laser capture microdissection (Sacci *et al.*, 2002). All these technologies will permit to down scale the genome to its expressed portion during a process of interest and to identify stage-specific genes. More focused screens (e.g. involving drug-treated or mutant parasites, or parasites in *ex* or *in vivo* conditions) may narrow down to smaller subsets of co-expressed genes and provide sharper leads to investigators. But it is likely that in most cases these global mRNA techniques will leave us with large numbers of differentially expressed genes. As for the entire genome, sequencing will hardly by itself constitute a rationale for further analysis, although sequence may occasionally suggest function. The challenge remains to translate the flow of expression data into biological activities.

# Functional Genomics in *Plasmodium*

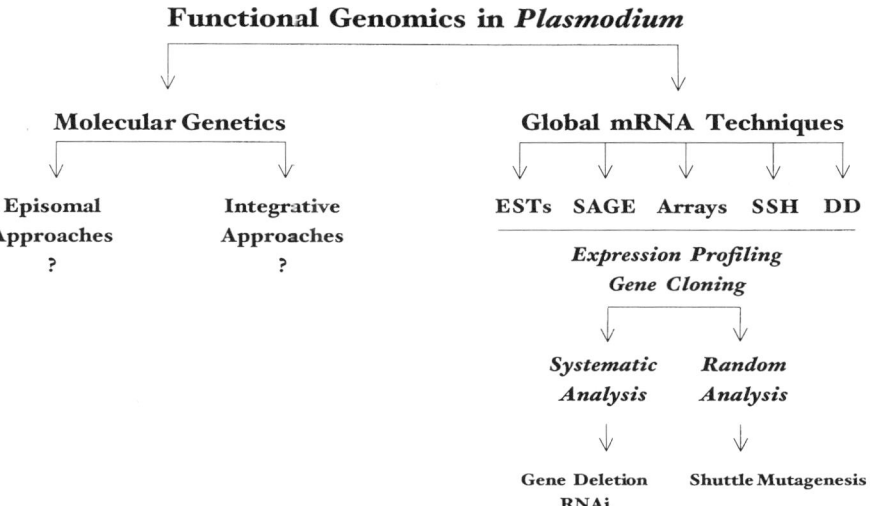

**Figure 6.** An outlook at functional genomics in *Plasmodium*.

Function could be addressed by a systematic, gene-by-gene approach. One possibility would be to systematically delete expressed genes by homologous (double crossover) recombination, and generate null mutants. However, the transition from gene sequence to parasite mutant takes at least 12 weeks in the relatively handy *P. berghei* system, making such large-scale functional studies impractical in most laboratories. If it proves reliable in *Plasmodium*, RNAi technology would be a more rapid method for testing the function of many genes. Systematic functional studies using RNAi have been performed against the products encoded by entire chromosomes in *Caenorhabditis elegans* (Fraser *et al.*, 2000; Gönczy *et al.*, 2000) or the components of complete pathways in *Drosophila* (Clemens *et al.*, 2000) (reviewed in Kuwabara and Coulson, 2000; Barstead, 2001). RNAi would be an efficient way to mine the *Plasmodium* genome for potential drug targets, or to screen for important genes that may deserve further analysis by homologous recombination.

Yet for most laboratories, a direct screen for genes of interest or essential genes would be a more appealing prospect than the gene-by-gene approach. New opportunities for generating random mutants in *Plasmodium* may arise from the construction of (differentially) expressed gene libraries, which reduces the initial pool of genes and allows their mutagenesis in other organisms. Cloned genes could for example be mutagenized in *E. coli* and mutated alleles subsequently introduced into *Plasmodium* for replacement of their chromosomal copy by homologous recombination. Such 'shuttle mutagenesis' has been used in yeast, after Tn*3* or Tn*7* mutagenesis in *E. coli* (Kumar and Snyder, 2001). There is now a wide choice of mutagenizing agents, including multifunctional transposons which use the same insertion event to determine (i) when the gene is expressed (via reporter fusion), (ii) where the product is localized in the cell (via formation of epitope-tagged products), and (iii) the consequence of the absence of the product (via gene inactivation) (Ross-Macdonald *et al.*, 1997; 1999). To avoid having to screen mutants individually, molecular barcodes have been developed for bacterial pathogenesis studies (Hensel *et al.*, 1995) and yeast functional genomics (Shoemaker *et al.*, 1996; Winzeler *et al.*, 1999). These short sequences serve as clone identifiers and allow large numbers of mutants to be pooled and analyzed simultaneously by comparative hybridization on filters or high-density arrays. In bacteria for example, these tags associated with classical transposition (signature-tagged transposition method, STM) have served to isolate mutants that were unable to survive in the host (Hensel *et al.*, 1995). Establishing such tools in *Plasmodium* would permit to envisage focused approaches to virtually any aspect of parasite biology.

Animal models of malaria should be particularly valuable for functional genomic studies and tackling basic aspects of parasite biology. Rodent *Plasmodium* species, including *P. berghei* and *P. yoelii* (Mota *et al.*, 2001), are practical because they can be studied routinely and safely in the laboratory, and *in vivo* throughout their life cycle. Double crossover recombination is readily obtained with linear DNA, allowing in principle shuttle strategies. An additional attractive feature of rodent systems is that the three actors (parasite, mosquito and vertebrate host) can be genetically manipulated, and the sequence of their genome is, or will soon be known. *P. knowlesi* and *P. cynomologi*, which infect primates and are closely related to the human parasite *P. vivax*, can also be manipulated by double crossover recombination (van der Wel, 1997; Kocken *et al.*, 1999; 2002). However, their use for large-scale studies is prohibited by ethical and practical reasons. *P. falciparum* remains the mandatory target for studying specific virulence traits, such as cytoadherence of infected erythrocytes. This system offers the advantage of an erythrocytic cycle that can be studied *in vitro* and synchronized, but is limited by the difficulty to produce mosquito stages of the parasite and the time consuming molecular genetic procedures. A precise understanding of malaria pathogenesis will necessitate that each system contributes its part.

# 6. Conclusion

The landscape of malaria research has changed dramatically in the last decade. The sequence of the genome of several *Plasmodium* species is now known, genomic techniques have been developed, and the parasite can be transfected. The molecular genetics toolbox, however, is far from complete. On the one hand, understanding the function of a given gene (reverse genetics) is straightforward, and we should soon have the tools for manipulating any gene. On the other hand, identification of genes based on their function (forward genetics) is still problematic. The powerful genomic techniques will continue to categorize the genome into subsets of interest, and may suggest function of groups of genes, but only constitute a first step. Molecular genetic methods must be adapted to translate the wealth of sequence and expression data into biological functions, and to link them to investigator-driven research addressing specific questions that can only be answered by a reductionist approach.

These are exciting times for the malaria research community. The blending of these new technologies will lead to an increasingly sophisticated view of parasite biology, and uncover the molecular details behind the unique features of this ancient eukaryote. More importantly, they hold great promise to help reducing the burden of malaria in allowing a systematic hunt for drug targets and a rational choice of vaccine candidates, and will certainly lead us to other intervention strategies that are now unforseeable.

# 7. Acknowledgements

We thank Alka Agrawal, Patricia Baldacci, Freddy Frischknecht, Hiroshi Sakamoto, and Sabine Thiberge for their review of the manuscript and many helpful comments.

# 8. References

Baldi, D.L., Andrews, K.T., Waller, R.F., Roos, D.S., Howard, R.F., Crabb, B.S., and Cowman, A.F. 2000. RAP1 controls rhoptry targeting of RAP2 in the malaria parasite *Plasmodium falciparum*. EMBO J. 19: 2435-2443.
Integration of an insertion plasmid at *RAP1* (rhoptry associated protein 1) generates a parasite clone that produces a RAP1 truncate. In these parasites, the components of the rhoptry low molecular weight complex (RAP1-3) fail to associate, indicating that the complex is not important for parasite invasion of red blood cells.
Barker, R.H.Jr., Metelev, V., Coakley, A., and Zamecnik, P. 1998. *Plasmodium falciparum*: effect of chemical structure on efficacy and specificity of antisense oligonucleotides against malaria *in vitro*. Exp. Parasitol. 88: 51-59.
Barker, R.H.Jr., Metelev, V., Rapaport, E., and Zamecnik, P. 1996. Inhibition of *P. falciparum* malaria using antisense oligodeoxynucleotides. Proc. Natl. Acad. Sci. USA. 93: 514-518.
Baron, U., Schnappinger, D., Helbl, V., Gossen, M., Hillen, W., and Bujard, H. 1999. Generation of conditional mutants in higher eukaryotes by switching between the expression of two genes. Proc. Natl. Acad. Sci. USA. 96: 1013-1018.
Barstead, R. 2001. Genome-wide RNAi. Current Opin. Chem. Biol. 5: 63-66.
Ben Mamoun, C., Gluzman, I.Y., Beverley, S.M., and Goldberg, D.E. 2000. Transposition of the *Drosophila* element *mariner* within the human malaria parasite *Plasmodium falciparum*. Mol. Biochem. Parasitol. 110: 405-407.
The *Drosophila* transposable element *mariner* is a member of the Tc1 family of transposons, which are ubiquitous elements of eukaryotic genomes capable of

mobilization independently of host factors. The transposon was introduced in *P. falciparum* by stable episomal transfection; few insertions in the parasite genome occurred, a number of which were in the same gene (protein kinase A).

Ben Mamoun, C., Gluzman, I.Y., Goyard, S., Beverley, S.M., and Goldberg, D.E. 1999. A set of independent selectable markers for transfection of the human malaria parasite *Plasmodium falciparum*. Proc. Natl. Acad. Sci. USA. 96: 8716-8720.

Describes episomal transfection in *P. falciparum* using *BSD* (blasticidin S deaminase) of *Aspergillus terreus*, which confers resistance to blasticidin, and *NEO* (neomycin phosphotransferase II) of transposon Tn5, which confers resistance to geneticin/G418.

Ben Mamoun, C., Gluzman, I.Y., Hott, C., MacMillan, S.K., Amarakone, A.S., Anderson, D.L., Carlton, J.M.-R., Dame, J.B., Chakrabarti, D., Martin, R.K., Brownstein, B.H., and Goldberg, D.E. 2001. Co-ordinated programme of gene expression during asexual intraerythrocytic development of the human malaria parasite *Plasmodium falciparum* revealed by microarray analysis. Mol. Microbiol. 39: 26-36.

Microarrays were constructed with 944 amplified inserts from *P. falciparum* cDNA expressed sequenced tags of known sequence (representing ~15% of the estimated number of parasite genes), and hybridized with Cy5-labelled cDNA from 5 intraerythrocytic stages of the parasite: rings, early and late trophozoites, and early and late schizonts. These stages express distinct sets of genes that can be clustered into functional groups: genes involved in protein synthesis, glucose metabolism and energy production, erythrocyte invasion.

Bischoff, E., Guillotte, M., Mercereau-Puijalon, O., and Bonnefoy, S. 2000. A member of the *Plasmodium falciparum* Pf60 multigene family codes for a nuclear protein expressed by readthrough of an internal stop codon. Mol. Microbiol. 35: 1005-1016.

Transient transfection is used to demonstrate that translation occurs through an internal ochre codon in the 6.1 member of the Pf60 gene family, as a luciferase reporter inserted downstream from the internal stop codon can be expressed.

Bowman, S., *et al.* 1999. The complete nucleotide sequence of chromosome 3 of *Plasmodium falciparum*. Nature. 400: 532-538.

Brecht, S., Erdhart, H., Soete, M., and Soldati, D. 1999. Genome engineering of *Toxoplasma gondii* using the site-specific recombinase Cre. Gene. 234: 239-247.

Carlton, J.M.R., *et al.* 2001. Profiling the malaria genome: a gene survey of three species of malaria parasite with comparison to other apicomplexan species. Mol. Biochem. Parasitol. 118: 201-210.

A comparative analysis of six datasets: 5482 *P. berghei* GSSs [genome survey sequences, obtained after mung bean treatment of genomic DNA], 5582 *P. berghei* ESTs [expressed sequenced tags, obtained from cDNA libraries], 10 874 GSSs in two *P. vivax* lines, and 2438 *P. falciparum* ESTs or GSSs present in GenBank. Approximately 1000 putative new *Plasmodium* genes are identified. Their functional categorization using InterPro (a database of protein domains and signatures) is presented.

Chiang, S.L., Mekalanos, J.J., and Holden, D.W. 1999. *In vivo* genetic analysis of bacterial virulence. Annu. Rev. Microbiol. 53: 129-154.

Chuang, C.-F., and Meyerowitz, E.M. 2000. Specific and heritable genetic interference by double-stranded RNA in *Arabidopsis thaliana*. Proc. Natl. Acad. Sci. USA. 97: 4985-4990.

Clemens, J.C., Worby, C.A., Simonson-Leff, N., Muda, M., Maehama, T., Hemmings, B.A., and Dixon, J.E. 2000. Use of double-stranded RNA interference in *Drosophila* cell lines to dissect signal transduction pathways. Proc. Natl. Acad. Sci. USA. 97: 6499-6503.

Cowman, A.F., Baldi, D.L., Healer, J., Mills, K.E., O'Donnell, R.A., Reed, M.B., Triglia, T., Wickham, M.E., and Crabb, B.S. 2000. Functional analysis of proteins involved in *Plasmodium falciparum* merozoite invasion of red blood cells. FEBS Letters. 476: 84-88.

A review that presents conclusions of gene disruption experiments at loci potentially involved in merozoite invasion of red blood cells: *MSP1-5, RAP1-2, RhopH3, Ag512, EBA175, AMA1, S-antigen*, and *ABRA*. They suggest that in the *P. falciparum* D10 line, all these genes except *EBA175* and *RAP* are essential for invasion of red blood cells in *vitro*.

Crabb, B.S., and Cowman, A.F. 1996. Characterization of promoters and stable transfection by homologous and nonhomologous recombination in *Plasmodium falciparum*. Proc. Natl. Acad. Sci. USA. 93: 7289-7294.

A second report of stable transfection and plasmid integration via homologous recombination in *P. falciparum* using a *Toxoplasma gondii* DHFR-TS pyrimethamine-resistance gene. The evidence for non homologous recombination is based on a single clone.

Crabb, B.S., Cooke, B.M., Reeder, J.C., Waller, R.F., Caruana, S.R., Davern, K.M., Wickham, M.E., Brown, G.V., Coppel, R.L., and Cowman, A.F. 1997a. Targeted gene disruption shows that knobs enable malaria-infected red cells to cytoadhere under physiological shear stress. Cell. 89: 287-296.

The *KAHRP* gene in *P. falciparum* is inactivated by insertion of a replacement plasmid via a single crossover. KAHRP(-) mutants do not correctly localize PfEMP1 and form knobs on the surface of infected erythrocytes, which do not adhere to CD36 under physiological flow conditions.

Crabb, B.S., Triglia, T., Waterkeyn, J.G., and Cowman, A.F. 1997b. Stable transgene expression in *Plasmodium falciparum*. Mol. Biochem. Parasitol. 90: 131-144.

Cui, L., Rzomp, K.A., Fan, Q., Martin, S.K., and Williams, J. 2001. *Plasmodium falciparum*: differential display analysis of gene expression during gametocytogenesis. Exp. Parasitol. 99: 244-254.

Dechering, K.J., Kaan, A.M., Mbacham, W., Wirth, D.F., Eling, W., Konings, R.N.H., and Stunnenberg, H.G. 1999. Isolation and functional characterization of two distinct sexual-stage-specific promoters of the human malaria parasite *Plasmodium falciparum*. Mol. Cell. Biol. 19: 967-978.

*Pfs16* and *Pfs25* are activated at the onset of gametocytogenesis (in erythrocytes) and gametogenesis (in mosquitoes) respectively. Their promoters are characterized by transient transfection in *P. falciparum* (blood stages) and *P. gallinaceum* (mosquito stages). An AAGGAATA sequence is identified that is present only in the *Pfs25* promoter region, binds a mosquito stage-specific transcription factor (PAF-1), and is important for *Pfs25* transcription.

Dechering, K.J., Thompson, J., Dodemont, H.J., Eling, W., and Konings, R.N. 1997. Developmentally regulated expression of *pfs16*, a marker for sexual differentiation of the human malaria parasite *Plasmodium falciparum*. Mol. Biochem. Parasitol. 89: 235-244.

Deitsch, K.W., Calderwood, M.S., and Wellems, T.E. 2001. Cooperative silencing elements in *var* genes. Nature. 412: 875-876.

Deitsch, K.W., del Pinal, A., and Wellems, T.E. 1999. Intra-cluster recombination and *var* transcription switches in the antigenic variation of *Plasmodium falciparum*. Mol. Biochem. Parasitol. 101: 107-116.

Transient transfection is used to analyze the promoter of a *var* gene (*var7b*). The promoter can be active when borne by an episome transfected in a clone where its chromosomal version is inactive. This favors a model of epigenetic regulation of *var* gene transcription, involving gene silencing through local changes in chromatin structure.

de Koning-Ward, T.F., Fidock, D.A., Thathy, V., Ménard, R., van Spaendonk, R.M.L., Waters, A.P., and Janse, C.J. 2000a. The selectable marker human dihydrofolate reductase enables sequential genetic manipulations of the *Plasmodium berghei* genome. Mol. Biochem. Parasitol. 106: 199-212.
Describes the selection of plasmid integration events using a WR99210-resistant form of human DHFR in a genome already manipulated using the pyrimethamine-resistant form of *P. berghei* DHFR-TS.

de Koning-Ward, T.F., Janse, C.J., and Waters, A.P. 2000b. The development of genetic tools for dissecting the biology of malaria parasites. Annu. Rev. Microbiol. 54: 157-185.

de Koning-Ward, T.F., Waters, A.P., and Crabb, B.S. 2001. Puromycin-N-acetyltransferase as a selectable marker for use in *Plasmodium falciparum*. Mol. Biochem. Parasitol. 117: 155-160.

Dessens, J.T., Beetsma, A.L., Dimopoulos, G., Wengelnik, K., Crisanti, A., Kafatos, F.C., and Sinden, R.E. 1999. CTRP is essential for mosquito infection by malaria ookinetes. EMBO J. 18: 6221-6227.
Ookinetes of *P. berghei* in which the *CTRP* gene is disrupted after a double crossover event are not motile, fail to invade the mosquito midgut epithelium and do not develop into oocysts.

Dessens, J.T., Margos, G., Rodriguez, M.C., and Sinden, R.E. 2000. Identification of differentially regulated genes of *Plasmodium* by suppression subtractive hybridization. Parasitol. Today. 16: 354-356.

Duraisingh, M.T., Triglia, T., and Cowman, A.F. 2002. Negative selection of *Plasmodium falciparum* reveals targeted gene deletion by double crossover recombination. Int. J. Parasitol. 32: 81-89.

Falkow, S. 1988. Molecular Koch's postulates applied to microbial pathogenicity. Rev. Infect. Dis. 10 Suppl. 2: S274-276.

Fidock, D.A., and Wellems, T.E. 1997. Transformation with human dihydrofolate reductase renders malaria parasites insensitive to WR99210 but does not affect the intrinsic activity of proguanil. Proc. Natl. Acad. Sci. USA. 94: 10931-10936.
*P. falciparum* transfected episomally with a human, methotrexate-resistant DHFR variant becomes resistant to the anti-folate compounds WR99210 and cycloguanil, but not the cycloguanil precursor proguanil, suggesting that the latter acts independently of DHFR.

Fidock, D.A., Nomura, T., Cooper, R.A., Su, X.-z., Talley, A.K., and Wellems, T.E. 2000. Allelic modifications of the *cg2* and *cg1* genes do not alter the chloroquine response of drug-resistant *Plasmodium falciparum*. Mol. Biochem. Parasitol. 110: 1-10.

Fidock, D.A., Nomura, T., Talley, A.K., Cooper, R.A., Dzekunov, S.M., Ferdig, M.T., Ursos, L.M.B., Sidhu, A., Naudé, B., Deitsch, K.W., Su, X.-z., Wootton, J.C., Roepe, P.D., and Wellems, T.E. 2000. Mutations in the *P. falciparum* digestive vacuole transmembrane protein PfCRT and evidence for their role in chloroquine resistance. Mol. Cell. 6: 861-871.

Fraser, A.G., Kamath, R.S., Zipperlen, P., Martinez-Campos, M., Sohrmann, M., and Ahringer, J. 2000. Functional genomic analysis of *C. elegans* chromosome I by systematic RNA interference. Nature. 408: 325-330.

Gardiner, D.L., Holt, D.C., Thomas, E.A., Kemp, D.J., and Trenholme, K.R. 2000. Inhibition of *Plasmodium falciparum clag9* gene function by antisense RNA. Mol. Biochem. Parasitol. 110: 33-41.

An internal fragment of *clag9* is cloned in an antisense direction under the control of the strong *calmodulin* promoter. Stable expression of the episome strongly decreases production of clag9 protein and reproduces the phenotype obtained after *clag9* disruption (decreased adherence of infected eythrocytes to CD36+ melanoma cells).

Gardner, M.J., *et al.* 1998. Chromosome 2 sequence of the human malaria parasite *Plasmodium falciparum*. Science. 282: 1126-1132.

Gueiros-Filho, F.J., and Beverley, S.M. 1997. Trans-kingdom transposition of the *Drosophila* element *mariner* within the protozoan *Leishmania*. Science. 276: 1716-1719.

Gönczy, P., *et al.* 2000. Functional genomic analysis of cell division in *C. elegans* using RNAi of genes on chromosome III. Nature. 408: 331-336.

Goonewardene, R., Daily, J., Kaslow, D., Sullivan, T.J., Duffy, P., Carter, R., Mendis, K., and Wirth, D. 1993. Transfection of the malaria parasite and expression of firefly luciferase. Proc. Natl. Acad. Sci. USA. 90: 5234-5236.

The first demonstration that *Plasmodium* can be transfected. *P. gallinaceum* gametes and zygotes (extracellular) are electroporated with a plasmid that contains the firefly luciferase gene inserted into the *Pgs28* gene and flanking DNA. After 24 hr, luciferase activity is transiently expressed in electroporated parasites.

Gossen, M., and Bujard, H. 1992. Tight control of gene expression in mammalian cells by tetracycline-responsive promoters. Proc. Natl. Acad. Sci. USA. 89: 5547-5551.

Hamann, L., Buss, H., and Tannich, E. 1997. Tetracycline-controlled gene expression in *Entamoeba histolytica*. Mol. Biochem. Parasitol. 84: 83-91.

Hammond, S.M., Caudy, A.A., and Hannon, G.J. 2001. Post-transcriptional gene silencing by double-stranded RNA. Nature Rev. Genet. 2: 110-119.

Hayward, R.E., DeRisi, J.L., Alfadhli, S., Kaslow, D.C., Brown, P.O., and Rathod, P.K. 2000. Shotgun DNA microarrays and stage-specific gene expression in *Plasmodium falciparum* malaria. Mol. Microbiol. 35: 6-14.

PCR-amplified inserts (3648 inserts of 1.5-kb average size, thought to represent ~40% of the parasite genes) from a *P. falciparum* mung bean nuclease genomic library (ORF- or exon-enriched) were printed on glass slides. These 'shotgun' microarrays were hybridized with a mixture of Cy3 or Cy5-labelled cDNA from the erythrocytic trophozoite and gametocyte stages, respectively. The 85 arrayed genes showing the highest differential fluorescence (red/green or green/red) were sequenced, and new genes were identified.

Hensel, M., Shea, J.E., Gleeson, C., Jones, M.D., Dalton, E., and Holden, D.W. 1995. Simultaneous identification of bacterial virulence genes by negative selection. Science. 269: 400-403.

Hoffman, S.L., Rogers, W.O., Carucci, D.J., and Venter, J.C. 1998. From genomics to vaccines: malaria as a model system. Nature Med. 4: 1351-1353.

Hoffman, S.L., Subramanian, G.M., Collins, F.H., and Venter, J.C. 2002. *Plasmodium*, human and *Anopheles* genomics and malaria. Nature. 415: 702-709.

Horrocks, P., and Kilbey, B.J. 1996. Physical and functional mapping of the transcriptional start sites of *Plasmodium falciparum* proliferating cell nuclear antigen. Mol. Biochem. Parasitol. 82: 207-215.

Horrocks, P., and Lanzer, M. 1999. Mutational analysis identifies a five base pair *cis*-acting sequence essential for *GBP130* promoter activity in *Plasmodium falciparum*. Mol. Biochem. Parasitol. 99: 77-87.

This work progressively narrows down a 5 bp-sequence, located within a 73 bp-enhancer element, which is crucial for efficient activity of the *GBP130* promoter and specifically binds nuclear factors.

Horrocks, P., Dechering, K., and Lanzer, M. 1998. Control of gene expression in *Plasmodium falciparum*. Mol. Biochem. Parasitol. 95: 171-181.

Jo, D., Nashabi, A., Doxsee, C., Lin, Q., Unutmaz, D., Chen, J., and Ruley, H.E. 2001. Epigenetic regulation of gene structure and function with a cell-permeable Cre recombinase. Nat. Biotech. 19: 929-933.

Jomaa, H. *et al.* 1999. Inhibitors of the nonmevalonate pathway of isoprenoid biosynthesis as antimalarial drugs. Science. 285: 1573-1576.

Kadekoppala, M., Cheresh, P., Catron, D., Ji, D., Deitsch, K., Wellems, T.E., Seifert, H.S., and Haldar, K. 2001. Rapid recombination among transfected plasmids, chimeric episome formation and *trans* gene expression in *Plasmodium falciparum*. Mol. Biochem. Parasitol. 112: 211-218.

Co-transfection with a plasmid that expresses GFP fluoresence and another that confers pyrimethamine resistance allows drug selection of a population where a third of the parasites stably fluoresce. These fluorescent parasites contain drug-selectable, chimeric concatemers of both plasmids, and can only be obtained when both plasmids share a large region of homology of *Plasmodium* DNA (5' *hrp3* UTR sequences).

Kadekoppala, M., Kline, K., Akompong, T., and Haldar, K. 2000. Stable expression of new chimeric fluorescent reporter in the human malaria parasite *Plasmodium falciparum*. Infect. Immun. 68: 2328-2332.

A DHFR-GFPmut2 fusion, which confers methotrexate-resistance and fluorescence, is used in *P. falciparum* episomal and integrative transfection.

Kaneko, O., Fidock, D.A., Schwartz, O.M., and Miller, L.H. 2000. Disruption of the C-terminal region of EBA-175 in the Dd2/Nm clone of *Plasmodium falciparum* does not affect erythrocyte invasion. Mol. Biochem. Parasitol. 110: 135-146.

EBA-175 (Erythrocyte-binding antigen) is localized in the merozoite micronemes and binds the red blood cell surface molecule glycophorin A in a sialic acid-dependent manner. Its encoding gene is targeted in the *P. falciparum* Dd2/Nm clone, whose merozoites invade erythrocytes mainly via a sialic acid-independent pathway. Parasites expressing a EBA-175 truncate lacking the cytoplasmic domain can be selected, implying that EBA-175 is not involved in the sialic acid-independent pathway of merozoite invasion.

Kappe, S., Bruderer, T., Gantt, S., Fujioka, H., Nussenzweig, V., and Ménard, R. 1999. Conservation of a gliding motility and cell invasion machinery in Apicomplexan parasites. J. Cell. Biol. 147: 937-943.

An ends-in strategy is used to demonstrate the role of the cytoplasmic tail of TRAP in sporozoite gliding motility and cell invasion. Exchanging the tail of TRAP by that of *Toxoplasma gondii* MIC2 does not affect sporozoite gliding or invasion, suggesting a functional conservation of the motor system that drives these processes in Apicomplexa. Mutations in the TRAP tail lead to a novel form of motility, termed pendulum motility.

Kocken, C.H.M., Ozwara, H., van der Wel, A., Beetsma, A.L., Mwenda, J.M., and Thomas, A.W. 2002. *Plasmodium knowlesi* provides a rapid *in vitro* and *in vivo* transfection system that enables double-crossover gene knockout studies. Infect. Immun. 70: 655-660.

Kocken, C.H.M., van der Wel, A.M., and Thomas, A.W. 1999. *Plasmodium cynomolgi*: transfection of blood-stage parasites using heterologous DNA constructs. Exp. Parasitol. 93: 58-60.

Kocken, C.H.M., van der Wel, A.M., Dubbeld, M.A., Narum, D.L., van de Rijke, F.M., van Gemert, G.J., van der Linde, X., Bannister, L.H., Janse, C., Waters, A.P., and Thomas, A.W. 1998. Precise timing of expression of a *Plasmodium falciparum*-derived transgene in *Plasmodium berghei* is a critical determinant of subsequent subcellular localization. J. Biol. Chem. 273: 15119-15124.

Kumar, A., and Snyder, M. 2001. Emerging technologies in yeast genomics. Nature Rev. Genet. 2: 302-312.

Kuwabara, P.E., and Coulson, A. 2000. RNAi – Prospects for a general technique for determining gene function. Parasitol. Today. 16: 347-349.

Lau, A.O.T., Sacci Jr, J.B., and Azad, A.F. 2000. Retrieving parasite specific liver stage gene products in *Plasmodium yoelii* infected livers using differential display. Mol. Biochem. Parasitol. 111: 143-151.

Lobo, C.-A., Fujioka, H., Aikawa, M., and Kumar, N. 1999. Disruption of the *Pfg27* locus by homologous recombination leads to loss of the sexual phenotype in *P. falciparum*. Mol. Cell. 3: 793-798.

*Pfg27* encoding a protein specifically produced at the onset of gametocytogenesis is targeted by single crossover integration of a replacement plasmid. Two mutants, in which full-length *Pfg27* lacks its promoter or terminator sequences, have reduced or no *Pfg27* transcription and display fewer or no typical gametocytes, respectively.

Matuschewski, K., Nunes, A., Nussenzweig, V., and Ménard, R. 2002. *Plasmodium* sporozoite invasion into insect and mammalian cells is directed by the same dual binding system. EMBO J. 21: 1597-1606.

This paper reports the effect of loss-of-function mutations in the extracellular domain of TRAP, a protein necessary for sporozoite gliding and cell invasion. Results indicate that neither the A-domain nor the thrombospondin type 1 repeat is necessary for gliding but that both are important for sporozoite invasion into mosquito salivary glands, the rodent liver, and cultured cells.

McRobert, L., and McConkey, G.A. 2002. RNA interference (RNAi) inhibits growth of *Plasmodium falciparum*. Mol. Biochem. Parasitol. 119: 273-278.

*P. falciparum*-infected red blood cells are electroporated with ~1-kb long double-stranded (ds) RNA corresponding to a segment of an essential gene (dihydroorotate dehydrogenase, DHODH). This causes a ~50% decrease in DHODH mRNA levels and in parasite growth, detected after 24 h and up to 72 h after electroporation. Electroporation with single-stranded RNA, or with dsRNA corresponding to nonessential genes, do not inhibit parasite growth in red blood cells.

Meissner, M., Brecht, S., Bujard, H., and Soldati, D. 2001. Modulation of myosin A expression by a newly established tetracycline repressor-based inducible system in *Toxoplasma gondii*. Nucleic Acids Res. 29: e115.

Ménard, R., and Janse, C.J. 1997. Gene targeting in malaria parasites. Methods. 13: 148-157.

Ménard, R., and Nussenzweig, V. 2000. Structure-function analysis of malaria proteins by gene targeting. Parasitol. Today. 16: 222-224.

Ménard, R., Sultan, A.A., Cortes, C., Altzuler, R., van Dijk, M.R., Janse, C.J., Waters, A.P., Nussenzweig, R.S., and Nussenzweig, V. 1997. Circumsporozoite protein is required for development of malaria sporozoites in mosquitoes. Nature. 385: 336-340.

The first *Plasmodium* mutant obtained by site-directed mutagenesis. Clones in which the single-copy *CS* gene is disrupted by allelic exchange do not produce sporozoites in the mosquito midgut.

Metzger, D., and Chambon, P. 2001. Site- and time-specific gene targeting in the mouse. Methods. 24: 71-80.

Mota, M.M., Thathy, V., Nussenzweig, R.S., and Nussenzweig, V. 2001. Gene targeting in the rodent malaria parasite *Plasmodium yoelii*. Mol. Biochem. Parasitol. 113: 271-278.

Munasinghe, A., Patankar, S., Cook, B.P., Madden, S.L., Martin, R.K., Kyle, D.E., Shoaibi, A., Cummings, L.M., and Wirth, D.F. 2000. Serial analysis of gene expression (SAGE) in *Plasmodium falciparum*. Application of the technique to A-T rich genomes. Mol. Biochem. Parasitol. 113: 23-34.

Natarajan, R., Thathy, V., Mota, M.M., Hafalla, J.C.R., Ménard, R., and Vernick, K.D. 2001. Fluorescent *Plasmodium berghei* sporozoites and pre-erythrocytic stages: a new tool to study mosquito and mammalian host interactions with malaria parasites. Cell. Microbiol. 3: 371-379.

The *gfp* gene is introduced at the *CS* locus and placed under the control of *CS* 5' and 3' expression sequences, rendering recombinant parasites fluorescent at the sporozoite and liver stages. This clone, which has a wild-type phenotype, can be used to separate infected cells by FACS.

Nunes, A., Thathy, V., Bruderer, T., Sultan, A.A., Nussenzweig, R.S., and Ménard, R. 1999. Subtle mutagenesis by ends-in recombination in malaria parasites. Mol. Cell. Biol. 19: 2895-2902.

This paper shows that in *P. berghei* insertion plasmids integrate at the homologous genomic locus via the double-strand gap repair model. Short lengths of gene conversion tracts are associated with plasmid integration at the *TRAP* locus, allowing various modifications to be introduced in the gene by a single-step ends-in strategy.

O'Donnell, R.A., Freitas-Junior, L.H., Preiser, P.R., Williamson, D.H., Duraisingh, M., McElwain, T.F., Scherf, A., Cowman, A.F., and Crabb, B.S. 2002. A genetic screen for improved plasmid segregation reveals a role for Rep20 in the interaction of *Plasmodium falciparum* chromosomes. EMBO J. 21: 1231-1239.

O'Donnell, R.A., Preiser, P.R., Williamson, D.H., Moore, P.W., Cowman, A.F., and Crabb, B.S. 2001. An alteration in concatemeric structure is associated with efficient segregation of plasmids in transfected *Plasmodium falciparum* parasites. Nucleic Acids Res. 29: 716-724.

In *P. falciparum*, transfected plasmids are arranged as head-to-tail concatemers. These episomes are first unstable (unstable replicating forms, URFs) and lost if selection is removed. After extended periods (>4 months), they become stable and segregate evenly between daughter merozoites, even in the absence of drug pressure (stably replicating forms, SRFs). These SRFs are more complex structures and may contain single-stranded DNA.

O'Donnell, R.A., Saul, A., Cowman, A.F., and Crabb, B.S. 2000. Functional conservation of the malaria vaccine antigen MSP-1$_{19}$ across distantly related *Plasmodium* species. Nature Med. 6: 91-95.

This study in *P. falciparum* shows that *MSP-1* (merozoite surface protein 1) can be targeted but not disrupted, and can be replaced by a gene encoding a hybrid MSP-1 whose C-terminus [MSP-1$_{19}$] is from *P. chabaudi*. Since Pf MSP-1$_{19}$ sequences are conserved among field isolates but highly divergent from that of Pc MSP-1$_{19}$, sequence conservation in *P. falciparum* cannot be explained by a functional constraint.

Pace, T., Scotti, R., Janse, C.J., Waters, A.P., Birago, C., and Ponzi, M. 2000. Targeted terminal deletions as a tool for functional genomics studies in *Plasmodium*. Genome Res. 10: 1414-1420.

This study describes an interesting system based on transfection of linear replacement molecules in *P. berghei* for introducing large (up to 60 kb) and defined deletions at chromosome ends. It provides a means to study the functional consequences of genomic rearrangments, or the loss of specific genes associated with chromosome ends.

Patankar, S., Munasinghe, A., Shoaibi, A., Cummings, L.M., and Wirth, D.F. 2001. Serial analysis of gene expression in *Plasmodium falciparum* reveals the global expression profile of erythrocytic stages and the presence of anti-sense transcripts in the malarial parasite. Mol. Biol. Cell. 12: 3114-3125.

A SAGE library of ~8,000 tags is generated from erythrocytic stages of *P. falciparum*. BLAST analysis of highly abundant tags documents the major metabolic pathways of the parasite cultured in normal conditions, and reveals the presence of antisense transcription in the parasite.

Porter, A. 1998. Controlling your losses: conditional gene silencing in mammals. Trends Genet. 14: 73-79.

Reed, M.B., Caruana, S.R., Batchelor, A.H., Thompson, J.K., Crabb, B.S., and Cowman, A.F. 2000. Targeted disruption of an erythrocyte binding antigen in *Plasmodium falciparum* is associated with a switch toward a sialic acid-independent pathway of invasion. Proc. Natl. Acad. Sci. USA. 97: 7509-7514.

An insertion plasmid is used to generate a *P. falciparum* clone that produces an EBA-175 truncate lacking the C-terminus including the cytoplasmic domain. Mutant merozoites still invade erythrocytes, indicating that EBA-175 is not essential for invasion. However, mutants have switched to a sialic acid-independent invasion pathway, showing that the parasite can utilize the sialic acid-independent pathway for invasion.

Reed, M.B., Saliba, K.J., Caruana, S.R., Kirk, K., and Cowman, A.F. 2000. Pgh1 modulates sensitivity and resistance to multiple antimalarials in *Plasmodium falciparum*. Nature. 403: 906-909.

Roos, D.S., Sullivan, W.J., Striepen, B., Bohne, W., and Donald, R.G.K. 1997. Tagging genes and trapping promoters in *Toxoplasma gondii* by insertional mutagenesis. Methods. 13: 112-122.

Ross-Macdonald, P., *et al.* 1999. Large-scale analysis of the yeast genome by transposon tagging and gene disruption. Nature. 402: 413-418.

Ross-Macdonald, P., Sheehan, A., Shirleen Roeder, G., and Snyder, M. 1997. A multipurpose transposon system for analyzing protein production, localization, and function in *Saccharomyces cerevisiae*. Proc. Natl. Acad. Sci. USA. 94: 190-195.

Sacci Jr, J.B., Aguiar, J.C., Lau, A.O., Hoffman, S.L. 2002. Laser capture microdissection and molecular analysis of *Plasmodium yoelii* liver-stage parasites. Mol. Biochem. Parasitol. 119: 285-289.

Sauer, B. 1998. Inducible gene targeting in mice using the Cre/*lox* system. Methods. 14: 381-392.

Sharp, P.A. 2001. RNA interference-2001. Genes & Dev. 15: 485-490.

Shi, H., Djikeng, A., Mark. T., Wirtz, E., Tschudi, C., and Ullu, E. 2000. Genetic interference in *Trypanosoma brucei* by heritable and inducible double-stranded RNA. RNA. 6: 1069-1076.

Shoemaker, D.D., Lashkari, D.A., Morris, D., Mittmann, M., and Davis, R.W. 1996. Quantitative phenotypic analysis of yeast deletion mutants using a highly parallel molecular bar-coding strategy. Nature Genet. 14: 450-456.

Strauss, E.J., and Falkow, S. 1997. Microbial pathogenesis: genomics and beyond. Science. 276: 707-712.

Sultan, A.A., de Koning-Ward, T.F., Fidock, D., Nussenzweig, V., and Ménard, R. 1999a. Complementation of TRAP knockout parasites of *Plasmodium berghei* using human dihydrofolate reductase gene as a selectable marker. Meeting Abstract. Woods Hole, MA, USA. September 1999.

Sultan, A.A., Thathy, V., de Koning-Ward, T.F., and Nussenzweig, V. 2001. Complementation of *Plasmodium berghei* TRAP knockout parasites using human dihydrofolate reductase gene as a selectable marker. Mol. Biochem. Parasitol. 113: 151-156.

Sultan, A.A., Thathy, V., Frevert, U., Robson, K., Crisanti, A., Nussenzweig, V., Nussenzweig, R.S., Ménard, R. 1997. TRAP is necessary for gliding motility and infectivity of *Plasmodium* sporozoites. Cell. 90: 511-522.
The sporozoite-specific *TRAP* gene is disrupted in *P. berghei* using insertion and replacement plasmids. TRAP appears to be essential for sporozoite infection of mosquito salivary glands and the rat liver, as well as for sporozoite gliding motility, suggesting that host cell invasion and gliding motility of sporozoites have a common molecular basis.

Sultan, A.A., Thathy, V., Nussenzweig, V., Ménard, R. 1999b. Green fluorescent protein as a marker in *Plasmodium berghei* transformation. Infect. Immun. 67: 2602-2606.
A DHFR-TS fusion to the rapidly folding GFPmut2 is used as a new marker for selecting integration events. Pyrimethamine selection is associated with flow cytometry, which can be used to eliminate drug resistant parasites that arise by spontaneous mutations or gene conversion events at the *DHFR-TS* locus, and clone integrants at day 10 post-electroporation.

Surolia, N., and Surolia, A. 2001. Triclosan offers protection against blood stages of malaria by inhibiting enoyl-ACP reductase of *Plasmodium falciparum*. Nature Med. 7: 167-173.

Tavernarakis, N., Wang, S.L., Dorovkov, M., Ryazanov, A., and Driscoll, M. 2000. Heritable and inducible genetic interference by double-stranded RNA encoded by transgenes. Nature Genet. 24: 180-183.

* Taylor, H.M., Triglia, T., Thompson, J., Sajid, M., Fowler, R., Wickham, M.E., Cowman, A.F., and Holder, A.A. 2001. *Plasmodium falciparum* homologue of the genes for *Plasmodium vivax* and *Plasmodium yoelii* adhesive proteins, which is transcribed but not translated. Infect. Immun. 69: 3635-3645.
Disruption in *P. falciparum* of a member of the *P. vivax* PvRBP and *P. yoelii* Py235 gene family, *PfRH3*, has no consequence on parasite replication in red cells. *PfRH3* appears to be a full-length and transcribed pseudogene.

Templeton, T.J., Kaslow, D.C., and Fidock, D.A. 2000. Developmental arrest of the human malaria parasite *Plasmodium falciparum* within the mosquito midgut via *CTRP* gene disruption. Mol. Microbiol. 36: 1-9.

Thathy, V., Fujioka, H., Gantt, S., Nussenzweig, R.S., Nussenzweig, V., and Ménard, R. 2002. Levels of CS protein in the *Plasmodium* oocyst control sporozoite budding and morphology. EMBO J. 21: 1586-1596.
*P. berghei* clones producing low amounts of CS are constructed by truncation of the 3' UTR of the *CS* gene. Comparison of oocyst differentiation between the wild-type, non CS producer, low CS producer and complemented clones indicate that CS has a specific role in establishing polarity in oocysts and in the formation of sporozoites.

Tomas, A.M., Margos, G., Dimopoulos, G., van Lin, L.H.M., de Koning-Ward, T.F., Sinha, R., Lupetti, P., Beetsma, A.L., Rodriguez, M.C., Karras, M., Hager, A., Mendoza, J., Butcher, G.A., Kafatos, F., Janse, C.J., Waters, A.P., and Sinden, R.E. 2001. P25 and P28 proteins of the malaria ookinete surface have multiple and partially redundant functions. EMBO J. 20: 3975-3983.

Parasites lacking either P25 or P28 generate normal numbers of ookinetes and mildly reduced numbers of oocysts, while parasites lacking both proteins (both genes are linked and were deleted by a single event) generate fewer ookinetes and greatly reduced numbers of oocysts. The functions of P25 and P28 are proposed to be multiple and partially redundant.

Tomas, A.M., van der Wel, A.M., Thomas, A.W., Janse, C.J., and Waters, A.P. 1998. Transfection systems for animal models of malaria. Parasitol. Today. 14: 245-249.

Trenholme, K.R., Gardiner, D.L., Holt, D.C., Thomas, E.A., Cowman, A.F., and Kemp, D.J. 2000. *Clag9*: a cytoadherence gene in *Plasmodium falciparum* essential for binding of parasitized erythrocytes to CD36. Proc. Natl. Acad. Sci. USA. 97: 4029-4033.

Triglia, T., Healer, J., Caruana, S.R., Hodder, A.N., Anders, R.F., Crabb, B.S., and Cowman, A.F. 2000. Apical membrane antigen 1 plays a central role in erythrocyte invasion by *Plasmodium* species. Mol. Microbiol. 38: 706-718.

The involvement in merozoite invasion of erythrocytes of AMA-1 (apical membrane antigen 1), a merozoite surface protein, is suggested by several lines of evidence. (i) Only single crossover events that do not disrupt the gene can be selected at the *PfAMA-1* locus. (ii) When *PcAMA-1* (P. chabaudii) is co-expressed with endogenous *PfAMA-1*, its contribution to merozoite invasion is shown by increased invasion in mouse red blood cells and using antibody inhibition assays. (iii) However, *PfAMA-1* cannot be disrupted despite simultaneous expression of *PcAMA-1*.

Triglia, T., Thompson, J.K., and Cowman, A.F. 2001. An EBA175 homologue which is transcribed but not translated in erythrocytic stages of *Plasmodium falciparum*. Mol. Biochem. Parasitol 116: 55-63.

Searches of the *P. falciparum* genome databases reveal the presence of an *EBA175* homologue, called *YEBA165*. This study shows that *YEBA165*, which can be disrupted, is a transcribed pseudogene.

Triglia, T., Wang, P., Sims, P.F.G., Hyde, J.E., and Cowman, A.F. 1998. Allelic exchange at the endogenous genomic locus in *Plasmodium falciparum* proves the role of dihydropteroate synthase in sulfadoxine-resistant malaria. EMBO J. 17: 3807-3815.

Single crossovers are used to generate a series of modified *dhps* genes to assess the contribution of mutations found in sulfadoxine-resistant field isolates of *P. falciparum*. It is shown that high levels of sulfadoxine resistance can be conferred by a stepwise accumulation of mutations around a primary A437G substitution in DHPS.

Tsai, Y.L., Hayward, R.E., Langer, R.C., Fidock, D.A., and Vinetz, J.M. 2001. Disruption of *Plasmodium falciparum* chitinase markedly impairs parasite invasion of mosquito midgut. Infect. Immun. 69: 4048-4054.

Urlinger, S., Baron, U., Thellmann, M., Hasan, M.T., Bujard, H., and Hillen, W. 2000. Exploring the sequence space for tetracycline-dependent transcriptional activators: novel mutations yield expanded range and sensitivity. Proc. Natl. Acad. Sci. USA. 97: 7963-7968.

Valdivia, R.H., and Falkow, S. 1997. Fluorescence-based isolation of bacterial genes expressed within host cells. Science. 277: 2007-2011.

van der Wel, A.M., Tomas, A.M., Kocken, C.H.M., Malhotra, P., Janse, C.J., Waters, A.P., and Thomas, A.W. 1997. Transfection of the primate malaria parasite *Plasmodium knowlesi* using entirely heterologous constructs. J. Exp.Med. 185, 1499-1503.

van Dijk, M.R., Janse, C.J., and Waters, A.P. 1996. Expression of a *Plasmodium* gene introduced into subtelomeric regions of *Plasmodium berghei* chromosomes. Science. 271: 662-665.

The first evidence of homologous recombination in *P. berghei*. A targeting plasmid, which is linearized in a 2.2 kb-subtelomeric fragment present as 200-300 copies in the parasite genome, integrates at several of the endogenous copies in three different chromosomes.

van Dijk, M.R., Janse, C.J., Thompson, J., Waters, A.P., Braks, J.A.M., Dodemont, H.J., Stunnenberg, H.G., van Gemert, G.-J., Sauerwein, R.W., and Eling, W. 2001. A central role for P48/45 in malaria parasite male gamete fertility. Cell. 104: 153-164.

This gene disruption study in both *P. berghei* and *P. falciparum* shows that P48/45, which is produced on the surface of male and female gametocytes and gametes, is important specifically for zygote formation. Elegant *in vitro* studies using *P. berghei* parasites indicate that only male null gametes are impaired in fertilization.

van Dijk, M.R., Vinkenoog, R., Ramesar, J., Vervenne, R.A.W., Waters, A.P., and Janse, C.J. 1997. Replication, expression and segregation of plasmid-borne DNA in genetically transformed malaria parasites. Mol. Biochem. Parasitol. 86: 155-162.

van Dijk, M.R., Waters, A.P., and Janse, C.J. 1995. Stable transfection of malaria parasite blood stages. Science. 268: 1358 –1362.

The first report of stable transfection in *Plasmodium*. A plasmid carrying a pyrimethamine-resistance *DHFR-TS* (dihydrofolate reductase thymidylate synthase) variant is electroporated into *P. berghei* merozoites, and erythrocytic stages of the parasites bearing replicating episomes are selected in pyrimethamine-treated rodents.

van Spaendonk, R.M.L., Ramesar, J., van Wigcheren, A., Eling, W., Beetsma, A.L., van Gemert, G.-J., Hooghof, J., Janse, C.J., and Waters, A.P. 2001. Functional equivalence of structurally distinct ribosomes in the malaria parasite, *Plasmodium berghei*. J. Biol. Chem. 276: 22638-22647.

Disruption of the C- or D- rRNA gene units (S-type, specifically expressed during the sporogonic cycle), using insertion and replacement plasmids, has little effect on parasite development in the mosquito. This excludes the view that *P. berghei* requires two functionally different S-type ribosomes (distinct from the A-type ribosomes of vertebrate stages) to complete its sporogonic cycle.

Waller, R.F., Reed, M.B., Cowman, A.F., and McFadden, G.I. 2000. Protein trafficking to the plastid of *Plasmodium falciparum* is via the secretory pathway. EMBO J. 19: 1794-1802.

In *P. falciparum*, protein targeting to the apicoplast (surrounded by four membranes) is analyzed using GFP fusions expressed from episomes. The bipartite N-terminal pre-sequence on apicoplast-targeted proteins is necessary and sufficient for apicoplast targeting: the signal peptide directs GFP entry into the secretory pathway, and the transit peptide further allows GFP import into the apicoplast.

Wang, Z., Morris, J.C., Drew, M.E., and Englund, P.T. 2000. Inhibition of *Trypanosoma brucei* gene expression by RNA interference using an integratable vector with opposing T7 promoters. J. Biol. Chem. 275: 40174-40179.

Waterkeyn, J.G., Wickham, M.E., Davern, K.M., Cooke, B.M., Coppel, R.L., Reeder, J.C., Culvenor, J.G., Waller, R.F., and Cowman, A.F. 2000. Targeted mutagenesis of *Plasmodium falciparum* erythrocyte membrane protein 3 (PfEMP3) disrupts cytoadherence of malaria-infected red blood cells. EMBO J. 19: 2813-2823.

Inactivation of *PfEMP3* using an insertion plasmid abolishes PfEMP3 production but does not alter cytoadherence of infected red cells to CD36. Production of truncated forms of PfEMP3 reduces cytoadherence, probably by a dominant negative action of the PfEMP3 truncates that block PfEMP1 transfer to the infected red cell surface.

Waters, A.P., Thomas, A.W., van Dijk, M.R., and Janse, C.J. 1997. Transfection of malaria parasites. Methods. 13: 134-147.

Wengelnik, K., Spaccapelo, R., Naitza, S., Robson, K.J.H., Janse, C.J., Bistoni, F., Waters, A.P., and Crisanti, A. 1999. The A-domain and the thrombospondin-related motif of *Plasmodium falciparum* TRAP are implicated in the invasion process of mosquito salivary glands. EMBO J. 18: 5195-5204.

A replacement strategy is used in *P. berghei* to exchange the endogenous TRAP by *P. falciparum* TRAP, but the resulting clone is severely impaired in all TRAP-dependent phenotypes.

Wickham, M.E., Rug, M., Ralph, S.A., Klonis, N., McFadden, G.I., Tilley, L., and Cowman, A.F. 2001. Trafficking and assembly of the cytoadherence complex in *Plasmodium falciparum*-infected human erythrocytes. EMBO J. 20: 5636-5649.

The use of two KAHRP-GFP fusions expressed from stably maintained episomes reveals that before being assembled into knobs on the erythrocyte surface, KAHRP is secreted in the parasitophorous vacuole and transiently associates with the Maurer's clefts in the erythrocyte cytoplasm.

Winzeler, E.A., *et al.* 1999. Functional characterization of the *S. cerevisiae* genome by gene deletion and parallel analysis. Science. 285: 901-906.

Wirtz, E., and Clayton, C. 1995. Inducible gene expression in trypanosomes mediated by a prokaryotic repressor. Science. 268: 1179-1183.

Wirtz, E., Leal, S., Ochatt, C., and Cross, G.A.M. 1999. A tightly regulated inducible expression system for conditional gene knock-outs and dominant-negative genetics in *Trypanosoma brucei*. Mol. Biochem. Parasitol. 99: 89-101.

Wren, B.W. 2000. Microbial genome analysis: insights into virulence, host adaptation and evolution. Nature Rev. Genet. 1: 30-39.

Wu, Y., Kirkman, L.A., and Wellems, T.E. 1996. Transformation of *Plasmodium falciparum* malaria parasites by homologous integration of plasmids that confer resistance to pyrimethamine. Proc. Natl. Acad. Sci. USA. 93: 1130-1134.

The first report of homologous integration of exogenous DNA into the *Plasmodium* genome. Selection of integrants with a pyrimethamine-resistant form of *DHFR-TS* from *P. falciparum* or *T. gondii* necessitates a complex procedure during which drug pressure is applied (selecting replicating episomes), removed (allowing plasmid loss), and reapplied (selecting parasites in which the plasmid integrated).

Wu Y., Sifri C.D., Lei H.-H., Su X.-Z., and Wellems, T.E. 1995. Transfection of *Plasmodium falciparum* within human red blood cells. Proc. Natl. Acad. Sci. USA. 92: 973-977.

The first report of transient transfection of red blood cell stages of *Plasmodium*. *P. falciparum*-infected red blood cells are electroporated with plasmids that contain *CAT* flanked by *P. falciparum* expression sequences.

Yan, S., Martinez-Calvillo, S., Schnaufer, A., Sunkin, S., Myler, P.J., and Stuart, K. 2002. A low-background inducible promoter system in *Leishmania donovani*. Mol. Biochem. Parasitol. 119: 217-223.

Yuda, M., Sakaida, H., and Chinzei, Y. 1999. Targeted disruption of the *Plasmodium berghei CTRP* gene reveals its essential role in malaria infection of the vector mosquito. J. Exp. Med. 190: 1711-1715.

Disruption of *CTRP*, which encodes an ookinete-specific product, blocks ookinete invasion and oocyst formation in the mosquito. CTRP(-) or (+) sporozoites (haploid) are generated by mating CTRP(-) and wild-type parasites, via ookinete diploidy and invasion of heterozygous ookinetes. As both sporozoite types are infective, CTRP is shown to be essential only for invasion of the ookinete stage.

From: Malaria Parasites: Genomes and Molecular Biology
Edited by: A.P. Waters and C.J. Janse

# Chapter 5

## *Toxoplasma gondii* a Model Organism for the Apicomplexans?

Dominique Soldati
and Markus Meissner

## Abstract

*Toxoplasma gondii* is closely related to the *Plasmodium* species sharing with them multiple structural and functional features but differing significantly in their mode of transmission and differentiation. The broad host range specificity and the robust nature of the extracellular form of *T. gondii* have facilitated its propagation in cell culture and led to the rapid development of genetic tools. Genetic manipulation has also revolutionized the research on Malaria and rather unexpectedly, some genetic properties differ significantly between *T. gondii* and *P. falciparum* imposing limits as to which strategies are specifically applicable to each of these parasites. Advantages, constraints are intrinsic to both systems and consequently our understanding of both parasites is not only complementary but also synergistic. The information obtained from the *Plasmodium* and several other Apicomplexa genomes have led to the identification of exciting new classes of genes that are restricted to the phylum and thus of considerable interest as potential targets for intervention. *T. gondii* is particularly amenable to ultrastructural and biochemical studies and some important findings regarding the composition and function of subcellular structures and organelles, unique metabolic pathways and mechanisms leading to the establishment of intracellular parasitism are of great significance for malaria research.

# 1. Introduction

*Plasmodium falciparum* is the most notorious member of the *Apicomplexa*, which include many thousands of obligate intracellular protozoan parasites sharing multiple structural and functional features. *Toxoplasma gondii*, is among the most successful parasites, with half of the human population chronically infected. Although the infection is generally asymptomatic, the development of toxoplasmosis is characterized by focal lesions in the central nervous system and thus represents a serious cause of neurological birth defects. This opportunistic parasite may also cause a fatal cerebral toxoplasmosis in association with a variety of immunosuppressive diseases and treatments.

Significant progress in the understanding of the biology and genetics of apicomplexans has been accomplished by taking advantage of diverse parasites and experimental systems and integrating the results. In this respect, *Toxoplasma* has provided a particularly attractive and effective experimental system. This ubiquitous parasite is capable of infecting virtually any nucleated cell from warm-blooded vertebrates. Consequently, the parasite is readily cultivated in any cell type including host cell mutants, allowing the determination of the host cell influence on parasite development. The robust nature of the haploid form has facilitated its propagation in cell culture and led to the rapid development of its reverse genetics. *T. gondii* is currently equipped with a wide array of experimental tools making it particularly suitable as a model parasite for genetic and biological studies (Figure 1). Genetic manipulation has also revolutionized research in *Plasmodium* but rather unexpectedly, some genetic properties differ significantly between *T. gondii* and *P. falciparum* imposing limits as to which strategies are specifically applicable and appropriate to one or the other parasite.

# 2. Accessibility to Experimental Approaches

All apicomplexans are obligate intracellular parasites and most of them grow and replicate within a parasitophorous vacuole, which is resistant to lysosomal fusion and segregated from most cellular trafficking pathways. Replication takes place until the host cell lyses, then the extracellular parasites must rapidly reinvade other host cells in order to survive. Many apicomplexans have not been easy to propagate and study in the laboratory and conditions to cultivate them anexically have not been established. Nevertheless, extracellular tachyzoites can survive several hours in the absence of host cells. In contrast, although large numbers of *P. falciparum*-infected red cells can be obtained, the extracellular merozoites die within minutes. What follows is a brief summary of the specific advantages that working with *T.gondii* may offer over *Plasmodium*.

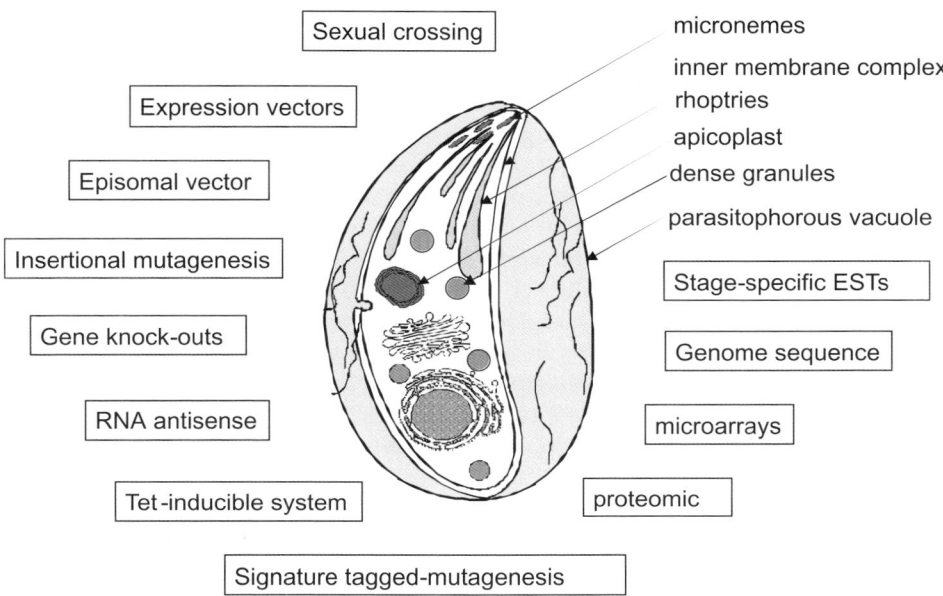

**Figure 1.** Genetic and genomic tools available to study intracellular parasitism by *Toxoplasma gondii*. Schematic representation of the subcellular organelles and compartments of an intracellular *T. gondii* tachyzoite.

## 2.1. Propagation in Cell Culture, Tissue Cyst Formation/ Animal Models

Apicomplexans replicate in haploid form and undergo sexual recombination in the definitive host, the mosquito vector in the case of *Plasmodium* and cats in the case of *Toxoplasma*. While the complete life cycle of *Plasmodium* is amenable to experimental procedures, studies on the sexual stages of *T. gondii,* which develop uniquely in felids are extremely limited. Systemic infection in a large variety of animals leads to the formation and persistence of tissue cysts, which are a source of transmission through various carnivores. The two life stages found in the intermediate host, the rapidly dividing tachyozoites and the dormant cyst forming bradyzoites can be propagated in cell culture. Many strains of *T. gondii* can produce tissue cysts in vitro. Strains less pathogenic for mice usually produce more tissue cysts than the more rapidly dividing, pathogenic strains (Bohne *et al.*, 1994; Gross and Bohne, 1994). Several stress conditions (pH, heat shock or chemical stress) have been reported to stimulate tissue cyst development *in vitro* (for a review see Dubey *et al.*, 1998). The slowly dividing encysted bradyzoites are the focus of great attention because they remain latent within the tissues for the life of the host and are the source of transmission and the source of reactivation in situations of immunodeficiency. Vero cells (a green monkey kidney cell line) are traditionally used to obtain higher titer of tachyzoites and can generate enough material for biochemical studies.

Members of the phylum Apicomplexa, replicate by simple binary division (endodyogeny in case of *Toxoplasma*) or by multinuclear processes (schizogony in the case of *Plasmodium*) that either involve multiple rounds of S and M phase prior to cytokinesis or the synthesis of polyploid chromatin followed by a single, multifocal mitosis and cytokinesis. One of the experimental limitations with Toxoplasma is the lack of cell cycle synchronicity within the populaion in culture. The introduction of herpes simplex thymidine kinase into the parasite allowed the synchronization of tachyzoite growth using the traditional exogenous supply of thymidine to block DNA synthesis and arrest cell growth (Radke and White, 1998).

## 2.2. Biochemical Studies

Biochemical studies on apicomplexan parasites have been limited by the difficulty of obtaining sufficient material for detailed analysis. Although standard biochemical methods can be used, to purify proteins from *T. gondii*, it is impractical to obtain milligram amounts of proteins. The purification of the functionally active myosin A from recombinant tachyzoites expressing Histidine-tagged TgMyoA required intensive cell culture work and use of a sophisticated stop flow system to measure enzymatic reactions in volumes as small as a few microliters (Herm-Götz *et al.*, 2002).

The extreme AT richness of *P. falciparum* genome introduces an additional limitation by compromising expression in heterologous systems. The strong codon bias hampers efficient translation in many heterologous systems and can only be circumvented by the generation of synthetic genes with a remodelled and higher GC content (Pan *et al.*, 1999; Withers-Martinez *et al.*, 1999). No such problems have been reported for the genes of *T. gondii*.

## 2.3. Light and Electron Microscopy Studies

Apicomplexans exhibit a conserved subcellular architecture, characterized by the apical complex. Because of its convenient size and its excellent preservation after fixation *T. gondii* tachyzoites and bradyzoites are very well suited to for high-resolution light and electron microscopy analyses (Figures 2 and 3), far more so than *Plasmodium*. They harbour distinct and morphologically well-defined intracellular organelles, including three distinct types of secretory vesicles (rhoptries, micronemes, and dense granules) and two endosymbiotic organelles (apicoplast and mitochondrion). *T. gondii* has been extensively used to characterize these subcellular structures and to

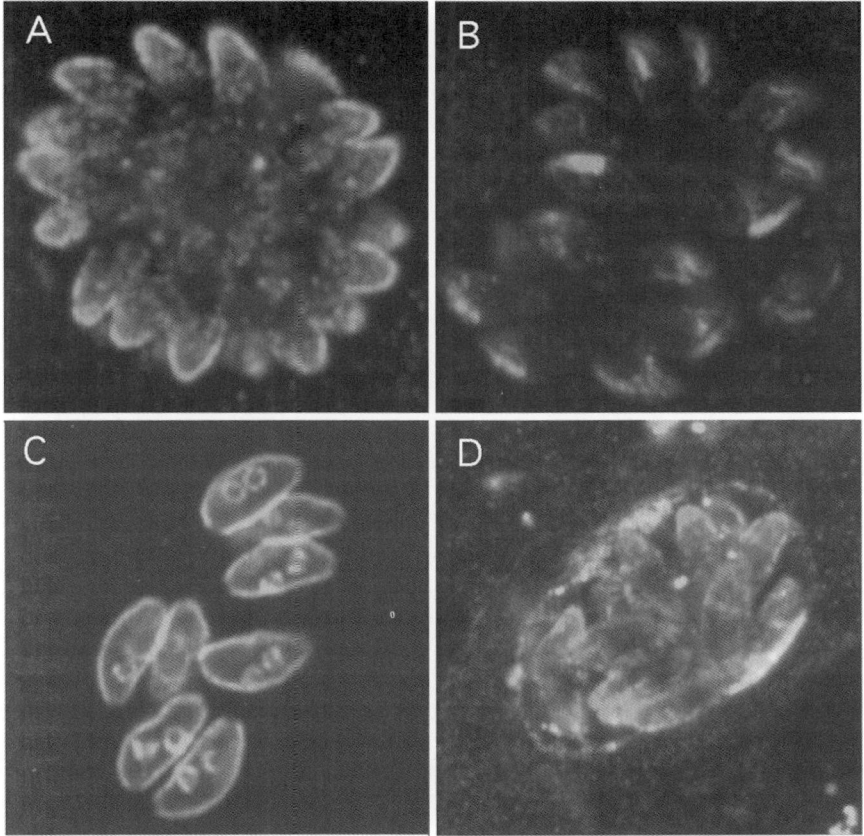

**Figure 2.** Indirect immunofluorescence confocal microscopy analysis of intracellular *T. gondii* tachyzoites. Panel **A** shows the micronemes stained with anti-MIC6. **B** shows the rhoptires stained with anti-ROP2. **C** shows the inner membrane complex and the nascent daughter cells stained with anti-IMC1. **D** shows the parasitophorous vacuole stained with anti-GRA3 in green and the micronemes are stained in red. **See Colour Plate at the back of the book.**

unambiguously determine the subcellular distribution and trafficking of an individual protein (Joiner and Roos 2002). As example, the apical membrane antigen (TgAMA-1) was unambiguously described as a microneme protein (Donahue *et al.*, 2000; Hehl *et al.*, 2000) while the homologue in Plasmodium species was first localized in the rhoptries and in part at the surface of the parasite (Crewther *et al.*, 1990), before being assigned to the micronemes (Healer *et al.*, 2002). The localization of numerous other proteins in *Plasmodium* is still matter of debate. *Toxoplasma* exhibits also a simple, polarized secretory apparatus with a well-defined endoplasmic reticulum and Golgi that can serve as a model for understanding secretion. It has been recently exploited to definitively establish how the Golgi apparatus segregates during cell division (Pelletier *et al.*, 2002).

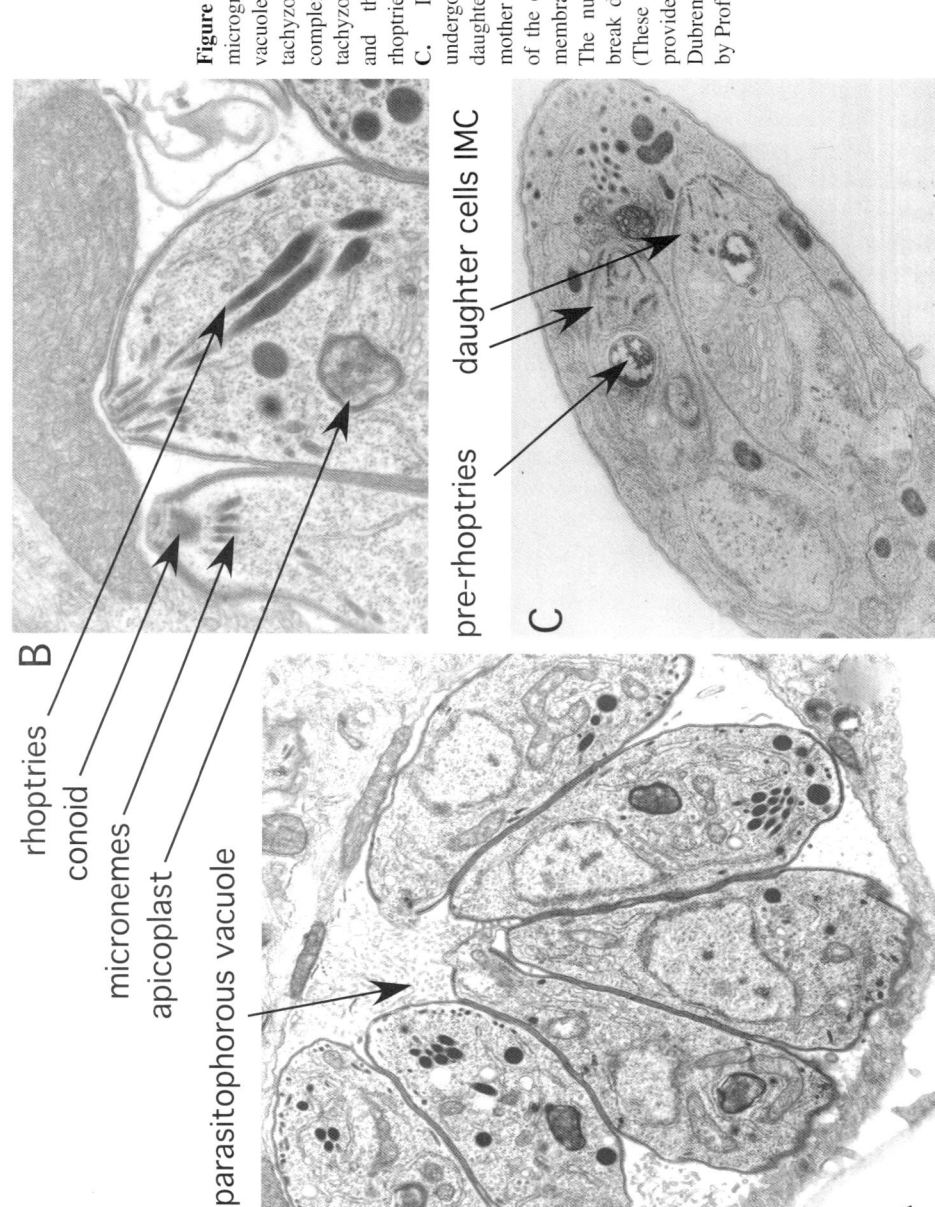

**Figure 3. A.** Electron microscopy micrographs of a parasitophrous vacuole containing *T. gondii* tachyzoites **B.** The apical complex apparatus of *T. gondii* tachyzoites showing the conoid and the secretory organelles rhoptries and micronemes. **C.** Intracellular tachyzoite undergoing endodyogeny. Two daughter cells develop into the mother cell with the formation of the conoid, the nascent inner membrane and the pre-rhoptries. The nuclear envelope does not break down during cell division. (These pictures were kindly provided by Dr. Jean Francois Dubremetz, the panel B was taken by Prof. E. Vivier).

# 3. Forward and Reverse Genetics: Old and New Tools

As already mentioned, *T. gondii* can be propagated in cell culture and in animals using conventional culture techniques, and this parasite is relatively accessible to biochemical and well suited for morphological studies. An additional reason why this parasite has emerged as a model system is its great flexibility and accessibility to genetic manipulation. The haploid genome of the asexual stages consists of $\sim 8.10^7$ bp, three times that of *Plasmodium*, although the number of anticipated genes is approximately the same (Nene *et al.*, 2000). The initial genetic manipulations were performed by chemical mutagenesis (Pfefferkorn and Pfefferkorn, 1979; Pfefferkorn and Kasper, 1983) and single clones were easily identified and isolated by limiting dilution and single plaque formation (Pfefferkorn and Pfefferkorn, 1976). Forward genetics has been applied to *T. gondii* by the generation of classical experimental crosses performed in cats (Pfefferkorn and Pfefferkorn, 1980). Such crosses can then be used to map and quantify the loci involved in a specific phenotype and they were used to generate a linkage map using restriction fragment length polymorphism (Sibley and Boothroyd, 1992). The establishment of transfection in *T. gondii* has rapidly led to the development of a broad range of tools associated with DNA transformation (Soldati, 1996). Importantly, the same protocol was applied to transfect *Plasmodium* species, thus opening the approaches of reverse genetic manipulation to the malarial parasite (Goonewardene *et al.*, 1993). However, the molecular genetics of *T. gondii* offers a wider range of experimental strategies including, positive and negative selectable markers, integrating and episomal vectors, homologous recombination to disrupt genes and non-homologous random integration as a strategy for insertional mutagenesis. These are described in more detail in the following sections

## 3.1. Expression Vectors

A large repertoire of selectable markers including chloramphenicol actetyltransferase (CAT), a mutated dihydrofolate-reductase (DHFR) conferring resistance to pyrimethamine, the nucleotide salvage pathways enzymes, uracil phosphoribosyl-transferase (UPRT) and hypoxanthine-xanthine-guanine phosphoribosyltransferase (HXGPRT) are available to generate multiple rounds of stable transformation of *T. gondii*. This system is all the more powerful given the particularly high frequency of stable transformation, of up to $10^{-2}$ observed. The monitoring of gene expression, promoter activity and subcellular distribution can be achieved by using several reporter genes such as *CAT* (Soldati and Boothroyd 1993), *lacZ* (Seeber and Boothroyd, 1996), GFP (Striepen *et al.*, 1998) and Luciferase (Matrajt *et al.*, 2002b). The expression of several heterologous proteins turned out to be problematic in tachyzoites and one way to circumvent this problem was to create either N-terminal fusions or synthetic genes with remodelled codon usage for expression in *T.gondii* (Striepen *et al.*, 1998; Meissner *et al.*, 2001). A more systematic analysis of the expression levels depending on the nature of the amino acid in position 2 allows now the optimisation of the steady-state level of expression of any transgene (Matrajt *et al.*, 2002b). Finally, a shuttle vector showing the characteristics of an episomal vector has been constructed for *T. gondii* (Black and Boothroyd 1998).

## 3.2. Insertional Mutagenesis

One of the most interesting and practically useful features of molecular genetic in *T. gondii* is the unusually high rate of stable integration in the genome. While the frequency of *Toxoplasma* transformation is about $10^{-2}$ if DHFR-TS sequence is present in the selectable vector (Donald and Roos, 1993) only a frequency between $10^{-5}$ -$10^{-6}$ has been reported for *P. berghei* (de Koning-Ward *et al.*, 2000). In *Plasmodium*, vectors integrate in the genome strictly by homologous recombination whereas integration occurs predominantly at random locations scattered throughout the genome of *Toxoplasma*. To favour homologous recombination, large contiguous genomic sequences are required. Approximately 8 kbp of sequences are necessary to obtain 50% of homologous integration (Donald and Roos, 1994). The combination of non-homologous recombination with high frequency of integration can be exploited to saturate the parasite genome with insertional transgenes. The tagged loci can readily be identified by plasmid rescue. This is an invaluable tool to identify non-essential genes for which a phenotypic selection exists to monitor its absence. As an example, mutants deficient in conversion from tachyzoites into bradyzoites have been generated by insertional mutagenesis (Matrajt *et al.*, 2002a). A similar approach based on chemical mutagenesis also led to the isolation mutants compromised in their ability to differentiate *in vitro* and *in vivo* (Singh *et al.*, 2002). In both studies, the pattern of gene expression in the mutants was subsequently investigated by microarrays as a novel approach to identify the sets of genes along the developmental pathways involved life stage differentiation.

An alternative strategy taking advantage of random integration in the parasite genome is signature tagged mutagenesis, which uses a unique DNA sequence to tag an individual mutant so that it can later be identified within a pool. Such a library of mutants has recently been generated in *T. gondii* and will be used to identify mutant parasites carrying mutations in genes that are essential for growth in restrictive but not permissive conditions (Knoll *et al.*, 2001).

## 3.3. Gene Knock-out; Gene Knock-down, and Conditional Knock-out

Whereas random integration occurs by using a DNA-fragment that either lacks or has only a little homology to the genome, targeted gene knockout by double homologous recombination can be favoured by providing sufficient contiguous homology to the targeted locus (Donald and Roos, 1994). Generation of knockout is only possible for non-essential gene in the haploid form of the parasite. A two-step hit and run strategy can be applied if it is suspected that the function of the target gene might be essential for the parasite survival (Donald and Roos, 1998).

Alternatively, if the gene of interest fulfils a critical role for survival, targeted gene disruption can be replaced by a knock down approach, which instead of eliminating only reduces the production of a protein. Use of antisense RNA, ribozyme or RNA interference (RNAi) can specifically lower the level of an mRNA and consequently the level of the corresponding protein. Previous studies report the successful use of antisense/ribozyme in *T. gondii* tachyzoites (Nakaar *et al.*, 1999; Nakaar *et al.*, 2000). An optimal strategy has been developed in *Trypanosoma brucei* where a powerful inducible system and RNAi are combined to routinely analyse gene function in vivo. Recent promising studies in *P. falciparum* suggest that RNAi mechanism can operate in apicomplexan parasites (Malhotra *et al.*, 2002; McRobert and McConkey, 2002). However, to date no report exists that describes the successful application of RNAi technology to *Toxoplasma*.

Instead of regulating gene expression at the mRNA level, the widely used tetracycline-based inducible system controls gene expression at the transcriptional level. The tetracycline-repressor system interferes with transcription and has been perfectly optimised to tightly regulate gene expression in *T. brucei* (Wirtz *et al.*, 1999). The same system has proved less successful when transferred to *T. gondii* (Meissner *et al.*, 2001). Although this system is suitable for the expression of toxic gene and dominant negative mutants, it is not appropriate to generate conditional knockouts. Indeed, the necessity to keep the parasites in presence of drug (anhydrotetracycline) to maintain the expression of the essential gene may lead to the creation of revertants that lose inducibility. In order to improve the system, a genetic screen based on random insertion was designed to identify a functional transcriptional activating domain in *T. gondii* and to establish a tetracycline transactivator-based inducible system similar to the one developed by Bujard and co-workers for higher eukaryotic cells (Gossen *et al.*, 1994). This new system permitted the creation of conditional knockout for an essential gene with no apparent reversion effect. Furthermore, the inducible system was shown to operate in the animal model (Meissner *et al.*, 2002b). The transactivator generated for *T. gondii* proved to be inactive in HeLa cells (Meissner *et al.*, unpublished). The activating domain appears to selectively interact with the transcription machinery of *T. gondii* and might possibly do so in other apicomplexan parasites. Using *Toxoplasma* as screening system, more transactivators can be identified, which might be also functional in *Plasmodium*.

# 4. Tools and Techniques to Exploit the Genome

Intracellular parasitism is governed by a complicated network of interactions between the host and the parasite cells. Dramatic changes occur within the infected cells and in the invading parasites and the monitoring and the significance of these changes is a key challenge of modern Parasitology. The resources provided by the genome information for both the host and the parasite put us in an unprecedented position to obtain rapidly global information about these changes.

## 4.1. Microarrays

High-density arrays of DNA fragments on a solid surface allow the expression of thousands of genes to be assessed in a single experiment. The *Plasmodium* genome sequence has considerably boosted this type of global analysis (for review see Rathod *et al.*, 2002). Similarly, in *Toxoplasma*, the ESTs projects have provided enough information to initiate this kind of approach. Two recent studies report the use of microarrays to analyse global gene expression in bradyzoite deficient mutants. Interestingly, a different application of this technology consists of investigating the host cell response to parasitic infection. It is known that cells infected with *T. gondii* undergo up-regulation of pro-inflammatory cytokines and become resistant to apoptosis. (Blader *et al.*, 2001; Gail *et al.*, 2001). Human cDNA microarrays consisting of 22,000 genes and ESTs are available and offer a unique chance to identify sets of genes modulated by the presence of the parasite and thus potentially implicated in host parasite interaction. Comparison studies report the transcriptional profiles of host cells upon infection with *T. gondii* tachyzoites.

## 4.2. Proteomics

The progress in two-dimensional electrophoresis, mass spectrometry and bioinformatics, combined to genome sequence information offer a unique opportunity to undertake global protein-expression analyses. The mapping and characterization of the protein population expressed in the tachyzoite stage of *T. gondii* has been initiated (Cohen *et al.*, 2002). Additional layers of information such as subcellular localization, post-translational modification and alternative splicing can be assessed in addition to the life stage specificity. Armed with the complete *P. falciparum* genome sequence, researchers are currently deciphering the repertoire of stage specifically expressed proteins in *P. falciparum*.

## 4.3. Sequence Polymorphisms and Population Studies

*T. gondii* has an usual clonal population structure with three lines originating from the genetic mixing of two discrete ancestral lines (Grigg *et al.*, 2001). High-resolution typing of *T. gondii* is essential to understand the effect of genetic differences among strains on the variation in disease manifestation and transmission patterns. In mice, the various strains of the parasite differ enormously in their virulence and disease presentation. In humans, disease manifestations are highly variable, ranging from asymptomatic to severe, especially in cases of brain and eye infection. The variability in human infection may be in part result from the type of strain that has caused the infection. Sexual recombination leads to a combination of polymorphisms and appears to contribute to virulence. The identification of the loci responsible for virulence and pathogenicity in the animals will be greatly accelerated by the knowledge of the complete *T. gondii* genome sequence. Currently, two loci contributing to the acute virulence in mice, located on different chromosomes have been mapped (Su *et al.*, 2002). Similar studies using fast congenic breeding techniques might be applied to mice host populations and used to map resistance/susceptibility loci in the host genome as has been performed for *Plasmodium* (Bagot *et al.*, 2002; Burt *et al.*, 2002).

# 5. Subcellular Structures and Organelles Common to the Apicomplexans

Apicomplexans exhibit a collection of unique organelles termed the apical complex (Figure 3B) These organelles include the rhoptries, the micronemes and cytoskeletal elements composing the apical polar ring that has a microtubule-organizing center, and the conoid, a small cone-shaped structure presumed to play a mechanical role in invasion. These parasites also harbour a variety of additional unusual organelles. The inner membrane complex, composed of flattened vesicles, is central to the peculiar mode of division observed in these parasites. Last but not least, most apicomplexan parasites contain a plastid, a chloroplast-like organelle acquired by secondary endosymbiosis of an algal ancestor (McFadden *et al.*, 1996; Kohler *et al.*, 1997). *T. gondii* is extremely well suited for cell biological studies because these subcellular organelles can be labelled with different fluorescent reporters for quantitative analysis in living cells and in combination with the molecular organellar targeting signals can be dissected (Joiner and Roos 2002). As an illustration, some of these distinctive organelles are revealed by indirect immunofluorescence (Figure 2). In contrast, it has been more difficult to recognise the mitochondria, Golgi apparatus, and the specialized apical complex organelles in *Plasmodium*.

## 5.1. Micronemes: Recognition, Attachment and Gliding Motion

Three sets of organelles sequentially secrete their content during host cell invasion (Carruthers and Sibley, 1997; Dubremetz *et al.*, 1998). Micronemes release adhesins early during the attachment-invasion process, followed by the rhoptries, which contribute to the creation of the parasitophorous vacuole, as invasion proceeds and finally the content of the dense granules constitutively accumulates in the parasitophorous vacuole. Upon contact with the host cell the content of the micronemes is discharged by stimulation of a yet unknown $Ca^{2+}$ signalling cascade (Carruthers *et al.*, 1999a; Carruthers *et al.*, 1999b). Although little is known about the molecular nature of this signalling pathway, recently two calcium dependent protein kinases (TgCDK1 and TgCDK2) have been identified and potentially linked to motility and invasion (Kieschnick *et al.*, 2001). After discharge, some micronemal proteins remain associated with the parasite's apical surface where they are thought to mediate attachment by binding to specific receptors on the target host cell (Carruthers and Sibley, 1997; Garcia-Reguet *et al.*, 2000). Many different micronemal proteins have been identified so far and they share high level of conservation throughout the Apicomplexans. Microneme proteins are likely to contribute to host cell binding and contain adhesive domains also found in vertebrates, like integrins, thrombospondin, kallikrein, lectin and epidermal growth factor (EGF) (Tomley and Soldati, 2001). The proteins carrying a membrane-spanning domain are likely candidates to be involved in gliding motility and previous studies on the micronemal protein TRAP have provided substantial evidence supporting such a role in *Plasmodium*. The destruction of the corresponding gene for TRAP resulted in a phenotype where the sporozoites were incapable to invade hepatocytes or to migrate into the salivary glands of the mosquito. Moreover the sporozoites without TRAP are not able to move (Sultan *et al.*, 1997). Homologues and orthologues to TRAP have been identified, like the ookinete specific protein of *P.falciparum* CTRP (Trottein *et al.*, 1995; Yuda *et al.*, 1999), and MIC2 in *T. gondii* (Wan *et al.*, 1996).

According to the current model of gliding motility, TRAP interacts via adhesive motifs present in the ectodomain with receptors and structures at the surface of the host cell, whereas the short cytoplasmic tail is presumed to establish a connection with the actomyosin system, enabling the parasite to glide. Studies have demonstrated that the cytoplasmic tail is essential for gliding motility and interchangeable between the homologous proteins of *T. gondii* and *P. berghei* (Kappe *et al.*, 1999) suggesting a highly conserved mechanism. A dozen of proteins have been localized to the micronemes of Toxoplasma so far. About half of these proteins are type I transmembrane proteins, with a transmembrane and cytoplasmic domain homologues to TRAP. Some conserved residues resembling the tyrosine-based sorting signal are present in the cytoplasmic domain of these proteins and play an essential role in targeting to the micronemes. Two conserved motifs have been identified within this domain, which are necessary for the transport of transmembrane proteins to the micronemes, whereas the ectopic domains appear to play no role in this process (Di Cristina *et al.*, 2000). The sorting of proteins to the micronemes appears to be conserved between apicomplexan parasite, since the heterologous expression of a chimeric protein between the transmembrane and cytoplasmic domain of TgMIC2 and the ectodomain of PbTRAP in *P.berghei* results in correct localisation to the *Plasmodium* micronemes (Kappe *et al.*, 1999).

A common theme for the soluble microneme proteins in *T. gondii* is that they are sorted to the micronemes due to formation of a complex with a transmembrane escorter protein. Up to now, three distinct complexes have been identified in *T. gondii*. The transmembrane protein TgMIC6 interacts with the soluble adhesins TgMIC1 and TgMIC4 (Reiss *et al.*, 2001), TgMIC8 interacts with the soluble adhesin TgMIC3 (Meissner *et al.*, 2002a) and TgMIC2 interacts with the soluble protein TgM2AP (Rabenau *et al.*, 2001) (for review see Soldati *et al.*, 2001; Carruthers, 2002). Another common theme of the micronemal proteins is the fact that they are processed on several occasions during their lifetime. In addition to processing during the transport to the micronemes, extensive cleavage occurs after release on the parasite surface. Two distinct parasite derived protease activities, called MPP1 and MPP2, have been identified that process MIC2 after release by the micronemes (Carruthers *et al.*, 2000). MPP1 activity appears to be conserved throughout the phylum of *Apicopmlexa* since PbTRAP expressed in *T. gondii* is correctly processed. The precise cleavage site was mapped by mass spectrometry in a conserved sequence within the transmembrane domain of numerous micronemal proteins (Opitz *et al.*, 2002). A candidate for some of these cleavage activities has been characterized recently. A subtilisin-type protease TgSUB1 cross reacts with antibodies specific for *Plasmodium* PfSUB1 and is transported to the micronemes (Miller *et al.*, 2001).

## 5.2. Rhoptries: Formation and Modification of the Parasitophorous Vacuole

Rhoptries are large, club shaped secretory organelles with an acidic pH (Shaw *et al.*, 1998). As in *Plasmodium*, rhoptries have two distinct regions with different protein content, the basal bulbous portion and a narrow apical duct. Rhoptries secretion occurs during invasion in a $Ca^{2+}$ independent manner (Carruthers and Sibley, 1997). Upon discharge, rhoptries change their appearance dramatically as they collapse (Nichols *et al.*, 1983). To date seven rhoptry proteins (ROP1-7) have been identified in *T.gondii* and unlike microneme proteins, these proteins do not exhibit any significant homology with the proteins present in the *Plasmodium* rhoptries. The rhoptry proteins are believed to modify the parasitophorous vacuole and although the majority of their actual function remains to be determined, a recent elegant study suggests that ROP2 mediates the association of host mitochondria with the parasitophorous vacuole (Sinai and Joiner, 2001). Furthermore another recent study has reported the formation of vacuoles devoided of parasites (called evacuoles) upon treatment with cytochalasin D (Hakansson *et al.*, 2001). In presence of this drug, the parasites were unable to invade but discharge of the rhoptries still occurred and led to the accumulation of vesicles, which exhibit the characteristics of the parasitophorous vacuole including association with the host mitochondria. In addition, these vesicles were able to fuse and deliver their content and thus contribute as a second step of secretion and fusion to the formation of the parasitophorous vacuole (Hakansson *et al.*, 2001).

*T. gondii* has also been exploited to identify the determinants necessary for rhoptry targeting. Here also tyrosine-based sorting signals appear to be involved and evolutionarily conserved (Hoppe *et al.*, 2000). With regard to the biogenesis of rhoptries it has been shown for *Toxoplasma* and *Plasmodium* that rhoptries mature from Golgi derived Pre-rhoptry bodies (Shaw *et al.*, 1998; Soldati *et al.*, 1998). Interestingly, recent studies established that rhoptry proteins RAP1 and RAP2 in *Plasmodium* form a complex, which plays a critical role in targeting to the rhoptries (Baldi *et al.*, 2000), very similar to the observations made on microneme proteins in *Toxoplasma*. Analytical procedures such as the systematic application of TAP tag technology could be usefully employed to investigate the extent of this phenomenon. This powerful purification procedure proved very useful in mammalian cell and might be applied also in protozoan organisms (Cox *et al.*, 2002)

## 5.3. Dense Granules

Dense granules (DG), the third class of secretory organelles in apicomplexans are localized throughout the parasite (for review see (Cesbron-Delauw 1994)). DGs secrete their contents at the end of the invasion process and continue to do so during the intracellular residence of the parasite (Dubremetz *et al.*, 1993) (Carruthers and Sibley, 1997). The proteins GRA1-8 show no homology to any other protein in other organisms, including *Plasmodium* and no significant homology between each other (for review see Carruthers, 1999). The different repertoire of dense granule proteins in different Apicomplexans might reflect differences in the organization of the parasitophorous vacuole. Although the function of these proteins is not yet understood, all of them are secreted and are thought to play a role in the maintenance and modification of the parasitophorous vacuole, by either remaining in the lumen or by integrating into the parasitophorous membrane. In addition, NTPases have been identified in dense granules, which are involved in purine salvage and may play a role in initiating the egression of the parasite from the host cell (Stommel *et al.*, 1997; Silverman *et al.*, 1998).

The non-fusogenic parasitophorous vacuole of *T. gondii* contains a network of tubules connected to the vacuolar membrane and proteins discharged from dense granules decorate this intravacuolar network after invasion. Recent studies have established that the secretory protein GRA2 and GRA6 are organized and stabilise this network (Figure 3A) (Mercier *et al.*, 2002).

## 5.4. The Inner Membrane Complex (IMC) and Cell Division

The pellicle of Apicomplexans is very unusual and characterized by three membrane layers composed of the plasma membrane and an inner membrane complex formed by juxtaposed flattened vesicles underneath the plasma membrane and covering the basket of sub-pellicular microtubules. Ultrastructural analyses by freeze fracture have illustrated the resemblance in the organization of the IMC between *Toxoplasma*, *Plasmodium* and other Apicomplexans (for review (Morrissette and Sibley, 2002a)). The membranes of the IMC are characterized by parallel alignment of intramembranous particles (IMPs) (Morrissette *et al.*, 1997). Some microtubule-associated proteins may confer great stability to the subpellicular microtubules and may play an important role in connecting them to the IMC (Morrissette and Roos, 1998). Recently, the inner membrane protein 1 IMC1 has been described as a marker of the IMC (Mann and Beckers, 2001). This protein is the major subunit composing a novel intermediate filament-like element in these parasites. The structural stability of the subpellicular network changes dramatically during the cell cycle and correlates with the proteolytic status of IMC1 (Mann *et al.*, 2002). This structure plays a key role during cell division and is likely to be part of the gliding motility machinery (see below).

The apicomplexans replicate by assembling progeny within the parental cell, using the inner membrane complex. *Plasmodium*, which package many progeny within a single parental cell during mitosis (called schizogony), while *T. gondii* typically divides by endodyogeny, which results in the formation of two daughter cells (Figure 4C). The simpler pattern of cell division in Toxoplasma has attracted scientists to analyse it in great detail by electron microscopy (van der Zypen and Piekarski, 1967) and more recently in combination with cell and appropriate molecular genetics techniques. Recently it was shown that also in the case of *T. gondii* more than two (up to 8) daughter cells can be produced within a single mother cell (Hu *et al.*, 2002), indicating that the machinery for cell division is conserved in all apicomplexans. Both in *Plasmodium and Toxoplasma*, microtubules play a similar, critical role in cellular division, since the treatment with oryzalin yields basically the same results for both parasites (Bejon *et al.*, 1997; Shaw *et al.*, 2000). Two microtubules organizing centers (MTOC) act at different steps during cell division: the subpellicular microtubules are organized by the apical ring (Nichols and Chiappino, 1987) and the spindle microtubules are organized by two spindle pole plaques. Whereas the apical polar ring MTOC drives daughter cell budding, the spindle microtubules are essential for nuclear division (Morrissette and Sibley, 2001). This organisation facilitates the maintenance of an intact cytoskeletal and organellar network throughout most of *Toxoplasma* endodyogeny such that the parasites are capable to invade the host cell at virtually any point during its cell cycle. It also allows the control of nuclear division independently from cell polarity and cytokinesis (Morrissette and Sibley, 2002b). The same arrangement allows other apicomplexan parasites, like *Plasmodium,* to accumulate multiple nuclei prior to budding and thus the production of 12-24 daughter parasites during schizogony. *T. gondii* expressing a YFP-tagged version of the IMC1 protein has been exploited to visualize and characterize the dynamics of the intramembranous complex in living cells during endodyogeny (Hu *et al.*, 2002). A similar approach was used to establish that the equal segregation of the apicoplast to each daughter cell is ensured by direct linkage of the plastid to the centrosomes (Striepen *et al.*, 2000).

## 5.5. The Apicoplast: A Unique Genome, Structure and Function

Like plants, most apicomplexans have two extrachromosomal genetic elements: the 6 kb mitochondrial genome and a 35 kb circular DNA contained within a non-photosynthetic plastid, the apicoplast (for review see Roos *et al.*, 2002). The origin and function of this plastid non-photosynthetic organelle has been the focus of great interest and a perfect illustration of successful genome database mining. The sequence of the 35 kb element in *T. gondii* closely related to green algal origin (Fichera and Roos, 1997; Kohler *et al.*, 1997), and the gene organization of the 35 kb genome is conserved between *Plasmodium* and *Toxoplasma*. Reminiscent of the chloroplast in plants, most proteins present in the apicoplast are encoded in the nuclear genome and transported post-translationally to the plastid. Translocation across the four membranes surrounding the apicoplast is mediated by an N-terminal bipartite targeting sequence (Waller *et al.*, 1998; Roos, 1999), which is interchangeable between *T. gondii* and *P. falciparum* (Waller *et al.*, 2000). The requirement for the bipartite leader sequence has been exploited bioinformatically; PlasmoDB has been extensively explored and scrutinized to identify metabolic pathways associated with the apicoplast (Roos *et al.*, 2002). Essential metabolic functions such as fatty acid synthesis (Waller *et al.*, 1998) and isoprenoids synthesis (Jomaa, 1999) have been assigned to this organelle. Several studies have demonstrated that this organelle is essential for long-term parasite survival and thus validate this organelle as an attractive new potential target for chemotherapy (McFadden and Roos, 1999; He *et al.*, 2001). However, because of the very distinct host cell metabolism in which the *Plasmodium* merozoites and *T. gondii* tachyzoites develop, the importance of apicoplast functions may vary significantly between the two parasites. For example, the mevalonate-independent pathway for isoprenoids synthesis is essential for the survival of blood stages forms of *Plasmodium* but not for *T. gondii* tachyzoites (Jomaa *et al.*, 1999).

# 6. Mechanisms Leading to the Establishment of Intracellular Parasitism

Most apicomplexan parasites actively penetrate the host cells forming a unique parasitophorous vacuole, biochemically and functionally distinct from a phagosome. Despite the lack of any locomotive organelles, like cilia or flagella, these parasites are highly motile by their ability to glide (Russell and Sinden ,1981). Like *Plasmodium, Toxoplasma* rapidly gains access to the intracellular environment of the host cell. This process involves gliding motility and the sequential secretion of the specialised secretory organelles, micronemes, rhoptries and dense granules (Morrissette and Sibley, 2002a; Opitz and Soldati, 2002). In *Plasmodium* it was shown that sporozoites are able to migrate through several host cells, without forming a parasitophorous vacuole, before they establish an infection. Furthermore this process appears to stimulate sporozoites in order to establish an infection (Mota *et al.*, 2001; 2002).

## 6.1. Host Cell Specificity, Recognition and Attachment

To accommodate for the broad host range specificity of *T. gondii*, adhesion might involve the recognition of ubiquitous surface-exposed host molecules or, alternatively, the presence of various parasite attachment molecules able to recognise different host cell receptors. SAG1 is a glycosyl-phosphatidyl-inositol (GPI) anchored protein at the surface of *T. gondii* tachyzoites, which could potentially be implicated in a first broad and low affinity interaction with the host cell membrane (He *et al.*, 2002). The stage-specific ESTs sequencing project has uncovered the existence of a large repertoire of surface antigens (SAGs). Most of these antigens belong to the developmentally regulated and distantly related SAG1 or SAG2 families. This family may group proteins that function during parasite attachment and play a crucial role in immune modulation or virulence attenuation (Lekutis *et al.*, 2001). Subsequently, micronemes release adhesins that might be involved in higher affinity and more specific host cell attachment.

## 6.2. Gliding Motility and Host Cell Invasion

Apicomplexan actin and myosins exhibit unusual properties and are utilized to generate atypical form of actin/myosin-dependent gliding motility, which propel parasites into host cell (Sibley and Andrews, 2000; Opitz and Soldati, 2002). According to the "capping model", the gliding motion is driven by the movement of transmembrane proteins from the apical pole along the subcortical actomyosin complex towards the posterior end of the parasite. The conditional knockout of myosin A established the crucial role of this

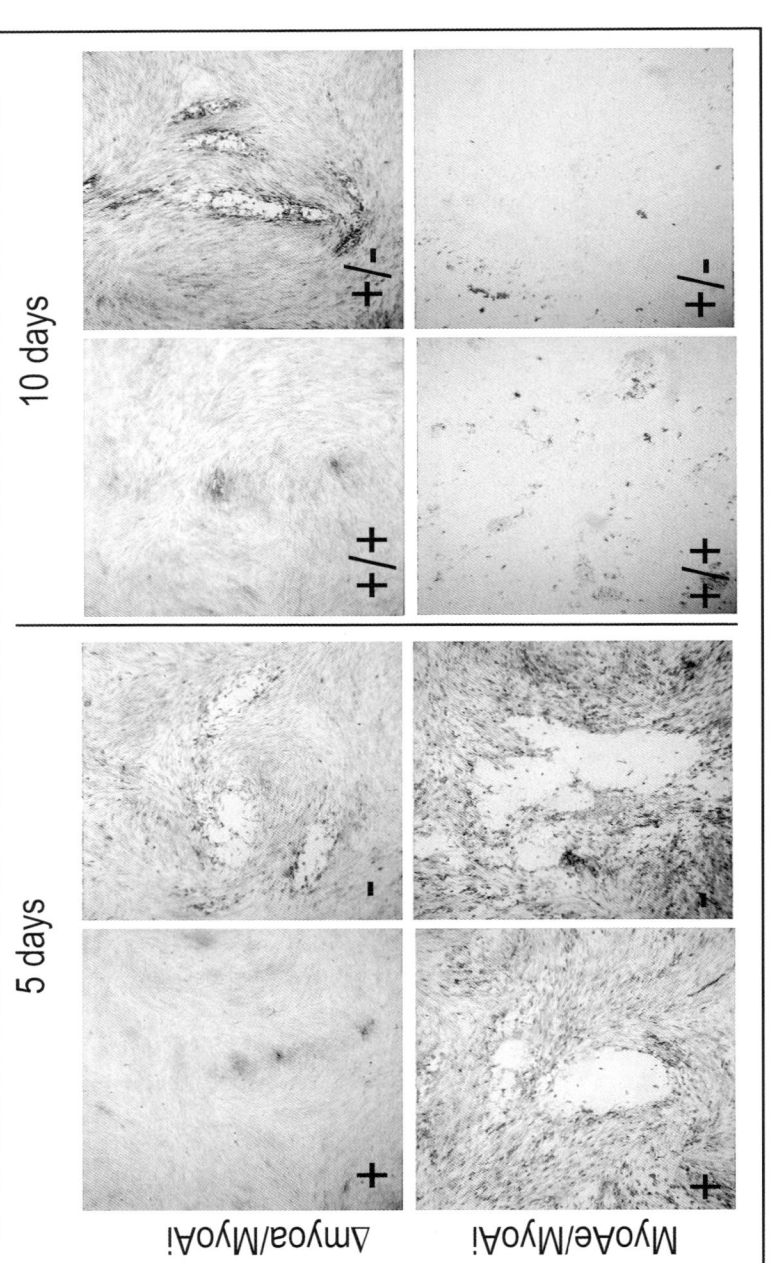

**Figure 4.** Conditional knockout of *T. gondii* myosin A using a newly established tetracycline inducible transactivator system. Reversible depletion of myosin A leads to the inhibition of plaque formation on monolayers of human foreskin fibroblasts. Parasites expressing the endogenous *MyoA* gene and a transgenic, inducible copy of MyoA (MyoAe/MyoAi) or parasites lacking the endogenous copy (disrupted by double homologous recombination) and expressing only the inducible copy (Δmyoa/MyoAi) were inoculated for five days or 10 days in presence (+) or absence (−) of ATc. In the case of MyoAe/MyoAi, MyoAe is constitutively expressed, down-regulation of MyoAi has no effect on plaque formation and after 10 days the host cells were completely lysed. In the case of Δmyoa/MyoAi, plaques are not formed after 5 days in presence of ATc (MyoAi switched off) and in absence of ATc, the size of the plaques appeared to be significantly smaller than the size of

actin-dependent motor in promoting gliding motility, host cell invasion and egress (Meissner *et al.*, 2002b). A plaque assay was used to demonstrate the inability of parasites depleted in MyoA to invade, to egress and consequently to propagate in tissue culture. This phenotype was reversible (Figure 4). The nature of the connection between the actomyosin system and the parasite ligands coupled to host cell receptors has not been elucidated yet. The precise topology of the actomyosin system between the plasma membrane and the inner membrane complex (IMC) is still a matter of debate, due to the technical difficulties in localizing proteins within this narrow space and added to the fact that actin filaments cannot be visualized in Apicomplexans presumably inherent to the instability of the filaments.

Currently three proteins have been identified as part of the myosin A motor complex present in the pellicule, the MyoA heavy chain, a myosin light chain and a putative docking protein (Herm-Gotz *et al.*, 2002). Two topologies can be envisioned. While a previous report suggested that actin was on the IMC side and myosin at the plasma membrane in *T. gondii* (Figure 5A), more recent work suggests that the *Plasmodium* homologue of TgMLC, called MTIP localizes to the IMC (Bergman *et al.*, 2002) (Figure 5B). This latter topology would imply that actin filaments are directly or indirectly contacting the tails of microneme proteins. The bulk of the actin pool in *T. gondii* appears to be maintained as monomeric or globular forms due to the presence of an actin depolymerising factor. Despite that, actin polymerisation is required for motility. It is intriguing to note that the conserved C-terminal acidic motif and tryptophan residue mapped in the tails of TRAP and conserved in several microneme protein in *T. gondii*, shows a significant level of similarity to a motif found in WASP/Scar proteins and several myosins in eukaryotes (Higgs and Pollard, 2001). This domain has been demonstrated to stimulate actin assembly by interacting with the Arp2/3 complex. Additionally, MyoA is a single-headed, non-processive enzyme, spending most of the time dissociated from the actin filaments implying that at least 100 molecules should bind the same actin filament at a time in order to generate movement (Herm-Gotz *et al.*, 2002). The presence of actin filaments on the plasma membrane side would more conveniently accommodate the existence of an array of myosins interacting with a limited number of microneme proteins. A recent study suggests that the glycolytic enzyme aldolase interacts with the cytoplasmic tail of MIC2 (D. Sibley and cowrokers, abstract MPMXIII, Woods Hole, 2002). In another system this enzyme has previously been shown to associate to atin filaments (Wang *et al.*, 1996).

## 6.3. Signal Transduction Cascades Triggering Invasion and Egress

Host cell invasion and egress are mechanistically intimately related (for review see Hoff and Carruthers, 2002) and likely to be regulated both by a calcium-dependent signal transduction pathway. The cascade of

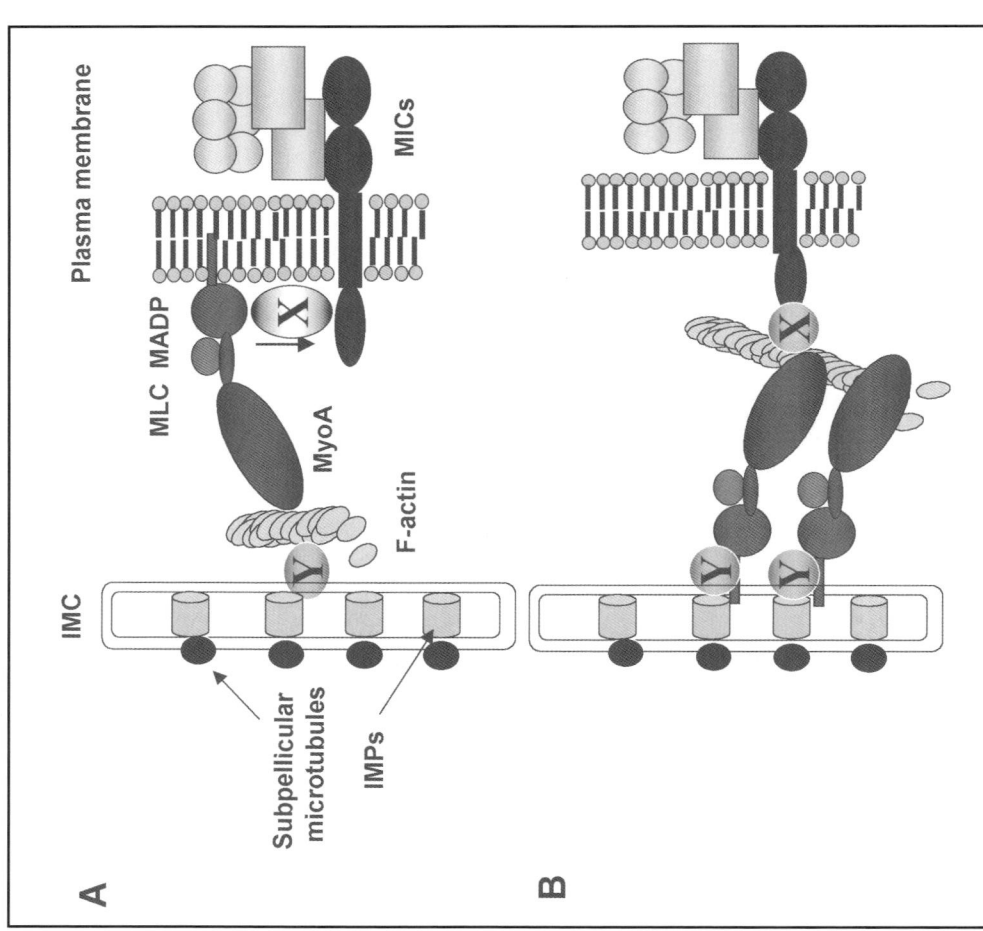

**Figure 5.** Schematic representation of the myosin motor complex's topology at the pellicle of *T. gondii* tachyzoites. The motor complex is composed of myosin A (MyoA), the myosin light chain 1 (MLC1) and the MyoA docking protein (MADP). It is permanently anchored either in the plasma membrane (**A**) or in the inner membrane complex (**B**). Actin monomers (globular actin, G-actin) are abundant but no actin filament can be visualized under physiological conditions. Upon stimulation, microneme protein complexes composed of adhesins and escorters are secreted and redistributed to the surface of the parasites. They presumably bind to host cell receptors or to components of the extracellular matrix and connect with the actomyosin system through their cytoplasmic tail. Unkown factors X and Y might be necessary to bridge the actomyosin system to the microneme proteins and to the subpellicular microtubules, possibly by the intermediate of the intramembranous particles (IMPs).

signals triggering these events has not been unravelled yet but some of the components are emerging. The discharge of adhesins by microneme is known to occur in response to a rise in intracellular calcium. Ethanol treatment has recently been shown to stimulate an increase in parasite inositol 1,4,5-triphosphate in tachyzoites, and thus this molecule might serve as second messenger to mediate intracellular calcium release (Lovett *et al.*, 2002). Beside the endoplasmic reticulum, the presence of acidocalcisomes (distinct and specialized organelles involved in the storage of polyphosphate) represents an unconventional storage compartment for calcium in *T. gondii* and other intracellular protozoan parasites (Rodrigues *et al.*, 2002).

# 7. Survival Within the Parasitophorous Vacuole

## 7.1. Autotroph or Auxotroph: Dependence on the Host Metabolism

Our knowledge of the metabolic requirement for intracellular replication of Apicomplexans is only partial. Two examples are discussed here which illustrates the balanced contribution of the host. All apicomplexan parasites synthesize pyrimidines *de novo* and are purine auxotrophs. *T. gondii* is partially capable of pyrimidine salvage by the action of uracil phosphoribos yltransferase and this gene is not essential. In contrast, the recent generation *T. gondii* mutants that lack carbamoyl phosphate synthetase II (uracil auxotrophs) unambiguously established the importance of parasite *de novo* pyrimidine biosynthesis *in vivo*. These uracil auxotrophs are completely avirulent in mice even in immuno-compromised mutant strains, illustrating the significance of the pyrimidine biosynthesis pathway for the virulence (Fox and Bzik 2002).

Recruitment of other nutrients such as chlolesterol has been studied in some details. *T. gondii* is unable to produce sterols via the mevalonate pathway and acquires cholesterol from the host. Intriguingly, intracellular parasites appear to be able to specifically increase the host cholesterol uptake by the low-density lipoprotein (LDL) pathway. By a yet unknown mechanism and pathway the parasite vacuole actively accumulates LDL-derived cholesterol that has transited through host lysosomes (Coppens *et al.*, 2000).

## 7.2. Evading the Immune Response

Imitating a number of intracellular pathogens, *T. gondii* has not too surprisingly also evolved strategies to interfere with the apoptotic program of the host cell (for review see Heussler *et al.*, 2001). The manipulation of host cellular functions by the parasite suggests that modulating factors are secreted beyond the parasitophorous vacuole. Although numerous proteins are known to be targeted in and to the surface of infected red blood by *P. falciparum*, nothing is known yet about colonization of host cell compartments by *T. gondii* proteins.

# 8. Summary

Advantages, constraints and limitations are intrinsic to any experimental system. Basic research performed on diverse apicomplexan parasites have vastly contributed to the current knowledge of the biology of malaria parasites. *T. gondii* has not only been used to study molecular genetic but also for cell biological investigation. This parasite serves as model system to elucidate parasite specific features and functions, including structures and organelles composing the invasion apparatus, unique metabolic pathways and common mechanisms leading to the establishment of intracellular parasitism. This parasite is not only an attractive model system to study conserved features among the apicomplexans, it is also a relevant pathogen to study and understand for its own sake.

The information gained from the complete sequence of the *Plasmodium* genome, combined to several ESTs and genome sequencing projects for other Apicomplexans have been critical for the identification of new classes of genes that are restricted to the phylum of *Apicomplexa* and thus of considerable interest as genes involved in parasitism and as potential targets for intervention. The initial EST sequencing project gene expression analysis and the current sequencing of the entire genome http://ToxoDB.org represents an invaluable complementary tool for cross-species comparisons microarrays, proteomic and genetic population studies.

# 9. Acknowledgement

We express our gratitude to Dr. J.F. Dubremetz and E. Vivier who kindly provided with the electron micrographs presented in the Figure 4. We thank also Dr. Dubremetz for supplying us with the anti-GRA3 and anti-ROP2 antibodies and Dr. G. Ward for the anti-IMC1 antibodies. Dr Markus Meissner was funded by a Feodor Lynen Fellowship of the Alexander von Hambolt Stiftung and by an HHMI International Scholar Grant.

# 10. References

Bagot, S., Campino, S., Penha-Goncalves, C., Pied, S., Cazenave, P.A., and Holmberg, D. 2002. Identification of two cerebral malaria resistance loci using an inbred wild-derived mouse strain. Proc. Natl. Acad. Sci. USA 99: 9919-23.

Baldi, D.L., Andrews, K.T., Waller, R.F., Roos, D., Howard, R.F., Crabb, B.S., and Cowman, A.F. 2000. RAP1 controls rhoptry targeting of RAP2 in the malaria parasite *Plasmodium falciparum*. EMBO J. 19: 2435-2443.

Bejon, P.A., Bannister, L.H., Fowler, R.E., Fookes, R.E., Webb, S.E., Wright, A., and Mitchell, G.H. 1997. A role for microtubules in *Plasmodium falciparum* merozoite invasion. Parasitology 114: 1-6.

Bergman, L.W., Kaiser, K., Fujioka, H., Coppens, I., Daly, T.M., Fox, S., Matuschewski, K., Nussenzweig, V., and Kappe, S.H. 2002. Myosin A tail domain interacting protein (MTIP) localizes to the inner membrane complex of *Plasmodium* sporozoites. J. Cell Sci. 116: 39-49.

Black, M.W., and Boothroyd, J.C. 1998. Development of a stable episomal shuttle vector for *Toxoplasma gondii*. J. Biol. Chem. 273: 3972-3979.

Blader, I.J., Manger, I.D., and Boothroyd, J.C. 2001. Microarray analysis reveals previously unknown changes in *Toxoplasma gondii*-infected human cells. J. Biol. Chem. 276: 24223-31.

Bohne, W., Heesemann, J., and Gross, U. 1994. Reduced replication of *Toxoplasma gondii* is necessary for induction of bradyzoite-specific antigens: a possible role for nitric oxide in triggering stage conversion. Infect. Immun. 62: 1761-7.

Burt, R.A., Marshall, V.M., Wagglen, J., Rodda, F.R., Senyschen, D., Baldwin, T.M., Buckingham, L.A., and Foote, S.J. 2002. Mice that are congenic for the char2 locus are susceptible to malaria. Infect. Immun. 70: 4750-3.

Carruthers, V.B., and Sibley, L.D. 1997. Sequential protein secretion from three distinct organelles of *Toxoplasma gondii* accompanies invasion of human fibroblasts. Eur. J. Cell Biol. 73: 114-23.

Carruthers, V.B. 1999. Armed and dangerous:*Toxoplasma gondii* uses an arsenal of secretory proteins to infect host cells. Parasitol. Int. 48: 1-10.

Carruthers, V.B., Giddings, O.K., and Sibley, L.D. 1999a. Secretion of micronemal proteins is associated with *Toxoplasma* invasion of host cells. Cell. Microbiol. 1: 225-35.

Carruthers, V.B., Moreno, S.N., and Sibley, L.D. 1999b. Ethanol and acetaldehyde elevate intracellular $[Ca^{2+}]$ and stimulate microneme discharge in *Toxoplasma gondii*. Biochem. J. 342: 379-86.

Carruthers, V.B., Sherman, G.D., and Sibley, L.D. 2000. The *Toxoplasma* adhesive protein MIC2 is proteolytically processed at multiple sites by two parasite-derived proteases. J. Biol. Chem. 275: 14346-53.

Carruthers, V.B. 2002. Host cell invasion by the opportunistic pathogen *Toxoplasma gondii*. Acta Trop. 81: 111-122.

Cesbron-Delauw, M.-F. 1994. Dense-granule organelles of *Toxoplasma gondii*: their role in the host-parasite relationship. Parasitol. Today 10: 293-296.

Cohen, A.M., Rumpel, K., Coombs, G.H., and Wastling, J.M. 2002. Characterisation of global protein expression by two-dimensional electrophoresis and mass spectrometry: proteomics of *Toxoplasma gondii*. Int. J. Parasitol. 32: 39-51.

Coppens, I., Sinai, A.P., and Joiner, K.A. 2000. *Toxoplasma gondii* exploits host low-density lipoprotein receptor- mediated endocytosis for cholesterol acquisition. J. Cell. Biol. 149: 167-80.

Cox, D.M., Du, M., Guo, X., Siu, K.W., and McDermott, J.C. 2002. Tandem affinity purification of protein complexes from mammalian cells. BioTechniques 33: 267-8, 270.

Crewther, P.E., Culvenor, J.G., Silva, A., Cooper, J.A., and Anders, R.F. 1990. *Plasmodium falciparum*: two antigens of similar size are located in different compartments of the rhoptry. Exp. Parasitol. 70: 193-206.

de Koning-Ward, T.F., Fidock, D.A., Thathy, V., Menard, R., van Spaendonk, R.M., Waters, A.P., and Janse, C.J. 2000. The selectable marker human dihydrofolate reductase enables sequential genetic manipulation of the *Plasmodium berghei* genome. Mol. Biochem. Parasitol. 106: 199-212.

Di Cristina, M., Spaccapelo, R., Soldati, D., Bistoni, B., and Crisanti, A. 2000. Two conserved amino acid motifs mediate protein targeting to the micronemes of the apicomplexan parasite *Toxoplasma gondii*. Mol. Cell. Biol.

Donahue, C.G., Carruthers, V.B., Gilk, S.D., and Ward, G.E. 2000. The *Toxoplasma* homolog of Plasmodium apical membrane antigen-1 (AMA-1) is a microneme protein secreted in response to elevated intracellular calcium levels. Mol. Biochem. Parasitol. 111: 15-30.

Donald, R.G., and Roos, D.S. 1994. Homologous recombination and gene replacement at the dihydrofolate reductase-thymidylate synthase locus in *Toxoplasma gondii*. Mol. Biochem. Parasitol. 63: 243-53.

Donald, R.G., and Roos, D.S. 1998. Gene knock-outs and allelic replacements in *Toxoplasma gondii*: HXGPRT as a selectable marker for hit-and-run mutagenesis. Mol. Biochem. Parasitol. 91: 295-305.

Donald, R.G.K., and Roos, D.S. 1993. Stable molecular transformation of *Toxoplasma gondii*: a selectable markerDHFR-TS marker based on drug resistance mutations in malaria. Proc. Natl. Acad. Sci. USA 90: 11703-11707.

Dubey, J.P., Lindsay, D.S., and Speer, C.A. 1998. Structures of *Toxoplasma gondii* Tachyzoites, Bradyzoites and Sporozoites and Biology and Development of Tissue Cysts. Clin. Microbiol. Rev. 11: 267-299.

Dubremetz, J.F., Achbarou, A., Bermudes, D., and Joiner, K.A. 1993. Kinetics and pattern of organelle exocytosis during *Toxoplasma gondii*/host-cell interaction. Parasitol. Res. 79: 402-8.

Dubremetz, J.F., Garcia-Reguet, N., Conseil, V., and Fourmaux, M.N. 1998. Apical organelles and host-cell invasion by *Apicomplexa*. Int. J. Parasitol. 28: 1007-13.

Fichera, M.E., and Roos, D.S. 1997. A plastid organelle as a drug target in apicomplexan parasites. Nature 390: 407-9.

Fox, B.A., and Bzik, D.J. 2002. De novo pyrimidine biosynthesis is required for virulence of *Toxoplasma gondii*. Nature 415: 926-929.

Gail, M., Gross, U., and Bohne, W. 2001. Transcriptional profile of *Toxoplasma gondii*-infected human fibroblasts as revealed by gene-array hybridization. Mol. Genet. Genomics 265: 905-12.

Garcia-Reguet, N., Lebrun, M., Fourmaux, M.N., Mercereau-Puijalon, O., Mann, T., Beckers, C.J., Samyn, B., Van Beeumen, J., Bout, D., and Dubremetz, J.F. 2000. The microneme protein MIC3 of *Toxoplasma gondii* is a secretory adhesin that binds to both the surface of the host cells and the surface of the parasite. Cell. Microbiol. 2: 353-64.

Goonewardene, R., Daily, J., Kaslow, D., Sullivan, T.J., Duffy, P., Carter, R., Mendis, K., and Wirth, D. 1993. Transfection of the malaria parasite and expression of firefly luciferase. Proc. Natl. Acad. Sci. USA. 90: 5234-6.

Gossen, M., Bonin, A.L., Freundlieb, S., and Bujard, H. 1994. Inducible gene expression systems for higher eukaryotic cells. Curr. Opin. Biotechnol. 5: 516-20.

Grigg, M.E., Bonnefoy, S., Hehl, A.B., Suzuki, Y., and Boothroyd, J.C. 2001. Success and virulence in *Toxoplasma* as the result of sexual recombination between two distinct ancestries. Science 294: 161-5.

Gross, U., and Bohne, W. 1994. *Toxoplasma gondii:* strain- and host cell-dependent induction of stage differentiation. J. Eukaryot. Microbiol. 43: 114-116.

Hakansson, S., Charron, A.J., and Sibley, L.D. 2001. *Toxoplasma* evacuoles: a two-step process of secretion and fusion forms the parasitophorous vacuole. EMBO J. 20: 3132-44.

He, C.Y., Shaw, M.K., Pletcher, C.H., Striepen, B., Tilney, L.G., and Roos, D.S. 2001. A plastid segregation defect in the protozoan parasite *Toxoplasma gondii*. EMBO J. 20: 330-9.

He, X.L., Grigg, M.E., Boothroyd, J.C., and Garcia, K.C. 2002. Structure of the immunodominant surface antigen from the *Toxoplasma gondii* SRS superfamily. Nat. Struct. Biol. 9: 606-11.

Healer, J., Crawford, S., Ralph, S., McFadden, G., and Cowman, A.F. 2002. Independent translocation of two micronemal proteins in developing *Plasmodium falciparum* merozoites. Infect. Immun. 70: 5751-8.

Hehl, A.B., Lekutis, C., Grigg, M.E., Bradley, P.J., Dubremetz, J.F., Ortega-Barria, E., and Boothroyd, J.C. 2000. *Toxoplasma gondii* homologue of plasmodium apical membrane antigen 1 is involved in invasion of host cells. Infect. Immun. 68: 7078-86.

Herm-Gotz, A., Weiss, S., Stratmann, R., Fujita-Becker, S., Ruff, C., Meyhofer, E., Soldati, T., Manstein, D.J., Geeves, M.A., and Soldati, D. 2002. *Toxoplasma gondii* myosin A and its light chain: a fast, single-headed, plus-end-directed motor. EMBO J. 21: 2149-58.

Heussler, V.T., Kuenzi, P., and Rottenberg, S. 2001. Inhibition of apoptosis by intracellular protozoan parasites. Int. J. Parasitol. 31: 1166-76.

Higgs, H.N., and Pollard, T.D. 2001. Regulation of actin filament network formation through ARP2/3 complex: activation by a diverse array of proteins. Annu. Rev. Biochem. 70: 649-76.

Hoff, E.F., and Carruthers, V.B. 2002. Is *Toxoplasma* egress the first step in invasion? Trends Parasitol. 18: 251-5.

Hoppe, H.C., Ngo, H.M., Yang, M., and Joiner, K.A. 2000. Targeting to rhoptry organelles of *Toxoplasma gondii* involves evolutionarily conserved mechanisms. Nat. Cell. Biol. 2: 449-56.

Hu, K., Mann, T., Striepen, B., Beckers, C.J., Roos, D.S., and Murray, J.M. 2002. Daughter Cell Assembly in the Protozoan Parasite *Toxoplasma gondii*. Mol. Biol. Cell 13: 593-606.

Joiner, K.A., and Roos, D.S. 2002. Secretory traffic in the eukaryotic parasite *Toxoplasma gondii*: less is more. J. Cell. Biol. 157: 557-63.

Jomaa, H. 1999. Inhibitors of the nonmevalonate pathway of isoprenoid biosynthesis as antimalarial drugs. Science. 285: 1573-1576.

Jomaa, H., Wiesner, J., Sanderbrand, S., Altincicek, B., Weidemeyer, C., Hintz, M., Turbachova, I., Eberl, M., Zeidler, J., Lichtenthaler, H.K., Soldati, D., and Beck, E. 1999. Inhibitors of the nonmevalonate pathway of isoprenoid biosynthesis as antimalarial drugs. Science 285: 1573-6.

Kappe, S., Bruderer, T., Gantt, S., Fujioka, H., Nussenzweig, V., and Menard, R. 1999. Conservation of a gliding motility and cell invasion machinery in Apicomplexan parasites. J. Cell. Biol. 147: 937-44.

Kieschnick, H., Wakefield, T., Narducci, C.A., and Beckers, C. 2001. *Toxoplasma gondii* attachment to host cells is regulated by a calmodulin-like domain protein kinase. J. Biol. Chem. 276: 12369-12377.

Knoll, L.J., Furie, G.L., and Boothroyd, J.C. 2001. Adaptation of signature-tagged mutagenesis for *Toxoplasma gondii*: a negative screening strategy to isolate genes that are essential in restrictive growth conditions. Mol. Biochem. Parasitol. 116: 11-6.

Kohler, S., Delwiche, C.F., Denny, P.W., Tilney, L.G., Webster, P., Wilson, R.J., Palmer, J.D., and Roos, D.S. 1997. A plastid of probable green algal origin in Apicomplexan parasites. Science 275: 1485-9.

Lekutis, C., Ferguson, D.J., Grigg, M.E., Camps, M., and Boothroyd, J.C. 2001. Surface antigens of *Toxoplasma gondii*: variations on a theme. Int. J. Parasitol. 31: 1285-92.

Lovett, J.L., Marchesini, N., Moreno, S.N., and Sibley, L.D. 2002. *Toxoplasma gondii* microneme secretion involves intracellular Ca(2+) release from inositol 1,4,5-triphosphate (IP(3))/ryanodine-sensitive stores. J. Biol. Chem. 277: 25870-6.

Malhotra, P., Dasaradhi, P.V., Kumar, A., Mohmmed, A., Agrawal, N., Bhatnagar, R.K., and Chauhan, V.S. 2002. Double-stranded RNA-mediatedgene silencing of cysteine proteases (falcipain-1 and -2) of *Plasmodium falciparum*. Mol. Microbiol. 45: 1245-54.

Mann, T., and Beckers, C. 2001. Characterization of the subpellicular network, a filamentous membrane skeletal component in the parasite *Toxoplasma gondii*. Mol. Biochem. Parasitol. 115: 257-268.

Mann, T., Gaskins, E., and Beckers, C.J. 2002. Proteolytic processing of TgIMC1 during maturation of the membrane skeleton of *Toxoplasma gondii*. J. Biol. Chem. 12: 12.

Matrajt, M., Donald, R.G., Singh, U., and Roos, D.S. 2002a. Identification and characterization of differentiation mutants in the protozoan parasite *Toxoplasma gondii*. Mol. Microbiol. 44: 735-47.

Matrajt, M., Nishi, M., Fraunholz, M.J., Peter, O., and Roos, D.S. 2002b. Amino-terminal control of transgenic protein expression levels in *Toxoplasma gondii*. Mol. Biochem. Parasitol. 120: 285-9.

McFadden, G.I., Reith, M.E., Munholland, J., and Lang-Unnasch, N. 1996. Plastid in human parasites. Nature 381: 482.

McFadden, G.I., and Roos, D.S. 1999. Apicomplexan plastids as drug targets. Trends Microbiol. 6: 328-333.

McRobert, L., and McConkey, G.A. 2002. RNA interference (RNAi) inhibits growth of *Plasmodium falciparum*. Mol. Biochem. Parasitol. 119: 273-278.

Meissner, M., Brecht, S., Bujard, H., and Soldati, D. 2001. Modulation of myosin A expression by a newly established tetracycline repressor-based inducible system in *Toxoplasma gondii*. Nucleic Acids Res. 29: E115.

Meissner, M., Reiss, M., Viebig, N., Carruthers, V.B., Toursel, C., Tomavo, S., Ajioka, J.W., and Soldati, D. 2002a. A family of transmembrane microneme proteins of *Toxoplasma gondii* contain EGF-like domains and function as escorters. J. Cell. Sci.: 563-574.

Meissner, M., Schluter, D., and Soldati, D. 2002b. *Toxoplasma gondii* myosin A: a virulence factor powering parasite gliding and host cells invasion. Science. 298: 837-840.

Mercier, C., Dubremetz, J.F., Rauscher, B., Lecordier, L., Sibley, L.D., and Cesbron-Delauw, M.F. 2002. Biogenesis of nanotubular network in toxoplasma parasitophorous vacuole induced by parasite proteins. Mol. Biol. Cell. 13: 2397-409.

Miller, S.A., Binder, E.M., Blackman, M.J., Carruthers, V.B., and Kim, K. 2001. A conserved subtilisin-like protein TgSUB1 in microneme organelles of *Toxoplasma gondii*. J. Biol. Chem. 276: 45341-8.

Morrissette, N.S., Murray, J.M., and Roos, D.S. 1997. Subpellicular microtubules associate with an intramembranous particle lattice in the protozoan parasite *Toxoplasma gondii*. J. Cell. Sci. 110: 35-42.

Morrissette, N.S., and Roos, D.S. 1998. *Toxoplasma gondii*: a family of apical antigens associated with the cytoskeleton. Exp. Parasitol. 89: 296-303.

Morrissette, N.S., and Sibley, L.D. 2002a. Cytoskeleton of apicomplexan parasites. Microbiol. Mol. Biol. Rev. 66: 21-38.

Morrissette, N.S., and Sibley, L.D. 2002b. Disruption of microtubules uncouples budding and nuclear division in *Toxoplasma gondii*. J. Cell. Sci. 115: 1017-25.

Mota, M.M., Pradel, G., Vanderberg, J.P., Hafalla, J.C., Frevert, U., Nussenzweig, R.S., Nussenzweig, V., and Rodriguez, A. 2001. Migration of *Plasmodium* sporozoites through cells before infection. Science. 291: 141-4.

Mota, M.M., Hafalla, J.C., and Rodriguez, A. 2002. Migration through host cells activates *Plasmodium* sporozoites for infection. Nat. Med. 8: 1318-22.

Nakaar, V., Samuel, B.U., Ngo, E.O., and Joiner, K.A. 1999. Targeted reduction of nucleoside triphosphate hydrolase by antisense RNA inhibits *Toxoplasma gondii* proliferation. J. Biol. Chem. 274: 5083-7.

Nakaar, V., Ngo, E.O., and Joiner, K.A. 2000. Selection based on the expression of antisense hypoxanthine-xanthine- guanine-phosphoribos yltransferase RNA in *Toxoplasma gondii*. Mol. Biochem. Parasitol. 110: 43-51.

Nene, V., Bishop, R., Morzaria, S., Gardner, M.J., Sugimoto, C., ole-MoiYoi, O.K., Fraser, C.M., and Irvin, A. 2000. *Theileria parva* genomics reveals an atypical apicomplexan genome. Int. J. Parasitol. 30: 465-474.

Nichols, B.A., Chappino, M.L., and O'Connors, G.R. 1983. Secretion from the rhoptries of *Toxoplasma gondii* during host-cell invasion. J. Ultrastruct. Res. 83: 85-98.

Nichols, B.A., and Chiappino, M.L. 1987. Cytoskeleton of *Toxoplasma gondii*. J. Protozool. 34: 217-226.

Opitz, C., Di Cristina, M., Reiss, M., Ruppert, T., Crisanti, A., and Soldati, D. 2002. Intramembrane cleavage of microneme proteins at the surface of the apicomplexan parasite *Toxoplasma gondii*. EMBO J. 21: 1577-85.

Opitz, C., and Soldati, D. 2002. 'The glideosome': a dynamic complex powering gliding motion and host cell invasion by *Toxoplasma gondii*. Mol. Microbiol. 45: 597-604.

Pan, W., Ravot, E., Tolle, R., Frank, R., Mosbach, R., Turbachova, I., and Bujard, H. 1999. Vaccine candidate MSP-1 from *Plasmodium falciparum*: a redesigned 4917 bp polynucleotide enables synthesis and isolation of full-length protein from *Escherichia coli* and mammalian cells. Nucleic Acids Res. 27: 1094-103.

Pelletier, L., Stern, C.A., Pypaert, M., Sheff, D., Ngo, H.M., Roper, N., He, C.Y., Hu, K., Toomre, D., Coppens, I., Roos, D.S., Joiner, K.A., and Warren, G. 2002. Golgi biogenesis in *Toxoplasma gondii*. Nature 418: 548-52.

Pfefferkorn, E.R., and Pfefferkorn, L.C. 1976. *Toxoplasma gondii*: isolation and preliminary characterization of temperature-sensitive mutants. Exp. Parasitol. 39: 365-76.

Pfefferkorn, E.R., and Pfefferkorn, L.C. 1979. Quantitative studies on the mutagenesis of *Tooxoplasma gondii*. J. Parasitol. 65: 207-218.

Pfefferkorn, E.R., and Pfefferkorn, L.C. 1980. *Toxoplasma gondii:* genetic recombination between drug resistant mutants. Exp. Parasitol. 39: 365-374.

Pfefferkorn, E.R., and Kasper, L.H. 1983. *Toxoplasma gondii*: genetic crosses reveal phenotypic suppression of hydroxyurea resistance by fluorodeoxyuridine resistance. Exp. Parasitol. 65: 207-218.

Rabenau, K.E., Sohrabi, A., Tripathy, A., Reitter, A., Ajioka, J.W., Tomley, F.M., and Carruthers, V.B. 2001. TgM2AP participates in *Toxoplasma gondii* invasion of host cells and is tightly associated with the adhesive protein TgMIC2. Mol. Microbiol. 41: 1-12.

Radke, J.R., and White, M.W. 1998. A cell cycle model for the tachyzoite of *Toxoplasma gondii* using the Herpes simplex virus thymidine kinase. Mol. Biochem. Parasitol. 94: 237-47.

Rathod, P.K., Ganesan, K., Hayward, R.E., Bozdech, Z., and DeRisi, J.L. 2002. DNA microarrays for malaria. Trends Parasitol. 18: 39-45.

Reiss, M., Viebig, N., Brecht, S., Fourmaux, M.N., Soete, M., Di Cristina, M., Dubremetz, J.F., and Soldati, D. 2001. Identification and characterization of an escorter for two secretory adhesins in *Toxoplasma gondii*. J. Cell. Biol. 152: 563-78.

Rodrigues, C.O., Ruiz, F.A., Rohloff, P., Scott, D.A., and Moreno, S.N. 2002. Characterization of isolated acidocalcisomes from *Toxoplasma gondii* Tachyzoites reveals a novel pool of hydrolysable polyphosphate. J. Biol. Chem. 11: 11.

Roos, D. 1999. Origins, targetings and function of the apicomplexan plastid. Curr. Opin. Microbiol. 2: 426-432.

Roos, D.S., Crawford, M.J., Donald, R.G., Fraunholz, M., Harb, O.S., He, C.Y., Kissinger, J.C., Shaw, M.K., and Striepen, B. 2002. Mining the *Plasmodium* genome database to define organellar function: what does the apicoplast do? Philos. Trans. R. Soc. Lond. B. Biol. Sci. 357: 35-46.

Russell, D.G., and Sinden, R.E. 1981. The role of the cytoskeleton in the motility of coccidian sporozoites. J. Cell. Sci. 50: 345-359.

Seeber, F., and Boothroyd, J.C. 1996. *Escherichia coli* beta-galactosidase as an *in vitro* and *in vivo* reporter enzyme and stable transfection marker in the intracellular protozoan parasite *Toxoplasma gondii*. Gene 169: 39-45.

Shaw, M.K., Roos, D.S., and Tilney, L.G. 1998. Acidic compartments and rhoptry formation in *Toxoplasma gondii*. Parasitology 117: 435-43.

Shaw, M.K., Compton, H.L., Roos, D.S., and Tilney, L.G. 2000. Microtubules, but not actin filaments, drive daughter cell budding and cell division in *Toxoplasma gondii*. J. Cell. Sci. 113: 1241-54.

Sibley, L.D., and Boothroyd, J.C. 1992. Construction of a molecular karyotype for *Toxoplasma gondii*. Mol. Biochem. Parasitol. 51: 291-300.

Sibley, L.D., and Andrews, N.W. 2000. Cell invasion by un-palatable parasites. Traffic 1: 100-6.

Silverman, J.A., Qi, H., Riehl, A., Beckers, C., Nakaar, V., and Joiner, K.A. 1998. Induced activation of the *Toxoplasma gondii* nucleoside triphosphate hydrolase leads to depletion of host cell ATP levels and rapid exit of intracellular parasites from infected cells. J. Biol. Chem. 273: 12352-12359.

Sinai, A.P., and Joiner, K.A. 2001. The *Toxoplasma gondii* protein ROP2 mediates host organelle association with the parasitophorous vacuole membrane. J. Cell. Biol. 154: 95-108.

Singh, U., Brewer, J.L., and Boothroyd, J.C. 2002. Genetic analysis of tachyzoite to bradyzoite differentiation mutants in *Toxoplasma gondii* reveals a hierarchy of gene induction. Mol. Microbiol. 44: 721-33.

Soldati, D., and Boothroyd, J.C. 1993. Transient transfection and expression in the obligate intracellular parasite *Toxoplasma gondii*. Science. 260: 349-352.

Soldati, D. 1996. Molecular genetic strategies in *Toxoplasma gondii*: close in on a successful invader. Febs Lett 389: 80-3.

Soldati, D., Lassen, A., Dubremetz, J.F., and Boothroyd, J.C. 1998. Processing of *Toxoplasma* ROP1 protein in nascent rhoptries. Mol. Biochem. Parasitol. 96: 37-48.

Soldati, D., Dubremetz, J.F., and Lebrun, M. 2001. Microneme proteins: structural and functional requirements to promote adhesion and invasion by the apicomplexan parasite *Toxoplasma gondii*. Int. J. Parasitol. 31: 1293-302.

Stommel, E.W., Ely, K.H., Schwartzman, J.D., and Kasper, L.H. 1997. *Toxoplasma gondii:* dithiol-induced $Ca^{2+}$ flux causes egress of parasites from the parasitophorous vacuole. Exp. Parasitol. 87: 88-97.

Striepen, B., He, C.Y., Matrajt, M., Soldati, D., and Roos, D.S. 1998. Expression , selection, and organellar targeting of the green fluorescent protein in *Toxoplasma gondii*. Mol. Biochem. Parasitol.

Striepen, B., Crawford, M.J., Shaw, M.K., Tilney, L.G., Seeber, F., and Roos, D.S. 2000. The plastid of *Toxoplasma gondii* is divided by association with the centrosomes. J. Cell. Biol. 151: 1423-34.

Su, C., Howe, D.K., Dubey, J.P., Ajioka, J.W., and Sibley, L.D. 2002. Identification of quantitative trait loci controlling acute virulence in *Toxoplasma gondii*. Proc. Natl. Acad. Sci. USA. 99: 10753-8.

Sultan, A.A., Thathy, V., Frevert, U., Robson, K.J., Crisanti, A., Nussenzweig, V., Nussenzweig, R.S., and Menard, R. 1997. TRAP is necessary for gliding motility and infectivity of plasmodium sporozoites. Cell 90: 511-22.

Tomley, M.F., and Soldati, D. 2001. Mix and match modules: structure and function of microneme proteins in apicomplexan parasites. Parasitol. Today 17: 81-88.

Trottein, F., Triglia, T., and Cowman, A.F. 1995. Molecular cloning of a gene from *Plasmodium falciparum* that codes for a protein sharing motifs found in adhesive molecules from mammals and plasmodia. Mol. Biochem. Parasitol. 74: 129-41.

van der Zypen, E., and Piekarski, G. 1967. Endodyogeny in *Toxoplasma gondii*. A morphological analysis. Z. Parasitnkd. 29: 15-35.

Waller, R.F., Keeling, P.J., Donald, R.G.K., Striepen, B., Handman, E., Lang-Unnasch, N., Cowman, A.F., Besra, G.S., Roos, D.S., and McFadden,

G.I. 1998. Nuclear-encoded proteins target to the plastid in *Toxoplasma gondii* and *Plasmodium falciparum*. Proc. Natl. Acad. Sci. USA. 95: 12352-7.

Waller, R.F., Reed, M.B., Cowman, A.F., and McFadden, G.I. 2000. Protein trafficking to the plastid of *Plasmodium falciparum* is via the secretory pathway. EMBO J. 19: 1794-1802.

Wan, K.L., Blackwell, J.M., and Ajioka, J.W. 1996. *Toxoplasma gondii* expressed sequence tags: insight into tachyzoite gene expression. Mol. Biochem. Parasitol. 75: 179-86.

Wang, J., Morris, A.J., Tolan, D.R., and Pagliaro, L. 1996. The molecular nature of the F-actin binding activity of aldolase revealed with site-directed mutants. J. Biol. Chem. 271: 6861-5.

Wirtz, E., Leal, S., Ochatt, C., and Cross, G.A. 1999. A tightly regulated inducible expression system for conditional gene knock-outs and dominant-negative genetics in *Trypanosoma brucei*. Mol. Biochem. Parasitol. 99: 89-101.

Withers-Martinez, C., Carpenter, E.P., Hackett, F., Ely, B., Sajid, M., Grainger, M., and Blackman, M.J. 1999. PCR-based gene synthesis as an efficient approach for expression of the A+T-rich malaria genome. Protein Eng. 12: 1113-20.

Yuda, M., Sawai, T., and Chinzei, Y. 1999. Structure and expression of an adhesive protein-like molecule of mosquito invasive stage malarial parasite. J. Exp. Med. 189: 1947-1952.

From: Malaria Parasites: Genomes and Molecular Biology
Edited by: A.P. Waters and C.J. Janse

# Chapter 6

# Microsatellite Markers and Population Genetics in *Plasmodium falciparum*

Deirdre Joy, Jianbing Mu,
and Xin-zhuan Su

## Abstract

Microsatellites are short (1-6 bp in length) tandemly repeated sequences. They have been shown to be abundant in the *Plasmodium falciparum* genome and as such will play a significant role in the study of malaria in the post-genome era. Microsatellites are useful for dissecting gene function, identifying vaccine candidates and drug targets, and elucidating the evolutionary history of the malaria parasite. With the availability of a high-density genetic marker map for *P. falciparum*, it is now possible to perform genome-wide allelic association studies for genes contribution to such specific phenotypes as drug resistance and virulence. Additionally, the application of a large number of microsatellite markers to field samples promises to reveal important information about population genetic structure, demographic history, and transmission dynamics.

## 1. Introduction

Publication of the genome of the human malaria parasite *Plasmodium falciparum* inaugurates a new era for malaria research and presents an unprecedented opportunity to study the biology and disease pathogenesis

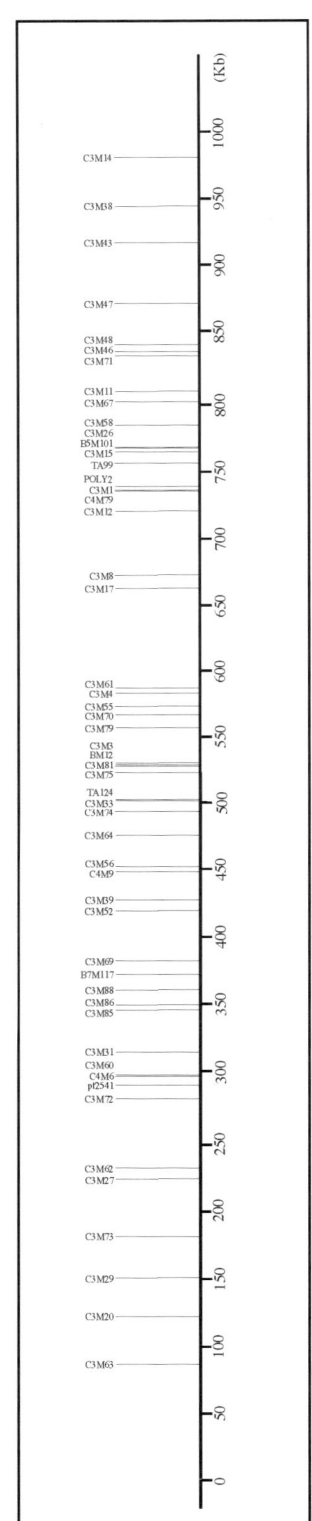

**Figure 1.** A physical map of *P. falciparum* chromosome 3 showing microsatellite loci (Su *et al.*, 1999). A high density map such as the one presented here has great utility for genetic mapping and population studies.

of the parasite. Completion of the genome sequence will allow large-scale characterization of genetic and phenotypic variations, in-depth analysis of parasite origin and evolutionary history, and construction of molecular and cellular circuitry, leading to improved parasite and disease control measures.

The parasite's ability to survive and prosper in hostile environments is imprinted in its relatively small genome and relies on rapid evolution in response to environmental variability. Knowing the DNA sequences is only the first step in understanding the molecular mechanisms of the parasite; to understand gene function, rigorous experiments are necessary. Microarray, genetic mapping, and proteomics are among the approaches that may shed light on gene function. Candidate genes can then be confirmed by other molecular, cellular, and biochemical methods. A parasite can be virulent, causing severe disease; non-virulent; or resistant to certain drugs. These phenotypic differences are all controlled by certain genetic elements (alleles). Catalogs of genome-wide polymorphisms—including single nucleotide polymorphism (SNP), microsatellite (MS), deletion/insertion, and gene conversion—will prove useful for dissecting gene function, identifying vaccine and drug targets, and elucidating the evolutionary history of the parasite (Figure 1).

MS have been shown to be abundant in the *P. falciparum* genome (Su and Wellems, 1996; Gardner *et al.*, 1998; Bowman *et al.*, 1999). Due to their genome-wide distribution and high level of polymorphism, MS will play a significant role in the study of malaria in the post-genome era. Large numbers of MS have been identified, and a high-resolution genetic map is now available (Su *et al.*, 1999). In the following sections we will describe MS markers and discuss their many applications, with particular attention to their usefulness in studying population genetics.

## 1.1. Microsatellites

MS are short (typically 1–6 base pairs in length) tandemly repeated sequences. They are abundant in most genomes studied, although their density varies among species. In addition to MS, various terms have been used to describe these repeats, including simple sequence repeat (SSR), simple sequence-length polymorphism (SSLP), and short tandem repeat polymorphism (STRP). MS alleles differ from each other in size as measured against size standards using electrophoresis; the copy number of the repeat motif characterizes individual alleles. Mutations typically involve the addition or loss of a single repeat unit, although larger mutations also occur (Weber and Wong, 1993; Di Rienzo *et al.*, 1994) (Figure 2).

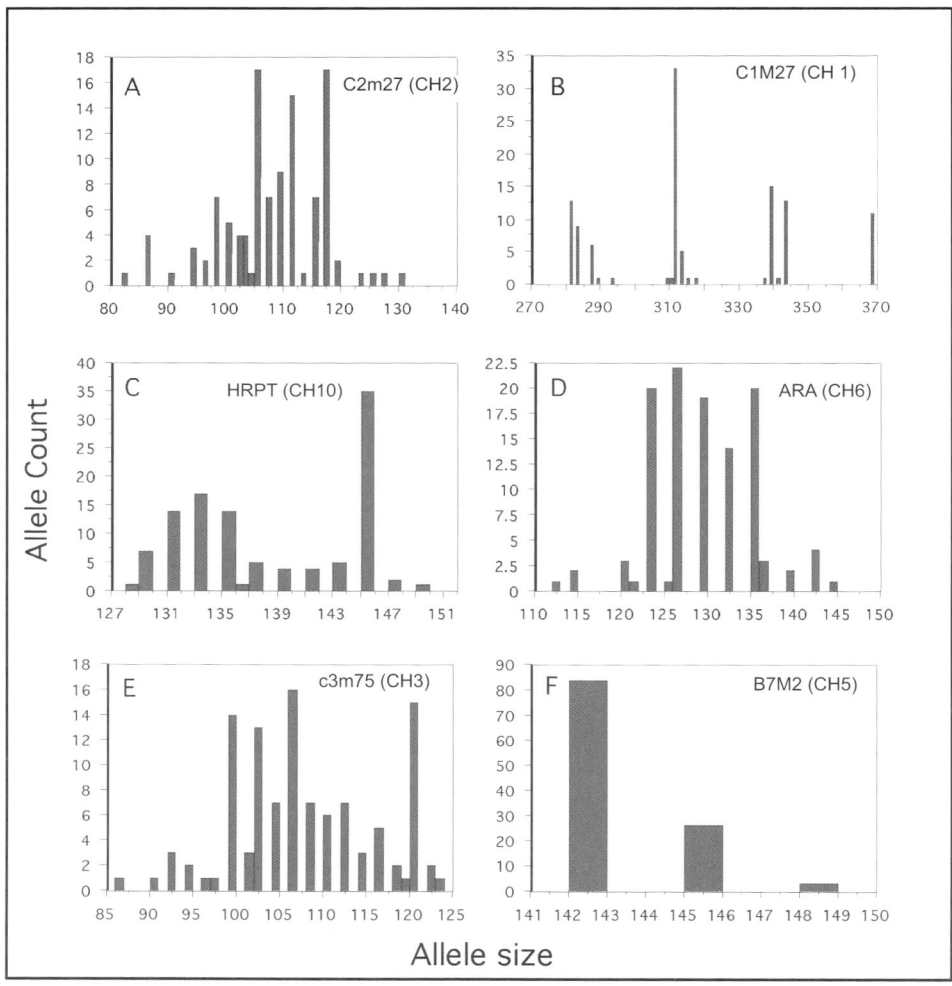

**Figure 2**. Distribution of allele sizes for six microsatellite markers from the *Plasmodium falciparum* genome. Allele counts based on 91 isolates. C2m27 appears to evolve via a stepwise mutation process (A) in contrast with C1M27, which shows evidence of complex mutations (B). HRPT consists of dinucleotide repeats (C) versus ARA which is made up of trinucleotide repeats (D). c3m75 is an example of a highly polymorphic satellite marker (E), whereas B7M2 shows very little variation in size (F). CH# refers to the chromosome where the microsatellite is located.

Rates of mutation in MS are high compared with point mutations, which range between $10^{-9}$ and $10^{-10}$. MS mutation rates have been estimated to be $10^{-2}$ in *E. coli* (Levinson and Gutman, 1987a), $10^{-3}$ in humans (Weber and Wong, 1993), and $10^{-3}$–$10^{-4}$ in mice (Dallas, 1992). In *P. falciparum*, our estimated mutation rate of $7.56 \times 10^{-5}$ to $1.51 \times 10^{-4}$ is based on nine polymorphisms in 238 MS loci for ten lines grown for 20-50 *in vitro* generations (X-z. Su, unpublished). A second estimate based on a genetic

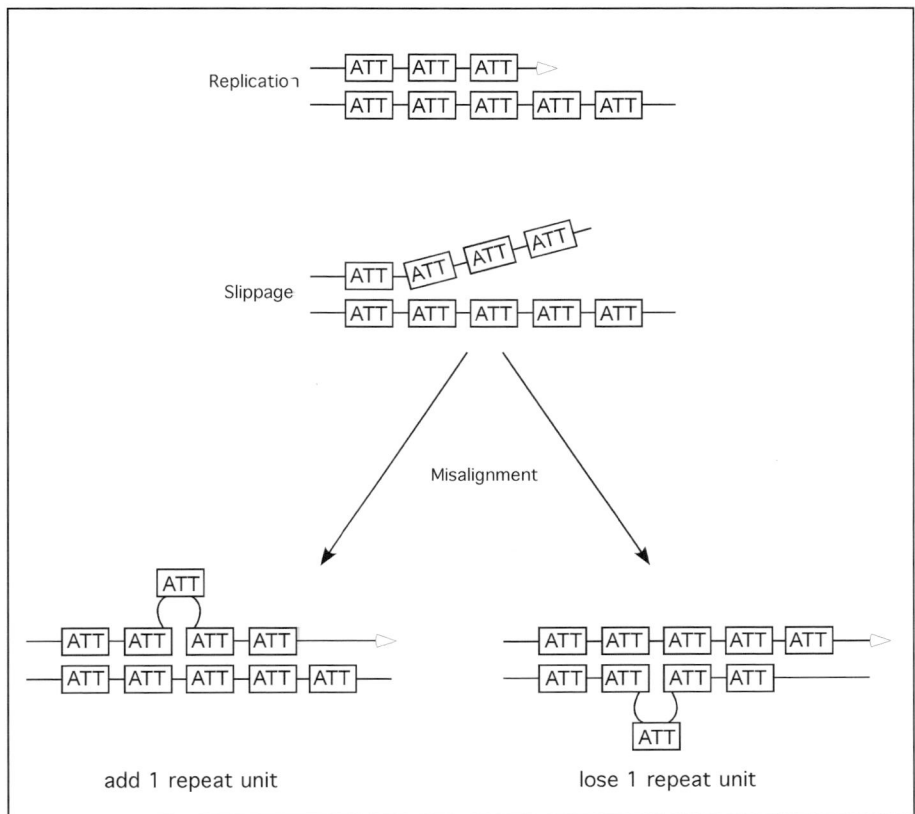

**Figure 3**. Cartoon of slipped-strand mispairing at microsatellite loci. Microsatellite repeat units are contained in boxes and the DNA strand is represented as a line with the arrow pointing in the direction of replication. Slippage followed by misalignment of the two DNA strands can lead to either the addition or the loss of a repeat unit, and rarely, more than one unit.

cross with 35 progeny genotyped for more than 800 MS makers puts the mutation rate between $6.98 \times 10^{-5}$ and $3.7 \times 10^{-4}$, with dinucleotide repeat loci evolving slightly faster than trinucleotide repeats (Anderson *et al.*, 2000a; 2000b). An inverse relationship between the size of the repeat motif and mutation rate has been observed in other organisms as well (Chakraborty *et al.*, 1997). For some organisms there is a mutational bias favoring repeat gains over losses, although the molecular mechanism responsible is not known (Amos and Rubinsztein, 1996; Primmer *et al.*, 1996). Despite this, most MS have a finite size that is, in general, shorter than a few tens of repeat units (Estoup and Cornuet, 1999).

The high mutation rate in MS has been attributed to two mechanisms: slipped strand mispairing during DNA replication (Levinson and Gutman, 1987b) and recombination (Jeffreys *et al.*, 1994), with the weight of the observational and experimental evidence supporting slippage as the dominant

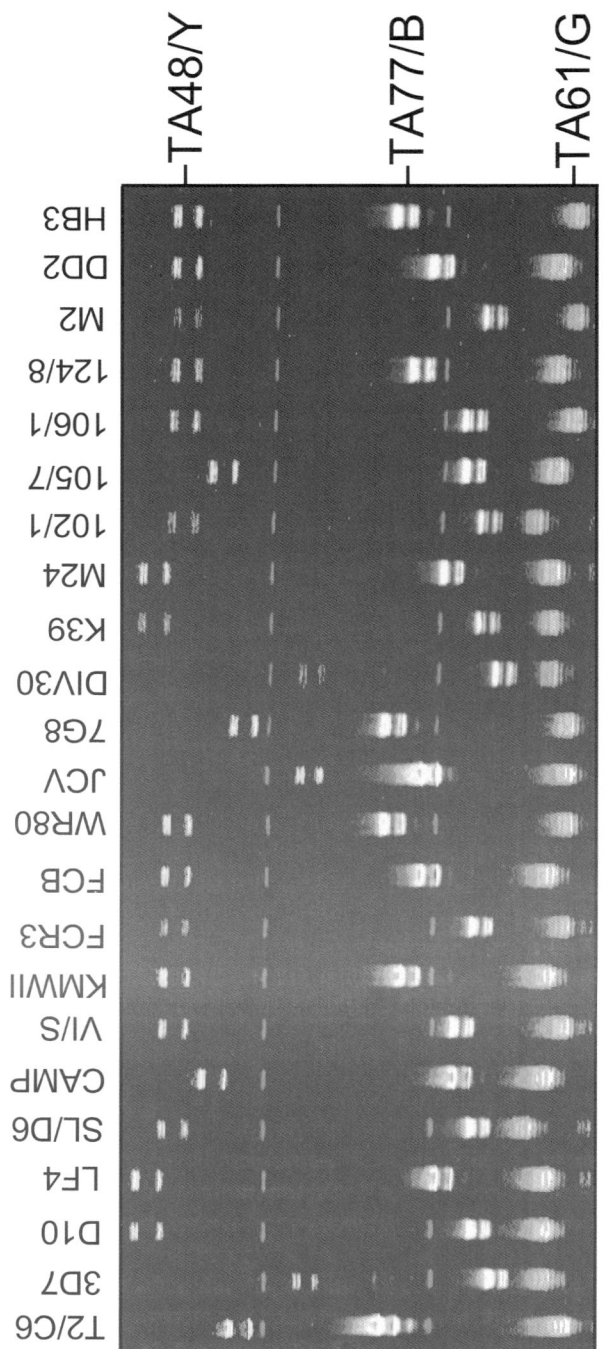

**Figure 4.** Microsatellite typing of *P. falciparum* isolates. Three microsatellite markers (right) were used to type 23 isolates (top). Most isolates exhibit an unique banding pattern. Red bands indicate size markers. Note: Different chromosomal strands migrate at slightly different rates, giving the appearance of double bands (Blue and yellow markers). **See Colour Plate at the back of the book.**

mutational mechanism. Slippage takes place when, during replication, the nascent strand dissociates from the template strand and reanneals out of frame because of repetitive sequences, as is the case with MS (Levinson and Gutman, 1987b). The resulting strand will be either shorter or longer than the template (Figure 3). Polymerase slippage generally produces alleles close in length to the original such that the mutant alleles differ by a single repeat unit. Point mutations are also believed to play a role in the generation and loss of MS repeat units (Kruglyak *et al.*, 1998).

MS typing relies on the polymerase chain reaction (PCR) and requires only a small amount of DNA, an advantage for studies involving malaria parasites, from which large amounts of DNA are often difficult to obtain. DNA can be labeled with radioactive materials such as $P^{32}$ or $S^{35}$ or with fluorescent dyes, and the labels can be incorporated through fluorescent dNTP or, alternatively, the 5' end phosphate group of one of the PCR primers can be labeled (Su and Ferdig, 2002). In the case of fluorescent labeling, different colors can be used to label different markers, and multiplex PCR can be employed to type several markers in a single reaction tube, dramatically increasing throughput (Figure 4).

In *P. falciparum* MS occur at a frequency of about 1-2 kb genome-wide (Su and Wellems, 1996; Gardner *et al.*, 1998; Bowman *et al.*, 1999). The abundance of MS in the *P. falciparum* genome has facilitated the development of a high-resolution linkage map (Su *et al.*, 1999), laying the foundation for the genetic characterization of the malaria parasite. Most MS in *P. falciparum* are $(TA)_n$, $(TAA)_n$, or $(T or A)_n$, reflecting the high AT content of the parasite genome. They have been isolated from introns, intergenic regions, and coding regions. Minisatellites, like MS are tandem repeat sequences, but unlike MS, the repeat motif is generally >14 bp. Minisatellites containing MS appear to be a common feature of the *P. falciparum* genome (Anderson *et al.*, 2000b). The high abundance of polymorphic MS suggests that they may be part of a gene regulation strategy employed by the parasite in response to the highly variable environments encountered in human and mosquito hosts.

Little is known about MS in other malaria species, and few MS have been reported from other human malaria parasites. A computer search of a yeast artificial chromosome (YAC) for *P. vivax* (# AY003872) produced only four short dinucleotide repeats: $(AT)_9$, $(AC)_5$, $(TC)_5$, and $(CG)_5$, although PolyT and polyA of 15 repeating units or longer are abundant (~one per 2 kb). Minisatellites and MS have been reported in several rodent malaria species, but in limited numbers and exhibiting low levels of polymorphism (van Belkum *et al.*, 1992); however, a recent computer search of the *P. yoelii* genome database (http://www.tigr.org/tdb/edb2/pya1/htmls/) identified numerous MS, most of which consisted of $(AT)_n$ (D. Joy, unpublished), suggesting an A-T content similar to that of *P. falciparum*.

# 2. Microsatellites as a Tool for Population Genetics

Applying large numbers of MS markers to field samples will reveal important information about population genetic structure and transmission dynamics of malaria parasites. Because of their high mutation rate, MS markers exhibit multiple alleles in population surveys, making them very informative for genetic studies. Other advantages of MS markers include their ubiquitousness and the ease with which they can be scored. With the completion of the genome sequence of the 3D7 line, the next phase of genome-wide characterization will include the generation of a database of cataloged variations within and among parasite populations to better understand parasite evolution, transmission dynamics, and the interactions of these factors with the more complex human genome. There are several important aspects of *P. falciparum* population genetics — including the effective rate of recombination in natural populations, the relationship between transmission rate and mating structure, the size and age of parasite populations, and the geographic origin and spread of drug resistance — for which MS data have the potential to shed new light.

Until quite recently, studies of genetic structure in *P. falciparum* have been limited to a small number of loci, namely, merozoite surface antigens *MSP*-1 and *MSP*-2, circumsporozoite surface antigens (*CSP*), and the S-antigen locus (Babiker *et al.*, 1991; Paul *et al.*, 1995; Daubersies *et al.*, 1996; Babiker *et al.*, 1997; Anderson and Day, 2000). Population studies of *P. falciparum* based on these markers have shown high levels of genetic diversity at the local level. Studies of worldwide genetic structure in *P. falciparum* show strong geographic structure (Drakeley *et al.*, 1996; Conway *et al.*, 2000); however, polymorphisms within one or a few genes are not likely to accurately record global relationships among parasite isolates. Additionally, there is evidence that these genes may be under strong immune pressure (Hughes, 1991; 1992). Population studies based on markers known to be under selection pressure, rather than neutral markers from non-coding regions, will produce biased results. For example, elevated levels of non-synonymous substitutions seen in some of these genes may be due to immune pressure for antigenic variants. Linkage disequilibrium (LD) based on these genes may reflect selective sweeps rather than actual recombination rates. Selection pressure is likely to result in an overestimation of genetic diversity within, and differentiation among, populations. Because most MS are in non-coding regions and therefore not expected to be under strong selection, they offer a neutral alternative to genetic markers used previously in population studies of *P. falciparum.*

## 2.1. Genetic Distance

Genetic distance measures observed genetic variation among individuals or populations and can be used to estimate the separation time between them. Several estimates of genetic distance have been developed specifically for use with MS. Such distances as Nei's standard genetic distance $D_s$ (Nei, 1972) and $D_{AS}$ (Stephens *et al.*, 1992) are based on allele frequency only. Others take into account the distribution of allele size as well as allele frequency and are therefore appropriate for use under the assumptions of a stepwise mutation model (SMM). These include $D_{SW}$ (Shriver *et al.*, 1995), $R_{st}$ (Slatkin, 1995), and $(\delta\mu)^2$ (Goldstein *et al.*, 1995). For example, the date of the split between African and non-African human populations was estimated to be approximately 156,000 years ago using $(\delta\mu)^2$ (Goldstein *et al.*, 1995), which is linear with regard to time (in the absence of allele size constraints), thus allowing divergence time to be estimated without reference to population size.

MS genetic distance measures were developed based on various assumptions about the nature of MS evolution. Because the mutational processes controlling MS evolution differ from those for classical genetic markers such as SNP, they provide a novel tool for measuring population parameters. However, models that can accurately describe MS evolution are required. One model that has been applied to MS is the infinite alleles model (IAM) (Kimura and Crow, 1964), in which every mutation is assumed to create a novel allele. Due to the high mutation rate and limited numbers of alleles observed for MS, this assumption is likely to be violated. In fact, MS evolution differs from that of classical genetic markers in that it more closely fits the SMM (Ohta and Kimura, 1973). The SMM assumes that mutations involve the loss or gain of a single tandem repeat, that gains and losses are equally likely, and that new mutations may produce alleles already present in the population. Under the SMM, allele size provides information about the evolutionary relationships among alleles.

In reality, MS do not appear to evolve in a uniform fashion and may often violate some of the more restrictive assumptions of the IAM and the SMM. For example, some MS in *P. falciparum* have been found to contain large indels and other complex mutations (Anderson *et al.*, 2000b). MS mutation rates can be affected by several factors including the type of repeat motif (e.g. CA *vs* TA), the length of the repeat motif, the length of the microsatellite array, and the step size. To address some of these issues, a third model, the two-phase model (TPM) incorporating features of the first two, was developed (Di Rienzo *et al.*, 1994). The TPM assumes that most mutations are stepwise, while allowing for a limited number of mutations that involve several repeat units. This model appears to produce more reliable genealogical estimates than models assuming a constant mutation rate (Stumpf and Goldstein, 2001).

## 2.2. Population Diversity, Genetic Structure, and Gene Flow

The application of MS markers to field samples is expected to yield in-depth information concerning parasite transmission and population structure. Because MS are generally assumed to be selectively neutral, individual profiles comprising multiple MS markers are ideal for characterizing fine-scale genetic structure in natural populations. Anderson *et al.* (1999) used MS to generate an estimate of heterozygosity and to identify malaria infections containing more than one clone in Papua New Guinea. Anderson *et al.* (2000a) have also conducted a worldwide survey of MS variation in parasite populations on three continents and have shown that genetic diversity at the local level differs among continents. Parasites from Africa had the highest genetic diversity when compared with those from Southeast Asia and South America, while South American parasites had the lowest diversity. The high diversity observed in Africa may be due to the older age of this population, providing support for the argument that malaria originated there. Or it may indicate more intense transmission, mediated by the vector species. The low diversity in South American populations could reflect a severe population bottleneck, perhaps brought about by the extensive use of DDT and chloroquine in the 1950s. On the other hand, a recent introductions of the parasite into South America by small founder populations could explain the observed low diversity. Lastly, epidemic populations also exhibit extremely low diversity.

The distribution of genetic variation among populations is called population structure. MS have been used in conjunction with Wright's $F_{st}$ (Wright, 1978) and Slatkin's $R_{st}$ (Slatkin, 1995) to measure population structure. Current population structure will reflect both ongoing gene flow and the effects of historical events such as recent demographic bottlenecks and expansions, population fragmentation, and long-distance dispersal. MS have recently provided new insight into the population structure of the malaria parasite. Population structure in *P. falciparum*, like genetic diversity, appears to differ markedly among continents. Whereas South American populations show evidence of isolation by distance, African populations separated by the same distances are genetically indistinguishable from each other (Anderson *et al.*, 2000a; Wootton *et al.*, 2002). The two regions differ in that most genetic variation is found within populations in Africa, but is distributed among populations in South America. This finding, which highlights regional differences in the population structure of *P. falciparum*, provides guidance for the design of future population studies of the malaria parasite. The dynamics of parasite populations on a local scale over several years can also be tracked with MS markers, due to their relatively high mutation rate, allowing for the study of changes in diversity within, and the pattern of gene flow among, local populations with time.

MS have been employed to study population structure and gene flow among sibling species of the malaria vector *Anopheles gambiae* (Lanzaro *et al.*, 1998). Estimates of gene flow differed depending on the location of the MS marker, with loci on the X chromosome and chromosome 3 showing high levels of gene flow and those on chromosome 2 showing low levels. However, when population size has not been stable for long periods of time, as is likely the case for *A. gambiae* and *A. arabiensis*, gene flow estimates based on $F_{st}$ may be inflated (Donnelly *et al.*, 1999). Additionally, translation of Wright's $F_{st}$ and Slatkin's $R_{st}$ into migration rates can be problematic because it assumes that all subpopulations are of equal size and that migration rates are symmetric between populations. Maximum likelihood (ML) estimates of migration rates based on coalescent theory (Kingman, 1982b, a) are free from these assumptions (Beerli and Felsenstein, 1999; 2001), and should be more efficient and accurate than methods using $F_{st}$, because they make use of the information contained in the underlining geneology (Felsenstein, 1992). Other population parameters such as population size, and growth rate (Kuhner *et al.*, 1995, 1998) can also be estimated using the coalescent in a maximum likelihood framework; however, given current methods, not all of these parameters can be estimated using MS. The accuracy of ML evolutionary inference methods increases with the number of independent loci examined because each locus represents a different realization of the evolutionary history of the organism (Rosenberg and Nordborg, 2002). For this reason multi-locus MS offer a potential advantage over single-locus sequence data for estimating demographic parameters.

## 2.3. Demographic History

Demographic history such as bottlenecks, population expansion or contraction, range expansion, and long-distance colonization all produce characteristic patterns in the genome. For example, population bottlenecks are associated with founder effects that reduce genetic variability and increase linkage among markers. Disease pathogens capable of epidemic growth may be particularly prone to bottlenecks and subsequent rapid expansion, especially following a recent introduction. MS were successfully used to detect a founder effect associated with the introduction of the pathogenic fungus *C. immitis* from North America into South America (Fisher *et al.*, 2001), and more recently, multiple chloroquine resistant founder mutations in *P. falciparum* (Wootton *et al.*, 2002).

MS are a particularly useful tool for detecting demographic change in the recent past, due to their high mutation rate. In addition, because they are distributed genome wide, MS avoid some of the pitfalls of single locus markers. For example, the reduced genetic variation and increased linkage associated with population bottlenecks can also be indicative of a selective sweep. Studies based on genetic markers tightly linked to a selected locus, such as those responsible for drug resistance, will incorrectly infer demographic expansion when none has occurred. Because multilocus MS data can be sampled genome wide, a distinction can be made between the effects of expansion, seen at all loci, and selection, seen only in those regions tightly linked to selected loci (Wootton *et al.*, 2002).

Statistical tests for discerning past demographic events have been developed specifically for MS markers. The within-locus *k*-test and the interlocus *g*-test (Reich and Goldstein, 1998) are two such tests for population expansion. The *k*-test uses the relationship between the branching pattern of gene genealogies and the distribution of MS allele lengths to infer demographic expansion events. A population of constant size will generally produce genealogies composed of a single deep split and multiple shallow splits. In contrast, under sudden expansion all the branches in a genealogy will tend to be of similar length. A ragged, multipeaked distribution of allele sizes is expected for a population growing at a constant size, whereas a single-peaked distribution suggests a past expansion event. The interlocus *g*-test has no analog among single-locus statistics. The pattern of variation across unlinked MS loci can be used to directly test whether population size has been constant or expanding in the recent past, because rapid population growth decreases the difference in total branch length across genealogies produced by multiple MS markers (Goldstein *et al.*, 1999). Using the *k*-test and the *g*-test, Reich and Goldstein (Reich and Goldstein, 1998) present evidence for a major Paleolithic human population expansion in Africa.

A population that has undergone a recent demographic expansion or decline may not be in mutation-drift equilibrium (MDE); as a result, measures of gene flow that are sensitive to departures from equilibrium may not reflect true values. It is therefore important to be able to assess the demographic stability of a population before estimating gene flow. Cornuet and Luikart (1996) have developed a multilocus homozygosity test for use with MS markers under all three evolution models (IAM, SMM, TPM) to test whether a population is at equilibrium. Results from this test as well as others suggest that the African malaria vectors *A. gambiae* and *A. arabiensis* have both undergone recent population expansions (Donnelly *et al.*, 1999).

## 2.4. Phylogenetic Relationships

Although they are a source of highly polymorphic genetic data, and are thus potentially highly informative, MS present specific challenges to genealogical inference because back mutations and homoplasy cannot be ruled out. MS may evolve too rapidly to draw reliable conclusions when genetic distances between taxonomic units are large, as is the case for most well-defined species. The limited utility of MS at higher taxonomic levels is due to possible constraints on the range of allele sizes, placing a ceiling on the amount of genetic distance that can accumulate between genetically isolated taxa (Garza *et al.*, 1995; Lehmann *et al.*, 1996; Kruglyak *et al.*, 1998).

On the other hand, MS are useful markers for reconstructing intraspecific relationships among populations as well as phylogenetic relationships among closely related species, including cryptic species.

The genetic distances $R_{st}$, $(\delta\mu)^2$, and allele frequencies can all be used to reconstruct evolutionary relationships among divergent populations; however, for highly divergent populations and cryptic species, a large number of MS is required to recover the correct phylogeny (Takezaki and Nei, 1996) because upper limits on allele size cause them to lose resolution at larger distances. Fisher *et al.* (2001) used MS to reconstruct genealogical relationships within the pathogenic fungus *Coccidioides immitis* and concluded that they were useful markers for delineating historical relationships but cautioned that genetic distances calculated from MS underestimate the length of long branches and that indels flanking MS can contribute considerable noise to the data. They also concluded that distances based on allele frequency alone performed better than those that included allele size. Other authors have also found that for large distances, genetic distances based on allele frequencies performed best. For example, in a study of human populations, trees based on allele frequencies were highly consistent with those obtained with other nuclear markers (Bowcock *et al.*, 1994). In another study, allele-sharing genetic distances provided high resolution in distinguishing among island fox populations (Goldstein *et al.*, 1999). Because of their high variability, MS are best able to recover geneologies when genetic distances are small, However, allele frequency alone will be the most accurate at recovering genealogies when large genetic distances exist.

## 2.5. Linkage Disequilibrium

Linkage disequilibrium (LD) refers to the non-random association of alleles and can arise between genetic markers that are tightly linked physically. It can occur in natural populations where asexual reproduction predominates to such an extent that the rate of recombination is not high enough to randomize genomes or break up clonal associations; however, LD can also arise in populations for which recombination is common. For example, LD can result when there is admixture of two or more subpopulations that differ in allele frequencies. LD seen in South American malaria parasite populations may be the result of recent admixture of genetically distinct populations (Anderson *et al.*, 2000a). Also, for disease organisms, an "epidemic" population structure, in which there has been a recent and rapid increase in a particular genotype, can result in temporary LD (Smith *et al.*, 1993).

The index of association ($I_A$) measures the strength of LD for multilocus data sets and can therefore be applied to MS. When values for $I_A$ differ significantly from zero, recombination is assumed to have been rare or absent. Smith *et al.* (1993) calculated $I_A$ for to a range of bacterial and parasite organisms using electrophoretic data and detected mating structures ranging from panmitic random mating to epidemic to exclusively clonal. The relative ease with which loci can be scored makes MS an attractive tool for use with this test. MS markers have revealed that levels of LD in *P. falciparum* vary among geographic regions and have demonstrated a link between transmission and effective recombination rates (Anderson *et al.*, 2000a). In fact, the parasite mating system appears to include both inbreeding and outbreeding, with the relative proportions varying in response to transmission levels. Extensive LD surrounding the chlorquine resistant gene *pfcrt* has been detected using >300 MS markers (Wootton *et al.*, 2002), demonstrating the utility of MS markers for studying drug resistance in the malaria parasite. MS can also be used to measure LD between polymorphic loci on a fine scale. Effective recombination rates for *P. falciparum* were found to be quite high based on LD within small chromosomal regions for a large number of isolates (Conway *et al.*, 1999). Recombination rates were also found to be high (17 Kb / cM) based on genome-wide markers typed for 35 isolates (Su *et al.*, 1999).

# 3. Trait Mapping

As genetic markers, MS can be used to map genes controlling various traits, especially drug resistance. Based on a genetic cross in *P. falciparum* (Wellems *et al.*, 1990), MS markers have been successfully used to map a locus associated with chloroquine resistance, leading to the identification of the *pfcrt* gene (Su *et al.*, 1997; Fidock *et al.*, 2000). The advantage of genetic mapping using progeny from a genetic cross is that the genetic backgrounds are similar for all progeny. On the other hand, by controlling the genetic background, elements that may be important in other parasites but not present in the parents of the cross will be missed completely. The high cost and complex procedures involved in setting up a genetic cross in *P. falciparum* have prevented the wider use of this technique.

With the availability of high-density genetic markers, including MS markers (Su *et al.*, 1999), it is now possible to perform genome-wide allelic association studies for genes contributing to such specific phenotypes as drug resistance. Genome-wide association depends on the presence of LD between the target gene(s) and the genetic markers used. LD, in turn, is affected by various factors such as the distance between the affected gene and the markers, the recombination rate, the age of the mutation, and population structure of the parasite. Although *P. falciparum* has been shown to have a relatively high recombination rate in a genetic cross at ~ 17 cM (Su *et al.*, 1999) and in the field (Conway *et al.*, 1999), there are many reasons why mapping drug-resistant genes in malaria parasites using field isolates remains feasible. First, despite a high recombination rate in Africa and Papua New Guinea, Asia and South America contain regions where transmission rates are low and, as a result, recombination rates are also low (Anderson *et al.*, 2000a). Second, most drug-resistant mutations are relatively recent events, having generally arisen since World War II. For such recent mutations, not enough time has passed to completely break down LD within local populations. One potentially fruitful approach would be to collect samples from areas representing "spreading front lines". For example, chloroquine-resistant parasites were reported in Southeast Asia and South America in the late 1950s but did not reach Africa until the late 1970s (Payne, 1987). Third, recombination rates may also be depressed because drug pressure can prevent or disrupt genetic recombination between resistant and sensitive parasites. If, for example, a patient is infected with a pool of parasites exhibiting a range of sensitivities to a particular drug, administering the drug would eliminate sensitive parasites before they could be taken up by a mosquito, thus preventing genetic exchange between resistant and sensitive strains. The confounding effects of population structure can be avoided by sampling parasites from the same location or nearby villages, obtaining parasites with similar genetic backgrounds.

MS markers are highly polymorphic with multiple alleles, making them more informative than markers such as SNP that can have at most four alleles, and typically have only two. Loci under LD due to drug selection will exhibit significantly reduced allelic diversity. The *pfcrt* locus, which contributes to chloroquine resistance, has recently been confirmed using this method (Wootton *et al.*, 2002).

High-throughput MS typing is now available, a key element for large-scale typing studies. Field isolates are relatively easy to obtain, and samples can be collected from regions of high or low transmission rates. These factors make it practical to conduct genome-wide searches for drug-resistant genes using field isolates. MS are certain to play a significant role in malaria genetics and disease control in the future.

# 4. Conclusions

There is ample evidence to suggest that the population genetics of *P. falciparum* is complex, incorporating a range of population structures and levels of recombination depending on the geographic location of the isolate. Drug pressure and other malaria control measures have left their mark on the demographic history of the parasite. Transmission dynamics vary regionally and depend on the vector as well as the parasite. Given this complexity, it is essential that we develop tools that can accurately describe parasite population parameters on local, regional, and global scales.

MS will likely shed new light on the recent debate concerning the age of the malaria parasite (Rich *et al.*, 1998; Volkman *et al.*, 2001; Mu *et al.*, 2002;

Joy *et al.*, 2003). SNP have thus far provided conflicting evidence, with the malaria parasite exhibiting lower than expected numbers of synonymous SNP (those not resulting in an amino acid change) and elevated numbers of non-synonymous SNP (those that alter the resulting amino acid). Low levels of synonymous SNP was one of the observations that led to the hypothesis of a recent origin for the parasite on the assumption that not enough time has passed to allow these largely neutral mutations to accumulate. On the other hand, high levels of non-synonymous SNP may suggest selection or other factors such as compositional constraint.

Some issues associated with MS remain to be addressed, in particular the suitability and generality of current models of microsatellite evolution. However the genome-wide distribution, high variability, and relatively neutral evolution of MS make them ideal markers for addressing questions of population structure and demographic history, in addition to genetic mapping and parasite typing.

# 5. References

Papers of particular interest have been highlighted as:
- of special interest
- - of outstanding interest

Amos, W. and Rubinsztein, D.C. 1996. Microsatellites are subject to directional evolution. Nature Genetics. 12: 13-14.

Anderson, T.J. and Day, K.P. 2000. Geographical structure and sequence evolution as inferred from the *Plasmodium falciparum* S-antigen locus. Mol. Biochem. Parasitol. 106: 321-326.

•• Anderson, T.J., Haubold, B., Williams, J.T., Estrada-Franco, J.G., Richardson, L., Mollinedo, R., Bockarie, M., Mokili, J., Mharakurwa, S., French, N., Whitworth, J., Velez, I.D., Brockman, A.H., Nosten, F., Ferreira, M.U. and Day, K.P. 2000a. Microsatellite markers reveal a spectrum of population structures in the malaria parasite *Plasmodium falciparum*. Mol. Biol. Evol. 17: 1467-1482.
A large-scale survey of parasite population structure using 465 infections from 9 worldwide locations. Revealed that population structure differs markedly among regions, reflecting regional differences in demographic and epidemiological histories.

Anderson, T.J., Su, X.-z., Bockarie, M., Lagog, M. and Day, K.P. 1999. Twelve microsatellite markers for characterization of *Plasmodium falciparum* from finger-prick blood samples. Parasitology. 119: 113-125.

Anderson, T.J., Su, X.-z., Roddam, A. and Day, K.P. 2000b. Complex mutations in a high proportion of microsatellite loci from the protozoan parasite *Plasmodium falciparum*. Mol. Ecol. 9: 1599-1608.

Babiker, H.A., Creasey, A.M., Fenton, B., Bayoumi, R.A., Arnot, D.E. and Walliker, D. 1991. Genetic diversity of *Plasmodium falciparum* in a village in eastern Sudan. 1. Diversity of enzymes, 2D-PAGE proteins and antigens. Trans. R. Soc. Trop. Med. Hyg. 85: 572-577.

Babiker, H.A., Lines, J., Hill, W.G. and Walliker, D. 1997. Population structure of *Plasmodium falciparum* in villages with different malaria endemicity in east Africa. Am. J. Trop. Med. Hyg. 56: 141-147.

Beerli, P. and Felsenstein, J. 1999. Maximum-likelihood estimation of migration rates and effective population numbers in two populations using a coalescent approach. Genetics. 152: 763-773.

Beerli, P. and Felsenstein, J. 2001. Maximum likelihood estimation of a migration matrix and effective population sizes in *n* subpopulations by using a coalescent approach. Proc. Natl. Acad. Sci. USA. 98: 4563-4568.

Bowcock, A.M., Ruiz-Linares, A., Tomfohrde, J., Minch, E., Kidd, J.R. and Cavalli-Sforza, L.L. 1994. High resolution of human evolutionary trees with polymorphic microsatellites. Nature. 368: 455-457.

Bowman, S., Lawson, D., Basham, D., Brown, D., Chillingworth, T., Church, D., Craig, A., Davies, R.M., Devlin, K., Feltwell, T. and Gentles, S. 1999. The complete nucleotide sequence of chromosome 3 of *Plasmodium falciparum*. Nature. 400: 532-538.

Chakraborty, R., Kimmel, M., Stivers, D.N., Davison, J. and Deka, R. 1997. Relative mutation rates at di-, tri-, and tetranucleotide microsatellite loci. Proc. Natl. Acad. Sci. USA. 94: 1041-1046.

Conway, D.J., Fanello, C., Lloyd, J.M., Al-Joubori, B.M., Baloch, A.H., Somanath, S.D., Roper, C., Oduola, A.M., Mulder, B., Povoa, M.M., Singh, B. and Thomas, A.W. 2000. Origin of *Plasmodium falciparum* malaria is traced by mitochondrial DNA. Mol. Biochem. Parasitol. 111: 163-171.

Conway, D.J., Roper, C., Oduola, A.M., Arnot, D.E., Kremsner, P.G., Grobusch, M.P., Curtis, C.F. and Greenwood, B.M. 1999. High recombination rate in natural populations of *Plasmodium falciparum*. Proc. Natl. Acad. Sci. USA. 96: 4506-4511.

Cornuet, J. and Luikart, G. 1996. Description and power analysis of two tests for detecting recent population bottlenecks from allele frequency data. Genetics. 144: 2001-2014.

Dallas, J.F. 1992. Estimation of microsatellite mutation rates in recombinant inbred strains of mouse. Mammalian Genome. 3: 452-456.

Daubersies, P., Sallenave-Saies, S., Magne, S., Trape, J.-F., Contamin, H., Fandeur, T., Rogier, C., Mercereau-Puijalon, O. and Druilhe, P. 1996. Rapid turnover on *Plasmodium falciparum* populations in asymptomatic individuals living in a high transmission area. Am. J. Trop. Med. Hyg. 54: 18-26.

Di Rienzo, A., Peterson, A.C., Garza, J., Valdes, A.M., Slatkin, M. and Freimer, N. 1994. Mutational processes of simple-sequence repeat loci in human populations. Proc. Natl. Acad. Sci. USA. 91: 3166-3170.

Donnelly, M.J., Cuamba, N., Charlwood, J.D., Collins, F.H. and Townson, H. 1999. Population structure in the malaria vectork, *Anopheles arabiensis* Patton, in East Africa. Heredity. 83: 408-417.

Drakeley, C.J., Duraisingh, M.T., Povoa, M., Conway, D.J., Targett, G.A. and Baker, D.A. 1996. Geographical distribution of a variant epitope of Pfs48/45, a *Plasmodium falciparum* transmission-blocking vaccine candidate. Mol. Biochem. Parasitol. 81: 253-257.

Estoup, A. and Cornuet, J.-M. 1999. Microsatellite evolution: inferences from population data. In: Microsatellites: Evolution and Applications. D.B. Goldstein and C. Schlotterer, eds. Oxford University Press, Oxford, p. 49-65.

Felsenstein, J. 1992. Estimating effective population size from samples of sequence: inefficiency of pairwise and segregating sites as compared to phylogenetic estimates. Genetical Research. 59: 139-147.

Fidock, D.A., Nomura, T., Cooper, R.A., Su, X., Talley, A.K. and Wellems, T.E. 2000. Allelic modifications of the cg2 and cg1 genes do not alter the chloroquine response of drug-resistant *Plasmodium falciparum*. Mol. Biochem. Parasitol. 110: 1-10.

Fisher, M.C., Koenig, G.L., White, T.J., San-Blas, G., Negroni, R., Alvarez, I.G., Wanke, B. and Taylor, J.W. 2001. Biogeographic range expansion into South America by *Coccidioides immitis* mirrors New World patterns of human migration. Proc. Natl. Acad. Sci. USA. 98: 4558-4562.

Gardner, M.J., Tettelin, H., Carucci, D.J., Cummings, L.M., Aravind, L., Koonin, E.V., Shallom, S., Mason, T., Yu, K., Fujii, C. and Pederson, J. 1998. Chromosome 2 sequence of the human malaria parasite *Plasmodium falciparum*. Science. 282: 1126-1132.

Garza, J., Slatkin, M. and Freimer, N. 1995. Microsatellite allele frequencies in humans and chimpanzees, with implications for constraints on allele size. Mol. Biol. Evol. 12: 594-603.

Goldstein, D., Linares, A., Cavalli-Sforza, L. and Feldman, M. 1995. Genetic absolute dating based on microsatellites and the origin of modern humans. Proc. Natl. Acad. Sci. USA. 92: 6723-6727.

Goldstein, D.B., Roemer, G.W., Smith, D.A., Reich, D.E., Bergman, A. and Wayne, R.K. 1999. The use of microsatellite variation to infer population structure and demographic history in a natural model system. Genetics. 151: 797-801.

Hughes, A.L. 1991. Circumsporozoite protein genes of malaria parasites (*Plasmodium* spp.): evidence for positive selection on immunogenic regions. Genetics. 127: 345-353.

Hughes, A.L. 1992. Positive selection and interallelic recombination at the merozoite surface antigen-1 (MSA-1) locus of *Plasmodium falciparum*. Mol. Biol. Evol. 9: 381-393.

Jeffreys, A.J., Tamaki, K., MacLeod, A., Monckton, D.G., Neil, D.L. and Armour, J.A.L. 1994. Complex gene conversion events in germline mutation at human minisatellites. Nature Genetics. 6: 136-145.

Joy, D.A., Feng, X., Mu, J., Furuya, T., Chotivanich, K., Krettli, A.U., Ho, M., Wang, A., White, N.J., Suh, E., Beerli, P. and Su, X.-z. 2003. Early origin and recent expansion of *Plasmodium falciparum*. Science. 300: 318-321.

Kimura, M. and Crow, J.F. 1964. The number of alleles that can be maintained in a finite population. Genetics. 49: 725-738.

Kingman, J.F.C. 1982a. The coalesent. Stochastic Processes and Their Applications. 13: 235-248.

Kingman, J.F.C. 1982b. On the geneology of large populations. In: Essays in Statistical Science. J. Gani and E. J. Hannan, eds. Applied Probability Trust, London, p. 27-33.

Kruglyak, S., Durrett, R.T., Schug, M.D. and Aquadro, C.F. 1998. Equilibrium distributions of microsatellite repeat length resulting from a balance between slippage events and point mutations. Proc. Natl. Acad. Sci. USA. 95: 10774-10778.

Kuhner, M.K., Yamato, J. and Felsenstein, J. 1995. Estimating effective population size and mutation rate from sequence data using Metropolis-Hastings sampling. Genetics. 140: 142-1430.

Kuhner, M.K., Yamato, J. and Felsenstein, J. 1998. Maximum likelihood estimation of population growth rates based on the coalescent. Genetics. 149: 429-434.

Lanzaro, G.C., Toure, Y.T., Carnahan, J., Zheng, L., Dolo, G., Traore, S., Petrarca, V., Vernick, K.D. and Taylor, C.E. 1998. Complexities in the genetic structure of

*Anopheles gambiae* populations in west Africa as revealed by microsatellite DNA analysis. Proc. Natl. Acad. Sci. USA. 95: 14260-14265.

Lehmann, T., Hawley, W. and Collins, F. 1996. An evaluation of evolutionary constraints on microsatellite loci using null alleles. Genetics. 144: 1155-1163.

Levinson, G. and Gutman, G.A. 1987a. High frequency of short frameshifts in poly-CA/TG tandem repeats borne by bacteriophage M13 in *Escherichia coli* K-12. Nucleic Acid Research. 15: 5323-5338.

Levinson, G. and Gutman, G.A. 1987b. Slipped-strand mispairing: a major mechanism for DNA sequence evolution. Mol. Biol. Evol. 4: 203-221.

Mu, J., Duan, J., Makova, K.D., Joy, D.A., Huynh, C.Q., Branch, O.H., Li, W.-H. and Su, X.-z. 2002. Chromosome-wide SNPs reveal an ancient origin for *Plasmodium falciparum*. Nature. 418: 323-325.

Nei, M. 1972. Genetic distances between populations. Am. Nat. 106: 283-292.

Ohta, T. and Kimura, M. 1973. A model of mutation appropriate to estimate the number of elecrophoretically detectable alleles in a finite population. Genet. Res. Camb. 22: 201-204.

Paul, R.E., Packer, M.J., Walmsley, M., Lagog, M., Ranford-Cartwright, L.C., Paru, R. and Day, K.P. 1995. Mating patterns in malaria parasite populations of Papua New Guinea. Science. 269: 1709-1711.

Payne, D. 1987. Spread of chloroquine resistance in *Plasmodium falciparum*. Parasitology Today. 3: 241-246.

Primmer, C.R., Ellegren, H., Saino, N. and Moller, A.P. 1996. Direcitional evolution in germline microsaellite mutations. Nature Genetics. 13: 391-393.

Reich, D.E. and Goldstein, D.B. 1998. Genetic evidence for a Paleolithic human population expansion in Africa. Proc. Natl. Acad. Sci. U S A. 95: 8119-8123.

Rich, S.M., Licht, M.C., Hudson, R.R. and Ayala, F.J. 1998. Malaria's eve: evidence of a recent population bottleneck throughout the world population of *Plasmodium falciparum*. Proc. Natl. Acad. Sci. USA. 95: 4425-4430.

Rosenberg, N.A. and Nordborg, M. 2002. Geneological trees, coalescent theory and the analysis of genetic polymorphisms. Nature. 3: 380-390.

Shriver, M., Jin, L., Boerwinkle, E., Deka, R., Ferrell, R. and Chakraborty, R. 1995. A novel measure of genetic distance for highly polymorphic tandem repeat loci. Mol. Biol. Evol. 12: 914-920.

Slatkin, M. 1995. A measure of population subdivision based on microsatellite allele frequencies. Genetics. 139: 457-462.

Smith, J.M., Smith, N.H., O'Rourke, M. and Spratt, B.G. 1993. How clonal are bacteria? Proc. Natl. Acad. Sci. USA. 90: 4384-4388.

Stephens, J., Gilbert, D., Yuhki, N. and O'Brien, S. 1992. Estimation of heterozygosity for single-probe multilocus DNA fingerprints. Mol. Biol. Evol. 9: 729-743.

• Stumpf, M.P.H. and Goldstein, D.B. 2001. Genealogical and evolutionary inference with the human Y chromosome. Science. 291: 1738-1742.
   Thorough review of methods available for geneological inference using microsatellites, with comparisons made between demographic model-based and model-free methods, and particular attention given to the difficulties associated with modeling the mutation process for microsatellites.

Su, X.-z. and Ferdig, M.T. 2002. Microsatellite analysis in human malaria *Plasmodium falciparum*. In: Methods in Molecular Medicine. D.L. Doolan, ed. Humana Press, Inc., Tottowa, NJ.

• Su, X.-z., Ferdig, M.T., Huang, Y., Huynh, C.Q., Liu, A., You, J., Wootton, J.C. and Wellems, T.E. 1999. A genetic map and recombination parameters of the human malaria parasite *Plasmodium falciparum*. Science. 286: 1351-1353.
   A genome-wide, high-resolution linkage map of *Plasmodium falciparum*

from 901 markers based on a genetic cross. Fourteen inferred linkage groups correspond to the 14 nuclear chromosomes. Demonstrated that meiotic crossover activity in the genome was high (17 kilobases per Centimorgan). The markers, map, and recombination parameters are facilitating genome sequence assembly, localization of determinants for such traits as virulence and drug resistance, and genetic studies of parasite field populations.

Su, X.-z., Kirkman, L.A., Fujioka, H. and Wellems, T.E. 1997. Complex polymorphisms in an approximately 330 kDa protein are linked to chloroquine-resistant *P. falciparum* in Southeast Asia and Africa. Cell. 91: 593-603.

Su, X.-z. and Wellems, T.E. 1996. Towards a high-resolution *Plasmodium falciparum* linkage map: polymorphic markers from hundreds of simple sequence repeats. Genomics. 33: 430-444.

Takezaki, N. and Nei, M. 1996. Genetic distances and reconstruction of phylogenetic trees from microsatellite DNA. Genetics. 144: 389-399.

van Belkum, A., Ramesar, J., Trommelen, G. and Uitterlinden, A. 1992. Mini- and micro-satellites in the genome of rodent malaria parasites. Gene. 118: 81-86.

Volkman, S.K., Barry, A.E., Lyons, E.J., Nielsen, K.M., Thomas, S.M., Choi, M., Thakore, S.S., Day, K.P., Wirth, D.F. and Hartl, D.L. 2001. Recent origin of *Plasmodium falciparum* from a single progenitor. Science. 293: 482-484.

Weber, J.L. and Wong, C. 1993. Mutation of human short tandem repeats. Human Molecular Genetics 2: 1123-1128.

Wellems, T.E., Panton, L.J., Gluzman, I.Y., do Rosario, V.E., Gwadz, R.W., Walker-Jonah, A. and Krogstad, D.J. 1990. Chloroquine resistance not linked to mdr-like genes in a *Plasmodium falciparum* cross. Nature. 345: 253-255.

Wootton, J.C., Feng, X., Ferdig, M.T., Cooper, R.A., Mu, J., Baruch, D., Magill, A.J. and Su, X.-z. 2002. Genetic diversity and chloroquine sweeps in the *Plasmodium falciparum*. Nature. 418: 320-322.

Wright, S. 1978. Variability among and within populations. University of Chicago Press, Chicago.

From: Malaria Parasites: Genomes and Molecular Biology
Edited by: A.P. Waters and C.J. Janse

# Chapter 7

## Chromosome Structure and Dynamics of Plasmodium Subtelomeres

Artur Scherf, Luisa M. Figueiredo
and Lúcio H. Freitas-Junior

## Abstract

Substantial gene synteny has been observed between malaria species. However, new data on the organization of plasmodial subtelomeres has recently become available and demonstrates that chromosome ends show, unlike the central region, a very dynamic evolution of its DNA sequence. Sequences of the subtelomere elements vary greatly among malaria species. Duplications among subtelomeres have created large families of expressed genes often encoding variable surface antigens. Fluorescence *in situ* hybridisation (FISH) combined with three-dimensional microscopy has demonstrated that chromosome ends in *Plasmodium* are not randomly arranged in the nucleus. Telomeres form clusters of 4 to 7 heterologous chromosome ends and are associated with the nuclear periphery. The physical alignment of subtelomeres promotes frequent recombination between members of telomere-associated virulence factor genes in heterologous chromosomes. This has important implications for the parasite survival and its adaptation to environmental stress.

**Abbreviations**:
TAS:    telomere-associated sequences
TARE:   telomere-associated repetitive element
ALT:    alternative lengthening of telomeres
FISH:   Fluorescence *in situ* Hybridisation

# 1. Introduction

Chromosome mapping studies and DNA sequence analysis has revealed that specific linkage groups are conserved between malaria species (Carlton *et al.*, 1998; Tchavtchitch *et al.*, 2001). However this conserved location of homologous genes and non-coding DNA sequence elements appears to be confined to the central chromosome part, whereas the subtelomeres consist of sequences that are often not found in other malaria species (Pace *et al.*, 1987; Dore *et al.*, 1990; Figueiredo *et al.*, 2000). Here we summarise our knowledge of a dynamic chromosome compartment, the subtelomeres of *Plasmodium* and discuss relevant aspects for the biology of plasmodial virulence factor genes. We also highlight the impact of novel findings on the nuclear organisation of chromosome extremities for DNA recombination and the regulation of transcription of telomere-associated genes.

# 2. Functional Consequences of Nuclear Organisation

## 2.1. The Far-ends of Eukaryotic Chromosomes: Structure and Function

Systematic genome analysis has provided precious information on the sequence and localisation of genes, which has improved our understanding of basic biological processes and has revealed that a significant part of a genome consists of domains devoid of genes. The extremities of chromosomes are one of these domains. Chromosome ends comprise a specialised nucleoprotein structure called the telomere and the adjacent region is called subtelomere, or Telomere Associated Sequence (TAS). For the vast majority of eukaryotes, telomeres consist of a tandem array of short G-rich repeats and specific associated proteins. They are responsible for protecting chromosome extremities from degradation, prevent end-to-end fusion by DNA repair mechanisms and solve the 'end replication problem' (the gradual loss of DNA at the chromosome end with each replication cycle) (for review see McEachern *et al.*, 2000). Telomeres are also implicated in transcriptional silencing (Gottschling *et al.*, 1990), chromosome positioning in the nucleus (Gotta *et al.*, 1996), as well as homologous and ectopic recombination during meiosis (for reviews see Cooper, 2000; Ishikawa and Naito, 1999).

Subtelomeric regions appear to be species-specific, extremely polymorphic in size and usually are composed of a mosaic of repetitive blocks. Each block is usually highly variable in location and copy number. In subtelomeres, it is common to find members of gene families, such as the human olfactory receptor gene family (Trask *et al.*, 1998). The role of subtelomeric regions in different organisms is still a matter of debate. It was hypothesised that subtelomeres act as a buffer in telomere silencing events (Levis *et al.*, 1985; Gottschling *et al.*, 1990; Nimmo *et al.*, 1994). More recently, it was shown that the high degree of homology between subtelomeric regions is important in the maintenance of telomeres, when telomerase is disrupted in yeast (Lundblad and Szostak, 1989; Teng and Zakian, 1999) and humans (Dunham *et al.*, 2000). Recent studies in *Plasmodium falciparum* support a role for the subtelomeric region in cluster arrangement of chromosome ends (Figueiredo *et al.*, 2002; O'Donnell *et al.*, 2002).

## 2.2. Plasmodial Chromosome Ends: A Specific Compartment for Multigene Families

All Plasmodial species display a canonical telomere structure: G-rich repeats in tandem arrays. Telomeric DNA was first cloned from *Plasmodium berghei* (Ponzi *et al.*, 1985) and later from *P. falciparum* (Vernick and McCutchan, 1988). It was shown to be a degenerate heptamere, in which GGGTT(T/C)A are the two type of repeats in *P. berghei* and the most frequent in ones observed in *P. falciparum* (including some more degenerate type of telomere repeats). Sequencing data from *Plasmodium yoelii, Plasmodium chabaudi* and *Plasmodium knowlesi* showed that telomeric DNA is composed approx. of 90% GGGTT(T/C)A repeats (approx. of 10% consists of degenerate G-rich repeats), which confirmed previous cross-hybridisation results obtained with *P. berghei* telomeric DNA (Dore *et al.*, 1986).

Data released from Plasmodial genome sequencing projects has revealed that the different Plasmodia species display a high degree of sequence conservation and gene synteny (Carlton *et al.*, 1998). Subtelomeres, however, diverge dramatically. Since the beginning of the Plasmodial spp genome project, our group has undertaken a broad approach, looking at the DNA composition of all available sequenced chromosome ends with the help of bioinformatic tools. In *P. falciparum* we observed a unique conservation of the higher-order structure of subtelomeres (Figueiredo *et al.*, 2000). Upstream of all sequenced *P. falciparum* telomeres (18 analysed till present), lies a non-coding region of 20-40kb composed of a mosaic of six different polymorphic repetitive elements (Telomere-Associated Repetitive Elements: TAREs1-6) (see Figure 1). In *P. falciparum*, four subtelomere blocks of complex repetitive DNA had previously been described (Oquendo *et al.*, 1986; Vernick and McCutchan, 1988; Dolan *et al.*, 1993; Patarapotikul and Langsley, 1988). The six elements are present in all intact chromosomes and are always positioned in the same order, in contrast to previously described subtelomeres from other organisms that display higher variability (reviewed in Scherf *et al.*, 2001; Mefford and Trask, 2002). The AT content of this region is ~70%, which is unusually high for a non-coding region (90% AT), suggesting that subtelomeres might originate from coding regions. Towards the centromere, TAREs are flanked by members of the *var* gene family coding for virulence factors (Rubio *et al.*, 1996; Hernandez-Rivas *et al.*, 1997). Most of the *var* genes located next to the TARE6 repeated element are

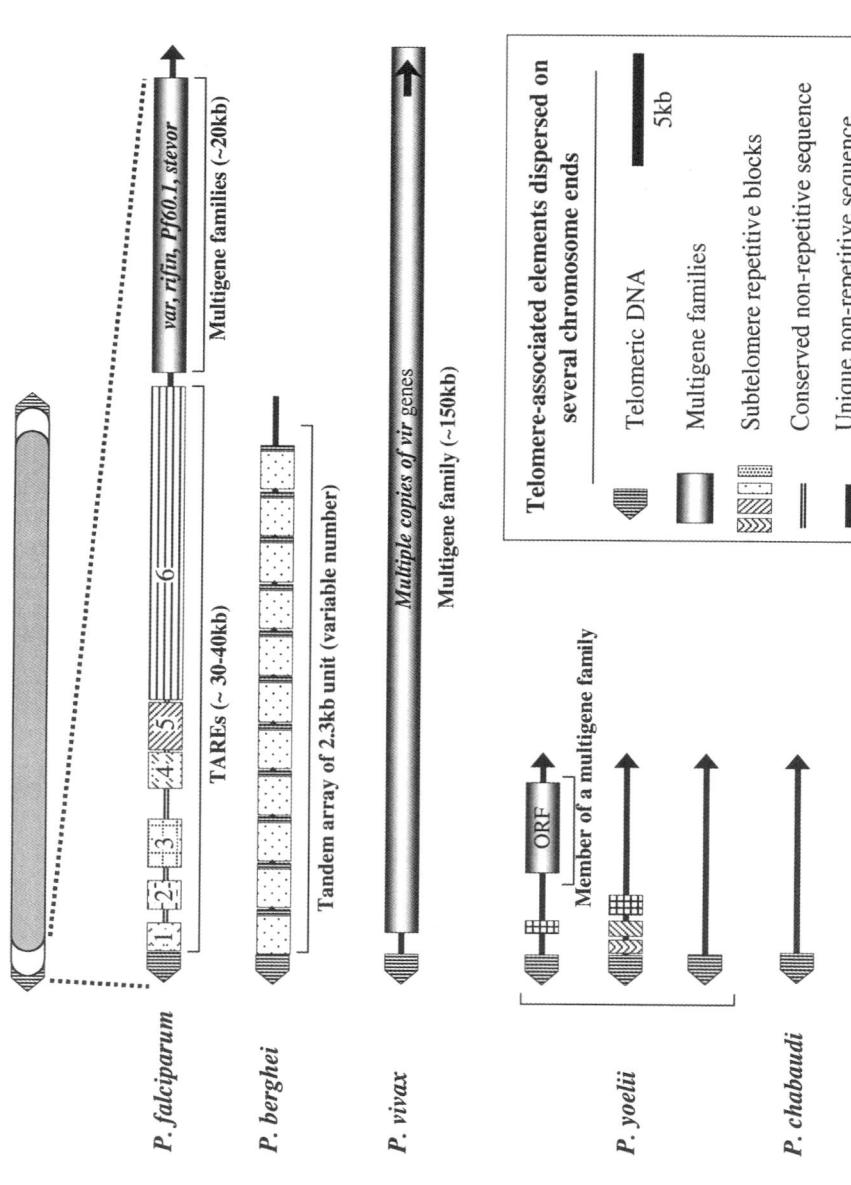

**Figure 1.** Model showing the organisation of subtelomeres in several *Plasmodium* species. The *Plasmodium* haploid nuclear genome is organised in 14 linear chromosomes. Each is composed of an internal region, where housekeeping genes are located, and a chromosome end. The terminus of the left arm of a chromosome is schematically represented as an example of a chromosome end organisation. Adjacent to *P. falciparum* telomeres, there is a highly polymorphic TAS, composed of two zones: a non-coding and a coding region. The non-coding region contains six TAREs, always positioned in the same order but of variable length. The adjacent coding region is the locus of several gene-families encoding important virulence factors, such as the *var* and *rifin* genes. The subtelomere organisation of four distinct *Plasmodium* species, (whose genome sequences are in the process of being determined) is also shown. See text for details.

transcribed toward the centromere (Gardner *et al.*, 2002). The *var* genes are followed by other multigene families such as *rif, stevor* and *Pf60.1* (Gardner *et al.*, 1998; Bowman *et al.*, 1999). Their precise role is described elsewhere in this book.

The analysis of *P. falciparum* chromosome mutants revealed that telomere repeats and TARE's do have distinct functions. Telomeres are absolutely essential for chromosomal function, whereas TARE's can be deleted without any direct effect on the viability of the parasite during mitotic and meiotic divisions. The application of FISH analysis revealed novel functions for these genetic elements. Chromosome ends still are able to attach to the nuclear periphery in the absence of TARE's. However, these mutant chromosome ends tend to delocalise from the clusters suggesting strongly their involvement in cluster formation and/or maintenance (Figueiredo *et al.*, 2002) (Figure 2A and B).

*P. berghei* subtelomeric regions were the first ones to be described: they consist of a tandem array of a 2.3 kb unit, clustered exclusively in a subtelomeric position on several chromosomes (Pace *et al.*, 1987; Dore *et al.*, 1990). The 2.3kb unit includes ~160bp stretch of telomere-related sequences, which appear to be important for recombinational events that contribute to chromosome-size polymorphism (Pace *et al.*, 1990). No gene families have been described so far adjacent to the 2.3kb unit.

At the time of this writing, *Plasmodium vivax* genome database contains a single hit for telomeric DNA: a YAC clone constructed by del Portillo *et al.* (2001). In contrast to *P. falciparum* and *P. berghei*, no large repetitive blocks exist in this *P. vivax* subtelomere. Instead, at ~1kb from telomere repeats there is an ordered array of members of the *vir* gene-family. They are present at about 600-1000 copies per haploid genome and encode proteins that are immunovariant in natural infections, suggesting a role in the establishment of a chronic infection through antigenic variation.

Analysis of the DNA sequence data available in the *P. yoelii* genome database (February 2002), suggests that subtelomere organisation is distinct among heterologous chromosomes. For example, some chromosome ends present short repetitive blocks (<1kb), but the type and number of these is highly variable. Another chromosome end carries a novel putative gene sequence, which is a member of a large gene-family (at least 50 distinct hits in the *P. yoelii* database) coding for putative proteins of small size (22-35kDa). Alignment of some of the members of this family reveals a great conservation of the N and C-termini. This adds another subtelomeric gene family to the one described recently, the Py235 gene-family (Khan *et al.*, 2001).

At this stage, the *P. chabaudi* genome project is not as advanced as the other rodent malaria species. Thus, only 3 hits were found containing

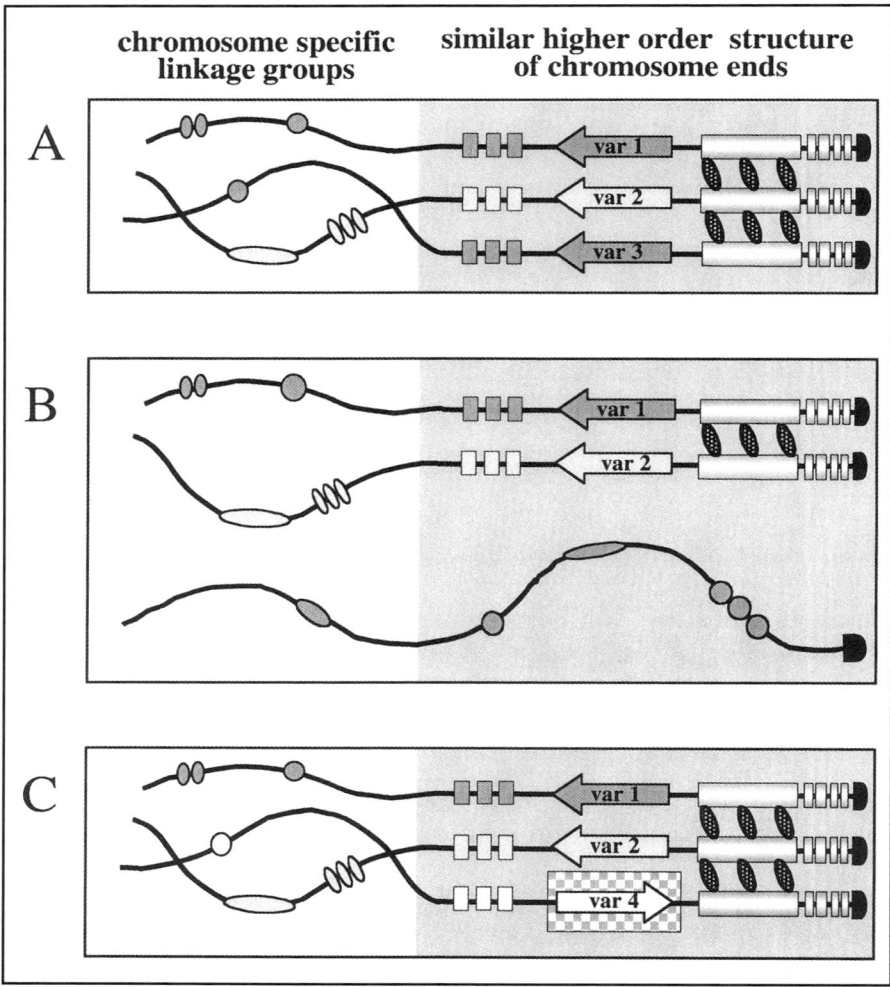

**Figure 2**. Chromosome clusters provide a nuclear environment for frequent recombination events. A similar subtelomere organisation is conserved among *P. falciparum* chromosome ends. A. The physical alignment of chromosome ends is shown where increased rates of recombination occur between homologous loci of heterologous chromosomes (ectopic recombination). *Var* genes adjacent to *TARE* are generally transcribed towards the centromere. TARE6 specific DNA binding proteins that might cross-link chromosome ends are indicated. B. The deletion of the *TARE* is associated with the delocalisation of this chromosome end from a cluster. C. An unusual orientation of a *var* gene within a chromosome end has been observed (Vázquez-Macías *et al.*, 2002). Ectopic recombination between opposite orientated *var* genes predicts that ectopic recombination events would lead to dicentric chromosomes (lethal event).

telomere repeats and their size was not larger than 5kb. Sequence analysis did not reveal any open-reading frames, nor repetitive elements upstream (up to 5 kb) from telomere repeats.

The restricted sequence data available for the subtelomeres of the rodent malaria species do allow drawing a number of important conclusions: 1) Each *Plasmodium* species analysed (except for *P. vivax*) contains distinct non-coding sequence elements adjacent to the same type of telomere repeats. In *P. falciparum*, a similar higher order structure is found on virtually all chromosome ends. This contrasts with *P. yoelii*, which displays distinct elements adjacent to the telomeres. 2) Chromosome ends harbour generally highly diverse gene families.

## 3. Non-Random Nuclear Organisation of Chromosomes

A prevailing view is that chromosomes are like spaghetti randomly floating around in the nucleoplasm. However, more and more data from different organisms point to non-random organisation of chromosomes in the nucleus (Marshall, 2002). We demonstrated, using FISH technology, that the 28 distinct chromosomes ends in *P. falciparum* are located at the nuclear periphery of asexual and sexual blood stage parasites. Several chromosome ends are physically grouped together and the number of clusters per haploid nuclei varies between 4 and 7. This observation implies that each cluster consists of 4 to 7 non-homologous chromosome ends (Figure 3A).

It has been a mystery as to how the heterologous subtelomeres are physically aligned within a cluster and what provides the anchor for telomeres to the nuclear envelope. Two independent studies point to the subtelomeres as a critical DNA region involved in clustering. One study showed that the deletion of the entire subtelomeric region delocalises the truncated chromosome end from the cluster (Figueiredo *et al.*, 2002) (see Figure 2B). However, this mutation apparently does not change the anchoring to the nuclear periphery. A second study demonstrated that a plasmid carrying a subtelomeric sequence called rep20 (TARE6) co-localises preferentially to the cluster in *P. falciparum* blood stage parasites, whereas a control plasmid showed a random distribution in the nucleus (O'Donnell *et al.*, 2002). These results suggest a function for the subtelomeric element rep20 in keeping together chromosome ends. How this is promoted is still unknown. Nonetheless, preliminary data from our laboratory indicate that specific proteins bind to rep20 repeats, raising the possibility that these molecules are involved in cross-linking *P. falciparum* chromosome ends (L.M. Figueiredo and A. Scherf, data not shown).

Is clustering of *P. falciparum* chromosome ends random or are the same ends always involved in cluster formation? In order to address this question,

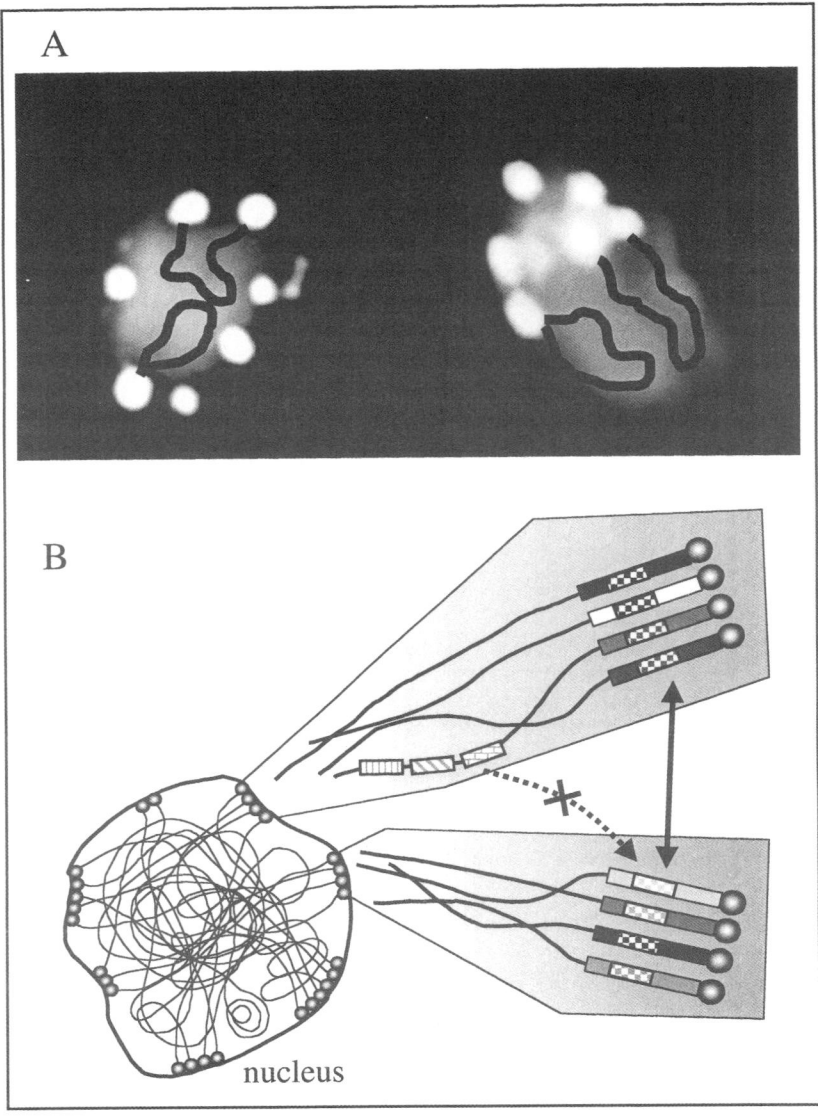

**Figure 3.** A. Nuclear architecture of *P. falciparum* chromosome ends. FISH analysis of a telomere specific DNA probe showing that the interphase chromosome termini form clusters at the nuclear periphery during trophozoite asexual blood stages (left) and a bouquet like structure in pre-meiotic gametocytes sexual blood stage (right). Parasite nuclei stained with DAPI (in light grey) and telomere clusters (in white). Two hypothetical chromosomes are shown schematically. Both chromosome ends are in the same cluster (white) or are located in distinct clusters (grey). B. *P. falciparum* cluster formation has a random component. Multicolour FISH analysis of individual chromosome ends has demonstrated that a given end can associate with different chromosome ends (Freitas-Junior *et al.*, 2000). A schematic of 2 clusters summarizing these results is shown.

the ends of three distinct chromosomes were analysed by multicolour FISH to determine the frequency of paired heterologuous telomeres. Promiscuous cluster formation was observed for the three chromosome ends studied (Freitas-Junior *et al.*, 2000). This observation implies that telomere clustering in *P. falciparum* appears to be random, suggesting that a given subtelomeric *var* gene might potentially recombine with *var* genes from any chromosome end. These results also imply that subtelomeric *var* loci will be close together in the nucleus whereas the constraint of diffusion might prevent interaction with other homologous *var* sequences being located in the central chromosome region (see Figure 3B). This indicates that the telomere associated *var* genes diverge more rapidly than members located in the central chromosome domain.

Can the subnuclear architecture of *P. falciparum* be considered to be a blueprint for other malaria species? Although the chromosome tips of different malaria species are bounded by the same type of telomere repeats, the telomere-associated sequences are surprisingly divergent from each other (see below). Telomere-FISH of evolutionary distinct malaria species revealed that chromosome ends are organized in clusters during interphase of blood stages, as first described for *P. falciparum*.

However, *P. vivax* and two of the rodent malaria parasites (*P. yoelii and P. chabaudi*) displayed significantly more fluorescent foci than *P. falciparum*, suggesting that the number of chromosomes per cluster is smaller or that several chromosomes are not located in clusters. It will be interesting to see if the heterogeneity of subtelomere organisation observed for some *Plasmodium* species does influence the cluster formation of chromosome ends. In summary, this analysis underlines that the nuclear architecture of chromosome ends is preserved between malaria species and might be related to an evolutionary advantage that this peculiar structure promotes.

## 3.1. Interaction Between Subtelomeric Loci (Recombination Frequencies)

What are the biological implications of the *P. falciparum* nuclear organisation on genes located close to telomeres? Does it advance our understanding of the genomic plasticity and recombination processes? It is now well established that plasmodial chromosome ends form a highly dynamic chromosome compartment and most of the chromosome polymorphism is due to DNA rearrangements occurring in the subtelomeric region (reviewed in Scherf *et al.*, 1999). The high recombination frequencies observed in subtelomeric regions seem to create an environment that allows the expansion and diversification of gene families located at chromosome ends. Thus, telomeres provide an ideal setting for genes involved in host parasite interactions such as parasite cytoadhesion and antigenic variation.

| *Plasmodium* species | Mammalian host | FISH | Number of clusters per nuclei (average) |
|---|---|---|---|
| *P. falciparum* | Human | 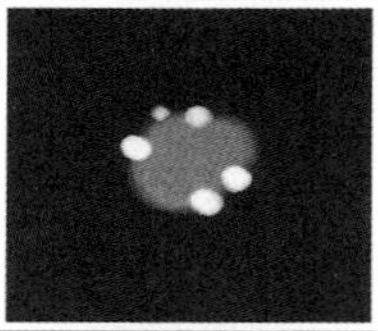 | 4-7 **(5)** |
| *P. vivax* | Human | 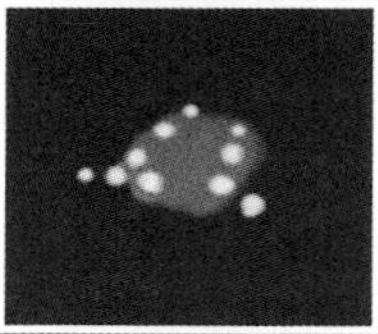 | 7-13 **(9)** |
| *P. gallinaceum* | Bird | 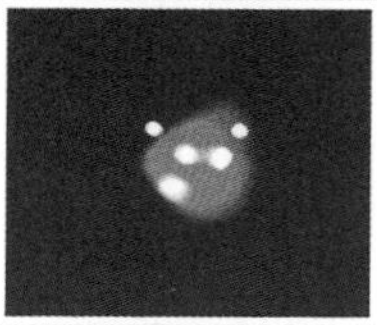 | 4-7 **(5)** |
| *P. berghei* | Rodent | 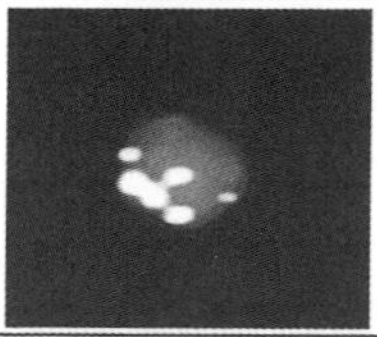 | 3-8 **(5)** |
| *P. yoelii* | Rodent | 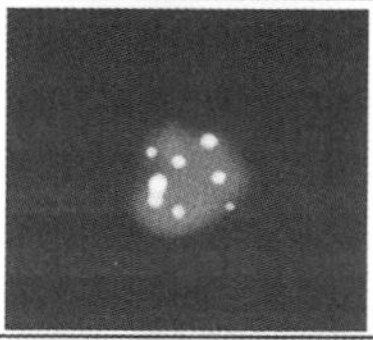 | 4-12 **(8,1)** |
| *P. chabaudi* | Rodent | 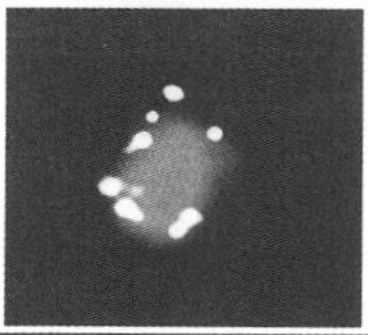 | 4-13 **(8,4)** |

**Figure 4**. Chromosome end clustering is a conserved feature of malarial species. FISH analysis of blood stage interphase parasites (young trophozoites) was performed using a fluorescent telomere probe (GGGTT$^{T}/_{C}$A). The mean number of clusters per nucleus is indicated in brackets.

Our new knowledge about the nuclear organisation points to a strong influence on chromosome interactions (Marshall, 2002). In *P. falciparum* the spatial proximity of heterologous chromosome ends combined with a common higher order structure of subtelomeres seems to be a critical factor in determining the high rate of DNA recombination observed in *var* loci in a genetic cross of *P. falciparum* (Freitas-Junior *et al.*, 2000) (see Figure 2A). It was shown that chimeric *var* forms were created by gene conversion involving two *var* members located at different chromosome ends. The marked diversity of *var* genes in genetically different laboratory strains and clinical isolates can lead to parasites with minimal overlapping *var* gene repertoires (Freitas-Junior *et al.*, 2000). This demonstrates the important role of the subtelomere in shaping genetically distinct parasite populations. Clusters are also observed during blood stage parasite proliferation and thus might enhance mitotic ectopic recombination events leading to a continual reorganisation of subtelomeric gene families during chronic patient infection.

Recently, a highly conserved subtelomere located *var* gene (*varCSA*), which is expressed during placental malaria, was discovered in parasite isolates from all over the world ((Vázquez-Macías *et al.*, 2002; Scherf *et al.*, data not shown). How does this 'non-variant' *varCSA* gene resist the diversification process? This might be explained by the exceptional organisation of this particular *var* gene, which is orientated in the opposite direction compared to other telomere-associated *var* genes (Figure 2C). Ectopic recombination events would lead to the formation of dicentric chromosomes that are generally lethal or would yield defective cells. Furthermore, being trapped in a cluster, would constrain the *varCSA* interaction with central *var* gene copies, making it relative inert to ectopic DNA recombination. Further analysis of the *P. falciparum* database suggests that an additional small number of subtelomeric *var* genes exist in the 3D7 genome being transcribed towards the telomere. For example, we identified a cluster of two subtelomeric *var* genes that are transcribed in opposite directions (data not shown). It will be interesting to investigate how this type of *var* genes relate to each other and if they are also conserved between distinct clinical isolates as the *varCSA* gene. In conclusion, this example illustrates the influence of nuclear organisation on important biological events related to parasite virulence.

### 3.2. Epigenetic Factors Involved in Antigenic Variation

The telomere position effect brings about a heritable yet variegated repression of genes placed near yeast telomere repeats (Gottschling *et al.*, 1990). This effect has been associated with chromatin organization, as proteins essential for telomeric silencing are found to be concentrated at the telomeres and/or subtelomeric regions (Kyrion *et al.*, 1992). In yeast, a number of telomere-associated proteins involved in telomere-length regulation, cluster formation, telomere anchoring to the periphery and gene silencing have been described

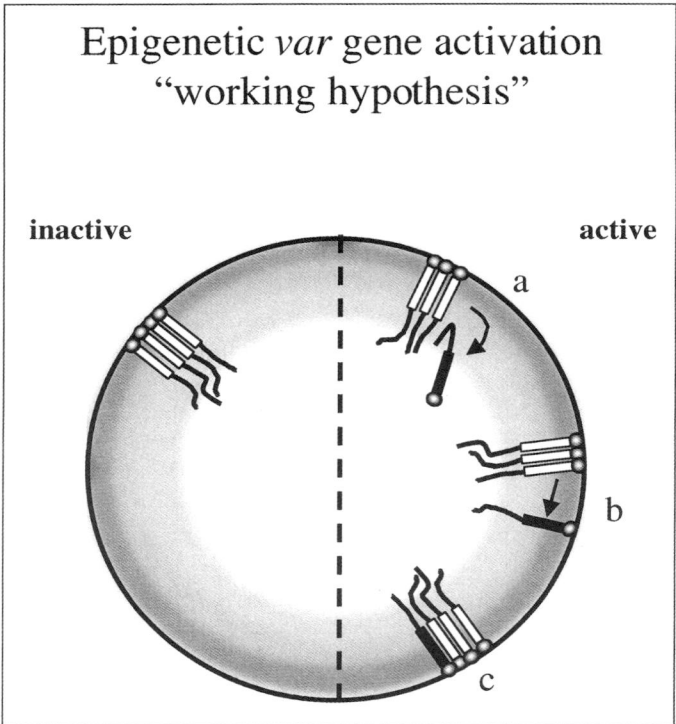

**Figure 5**. Model of the epigenetic regulation of telomere-associated *var* genes. A cluster of chromosome ends with silent *var* genes is shown (white). Three alternative activation processes are proposed (subtelomere carrying an activated *var* gene is shown in black). The first two involve chromosome end movements outside the cluster to either a specific chromosome compartment in a more internal region of the nucleus (a) or in the nuclear periphery (b). Alternatively, the activation could occur within a cluster due to local changes in the chromatin organisation of the *var* gene (c).

and characterized (reviewed in Gasser, 2001). The mechanisms that control the regulation of *var* gene expression in *P. falciparum* are still unresolved but many indications point to epigenetic factors being involved in the regulation of *var* gene transcription (Scherf *et al.*, 1998; Freitas-Junior *et al.*, 2000; Deitsch *et al.*, 2001). *P. falciparum* orthologues to most yeast telomere-associated proteins have been identified including the genes coding for the "silent information regulators 3 and 4" (LH Freitas-Junior, A Scherf, unpublished data), This finding implies that *P. falciparum* might use a similar mechanism to silence telomere-associated *var* genes. Gene inactivation studies are in progress to analyse the role of these telomere-associated molecules in *P. falciparum*.

Changes in the intra-nuclear chromosome positioning have been associated with transcriptional activation in mammalian cells and African trypanosomes (Tumbar and Belmont, 2001; Navarro and Gull, 2001). Thus, a second attractive working hypothesis is that telomere clusters form a repression zone for gene expression. In order to overcome epigenetic *var* gene silencing, the corresponding subtelomere needs to move to a distinct nuclear compartment that would be compatible with gene expression. Figure 5 shows different hypothetical models for *var* gene activation.

# 4. Conclusions

The recent discovery of telomere clustering at the nuclear periphery in *P. falciparum* and its role in ectopic recombination between members of the *var* gene families has generated more appeal to a field which was before considered to be a domain of a handful of specialists. Similar nuclear organisation of subtelomeres in clusters has now been observed in other human and rodent malaria species. Other malaria pathogens seem to follow the same strategy in order to create high levels of continual diversity in subtelomeric gene families. Yet, it is clear form the preliminary genome analysis, that each species has developed its own strategy to organize the chromosome ends.

Another important feature of the telomere is its potential role in epigenetic gene regulation. A better understanding of the biology of this specific nuclear region is necessary and might lead to a better comprehension of the chromatin factors necessary to establish repression and the mutually exclusive expression of genes involved in the antigenic variation in *Plasmodium*.

# 5. Acknowledgments

We thank L. Pirrit for helpful comments and discussion, J. Gysin for the *P. vivax*, R. Paul for *P. gallinaceum*, P. Uzureau for *P. berghei* and *P. yoelii* and T. Blisnik for *P. chabaudi* blood stage parasites. This work was supported by an EC grant QLK-CT-2000-00109, a grant VIH-PAL from the French Ministry and by the fellowships from Fundação para a Ciência e a Tecnologia, PRAXIS XXI/BD/16020/98, Portugal (L.M.F.) and Fundação de Amparo à Pesquisa do Estado de São Paulo (FAPESP) (L.H.F.J.). *P. vivax* YAC data was obtained from the Sequencing Group at the Sanger Institute and can be retrieved from ftp://ftp.sanger.ac.uk/pub/pathogens/vivax. *P. berghei* sequence data was obtained from the University of Florida, Malaria Genome Tag Sequencing Project (Carlton and Dame, 2000). Data produced by the *P. knowlesi* Partial Genome Shotgun Sequencing Group at the Sanger Institute can be obtained from ftp://ftp.sanger.ac.uk/pub/pathogens/knowlesi. Data produced by the *P. chabaudi* Partial Genome Shotgun Sequencing Group at the Sanger Institute can be obtained from ftp://ftp.sanger.ac.uk/pub/pathogens/chabaudi. Preliminary sequence and/or preliminary annotated sequence data from the *P. yoelii* genome was obtained from The Institute for

Genomic Research website (www.tigr.org). We wish to thank the scientists and funding agencies comprising the international Malaria Genome Project for making sequence data from the genome of *P. falciparum* (3D7) public prior to publication of the completed sequence. The Sanger Centre (UK) provided sequence for chromosomes 1, 3-9, & 13, with financial support from the Wellcome Trust. A consortium composed of The Institute for Genome Research, along with the Naval Medical Research Center (USA), sequenced chromosomes 2, 10, 11 & 14, with support from NIAID/NIH, the Burroughs Wellcome Fund, and the Department of Defense. The Stanford Genome Technology Center (USA) sequenced chromosome 12, with support from the Burroughs Wellcome Fund. The Plasmodium Genome Database is a collaborative effort of investigators at the University of Pennsylvania (USA) and Monash University Melbourne, Australia), supported by the Burroughs Wellcome Fund.

# 6. References

*    indicates a paper of special interest

Bowman, S., Lawson, D., Basham, D., Brown, D., Chillingworth, T. and Churcher, C.M. 1999. The complete nucleotide sequence of chromosome 3 of *Plasmodium falciparum*. Nature. 400: 532-538.

Carlton, J.M. and Dame, J.B. 2000. The *Plasmodium vivax* and *P. berghei* gene sequence tag projects. Parasitol. Today. 16: 409.

Carlton, J.M., Vinkenoog, R., Waters, A.P. and Walliker, D. 1998. Gene synteny in species of Plasmodium. Mol. Biochem. Parasitol. 93: 285-94.

Cooper, J.P. 2000. Telomere transitions in yeast: the end of the chromosome as we know it. Curr. Opin. Genet. Dev. 10: 169-177.

Deitsch, K.W., Calderwood, M.S. and Wellems, T.E. 2001. Malaria. Cooperative silencing elements in *var* genes. Nature. 412: 875-6.

del Portillo, H.A., Fernandez-Becerra, C. and et al. 2001. A superfamily of variant genes encoded in the subtelomeric region of *Plasmodium vivax*. Nature. 410: 839-42.

Dolan, S.A., Herrfeldt, J.A. and Wellems, T.E. 1993. Restriction polymorphisms and fingerprint patterns from an interspersed repetitive element of *Plasmodium falciparum* DNA. Mol. Biochem. Parasitol. 61: 137-42.

Dore, E., Pace, T., Ponzi, M., Picci, L. and Frontali, C. 1990. Organization of subtelomeric repeats in *Plasmodium berghei*. Mol. Cell. Biol. 10: 2423-7.

Dore, E., Pace, T., Ponzi, M., Scotti, R. and Frontali, C. 1986. Homologous telomeric sequences are present in different species of the genus Plasmodium. Mol. Biochem. Parasitol. 21: 121-7.

* Dunham, M.A., Neumann, A.A., Fasching, C.L. and Reddel, R.R. 2000. Telomere maintenance by recombination in human cells. [see comments]. Nat. Genet. 26: 447-50.
    This report provides direct evidence that recombination between telomeres is responsible for the telomerase-independent mechanism of telomere maintenance in ALT cells. As telomerase is a potential target for cancer therapy, an understanding of this alternative pathway is crucial.

* Figueiredo, L.M., Freitas-Junior, L.H., Bottius, E., Olivo-Marin, J.C. and Scherf, A. 2002. A central role for *Plasmodium falciparum* subtelomeric regions in spatial positioning and telomere length regulation. EMBO J. 21: 815-824.

By studying natural mutants in *P. falciparum*, in which some chromosome ends are truncated, the authors suggest two distinct roles for the subtelomeric region of chromosomes: modulating *in cis* the average length of telomeric tracts and defining the spatial organisation of the chromosome end in the nucleus. Subtelomeric regions are shown to be important for chromosome end clustering, but not for localisation at the perinuclear space.

Figueiredo, L.M., Pirrit, L.A. and Scherf, A. 2000. Genomic organisation and chromatin structure of *Plasmodium falciparum* chromosome ends. Mol. Biochem. Parasitol. 106: 169-174.

* Freitas-Junior, L.H., Bottius, E., Pirrit, L.A., Deitsch, K.W., Scheidig, C., Guinet, F., Nehrbass, U., Wellems, T.E. and Scherf, A. 2000. Frequent ectopic recombination of virulence factor genes in telomeric chromosome clusters of *P. falciparum*. Nature. 407: 1018-1022.

The authors demonstrated for the first time that telomeres from *P. falciparum* form clusters containing 4 to 7 heterologous chromosome ends at the nuclear periphery of asexual and sexual stages. This spatial organisation creates a highly recombinogenic compartment, where the genes coding for important virulence factors (such as the *var* genes) have their locus. Cluster formation is proposed to be important for generating diversity in molecules responsible for tissue tropism and pathogenesis of malaria in the human host.

Gardner, M.J., Tettelin, H., Carucci, D.J. *et al.* 1998. Chromosome 2 sequence of the human malaria parasite *Plasmodium falciparum* [published erratum appears in Science 1998 Dec 4;282(5395):1827]. Science. 282: 1126-32.

Gardner, M.J., Shallom, S.J., Carlton, J.M., *et al.* 2002. Sequence of *Plasmodium falciparium* chromosomes 2, 10, 11, and 14. Nature. 419: 498-511.

Gasser, S.M. 2001. Positions of potential: Nuclear organization and gene expression [Review]. Cell. 104: 639-642.

Gotta, M., Laroche, T., Formenton, A., Maillet, L., Scherthan, H. and Gasser, S.M. 1996. The Clustering of telomeres and colocalization with Rap1, Sir3, and Sir4 Proteins in wild-type *Saccharomyces cerevisiae*. J. Cell Biol. 134: 1349-1363.

Gottschling, D.E., Aparicio, O.M., Billington, B.L. and Zakian, V.A. 1990. Position effect at *S. cerevisiae* telomeres: reversible repression of Pol II transcription. Cell. 63: 751-62.

Hernandez-Rivas, R., Mattei, D., Sterkers, Y., Peterson, D.S., Wellems, T.E. and Scherf, A. 1997. Expressed *var* genes are found in *Plasmodium falciparum* subtelomeric regions. Mol. Cell Biol. 17: 604-11.

Ishikawa, F. and Naito, T. 1999. Why do we have linear chromosomes? A matter of Adam and Eve [Review]. Mutat. Res. 434: 99-107.

Khan, S.M., Jarra, W., Bayele, H. and Preiser, P.R. 2001. Distribution and characterisation of the 235 kDa rhoptry multigene family within the genomes of virulent and avirulent lines of *Plasmodium yoelii*. Mol. Biochem. Parasitol., 114: 197-208.

Kyrion, G., Boakye, K.A. and Lustig, A.J. 1992. C-terminal truncation of RAP1 results in the deregulation of telomere size, stability, and function in *Saccharomyces cerevisiae*. Mol. Cell Biol. 12: 5159-73.

Levis, R., Hazelrigg, T. and Rubin, G.M. 1985. Effects of genomic position on the expression of transduced copies of the white gene of Drosophila. Science. 229: 558-61.

Lundblad, V. and Szostak, J.W. 1989. A mutant with a defect in telomere elongation leads to senescence in yeast. Cell. 57: 633-43.

Marshall, W.F. 2002. Order and disorder in the nucleus. Curr Biol, 12, R185-92.

McEachern, M.J., Krauskopf, A. and Blackburn, E.H. 2000. Telomeres and their control. Annu. Rev. Genet. 34: 331-358.

Mefford, H.C. and Trask, B.J. 2002. The complex structure and dynamic evolution of human subtelomeres. Nat. Rev. Genet. 3: 91-102.

* Navarro, M. and Gull, K. 2001. A pol I transcriptional body associated with VSG mono-allelic expression in *Trypanosoma brucei*. Nature 414: 759-763.

    A connection has been established between nuclear architecture and expression of a single VSG gene (an important multigene family that codes for *Trypanosoma brucei* coat protein). VSG expression site is present in a nuclear body called the Expression Site Body (ESB), which is independent from the nucleolus. The authors propose a "privileged location" model, in which monoallelic expression results from the fact that there is only space for one ES in each ESB.

Nimmo, E.R., Cranston, G. and Allshire, R.C. 1994. Telomere-associated chromosome breakage in fission yeast results in variegated expression of adjacent genes. EMBO J. 13: 3801-11.

* O'Donnell, R.A., Freitas-Junior, L.H., Preiser, P.R., Williamson, D.H., Duraisingh, M., McElwain, T.F., Scherf, A., Cowman, A.F. and Crabb, B.S. 2002. A genetic screen for improved plasmid segregation reveals a role for Rep20 in the interaction of *Plasmodium falciparum* chromosomes. EMBO J. 21: 1231-1239.

    In a genetic screen for stable plasmids in *P. falciparum*, the authors revealed that Rep20 (or TARE6) promotes plasmid segregation between daughter merozoites. The authors suggest that Rep20 is directly implicated in the physical association of chromosome ends, which is a process that facilitates the generation of diversity in the terminally located *P. falciparum* virulence genes.

Oquendo, P., Goman, M., Mackay, M., Langsley, G., Walliker, D. and Scaife, J. 1986. Characterisation of a repetitive DNA sequence from the malaria parasite, *Plasmodium falciparum*. Mol. Biochem. Parasitol. 18: 89-101.

Pace, T., Ponzi, M., Dore, E. and Frontali, C. 1987. Telomeric motifs are present in a highly repetitive element in the *Plasmodium berghei* genome. Mol. Biochem. Parasitol. 24: 193-202.

Pace, T., Ponzi, M., Dore, E., Janse, C., Mons, B. and Frontali, C. 1990. Long insertions within telomeres contribute to chromosome size polymorphism in *Plasmodium berghei*. Mol. Cell Biol. 10: 6759-64.

Patarapotikul, J. and Langsley, G. 1988. Chromosome size polymorphism in *Plasmodium falciparum* can involve deletions of the subtelomeric pPFrep20 sequence. Nucleic Acids Res. 16: 4331-40.

Ponzi, M., Pace, T., Dore, E. and Frontali, C. 1985. Identification of a telomeric DNA sequence in *Plasmodium berghei*. EMBO J. 4: 2991-5.

Rubio, J.P., Thompson, J.K. and Cowman, A.F. 1996. The var genes of *Plasmodium falciparum* are located in the subtelomeric region of most chromosomes. EMBO J. 15: 4069-77.

Scherf, A., Bottius, E. and Hernandez-Rivas, R. 1999. The malaria genome. In: Malaria: Molecular and Clinical Aspects. Wahlgren, M. and Perlmann, P., eds, Harwood Academic Publishers, Amsterdam. p. 153-179.

Scherf, A., Figueiredo, L.M. and Freitas-Junior, L.H. 2001. Plasmodium telomeres: a pathogen's perspective. Curr. Opin. Microbiol. 4: 409-414.

Scherf, A., Hernandez-Rivas, R., Buffet, P., Bottius, E., Benatar, C., Pouvelle, B., Gysin, J. and Lanzer, M. 1998. Antigenic variation in malaria: *in situ* switching, relaxed and mutually exclusive transcription of *var* genes during intra-erythrocytic development in *Plasmodium falciparum*. EMBO J. 17: 5418-5426.

Tchavtchitch, M., Fischer, K., Huestis, R. and Saul, A. 2001. The sequence of a 200 kb portion of a Plasmodium vivax chromosome reveals a high degree of conservation with *Plasmodium falciparum* chromosome 3. Mol. Biochem. Parasitol. 118: 211-22.

Teng, S.C. and Zakian, V.A. 1999. Telomere-telomere recombination is an efficient bypass pathway for telomere maintenance in *Saccharomyces cerevisiae*. Mol. Cell Biol. 19: 8083-93.

Trask, B.J., Friedman, C., Martin-Gallardo, A., Rowen, L., Akinbami, C., Blankenship, J., Collins, C., Giorgi, D., Iadonato, S., Johnson, F., Kuo, W.L., Massa, H., Morrish, T., Naylor, S., Nguyen, O.T., Rouquier, S., Smith, T., Wong, D.J., Youngblom, J. and van den Engh, G. 1998. Members of the olfactory receptor gene family are contained in large blocks of DNA duplicated polymorphically near the ends of human chromosomes. Hum. Mol. Genet. 7: 13-26.

* Tumbar, T. and Belmont, A.S. 2001. Interphase movements of a DNA chromosome region modulated by VP16 transcriptional activator. Nat. Cell Biol. 3: 134-139.

Using a simplified, flexible system to study the cell-cycle-dependent localization of a chromosome site, the authors showed that the binding of a single transcriptional activator can induce a redistribution of a specific chromosome site from a predominantly peripheral to a more interior nuclear localization, in mammalian cells. This movement is associated with transcriptional activation.

* Vázquez-Macías, A., Martínez-Cruz, P., Castañeda-Patlán, M.C., Scheidig, C., Gysin, J., Scherf, A. and Hernández-Rivas, R. 2002. A distinct 5' flanking var gene region regulates *P. falciparum* variant erythrocyte surface antigen expression in placental malaria. Mol. Microbiol. 45: 155-167.

Telomere-associated *var* genes are generally very diverse among clinical isolates. Surprisingly, a *var* gene that is conserved among parasites isolates from different parts of the world has been located to the subtelomere by these authors. This *var* gene, which mediates binding to the placental adhesion receptor chondroitin sulfate A, is transcribed in the opposite direction than those commonly found at chromosome ends. The authors suggest that this location could account for reduced ectopic recombination frequencies with other subtelomeric *var* genes.

Vernick, K.D. and McCutchan, T.F. 1988. Sequence and structure of a *Plasmodium falciparum* telomere. Mol. Biochem. Parasitol. 28: 85-94.

From: Malaria Parasites: Genomes and Molecular Biology
Edited by: A.P. Waters and C.J. Janse

# Chapter 8

## Gene Expression

## Kirk W. Deitsch

### Abstract

The study of gene expression in malaria parasites has blossomed in recent years with the development and use of reporter gene constructs in transfection experiments and the recent release of the annotated sequence of the *P. falciparum* genome. The promoters and regulatory regions of many genes from several *Plasmodium* species have been characterized, however the exact nature of the transcription factors that bind them remain ill-defined. In addition, how expression of the large multigene families found in all *Plasmodium* genomes is regulated presents a particularly interesting problem. Recent findings with regard to the *var* gene family of *P. falciparum* may shed light on the topics of allelic exclusion and transcription switching that are important for understanding how expression of these gene families is controlled.

### 1. Introduction

Malaria parasites, like many other vector-borne pathogens, have evolved very complex life cycles where, during the course of natural transmission, they must invade and multiply within several distinctly different cell types. These include the hepatocytes and circulating erythrocytes of their vertebrate hosts as well as the cells of the gut and salivary glands of the mosquito vector. As they develop, these parasites go through drastic changes in morphology and undergo asexual replication as well as sexual differentiation and recombination. The ability to complete such complex life cycles is dependent upon the parasites' ability to control a tightly regulated pattern of gene expression. The correctly regulated expression of the large complement of genes found in their genomes is determined by intricate interactions between DNA regulatory elements and the transcription factors that bind them as well as the cellular machinery responsible for DNA replication, RNA transcription and chromatin assembly. Regulation of gene expression in *Plasmodium* is still poorly understood and represents a field of study that is only just beginning to yield significant advances.

Studies of transcription and gene expression in *Plasmodium* have been hampered by a lack of many of the tools that have proven so valuable in studying gene expression in other "model" organisms including such things as inducible promoters and temperature sensitive mutants. Indeed, such basic techniques as transfection and genetic modification were only developed relatively recently (Koning-Ward *et al.*, 2000). In *P. falciparum*, while gene mapping through the use of genetic crosses has been extremely valuable in identifying genes responsible for a few phenotypes (Wellems *et al.*, 1991; Vaidya *et al.*, 1995; Su *et al.*, 1997), performing crosses is time consuming and extremely expensive, making the technique impractical for routine genetic analysis. Recently the use of RNAi has proven very valuable for generating gene knockouts in other parasites (Ngo *et al.*, 1998), however, despite a preliminary report of its application in *P. falciparum* (McRobert and McConkey, 2002) the technique has not yet been convincingly demonstrated to work efficiently in any species of *Plasmodium*.

However, the release of the completed sequence of the *P. falciparum* genome is proving to be extremely valuable to all molecular studies of malaria parasites, including studies of gene expression (see Chapters 1 and 2). For example, it became obvious, even from the early gene sequences that became available prior to the genome projects, that regulatory regions of *P. falciparum* genes lack any similarity to known eukaryotic enhancers or repressors (Gardner *et al.*, 1998; Bowman *et al.*, 1999). This observation led to speculation that malaria parasites may have developed their own unique set of transcription factors distinct from those found in yeast or higher eukaryotes (Horrocks *et al.*, 1998). This idea is now supported by the fact that it is has not been possible to identify homologues to many of the known eukaryotic transcription factors using homology searches of the *P. falciparum* genome database. Even the TATA binding protein, one of the most conserved eukaryotic transcription factors, is substantially diverged (McAndrew *et al.*, 1993). Such extensive divergence might render less conserved transcription factors nearly unrecognizable by standard annotation methodologies. Thus, it is not yet clear if malaria parasites have developed their own set of transcription factors, or if they rely on mechanisms of gene regulation that depend on something other than the transcription factor dependent regulation paradigm that has been used to describe gene regulation in other eukaryotes.

This chapter aims not to provide a comprehensive review of all the work on gene expression that has been conducted in the various species of *Plasmodium*. Rather, the intention is to provide a concise introduction that highlights some of the basic findings regarding *Plasmodium* promoters and regulatory regions, a brief mention of some of the methods currently being used to investigate entire genome expression patterns (and dealt with in detail in other chapters in this book), followed by a more complete discussion of the complex topics of allelic exclusion and chromatin organization as they apply to *var* gene regulation. In this way it is hoped that the reader will gain an overall understanding of the state of what is known about gene expression in malaria parasites without being inundated with details.

## 2. Studies on *Plasmodium* Promoters and Regulatory Regions

Using simple sequence comparisons, transfection and electromobility shift assays, a number of promoters and upstream regulatory regions have been characterized from several *Plasmodium* species leading to some general conclusions about the nature of these promoters. Simple sequence comparisons of a large number of promoters have shown that they lack any definable similarity with one another, with the exception that most contain an over-representation of homopolymeric tracts (dA:dT) (Horrocks *et al.*, 1998; Porter, 2001). These tracts may be significant in that such sequences have been shown to adopt a rigid structure that can contribute to the overall topology of the DNA strand (Nelson *et al.*, 1987). It is conceivable that these elements could thus contribute to transcription factor binding and transcriptional regulation as has been demonstrated in other organisms (Struhl, 1985; Winter and Varshavsky, 1989; Hori and Firtel, 1994).

Studies using reporter gene constructs and transfection have yielded perhaps the greatest amount of information with regard to *Plasmodium* promoters. Firstly, it was established that assays done using transfected episomal constructs accurately reflected the activity of promoters in their endogenous location on the chromosome, both in terms of transcriptional activity and even at the level of chromatin structure (Koning-Ward *et al.*, 1999; Dechering *et al.*, 1999; Horrocks *et al.*, 2002). These studies showed that cloned promoters are correctly regulated both in terms of their stage specific expression (i.e. blood versus insect stages) (Dechering *et al.*, 1999) as well as their expression at specific points within the cell cycle (Koning-Ward *et al.*, 1999). Cell cycle specificity of promoter activity has important implications as demonstrated by the fact that expression of genes at an inappropriate point in the cell cycle can lead to accumulation of the encoded protein in an incorrect subcellular localization (Kocken *et al.*, 1998). While the sequences of *Plasmodium* promoters don't resemble those described in other eukaryotes, standard deletion analysis of upstream regulatory regions indicate that they seem to conform to the typical bipartite structure consisting of a basal promoter regulated by upstream enhancer and silencer elements (Crabb and Cowman, 1996; Horrocks and Kilbey, 1996; Mbacham *et al.*, 2001). In addition, promoters from one species of *Plasmodium* appear to be correctly regulated when transfected into other species indicating that there is some level of conservation of transcriptional control throughout the genus (van der Wel *et al.*, 1997).

Electromobility shift assays performed with *Plasmodium* nuclear extracts and portions of promoters have identified several specific interactions between DNA elements and trans-acting proteins or protein complexes (Dechering *et al.*, 1999; Horrocks and Lanzer, 1999). While the sequences of a couple of cis-acting DNA elements have been identified, the exact nature of the proteins that bind to them remain ill-defined. Nonetheless, these experiments support the conclusion that *Plasmodium* genes consist of a bipartite structure and that the upstream enhancers/repressor elements are bound by regulatory complexes in a sequence dependent fashion.

Some initial work has also been done to identify possible roles of 5' and 3' UTRs of mRNAs in gene expression. In the *P. gallinaceum Pgs28* transcript it was found that alteration of the 3' UTR sequence greatly reduces expression levels of the protein (Golightly *et al.*, 2000). Both a U-rich element and one of five consensus eukaryotic polyadenylation signals were required for efficient protein expression. The B7 gene of *P. berghei*, on the other hand, has been shown to utilize alternative transcriptional start sites, generating transcripts with different 5' UTRs in sexual versus asexual stage parasites (Pace *et al.*, 1998). While the data is not extensive, these studies imply that the flanking regions of mRNA molecules can be important in determining either mRNA stability or translational efficiency and as a result contribute to levels of gene expression.

## 2.1. "Global" Analysis of Gene Expression

While the vast majority of studies on gene expression in malaria parasites have been directed toward the analysis of expression of individual genes, new techniques are being developed that allow the analysis of the expression patterns of very large families of genes or even complete genomes. The sequencing of the entire *P. falciparum* genome has greatly facilitated the development of these techniques. Projects directed at investigating expression of large groups of genes range in technical sophistication from the simple random cloning and sequencing of large numbers of messages to yield libraries of expressed tags (often referred to as a transcriptome) (Chakrabarti *et al.*, 1994; Kappe *et al.*, 2001; Carlton *et al.*, 2001), to the more technically advanced projects like serial analysis of gene expression (SAGE) (Patankar *et al.*, 2001; Munasinghe *et al.*, 2001), microarrays and proteomics. These latter two techniques are described in detail in elsewhere and won't be dealt with further here. However, one cannot overstate the impact that the *P. falciparum* genome project and these "global" expression projects are having on the study of gene expression in malaria parasites.

An interesting observation that has come from a recent SAGE analysis of asexual stage *P. falciparum* parasites was that 17% of highly abundant tags corresponded to antisense transcripts of annotated genes (Patankar *et al.*, 2001). This obviously has led to speculation that these transcripts may play a role in regulating gene expression. Non-translated RNAs have been previously observed in malaria parasites as transcripts from apparent pseudogenes (Triglia *et al.*, 2001; Taylor *et al.*, 2001). Two additional examples are from the multicopy *var* gene family that plays a role in the antigenic variation of asexual *P. falciparum* parasites. The first refers to "sterile RNAs" that encode the second exon of *var* genes but lack a start methionine and yield no detectable protein (Su *et al.*, 1995). In the second example, transcripts are detectable in early ring stage parasites from the entire *var* gene family, while mRNA from only a single gene ultimately gets translated (Chen *et al.*, 1998; Scherf *et al.*, 1998). Both of these observations will be discussed in greater detail below.

A great deal of work remains to be done to gain a more complete understanding of *Plasmodium* promoters and regulatory regions as well as the protein complexes that bind to them. However, while individual transcription factors have been difficult to identify, the basic concept of a bipartite promoter bound by activators or repressors seems to apply to most *Plasmodium* genes. Two other problems of gene regulation remain largely a mystery. The first regards the changes in gene expression that lead to development of the alternative forms of the parasites (i.e. the various forms that infect the vertebrate liver, red blood cells, and the tissues within the insect vector). The details of how these developmental programs are initiated and controlled are not understood, however it is known that alternative ribosomal RNA expression plays an important role in this process, as is discussed in Chapter 9. The second concerns the regulation of the large, multicopy gene families that have been discovered in all *Plasmodium* species investigated. Often individual copies of the family are expressed one at a time in a mutually exclusive fashion, a process called allelic exclusion. Transcriptional regulation of these multicopy families and the process of allelic exclusion are perhaps most clearly exemplified by the expansive gene families whose expression patterns result in the antigenic variation of parasite populations during vertebrate infection.

**Table 1.** Multicopy gene families identified in several species of *Plasmodium*

|  | **Function of protein** | **Copy number** | **Reference number** |
|---|---|---|---|
| ***P. falciparum*** | | | |
| *var* | Antigenic variation, cytoadherence and sequestration | ~60 | 32, 38, 39 |
| *rif/stevor* | Antigenic variation, rosetting? | ~200 | 44, 85-87 |
| *Pf60* | Unknown, one member localized to nucleus | 10-80 | 51, 52 |
| ***P. vivax*** | | | |
| *vir* | Antigenic variation | 600-1000 | 36 |
| ***P. knowlesi*** | | | |
| *SICAvar* | Antigenic variation | Unknown | 37 |
| ***P. yoellii*** | | | |
| *Py235* | Unknown, localized to the merozoite surface | ~35 | 88 |

## 2.2. Antigenic Variation in Malaria- Escaping the Host's Assault

Most if not all species of malaria appear to undergo a process of antigenic variation during asexual replication whereby they avoid the antibody response of their vertebrate hosts by altering the antigenic phenotype of the infected red blood cells (Kyes *et al.*, 2001). The gene family encoding the major antigenic determinant in *P. falciparum* has been identified and extensively characterized (Baruch *et al.*, 1995; Su *et al.*, 1995; Smith *et al.*, 1995), and multicopy gene families presumed to be responsible for antigenic variation have also be identified in *P. vivax* (del Portillo *et al.*, 2001) and *P. knowlesi* (al Khedery *et al.*, 1999). These gene families range in size from the ~40 to 60 genes seen in *P. falciparum* to an estimated 600-1000 copies in *P. vivax* (Table 1). The genes are expressed singularly and over the course of an infection small sub-populations of parasites arise that have switched to expressing a different copy, thus avoiding the antibody response of the host to the previously expressed copy. This process is thought to be responsible for the persistent nature of malaria infections as well as the waves of parasitemia frequently observed in *P. falciparum*. The changes in antigenic phenotype in *P. falciparum* are accompanied by changes in virulence as a result of the cytoadherent properties determined by the expressed copy of the gene family (see Chapter 12). Thus, at least in the case of this human parasite, antigenic variation is also tightly linked to virulence and pathogenicity.

Antigenic variation in malaria parasites poses a particularly vexing yet fascinating problem from the perspective of understanding the regulation of gene expression. In each case the parasite possesses a large family of equivalent genes whose expression must be tightly regulated. Expressing more than one copy at a time would result in premature expenditure of the antigenic repertoire and therefore strict allelic exclusion must be imposed. In addition, the rate of expression switching must be maintained sufficiently high to stay ahead of the corresponding antibody response of the host, yet not so fast as to have expressed the entire family sooner than absolutely necessary. Elucidation of the mechanisms responsible for this phenomenon will undoubtedly prove to be a major milestone in the understanding of the biological relationship between the parasites and their hosts, but will also shed light onto the ways in which these organisms maintain, organize and control the expression of their genomes.

## 2.2.1. var Genes- A Problem of Transcriptional Control

As mentioned above, the most extensively studied family of genes responsible for antigenic variation in malaria parasites are the *var* genes of *P. falciparum*. The *var* gene family consists of approximately 40-60 genes found either as single copies in subtelomeric regions or as clusters of tandemly repeated genes in the central regions of chromosomes. Different parasite isolates typically have completely different *var* complements and frequent duplications, deletions and recombinations between genes provide parasites with a virtually limitless repertoire of antigenic determinants (Kyes *et al.*, 1997; Taylor *et al.*, 2000a). The genes have been shown to be expressed from either subtelomeric locations or from sites within the internal regions of the chromosomes without translocations or gene duplications as has been observed for the genes involved in antigenic variation in African trypanosomes (Su *et al.*, 1995; Hernandez-Rivas *et al.*, 1997; Fischer *et al.*, 1997). The active copy is expressed in the ring stage of the asexual cell cycle (3-18 hours post-invasion), with transcripts no longer detectable in late trophozoites or schizonts (Kyes *et al.*, 2000).

Most models of antigenic variation predicted that *var* genes would be expressed one at a time and that switches in expression would involve the coordinated silencing of the previously active gene in conjunction with the activation of a newly expressed copy. However, while widely believed, direct evidence that *var* genes were singularly transcribed was difficult to prove experimentally because the switching rate was sufficiently high (estimated to be up to 2% per generation (Roberts *et al.*, 1992)) that once a population of parasites was large enough to study, multiple expressed *var* genes were detectable. This problem was first solved through the use of single cell RT-PCR on individual infected red blood cells (Chen *et al.*, 1998). These studies did indeed verify that only a single *var* gene is expressed in any given parasite, however it was also observed that in very early ring stage parasites (1-3 hours after red cell invasion) many, if not all *var* genes are detectably transcribed. These early transcripts are not translated and appear

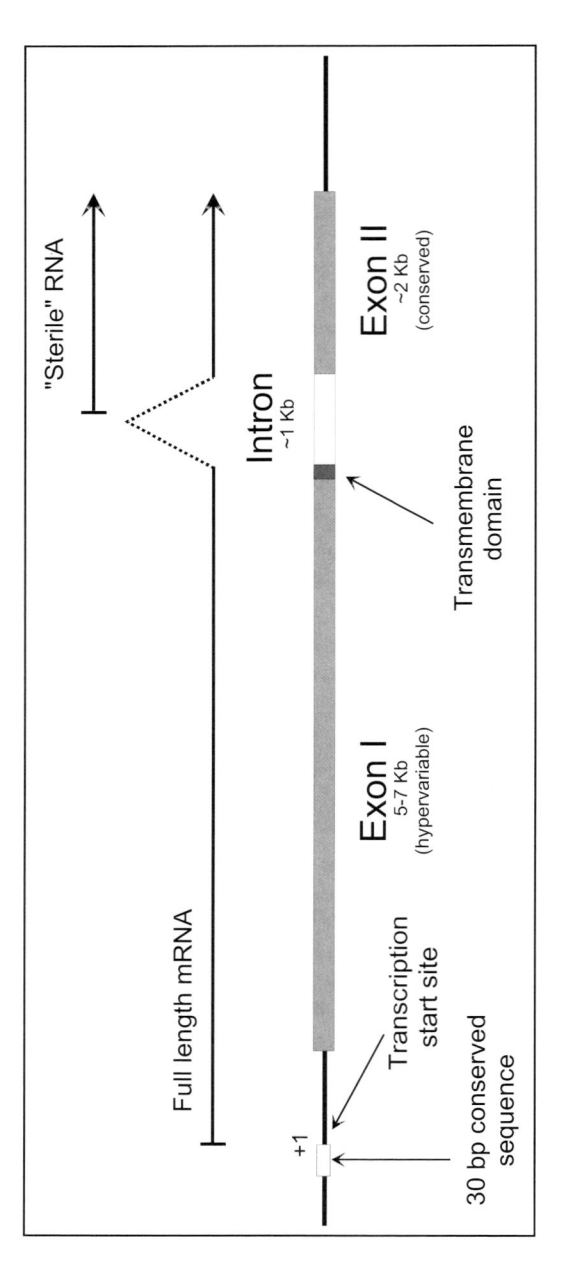

**Figure 1.** Schematic diagram showing the structure of *var* genes. The shaded boxes identify the protein coding regions of the two exons, including the transmembrane domain (darker shade). The positions of the transcription start site and 30 bp degenerate sequences are identified. The intron contains typical eukaryotic splice donor and acceptor sites.

to be truncated (Taylor *et al.*, 2000b). As the parasite grows through the ring stage of development, transcription becomes confined to a single copy and it is these transcripts that are translated and determine the antigenic phenotype of the infected red blood cell. Both the exclusive expression of a single *var* gene at a time and the presence of non-translated transcripts from the entire *var* complement during early ring stages were verified using nuclear run-on analysis from parasites selected for a particular cytoadherent phenotype (Scherf *et al.*, 1998). The significance of the early, untranslated transcripts remains unclear, however, as experiments have also detected early ring stage, non-translated transcripts originating from genes that were specific to other life stages of the parasite, indicating that the phenomenon may not be unique to the *var* gene family (Chen *et al.*, 1998).

### 2.2.2. *The Structure of var Genes*

All *var* genes thus far identified, either through direct experiments or through the *P. falciparum* genome project, have an identical structure (Figure 1) (Su *et al.*, 1995; Gardner *et al.*, 1998; Bowman *et al.*, 1999). The genes consist of two exons, the first encoding the large, polymorphic extracellular portion of the protein while the smaller second exon encodes the more highly conserved part that is thought to anchor the protein into the membrane of the infected cell. The two exons are separated by a relatively conserved intron that ranges in size from 0.8 to 1.2 kb. The single transmembrane domain of the protein is located just upstream of the intron/exon boundary at the very 3' end of exon I (Su *et al.*, 1995). As mentioned above, the genes are located either as single copies in the subtelomeric regions of most chromosomes or as tandemly repeated genes found in clusters on chromosomes 4, 7, 8 and 12 (Rubio *et al.*, 1996).

While the overall structure of *var* genes is identical regardless of their chromosomal location, the sequences of their promoter and upstream regions differ depending on whether the genes are located subtelomerically or within the internal chromosomal clusters (Voss *et al.*, 2000). Sequence identities can be as high as 80% between promoters from genes found in the same portion of the chromosome and a degenerate 30 bp sequence is shared between genes from both classes. The significance of the two different promoter types or the 30 bp conserved sequence remains unknown. The start site of transcription was identified for the promoter of an internal *var* gene (Deitsch *et al.*, 1999), and was found to be nearly 1 kb from the start methionine indicating that the mRNA includes a relatively long 5' UTR, again of unknown significance.

## 2.2.3. var Gene Transcripts

Unlike the variant antigen genes in African trypanosomes, *var* genes appear to be transcribed monocistronically and are presumably capped, spliced and polyadenylated like typical RNA pol II messages. The introns also contain consensus splice sites in the appropriate positions with regard to the splice junction (Su *et al.*, 1995). Full-length *var* transcripts are detected as early as 3 hours after red cell invasion and remain detectable through about 18 hours, after which transcript levels rapidly fall (Kyes *et al.*, 2000). The presence of the full-length transcripts precedes the appearance of the protein on the infected cell surface by many hours, with protein first being detected around 16 hours after invasion (Gardner *et al.*, 1996). This delay is thought to reflect a rather lengthy process that includes transporting the protein across the parasite plasma membrane and the parasitophorous vacuolar membrane followed by anchoring it into the red cell plasma membrane. The mechanisms involved in the targeting and transport of the protein are not understood.

In addition to the mature, full-length transcript, an additional family of RNA molecules are detectable that seem to be derived from the *var* gene family. Northern blots probed with DNA sequences from exon II identified not only the full-length 7-10 kb message, but also detected a highly abundant group of RNAs of 1.8-2.4 kb (Su *et al.*, 1995). The sequences of several of these molecules were determined from a cDNA library and found to contain sequences starting from within the *var* intron and extending through exon II. These cDNAs lacked consensus start codons preceded by AT-rich noncoding DNA as is typical for translation start sites in *P. falciparum* and thus were presumed not to encode protein. They have therefore been referred to as *var* "sterile" transcripts. Complicating this analysis was the discovery of the Pf60 multigene family (Carcy *et al.*, 1994; Bischoff *et al.*, 2000). This gene family consists of at least 10 and potentially up to 90 copies whose 3' ends share significant identity (73% in one case) with *var* exon II. This extremely close identity has led to the suggestion that all of these genes be included in a single gene family called Pf60/*var*, although this name has yet to gain broad acceptance. Because of their close identity and similar size, it has been difficult to distinguish between *var* sterile transcripts and Pf60 transcripts by Northern analysis, thus the expression profile for the sterile transcripts remains poorly defined. It has been speculated that the sterile transcripts could play a role in *var* gene recombination or expression switching, however no definitive role for these RNA molecules has yet been determined.

An interesting aspect of at least one member of the Pf60 family is the finding of read through of an internal stop codon (Bischoff *et al.*, 2000). The exact function of the internal stop codon is unknown, however reporter gene assays have demonstrated that translational efficiency was ~10% of that of constructs with no internal stop codon. This implies that the presence of such codons may play a role in regulating protein production. The discovery of stop codon read through may be very important for annotating the *P. falciparum* genome sequence as genes initially annotated as pseudogenes due to the presence of one or two internal stops may subsequently be found to be fully functional.

### 2.2.4. var Gene Silencing

As mentioned earlier, studies have shown that *var* genes can be expressed regardless of whether they are located in the subtelomeric portions of the chromosomes or within internal chromosomal clusters. In addition, examination of an individual *var* gene before and after a switch in gene expression (i.e. in its transcriptionally active and silent state) identified no alterations at the DNA sequence level (Deitsch *et al.*, 1999). This implied that while all *var* genes are probably equally capable of being expressed, some ill-defined mechanism of allelic exclusion was acting to limit expression to a single copy. To pursue this topic further, the promoter and 5′ regulatory region from a *var* gene were cloned upstream of a luciferase reporter gene within a plasmid construct. Transfection of this plasmid into malaria parasites gave appropriate levels of reporter gene activity at the correct stage of the cell cycle when *var* genes are transcribed (Deitsch *et al.*, 1999). However, in parasites that carried this construct, the *var* promoter on the episomal construct appeared to always be transcriptionally active and never silent. This indicated that the *var* upstream regulatory region included in the construct was capable of mediating *var* gene transcription, however, the ability to silence the promoter resulting in mutually exclusive transcription of only a single *var* promoter at a time was lost. In other words, it appeared that silent *var* promoters become transcriptionally active when removed from their chromosomal context and placed on transfected episomes.

Several conclusions can be drawn from these experiments. First, switches in *var* gene expression are likely not dependent on changes in the DNA sequence of the genes or their regulatory elements. Second, removal of a silent *var* promoter from the chromosome and placement on an episome renders it constitutively active, demonstrating that even though a particular gene may be silent, the transcription factors necessary for expression are present within the cell. Thus changes in expression are also not likely the result of changes in the presence or absence of individual transcription factors. Third, proper regulation of *var* promoter activity is dependent on chromosomal context. This implies that some type of silencing mechanism is responsible for maintaining all but a single *var* promoter in a transcriptionally silent state. In addition, episomal constructs that contained isolated *var* promoters were missing an important element necessary for assembly of this silent state. Regulation of transcription that is not based on either changes in DNA sequence or on the presence or absence of transcription factors is referred to as "epigenetic", a term now frequently applied to the control of *var* gene transcription.

# 3. "Epigenetic" Gene Regulation

The molecular basis of epigenetic transcriptional regulation has been studied extensively in many model systems including yeast, *Drosophila*, mammalian cells and several species of plants. In most instances it involves the silencing of individual genes through the condensation of the surrounding chromatin fiber (Moazed, 2001). Several examples exist in the literature of chromatin mediated gene silencing. Frequently, silenced promoters are flanked by binding sites for chromatin modifying protein complexes and many of these binding sites resemble chromosomal origins of replication (Abraham *et al.*, 1984; Feldman *et al.*, 1984) implicating a relationship between DNA replication and chromatin assembly (Abraham *et al.*, 1984). In addition, inclusion of these binding sites on episomal constructs allows proper chromatin formation and gene silencing in a non-chromosomal context. In the yeast *Saccharomyces cerevisiae*, assembly of a silent chromatin structure on episomes has been shown to be an accurate reflection of what happens at silent loci on the chromosome (Shei and Broach, 1995), providing a convenient system for studying chromatin formation and its effects on transcription.

The use of this system and others like it has allowed many basic characteristics of this mechanism of gene silencing to be determined. The DNA of eukaryotes is wrapped around nucleosomes to form a compact, highly organized fiber. In silent chromatin, the histones that make up the nucleosomes are frequently modified through the addition or subtraction of acetyl, methyl, and/or phosphoryl groups (Grant, 2001). These modifications then play a role in the assembly of large protein complexes along the length of the DNA strand that organize the chromatin fiber into a condensed structure that is largely inaccessible to most transcription factors, thereby effectively silencing the gene. Many of the proteins involved in this process have been identified in yeast and include the Swi/Snf and Sir families of proteins as well as proteins that contain Chromo, Bromo and SET domains (Jenuwein *et al.*, 1998; Jones *et al.*, 2000; Marmorstein and Berger, 2001). The physical condensation of the DNA can be detected by its inaccessibility to restriction endonucleases or DNAses in experimental assays (Loo and Rine, 1994; Donze *et al.*, 1999). This condensed DNA is then moved away from the center of the nucleus and localized near the nuclear membrane in an area referred to as the perinuclear space (Cockell and Gasser, 1999; Gasser, 2001). When plasmid DNA is used to follow this process, transition of the transfected cell through S-phase of the cell cycle is required before either transcriptional silencing or chromatin condensation on the transfected DNA can occur (Miller and Nasmyth, 1984). While this might lead one to believe that DNA replication is therefore required for silencing, it has recently been shown that while S-phase transition is required, actual DNA replication is not (Li *et al.*, 2001; Kirchmaier and Rine, 2001). The exact role that S-phase transition plays is not yet understood.

## 3.1. Assembly of a Silent State on Transfected *Var* Promoters

As described above, transfected constructs containing a *var* promoter were constitutively expressed and failed to replicate the silencing phenomenon seen of chromosomal *var* promoters. They were thus not useful for studying epigenetic gene regulation and the properties of mutually exclusive *var* gene expression and the mechanisms behind expression switching. If the mechanisms responsible for gene silencing in yeast applied to *Plasmodium*, then it should have been possible to assemble a silent chromatin structure on *var* promoter containing episomes if all the necessary DNA components were present on the plasmid. The original *var* promoter constructs contained approximately 2.5 kb of DNA from upstream of the start methionine (Deitsch *et al.*, 1999). This included about 1.5 kb upstream of the transcriptional start site and an additional 1 kb between +1 of the transcript and the start site of translation. It was conceivable that simply additional sequence from further upstream of the gene was required, however it was also noted that silencing elements in yeast frequently are located in such a way that they flank the promoters that they regulate. They act as cooperative elements, incapable of silencing transcription when not paired with one another (Boscheron *et al.*, 1996). When properly paired, a silent chromatin structure is assembled between them, extending across the promoter and silencing transcription. These silencers also contain boundary elements that constrain the spread of the silent chromatin to the area surrounding the targeted promoters (Donze *et al.*, 1999). This leads to a partitioning of the chromosome, and ultimately the genome, into regions of transcriptionally active chromatin and chromatin containing silenced genes. In higher eukaryotes, this mechanism is thought to play a prominent role in the "programming" of the genome that results in cell differentiation during development (Mechali, 2001). With this in mind, the possibility of a regulatory element responsible for transcriptional silencing existing downstream of *var* promoters was considered.

Examination of the sequences of *var* genes in the regions downstream of the promoters rapidly leads one to consider the single intron found in all *var* genes. Comparison of the sequences of introns from numerous *var* genes quickly determined that they are highly conserved, both in terms of sequence identity and structure. Their length varies between 800 and 1200 bp, and they can be divided into three regions. Regions 1 and 3 are each about 150 bp long. Region 1 includes a number of repeats of the sequence $5'$-TGT$\binom{G}{A}$T$\binom{G}{A}$TG-$3'$, while region 3 contains several copies of the complementary repeat $5'$-ACA$\binom{C}{T}$A$\binom{C}{T}$AC-$3'$. Region 2 does not contain these repeats, but rather consists primarily of long stretches of As. In fact, adenine residues make up greater than >70% of the region 2 sequence. The presence of these repeats and the composition of the DNA surrounding them results in stretches of DNA with prominent asymmetry in the G *vs.* C and A *vs.* T contents within regions 1 and 3. The features of strand asymmetry and frequent tracts of alternating purines and pyrimidines are thought to form atypical DNA structures that may play a role in chromatin assembly (Wells *et al.*, 1988; van Holde and Zlatanova, 1994). However, the role that these sequence features play in the case of *var* introns remains unexplored. The conserved nature of these introns, the interesting structure that they display and the presence of the previously mentioned "sterile" transcripts originating from within the introns all suggested that perhaps these DNA elements play a role in *var* promoter regulation.

To investigate this possibility, a representative *var* intron was inserted downstream of a *var* promoter/luciferase transcription unit in a plasmid construct. The intron was placed downstream of the stop codon and polyadenylation signal of the transcript and thus was no longer functioning as an intron but rather simply as a DNA element in a position flanking the promoter with respect to the 5′ regulatory region of the *var* gene. It was hoped that this position would approximate the position that the intron occupies within a chromosomal *var* gene. Transfection of these plasmids into cultured *P. falciparum* parasites resulted in virtually complete *var* promoter silencing (Deitsch *et al.*, 2001). The silencing effect of the intron was specific to transcription initiated from a *var* promoter as the intron had no effect on transcription from a control *P. falciparum* promoter. This led to the conclusion that this was a cooperative effect requiring both the intron and an additional element that exists somewhere in the 5′ regulatory region of the *var* gene. In addition, this silencing required transition of the transfected parasites through S-phase of the cell cycle in much the same way silencing in yeast is S-phase dependent. Expression assays performed after the parasites had taken up the plasmid DNA but prior to the onset of S-phase resulted in no detectable promoter silencing. The similarities between the silencing observed here and that previously described in yeast is compelling and provides encouragement that many of the concepts that have been developed to describe yeast chromatin assembly and gene regulation may also apply to *Plasmodium*. The ability to assemble a silent transcriptional state on plasmid constructs outside of the normal chromosomal context, whether that state be based on chromatin structure or some other mechanism, should also provide a useful tool to further explore the mechanisms of exclusive *var* transcription in *P. falciparum*.

## 3.2. How Could This All Work?

While the discovery that the *var* intron acts as one component of a cooperative silencing mechanism provides some insight into how most of the *var* gene family is kept transcriptionally silent, it provides little help in attempting to understand how an individual *var* gene is activated or how the parasites are able to switch expression between members of the family while maintaining mutually exclusive transcription. It may be fruitful to look at what has been accomplished in other organisms for concepts that might apply to regulation of the *var* gene family. Chromatin mediated gene regulation and epigenetics is one of the most rapidly developing topics within the broad field of gene expression. A great deal of data is accumulating that may be relevant to the design of informative experiments with malaria parasites. Several specific examples include the concept of cellular memory modules, other models of allelic exclusion, and several defined "epigenetic marks" including DNA modification and the so called "histone code".

Cellular memory module is a term that refers to chromosomal elements that confer epigenetic inheritance. These are defined as sequences that play a role in stable differential expression of identical genetic information within a cell population (Lyko and Paro, 1999). Typically these elements have been identified in multi-cellular organisms where they play a role in establishing patterns of gene expression that govern cellular differentiation and development (Francis and Kingston, 2001). Essentially, these modules act as switch elements that can be acted on by either positive or negative transcriptional regulators (Cavalli and Paro, 1998). The action of these regulators effectively "set" the switch to either the on or off state and this state is inherited for many cell generations. The molecular mechanisms that control cellular memory appear to depend on chromatin mediated gene silencing or activation and are likely to be conserved in most higher eukaryotic organisms (Francis and Kingston, 2001). The concept of establishing a gene in its on or off state, then maintaining this state for multiple generations could easily be applied to *var* gene expression in *P. falciparum*. Perhaps there exists an element within the upstream regulatory region or intron of *var* genes that acts as a memory module to maintain expression of a particular *var* gene for several generations before a switch is made.

Other examples of allelic exclusion that have been described in mammalian cells including paternal/maternal imprinting (Arney *et al.*, 2001), X inactivation in females (Boumil and Lee, 2001) and the mutually exclusive expression of antigen-receptor genes of the immune system (Nemazee, 2000) and odorant receptor genes within the cells of the olfactory system (Serizawa *et al.*, 2000). In all these cases the structure of chromatin surrounding the genes has been shown to be important, although the exact mechanisms remain largely a mystery. In many cases, once the transcriptional state of the gene in question has been set, an "epigenetic mark" can be detected (Lyko and Paro, 1999). This can be a covalent modification to the DNA itself, the most common of which is DNA methylation (Meehan *et al.*, 2001; Nakao, 2001), or it can involve modifications to the histone proteins that make up the nucleosomes on which the DNA is wound. The histone modifications that have been demonstrated to affect the transcriptional state of the associated DNA include acetylation, phosphorylation and methylation (Grant, 2001). These modifications can make up a pattern that has been referred to as a "histone code" (Jenuwein and Allis, 2001), which then is thought to help determine the structure of the chromatin that assembles along the stretch of DNA and affects the transcriptional status of the gene. While malaria parasites are not thought to methylate their DNA extensively, some methylation has been reported (Pollack *et al.*, 1991). Histone acetylation has been shown to occur (Darkin-Rattray *et al.*, 1996) and other nucleosomal or DNA modifications are likely to be found. The possible role of such modifications in *var* gene regulation awaits investigation.

In addition to *var*, at least two other large, multicopy gene families in the *P. falciparum* genome have been extensively described. These include the *rifin/stevor* family (Cheng *et al.*, 1998; Kyes *et al.*, 1999; Fernandez *et al.*, 1999) and the *Pf60* genes (Carcy *et al.*, 1994; Bischoff *et al.*, 2000). Besides the gene families responsible for antigenic variation in *P. vivax* and *P. knowlesi* mentioned earlier, a gene family encoding a protein called Py235 has been identified in *P. yoelii*. This family consists of at least 35 members with each merozoite that is released from a mature schizont expressing a different copy (Preiser *et al.*, 1999). Undoubtedly, as more sequence data becomes available from more *Plasmodium* species, other multicopy families of genes will be identified and investigated. All of these gene families have interesting and complex expression patterns and presumably play important roles in the survival of the parasites. Determining the molecular mechanisms responsible for controlling these expression patterns poses a difficult and fascinating problem that extends well beyond the control of *var* genes.

# 4. Conclusions and Future Directions

The study of gene expression in malaria parasites remains in a relatively young state and is now beginning to take advantage of some of the recent advancements in *Plasmodium* molecular biology. The development of transfection several years ago allowed the use of the reporter gene constructs that led to many of the recent advancements in the field as well as the ability to genetically modify the parasite. Hopefully, as more discoveries are made from the data coming from the genome project similar advancements will result. Examination of the sequences already available has allowed the identification of probable histone acetyltransferases and deacetylases, several Chromo (Figure 2) and Bromo containing proteins as well as SET methylases and members of the Swi/Snf ATPase family. In addition, the application of techniques like SAGE and microarrays should also begin to contribute to the understanding of patterns of gene expression and development of these parasites. The challenge now appears to be to efficiently utilize the large amounts of data that are rapidly becoming available.

Understanding *var* gene regulation and the mechanisms of allelic exclusion and expression switching continue to pose a rather imposing challenge. While much has been learned about similar topics in other model organisms, a true understanding of this process has not yet been achieved.

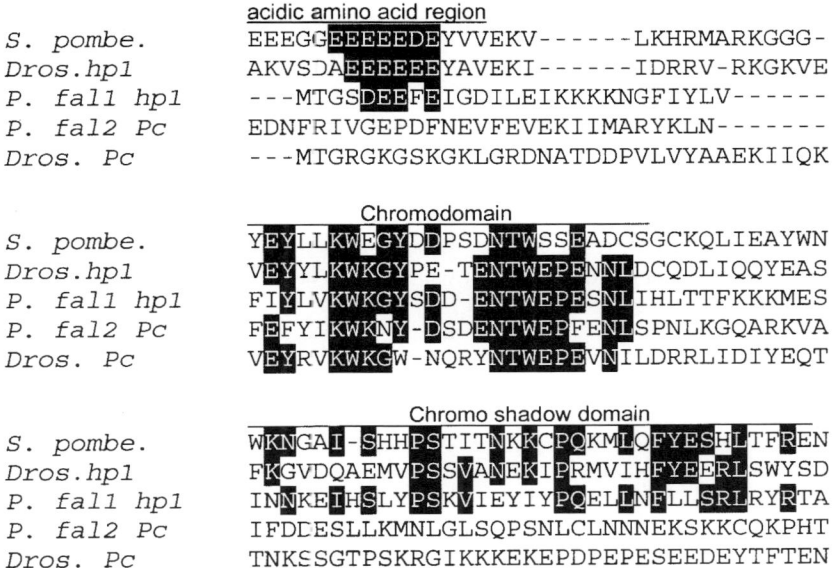

**Figure 2.** Alignment showing the identification of 2 Chromodomain containing genes from the *P. falciparum* genome. Such genes are often classified as either hp1-like or polycomb-like based on the presence or absence of the acidic amino acid region and the chromo shadow domain. As can be seen in the figure, representatives of both classes were found within the *P. falciparum* genome. *S. pombe*: swi6 gene from *Schizosaccharomyces pombe*; *Dros. hp1*: heterchromatin protein 1 (hp1) from *Drosophila melanogaster*; *Dros. Pc*: polycomb gene from *Drosophila melanogaster*; *P. fal1 hp1*: hp1 like gene from *P. falciparum*; *P. fal2 Pc*: polycomb like gene from *P. falciparum*.

While this may make the study of *var* gene regulation seem daunting, it also provides an opportunity for experiments with malaria parasites to make a substantial contribution to the broader field of epigenetic control of gene expression. Gaining a greater understanding of this process is also likely to be of enormous value to the development of alternative and innovative strategies for malaria treatment and control.

# 5. Acknowledgments

KWD is a Stavros S. Niarchos Scholar. The Department of Microbiology & Immunology at Weill Medical College of Cornell University acknowledges the support of the William Randolph Hearst Foundation.

# 6. References

Abraham, J., Nasmyth, K.A., Strathern, J.N., Klar, A.J., and Hicks, J.B. 1984. Regulation of mating-type information in yeast. Negative control requiring sequences both 5' and 3' to the regulated region. J. Mol. Biol. 176: 307-331.

al Khedery, B., Barnwell, J.W., and Galinski, M.R. 1999. Antigenic variation in malaria: a 3' genomic alteration associated with the expression of a *P. knowlesi* variant antigen. Mol. Cell. 3: 131-141.

Arney, K.L., Erhardt, S., Drewell, R.A., and Surani, M.A. 2001. Epigenetic reprogramming of the genome--from the germ line to the embryo and back again. Int. J. Dev. Biol. 45: 533-540.

Baruch, D.I., Pasloske, B.L., Singh, H.B., Bi, X., Ma, X.C., Feldman, M., Taraschi, T.F., and Howard, R.J. 1995. Cloning the *P. falciparum* gene encoding PfEMP1, a malarial variant antigen and adherence receptor on the surface of parasitized human erythrocytes. Cell. 82: 77-87.

Bischoff, E., Guillotte, M., Mercereau-Puijalon, O., and Bonnefoy, S. 2000. A member of the *Plasmodium falciparum* Pf60 multigene family codes for a nuclear protein expressed by readthrough of an internal stop codon. Mol. Microbiol. 35: 1005-1016.

Boscheron, C., Maillet, L., Marcand, S., Tsai-Pflugfelder, M., Gasser, S.M., and Gilson, E. 1996. Cooperation at a distance between silencers and proto-silencers at the yeast *HML* locus. EMBO J. 15: 2184-2195.

Boumil, R.M. and Lee, J.T. 2001. Forty years of decoding the silence in X-chromosome inactivation. Hum. Mol. Genet. 10: 2225-2232.

Bowman, S., Lawson, D., Basham, D., Brown, D., Chillingworth, T., Churcher, C.M., Craig, A., Davies, R.M., Devlin, K., Feltwell, T., Gentles, S., Gwilliam, R., Hamlin, N., Harris, D., Holroyd, S., Hornsby, T., Horrocks, P., Jagels, K., Jassal, B., Kyes, S., McLean, J., Moule, S., Mungall, K., Murphy, L., and Barrell, B.G. 1999. The complete nucleotide sequence of chromosome 3 of *Plasmodium falciparum*. Nature. 400: 532-538.

Carcy, B., Bonnefoy, S., Guillotte, M., Le Scanf, C., Grellier, P., Schrevel, J., Fandeur, T., and Mercereau-Puijalon, O. 1994. A large multigene family expressed during the erythrocytic schizogony of *Plasmodium falciparum*. Mol. Biochem. Parasitol. 68: 221-233.

Carlton, J.M., Muller, R., Yowell, C.A., Fluegge, M.R., Sturrock, K.A., Pritt, J.R., Vargas-Serrato, E., Galinski, M.R., Barnwell, J.W., Mulder, N., Kanapin, A., Cawley, S.E., Hide, W.A., and Dame, J.B. 2001. Profiling the malaria genome: a gene survey of three species of malaria parasite with comparison to other apicomplexan species. Mol. Biochem. Parasitol. 118: 201-210.

Cavalli, G. and Paro, R. 1998. The Drosophila Fab-7 chromosomal element conveys epigenetic inheritance during mitosis and meiosis. Cell. 93: 505-518.

Chakrabarti, D., Reddy, G.R., Dame, J.B., Almira, E.C., Laipis, P.J., Ferl, R.J., Yang, T.P., Rowe, T.C., and Schuster, S.M. 1994. Analysis of expressed sequence tags from *Plasmodium falciparum*. Mol. Biochem. Parasitol. 66: 97-104.

Chen, Q., Fernandez, V., Sundstrom, A., Schlichtherle, M., Datta, S., Hagblom, P., and Wahlgren, M. 1998. Developmental selection of *var* gene expression in *Plasmodium falciparum*. Nature. 394: 392-395.

Cheng, Q., Cloonan, N., Fischer, K., Thompson, J., Waine, G., Lanzer, M., and Saul, A. 1998. stevor and rif are *Plasmodium falciparum* multicopy gene families which potentially encode variant antigens. Mol. Biochem. Parasitol. 97: 161-176.

Cockell, M. and Gasser, S.M. 1999. Nuclear compartments and gene regulation. Curr. Opin. Genet. Dev. 9: 199-205.

Crabb, B.S. and Cowman, A.F. 1996. Characterization of promoters and stable transfection by homologous and nonhomologous recombination in *Plasmodium falciparum*. Proc. Natl. Acad. of Sci. USA. 93: 7289-7294.

Darkin-Rattray, S.J., Gurnett, A.M., Myers, R.W., Dulski, P.M., Crumley, T.M., Allocco, J.J., Cannova, C., Meinke, P.T., Colletti, S.L., Bednarek, M.A., Singh, S.B., Goetz, M.A., Dombrowski, A.W., Polishook, J.D., and Schmatz, D.M. 1996. Apicidin: a novel antiprotozoal agent that inhibits parasite histone deacetylase. Proc. Natl. Acad. Sci. USA. 93: 13143-13147.

Dechering, K.J., Kaan, A.M., Mbacham, W., Wirth, D.F., Eling, W., Konings, R.N., and Stunnenberg, H.G. 1999. Isolation and functional characterization of two distinct sexual-stage- specific promoters of the human malaria parasite *Plasmodium falciparum*. Mol. Cell Biol. 19: 967-978.

Deitsch, K.W., Calderwood, M.S., and Wellems, T.E. 2001. Malaria. Cooperative silencing elements in *var* genes. Nature. 412: 875-876.

Deitsch, K.W., del Pinal, A., and Wellems, T.E. 1999. Intra-cluster recombination and var transcription switches in the antigenic variation of *Plasmodium falciparum*. Mol. Biochem. Parasitol. 101: 107-116.

del Portillo, H.A., Fernandez-Becerra, C., Bowman, S., Oliver, K., Preuss, M., Sanchez, C.P., Schneider, N.K., Villalobos, J.M., Rajandream, M.A., Harris, D., Pereira da Silva, L.H., Barrell, B., and Lanzer, M. 2001. A superfamily of variant genes encoded in the subtelomeric region of *Plasmodium vivax*. Nature. 410: 839-842.

Donze, D., Adams, C.R., Rine, J., and Kamakaka, R.T. 1999. The boundaries of the silenced HMR domain in *Saccharomyces cerevisiae*. Genes Dev. 13: 698-708.

Feldman, J.B., Hicks, J.B., and Broach, J.R. 1984. Identification of sites required for repression of a silent mating type locus in yeast. J. Mol. Biol. 178: 815-834.

Fernandez, V., Hommel, M., Chen, Q., Hagblom, P., and Wahlgren, M. 1999. Small, clonally variant antigens expressed on the surface of the *Plasmodium falciparum*-infected erythrocyte are encoded by the *rif* gene family and are the target of human immune responses. J. Exp. Med. 190: 1393-1404.

Fischer, K., Horrocks, P., Preub, M., Wiesner, J., Wunsch, S., Camargo, A., and Lanzer, M. 1997. Expression of *var* genes located within polymorphic subtelomeric domains of *Plasmodium falicparum* chromosomes. Molecular and Cellular Biology. 17: 3679-3686.

Francis, N.J. and Kingston, R.E. 2001. Mechanisms of transcriptional memory. Nat. Rev. Mol. Cell Biol. 2: 409-421.

Gardner, J.P., Pinches, R.A., Roberts, D.J., and Newbold, C.I. 1996. Variant antigens and endothelial receptor adhesion in *Plasmodium falciparum*. Proc. Natl. Acad. of Sci. USA. 93: 3503-3508.

Gardner, M.J., Tettelin, H., Carucci, D.J., Cummings, L.M., Aravind, L., Koonin, E.V., Shallom, S., Mason, T., Yu, K., Fujii, C., Pederson, J., Shen, K., Jing, J., Aston, C., Lai, Z., Schwartz, D.C., Pertea, M., Salzberg, S., Zhou, L., Sutton, G.G., Clayton, R., White, O., Smith, H.O., Fraser, C.M., and Hoffman, S.L. 1998. Chromosome 2 sequence of the human malaria parasite *Plasmodium falciparum*. Science. 282: 1126-1132.

Gasser, S.M. 2001. Positions of potential: nuclear organization and gene expression. Cell. 104: 639-642.

Golightly, L.M., Mbacham, W., Daily, J., and Wirth, D.F. 2000. 3' UTR elements enhance expression of Pgs28, an ookinete protein of *Plasmodium gallinaceum*. Mol. Biochem. Parasitol. 105: 61-70.

Grant, P.A. 2001. A tale of histone modifications. Genome Biol. 2: REVIEWS0003.

Hernandez-Rivas, R., Mattei, D., Sterkers, Y., Peterson, D.S., Wellems, T.E., and Scherf, A. 1997. Evidence for mobile and differentially expressed *var* genes in subtelomeric regions of *Plasmodium falciparum* chromosomes. Molecular and Cellular Biology. 17: 604-611.

Hori, R. and Firtel, R.A. 1994. Identification and characterization of multiple A/T-rich cis-acting elements that control expression from Dictyostelium actin promoters: the Dictyostelium actin upstream activating sequence confers growth phase expression and has enhancer-like properties. Nucleic Acids Res. 22: 5099-11.

Horrocks, P., Dechering, K., and Lanzer, M. 1998. Control of gene expression in *Plasmodium falciparum*. Mol. Biochem. Parasitol. 95: 171-181.

Horrocks, P. and Kilbey, B.J. 1996. Physical and functional mapping of the transcriptional start sites of *Plasmodium falciparum* proliferating cell nuclear antigen. Mol. Biochem. Parasitol. 82: 207-215.

Horrocks, P. and Lanzer, M. 1999. Mutational analysis identifies a five base pair cis-acting sequence essential for GBP130 promoter activity in *Plasmodium falciparum*. Mol. Biochem. Parasitol. 99: 77-87.

Horrocks, P., Pinches, R., Kriek, N, and Newbold, C. 2002. Stage-specific promoter activity from stably maintained episomes in *Plasmodium falciaprum*. Int. J. Parasitol. 32:1203-1206.

Jenuwein, T. and Allis, C.D. 2001. Translating the histone code. Science. 293: 1074-1080.

Jenuwein, T., Laible, G., Dorn, R., and Reuter, G. 1998. SET domain proteins modulate chromatin domains in eu- and heterochromatin. Cell Mol. Life Sci. 54: 80-93.

Jones, D.O., Cowell, I.G., and Singh, P.B. 2000. Mammalian chromodomain proteins: their role in genome organisation and expression. Bioessays. 22: 124-137.

Kappe, S.H., Gardner, M.J., Brown, S.M., Ross, J., Matuschewski, K., Ribeiro, J.M., Adams, J.H., Quackenbush, J., Cho, J., Carucci, D.J., Hoffman, S.L., and Nussenzweig, V. 2001. Exploring the transcriptome of the malaria sporozoite stage. Proc. Natl. Acad. Sci. USA. 98: 9895-9900.

Kirchmaier, A.L. and Rine, J. 2001. DNA replication-independent silencing in *S. cerevisiae*. Science. 291: 646-650.

Kocken, C.H., van der Wel, A.M., Dubbeld, M.A., Narum, D.L., van de Rijke, F.M., van Gemert, G.J., van, d.L., X, Bannister, L.H., Janse, C., Waters, A.P., and Thomas, A.W. 1998. Precise timing of expression of a *Plasmodium falciparum*-derived transgene in *Plasmodium berghei* is a critical determinant of subsequent subcellular localization. J. Biol. Chem. 273: 15119-15124.

Koning-Ward, T.F., Janse, C.J., and Waters, A.P. 2000. The development of genetic tools for dissecting the biology of malaria parasites. Annu. Rev. Microbiol. 54: 157-185.

Koning-Ward, T.F., Speranca, M.A., Waters, A.P., and Janse, C.J. 1999. Analysis of stage specificity of promoters in *Plasmodium berghei* using luciferase as a reporter. Mol. Biochem. Parasitol. 100: 141-146.

Kyes, S., Horrocks, P., and Newbold, C. 2001. Antigenic variation at the infected red cell surface in malaria. Annu. Rev. Microbiol. 55: 673-707.

Kyes, S., Pinches, R., and Newbold, C. 2000. A simple RNA analysis method shows var and rif multigene family expression patterns in *Plasmodium falciparum*. Mol. Biochem. Parasitol. 105: 311-315.

Kyes, S., Taylor, H., Craig, A.G., Marsh, K., and Newbold, C.I. 1997. Genomic representation of *var* gene sequences in *Plasmodium falciparum* field isolates from different geographic regions. Mol. Biochem. Parasitol. 87: 235-238.

Kyes, S.A., Rowe, J.A., Kriek. N., and Newbold, C.I. 1999. Rifins: a second family of clonally variant proteins expressed on the surface of red cells infected with *Plasmodium falciparum*. Proc. Natl. Acad. Sci. USA. 96: 9333-9338.

Li, Y.C., Cheng, T.H., and Gartenberg, M.R. 2001. Establishment of transcriptional silencing in the absence of DNA replication. Science. 291: 650-653.

Loo, S. and Rine, J. 1994. Silencers and domains of generalized repression. Science. 264: 1768-1771.

Lyko, F. and Paro, R. 1999. Chromosomal elements conferring epigenetic inheritance. Bioessays. 21: 824-832.

Marmorstein, R. and Berger, S.L. 2001. Structure and function of bromodomains in chromatin-regulating complexes. Gene. 272: 1-9.

Mbacham, W.F., Chow, C.S., Daily, J., Golightly, L.M., and Wirth, D.F. 2001. Deletion analysis of the 5' flanking sequence of the *Plasmodium gallinaceum* sexual stage specific gene *pgs28* suggests a bipartite arrangement of cis-control elements. Mol. Biochem. Parasitol. 113: 183-187.

McAndrew, M.B., Read, M., Sims, P.F., and Hyde, J.E. 1993. Characterisation of the gene encoding an unusually divergent TATA- binding protein (TBP) from the extremely A+T-rich human malaria parasite *Plasmodium falciparum*. Gene. 124: 165-171.

McRobert, L. and McConkey, G.A. 2002. RNA interference (RNAi) inhibits growth of *Plasmodium falciparum*. Mol. Biochem. Parasitol. 119: 273-278.

Mechali, M. 2001. DNA replication origins: from sequence specificity to epigenetics. Nat. Rev. Genet. 2: 640-645.

Meehan, R.R., Pennings, S., and Stancheva, I. 2001. Lashings of DNA methylation, forkfuls of chromatin remodeling. Genes Dev. 15: 3231-3236.

Miller, A.M. and Nasmyth, K.A. 1984. Role of DNA replication in the repression of silent mating type loci in yeast. Nature. 312: 247-251.

Moazed, D. 2001. Common themes in mechanisms of gene silencing. Mol. Cell. 8: 489-498.

Munasinghe, A., Patankar, S., Cook, B.P., Madden, S.L., Martin, R.K., Kyle, D.E., Shoaibi, A., Cummings, L.M., and Wirth, D.F. 2001. Serial analysis of gene expression (SAGE) in *Plasmodium falciparum*: application of the technique to A-T rich genomes. Mol. Biochem. Parasitol. 113: 23-34.

Nakao, M. 2001. Epigenetics: interaction of DNA methylation and chromatin. Gene. 278: 25-31.

Nelson, H.C., Finch, J.T., Luisi, B.F., and Klug, A. 1987. The structure of an oligo(dA).oligo(dT) tract and its biological implications. Nature. 330: 221-226.

Nemazee, D. 2000. Receptor selection in B and T lymphocytes. Annu. Rev. Immunol. 18: 19-51.

Ngo, H., Tschudi, C., Gull, K., and Ullu, E. 1998. Double-stranded RNA induces mRNA degradation in Trypanosoma brucei. Proc. Natl. Acad. Sci. USA. 95: 14687-14692.

Pace, T., Birago, C., Janse, C.J., Picci, L., and Ponzi, M. 1998. Developmental regulation of a Plasmodium gene involves the generation of stage-specific 5' untranslated sequences. Mol. Biochem. Parasitol. 97: 45-53.

Patankar, S., Munasinghe, A., Shoaibi, A., Cummings, L.M., and Wirth, D.F. 2001. Serial analysis of gene expression in *Plasmodium falciparum* reveals the global expression profile of erythrocytic stages and the presence of anti-sense transcripts in the malarial parasite. Mol. Biol. Cell. 12: 3114-3125.

Pollack, Y., Kogan, N., and Golenser, J. 1991. *Plasmodium falciparum*: evidence for a DNA methylation pattern. Exp. Parasitol. 72: 339-344.

Porter, M.E. 2001. The DNA polymerase delta promoter from *Plasmodium falciparum* contains an unusually long 5' untranslated region and intrinsic DNA curvature. Mol. Biochem. Parasitol. 114: 249-255.

Preiser, P.R., Jarra, W., Capiod, T., and Snounou, G. 1999. A rhoptry-protein-associated mechanism of clonal phenotypic variation in rodent malaria. Nature. 398: 618-622.

Roberts, D.J., Craig, A.G., Berendt, A.R., Pinches, R., Nash, G., Marsh, K., and Newbold, C.I. 1992. Rapid switching to multiple antigenic and adhesive phenotypes in malaria. Nature. 357: 689-692.

Rubio, J.P., Thompson, J.K., and Cowman, A.F. 1996. The var genes of *Plasmodium falciparum* are located in the subtelomeric region of most chromosomes. EMBO J. 15: 4069-4077.

Scherf, A., Hernandez-Rivas, R., Buffet, P., Bottius, E., Benatar, C., Pouvelle, B., Gysin, J., and Lanzer, M. 1998. Antigenic variation in malaria: *in situ* switching, relaxed and mutually exclusive transcription of *var* genes during intra-erythrocytic development in *Plasmodium falciparum*. EMBO J. 17: 5418-5426.

Serizawa, S., Ishii, T., Nakatani, H., Tsuboi, A., Nagawa, F., Asano, M., Sudo, K., Sakagami, J., Sakano, H., Ijiri, T., Matsuda, Y., Suzuki, M., Yamamori, T., Iwakura, Y., and Sakano, H. 2000. Mutually exclusive expression of odorant receptor transgenes. Nat. Neurosci. 3: 687-693.

Shei, G.J. and Broach, J.R. 1995. Yeast silencers can act as orientation-dependent gene inactivation centers that respond to environmental signals. Mol. Cell Biol. 15: 3496-3506.

Smith, J.D., Chitnis, C.E., Craig, A.G., Roberts, D.J., Hudson-Taylor, D.E., Peterson, D.S., Pinches, R., Newbold, C.I., and Miller, L.H. 1995. Switches in expression of *Plasmodium falciparum var* genes correlate with changes in antigenic and cytoadherent phenotypes of infected erythrocytes. Cell. 82: 101-110.

Struhl, K. 1985. Naturally occurring poly(dA-dT) sequences are upstream promoter elements for constitutive transcription in yeast. Proc. Natl. Acad. Sci. USA. 82: 8419-8423.

Su, X., Heatwole, V.M., Wertheimer, S.P., Guinet, F., Herrfeldt, J.V., Peterson, D.S., Ravetch, J.V., and Wellems, T.E. 1995. A large and diverse gene family (*var*) encodes 200-350 kD proteins implicated in the antigenic variation and cytoadherence of *Plasmodium falciparum*-infected erythrocytes. Cell. 82: 89-100.

Su, X., Kirkman, L.A., Fujioka, H., and Wellems, T.E. 1997. Complex polymorphisms in an approximately 330 kDa protein are linked to chloroquine-resistant *P. falciparum* in Southeast Asia and Africa. Cell. 91: 593-603.

Taylor, H.M., Kyes, S.A., Harris, D., Kriek, N., and Newbold, C.I. 2000b. A study of var gene transcription in vitro using universal var gene primers. Mol. Biochem. Parasitol. 105: 13-23.

Taylor, H.M., Kyes, S.A., and Newbold, C.I. 2000a. *Var* gene diversity in *Plasmodium falciparum* is generated by frequent recombination events. Mol. Biochem. Parasitol. 110: 391-397.

Taylor, H.M., Triglia, T., Thompson, J., Sajid, M., Fowler, R., Wickham, M.E., Cowman, A.F., and Holder, A.A. 2001. *Plasmodium falciparum* homologue of the genes for *Plasmodium vivax* and *Plasmodium yoelii* adhesive proteins, which is transcribed but not translated. Infect. Immun. 69: 3635-3645.

Triglia, T., Thompson, J.K., and Cowman, A.F. 2001. An EBA175 homologue which is transcribed but not translated in erythrocytic stages of *Plasmodium falciparum*. Mol. Biochem. Parasitol. 116: 55-63.

Vaidya, A.B., Muratova, O., Guinet, F., Keister, D., Wellems, T.E., and Kaslow, D.C. 1995. A genetic locus on *Plasmodium falciparum* chromosome 12 linked to a defect in mosquito-infectivity and male gametogenesis. Mol. Biochem. Parasitol. 69: 65-71.

van der Wel, A.M., Tomas, A.M., Kocken, C.H., Malhotra, P., Janse, C.J., Waters, A.P., and Thomas, A.W. 1997. Transfection of the primate malaria parasite *Plasmodium knowlesi* using entirely heterologous constructs. J. Exp. Med. 185: 1499-1503.

van Holde, K. and Zlatanova, J. 1994. Unusual DNA structures, chromatin and transcription. Bioessays. 16: 59-68.

Voss, T.S., Thompson, J.K., Waterkeyn, J., Felger, I., Weiss, N., Cowman, A.F., and Beck, H.P. 2000. Genomic distribution and functional characterisation of two distinct and conserved *Plasmodium falciparum var* gene 5' flanking sequences. Mol. Biochem. Parasitol. 107: 103-115.

Wellems, T.E., Walker-Jonah, A., and Panton, L.J. 1991. Genetic mapping of the chloroquine-resistance locus on *Plasmodium falciparum* chromosome 7. Proc. Natl. Acad. Sci. USA. 88: 3382-3386.

Wells, R.D., Collier, D.A., Harvey, J.C., Shimizu, M., and Wohlrab, F. 1988. The chemistry and biology of unusual DNA structures adopted by oligopurine.oligop yrimidine sequences. FASEB J. 2: 2939-2949.

Winter, E. and Varshavsky, A. 1989. A DNA binding protein that recognizes oligo(dA).oligo(dT) tracts. EMBO J. 8: 1867-1877.

From: Malaria Parasites: Genomes and Molecular Biology
Edited by: A.P. Waters and C.J. Janse

# Chapter 9

# Regulation of rRNA Transcription and Processing During the *Plasmodium* Life Cycle

Thomas F. McCutchan, Rosalinda van Spaendonk, Chris Janse, Jun Fang, and Andrew P. Waters

## Abstract

A remarkable feature of *Plasmodium* parasites is that they express structurally distinct sets of rRNA in a developmentally specific manner. The expression of three different rRNAs parallels the developmental cycle of the parasite. Parasites missing the rRNA genes predominantly expressed in the mosquito develop at a slower rate than wild type parasites but never the less develop fully. The normal cyclical pattern of expression is influenced by temperature and ambient glucose concentrations and changes in these factors can result in dramatic changes in the parasite's physiology. Variable forms of ribosomes are not unique to Plasmodium species. They are often found in plants and bacteria but alteration in these cases result from changes in the protein complement of the ribosome. Alteration of the rRNA itself occurs in Plasmodium species and presents opportunities to study a novel biochemical property of the parasite and to gain insights into RNA as a functional moiety.

# 1. Ribosomes

Ribosomes are small but complex structures, roughly 20 to 30 nm in diameter, which serve as the center for protein synthesis in all cells (for recent review see Doudna and Rath, 2002). Herein lies the machinery that decodes information from the messenger RNA and catalyzes the ordered assembly of amino acids into proteins. The design is ancient and has changed minimally throughout evolution. For example, eukaryotic and prokaryotic ribosomes are very similar. Both are asymmetric and composed of one small and one large subunit. Each ribosomal subunit consists of at least one ribosomal RNA (rRNA) molecule bound by a complex of ribosomal proteins (r-proteins), assembled in a precise fashion. Decoding messenger RNA and assembly of protein occurs in a complex, the polysome, that has a number of ribosomes all travelling along a message at the same time, each making a separate but identical polypeptide chain. Polysomes, as well as single ribosomes, are present in the cytosol of the cell in two different forms: free or membrane-bound. The type of protein being synthesised tends to segregate, depending on its origin, the free or the attached ribosome. The free ribosomes synthesise proteins that are retained in the cell. Free ribosomes frequently bind to cytoskeletal fibres in the cytosol and are responsible for the synthesis of proteins that remain in solution in the cytoplasm or form important cytoplasmic structures such as the cytoskeleton. Other ribosomes are bound to the cytosolic side of the membranes and constitute the rough endoplasmic reticulum (ER). After synthesis, proteins synthesised on the rough ER are inserted through the cisternal membrane and released into the ER lumen. These proteins become part of membranes, are packaged into vesicles for storage in the cytoplasm, trafficking to subcellular organelles or are exported to the cell exterior.

In order to form a single active ribosome, the r-proteins must associate with the rRNA. Nomura and colleagues were able to reconstitute the ribosome *in vitro* (Nomura *et al.*, 1984), opening the door to a functional understanding of the component parts of the ribosome. The process of reconstitution of component parts sheds light on the role of a number of r-proteins in protein synthesis. Further advances in understanding the higher order conformation of the rRNA has led to defining its role in catalysis (Allen and Noller, 1989; Noller *et al.*, 1992; Green and Noller, 1997; Wilson and Noller, 1998; Maguire and Zimmerman, 2001).

Below we describe some general principles of control over ribosome synthesis gathered from a variety of organisms, both eukaryotic and prokaryotic. Mechanisms that to date have been found only in prokaryotic systems are intentionally included as they lend background and precedent to the emerging information about protein synthesis in *Plasmodium*.

## 1.1. Regulation of Ribosome Production is Essential to all Cells

The number of ribosomes present in a cell is directly related to the protein-synthesising activity and size of the cell. In an actively growing bacterium, there may be roughly 20,000 ribosomes per cell; whereas in some eukaryotic cells, 10 million ribosomes are required to keep up with the cell's need for protein synthesis (Alberts *et al.*, 1994). The ribosomes may make up 45% of the dry weight of an actively growing cell; thus, regulation that balances the needs of the cell with its external environment are essential. Control of growth and function of a cell by regulation of ribosome production is somewhat analogous to controlling a vehicle with accelerator and brakes. Speed has to be monitored and regulated in conjunction with driving conditions for the sake of both fuel efficiency and safety. Likewise, the organism must have the means to speed up, slow down or remain stationary in response to its environment. Any of these states may have an adaptive advantage depending on the situation. A number of regulation strategies for ribosome syntheses have been characterised and will be briefly described here because some will apply to *Plasmodium*.

One mechanism of control of the rate of rRNA (and hence, ribosome) synthesis is the "stringent response" that occurs in both eukaryotic and prokaryotic organisms (Cashel *et al.*, 1996). The presence of external signals (e.g., glucose or amino acid concentrations) are sensed by receptors on the cell surface leading to adaptive changes to the cell's interior. In this case, this response is directly related to the available carbon source for the organism. Another regulator of rRNA synthesis is called growth-rate-dependent control (GRDC). It ensures that rRNA synthesis relates to the growth rate of the cell, perhaps by sensing nucleotide triphosphate levels in the cytosol (Barker and Gourse, 2001), and not directly to the availability of amino acids or glucose. Often, both stringent and GRDC mechanisms work in the same organism. A number of eukaryotic organisms, such as yeast, respond to their environment with both a stringent and growth-rate response.

Syntheses of ribosomal r-proteins are coordinated with the synthesis of the rRNA. In many microorganisms, both eukaryote and prokaryote ribosomal protein production can be controlled by a feedback mechanism at the translational level. For example, r-protein that is produced in excess of that needed to form ribosomes may play the role of repressor. A balance is achieved because it binds to the rRNA in an active ribosome with higher affinity than to its own mRNA. Translation is only repressed when the level of the r-protein exceeds that of the rRNA. In eukaryotic organisms translational repression is more often related to a balance between phosphorylation and dephosphorylation of initiation factors; however, in some cases the r-proteins directly bind to the mRNA and act as translational repressors as described above. This type of feedback translational repression has been demonstrated in *Plasmodium* for the dihydrofolate reductase-thymidylate synthetase (dhfr-ts) proteins and its cognate mRNA (Zhang and Rathod, 2002).

In eukaryotic organisms, gene regulation may also directly relate to chromatin structure. Most of these organisms probably do not have a mechanism for selectively affecting individual rDNA units. Little sequence heterogeneity is found in rRNA sequences within organisms. In *Plasmodium* species, however, there are four to eight rDNA copies dispersed to as many chromosomes (Wellems *et al.*, 1987; Janse *et al.*, 1994). Although gene conversion affecting different loci affected rDNA evolution (Corredor and Enea, 1993), each member of the genus that we have studied, with the exception of *P. berghei,* maintains at least these three distinct sets of ribosomal rRNA genes. One must thus consider chromatin structure as a possible transcription regulator. For example, regions of DNA alternatively are found in open or closed (i.e., condensed) regions within the chromatin. Because packaging or condensing of chromatin structure has the effect of strongly repressing transcription, regulation of chromatin packaging could act to regulate rRNA transcription.

## 2. The Plasmodium Ribosome Genes

The ribosomal gene (rDNA unit) of *Plasmodium* is similar to other eukaryotic organisms. It is arranged in the standard fashion with a copy of the small subunit rRNA (SSU rRNA), an internal transcribed spacer (ITS1), the 5.8S rRNA, another internal transcribed spacer (ITS2), and the large subunit rRNA (LSU rRNA) genes proceeding in a 5' to 3' direction. Usually, these genes are referred to as "dosage-response" genes, in that deletion of copies has serious phenotypic consequences for the organism, presumably because cellular requirements for rRNA are not being met (Mitchell *et al.*, 1992). In contrast, *Plasmodium* species have only four to eight genes (Dame and McCutchan, 1983 Langsley *et al.*, 1983). It is unlikely that more than one or two of the genes are transcribed at any one-time (Rogers *et al.*, 1995). Hence, mutations that occur in rRNA genes of *Plasmodium* have a leveraged effect not seen in most other organisms: mutation in a single gene may affect up to fifty percent of the transcripts and result in phenotypic changes of the magnitude seen in the "minute mutants" of *Drosophila* caused by large deletions in the rDNA complex (reviewed in Mitchell *et al.*, 1992 ).

## 3. Correlation of rRNA Expression and the Life Cycle

*Plasmodium* rRNA genes have been reviewed at length (Gunderson *et al.*, 1987; McCutchan *et al.*, 1988; 1995; Rogers *et al.*, 1998) and the reader is encouraged to refer to those papers with regard to developmental regulation of stage-specific rRNA transcription, targeting of drugs to the ribosome, and use of rRNA as a diagnostic tool.

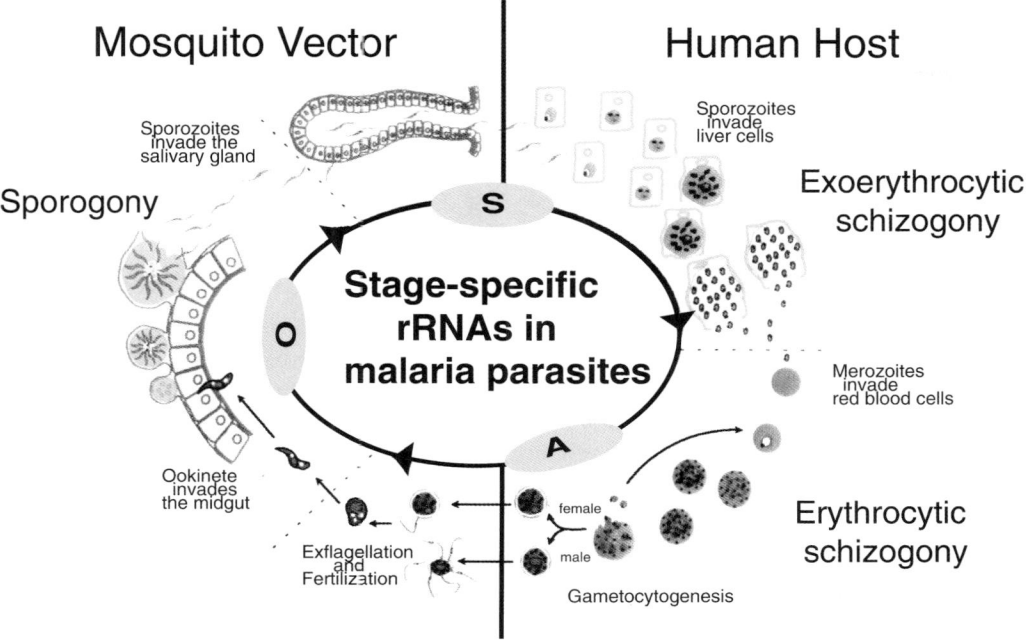

**Figure 1.** Life cycle of *Plasmodium vivax*. Changes in rRNA transcription correlate with the developmental cycle. Active transcription periods for three different rDNA units are indicated.

The life cycle of *Plasmodium vivax* with corresponding changes in ribosomal type is shown in Figure 1. The remarkable feature of *Plasmodium* parasites of differentially expressing structurally distinct sets of rRNA gene units in a stage-specific manner leads to a unique opportunity to study growth-dependent regulation in *Plasmodium* and the effects of sequence heterogeneity on ribosome function and evolution. Classification of the different rRNA units is based to some extent upon their structural differences but principally upon the point in the life cycle at which they are maximally expressed. To date, three classes have been identified: A-type (maximally expressed in blood-stage parasites), S-type (maximally expressed in oocysts and sporozoites), and O-type (uniquely expressed and retained with the oocyst). The A- and S-types are found in all *Plasmodium* species examined; whereas, to date, the O-type has only been fully characterised in *P. vivax* (Li *et al.*, 1997) and does not exist in *P. berghei* (van Spaendonk *et al.*, 2000).

Most of the investigation of this phenomenon has been done using *P. falciparum* and *P. berghei*. Incorporated into the description of data is an important distinction that could easily be missed: the difference between new transcription and the presence of mature transcripts. Active transcription initiation is assessed by determining sequences specific to the different unprocessed precursor rRNA. Further, some characteristics of ribosomal transcription and processing appear to be characteristic of the genus, while others appear to be species-specific. The extent to which this is the result of the same general mechanisms of ribosomal control being molded to fit the framework of a species' development process, versus there being characteristics that are genuinely species-specific, is unknown.

### 3.1. Sporozoite Development Into Exoerythrocytic Forms

The changes in relative abundance and transcription of the different rRNA types appear to correlate with discrete developmental stages of the parasite. Host transitions occur that influence these changes but they are not sufficient to drive the mechanisms involved in ribosomal control. The process of infection of the vertebrate by parasites from the mosquito is an example. The transition from the S ribosome to the A ribosome occurs during development of the malaria parasite in the liver (Zhu *et al.*, 1990; Figure 1). Infection of the vertebrate and even invasion of liver cells is not enough, however, to trigger the event. For example, observing invasion of sporozoites into cultured mouse liver cells has been informative. Sporozoite-specific antibody or cytochalasin B, which also prevents invasion of liver cells, prevents the switch of ribosomal types.

From the above, it was clear that some measure of development was necessary prior to the rRNA transition. It is known that increasing doses of irradiation progressively disables the parasite's potential to develop. The appropriate dose prevents development beyond a certain point, yielding a sporozoite that is effectively attenuated. Further, when such sporozoites are used as a vaccine, the individual receiving them is protected from future infection with live, fully competent sporozoites. Irradiation beyond the critical point generates sporozoites that cannot develop in the hepatocyte and fail to protect immunised individuals. Parenthetically, the same dose of irradiation that is optimal for producing vaccine-grade attenuated parasites still permits the rRNA transition, although development ceases at some point thereafter. Lethally irradiated sporozoites do not switch from S- to A-type rRNA. Therefore, by comparing the efficacy of vaccination with the ribosomal transition it was shown that the switch is an essential component of the developmental process of the attenuated parasite.

### 3.2. Schizogony and Gametocytogenesis

Two to three virtually identical A genes are the predominant source of rRNA during blood-stage development in all species (Figure 2). A-type units are transcribed in direct response to cell growth in blood-stage asexual parasites. A-type units are transcribed in direct response to cell growth

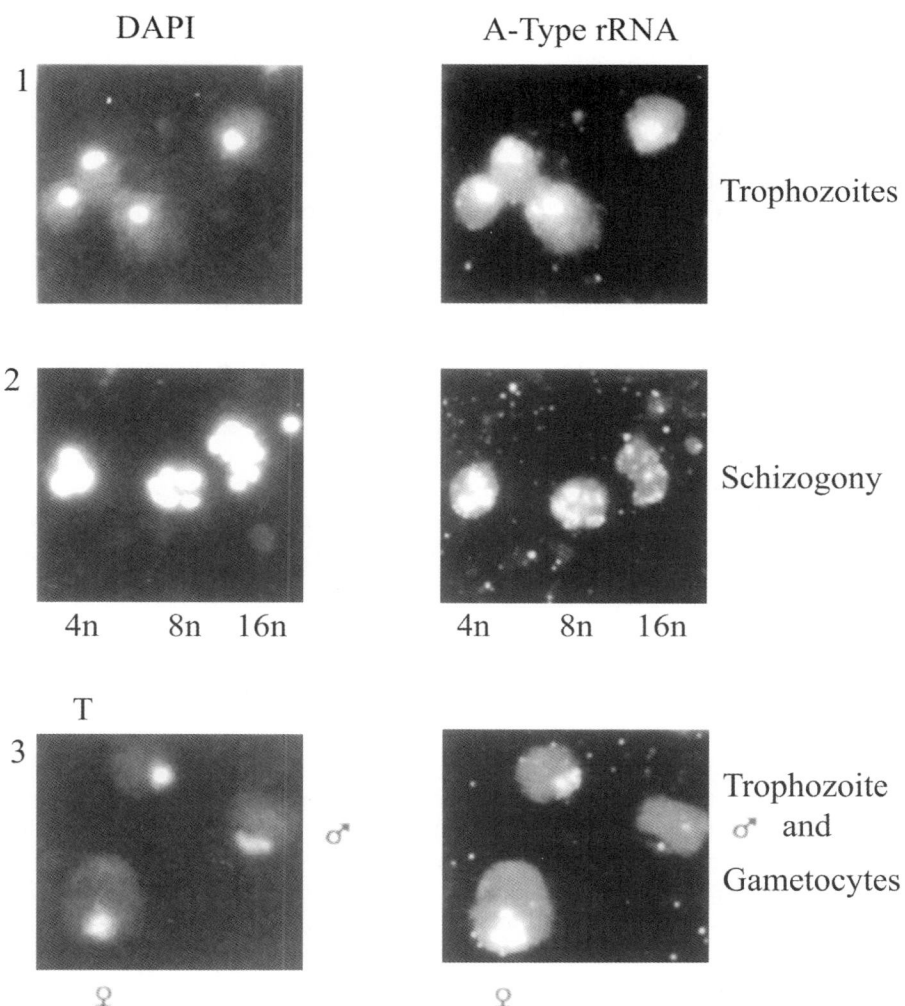

**Figure 2.** *In situ* localization of A transcription during the parasites 24-hr cycle of asexual development in the blood. Note the suppression of transcription in male gametocytes (Thompson *et al.*, 1999).

in blood-stage asexual parasites. A rapid downregulation of transcription is the result of glucose starvation, as would be expected for an organism displaying a stringency response. Curiously, only the A genes are subject to this downregulation, while the S gene transcription is actually upregulated in response to carbon source starvation—indicating that they are not influenced by the stringency mechanism (Fang and McCutchan, unpublished). In support of the idea of a selective stringency response is the identification of a *P. falciparum* gene with 40% identity to the stringency factor (RelA protein). This protein initiates the response, has no other known function, and is found in bacteria and plants that display the stringency response. The copy number of A-type genes may indeed represent the classic gene-dosage effect noted earlier and as seen for a small number of other housekeeping genes (Vinkenoog *et al.*, 1998). Whereas it has proved possible to knock out individual members of duplicated housekeeping genes, to date there has been no success in producing parasites with deleted A genes (Waters and Janse, unpublished). Based on this negative result, one would be inclined to believe that, unlike the S genes (see below), all A genes are essential and the dose dependence is strict.

Transcription of the A units in *P. berghei* is differentially regulated during male (inactive) and female (active) gametocytogenesis, as shown in Figure 2 (Waters *et al.*, 1997). The same may well be true of other species but has not yet been tested. Small amounts of S-type rRNA precursors are also observed during the development of parasites for *P. berghei*. In *P. berghei*, this occurs at about 20 hours after merozoite invasion and, as determined by Northern blot analysis, the S precursors are transcribed but not processed into mature S-type rRNA molecules (Waters *et al.*, 1997). Productive S-type transcription is clearly related to only a subset of cells (probably female gametocytes) and is not seen in mature schizonts or male gametocytes (Waters *et al.*, 1997). Its transcription is elevated in the gametocyte-producing strains of *P. berghei* versus non-producing strains of *P. berghei*. Purification of the gametocytes from *P. falciparum* show very significant increases in S-type transcription (Fang and McCutchan, unpublished). It seems likely that S-type transcription in gametocytes is characteristic of the genus.

In both *P. falciparum* and *P. berghei*, S-type precursors appear to have an extended life in the gametocyte reminiscent of 5S precursor storage in oocytes of *Xenopus laevis*. The accumulation of S-type precursor in *P. falciparum* continues until zygote formation (Waters *et al.*, 1988). This does not appear to be the case in either *P. berghei* (Waters *et al.*, 1997) or *P. vivax* (Li *et al.*, 1997), both of which produce gametocytes much more rapidly (28 and 50 hours, respectively) than *P. falciparum* (8 days). Storage of both unprocessed rRNA and messenger RNA (e.g., Pbs21; Thompson and Sinden, 1994) is an interesting phenomenon that does occur during the sexual stages of parasite development and is under further investigation (Waters and Janse, unpublished data).

## 3.3. The Sexual Phase of Development

Major morphologic changes occur in malaria parasites as they are transmitted between human and mosquito. In parallel with parasite development are changes in the surrounding temperature and metabolic environment. To respond to the dramatic environmental shifts that occur during these periods, cells must have sensors to monitor the environment and, in turn, to initiate appropriate metabolic changes in the cell. It has been shown that thermoregulation—in the form of cold-induced upregulation of the transcription of a single rDNA unit—is a component of *P. falciparum* transmission to the mosquito.

The two types of A genes located on different *P. falciparum* chromosomes are identical, at least with regard to mature SSU rRNA. The sequences preceding the two mature A-type RNA (A1, A2) and two mature S-type RNA (S1, S2) are, however, all different. The differences in upstream regions lead to differential control over transcription (Fang and McCutchan, 2002). The full-length transcripts that contain the mature A-type rRNA also contain a 5' leader region, the external transcribed spacer (ETS). The ETS region is similar over a 530-nucleotide region preceding the mature 18S rRNA, but no other similarities were found in the approximately 1-kb preceding the start of the mature transcript. The only similarity preceding the mature S-type rRNA is a 20-bp region common to all the rRNA genes. Polymorphisms in the RNA sequence of the precursor 5' to the mature 18S rRNA gene allowed us to follow the transcriptional regulation of each of the four genes, because the transient precursor region could be quantified using real-time PCR. The numbers of each rRNA precursor were measured from each stage of the life cycle. The amounts of the freshly transcribed precursor rRNAs in every stage were compared (Fang and McCutchan, 2002). S1 transcripts, dominant in gametocytes, decrease in relative amounts as the parasite develops through the mosquito stages, while S2 transcripts, first seen in overwhelming amounts in oocysts, are the dominant gene product in sporozoites. The burst of S1 transcription is gametocyte specific and may be preparatory to transmission to the mosquito; i.e., it may play a "bridge-like" role, allowing the S gene product, or precursor, to be already in place when the human-to-mosquito switch occurs.

The intrinsic responsiveness of S-type promoters to temperature fluctuates in blood-stage parasites (Figure 3). Transcription of the asexual blood-stage parasite genes A1 and A2 remained relatively constant at different temperatures, whereas the rate of S gene transcription was sensitive to temperature: at 42°C, neither S gene was transcribed. Further, upon decreasing temperature, transcription of S2 increased 4.4 fold at 31°C and 15 fold at 26°C. The S1 gene was dramatically less affected by the decrease in temperature. In a similar fashion, we have now measured the steady-state transcription of other messenger RNAs and found that about 250 other RNA were also differentially regulated by temperature (Fang and McCutchan, unpublished).

Downregulation of S2 transcription in mosquitoes maintained at high temperature (the converse of the above experiment) supports the idea that transcriptional control is directly affected by temperature. Mosquitoes were allowed to blood feed at 26°C and were maintained at 26°C for 8 days. Development proceeded naturally into oocysts on the mosquito midgut.

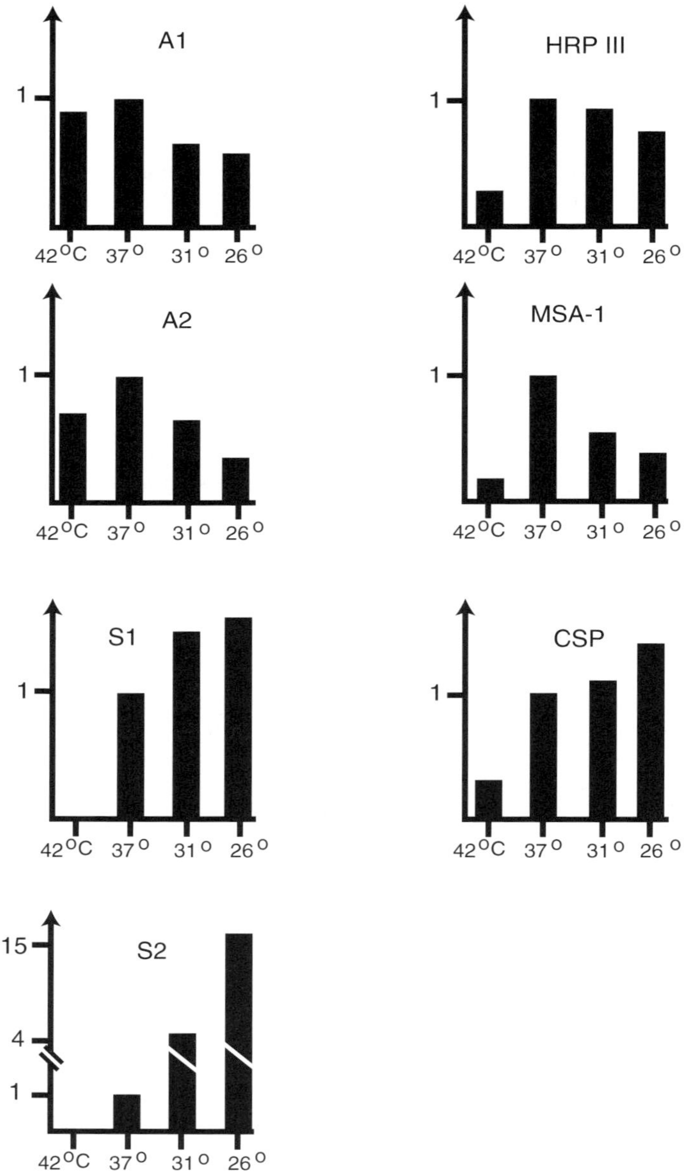

**Figure 3.** Transcriptional sensitivity of two A-type and two S-type *P. falciparum* rDNA units to changes in temperature. Three protein encoding RNAs were compared using the same temperature shifts (right column).

Eight days after the feed, mosquitoes were separated into two batches, one maintained at 26°C, the other at 37°C. The results showed that S2 gene transcription was inhibited by temperature increase, while transcription of the other genes was not.

Recently, a third type of rRNA gene unit, the O-type, was isolated and cloned (Li *et al.*, 1997). O-type genes are significantly different structurally from the A- and S-types, containing sequence insertions in regions that have hitherto been considered almost universally conserved. In *P. vivax*, transcripts of the O-type can be detected in ookinetes and oocysts, starting from two days after fertilisation (Figure 4).

### 3.4. Transition Between Types

Additional forms of control of ribosomal number may act on mature ribosomes. The mechanisms of discrimination between "incoming and outgoing" forms of the ribosomes may be a controlled process that merits further investigation. A potentially interesting example is shown in Figure 4. The O-type of rRNA is first seen at about two days after fertilisation in *P. vivax*. After approximately six days, an upregulation of S gene transcription begins. Both types of ribosomes are clearly present in the mature oocyst and yet only the S-type rRNA is contained in the parasite as it enters the salivary gland (Figure 4). The two types of ribosome are somehow physically separated in the oocyst (Li *et al.*, 1997). How broadly this applies to the genus is unknown but the potential for the ribosome types being physically separated in the cell exists. It is clear from numerous studies that ribosomes of different types exist in the same cell at the same time. Therefore, the potential variety of the *Plasmodium* ribosome complement could be extended by the formation of hybrid ribosomes combining subunits of different rRNA types. However, the clear physical segregation exhibited by the O- and S-type in *P. vivax* may indicate that this opportunity is severely limited.

## 4. The Search for Functional Differences Between Ribosomes

The primary function of all *Plasmodium* ribosomes is the same: protein synthesis. Subtle differences in the assembled components of the ribosome, however, can have an essential role to play in the survival of an organism. Dame and McCutchan (1983) initially proposed two possible reasons for sequence heterogeneity in the *Plasmodium* ribosomes. One was that different regions of the chromosome were open to transcription at transition points in the parasite's developmental cycle. The dispersal of the genes into regions that are open or closed to transcription would explain the developmental

# Development of *P. gallinaceum* in *Aedes aegypti*

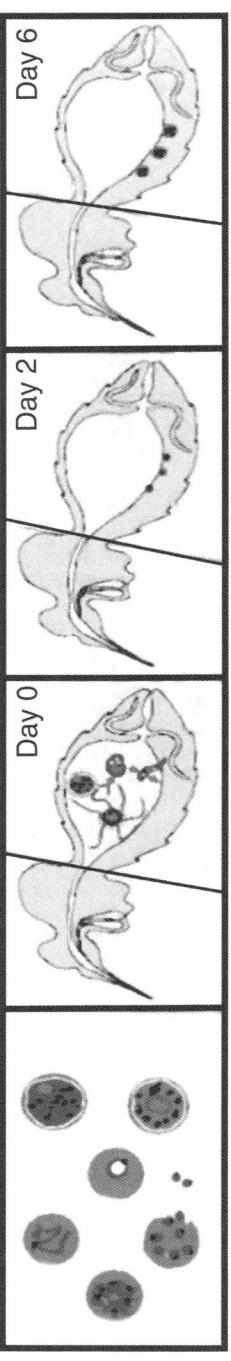
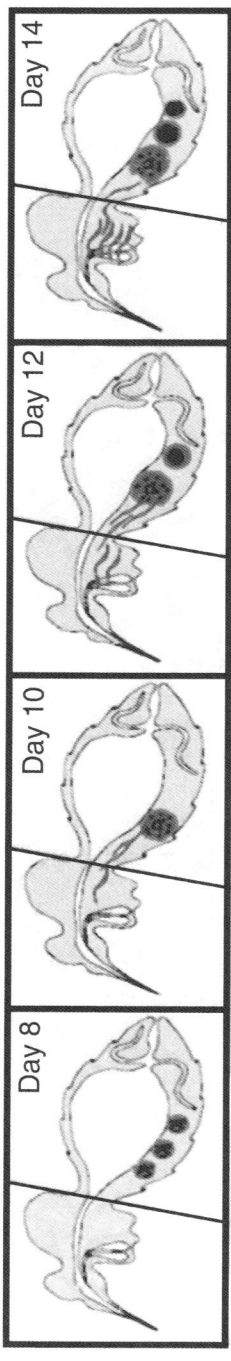

**Figure 4.** Schematic view of parasite development in mosquitoes following a blood meal. S- and O-type RNAs are physically separated during sporogony. Developmental progression is correlated with the type of rRNA found in either head or thorax. The progressive course of differentiation relates to days following the blood meal. A line indicates the separation of thorax and abdomen.

GTPase Site of *P. falciparum* rRNA

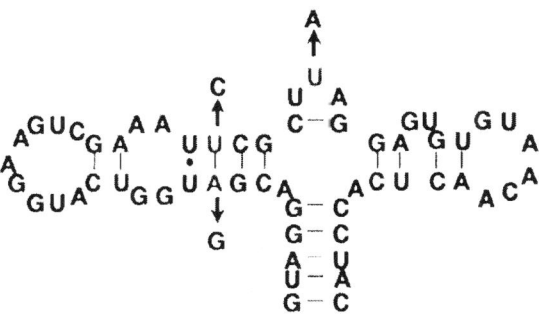

**Figure 5.** Two-dimensional structure of the *P. falciparum* rRNA GTPase site, showing the location of sequence differences between the A- and S-type RNAs.

specificity of expression. The second possibility noted was that ribosomes could have functional differences. For example, the transition between ribosome forms might serve to optimise the parasites relative fitness in a changing environment.

Initially, it was thought that one might gain insight into functional differences among heterologous ribosomes simply by comparing the primary sequences of gene types within a species and among homologues found of different species. The position of sequence differences between rRNA types is of interest because the increasing body of evidence supports a functional role for rRNA in protein synthesis (Noller *et al.*, 1992). The majority of *Plasmodium* rRNA nucleotide variation is located in regions that are normally variable and probably have little effect on function. Some differences, however, occur in conserved 'core regions' of the rRNA where structure–function relationships have been established. Structural analysis of *P. falciparum* A and S genes was particularly curious because it revealed differences in the GTPase region of the molecule (Figure 5 and Rogers *et al.*, 1996). While the sequence of the A-type 28S RNA resembles that found in other eukaryotes, the S-type contains a compensatory base pair change in the loop that joins two helices (Figure 5). This S-type variation is not seen in other eukaryotes (Gutell *et cl.*, 1993).

The extreme functional conservation of the GTPase region shown previously by replacement of the *E. coli* sequence with the corresponding one from yeast, had no effect on growth. In a parallel experiment, the A- and S-type GTPase sites were engineered to replace the GTPase region of a complete yeast rDNA unit and the mosaic genes were tested in yeast. It was demonstrated that the *P. falciparum* S-type GTPase cannot complement the yeast rDNA unit while the A-type gene can (Velichutina *et al.*, 1998). Clearly, there is a functional difference between the ribosomes carrying the different GTPase sites but the significance of this finding is yet unknown.

To obtain more insight into possible functional differences between the rRNA molecules of the different rDNA units, van Spaendonk *et al.* investigated development in another species (2001). It was demonstrated that the distinct *P. berghei* rRNA molecules did not differ in their core regions in a manner similar to *P. falciparum* (van Spaendonk *et al.*, 2001). Thus, the structural differences between *Plasmodium* rRNA molecules are not totally conserved among members of the genus, although *P. berghei* is a glaring exception. Based on this observation, it seems unlikely that functional differences between ribosomes will be established on the basis of sequence analysis alone.

In experiments designed to determine whether each rRNA gene was essential to development, van Spaendonk *et al.* (2001) disrupted the C (S-type) rRNA gene units of *P. berghei* and analysed the resulting phenotype. They demonstrated that the presence of only one of the two S-type rRNA gene units was sufficient to allow the parasite to undergo full development. There is, however, a growth retardation of the oocysts in parasites with either a C or D (S-type) deletion. The degree of retardation in the C- or D-knockout parasites is proportional to the level of expression of these units in the wild-type parasite, again pointing to gene-dosage phenomena.

Population data showing the fixation of crossovers between A- and S-types in the field would at least tend to support the contention that GTPase site variation was not a factor in parasite survival, even in the environment. Crossover between different chromosomes could result in natural generation of A-S mosaic transcripts. Mosaic A-S genes have been seen by a number of other groups. Data showing a crossover between A and S that has been fixed in the population was shown by Li *et al.* (2000). They found that *P. vivax* from the New World could be distinguished from Old World *P. vivax* on the basis of a crossover between A and S genes in variable region 7 of the SSU RNA (Figure 6). This presumably leaves the altered GTPase site of one of the S genes under the control of an A gene promoter. The parasites are phenotypically different with regard to successful infection of the *Anopheles albimanus* mosquito. What role the ribosome might play in this, if any, is not fully understood. The mosaic S-gene is expressed in the mosquito and thus the crossover between the two genes is expressed, posing no problem for the survival of the organism in its natural environment.

# 5. Future Directions

The availability of the nearly complete genome sequence for *P. falciparum* was anticipated to give the first glimpse of the full rRNA gene complement of a *Plasmodium* parasite. As expected, previously characterised pairs of A and S genes are found on different chromosomes (A1 and A2 on chromosomes 5 and 7; S1 and S2 on 11 and 13, respectively). In addition, a third type of rRNA unit is found on chromosome 1 that contains a hybrid unit appearing to have S-type 18 and 5.8S genes but a unique 28S gene (65% identical to the A-type and 75% to the S-type) which, unlike other characterised units, contains

# a. rRNA Gene Conversion

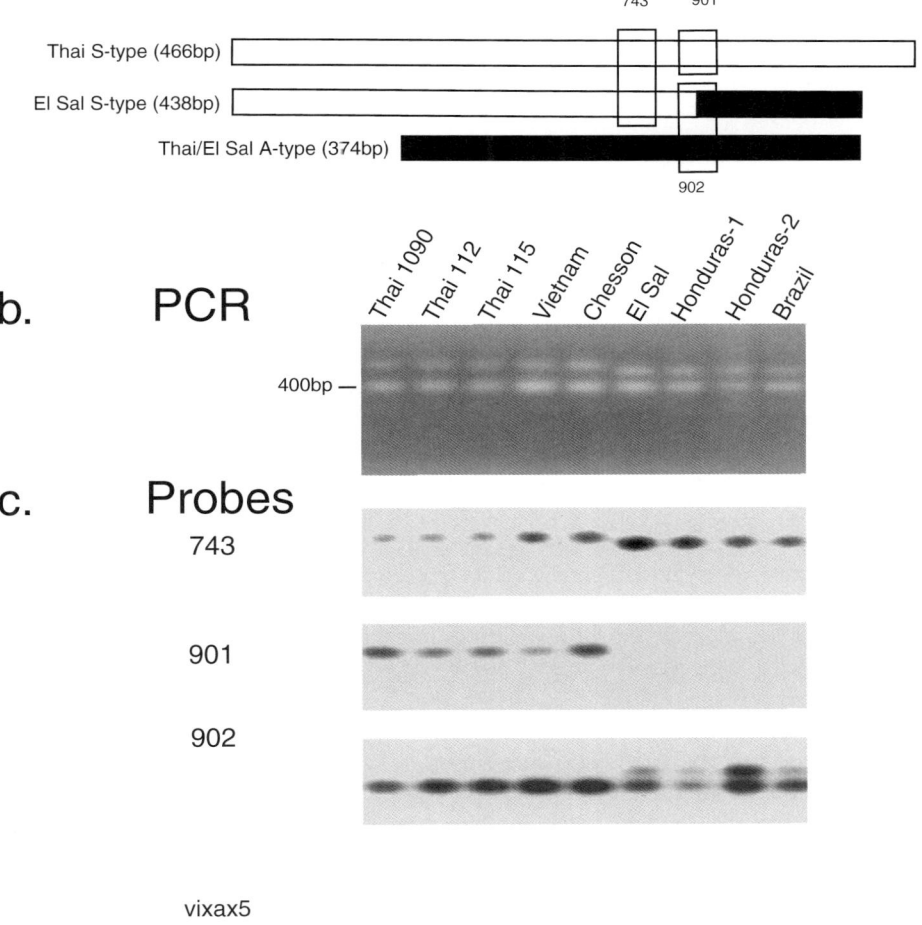

vixax5

**Figure 6.** Cross-over event between A and S genes has become fixed in *P. vivax* from Central and South America. *a*) schematic of a representative type of the S gene from the Old World (Thai S-type), the S gene from Central America (El Sal S-type) and A-type gene that both share. Position and number of gene-specific probes are indicated. *b*) separation and comparison of PCR products resulting from amplification of a variable region of the rRNA genes of *P. vivax* from different sources. *c*) Southern blot analysis of amplification products from New and Old World *P. vivax*, using the probes shown in Part (*a*) of this figure.

a further variation of the GTPase site, again conserving secondary structure predictions. It does not possess the insertions that characterise the O-type 18S gene and, to date, expression pattern is unknown and will clearly be the subject of further research. The last two units are intriguing; they consist of intact 5.8S and 28S genes and lack the 18S unit. They are symmetrically arranged at each end of chromosome 8, which may indicate they have recently been silenced by a mutual deletion event and represent holes in the current genome sequence. Their expression pattern, if any, is unknown and it may not be possible to study these particular units in the 3D7 parasite clone that was used for the genome sequencing project.

Once the full complement and expression patterns of the rRNA genes of *Plasmodium* parasites are entirely clear, it will be possible to examine the individual roles of the different classes and units. Detailed studies on the growth characteristics of parasites with altered S-type rRNA molecules are required to gain insight into circumstances surrounding their introduction into, and their maintainence in, this genus. This can be most readily achieved in the relatively simple *P. berghei* system as single S-type knockout lines exist and the potential for multiple rounds of genetic manipulation has been realised in this system. As mentioned above, the rate of development in the mosquito may be a key factor and diurnal environmental temperature fluctuation may have a greater effect on parasite lines lacking the appropriate number and type of rRNA genes. Alternatively, the three ribosome system could be vestigial; like the appendix, a remaining product of the evolutionary history of the organism.

Interest in cold-stimulated promoters is significant because it may offer a versatile tool for experimentation on gene function in the parasite, such as a way to control the production of a particular transcript. A cold-induced promoter that worked in *Plasmodium* would be of interest for studying the function of a gene, eliminating the presence of a messenger RNA with a ribozyme or preventing translation with antisense messages. A full range of control appears to be accessible within the temperature ranges at which *P. falciparum* is viable in culture. This is particularly appealing because, to date, there have been no reports of the successful introduction into *Plasmodium* of tetracycline-inducible expression systems. Set against the value of cold-inducible specific transcription is the fact that it will take place in a background of at least 250 blood-stage genes whose transcription profile is sensitive to lowered environmental temperature.

It has been shown that temperature-sensitive promoters play a role in development but the effects of factors such as oxygen tension and carbon source availability have not yet been reported. It is also possible that ribosome function is not the reason for variation between the types. For example, it has been proposed that variation in human globin genes is not totally a product of evolutionary selection for gene function. This variation may also ensure that genes with great sequence similarity remain in their appropriate chromosomal location and not altered or deleted by homologous recombination.

One question that arises is why other vector-borne diseases do not also maintain sequence variability in their rRNAs. In fact, variable forms of ribosomes are often found in plants and bacteria but variation in these organisms is predominantly the product of changes in associated r-proteins. These changes act to facilitate translation of select messenger RNAs during dramatic environmental changes. For example, enteric bacteria undergo a regular temperature drop upon vacating their vertebrate host, similar to that occurring for *Plasmodium falciparum* during transmission. In response to their new environment, the bacteria undergo changes involving new transcription factors and ribosomal proteins that facilitate translation at the lower temperature. These changes have been shown to be a biologically advantageous response to environmental change and are thought to serve to make the bacteria competitive in two dramatically different environments. Reassembly of existing ribosomes occurs ostensibly as an adaptation to a new environment. It would not be surprising to find that thermally controlled transitions occur in other vector-borne diseases, as they do in bacteria and *Plasmodium*.

Reference to the genome sequence does not reveal obvious sets of duplicated or related r-protein encoding genes, which might offer an indication that the protein complement of the ribosomes assembled on the different rRNA classes might differ. However, genome-wide analyses such as transcriptome and possibly proteome studies, coupled with bioinformatics, are required to address this issue in greater depth. What is clear, is that there are still many mysteries surrounding the stage-specific ribosomes of *Plasmodium* and that continued research is essential to uncover the precise biology of what are traditionally lucrative targets for the development of effective chemotherapeutics.

# 6. References

Alberts, B., Bray, D., Lewis, J., Raff, M., Roberts, K., and Watson, J.D. 1994. The cell nucleus. In: Molecular Biology of the Cell. S. Smart, ed. Garland Publishing, Inc., New York. p. 335–399.

Allen, P.N. and Noller H.F. 1989. Mutations in ribosomal protein S4 and S12 influence the higher order structure of 16S ribosomal RNA. J. Mol. Biol. 208: 457–468.

Barker, M.M. and Gourse, R.L. 2001. Regulation of rRNA transcription correlates with nucleoside triphosphate sensing. J. Bacteriol. 183: 6315–6323.

Cashel, M., Gentry, D.G., Hernandez, V.H. and Vinella, D. 1996. The stringent response. In: *Escherichia coli* and *Salmonella*. F.C. Neidhardt, R. Curtis III, J.L. Ingraham, E.C.C. Lin, K.B. Low, B. Hagasanik, W.S. Reznikoff, M. Riley, M. Schaechter, and H.E. Umbarger, eds. Publisher, City. p. 1458–1496.

Corredor, V., and Enea, V. 2003. Plasmodium ribosomal RNA as phylogenetic probe: a cautionary note. Mol. Biol. Evol. 10: 924-926.

Dame, J.B. and McCutchan, T.F. 1983. The four ribosomal DNA units of the malaria parasite *Plasmodium berghei*. Identification, restriction map, and copy number analysis. J. Biol. Chem. 258: 6984–6990.

Doudna, J.A., and Rath, V.L. 2002. Structure and function of the eukaryotic ribosome: the next frontier. Cell. 109(2): 153–156.

Fang, J., and McCutchan, T.F. 2002. Thermoregulation in a parasite life cycle. Nature. 418: 742.

Green, R. and Noller, H.F. 1997. Ribosomes and translation. Annu. Rev. Biochem. 66: 679–716.

Gunderson, J.H., Sogin, M.L., and Wollett, G., Hollingdale, M., de la Cruz, V.F., Waters, A.P., and McCutchan, T.F. 1987. Structurally distinct, stage-specific ribosomes occur in *Plasmodium*. Science. 238: 933–937.

Gutell, R.R., Gray, M.W., and Schnare, M.N. 1993. A compilation of large subunit (23S and 23S-like) ribosomal RNA structures. Nucleic Acids Res. 21: 3055–3074.

Janse, C.J., Carlton, J.M.R., Walliker, D., and Waters, A.P. 1994. Conserved location of genes on polymorphic chromosomes of four species of malaria parasites. Mol. Biochem. Parasitol. 68: 285–296.

Langsley, G., Hyde, J.E., Goman, M., and Scaife, J.G. 1983. Cloning and characterisation of the rRNA genes from the human malaria parasite *Plasmodium falciparum*. Nucleic Acids Res. 11(24): 8703–8717.

Li, J., Gutell, R.R., Damberger, S.H., Wirtz, R.A., Kissinger, J.C., Rogers, M.J., Sattabongkot, J., and McCutchan, T.F. 1997. Regulation and trafficking of three distinct 18S ribosomal RNAs during development of the malaria parasite. J. Mol. Biol. 269: 203–213.

Li, J., Collins, W.E., Wirtz, R.A., Rathore, D., Lal, A., and McCutchan, T.F. 2000. Synergy between *Plasmodium vivax* and mosquito vectors can result in the genetic isolation of parasite populations. Emerg. Infect. Dis. 7: 35–42.

Maguire, B.A. and Zimmermann, R.A. 2001. The ribosome in focus. Cell. 104: 813–816.

McCutchan, T.F., de la Cruz, V.F., Lal, A.A., Gunderson, J.H., Elwood, H.L., and Sogin, M.L. 1988. Primary sequences of two small subunit ribosomal RNA genes from *Plasmodium falciparum*. Mol. Biochem. Parasitol. 28: 63–68.

McCutchan, T.F., Li, J., McConkey, G.A., Rogers, M.J., and Waters, A.P. 1995. The cytoplasmic ribosomal RNAs of *Plasmodium* spp. Parasitol. Today. 11: 134–138.

Mitchell, P., Osswald, M., and Brimacombe, R. 1992. Identification of intermolecular RNA cross-links at the subunit interface of the *Escherichia coli* ribosome. Biochemistry. 31: 3004–3011.

Noller, H.F., Hoffarth, V., and Zimniak, L. 1992. Unusual resistance of peptidyl transferase to protein extraction procedures. Science. 256: 1416–1419.

Nomura, M., Gourse, R., and Baughman, G. 1984. Regulation of the synthesis of ribosomes and ribosomal components. Annu. Rev. Biochem. 53: 75–115.

Rogers, M.J., Gutell, R.R., Damberger, S.H., Li, J., McConkey, G.A., Waters, A.P., and McCutchan, T.F. 1996. Structural features of the large subunit rRNA expressed in *Plasmodium falciparum* sporozoites that distinguish it from the asexually expressed large subunit rRNA. RNA. 2: 134–145.

Rogers, M.J., Li, J., and McCutchan, T.F. 1998. The *Plasmodium* rRNA genes: developmental regulation and drug target. In: Malaria: Parasite Biology, Pathogenesis, and Protection. I.W. Sherman, ed. ASM Press, Washington D.C. p. 203–217.

Rogers, M.J., McConkey, G.A., Li, J., and McCutchan, T.F. 1995. The ribosomal DNA loci in *Plasmodium falciparum* accumulate mutations independently. J. Mol. Biol. 254: 881–891.

Thompson, J., van Spaendonk R.M.L., Choudhuri, R., Sinden, R.E., Janse, C.J., Waters, A.P. 1999. Heterogeneous ribosome populations are present in *Plasmodium berghei* during development in its vector. Mol. Microbiol. 31: 253–260.

Thompson, J., and Sinden, R.E. 1994. *In situ* detection of Pbs21 mRNA during sexual development of *Plasmodium berghei*. Mol. Biochem. Parasitol. 68: 189–196.

van Spaendonk, R.M.L., Ramesar, J., Janse, C.J., and Waters, A.P. 2000. The rodent malaria parasite *Plasmodium berghei* does not contain a typical O-type small subunit ribosomal RNA gene. Mol. Biochem. Parasitol. 105: 169–174.

van Spaendonk, R.M.L., Ramesar, J., Wigcheren, A., Eling, W., Beetsma, A., Gemert, G.J., Hooghof, J., Janse, C.J., and Waters, A.P. 2001. Functional equivalence of structurally distinct ribosomes in the malaria parasite *Plasmodium berghei*. J. Biol. Chem. 276: 22638–22647.

Velichutina, I.V., Rogers, M.J., McCutchan, T.F., and Liebman, S.W. 1998. Chimeric rRNAs containing the GTPase centers of the developmentally regulated ribosomal rRNAs of *Plasmodium falciparum* are functionally distinct. RNA. 4: 594–602.

Vinkenoog, R., Speranca, M.A., van Breemen, O., Ramesar, J., Williamson, D.H., Ross-MacDonald, P.B., Thomas, A.W., Janse, C.J., del Portillo, H.A., and Waters, A.P. 1998. Malaria parasites contain two identical copies of an elongation factor-1 alpha gene. Mol. Biochem. Parasitol. 94(1): 1–12.

Waters, A.P., van Spaendonk, R.M.L., Ramesar, J., Vervenne, H.A.W., Dirks, R.W., Thompson, J., and Janse, C.J. 1997. Species-specific regulation and switching of transcription between stage-specific ribosomal RNA genes in *Plasmodium berghei*. J. Biol. Chem. 272: 3583–3589.

Waters, A.P., Syin, C., and McCutchan, T.F. 1988. Developmental regulation of stage-specific ribosome populations in *Plasmodium*. Nature. 333: 74–76.

Wellems, T.E., Walliker, D., Smith, C.L., do Rosario, V.E., Maloy, W.L., Howard, R.J., Carter, R., and McCutchan, T.F. 1987. A histidine-rich protein gene marks a linkage group favoured strongly in a genetic cross of *Plasmodium falciparum*. Cell. 49: 633–642.

Wilson, K.S. and Noller, H.F. 1998. Molecular movement inside the translational engine. Cell. 92: 337–349.

Zhang, K. and Rathod, P.K. 2002. Divergent regulation of dihydrofolate reductase between malaria parasite and human host. Science. 296: 545–547.

Zhu, J.D., Waters, A.P., Appiah, A., McCutchan, T.F., Lal, A.A., and Hollongdale, M.R. 1990. Stage-specific ribosomal expression switches during sporozoite invasion of hepatocytes. J. Biol. Chem. 265: 12740–12744.

From: Malaria Parasites: Genomes and Molecular Biology
Edited by: A.P. Waters and C.J. Janse

# Chapter 10

## Cell Cycle Control in *Plasmodium falciparum*: A Genomics Perspective

Christian Doerig
and Debopam Chakrabarti

## Abstract

The molecular mechanisms regulating cell proliferation and development in malaria parasites are still largely unknown. Phenomenological observations, pertaining to the organisation of the cell cycle during schizogony or to the signal transduction pathways whose activation is responsible for the developmental stage transitions, can now be complemented with information gathered from genomic databases. The PlasmoDB database has been used extensively to identify putative homologues of a number of eukaryotic cell cycle regulators such as cyclins, cyclin-dependent kinases, factors involved in the control of DNA synthesis, and components of signal transduction pathways. However, gene identification based on sequence homology is limited by the fact that any *Plasmodium*-specific functional homologue will be missed by this approach. Furthermore, experimental data indicate that the structure of some regulatory pathways (unlike that of metabolic pathways) cannot be deducted directly from database mining. Because of these limitations in the direct exploitation of genomic database, elucidation of the organisation of the signal transduction and cell cycle machineries requires experimental, proteomics-based approaches such as the characterisation of protein complexes containing cell cycle regulators and the establishment of a map of protein-protein interactions involving these elements.

# 1. Introduction

## 1.1. The Cell Cycle and the Life Cycle of Malaria Parasites

Like all living organisms, malaria parasites must tightly control cell proliferation and differentiation. Completion of the *Plasmodium* life cycle requires the succession of a large number of different developmental stages, which vary greatly with respect to proliferation status. Hence, the sporozoites accumulated in the mosquito's salivary gland have arrested their cell cycle, and maintain it in this state until they have found their way into a hepatocyte. Thereafter, schizogony is initiated, and the resulting merozoites withdraw from proliferation until they have invaded an erythrocyte, where schizogony resumes. The merozoites which are committed to sexual differentiation (gametocytogenesis) do not undergo a full division cycle while in the bloodstream of the vertebrate host; male gametocytes require ingestion by the mosquito to complete three rounds of cell division and produce eight gametes, each of which can fuse with a female gamete to yield a zygote (the only diploid stage in the entire life cycle). Meiosis rapidly ensues, and the ookinete is quiescent (in terms of cell cycle!) until after it has crossed the midgut epithelium. Then intense asexual division occurs again, generating several thousand sporozoites, the cell cycle-arrested form accumulating in salivary glands (see http://www.malaria.org and links therein for further information on the malaria life cycle).

Cell cycle control elements are presumably involved not only in cell multiplication processes themselves, e.g. during schizogony, but also in the transitions between developmental stages differing in their proliferation status. Their fundamental importance in cellular regulation makes them attractive targets for intervention. As will be discussed below, it is becoming clear that the parasite cell cycle control machinery diverges both structurally and functionally from that of mammalian cells, lending support to the idea that specific interference can be achieved.

This chapter starts with a review of essential features of eukaryotic cell cycle regulation (section 1). Section 2 presents the insight into the *Plasmodium* cell cycle that was gained form observational and phenomenological approaches, while section 3 focuses on specific genes which are likely to play a part in the control of cell division, a line of investigation that has witnessed a tremendous flourish since the advent of malaria parasite genomics. Finally, section 4 consists of a discussion of the benefits and limitation of genomics in the context of cell cycle research.

## 1.2. The Eukaryotic Cell Cycle Machinery

Chromosomal replication followed by mitosis and cytokinesis is central to proliferation of any cell. These events are coordinately regulated by intricate mechanisms to maintain accuracy in genetic information and constant chromosome numbers through successive generations. The eukaryotic cell cycle comprises two major phases: S phase (synthesis), in which the genome is duplicated, and M phase (mitosis), during which the replicated genome is segregated to the daughter cells. There are two intervening gap phases, G1 before S phase and G2 before M phase. Progression through the phases of the cycle is controlled by a family of protein kinases, the cyclin-dependent kinases (CDKs). These enzymes are inactive as monomers: activation requires binding of a positive regulatory subunit (cyclin). Cyclin binding causes a structural rearrangement of the kinase subunit that allows access of the substrates to the catalytic cleft of the enzyme. Several cyclins, each with a defined specificity for a subset of CDKs, are expressed at different phases of the cell cycle, and their abundance is tightly regulated by transcriptional control and ubiquitin-dependent proteolysis. Hence, the temporary association of particular kinase and cyclin subunits at the appropriate phase of the cell cycle ensures progression of the cycle to the next phase. In yeast a single CDK (cdc2 in *Schizosaccharomyces pombe*) associates with a wide variety of S-phase and mitotic cyclins to control the entire cell cycle, whereas in higher eukaryotes, several CDKs coexist in the cell. The CDK-cyclin complexes are regulated by their own phosphorylation status: kinases such as Wee1 and Myt1 inactivate CDKs by phosphorylating residues in the ATP binding pocket of the enzymes; this inhibition is relieved by the activity of phosphatases of the CDC25 family. Thus, human CDKs 1, 2, 4 and 6 are inhibited by active-site phosphorylation (Tyr15 in CDK2) as a response to activation of various checkpoint signaling pathways. Some CDKs are also substrates for the «CDK-activating kinases» (CAKs), whose activity results in the phosphorylation of a conserved threonine residue (Thr160 in CDK2) and increased activity of the target kinase subunit. Cyclin-dependent kinase inhibitors (CKIs) of the p21/p27 and p16 families provide additional means of negatively regulating the activity of CDKs and can lead to cell cycle arrest once activated by upstream signals (for recent reviews on different aspects of the eukaryotic cell cycle machinery, see (Morgan, 1997; Solomon and Kaldis, 1998; Sherr and Roberts, 1999; Harper and Adams, 2001).

The cell cycle control machinery is itself an effector that is placed under the control of a variety of signal transduction pathways. These pathways are responsible for transmitting intra- or extracellular signals to effectors such as the transcription and cell cycle control machineries, so that appropriate adaptative responses are activated. Activation of transduction pathways can result either in triggering proliferation of resting cells, or in induction of cell cycle arrest of proliferating cells; these effects can be mediated by a wide variety of mechanisms, including transcriptional regulation of cyclins and other cell cycle control elements, and direct effect (e.g. through phosphorylation) on molecules such as CKIs, kinases or phosphatases. A well understood example is cell cycle arrest during yeast mating-type differentiation: pheromone-dependent activation of a mitogen-activated protein kinase (MAPK) pathway results in the phosphorylation (and concomitant activation) of a CKI, which in turns interferes with CDK-cyclin activity, thereby arresting the cell cycle prior entry into S-phase (reviewed in (Bardwell *et al.*, 1994). In all likelihood, similar mechanisms operate at many points in the life cycle of malaria parasites.

## 1.3. The Control of Entry into S-Phase

Current models of S phase regulation, mainly based on studies in the yeast system, focus on two stages: (1°) assembly of the pre-replicative complex (pre-RC) at DNA replication origins, and (2°) initiation of DNA synthesis. A group of high molecular weight proteins, referred to as the origin recognition complex (ORC), is bound to origins of replication *in vivo* throughout the cell cycle and is required for the initiation of DNA synthesis. The ORC consists of six polypeptides, ORC1 to 6, with molecular masses, in yeast, ranging from 50 to 120 kDa. Most likely ORC1, ORC2, and ORC4 determine the specificity of binding, as these subunits were shown by protein-DNA cross-linking to interact with the major groove of the DNA and bind within 10 Å of the origin of replication. The binding of ORC to DNA is dependent on hydrolysis of ATP. ORC1 and ORC5 subunits contain purine nucleotide binding motifs, the mutation of which resulted in non-viable or slow-growing cells in *S. cerevisiae*. It is believed that the ORC acts as a landing pad to nucleate the loading of initiation proteins onto replication origins during the transition from late mitosis to G1 phase. During G0/G1, an unstable, cell cycle-regulated regulatory factor (CDC6 in *S. cerevisiae*, cdc18 in *S. pombe*) is synthesized and associates rapidly with ORC-bound origins. Cdc6 also contains a purine nucleotide-binding motif and the region carrying this motif exhibits similarity to that of ORC1. Cdc6 is now recognized as one of the factors (together with Cdt1, a protein recently identified in fission yeast) responsible for loading of the MCM complex (for mini-chromosome maintenance), another six-member family of related proteins, which contain nucleotide triphosphate-binding domain, onto the origin of replication. The chromatin binding of MCM proteins is stringently cell cycle regulated and is dependent on the replication status of the cell. MCM, Cdc6 and Cdt1 have been often referred to as one of the hypothetical "replication licensing factor" ("licensing" refers to the chromosome's acquisition of competence to replicate only once per cell cycle). In *S. cerevisiae*, localization of MCM proteins is similar to that predicted for a "licensing factor": They translocate to the nucleus at late M phase, remain there during G1 and disappear from nucleus at the onset of S phase. The activity of Cdc6 and Cdt1 is downregulated in G2 to prevent re-replication. The chromatin structure at the origin, therefore, oscillates between a pre-replicative (pre-RC) state in the G1 and post-replicative (post-RC) in S, G2, and M cell cycle stages. The pre-RC is disassembled at the G1/S boundary, allowing the replication origins to fire (reviewed in (Nasmyth, 1996; Dutta and Bell, 1997; Kelly and Brown, 2000; Lei and Tye, 2001).

## 1.4. Replication Complex and Onset of S Phase

Several lines of evidence indicate that the transition from G1 to S correlates with an increase in S phase specific CDK activity. CDK not only promotes the initiation of DNA replication but also prevents reinitiation before onset of mitosis. It has been proposed that proteins phosphorylated by S-phase CDKs are incapable of re-associating with the chromatin, thereby restricting firing of origins to only once per cell cycle. Recent studies have identified the targets for S-phase CDKs as being components of pre-RC. The fission yeast homologue of Cdc6 is highly phosphorylated by S-phase CDK. This phosphorylation is not essential for promoting DNA replication, but is needed for proteolysis of Cdc6; this phosphorylation-dependent proteolysis ensures that the formation a new pre-RC requires synthesis of new Cdc6 molecules. Hyperphosphorylation of MCM proteins has also been associated with their dissociation from the pre-RC. Likewise, the *Xenopus* homologue of MCM4 is specifically phosphorylated by a CDK (cdc2/cyclinB), which causes a substantial decrease in MCM4 affinity for chromatin. In addition to CDK, another protein kinase with homologues in fission yeast and metazoans, CDC7, is essential for the S-phase progression. CDC7 is also referred to as DDK (Dbf4-dependent kinase) because of its positive regulator Dbf4. DDK has been shown recently to phosphorylate MCM2, MCM3, MCM4, and MCM6 *in vitro*, and blocking this kinase at the G1/S boundary also inhibits the phosphorylation of MCM2 at this point *in vivo*. DDK is necessary for firing of different replication origins throughout the S phase. As a consequence of CDK and DDK activity, another protein, Cdc45, associates with MCM at the replication origin at the onset of S phase (reviewed in (Morgan, 1997). Cdc45 appears to serve as an initiation factor to transform the pre-RC to RC (Zou and Stillman, 1998), and is important for recruitment of proteins involved in DNA replication such as replication protein A (RPA), DNA polymerase $\alpha$ and $\epsilon$, and proliferating cell nuclear antigen (PCNA) (Schwacha and Bell, 2001).

## 1.5. Mitosis and Exit from Mitosis

Upon completion of DNA replication the cell enters into G2 phase, and refiring of replication origins is prevented until the next cycle. This is achieved, at least in part, by a decrease in Cdc6 and Cdt1 proteins levels. Recently it has been shown that expression of Cdc6 and/or Cdt1 in G2 induces premature initiation of DNA synthesis (Yanow *et al.*, 2001). In G2, mitotic CDK level rises slowly but the enzymes are kept inactive by inhibitory phosphorylation at Y15 and T14. Wee1 kinase is responsible for Y15 phosphorylation (Borgne and Meijer, 1996). The quiescence in G2 is maintained by the inactivity of CDK. Two checkpoint proteins, the Chk1 and Cds1 kinases, are activated by damaged DNA or unreplicated DNA. Activated Chk1 or Cds1 keep mitotic CDK inactive by phosphorylating Wee1 kinase (which is activated by phosphorylation) or Cdc25 phoshphatase (which is inactivated by phosphorylation). Entry into mitosis occurs once CDC25 inhibition is released and hence can dephosphorylate the CDK (Morgan, 1997; Solomon and Kaldis, 1998; Sherr and Roberts, 1999; Harper and Adams, 2001).

In mitosis, chromosome condensation is followed by alignment of condensed chromosomes at the metaphase plate. Many proteins belonging to the "structural maintenance of chromosomes (SMC)" gene family and others such as Scc1 and Scc3 are responsible for promoting chromosome condensation. The condensed chromosomes then become attached to the mitotic spindle, leading to the initiation of segregation of chromosomes in anaphase. The segregation of chromosomes is initiated by proteolytic degradation of Scc1. Separin proteins such as Esp1 or Cut1 regulate the cleavage of Scc1. Securin proteins such as Pds1 and Cut2 associates tightly with Separins to prevent their activity until the proper time. The metaphase to anaphase transition is triggered by the anaphase-promoting complex/ cyclosome (APC/C). The APC/C in association with Cdc20 targets the Esp1 inhibitor Pds1 (a securin) for ubiquitin-dependent proteolysis, which results in the cleavage of Scc1 by Esp1 and initiation of chromosome segregation. The chromosomes are then pulled towards the centrosome. Upon completion of chromosome segregation, exit from mitosis ensues due to inactivation of mitotic CDK. In *S cerevisiae*, a set of proteins, known as 'mitotic exit network' or MEN, are involved in the process of exit from mitosis. The MEN includes the protein phosphatase Cdc14, whose activation results (indirectly) in the inhibition of mitotic CDK (reviewed in Gardner and Burke, 2000; Hixon and Gualberto, 2000; Adams *et al.*, 2001; Losada and Hirano, 2001; McCollum and Gould, 2001).

## 1.6. Nucleocytoplasmic Transport

In other eukaryotes, this process plays a crucial role in transducing signals to nuclear effectors, and importing regulatory elements such as cyclins and kinases to the nucleus. It is controlled by sets of proteins functioning as transporters (e.g. importins), indicator proteins, or recycling factors. The small GTP-binding protein Ran (ras-related nuclear protein) is the principal indicator protein, whose function is to distinguish cytoplasmic versus nuclear environments (reviewed in Sazer and Dasso, 2000). The other core components of the Ran pathway are (i) RanGAP (GTPase activating protein), which is cytoplasmic and stimulates hydrolysis of GTP to GDP by Ran, (ii) RanGEF (guanine–nucleotide exchange factor), also called RCC1 (regulator of chromosome condensation), which is nuclear and ensures GTP binding to Ran, and (iii) RanBP1 (Ran Binding Protein 1), which interacts specifically with the GTP-bound form of Ran and acts as a co-activator of RanGAP. The GTP-bound form of Ran is present in the nucleus and the GDP-bound form is found predominantly in the cytoplasm, a consequence of the presence of RanGAP in the cytoplasm and RCC1 in the nucleus. In the case of importin-mediated transport, importins (karyopherins) α and β form a heterodimer able to bind to proteins possessing a nuclear import signal. The heterotrimer is then transloctaed to the nucleus, and direct interaction by Ran-GTP to karyopherin β releases karyopherin α and the imported protein into the nucleus. After export from the nucleus, transporters such as karyopherin β are bound to RanGTP and need to be released from it. The dissociation of transporter–RanGTP complexes is accomplished by RanBP1 (Coutavas *et al.*, 1993), and GTP hydrolysis is triggered by RanGAP on the transient complex RanGTP-RanBP1. At this point another round of transport can be initiated (reviewed in (Gorlich and Kutay, 1999)).

Homologues of many of the proteins discussed in this section have been identified in the PlasmoDB database, or have already been characterised at the biochemical level (see section 3 and Table 1).

# 2. Cell Cycle Control in Malaria Parasites: Lessons From Phenomenological Approaches

This section covers a topic that has been reviewed elsewhere (Sinden *et al.*, 1996; Doerig, 1997; 2000). Therefore, we will only briefly review data on cell division that have been obtained by direct structural, biochemical or pharmacological studies of parasites at different developmental stages, and of cellular events occurring during transition between stage (by opposition to the targeted study of specific genes, which is the subject of section 3).

## 2.1. Erythrocytic Schizogony and Sexual Development

The stage of the life cycle that has received most attention is erythrocytic schizogony, because it is responsible for pathology; it is also the most amenable to study, as it is comparatively easy to obtain biological material from blood stage cultures. The process of cell division in Apicomplexa is very different from the binary fission observed in most other Eukaryotes. Daughter cells are assembled within the mother, using the so-called "inner membrane complex", a scaffold of cytoskeletal and membranous components; this process, called endodyogeny if there are just two daughter cells, has

**Table 1.** *P. falciparum* gene products presumably belonging to transduction pathways and regulatory networks.

| | Maximal Homology To: | Remarks | References |
|---|---|---|---|
| **cAMP Pathway** | | | |
| PfACα (AY191005) | prokaryotic adenylyl cyclases | Complements *Dictyostelium* AC⁻ mutant | Muhia *et al.*, 2003 |
| PfACβ (NP_704518) | soluble cyclases | Possesses a twin catalytic domain. | Muhia *et al.*, 2003 |
| PfPKAc (CAA11945) | PKA family | cAMP-dependent kinase activity detected in parasite extracts | Syin *et al.*, 2001 |
| PfPKAr (CAD29699) | PKAr, RII subfamily | Recombinant PfPKAr associates with a cAMP-dependent kinase activity | Merckx and Doerig, unpublished |
| PFL0475w MAL13P1.118 PF14_0672 MAL13P1.119 | cyclic nucleotide phosphodiesterases | No data other than sequence and expression profiles in PlasmoDB | PlasmoDB |
| PfAKAP (PF10_0092) | A-Kinase anchoring protein | No data other than sequence and expression profiles in PlasmoDB | PlasmoDB |
| **cGMP Pathway** | | | |
| PfGCα (CAC00546) | G-protein-dependent adenylyl cyclases | No demonstrated activity of recombinant protein. | Carrucci *et al.*, 2000 |
| PfGCβ (CAC05389) | G-protein-dependent adenylyl cyclases | Guanylyl cyclase activity demonstrated for recombinant protein, despite structural relatedness to adenylyl cyclases. | Carrucci *et al.*, 2000 |
| PfPKG (PF14_0346) | PKG family | cGMP-dependent kinase activity demonstrated for recombinant protein | Deng and Baker, 2002 |
| **MAPK Pathway** | | | |
| Pfmap-1 (CAA57972) | ERK1/ERK2 subfamily | Typical ERK1/2 activation site (TXY); demonstrated kinase activity for recombinant enzyme. | Doerig *et al.*, 1996 Lin *et al.*, 1996 Graeser *et al.*, 1997 |

| | | | |
|---|---|---|---|
| Pfmap-2 (CAA67247) | ERK1/ERK2 subfamily | Atypical activation site (TSH); demonstrated kinase activity for recombinant enzyme. Expressed specifically in gametocytes. | Dorin *et al.*, 1999 |
| Pfnek-1 (CAB76949) | NIMA/Nek family | MAPKK-like (instead of NIMA-like) activation site. Phosphorylates Pfmap-2 *in vitro*. | Dorin *et al.*, 2001 |
| PfPK7 (CAD12771) | Similar levels of homology to MAPKK and PKA | Autophosphorylation (but not exogenous substrate phosphorylation) demonstrated for recombinant protein. | Dorin and Doerig, unpublished |
| Pfraf (PFB0520w) | RafB family | No activity of recombinant enzyme. Expressed in blood stages. | Nivez and Doerig, unpublished |
| Missing or undetected (by comparison with MAPK pathways in other eukaryotes): "true" MAPKK, ras, p38, Junk, MAP4K | | | |
| **Phosphoinositide Pathway** | | | |
| Akt/PKB (PFL2250c) | Mouse AKT1 kinase | No data other than sequence and expression profiles in PlasmoDB | PlasmoDB |
| MAL13P1.82 | phosphatidylinositol synthase | No data other than sequence and expression profiles in PlasmoDB | |
| PFE0765w | phosphatidylinositol 3-kinase | No data other than sequence and expression profiles in PlasmoDB | PlasmoDB |
| PFE0485w | phosphatidylinositol 4-kinase | No data other than sequence and expression profiles in PlasmoDB | PlasmoDB |
| PFA0515w | phosphatidylinositol-4-phosphate 5-kinase | No data other than sequence and expression profiles in PlasmoDB | PlasmoDB |
| PFB0410c | phospholipase | No data other than sequence and expression profiles in PlasmoDB | PlasmoDB |
| PF10_0132 | phospholipase C-like | No data other than sequence and expression profiles in PlasmoDB | PlasmoDB |
| PF10_0379 | phospholipase | No data other than sequence and expression profiles in PlasmoDB | PlasmoDB |
| Missing or undetected: Heterotrimeric G protein | | | |

*Table 1, Continued*

| | Maximal Homology To: | Remarks | References |
|---|---|---|---|
| **Calcium Signalling** | | | |
| PfCDPK1 (X67288) | Plant/Ciliate CDPKs | Demonstrated calcium-dependent kinase activity of recombinant protein | Zhao et al., 1994 |
| PfCDPK2 (X99763) | Plant/Ciliate CDPKs | Demonstrated calcium-dependent kinase activity of recombinant protein | Farber et al., 1997 |
| PfCDPK3 (AAF63154) | Plant/Ciliate CDPKs | Gametocyte-specific accumulation of mRNA | Li et al., 2000 |
| calmodulin (X56950) | calmodulin from a variety of species | Expression in schizonts higher than in earlier stages | Robson et al., 1991<br>Calvo et al., 2002 |
| Missing or not unambiguously identified: PKC | | | |
| **Cell Cycle Control Machinery** | | | |
| PfPK5 (CAA43923) | Similar levels of homology (60% identity) to p34cdc2 (CDK1) and CDK5. | Cyclin-dependent histone H1 kinase activity demonstrated for the recombinant enzyme. Promiscuity towards activating cyclin partner *in vitro.* 3D structure solved experimentally. | Ross-McDonald et al., 1994<br>Le Roch et al., 2000<br>Holton et al., 2003 |
| PfPK6 (AAC61592) | Similar levels of homology to CDKs and MAPKs | Cyclin-independent histone H1 kinase activity demonstrated for the recombinant enzyme | Bracchi-Ricard et al., 2000 |
| Pfcrk-1 (CAA56732) | p58-GTA/PISTLRE family | No activity detected for recombinant enzyme (even in the presence of available cyclins). mRNA detected in gametocytes but not asexual parasites. | Doerig et al., 1995 |
| Pfcrk-3 (PFD0740w) | p34-cdc2 (CDK1) (39% identity over the catalytic domain [excluding insertions]) | Large insertions and N-terminal extension in addition to the kinase catalytic domain. No activity demonstrated for recombinant protein (catalytic domain). | Equinet and Doerig, unpublished |
| Pfcrk-4 (CAB1114) | Similar levels of homology to CDKs and MAPKs | Large insertions and N-terminal extension in addition to the kinase catalytic domain. No activity demonstrated for recombinant protein (catalytic domain) | Equinet and Doerig, unpublished |

| | | | |
|---|---|---|---|
| Pfcrk-5 (MAL6P1.271) | CDK1/CDK2 | No data other than sequence and expression profiles in PlasmoDB | Reiniger, Nivez, and Doerig, unpublished |
| Pfmrk (AAC72269) | CDK7 (CDK-activating kinase) | Cyclin-dependent histone H1 kinase activity demonstrated for the recombinant enzyme | Li *et al.*, 1996<br>Li *et al.*, 2001 |
| Pfcyc-1 (AJ245852) | Cyclin H | Recombinant protein activates PfPK5 and Pfmrk (but no other malarial kinase) *in vitro*. Associates with histone H1 kinase activity in parasite extracts. | Le Roch *et al.*, 2000 |
| Pfcyc-2 (PFL1330c) | Cyclin A/B | Associates with histone H1 kinase activity in parasite extracts. | Merckx *et al.*, 2003 |
| Pfcyc-3 (PFE0920c) | Trypanosome CYC2 | Associates with histone H1 kinase activity in parasite extracts. Activates PfPK5 *in vitro*. | Merckx *et al.*, 2003 |
| Pfcyc-4 (PF13_0022) | Cyclin Ania-6 | Associates with histone H1 kinase activity in parasite extracts. | Merckx *et al.*, 2003 |
| PfGSK-3 (PFC0525) | GSK3β | Demonstrated kinase activity of recombinant enzyme. Localised in Maurer's Clefts. | Droucheau *et al.*, 2003 |

Missing or undetected: Myt1, CDC25, CKIs, CKS Heterotrimeric G protein, Wee1

### Pre-Replication & Replication Complex

| | | | |
|---|---|---|---|
| ORC1 (PFL0150w) | *D. melanogaster* | Expression in sexual stages detected | Li and Cox, 2003 |
| ORC2 (MAL7P1.21) | *D. melanogaster* | No data other than sequence and expression profiles in PlasmoDB | PlasmoDB |
| ORC5 (PFB0720c) | *S. pombe* | No data other than sequence and expression profiles in PlasmoDB | PlasmoDB |
| MCM2 (PF14_0177) | *S. cerevisiae* | Protein expression peaks in early schizonts. Interacts with MCM6 | Patterson and Chakrabarti, unpublished |
| MCM3 (PFE1345c) | *A. thaliana* | No data other than sequence and expression profiles in PlasmoDB | PlasmoDB |

*Table 1, Continued*

| | Maximal Homology To: | Remarks | References |
|---|---|---|---|
| MCM4 (PF13_0095) | *X. laevis* | Expression in sexual stages detected | Li and Cox, 2001 |
| MCM5 (PFD0790c) | *D. melanogaster* | No data other than sequence and expression profiles in PlasmoDB | PlasmoDB |
| MCM6 (PF13_0291) | *D. melanogaster* | Protein expression peaks in early schizonts. Interacts with MCM2 | Patterson and Chakrabarti, unpublished |
| MCM7 (PF07_0023) | *S.pombe* | No data other than sequence and expression profiles in PlasmoDB | PlasmoDB |
| MCM8 (PFL0560c) | Human | No data other than sequence and expression profiles in PlasmoDB | PlasmoDB |
| DNA polymerase α (AAK14825) | Mammalian | Native enzyme activity purified. | White *et al.*, 1993 |
| DNA polymerase δ (P30315) | *D. melanogaster* | Gene sequence and RNA expression analysed, Epstein-Barr virus-like | Ridley *et al.*, 1991 |
| RFC3 (AAG37985) | *S. pombe* | Other RFC subunits are also present. No data other than sequence and expression profiles in PlasmoDB | PlasmoDB |
| RPA1 (AL035475) | *S. pombe* | Single-stranded DNA-binding activity | Voss *et al.*, 2002 |
| PCNA1 (P31008) | *O. sativa* | Cloned and expression analysed | Kilbey *et al.*, 1993 |
| PCNA2 (AAN34972, AAG37983) | *T. gondii* | Differential expression of PCNA1 and PCNA2 in erythrocytic stages, forms oligomer | Patterson *et al.*, 2002, Li *et al.*, 2003. |
| Missing or undetected: Cdc6, Cdt1, Cdc45, DDK | | | |
| **Mitotic Exit Network** | | | |
| Cdc5 (PF10_0327) | *S.pombe* | No data other than sequence and expression profiles in PlasmoDB | PlasmoDB |
| Cdc14 (PF11_0139) | *S.cerevisiae* | No data other than sequence and expression profiles in PlasmoDB | PlasmoDB |

| | | | |
|---|---|---|---|
| APC10 (PFL0850w) | *S. pombe* | No data other than sequence and expression profiles in PlasmoDB | PlasmoDB |
| Cdc20 (PF10_0261) | *S. pombe* | No data other than sequence and expression profiles in PlasmoDB | PlasmoDB |
| Missing or undetected: Separin and Securin proteins | | | |
| **Nucleocytoplamsic Transport** | | | |
| PfRan (CAA52140) | Ras-related nuclear antigen Ran/TC4 | Demonstrated GTP binding property of the recombinant protein; binds to PfRanBP1 *in vitro* | Dontfraid *et al.*, 1994 Sultan *et al.*, 1994 |
| PfRCC1 (AAC61678) | Regulator of Chromosome Condensation RCC1 (guanidine nucleotide exchange factor) | Demonstrated nuclear localisation during schizogony | Ji *et al.*, 1998 |
| PfRanBP1 (CAD12772) | Ran-Binding protein 1 family | Recombinant protein binds to PfRan *in vitro* | Ramachandran *et al.*, 2002 |
| PfRanBP7 (chr2_11953) | Ran-Binding protein 7 family (importin 7) | No data other than sequence and expression profiles in PlasmoDB | PlasmoDB |
| PfKarα (AF529881) | Importin/karyopherin α | Recombinant PfKarα and PfKarβ form a heterodimer *in vitro*. | Mohmmed *et al.*, 2002 |
| PfKarβ (AF539437) | Importin/karyopherin β | | |

The name of the gene is indicated in the first (left) column, with the accession number in brackets. Where no accession number is available, the identifier of the gene in the PlasmoDB database is provided. The second column indicates the protein family to which the gene of interest shows maximal homology. It must be kept in mind that this has no absolute predictive value with respect to the actual function of the gene product. Where experimental data are available to ascertain the function of the gene product, this is indicated in the third column. The expression profiles in PlasmoDB mentioned in column 3, refer to two sets of microarray data (Le Roch *et al.*, 2003; Bozdech *et al.*, 2003).

This table essentially summarizes existing data, and does not pretend to be exhaustive in terms of database mining. However, a few genes that have been looked for, but for which no obvious homologue has been found in PlasmoDB, are indicated as "missing", because of their particular interest in the context of cell cycle control. Evidently, it is possible that functional homologues of these elements may be found in the future, either experimentally or by a more thorough exploitation of the database.

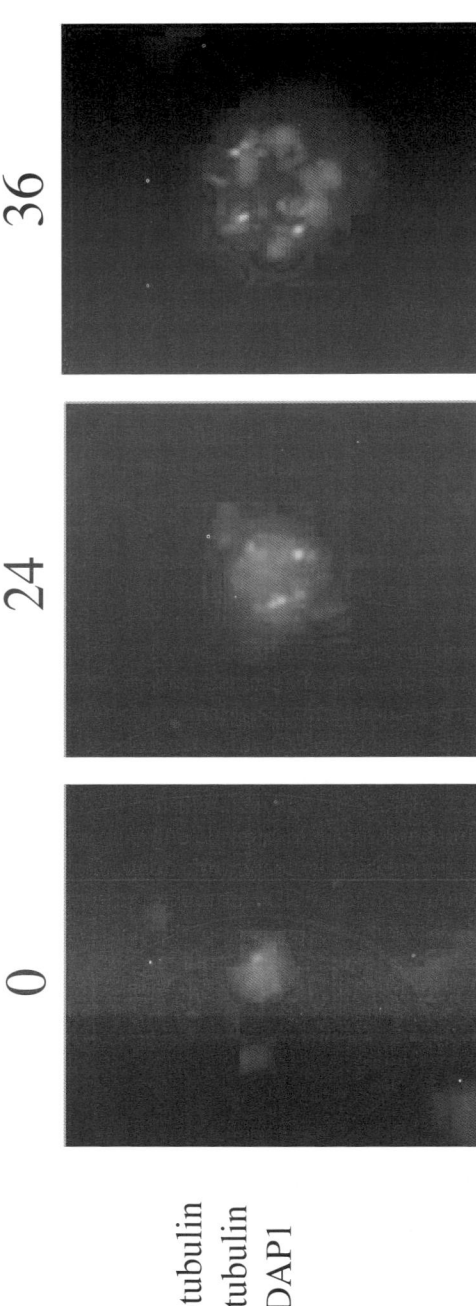

α-tubulin
γ-tubulin
DAP1

0    24    36

**Figure 1.** DeltaVision deconvolution fluorescence micrograph of cells stained with alpha-tubulin (chicken, red) and gamma-tubulin (human, green) antibodies, illustrating microtubule structure during *Plasmodium falciparum* asexual development. The use of these antibodies to label the *P. falciparum* microtubule-organizing center (MTOC) and microtubules was first reported by Fowler *et al.* (2001). Numbers on top indicate the number of hours post synchronisation. (0) Ring, (24), Mature Trophozoite, (36) Segmenter/schizont. In animal cells, gamma-tubulin is associated with the MTOC in the centrosome, from which microtubules (consisting of alpha- and beta-tubulins) spokes radiate. Here, alpha-tubulin can be seen to become organised into spindles as cell division progresses, while gamma-tubulin staining remains concentrated in a small number of punctuations, suggesting that the parasite utilizes the tubulins similarly to higher eukaryotes. As reported in Read *et al.*, 1993, spindles at different stages of development appear to be present in the schizont. This experimental approach does not yield conclusive information on the subcelluar location of the MTOC; it would be of interest to determine by immuno-electron microscopy what the relation is between the tubulins and the intranuclear hemicentriolar plaques referred to in the text. For a thorough discussion of microtubules in *Plasmodium* merozoites, see Fowler *et al.*, 2001. **See Colour Plate at the back of the book.**

been studied extensively in *Toxoplasma gondii* (Hu *et al.*, 2002). Another peculiarity of cell division in *Toxoplasma gondii* is that the segregation of a single apicoplast per daughter cell is linked to the mitotic spindle used for nuclear division (Striepen *et al.*, 2000). Hence, the overall organisation of cell division appears to be unique in these organisms. Accordingly, a clear correlation between the G1, S, G2 and M phases of the typical eukaryotic cell cycle and those of the schizogonic cycle of *Plasmodium* has yet to be established. It is generally accepted that merozoites and rings are in G1, and S phase is initiated in trophozoites, around 18 hours post invasion.

The processes involved in the ensuing S phases and nuclear divisions are not understood. Is there a clear succession of alternating S phases and nuclear divisions, or a burst of DNA synthesis followed by segregation of the genomes in different nuclei? Is there an equivalent to gap phases? Attempts to solve such questions using biochemical approaches are hampered by the difficulty in obtaining sufficiently finely synchronised populations, and by the apparent asynchronicity of nuclear division in a given schizont (see below) [reviewed in (Leete and Rubin, 1996; Arnot and Gull, 1998; Doerig *et al.*, 2000)]. Chromosomes do not condense, and electron microscopy examination of schizonts revealed that the nuclear membrane is maintained throughout the nuclear division process (endomitosis), a peculiarity that is shared by other lower eukaryotes such as yeast (Aikawa, 1966; Rudzinska, 1969). The mitotic spindle does not originate at a typical cytoplasmic centriole like in higher eukaryotes. Instead, microtubules of the mitotic spindle are generated from a centriolar plaque located at the inner side of the nuclear membrane; the plaque splits in two parts (the so-called hemicentriolar plaques) which migrate to opposite sides of the nucleus, forming an intranuclear spindle in the process (reviewed by Bannister *et al.*, 2000). The use of anti-tubulin antibodies in immunofluorescence staining allowed Read *et al.* (1993) to monitor the development of the spindle in individual nuclei, from the dot-like centriolar plaque to a full spindle. This yielded the surprising result that nuclei in a given schizont appear to divide independently from each other, as spindles at all stage of development could be observed in the same schizont (see Figure 1). This was confirmed by a quantitative analysis of the number of nuclei per schizont: if nuclear division was synchronous, one would expect to see a majority of schizonts distributed into cells with 1, 2, 4, 8 or 16 nuclei. Instead, the distribution curve was very smooth, almost Gaussian in appearance, with schizonts containing odd and even numbers (between 2 and 20) of nuclei being observed with equivalent frequency. Parasites with three nuclei and/or triploid DNA content have also been observed in *T. gondii*, indicating that in this system also individual nuclei seem to be autonomous with respect to at least some cell cycle progression events (Hu *et al.*, 2002).

The apparent asynchrony in nuclear divisions in a given schizont raises interesting questions with regard to the regulation of the process: it is generally accepted that progression of the cell cycle is driven, in part, by fluctuations in the cytoplasmic abundance of cyclins resulting from regulated gene expression; nuclear cell cycle events occur following transport of these elements into the nucleus. In *Plasmodium*, there must be another layer of regulation, since all nuclei in the same cytoplasm do not behave homogeneously. This peculiarity may be linked to endomitosis: if a nucleus is to complete its division cycle independently from other nuclei in the schizont, it has to maintain physical isolation for the entire duration of the process. In this context, investigations on the modalities of nucleocytoplasmic transport may be of particular interest (see Section 3).

It is thus apparent, despite the present lack of understanding of the molecular mechanisms underlying the process of *Plasmodium* asexual cell division, that the malaria parasite cell cycle during schizogony deviates from the canonical eukaryotic cell cycle. The malaria cell cycle resembles to some extent to the endoreduplication cell cycles in many insects and plants where nuclear polyploidization is observed (Edgar and Orr-Weaver, 2001). In contrast to the usual eukaryotic cell cycle, in endoreduplication the onset of S phase is uncoupled from completion of M phase. This may be achieved through an alteration of CDK activity. Support for this idea come from the observation that deletion of B-type cyclin in fission yeast causes multiple rounds of DNA replication. Likewise, overexpression of CDK inhibitors such as Rum1 also leads to repeated S phases. A simplified model for the mechanism of endoreduplication is that in which M-CDK activity is inhibited and the S-CDK activity fluctuates; in *Plasmodium* schizonts, however, the situation is likely to be much more complex than that. Studies of endodyogeny in *T. gondii* has showed that tachyzoites have long G1 and S phases (60% and 30% of the cell cycle, respectively), and that a G2 phase preceding the short M-phase is either very short or non-existent (Radke *et al.*, 2001). The relevance of these interesting findings to *Plasmodium* schizogony remains to be determined.

Another peculiarity of *Plasmodium* schizogony is the apparent absence of a checkpoint coupling exit from (nuclear) mitosis with proper spindle assembly: the use docetaxel, a microtubule-stabilising agent interfering with spindle dynamics, did not prevent further parasite DNA synthesis (Sinou *et al.*, 1998). Similar observations were reported with other agents interfering with spindle formation, such as taxol and vinblastine. Furthermore, γ-irradiation of schizonts had no effect on total DNA synthesis despite lethal chromosomal damage (R. Graeser, unpublished observations reported in Doerig *et al.*, 2000), suggesting that the checkpoint control pathway for DNA damage, like that for spindle assembly, may not be operating in malaria parasites. It is relevant to mention here a study of *T. gondii* tachyzoite division in which Shaw *et al.* demonstrated that daugther cell assembly is not tightly linked to completion of DNA replication, pointing to a down-regulation or absence of the DNA replication checkpoint in this stage of the parasite life cycle (Shaw *et al.*, 2001).

Differentiating male or female gametocytes withdraw from proliferation, and their cell cycle is arrested while the cells are in the vertebrate host; ingestion by the mosquito results in activation and release from the block in cell cycle (Janse *et al.* 1988; Sinden *et al.*, 1996). The DNA content of gametocytes has been shown to be higher than that corresponding to a haploid genome (1C), without however reaching 2C (Janse *et al.*, 1986). This suggests that the cell cycle is arrested at some point after S-phase has started, although here again a clear correlation with typical eukaryotic cell cycle phases has yet to be established. In agreement with the idea that gametocytes are engaged in at least the first stages of cell division (rather than being arrested at a phase analogous to the G1 phase of other eukaryotes) is the ultrastructural observation that gametocytes contain elongated nuclei with apparent spindles (Sinden, 1982). The very rapid formation of eight gamete cells from each male gametocyte upon exflagellation (Janse *et al.*, 1986) demonstrates that all the components required for immediate cell division are in place in the gametocyte, awaiting the activation signal(s) provided by the mosquito midgut environment (Sinden *et al.*, 1996). Virtually nothing is known, from a cell cycle control perspective, of the molecular mechanisms regulating meiosis and sporogony occurring in the insect vector.

## 2.2. Transduction Pathways Regulating Parasite Proliferation and Differentiation

Phenomenological approaches have implicated second messengers and enzymes functioning in a variety of transduction pathways in the control of parasite multiplication and development (see Table 1 and (Sinden *et al.*, 1996; Doerig, 1997; Garcia, 1999) for reviews on this topic, which will not be covered exhaustively here as the scope of the present article focuses on genomics aspects). Selected examples of such findings include (i) sudden production of cGMP, diacylglycerol and phosphoinositide second messengers during gametocyte activation (Martin *et al.*, 1994), with xanthurenic acid playing an important role as a triggering signal (Billker *et al.*, 1998; Garcia *et al.*, 1998), (ii) implication of cAMP in gametocytogenesis (a controversial finding) (Kaushal *et al.*, 1980), (iii) the block of morphological differentiation of *P. gallinaceum* zygotes to ookinetes by inhibition of Ca2+/calmodulin-dependent protein kinase activity (Silva-Neto *et al.*, 2002), and (iv) inhibition of schizogony by an inhibitor (H-89) with some specificity against the cAMP-dependent protein kinase (Syin *et al.*, 2001). The latter studies are two of many investigations using inhibitors of specific enzymes families to explore signalling pathways in the parasite. It must be kept in mind, however, that data generated from the use of inhibitors against enzymes from other systems have to be taken with caution, as (i) such molecules often do not have the narrow specificity they are advertised for (Davies *et al.*, 2000), and (ii) the effects of a given molecule on the homologues in different organisms can be very divergent (see, for example, Gray *et al.*, 1998). Therefore, the possibility, now offered by the availability of genomic databases, of identifying putative targets and monitoring the effect of inhibitors on individual enzymes, is of great interest.

# 3. Cell Cycle Control in Malaria Parasites: Lessons from the Study of Specific Genes and Database Mining

Identification and cloning of targeted *Plasmodium* genes was first achieved by "homology cloning" using PCR with degenerate oligonucleotides designed to hybridise to conserved regions. More recently, the availability of EST and genomic databases has rendered this approach unnecessary, as putative homologues can now be identified *in silico*. Many genes involved in the control of cell proliferation and/or differentiation have thus been cloned, expressed, and their products characterised. We will start by considering CDKs and cyclins, which constitute the core machinery of cell cycle control; then we will review the elements involved in the control of DNA replication; and finally we will consider regulatory pathways operating upstream of these effectors. Figure 2 summarises part of the findings discussed in this section.

## 3.1. CDKs and Cyclins

Several kinases with maximal homology to the CDK family have been identified in *P. falciparum* (Doerig *et al.*, 2002), some of which have only recently been found in PlasmoDB and still await characterisation (Table 1). Hence, the complement of CDK-related enzymes in *Plasmodium* appears to be more complex than that of yeast (Goffeau *et al.*, 1996). It is far from certain, however, that all these enzymes are cell cycle regulators (Kinnaird and Mottram, 1997), and it is difficult to predict their function from sequence data alone (see section 4). So far, cyclin-dependent activity has been demonstrated only for PfPK5, the kinase with highest homology to CDK1, and hence a prime candidate for cell cycle regulation (Ross-Macdonald *et al.*, 1994; Le Roch *et al.*, 2000), and for Pfmrk, a putative CDK-activating kinase (Li *et al.*, 1996; Waters *et al.*, 2000). Pfcyc-1 was the first *Plasmodium* functional

**Figure 2.** Tentative regulatory map of *P. falciparum* gene products potentially involved in the control of cell proliferation and/or development. All elements showed on the figure have been identified in the *P. falciparum* genome, and are arranged following guidance from other eukaryotic systems in which the pathways are well understood. Those labelled with an asterisk have been tentatively identified on the sole basis of database sequence information. It must be kept in mind that no clear-cut "ultimate" functional data are available for any of the gene products appearing in this figure, which should be taken more as a series of working hypotheses than as a representation of existing pathways (see the text for details). The proteins labelled as "receptors" can be any of the numerous membrane proteins identified at the surface of the parasite. For the sake of clarity, the parasitophorous and erythrocyte membranes have not been included in the picture, but their components may obviously have a potentially important role to play in signal transduction. Note that the depiction of the cell cycle phases at the bottom of the picture is an oversimplification (see text for details). **See Colour Plate at the back of the book.**

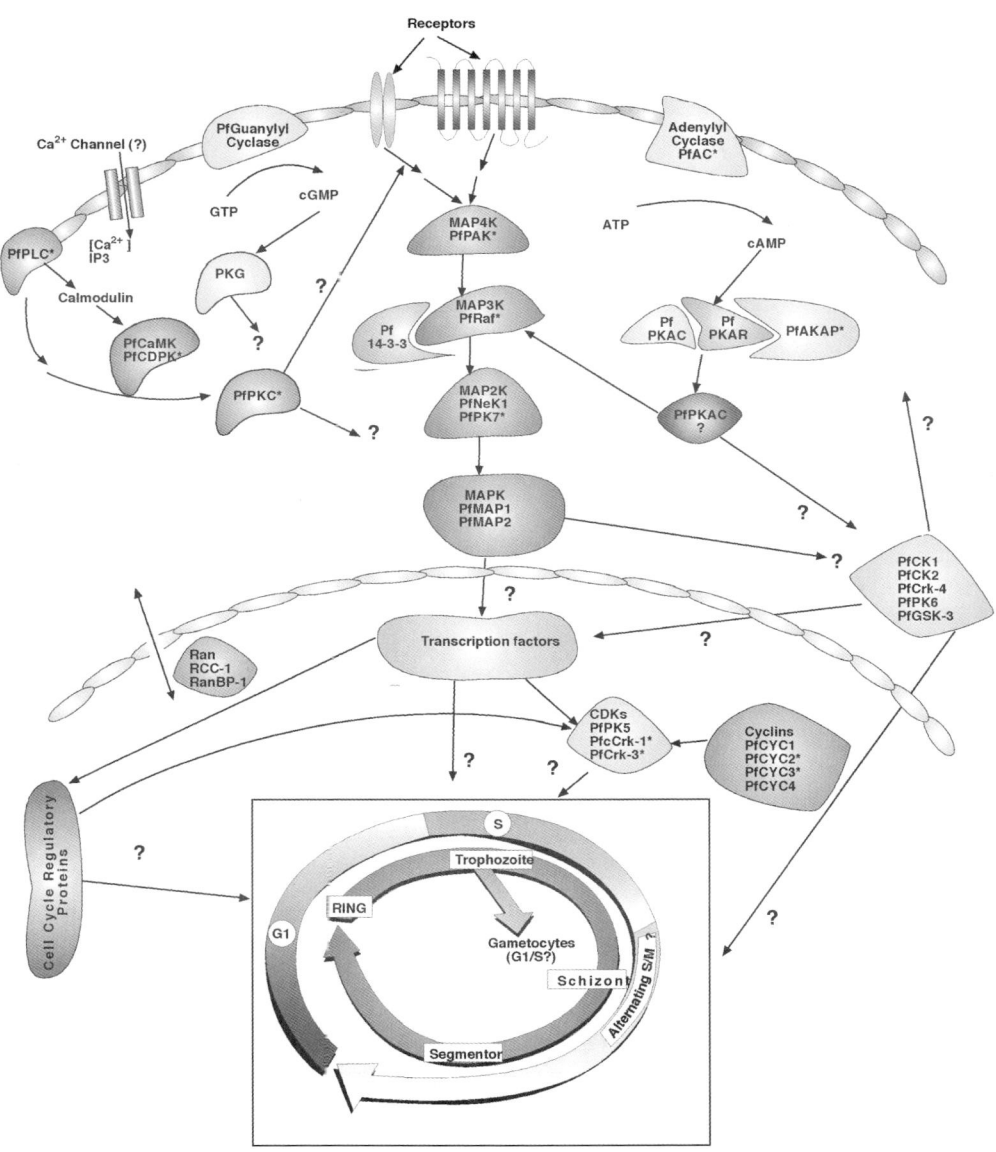

cyclin to be reported and is capable of activating PfPK5 *in vitro*, despite maximal homology to the cyclin H family (this is unexpected, since cyclin H homologues are specific activators of CDK7 subfamily members, and PfPK5 is much more closely related to CDK1 than to CDK7) (Le Roch *et al.*, 2000). Furthermore, PfPK5 responds to mammalian cyclin A, and also to p25 and RINGO (Le Roch *et al.*, 2000; Holton *et al.*, 2003; Merckx *et al.*, 2003), which are non-cyclin CDK activators from vertebrates, and autophosphorylates in the presence of some of these molecules. Autophosphorylation has to our knowledge not been documented in CDKs from other organisms, even *in vitro*. Taken together, autophosphorylation and promiscuity with respect to activators suggest that the regulation of PfPK5 activity differs considerably from that of CDKs of model eukaryotes, at least *in vitro*. Three additional proteins with homology to cyclins have been identified in PlasmoDB. These associate with a histone H1 kinase activity in parasite extracts, as shown by pull-down and immunoprecipitation experiments, and one of them (Pfcyc-3) is able to activate recombinant PfPK5 *in vitro* like Pfcyc-1 (Merckx *et al.*, 2003). Here again, predicting "ultimate" functions (such as a precise role in cell cycle progression, or interaction with a given catalytic subunit *in vivo*) from sequence homologies is not feasible (see section 4).

In other eukaryotes the activity of CDKs is regulated by two other key molecules: Cdc25 phosphatase and Wee1 kinase, which act antagonistically to modulate phosphorylation of a conserved residue (Y15 in CDK2) in the ATP binding site of the CDK (Borgne and Meijer, 1996; Morgan, 1997; see Section 1). All CDK-related enzymes of *P. falciparum* (Table 1) possess the conserved target Y residue in the ATP binding site, which suggests that the CDKs of the parasite may be subject to this mode of regulation. However, the sequence database lacks clear homologues of Cdc25 and Wee1, although some of the numerous kinase and phosphatase sequences in the genome are able to match with these elements (with low scores!) upon BLAST analysis. Thus, if CDK activity is regulated through phosphorylation/ dephosphorylation, this must be mediated through as yet uncharacterised kinase and phosphatase enzymes that would be functional homologues of Wee1 and Cdc25, despite a lack of obvious homology at the level of primary structure. It is worth mentioning here that homologues of Cdc25 (and many other proteins) have been identified much more easily in most other eukaryotes than in *Plasmodium*, which presumably reflects the phylogenetic isolation of Apicomplexa. Whether Pfmrk, displaying maximal homology to the CDK-activating kinase CDK7 which activates a subset of CDKs by phosphorylation of the Thr residue in the T-loop (T160 in CDK2, which is conserved as well in all malarial CDK-related enzymes), actually plays a role in cell cycle progression, remains to be determined.

The primary structure of another CDK-related kinase, PfPK6, displays characteristics of both CDKs and MAPKs (Bracchi-Ricard *et al.*, 2000). It is not the only example of a "patchwork" protein kinase in *Plasmodium* (see below). PfPK6 is active *in vitro* in the absence of cyclin, and localizes predominantly to cytoplasm in trophozoites and schizonts. Recently PfPK6 has been shown to interact with and phosphorylate elongation factor-1α (V. Bracchi-Ricard and D. Chakrabarti, unpublished observation). PfPK6 may play an indirect role in cell cycle regulation, through regulating other cellular activities. Pfcrk-1 is a gene encoding a CDK-related protein with highest homology to the p58-GTA (PISTLRE) family, which is represented in mammals by several isoforms and may play a role in cell differentiation and signal-dependent regulation of transcription (Lahti *et al.*, 1995; Trembley *et al.*, 2002). Pfcrk-1 mRNA has been found by Northern blot analysis to be expressed at detectable levels only in gametocytes (Doerig *et al.*, 1995), but microarray data on PlasmcDB indicate that the gene is expressed in asexual parasites as well. Discrepancy between microarray and Northern blot data occur in a few other instances as well.

The homologues of a number of CDK regulatory proteins, for example cyclin-dependent kinase inhibitors (CKIs) or Cks proteins, have not been identified through genome-wide sequence-based searches, either because functional homologues are not sufficiently conserved at the primary sequence level to allow detection, or because the *Plasmodium* machinery lacks such elements; it is worth mentioning here that a heterologous (mammalian) p21 CKI is able to inhibit PfPK5-Pfcyc-1 activity (Li *et al.*, 2001), indicating that the kinase possesses the structural features required for CKI-dependent regulation.

The *P. yoelii* orthologues of several of the CDKs and other kinases discussed here have been shown to be expressed (at the RNA level) during liver stages in infected mice (Nivez *et al.*, 2000), suggesting that the parasite may use (at least in part) the same machinery to regulate cell division in the context of both exoerythrocytic and erythrocytic schizogonies.

### 3.2. Entry into Schizogony and Nuclear Division

Several components of the two complexes that regulate initiation of DNA synthesis, the origin recognition complex (ORC) and mini chromosome maintenance complex (MCM), have been identified in *P. falciparum*. Expression of subunits of these complexes varies during the asexual erythrocytic developmental cycle (S. Patterson and D. Chakrabarti, unpublished observation) and a recent report indicates that the MCM4 transcript is also detected in the sexual erythrocytic stage (Li and Cox, 2001). Other key regulators of pre-replication complex formation, CDC6, Cdt1, and DDK are yet to be identified in PlasmoDB, possibly because functional homologues are not sufficiently conserved at the primary sequence level to allow detection. The genes encoding DNA polymerase α and δ and the polymerase δ accessory factor PCNA have been isolated (Ridley *et al.*, 1991;

Kilbey *et al.*, 1993; White *et al.*, 1993)). Recently, a gene encoding a second PCNA (PCNA2) has been identified, similar to that in *Toxoplasma* (Guerini *et al.*, 2000). PCNA1 and PCNA2 have been found to be differentially expressed, suggesting that the two PCNAs in *P. falciparum* may have distinct physiological roles (Patterson *et al.*, 2002). The large subunit of replication protein A (PfRPA1) has been characterized recently (Voss *et al.*, 2002). Although the 6.5 kb PfRPA1 transcript has the potential to encode a 134 kDa polypeptide, the mature protein appears, from mass spectrometry data, to contain only the COOH-terminus region of 55 kDa. PfRPA1 is expressed in late trophozoites and schizonts, in line with the observed DNA replication activity during these periods of development.

The detailed molecular mechanisms of nuclear division (which may in part be analogous to M-phase of higher eukaryotes) have yet to be elucidated. Nevertheless, putative orthologues of several components of the eukaryotic mitotic exit and cytokinesis machinery such as CDC14, subunits of anaphase promoting complex and CDC20 are present in the PlasmoDB database, suggesting that the malaria parasite utilizes at least some conserved components of the signalling network for coordination of late mitotic events (this, obviously, remains to be demonstrated experimentally).

### 3.3. Signal Transduction Pathways Controlling Proliferation and Differentiation

*3.3.1. cAMP Pathway*

In eukaryotes, cAMP is synthesised from ATP by adenylate cyclases in response to a variety of extracellular signals. One of the major roles of cAMP is to activate the cAMP-dependent kinase (PKA), by binding to the regulatory subunits of a complex constituted of two regulatory and two catalytic subunits. cAMP binding to the regulatory subunits causes dissociation of the complex and release of active catalytic subunits (reviewed in (Skalhegg and Tasken, 2000)). *P. falciparum* genes encoding PKA catalytic (PKAc) (Syin *et al.*, 2001) and regulatory (PKAr) (A. Merckx and C. Doerig, unpublished) subunits have been identified and characterised. The catalytic subunit is expressed in asexual blood stages and to a lesser extent in gametocytes. Although H-89, an inhibitor with some specificity against PKAs but able to inhibit other kinases as well (Davies *et al.*, 2000), has parasitocidal activity, it has not been clearly established that PfPKAc is its actual target; this issue is further complicated by the fact that erythrocytes contain an endogenous PKAc, which appears to be modified by infection with *P. falciparum* (Syin *et al.*, 2001). Recombinant PfPKAr is able to pull down a cAMP-dependent kinase activity from parasite extracts, suggesting that it functions as a true PKAr as predicted from its sequence (A. Merckx and C. Doerig, unpublished). Genomic database searches reveal that there presumably is only one member each in the PKAc and PKAr families in the parasite's genome. Two genes encoding proteins with maximal homology to adenylate cyclases have been identified in the *P. falciparum* genome (Muhia *et al.*, 2003). One of these (PfACα) is able to complement a *Dyctiostelium* null mutant, demonstrating genuine adenylate cyclase function. Interestingly, both PfACα and PfACβ appear to be related to prokaryotic cyclases.

### 3.3.2. cGMP Pathway

Two *P. falciparum* proteins identified on the basis of maximal homology to cyclases were expressed in *E. coli*, and one of these (PfGCβ) was demonstrated to possess guanylate cyclase (but no adenylate cyclase) activity *in vitro*. The other one, PfGCα, did not display any activity as a recombinant protein (Carucci *et al.*, 2000). Both genes encode potentially bifunctional polypeptides, as in both cases a domain related to P-ATPases is present as a N-terminal extension to the cyclase domain. Surprisingly, the PfGCβ cyclase domain possessing guanylate cyclase activity conforms to a structure that is usually associated with adenylate cyclases. Xanthurenic acid, an inducer of gametocyte activation present in the mosquito gut, is able to stimulate guanylate cyclase activity in purified gametocyte membranes; although this has not been formally demonstrated, it is likely that the gene involved is one (or both) of the PfGCs, both of which are expressed specifically in gametocytes (Muhia *et al.*, 2001). A homologue of the cGMP-dependent kinase (PfPKG) has also been characterised, and possesses cGMP-stimulated kinase activity as a recombinant protein. This gene is predominantly expressed in rings, and PfPKG protein is not detectable in gametocytes (Deng and Baker, 2002). The apparent discrepancy between the expression patterns of PfGCs and PfPKG is currently unexplained.

### 3.3.3. MAPK Pathways

Two *P. falciparum* enzymes displaying maximal homology to the ERK1/ERK2 family of MAPKs have been characterised. The Pfmap-1 catalytic domain carries the T-X-Y motif that is used as an activation site by all MAPKs characterised so far, and a C-terminal extension that is not shared by other members of the ERK1/ERK2 family (Doerig *et al.*, 1996; Lin *et al.*, 1996). Western blot analysis of parasite extracts suggests that this extension may be processed *in vivo* (Graeser *et al.*, 1997). The second plasmodial MAPK, Pfmap-2, is very unusual in that it does not possess a classical T-X-Y activation motif: the motif at this site is T-S-H (Dorin *et al.*, 1999). Site-directed mutagenesis has shown that both the T and the H in this motif are important for *in vitro* activity of the recombinant enzyme. Furthermore, three small insertions are found in the vicinity of this atypical activation site. Taken together, these observations suggest that the structure of the activation domain, and hence the mechanism of activation, of Pfmap-2, may differ from those of "regular" ERK1/ERK2 kinases. Interestingly, the PlasmoDB database appears not to contain any gene that unambiguously belongs to the MAPKK family. One open reading frame, PfPK7, displays maximal (but limited --approximately 34% identity) homology to vertebrate MAPKK3/6, but does not possess the two otherwise very conserved S or T residues used in MAPKK activation in other systems; it is therefore unlikely that PfPK7 functions in a classical MAPK module. PfPK7 is another example of a "patchwork" enzyme, with some regions closely related to the PKA family despite an overall maximal homology to MAPKK3/6. Recombinant PfPK7 autophosphorylates *in vitro* and phosphorylates exogenous substrates such as myelin basic protein and histone H2, but so far no activity towards MAPK pathway components, including Pfmap-1 and Pfmap-2, has been detected (D. Dorin and C. Doerig, unpublished). Another kinase was identified as carrying an activation site that is closely related to that found in MAPKKs, despite an overall maximal homology to the NIMA/Nek family of kinases. This enzyme, called Pfnek-1, was shown to specifically phosphorylate Pfmap-2, but not mouse ERK2; furthermore, Pfnek-1 and Pfmap-2 act synergistically to phosphorylate an exogenous substrate, suggesting (but not formally demonstrating) that one of the enzymes is able to activate the other (Dorin *et al.*, 2001). The PlasmoDB database contains sequences with maximal homology to MAPKKKs, but these enzymes still await functional characterisation. The small GTPase Ras plays a crucial role in MAPK pathway activation in other eukaryotes from yeast to mammals, but no clear homologue has been identified among the numerous small GTPases present in the *Plasmodium* genome. Interestingly, the ERK1/ERK2-related Pfmap-1 and Pfmap-2 seem to be the only members of the MAPK superfamily to exist in *P. falciparum*, in contrast with most eukaryotes, which possess several MAPK modules including those involving p38/stress-activated and Jun-kinases (Garrington and Johnson, 1999). It will be interesting to determine which molecular tools the parasite uses to fulfil the essential functions for which these pathways are responsible in other eukaryotes.

Thus, genomic database content and available experimental data concur in suggesting that the organisation of malarial MAPK pathways diverge considerably from that found in other eukaryotes.

### 3.3.4. Phosphoinositide (PI) Pathways and Calcium Signalling

In many eukaryotic systems, stimulation of the inositol-lipid pathway following receptor activation is relayed by heterotrimeric G-proteins and causes inositol 1,4,5-triphosphate ($IP_3$)-mediated release of $Ca^{2+}$. This also leads to the activation of protein kinase C by diacylglycerol (DAG). Two enzymes, phospholipase C (PLC) and PI-3 kinase, are critical for the generation of DAG and phosphoinositide second messengers (reviewed in (Leslie *et al.*, 2001; Newton, 2001).

High levels of phosphotidylinositol bisphosphate (PIP$_2$) have been detected in the late asexual stages of *Plasmodium*, and an increase in IP$_3$ level has been correlated with exflagellation during gametogenesis (Martin *et al.*, 1994). Putative homologues of PLC and PI-3 kinase are present in PlasmoDB. Likewise, protein kinases with some homology to the PKC family (but here again, functional relevance is highly speculative) and putative PKC-anchoring protein RACK-1 (Receptor for Activated C-kinase) have also been identified in the sequence database. Proteins with biochemical properties similar to those of heterotrimeric G-proteins have been detected in merozoites, and treatment of cultured parasites with cholera toxin, an activator heterotrimeric G-proteins, stimulates gametocytogenesis (Dyer and Day, 2000). However, no clear homologues of heterotrimeric G-proteins have to our knowledge been identified in genomic database. The reason for this apparent discrepancy remains to be determined.

The parasite appears to possess an extensive complement of proteins involved in calcium signaling (reviewed in Garcia *et al.*, 1999). Calmodulin, which regulates a number of biological activities in a calcium-dependent manner, is present in *P. falciparum* (Robson and Jennings, 1991), and appears to be developmentally regulated, with higher levels in schizonts than in early stages (Calvo *et al.*, 2002; Orfa Rojas and Wasserman, 1995). The EF-hand motif (two a helices separated by a Ca$^{2+}$ binding loop) has been found in several *P. falciparum* proteins, including at least three calcium-dependent protein kinases (CDPKs), a family of enzymes found in plant and ciliates but not animal cells (Zhao *et al.*, 1994; Farber *et al.*, 1997; Li *et al.*, 2000); reviewed in (Kappes *et al.*, 1999)). CDPK1 has been suggested to have a role during invasion and/or membrane biogenesis, and is expressed in asexual parasites, like CDPK2. PfCDKP3 mRNA, in contrast, is found predominantly in sexual stages.

### 3.3.5. Other Molecules Involved in Signalling

A preliminary count of genes encoding serine/threonine protein kinases in the *P. falciparum* genome yields a lower limit of approximately 70 easily recognisable members of this family (but apparently no true member of the tyrosine kinase family) (P. Refour and C. Doerig, unpublished). Several of these enzymes, in addition to those mentioned in the sections above, have been investigated, but a general discussion of this topic lies outside the scope of the present chapter (see Kappes *et al.*, 1999 for a review on *Plasmodium* protein kinases). It is nevertheless of interest to briefly mention here kinases which are likely to be involved in signalling pathways, like casein kinases 1 and 2 (CK1, CK2), and Glycogen-synthase kinase 3 (GSK3). The malarial casein kinase 1 (PfCK1) homologue that shows peak activity in ring and trophozoite stages is one of the smallest CK1 enzyme known (Barik *et al.*, 1997). The recombinant PfCK1 exhibits CK1 characteristic properties such as inhibition by CK1-7. Interestingly, PfCK1 and the CK1 homologues from trypanosomatids show high affinity for purvalanol B, an inhibitor of mammalian CDKs (Knockaert *et al.*, 2000). The malarial CK2 showed peak expression in the schizont stage. Unlike CK2 enzymes from other systems, malaria CK2, similar to *T. brucei* CK2, prefers ATP over GTP as the phosphate donor (V. Bracchi-Ricard and D. Chakrabarti, unpublished observation). Glycogen synthase kinase 3 (GSK-3) represents a highly conserved family of protein kinases involved in many cellular processes. In higher eukaryotes, GSK-3 plays a role in the Wnt and insulin signalling pathway, glycogen and protein synthesis, regulation of transcription factors, embryonic development, apoptosis and cell proliferation (reviewed in Grimes and Jope, 2001). A GSK-3 homologue has been identified in PlasmoDB, expressed in bacteria and shown to possess kinase activity *in vitro* (Droucheau *et al.*, 2003). Immunofluorescence experiments have located the PfGSK3 protein in the Maurer clefts, a structure found in the cytoplasm of the infected erythrocyte. Genomic databases also appear to contain genes related to histidine kinases which may play important roles in signal transduction, but this has not yet received experimental attention.

14-3-3 proteins are phosphoserine/threonine-dependent chaperones and have an important role in cellular signaling by mediating interactions between various proteins. A *Plasmodium* homologue of 14-3-3 has been identified, and its transcript shows peak expression in early trophozoites (Al-Khedery *et al.*, 1999).

### 3.3.6. Checkpoints

In the eukaryotic cell cycle, checkpoints at phase transitions ensure that previous phases have been completed before the cell moves on to the next phase. The checkpoint proteins are key regulators to maintain genomic stability, and function at the transition of G1 to S, G2 to M, and at the formation of mitotic spindle. The cell cycle transition from one stage to another depends on (1) formation of CDK-cyclin complexes, (2) phosphorylation events that either activate or inhibit CDK activity, and (3) CDK inhibitory molecules (see above). The spindle checkpoint in the yeast cell is regulated by the anaphase promoting complex (APC), MAD (mitotic arrest deficient) and BUB (budding uninhibited by benomyl) gene products. Only components of the APC complex have been detected in *Plasmodium*. Interestingly, possible homologues of two kinases Chk1 and Chk2 that participate in cell cycle arrest following DNA damage through stabilization of p53 are present in the malaria genome, despite an apparent lack of this checkpoint in the parasite (see above).

*3.3.7. Nucleocytoplasmic Transport*

A possible mechanism for the apparent asynchronous division of nuclei in a given schizont may implicate nucleocytoplasmic transport (see Section 2). *P. falciparum* genes encoding homologues of the Ran GTPase (Dontfraid and Chakrabarti, 1994; Sultan *et al.*, 1994), RCC1 (Ji *et al.*, 1998) and RanBP1 (Ramachandran *et al.*, 2002) proteins have been identified and shown to be expressed in blood stages; PfRCC1 was demonstrated to be localised predominantly in the nucleus, and PfRanBP1 was shown to be able to bind to PfRanGTP. Furthermore, homologues of the transporter proteins karyopherin α and β have been identified, and their expression shown to peak in trophozoites. As expected by analogy with data from other systems, these two proteins do heterodimerize *in vitro* (Mohmmed *et al.*, 2003). Hence, it appears that *P. falciparum* possesses several of the major elements which in other eukaryotes control nucleocytoplasmic transport. However, in model organisms some of these molecules function in other processes as well (Sazer and Dasso, 2000); whether PfRan, PfRanBP1 and PfRCC1 mediate transport, and whether this plays a direct role in the control of cell cycle during schizogony, remains to be determined experimentally.

# 4. Benefits, Limitations and Perspectives of the Use of Genomic Databases in the Study of Cell Cycle Control

## 4.1. Benefits

The most direct benefit of the availability of genomic database lies in the possibility of identifying targeted genes *in silico* on the basis of sequence homology. This is in essence no different from the *in vitro* procedures used prior to the advent of genomic databases (in particular those involving degenerate primers designed to hybridise to conserved regions), but allows much faster progress in gene identification. For example, all the cyclin genes discussed above have been identified through direct mining of PlasmoDB, but none of them was detected through numerous previous *in vitro* attempts using PCR with degenerate primers, presumably because of the low level of conservation displayed by the malarial cyclins relative to cyclins of other eukaryotes.

Database-dependent microarray and proteomics analyses of parasites at different developmental stages will provide information on which pathways and regulatory elements are likely to be operating in those stages --this avenue has already begun to be explored (Hayward *et al.*, 2000; Ben Mamoun *et al.*, 2001; Bozdech *et al.*, 2003; Le Roch *et al.*, 2003). Microarray analysis will be of particular interest with regard to those stages of the parasite which are not (or with difficulty) amenable to classical biochemical analysis, such as mosquito or liver stages. However, presence of a given mRNA at a given stage does not necessarily imply that the corresponding protein is produced or active. Since many of the components of transduction pathways are regulated by phosphorylation, the activity of specific pathways could in principle be detected by proteomic analysis of phosphoproteins (Sickmann and Meyer, 2001; Conrads *et al.*, 2002). Microarray and proteomics approaches, if used in conjunction with triggering signals (e.g. treatment of gametocytes with xanthurenic acid) or mutant lines defective is given transduction elements (see below), are likely to provide important information on the consequences of the activation or inhibition of given transduction pathways on the repertoire of expressed genes. They may also help to resolve current discrepancies over the expression patterns of different components predicted to function in the same pathways (such as that observed, for example, between the expression of PKG and guanylyl cyclase).

Another, more indirect use of databases, is that associated with peptide identification by mass spectrometry. This is likely to prove extremely useful in the context of cell cycle research, through the analysis of protein complexes such as CDK/cyclin or signal transduction modules, using affinity chromatography on given gene products. This can be achieved by examining the proteins in the cell extract that bind to a column on which the target recombinant protein is immobilised. A less straightforward but more potent approach would be to replace the wild-type allele of the target protein with a modified allele where the "bait"protein open reading frame is fused to a tag that allows direct purification of complexes from parasite extracts, as has been done in a yeast genome-wide analysis of protein-protein interactions (Gavin *et al.*, 2002) and in the study of trypanosomatid CDKs (Mottram *et al.*, 1996). The advantage of the latter approach over the former one is that it does not depend on *de novo* complex formation with recombinant proteins, and thus is more likely to detect tightly interacting proteins that are sequestered in complexes at the time of extract preparation. Likewise, affinity chromatography on immobilised kinase inhibitors can allow the identification of potential targets, as has been shown with Purvalanol B and the malarial casein kinase 1 homologue PfCK1 (Knockaert *et al.*, 2000). This will be useful not only in the context of drug discovery research, but also as a way to characterise inhibitors and their targets as tools for fundamental cell cycle and signal transduction research.

Finally, the availability of genomic databases allows genome-wide investigations of targeted gene families. An example that is directly relevant to cell proliferation research is that of protein kinases. As discussed above, protein kinases play essential roles in signal transduction and cell cycle control, and preliminary database mining indicated that there are approximately 70 easily recognisable serine/threonine -presumably including some "dual specificity"- protein kinases in the *Plasmodium* genome (but apparently, like in yeast, no true tryosine kinase). Identification of all these enzymes opens the way to systematic, genome-wide, mutagenesis in order to generate hypersensitive mutants in the context of a chemical genetics approach (Specht and Shokat, 2002), to which protein kinases are particularly well suited (Bishop *et al.*, 2001). This approach is currently being undertaken for the functional study of malarial kinases (Doerig *et al.*, 2002). This should allow the constitution of a library of parasite clones, each of which would carry a hypersensitive allele for one specific protein kinase. Such a library would represent an immensely useful tool not only for investigations into cell cycle and signal transduction, but for the study of most biological processes operating during the life cycle of the parasite.

## 4.2. Limitations: Problems in Reconstituting Regulatory Networks and Pathways

Available functional data concerning cell cycle elements (reviewed in section 3) tend to show that the function of given elements (e.g. kinases or cyclins) is not necessarily in line with that expected from sequence homology with putative orthologues in other systems. It is useful in this context to make a distinction between what can be called "proximate" and "ultimate" levels of function. The "proximate" (or "immediate") function of many of these elements can be predicted from sequence data and verified experimentally. For example, most of the ORFs encoding predicted protein kinases possess demonstrated phosphotransferase activity, and at least some of the genes examined on the basis of their homology to cyclins encode proteins that are able to activate CDKs. Hence, their "proximate" function as protein kinases or cyclins, respectively, can be established relatively easily. The "ultimate" function, however, which refers to the precise role of the protein in cell development or cell cycle progression and to its location regulatory networks, cannot be predicted from sequence alone; telling examples which are detailed in the sections above are (i) the organisation of MAPK pathways (where no clear orthologue of MAPKK can be identified in genomic databases), (ii) the apparent promiscuity of PfPK5 with regards to cyclins, (iii) the difficulty in identifying Wee1 and Cdc25 functional homologues among the numerous kinase- and phosphatase-encoding genes in the database (or a ras functional homologue among the small G-proteins), and (iv) the presence of protein kinases with "mixed" relatedness to subfamilies that are distinct in other eukaryotes, such as the CDK-MAPK "hybrid" kinases PfPK6 and Pfcrk-4, the NIMA/Nek-like Pfnek-1 which possesses a MAPKK-like activation site, or the dual relatedness of PfPK7 to MAPKKs and PKAs (see above and Table 1). Hence, it appears that regulatory systems like the cell cycle machinery or complex transduction pathways are proving very difficult to reconstitute from genomic databases. This is in sharp contrast to metabolic pathways, many of which can be reconstituted in their entirety from the content of databases (see http://sites.huji.ac.il/malaria/, a site which displays *P. falciparum* metabolic pathways reconstituted by H. Ginsburg).

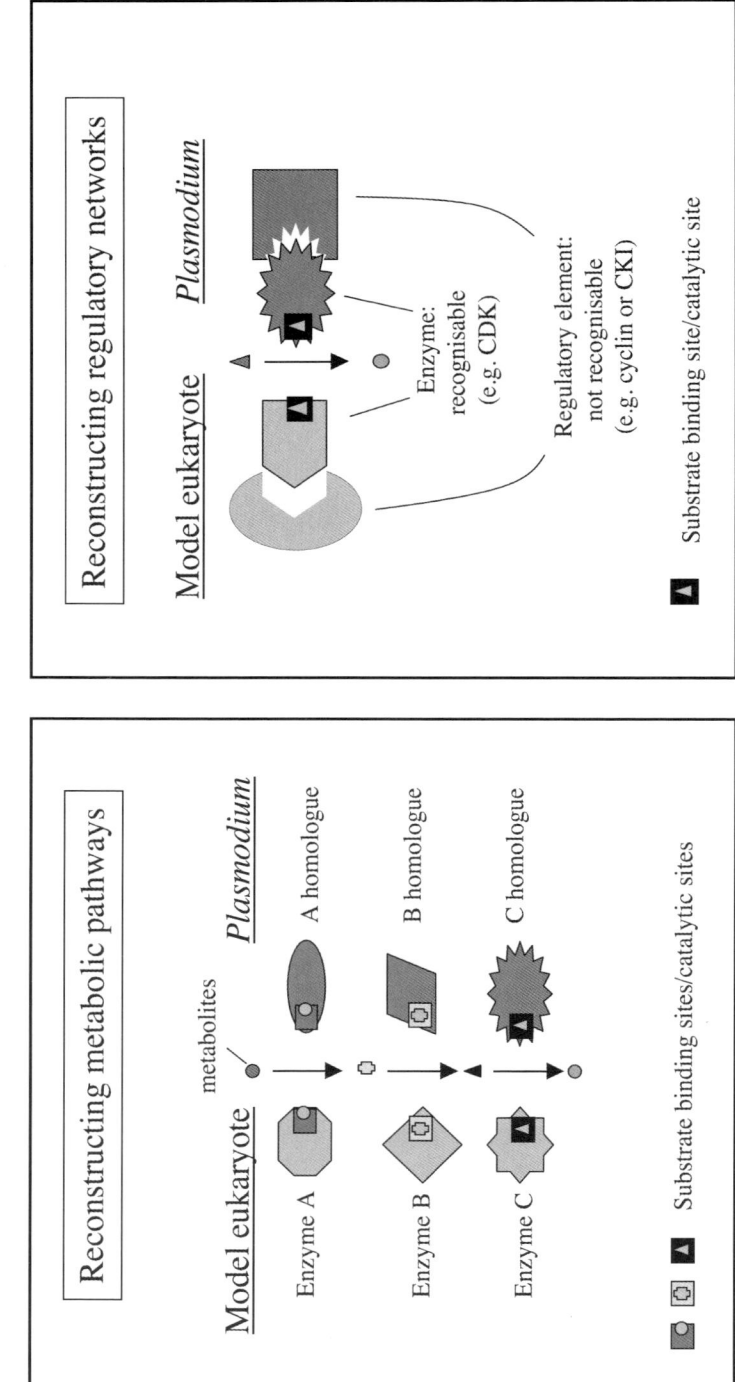

**Figure 3.** A scheme illustrating the difficulty of reconstituting *Plasmodium* regulatory pathways from sequence databases. Left Panel: Reconstruction of metabolic pathways. Presence of substrate binding motifs and catalytic sites allows to position the *Plasmodium* orthologue on the metabolic map, even though the molecule may be overall divergent from those of model organisms (as schematised by different shapes and colors). This is because the metabolite substrates are identical (species-independent) for all members of a given family of orthologues. Right panel: Regulatory elements (shown at the extreme left and right of the panel) have to interact not with invariant substrates, but with species-specific proteins (shown at the center), which are divergent from their orthologues in other systems. **See Colour Plate at the back of the book.**

Why such a difference, in the possibilities offered by exploitation of genomic sequences, between regulatory and metabolic pathways? The enzymes functioning in metabolic pathways usually display highly conserved domains involved in substrate binding and enzymatic activity, largely because the substrate is an invariant molecule. Hence, even if overall homology levels between a given metabolic enzyme in *Plasmodium* and its orthologue in other organisms are not particularly high, the presence of such "signature" motifs are often sufficient to assign them a function. Furthermore, for metabolic enzymes "ultimate" function and location within pathways (e.g. energy production through glycolysis for hexokinase) can in general be immediately deduced from "proximate" function (e.g. phosphorylation of glucose for the same enzyme), which in turn can be predicted directly from sequence motifs. In contrast, the substrates (or binding partners) of elements involved in regulatory networks are often not invariant metabolites, but other *Plasmodium* proteins, which themselves show divergences form their own homologues in other systems (see Figure 3). Therefore the "substrate" recognition site cannot be utilised to predict which partner the protein interacts with, and hence its precise position in a regulatory network and its ultimate function -these need to be investigated through the functional analysis of individual gene products (for a further discussion of the problem of assigning function from sequence data, even with respect to metabolic enzymes, see, Gerlt and Babbitt, 2000).

The limitations of the use of sequence databases discussed above are for a large part consequences of the considerable phylogenetic distance between Apicomplexa and model organisms such as yeast and metazoan cells (which, being members of the same phylogenetic lineage Opisthokonta, are much more closely related to each other than to malaria parasites) (Baldauf, 2003). Nonetheless, it is clear that the availability of *Plasmodium* genomic databases has revolutionised the study of the cell cycle in malaria parasites, and will undoubtedly allow much progress to be made in the next few years. In view of the central role played by cell cycle control elements in cell survival, this in turn will provide valuable information regarding potential targets for novel chemotherapeutic approaches. Eukaryotic proteins kinases, including CDKs, are now considered as promising targets for intervention against a variety of pathologies, with some kinase inhibitors already on the market or in clinical trial (Cohen, 2002; Meijer and Raymond, 2003). The possibility of specifically interfering with the parasite's cell cycle regulators such as CDKs or other protein kinases is now being explored with respect to malaria and other parasitic diseases (Doerig *et al.*, 2002; Doerig, 2003), but this is another story.

# 5. Acknowledgements

We wish to thank the scientists and funding agencies comprising the international Malaria Genome Project for making sequence data from the genome of *P. falciparum* (3D7) public prior to publication of the completed sequence. The Sanger Centre (UK) provided sequence for chromosomes 1, 3-9, & 13, with financial support from the Welcome Trust. A consortium composed of The Institute for Genome Research, along with the Naval Medical Research Center (USA), sequenced chromosomes 2, 10, 11 & 14, with support from NIAID/NIH, the Burroughs Wellcome Fund, and the Department of Defense. The Stanford Genome Technology Center (USA) sequenced chromosome 12, with support from the Burroughs Welcome Fund. The *Plasmodium* Genome Database is a collaborative effort of investigators at the University of Pennsylvania (USA) and Monash University (Melbourne, Australia), supported by the Burroughs Wellcome Fund.

Work in the C.D. laboratory is supported by INSERM, The French Ministries of Research (PAL+ and PRFMMIP programmes), Defence (Délégation Générale pour l'Armement), Education, and Foreign Affairs (French-South African Programme for Cooperation in Science and Technology), the Welcome Trust, and by the UNDP/World Bank/WHO Special Program for Research and Training in Tropical Diseases (TDR). The research in D.C. laboratory is supported by a grant from National Institutes of Health (AI48036). The authors wish to thank Shelly Patterson, University of Central Florida for her excellent assistance in fluorescence microscopy and Siglinde Quirk, University of Central Florida, for Figure 2 graphics.

# 6. References

Adams, R.R., Carmena, M. and Earnshaw, W.C. 2001. Chromosomal passengers and the (aurora) ABCs of mitosis. Trends Cell Biol 11: 49-54.

Aikawa, M. 1966. The fine structure of the erythrocytic stages of three avian malarial parasites, *Plasmodium fallax*, *P. lophurae*, and *P. cathemerium*. Am. J. Trop. Med. Hyg. 15: 449-471.

Al-Khedery, B., Barnwell, J.W. and Galinski, M.R. 1999. Stage-specific expression of 14-3-3 in asexual blood-stage Plasmodium. Mol. Biochem. Parasitol. 102: 117-130.

Arnot, D.E. and Gull, K. 1998. The Plasmodium cell-cycle: facts and questions. Ann Trop Med Parasitol 92: 361-365.

Baldauf, S.L. 2003. The deep roots of eukaryotes. Science 300: 1703-6.

Bannister, L.H., Hopkins, J.M., Fowler, R.E., Krishna, S. and Mitchell, G.H. 2000. A brief illustrated guide to the ultrastructure of *Plasmodium falciparum* asexual blood stages. Parasitol Today 16: 427-433.

Bardwell, L., Cook, J.G., Inouye, C.J. and Thorner, J. 1994. Signal propagation and regulation in the mating pheromone response pathway of the yeast *Saccharomyces cerevisiae*. Dev. Biol. 166: 363-379.

Barik, S., Taylor, R.E. and Chakrabarti, D. 1997. Identification, cloning, and mutational analysis of the casein kinase 1 cDNA of the malaria parasite, *Plasmodium falciparum*. Stage-specific expression of the gene. J. Biol. Chem. 272: 26132-26138.

Ben Mamoun, C., Gluzman, I.Y., Hott, C., MacMillan, S.K., Amarakone, A.S., Anderson, D.L., Carlton, J.M., Dame, J.B., Chakrabarti, D., Martin, R.K., Brownstein, B.H. and Goldberg, D.E. 2001. Co-ordinated programme of gene expression during asexual intraerythrocytic development of the human malaria parasite *Plasmodium falciparum* revealed by microarray analysis. Mol. Microbiol. 39: 26-36.

Billker, O., Lindo, V., Panico, M., Etienne, A.E., Paxton, T., Dell, A., Rogers, M., Sinden, R.E. and Morris, H.R. 1998. Identification of xanthurenic acid as the putative inducer of malaria development in the mosquito. Nature 392: 289-92.

Bishop, A.C., Buzko, O. and Shokat, K.M. 2001. Magic bullets for protein kinases. Trends Cell Biol. 11: 167-172.

Borgne, A. and Meijer, L. 1996. Sequential dephosphorylation of p34(cdc2) on Thr-14 and Tyr-15 at the prophase/metaphase transition. J. Biol. Chem. 271: 27847-27854.

Bozdech, Z., Zhu, J., Joachimiak, M.P., Cohen, F.E., Pulliam, B. and DeRisi, J.L. 2003. Expression profiling of the schizont and trophozoite stages of *Plasmodium falciparum* with a long-oligonucleotide microarray. Genome Biol. 4: R9.

Bracchi-Ricard, V., Barik, S., Delvecchio, C., Doerig, C., Chakrabarti, R. and Chakrabarti, D. 2000. PfPK6, a novel cyclin-dependent kinase/mitogen-activated protein kinase- related protein kinase from *Plasmodium falciparum*. Biochem. J. 347 (1): 255-263.

Calvo, E., Rubiano, C., Vargas, A. and Wasserman, M. 2002. Expression of housekeeping genes during the asexual cell cycle of *Plasmodium falciparum*. Parasitol. Res. 88: 267-271.

Carucci, D.J., Witney, A.A., Muhia, D.K., Warhurst, D.C., Schaap, P., Meima, M., Li, J.L., Taylor, M.C., Kelly, J.M. and Baker, D.A. 2000. Guanylyl cyclase activity associated with putative bifunctional integral membrane proteins in *Plasmodium falciparum*. J. Biol. Chem. 275: 22147-22156.

Cohen, P. 2002. Protein kinases--the major drug targets of the twenty-first century? Nat. Rev. Drug. Discov. 1: 309-315.

Conrads, T.P., Issaq, H.J. and Veenstra, T.D. 2002. New tools for quantitative phosphoproteome analysis. Biochem. Biophys. Res. Commun. 290: 885-890.

Coutavas, E., Ren, M., Oppenheim, J.D., D'Eustachio, P. and Rush, M.G. 1993. Characterization of proteins that interact with the cell-cycle regulatory protein Ran/TC4. Nature 366: 585-587.

Davies, S.P., Reddy, H., Caivano, M. and Cohen, P. 2000. Specificity and mechanism of action of some commonly used protein kinase inhibitors. Biochem. J. 351: 95-105.

Deng, W. and Baker, D.A. 2002. A novel cyclic GMP-dependent protein kinase is expressed in the ring stage of the *Plasmodium falciparum* life cycle. Mol. Microbiol. 44: 1141-1151.

Doerig, C. 1997. Signal transduction in malaria parasites. Parasitol Today 13: 307-313.

Doerig, C. 2003. Protein kinases as targets for anti-parasitic chemotherapy. Biochim. Biophys. Acta.

Doerig, C., Chakrabarti, D., Kappes, B. and Matthews, K. 2000. The cell cycle in protozoan parasites. Prog. Cell Cycle Re.s 4: 163-183.

Doerig, C., Endicott, J. and Chakrabarti, D. 2002. Cyclin-dependent kinase homologues of *Plasmodium falciparum*. Int. J. Parasitol. 32: 1575-1585.

Doerig, C., Horrocks, P., Coyle, J., Carlton, J., Sultan, A., Arnot, D. and Carter, R. 1995. Pfcrk-1, a developmentally regulated cdc2-related protein kinase of *Plasmodium falciparum*. Mol. Biochem. Parasitol. 70: 167-174.

Doerig, C., Meijer, L. and Mottram, J.C. 2002. Protein kinases as drug targets in parasitic protozoa. Trends Parasitol.18: 366-371.

Doerig, C.M., Parzy, D., Langsley, G., Horrocks, P., Carter, R. and Doerig, C.D. 1996. A MAP kinase homologue from the human malaria parasite, *Plasmodium falciparum*. Gene. 177: 1-6.

Dontfraid, F.F. and Chakrabarti, D. 1994. Cloning and expression of a cDNA encoding the homologue of Ran/TC4 GTP- binding protein from *Plasmodium falciparum*. Biochem. Biophys. Res. Commun. 201: 423-429.

Dorin, D., Alano, P., Boccaccio, I., Ciceron, L., Doerig, C., Sulpice, R. and Parzy, D. 1999. An atypical mitogen-activated protein kinase (MAPK) homologue expressed in gametocytes of the human malaria parasite *Plasmodium falciparum*. Identification of a MAPK signature. J. Biol. Chem. 274: 29912-29920.

Dorin, D., Le Roch, K., Sallicandro, P., Alano, P., Parzy, D., Poullet, P., Meijer, L. and Doerig, C. 2001. Pfnek-1, a NIMA-related kinase from the human malaria parasite *Plasmodium falciparum* Biochemical properties and possible involvement in MAPK regulation. Eur. J. Biochem. 268: 2600-2608.

Droucheau, E., Primot, A., Mattei, D., Knockaert, M., Richardson, C., Sallicandro, P., Alano, P., Jafarshad, A., Baratte, B., Kunick, C., Parzy, D., Pearl, L., Doerig, C. and Meijer, L. 2003. *Plasmodium falciparum* glycogen synthase kinase -3 (PfGSK-3): molecular model, expression, intracellular localisation and selective inhibitors. Biochim. Biophys. Acta in press.

Dutta, A. and Bell, S.P. 1997. Initiation of DNA replication in eukaryotic cells. Annu. Rev. Cell. Dev. Biol. 13: 293-332.

Dyer, M. and Day, K. 2000. Expression of *Plasmodium falciparum* trimeric G proteins and their involvement in switching to sexual development. Mol. Biochem. Parasitol. 110: 437-448.

Edgar, B.A. and Orr-Weaver, T.L. 2001. Endoreplication cell cycles: more for less. Cell. 105: 297-306.

Farber, P.M., Graeser, R., Franklin, R.M. and Kappes, B. 1997. Molecular cloning and characterization of a second calcium-dependent protein kinase of *Plasmodium falciparum*. Mol. Biochem. Parasitol. 87: 211-216.

Fowler, R.E., Smith, A.M., Whitehorn, J., Williams, I.T., Bannister, L.H. and Mitchell, G.H. 2001. Microtubule associated motor proteins of *Plasmodium falciparum* merozoites. Mol. Biochem. Parasitol. 117: 187-200.

Garcia, C.R. 1999. Calcium homeostasis and signaling in the blood-stage malaria parasite. Parasitol. Today 15: 488-491.

Garcia, G.E., Wirtz, R.A., Barr, J.R., Woolfitt, A. and Rosenberg, R. 1998. Xanthurenic acid induces gametogenesis in Plasmodium, the malaria parasite. J. Biol. Chem. 273: 12003-12005.

Gardner, R.D. and Burke, D.J. 2000. The spindle checkpoint: two transitions, two pathways. Trends Cell Biol 10: 154-8.

Garrington, T.P. and Johnson, G.L. 1999. Organization and regulation of mitogen-activated protein kinase signaling pathways. Curr. Opin. Cell. Biol. 11: 211-218.

Gavin, A.C., Bosche, M., Krause, R., Grandi, P., Marzioch, M., Bauer, A., Schultz, J., Rick, J.M., Michon, A.M., Cruciat, C.M., Remor, M., Hofert, C., Schelder, M., Brajenovic, M., Ruffner, H., Merino, A., Klein, K., Hudak, M., Dickson, D., Rudi, T., Gnau, V., Bauch, A., Bastuck, S., Huhse, B., Leutwein, C., Heurtier, M.A., Copley, R.R., Edelmann, A., Querfurth, E., Rybin, V., Drewes, G., Raida, M., Bouwmeester, T., Bork, P., Seraphin, B., Kuster, B., Neubauer, G. and Superti-Furga, G. 2002. Functional organization of the yeast proteome by systematic analysis of protein complexes. Nature. 415: 141-147.

Gerlt, J.A. and Babbitt, P.C. 2000. Can sequence determine function? Genome Biol. 1.

Goffeau, A., Barrell, B.G., Bussey, H., Davis, R.W., Dujon, B., Feldmann, H., Galibert, F., Hoheisel, J.D., Jacq, C., Johnston, M., Louis, E.J., Mewes, H.W., Murakami, Y., Philippsen, P., Tettelin, H. and Oliver, S.G. 1996. Life with 6000 genes. Science. 274: 546, 563-567.

Gorlich, D. and Kutay, U. 1999 Transport between the cell nucleus and the cytoplasm. Annu. Rev. Cell. Dev. Biol. 15: 607-660.

Graeser, R., Kury, P., Franklin, R.M. and Kappes, B. 1997. Characterization of a mitogen-activated protein (MAP) kinase from *Plasmodium falciparum*. Mol. Microbiol. 23: 151-159.

Gray, N.S., Wodicka, L., Thunnissen, A.M., Norman, T.C., Kwon, S., Espinoza, F.H., Morgan, D.O., Barnes, G., LeClerc, S., Meijer, L., Kim, S.H., Lockhart, D.J. and Schultz, P.G. 1998. Exploiting chemical libraries, structure, and genomics in the search for kinase inhibitors. Science. 281: 533-538.

Grimes, C.A. and Jope, R.S. 2001. The multifaceted roles of glycogen synthase kinase 3beta in cellular signaling. Prog. Neurobiol. 65: 391-426.

Guerini, M.N., Que, X., Reed, S.L. and White, M.W. 2000. Two genes encoding unique proliferating-cell-nuclear-antigens are expressed in *Toxoplasma gondii*. Mol. Biochem. Parasitol. 109: 121-31.

Harper, J.W. and Adams, P.D. 2001. Cyclin-dependent kinases. Chem. Rev. 101: 2511-2526.

Hayward, R.E., Derisi, J.L., Alfadhli, S., Kaslow, D.C., Brown, P.O. and Rathod, P.K. 2000. Shotgun DNA microarrays and stage-specific gene expression in *Plasmodium falciparum* malaria. Mol. Microbiol. 35: 6-14.

Hixon, M.L. and Gualberto, A. 2000. The control of mitosis. Front Biosci 5: D50-7.

Holton, S., Merckx, A., Burgess, D., Doerig, C., Noble, M. and Endicott, J. 2003. Structures of *P. falciparum* PfPK5 test the CDK regulation paradigm and suggest mechanisms of small molecule inhibition. Structure 11: 1329-1337.

Hu, K., Mann, T., Striepen, B., Beckers, C.J., Roos, D.S. and Murray, J.M. 2002. Daughter cell assembly in the protozoan parasite *Toxoplasma gondii*. Mol. Biol. Cell. 13: 593-606.

Janse, C.J., Ponnudurai, T., Lensen, A.H., Meuwissen, J.H., Ramesar, J., Van der Ploeg, M. and Overdulve, J.P. 1988. DNA synthesis in gametocytes of *Plasmodium falciparum*. Parasitol. 96: 1-7.

Janse, C.J., van der Klooster, P.F., van der Kaay, H.J., van der Ploeg, M. and Overdulve, J.P. 1986. DNA synthesis in *Plasmodium berghei* during asexual and sexual development. Mol. Biochem. Parasitol. 20: 173-182.

Janse, C.J., Van der Klooster, P.F., Van der Kaay, H.J., Van der Ploeg, M. and Overdulve, J.P. 1986. Rapid repeated DNA replication during microgametogenesis and DNA synthesis in young zygotes of *Plasmodium berghei*. Trans. R. Soc. Trop. Med. Hyg. 80: 154-157.

Ji, D.D., Sultan, A.A., Chakrabarti, D., Horrocks, P., Doerig, C. and Arnot, D.E. 1998. An RCC1-type guanidine exchange factor for the Ran G protein is found in the *Plasmodium falciparum* nucleus. Mol. Biochem. Parasitol. 95: 165-70.

Kappes, B., Doerig, C.D. and Graeser, R. 1999. An overview of Plasmodium protein kinases. Parasitol Today 15: 449-454.

Kaushal, D.C., Carter, R., Miller, L.H. and Krishna, G. 1980. Gametocytogenesis by malaria parasites in continuous culture. Nature 286: 490-492.

Kelly, T.J. and Brown, G.W. 2000. Regulation of chromosome replication. Annu. Rev. Biochem. 69: 829-80.

Kilbey, B.J., Fraser, I., McAleese, S., Goman, M. and Ridley, R.G. 1993. Molecular characterisation and stage-specific expression of proliferating cell nuclear antigen (PCNA) from the malarial parasite, *Plasmodium falciparum*. Nucleic Acids Res 21: 239-243.

Kinnaird, J. and Mothram, J. 1997. Plasmodium cdc2-related kinases: Do they regulate stage differentiation? Parasitol Today. 13: 7-8.

Knockaert, M., Gray, N., Damiens, E., Chang, Y.T., Grellier, P., Grant, K., Fergusson, D., Mottram, J., Soete, M., Dubremetz, J.F., Le Roch, K., Doerig, C., Schultz, P. and Meijer, L. 2000. Intracellular targets of cyclin-dependent kinase inhibitors: identification by affinity chromatography using immobilised inhibitors. Chem. Biol. 7: 411-422.

Lahti, J.M., Xiang, J. and Kidd, V.J. 1995. The PITSLRE protein kinase family. Prog Cell Cycle Res 1: 329-338.

Le Roch, K., Sestier, C., Dorin, D., Waters, N., Kappes, B., Chakrabarti, D., Meijer, L. and Doerig, C. 2000. Activation of a *Plasmodium falciparum* cdc2-related kinase by heterologous p25 and cyclin H. Functional characterization of a *P. falciparum* cyclin homologue. J. Biol. Chem. 275: 8952-8958.

Le Roch, K.G., Zhou, Y., Blair, P.L., Grainger, M., Moch, J.K., Haynes, J.D., De La Vega, P., Holder, A.A., Batalov, S., Carucci, D.J. and Winzeler, E.A. 2003. Discovery of gene function by expression profiling of the malaria parasite life cycle. Science. 301: 1503-1508.

Leete, T. and Rubin, H. 1996. Malaria and the cell cycle. Parasitol. Today 12: 442-444.

Lei, M. and Tye, B.K. 2001. Initiating DNA synthesis: from recruiting to activating the MCM complex. J. Cell. Sci. 114: 1447-1454.

Leslie, N.R., Biondi, R.M. and Alessi, D.R. 2001. Phosphoinositide-regulated kinases and phosphoinositide phosphatases. Chem. Rev. 101: 2365-80.

Li, J.L., Baker, D.A. and Cox, L.S. 2000. Sexual stage-specific expression of a third calcium-dependent protein kinase from *Plasmodium falciparum*. Biochim. Biophys. Acta. 1491: 341-349.

Li, J.L. and Cox, L.S. 2003. Characterisation of a sexual stage-specific gene encoding ORC1 homologue in the human malaria parasite *Plasmodium falciparum*. Parasitol Int. 52: 41-52.

Li, J.L. and Cox, L.S. 2001. Identification of an MCM4 homologue expressed specifically in the sexual stage of *Plasmodium falciparum*. Int. J. Parasitol. 31: 1246-1252.

Li, L., Brunk, B.P., Kissinger, J.C., Pape, D., Tang, K., Cole, R.H., Martin, J., Wylie, T., Dante, M., Fogarty, S.J., Howe, D.K., Liberator, P., Diaz, C., Anderson, J., White, M., Jerome, M.E., Johnson, E.A., Radke, J.A., Stoeckert, C.J. Jr,

Waterston, R.H., Clifton, S.W., Roos, D.S. and Sibley, L.D. 2003. Gene discovery in the apicomplexa as revealed by EST sequencing and assembly of a comparative gene database. Genome Res. 13: 443-54.

Li, J.L., Robson, K.J., Chen, J.L., Targett, G.A. and Baker, D.A. 1996. Pfmrk, a MO15-related protein kinase from *Plasmodium falciparum*. Gene cloning, sequence, stage-specific expression and chromosome localization. Eur. J. Biochem. 241: 805-813.

Li, Z., Le Roch, K., Geyer, J.A., Woodard, C.L., Prigge, S.T., Koh, J., Doerig, C. and Waters, N.C. 2001. Influence of human p16(INK4) and p21(CIP1) on the *in vitro* activity of recombinant *Plasmodium falciparum* cyclin-dependent protein kinases. Biochem. Biophys. Res. Commun. 288: 1207-1211.

Lin, D.T., Goldman, N.D. and Syin, C. 1996. Stage-specific expression of a *Plasmodium falciparum* protein related to the eukaryotic mitogen-activated protein kinases. Mol. Biochem. Parasitol. 78: 67-77.

Losada, A. and Hirano, T. 2001. Shaping the metaphase chromosome: coordination of cohesion and condensation. Bioessays. 23: 924-935.

Martin, S.K., Jett, M. and Schneider, I. 1994. Correlation of phosphoinositide hydrolysis with exflagellation in the malaria microgametocyte. J. Parasitol. 80: 371-378.

McCollum, D. and Gould, K.L. 2001. Timing is everything: regulation of mitotic exit and cytokinesis by the MEN and SIN. Trends Cell Biol. 11: 89-95.

Meijer, L. and Raymond, E. 2003. Roscovitine and other purines as kinase inhibitors. From starfish oocytes to clinical trials. Acc. Chem. Res. 36: 417-425.

Merckx, A., Le Roch, K., Nivez, M.P., Dorin, D., Alano, P., Guiterrez, G.J., Nebreda, A.R., Goldring, D., Whittle, C., Patterson, S., Chakrabarti, D. and Doerig, C. 2003. Identification and initial characterization of three novel cyclin-related proteins of the human malaria parasite *Plasmodium falciparum*u. J. Biol. Chem. 278: 39839-39850.

Mohmmed, A., Kishore, S., Dasaradhi, P., Patra, K., Malhotra, P. and Chauhan, V. 2003. Cloning and characterization of *Plasmodium falciparum* homologues of nuclear import factors, karyopherin α and karyopherin β. Mol. Biochem. Parasitol. 127: 199-203.

Morgan, D.O. 1997. Cyclin-dependent kinases: engines, clocks, and microprocessors. Annu Rev Cell Dev. Biol. 13: 261-291.

Mottram, J.C., McCready, B.P., Brown, K.G. and Grant, K.M. 1996. Gene disruptions indicate an essential function for the LmmCRK1 cdc2- related kinase of *Leishmania mexicana*. Mol. Microbiol. 22: 573-583.

Muhia, D.K., Swales, C.A., Deng, W., Kelly, J.M. and Baker, D.A. 2001. The gametocyte-activating factor xanthurenic acid stimulates an increase in membrane-associated guanylyl cyclase activity in the human malaria parasite *Plasmodium falciparum*. Mol. Microbiol. 42: 553-560.

Muhia, D.K., Swales, C.A., Eckstein-Ludwig, U., Saran, S., Polley, S.D., Kelly, J.M., Schaap, P., Krishna, S. and Baker, D.A. 2003. Multiple splice variants encode a novel adenylyl cyclase of possible plastid origin expressed in the sexual stage of the malaria parasite *Plasmodium falciparum*. J. Biol. Chem. 278: 22014-22022.

Nasmyth, K. 1996. Viewpoint: putting the cell cycle in order. Science. 274: 1643-5.

Newton, A.C. 2001. Protein kinase C: structural and spatial regulation by phosphorylation, cofactors, and macromolecular interactions. Chem. Rev. 101: 2353-2364.

Nivez, M., Achbarou, A., Bienvenu, J.D., Mazier, D., Doerig, C. and Vaquero, C. 2000. A study of selected *Plasmodium yoelii* messenger RNAs during hepatocyte infection. Mol. Biochem. Parasitol. 111: 31-39.

Orfa Rojas, M. and Wasserman, M. 1995. Stage-specific expression of the calmodulin gene in *Plasmodium falciparum*. J. Biochem. (Tokyo) 118: 1118-1123.

Patterson, S., Whittle, C., Robert, C. and Chakrabarti, D. 2002. Molecular characterization and expression of an alternate proliferating cell nuclear antigen homologue, PfPCNA2, in *Plasmodium falciparum*. Biochem. Biophys. Res. Commun. 298: 371-6.

Radke, J.R., Striepen, B., Guerini, M.N., Jerome, M.E., Roos, D.S. and White, M.W. 2001. Defining the cell cycle for the tachyzoite stage of *Toxoplasma gondii*. Mol. Biochem. Parasitol. 115: 165-175.

Ramachandran, V., Dorin, D., Khiong, C.W., Kara, U.A. and Doerig, C. 2002. A *Plasmodium falciparum* homologue of the Ran binding protein 1, a protein involved in nucleocytoplasmic transport. Mol. Biochem. Parasitol. 123: 67-71.

Read, M., Sherwin, T., Holloway, S.P., Gull, K. and Hyde, J.E. 1993. Microtubular organization visualized by immunofluorescence microscopy during erythrocytic schizogony in *Plasmodium falciparum* and investigation of post-translational modifications of parasite tubulin. Parasitol. 106: 223-232.

Ridley, R.G., White, J.H., McAleese, S.M., Goman, M., Alano, P., de Vries, E. and Kilbey, B.J. 1991. DNA polymerase delta: gene sequences from *Plasmodium falciparum* indicate that this enzyme is more highly conserved than DNA polymerase alpha. Nucleic Acids Res. 19: 6731-6736.

Robson, K.J. and Jennings, M.W. 1991. The structure of the calmodulin gene of *Plasmodium falciparum*. Mol. Biochem. Parasitol. 46: 19-34.

Ross-Macdonald, P.B., Graeser, R., Kappes, B., Franklin, R. and Williamson, D.H. 1994. Isolation and expression of a gene specifying a cdc2-like protein kinase from the human malaria parasite *Plasmodium falciparum*. Eur. J. Biochem. 220: 693-701.

Rudzinska, M.A. 1969. The fine structure of malaria parasites. Int. Rev. Cytol. 25: 161-199.

Sazer, S. and Dasso, M. 2000. The ran decathlon: multiple roles of Ran. J. Cell. Sci. 113: 1111-1118.

Schwacha, A. and Bell, S.P. 2001. Interactions between two catalytically distinct MCM subgroups are essential for coordinated ATP hydrolysis and DNA replication. Mol. Cell 8: 1093-1104.

Shaw, M.K., Roos, D.S. and Tilney, L.G. 2001. DNA replication and daughter cell budding are not tightly linked in the protozoan parasite *Toxoplasma gondii*. Microbes Infect. 3: 351-362.

Sherr, C.J. and Roberts, J.M. 1999. CDK inhibitors: positive and negative regulators of G1-phase progression. Genes Dev. 13: 1501-1512.

Sickmann, A. and Meyer, H.E. 2001. Phosphoamino acid analysis. Proteomics. 1: 200-206.

Silva-Neto, M.A., Atella, G.C. and Shahabuddin, M. 2002. Inhibition of Ca2+/calmodulin-dependent protein kinase blocks morphological differentiation of plasmodium gallinaceum zygotes to ookinetes. J. Biol. Chem. 277: 14085-14091.

Sinden, R.E. 1982. Gametocytogenesis of *Plasmodium falciparum in vitro*: ultrastructural observations on the lethal action of chloroquine. Ann. Trop. Med. Parasitol. 76: 15-23.

Sinden, R.E., Butcher, G.A., Billker, O. and Fleck, S.L. 1996. Regulation of infectivity of Plasmodium to the mosquito vector. Adv. Parasitol. 38: 53-117.

Sinou, V., Boulard, Y., Grellier, P. and Schrevel, J. 1998. Host cell and malarial targets for docetaxel (Taxotere) during the erythrocytic development of *Plasmodium falciparum*. J. Eukaryot. Microbiol. 45: 171-183.

Skalhegg, B.S. and Tasken, K. 2000. Specificity in the cAMP/PKA signaling pathway. Differential expression,regulation, and subcellular localization of subunits of PKA. Front. Biosci. 5: D678-693.

Solomon, M.J. and Kaldis, P. 1998. Regulation of CDKs by phosphorylation. Results Probl. Cell. Differ. 22: 79-109.

Specht, K.M. and Shokat, K.M. 2002. The emerging power of chemical genetics. Curr. Opin. Cell Biol. 14: 155-159.

Striepen, B., Crawford, M.J., Shaw, M.K., Tilney, L.G., Seeber, F. and Roos, D.S. 2000. The plastid of *Toxoplasma gondii* is divided by association with the centrosomes. J. Cell. Biol. 151: 1423-1434.

Sultan, A.A., Richardson, W.A., Alano, P., Arnot, D.E. and Doerig, C. 1994. Cloning and characterisation of a *Plasmodium falciparum* homologue of the Ran/TC4 signal transducing GTPase involved in cell cycle control. Mol. Biochem. Parasitol. 65: 331-338.

Syin, C., Parzy, D., Traincard, F., Boccaccio, I., Joshi, M.B., Lin, D.T., Yang, X.M., Assemat, K., Doerig, C. and Langsley, G. 2001. The H89 cAMP-dependent protein kinase inhibitor blocks *Plasmodium falciparum* development in infected erythrocytes. Eur. J. Biochem. 268: 4842-9484.

Trembley, J.H., Hu, D., Hsu, L.C., Yeung, C.Y., Slaughter, C., Lahti, J.M. and Kidd, V.J. 2002. PITSLRE p110 protein kinases associate with transcription complexes and affect their activity. J. Biol. Chem. 277: 2589-2596.

Voss, T.S., Mini, T., Jenoe, P. and Beck, H.P. 2002. *Plasmodium falciparum* possesses a cell-cycle regulated short-type replication protein A large subunit encoded by an unusual long transcript. J. Biol. Chem. 5: 5.

Waters, N.C., Woodard, C.L. and Prigge, S.T. 2000. Cyclin H activation and drug susceptibility of the Pfmrk cyclin dependent protein kinase from *Plasmodium falciparum*. Mol. Biochem. Parasitol. 107: 45-55.

White, J.H., Kilbey, B.J., de Vries, E., Goman, M., Alano, P., Cheesman, S., McAleese, S. and Ridley, R.G. 1993. The gene encoding DNA polymerase alpha from *Plasmodium falciparum*. Nucleic Acids Res. 21: 3643-3646.

Yanow, S.K., Lygerou, Z. and Nurse, P. 2001. Expression of Cdc18/Cdc6 and Cdt1 during G2 phase induces initiation of DNA replication. EMBO J. 20: 4648-4656.

Zhao, Y., Pokutta, S., Maurer, P., Lindt, M., Franklin, R.M. and Kappes, B. 1994. Calcium-binding properties of a calcium-dependent protein kinase from *Plasmodium falciparum* and the significance of individual calcium- binding sites for kinase activation. Biochem. 33: 3714-3721.

Zou, L. and Stillman, B. 1998. Formation of a preinitiation complex by S-phase cyclin CDK-dependent loading of Cdc45p onto chromatin. Science 280: 593-596.

From: Malaria Parasites: Genomes and Molecular Biology
Edited by: A.P. Waters and C.J. Janse

# Chapter 11

## The Apicoplast

## Ross F. Waller
## and Geoffrey I. McFadden

## Abstract

The apicoplast is a plastic organelle that is homologous to chloroplasts of plants. It occurs throughout the Apicomplexa and is an ancient feature of this group acquired by the process of endosymbiosis. Like plant chloroplasts, apicoplasts are semi-autonomous with their own genome and expression machinery. In addition, apicoplasts import numerous proteins encoded by nuclear genes. These nuclear genes largely derive from the endosymbiont through a process of intracellular gene relocation. The exact role of a plastid in parasites is uncertain but early clues indicate synthesis of lipids, heme and isoprenoids as possibilities. The various metabolic processes of the apicoplast are potentially excellent targets for drug therapy.

## 1. Introduction

The apicomplexan plastid was a discovery waiting to be made. With hindsight it seems incredible that such an organelle could have so long concealed its identity in cells that have received as much scientific attention as *Plasmodium*. In fact these plastids, referred to as apicoplasts in apicomplexan parasites, had been observed but not recognized as plastids. As early as the 1960s electron microscopists had described the ultrastructure of a mysterious multi-membrane-bound organelle from *Plasmodium* spp. as well as from all other major groups of apicomplexan parasites (see McFadden *et al.*, 1997 for review). Moreover, circular organellar DNAs, that we now know belong to the apicoplast, were observed, measured, and cruciform secondary structures noted (formed by inverted repeats of rRNA genes, a classic feature of plastid genomes) (Kilejian, 1975; Dore *et al.*, 1983; Borst *et al.*, 1984). We were even aware of the sensitivity of *Plasmodium* to a range of plastid-targeting antibiotics (though these were assumed to target the parasite mitochondrion). However, none of these observations could have reasonably inspired anyone to claim that *Plasmodium* is descended from photosynthetic stock and that it still maintains a plastid organelle.

It required the depth of genomic data to awaken us to the strange possibility of *Plasmodium* being a highly derived alga. Iain Wilson's group paved the way to apicoplast discovery with studies of extra-chromosomal DNAs recovered from isopycnic density gradient fractionation of total *Plasmodium* DNA. This group recovered two DNA forms; one a 6kb tandemly repeated element that was later identified as the mitochondrial genome, and a second, 35kb circle that was supposed to represent the DNA circles previously observed by microscopists (Wilson *et al.*, 1996b; Wilson and Williamson, 1997). This molecule was also thought to be mitochondrial DNA, and early sequence data of eubacterial-like rRNA genes supported this organellar conclusion. However, as the sequencing effort continued a new conclusion, that was originally embraced with some awkwardness ("Have malaria parasites three genomes?", Wilson *et al.*, 1991), began to emerge. Gradually, evermore convincing character traits of a plastid genome were uncovered, and strong parallels with plastid genomes from non-photosynthetic plants (*Epifagus virginiana*) and algae (*Astasia longa*) became clear. These studies culminated with the completion of the 35kb sequence, by which time the chromosome architecture, gene content and gene arrangement were undeniably plastid-like (Wilson *et al.*, 1996b; McFadden and Waller, 1997; Wilson and Williamson, 1997; Gleeson, 2000).

Final proof of a plastid organelle awaited defining the location of this plastid-like DNA. Maternal inheritance of this DNA circle strongly argued that it was located in a non-nuclear organelle (Creasey *et al.*, 1994). A lack of co-purification with the mitochondrion (converse to the 6kb extra-chromosomal DNA) told us that this organelle was separate to the mitochondrion (Wilson *et al.*, 1992). Therefore a relict plastid seemed most likely. Suddenly plastid spotting in the old ultrastructural literature became a favored pastime of many, and the mysterious multi-membrane organelles quickly became the 'odds-on bet'. *In situ* hybridization experiments in *Toxoplasma gondii* using probes complementary to the plastid genome or its gene transcripts offered the final nail in the coffin by defining the locality of these molecules as a multi-membrane-bound compartment distinct from the mitochondrion or other organelles (McFadden *et al.*, 1996; Köhler *et al.*, 1997).

Genomic data had provided the first hint of a plastid in Apicomplexa, and also provided the tools for its identification at the cell level. Two immediate questions were raised by this discovery. (1) What is the origin of the apicoplast? (2) What is the function of the apicoplast in parasites? Again genomic data provided the major springboard for tackling these questions. Addressing the first question, phlylogeneticist have continuously poured over apicoplast molecular data for hints of the source of this plastid ever since the Wilson group's first partial sequences of the 35kb circular DNA were produced. These studies have matured substantially as the body of genomic data has expanded from parasites and other protists alike and their somewhat alarming findings will be discussed below. The second question – apicoplast function – has also provided an interesting journey. The sequencing of the apicoplast genome promised to offer insights into apicoplast metabolism, but instead revealed an extremely streamlined genome principally dedicated to its own expression plus that of a small handful of unidentified open reading frames. Direct biochemical or proteomic studies of the apicoplast have been obstructed by the lack of effective apicoplast purification techniques. So again, genomics have been called upon, this time to paint a picture of apicoplast metabolism from nucleus-encoded genes whose products are targeted to the apicoplast. Through the characterization of the apicoplast-targeting signals in nucleus-encoded genes an ever-growing list of putative apicoplast genes in the nucleus is being assembled. This list enables a 'virtual tour' of apicoplast metabolism from genome data. To date this has provided our greatest insights into apicoplast function. Furthermore, these studies have lead directly to the identification of novel anti-malarials. This is undoubtedly the most exciting aspect of apicoplast research because the apicoplast's plant-like heritage offers a broad range of unique targets for chemotherapeutics. The completion of the *Plasmodium* genome potentially offers the identification of the complete protein content of the apicoplast, from which an extensive picture of apicoplast metabolism can be made.

Few areas of malarial research can boast gaining so much from genomic studies. This chapter presents a summary of our current understanding of the apicoplast based both on genomic and other data.

# 2. Origin of the Apicoplast

All plastids are derived from a prokaryote that entered into an endosymbiotic partnership with a eukaryotic host cell (Cavalier-Smith, 1982). In fact, it is widely believed that all modern plastids are derived from a single symbiotic event where the endosymbiont was a cyanobacterial-like organism (Delwiche and Palmer, 1997). The apicoplast is, without doubt, a *bona fide* member of this family of organelles because of its plastid-type genome. However, there are two main categories of plastids – primary plastids and secondary (or complex) plastids. Primary plastids are the products of a symbiosis between

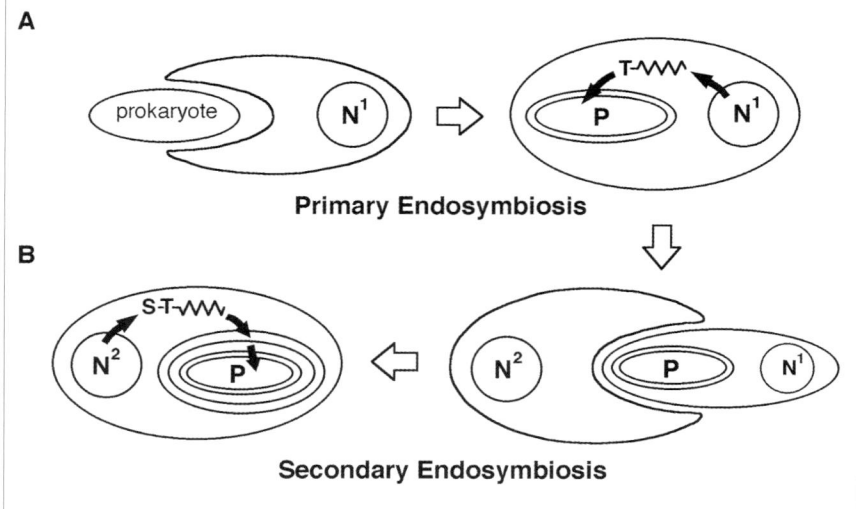

**Figure 1**. Plastid origins and protein targeting. (A) Primary endosymbiosis describes the uptake of a prokaryote by a eukaryote. Plastids derived by primary endosymbiosis are surrounded by two membranes and targeting of nucleus-encoded gene products to the endosymbiont is affected by an N-terminal transit peptide (T). (B) Secondary endosymbiotic plastid origin involves a heterotrophic eukaryote phagocytosing a photosynthetic eukaryote possessing a primary endosymbiont. The secondary endosymbiont's cytoplasm and nucleus ($N^1$) are typically lost and the resulting plastid is surrounded by four membranes. Sometimes one of the two outer membranes is lost at this point, resulting in a total of three. Targeting of nucleus-encoded ($N^2$) gene products to secondary plastids requires a signal peptide (S) to mediate protein passage across the outer membrane followed by a transit peptide (T) for import across the inner membranes.

a prokaryote and a eukaryote (see Figure 1A). These plastids are surrounded by two membranes and are represented in three major groups; plants and green algae, red algae, and an unusual algal group called glaucophytes (Delwiche, 1999). Secondary plastids are more complicated. These are derived from symbioses between two eukaryotes where a primary plastid-containing eukaryote is engulfed by a second eukaryote (Figure 1B). With time this engulfed eukaryotic is stripped of its now redundant features, and many of its genes are transferred to the host nucleus, until essentially only a plastid remains (exceptions do exist where a miniature nucleus and some cytosol persist). However, this plastid remains bound by either one or two extra membranes derived from its former host cell's plasma membrane and/or the food vacuole of its new host (Figure 1B). These extra membranes are diagnostic of secondary plastids that are found in the following diverse algal groups; dinoflagellates, heterokont algae (including kelps and other brown algae), haptophytes, cryptomonads, chlorarachniophytes and euglenoids (Delwiche, 1999). Apicoplasts are also surrounded by extra membranes (see following section for discussion of how many) which is evidence for them being secondary plastids. The route of protein trafficking to plastids is also strongly discriminating between primary and secondary plastids and in apicomplexans provides a compelling case for the apicoplast being a secondary plastid (protein trafficking is discussed in detail in a following section)

The identification of the apicoplast as a secondary plastid tells us the method by which the progenitor of apicomplexans acquired a plastid, but key questions remains unanswered. What type of plastid-containing organism (alga) was the endosymbiont? Was this plastid acquired before apicomplexans diverged from related protists or after? If the answer is before, do other groups share this plastid? If the answer is after, have different apicomplexans independently acquired their own plastids? It is important to note that the sister taxon to the Apicomplexa is the dinoflagellate algae, and the nearest relatives of this pair are the ciliates. This grouping is well supported by both molecular and ultrastructural data, and is referred to as the Alveolata (Vivier and Desportes, 1990; Baldauf *et al.*, 2000). Most dinoflagellates possess a secondary plastid (containing peridinin pigment) (Delwiche, 1999) which has lead to the enticing hypothesis that apicomplexans and dinoflagellates share a common plastid. However, this scenario is confounded by numerous of the dinoflagellates having acquired further secondary plastids independently (Delwiche, 1999), raising the possibility that either the dinoflagellates, or apicomplexans or both may have acquired their plastids separately. Ciliates, on the other hand, typically lack a plastid.

Traditional features that help define plastid lineages such as pigments (chlorophyll *b* versus chlorophyll *c*) and genes associated with photosynthesis (Rubisco forms) are absent in the non-photosynthetic apicoplast, so other molecular markers must be sought for the task of tracing apicoplast history. Genes encoded on the plastid genome are one obvious class of useful markers of plastid lineage. Nucleus-encoded genes, however, can also be very useful markers particularly as they are independent of any evolutionary constraints placed on plastid-encoded genes such as strong AT content or codon usage biases. Thus, genes that originated with the plastid (either from the plastid DNA or, in the case of secondary plastids, the endosymbiont's nucleus) but have relocated to the nucleus can also serve as useful markers of plastid heritage. Most of these genes traffic their products back to the plastid, although, some may come to fulfill functions in the host cytosol. In either case, such genes can offer phylogenetic information of the endosymbiont from which they derive. Genes that derive from the host cell rather than the endosymbiont, but become associated with plastid function, can also be useful markers. An example is a host gene whose protein product has become targeted to the plastid to fulfill an organellar role. If its role is stable, it can continue to be a useful marker for all plastid descendants subsequent to its adoption. Attempts to identify the source and history of the apicoplast have utilized all of these different classes of molecular markers.

## 2.1. Plastid-Encoded Markers

Apicoplast-encoded sequences have received the greatest attention in attempts to identify apicoplast affinities with other plastids. Early studies, conducted as the first apicoplast sequences were becoming available, were encumbered by poor taxa representation in the data sets and will not be discussed here (see Jeffries and Johnson, 1996, for a comprehensive review). With the completion of several plastid genome sequences, including two from the Apicomplexa (*Plasmodium falciparum* and *Toxoplasma gondii*) (Wilson *et al.*, 1996b; Köhler *et al.*, 1997) and partial genome sequences from a number of others (Denny *et al.*, 1998), these studies have become much more potent.

*2.1.1. Apicomplexa Share a Common Plastid*

Plastid sequence data strongly support that all apicoplasts share a common origin. Lang-Unnasch and co-workers constructed phylogenetic trees from apicoplast-encoded rRNA genes representing much of the diversity of the Apicomplexa (Lang-Unnasch *et al.*, 1998). These tree topologies basically match those derived from nuclear sequences suggesting that the apicoplast has been co-evolving with the host cell throughout apicomplexan diversification. Further evidence for all apicoplasts sharing a common origin comes from the striking similarities of apicoplast genomes. *Plasmodium falciparum* and *Toxoplasma gondii*, representing haemosporins and coccidians respectively, have near identical plastid genomes in terms of gene content and order (Wilson *et al.* 1996b; Köhler *et al.*, 1997; http: //www.sas.upenn.edu/~jkissing/toxomap.html). Denny *et al.* (1999) have added further partial genome sequences from another coccidian, *Eimeria tenella,* and the piroplasm *Theileria annulata*. These sequences also show a remarkable level of genome conservation. This uniformity of the genome not only supports a common origin of the apicoplast, but shows that its genome had reached a stable evolutionary state prior to apicomplexan diversification. This suggests that the acquisition of this secondary plastid may predate the divergence of the Apicomplexa from related protists.

This conclusion implies that all apicomplexa should contain a plastid, unless members have subsequently lost this organelle. Indeed, molecular and/or ultrastructural data indicates that an apicoplast is present in all major apicomplexan groups (McFadden *et al.*, 1997; Lang-Unnasch *et al.*, 1998). One possible exception is *Cryptosporidium* for which extensive searching for evidence of an apicoplast has failed to find any (Zhu *et al.*, 2000a). This member may, therefore, have lost its apicoplast since its divergence from other apicomplexa.

## 2.1.2. Apicoplast-Encoded Genes Support a Non-Green Origin?

If the apicoplast is monophyletic (of single origin) and possibly more ancient than apicomplexan diversification, then this raises the possibility that it may be the homologue of other algal secondary plastids. Again the spot-light falls upon the Apicomplexa's sister taxon the dinoflagellates. For a long time no plastid DNA could be recovered from dinoflagellates, preventing any comparison of heritage with the apicoplast. In 1999 two groups broke this impasse. However, what they found was highly divergent genes each arranged individually on DNA mini-circles (Takishita and Uchida, 1999; Zhang *et al.*, 1999; 2001). This radical reorganization of their plastid genomes is perhaps consistent with other aberrant features of dinoflagellate genetic organization (such as the nucleus). Again, comparisons to the apicoplast were frustrated.

Earlier studies argued for a link between apicoplasts and green (chlorophyll *b*-containing) plastids. rRNA and RNA Polymerase gene sequences often grouped the apicoplast with euglenoid algae (these contain green secondary plastids) (Howe, 1992; Gardner *et al.*, 1994; Egea and Lang-Unnasch, 1995), while TufA gene sequences suggested a direct link with green algae (implying that the apicoplast is derived from an green algal endosymbiont acquired independently from dinoflagellate plastids) (Köhler *et al.*, 1997). However, limited taxon sampling, poor statistical support for these alliances, and the artifactual phenomenon of divergent sequences (often with strong AT biases) attracting in phylogenies, all cast a shadow over these conclusions (Blanchard and Hicks, 1999).

Analyses of plastid gene loss and rearrangement on plastid genomes argues more convincingly for apicoplast affinities with non-green plastids. Plastid ribosomal protein genes are organized into two gene clusters in cyanobacteria, plants and green algae, whereas non-green plastids have combined these into a single "super operon". Apicoplasts share this "super operon" gene organization (Wilson *et al.*, 1996b; McFadden *et al.*, 1997). Furthermore, while gene loss from the plastid genome has occurred in all plastids since their endosymbiotic origin, the apicoplast retains more of these ribosomal protein genes in common with non-green plastids compared to the green counter-parts (McFadden *et al.*, 1997; Blanchard and Hicks, 1999).

## 2.2. Nucleus-Encoded Markers

Markers for the apicoplast that are encoded in the nucleus didn't become available for some time after the discovery of this organelle. The first to be analyzed were ribosomal protein genes *rps*9 and *rpl*28, and fatty acid biosynthesis gene *fab*H, all of which encode proteins that are trafficked into the apicoplast (Waller *et al.*, 1998). Phylogenies with these sequences supported their plastid origin but again limited taxon sampling prevented any meaningful resolution of the order amongst the plastids (Waller *et al.*, 1998). Nevertheless, these sequences offered a ray of hope given that many more (most probably hundreds) of other apicoplast genes promised to be found in the nuclear genome. This promise has indeed been fulfilled and this resource will undoubtedly provide the greatest power for resolving this issue of apicoplast origin. Amongst these genes, Sato *et al*, (2000) have even apparently identified genes that derive from the apicoplast, but now fulfill cytoplasmic functions. Useful phylogenetic markers still await identification from these genes. However, one group (Fast *et al.*, 2001) have made a promising break through from an unexpected quarter.

Fast *et al.* (2001) noted that plastid-targeted GAPDH (glyceraldehyde-3-phosphate dehydrogenase) in dinoflagellates was distinct from that of plants, green aglae and red algae. The later all possess a cyanobacterium-type version of this enzyme that is nucleus-encoded and plastid-targeted (this is in addition to a eukaryotic-type cytosolic version of this enzyme that all eukaryotes possess). Dinoflagellates, on the other hand, target a eukaryotic version of this enzyme to their plastids (Fagan *et al.*, 1998). This suggests that a gene duplication of the cytosolic form has lead to one copy displacing the cyanobacterial version at some point in dinoflagellate evolution (endosymbiotic gene displacement is not uncommon in plastids) (Liaud *et al.*, 1997; Fagan *et al.*, 1998). Fast *et al.* (2001) recognized that this dichotomy amongst plastids offered an excellent opportunity to test the relationship of apicoplasts to dinoflagellate plastids. They sequenced several plastid GAPDH genes including one from *Toxoplasma gondii*, and several from non-green algae, and revealed that apicoplasts possess a eukaryotic GAPDH similar to dinoflagellates. Moreover, other non-green algae (heterokont algae and cryptomonads) also possess this eukaryotic version, and in GAPDH phylogenies these group tightly together with the Apicomplexa and dinoflagellates. Interestingly, the cytosolic version of GAPDH from all of these taxa (with the exception of cryptomonads) also group together and form a sister group to the plastid targeted form. This is consistent with a gene duplication event prior to the divergence of these groups.

This stunning result has the profound consequence that the apicoplast is derived from an extremely ancient secondary endosymbiotic event, and is shared not only by dinoflagellates, but also heterokont algae, and probably cryptomonads (notable plastid losses are also predicted, see Figure 2). The original endosymbiont was almost certainly a red alga based on strong pigmentation and plastid gene data that links the plastids of heterokont algae and cryptomonads with those of red algae (Delwiche, 1999). This result also lends support to heterokonts being the sister to the Alveolata, a notion with increasing support (Baldauf *et al.*, 2000; Van de Peer *et al.*, 2000). Further analysis of the *Plasmodium* genome, in conjunction with expanding genome data from other protists, will offer many more opportunities to test the hypothesis of Fast *et al.* (2001). However for now, we are enjoying

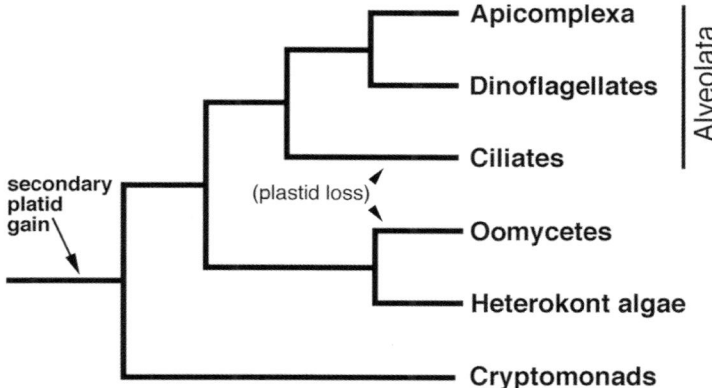

**Figure 2.** Hypothesised evolution of protists sharing a common ancient secondary plastid. Apicomplexans, dinoflagellates and ciliates form a well supported grouping known as the Alveolata. GAPDH phylogenies of Fast *et al.*, (2001) support that apicomplexans, dinoflagellates, heterokont algae and cryptomonad algae share a secondary plastid derived from a single endosymbiotic event (arrow). This implies that ciliates and oomyctes (well established sisters to the heterokont algae) are derived from plastid containing ancestors that have lost their plastid (arrowheads).

the clearest picture that we have had yet of the apicoplast, as a formerly photosynthetic, ancient secondary plastid that has played a major role in the evolution of protistan life.

# 3. Apicoplast Morphology and Division

Apicoplast ultrastructural studies enjoy the unusual distinction that much of this work was done decades before the discovery of this organelle's true identity. These early apicoplast studies span much of the diversity of the Apicomplexa and, because no one knew they were observing a plastid, the apicoplast acquired many different labels (see McFadden *et al.*, 1997 for summary of early microscopy). Amongst *Plasmodium* spp., "spherical body", "double walled organelle" and simply "vacuole" were used. These studies, in conjunction with more recent work, still contribute substantially to our understanding of apicoplast structure.

## 3.1. Number of Surrounding Membranes

The apicoplast is a multi-membrane-bound compartment that can vary in size and shape during the cell cycle. The number of membranes surrounding apicoplasts has been a matter of some debate due to the required exceptional levels of membrane preservation for this to be accurately defined. Consequently, in the past the apicoplast was often simply described as a multi-membrane compartment without any commitment to an actually

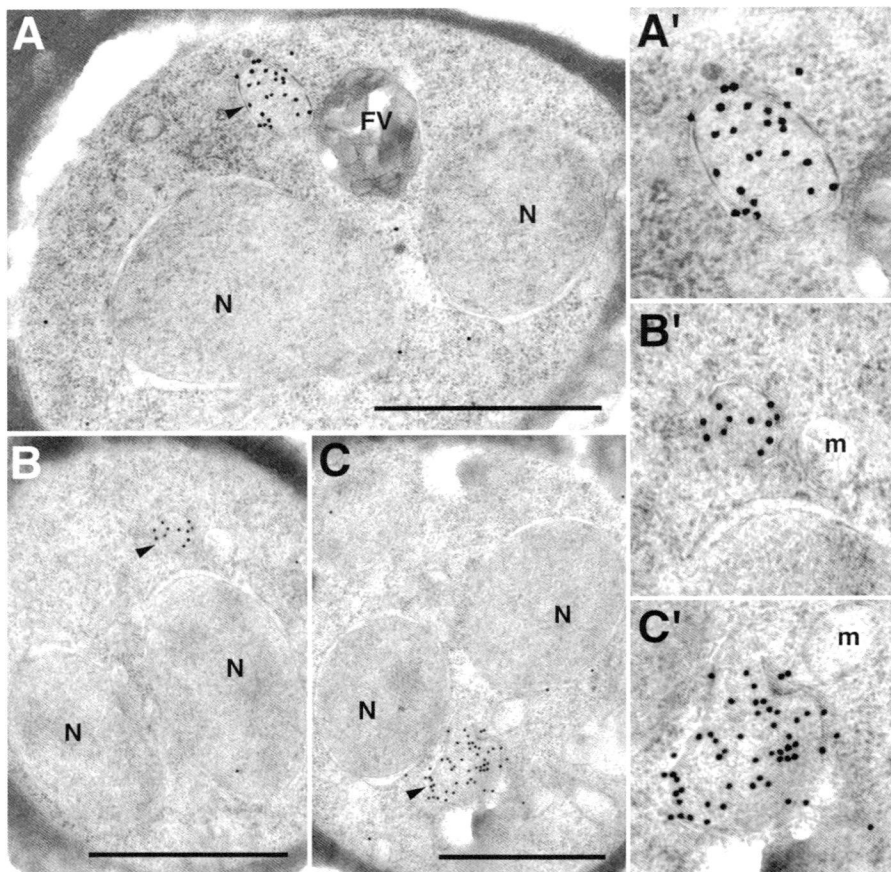

**Figure 3**. Apicoplast ultrastructure in *Plasmodium falciparum*. Parasites represent schizont stages containing multiple nuclei. GFP is targeted to the apicoplast via a transgene (encoding ACP$_{presequence}$-GFP, see 'Protein targeting to apicoplasts' section) and is visualized by immuno-gold detection of GFP. Mitochondrion profiles are seen in close proximity to apicoplasts in B' and C'. Sections are made from cryopreserved cells. N, nucleus; FV, food vacuole; arrowhead, apicoplast; m, mitochondrion. Scale bar = 1$\mu$m.

membrane number. Some exceptions, however, do exist, and *Toxoplasma gondii* is the most noteworthy having dazzled electron microscopists for years with superb ultrastructural preservation. In this case it is clear that four membranes surround the apicoplast (Dubremetz, 1995; Köhler *et al.*, 1997). Other apicomplexa have also lent themselves to answering this question of membrane number. Another coccidian, *Hepatozoon domerguei*, is identified as having four membranes around its apicoplast (Vivier *et al.*, 1972), as does *Coelotropha durchoni* (Viviere and Hennere, 1965), a further coccidian, and the gregarine *Selenidiidae pendula* (Schrevel, 1971). More recently, a study of the haemosporin, *Garnia gonandati*, clearly identifies a four membrane apicoplast in this close relative of *Plasmodium* spp. (Diniz *et al.*, 2000).

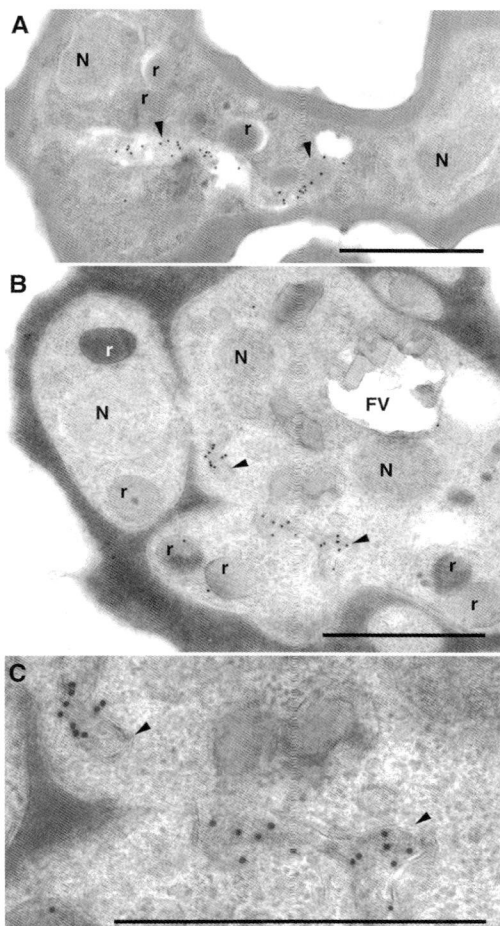

**Figure 4**. Apicoplast ultrastructure in late schizont-stage *Plasmodium falciparum* parasites expressing ACP$_{presequence}$-GFP (see Figure 3). Immuno-gold detection of GFP labels the apicoplast. (A) Multi-nucleate stage with a long apicoplast profile. (B and C) Segregating parasite with profiles of branched forms of the apicoplast. (C) Enlarged branched apicoplast showing multiple membranes surrounding the apicoplast. Panel A represents standard chemical fixation whilst panels B and C represent cyropreservation. Arrowheads, apicoplast; N, nucleus; FV, food vacuole; r, rhoptry. Scale bar = 1μm.

From this broad representation of apicomplexa it is probable that most, if not all, apicoplasts will be bounded by four membranes. However, one group (Hopkins *et al.*, 1999) have argued that *Plasmodium* spp. are exceptions, having three membranes around their apicoplasts. Hopkins *et al.* (1999) have made a detailed microscopic study, including the use of serial sections, of apicoplasts from *P. falciparum*. *P. falciparum* in notorious for poor ultrastructural preservation by chemical fixation, and this study also is confounded by wavy and folded membrane profiles that are often dilated, discontinuous and merge into one another. Comparisons with smoother membrane profiles from cryopreserved material (Figures 3 and 4) suggest that this waviness is likely the result of chemical fixation artifacts. Despite recognizing that different apicoplast profiles showed three, four and five bounding membranes, Hopkins and co-workers are convinced that three is the true number for *P. falciparum*. It is interesting to note that if the apicoplast has a common ancestry with dinoflagellate, heterokont and cryptomonad plastids (see previous section), then the loss (or gain) of plastid membranes has occurred before in this plastid lineage. Heterokont and cryptomonad plastids are surrounded by four membranes (most likely the ancestral state) while this plastid in dinoflagellates has only three (Delwiche, 1999). This suggests that one membrane has been lost during dinoflagellate evolution, and an equivalent event could have happened much more recently in *Plasmodium* evolution (since the two haemosporins, *Garnia* and *Plasmodium*, diverged). This membrane loss is not evident, however, by protein-targeting to apicoplasts, either in the route or interchangeability of protein-targeting motifs between *P. falciparum* and *Toxoplasma gondii* (discussed in the following section). Furthermore, one image from a *P. falciparum* gametocyte (for which ultrastructural preservation is typically better than for the asexual blood stages) shows a likely apicoplast clearly with four surrounding membranes (McFadden and Roos, 1999). Therefore, we await further confirmation of membrane number from *Plasmodium* spp..

## 3.2. Apicoplast Contents

Throughout the apicomplexa apicoplast contents are typically homogenous with a finely granular and sometimes fibrous texture (Figures 3 and 4, and McFadden *et al.*, 1997 for references). These fine granules are consistent in size with plastid-type 70S ribosomes (McFadden *et al.*, 1996; Hopkins *et al.*, 1999). Occasionally, some differentiation of the lumenal contents has been noted where more darkly staining regions in electron micrographs co-occur with pools of rRNAs (McFadden *et al.*, 1996; McFadden *et al.*, 1997). The fibrous nature of the apicoplast lumen has been suggested to represent the DNA content of these organelles. Apicoplast genome copy number is another contentious issue, with earlier estimates of average apicoplast content based on Southern blot analysis suggesting extremely low counts of six for *Toxoplasma gondii* and only one for *Plasmodium falciparum* (Fichera and Roos, 1997; Köhler *et al.*, 1997). More recently, measurements have been made on individual cells using video-intensified microscope photon counting, and less extreme values were determined with approximately 25 copies for *T. gondii* and 15 for *P. falciparum* (Matsuzaki *et al.*, 2001). Interestingly, in *T. gondii*, the apicoplast DNA content showed significant variation, with up to 80 copies evident in some apicoplasts (Matsuzaki *et al.*, 2001). Apicoplast DNA has been observed as a discrete nucleoid concentrated in a small portion of the *T. gondii* apicoplast by fluorescence microscopy (Striepen *et al.*, 2000). At the electron microscope level, a similar result is seen by immuno-detection of DNA, with most staining in concentrated foci (Matsuzaki *et al.*, 2001). The later authors, however, suggest that this is artifactual based on seeing more homogenous DNA staining in their fluorescent studies. Yet in the absence of an independent apicoplast stain for these fluorescent images, they may have overlooked any segregation of DNA staining within apicoplasts.

The Hopkins study (Hopkins *et al.*, 1999) of *Plasmodium falciparum* apicoplasts report extra membranous structures both inside the apicoplast lumen as well as on the outside surfaces. These internal structures are unique to *Plasmodium* and consist of infolding waves and whorls of membrane that are apparently continuous with the bounding plastid membranes (the nature of the external structures is less clear). Hopkins *et al.* (1999) observe these structures to become more elaborate as the parasites develop right through to schizonts and early merozoites. It is suggested that these membranes may indicate mass lipid transport out of the apicoplast to the rest of the parasite (Hopkins *et al.*, 1999). However, at this stage we cannot rule out that these structures represent artifacts of chemical fixation, and we note that we have not observed any equivalent structures in cryopreserved material representing late trophozoites and schizonts (Figures 3 and 4).

## 3.3. Apicoplast Lifecycle In Asexually Reproducing *Plasmodium falciparum*

In *Plasmodium falciparum* each parasite contains only one apicoplast. This is clearly evident from studies of live parasites using targeted GFP as a marker for the plastid (Figures 5 and 6A) (Waller *et al.*, 2000). Apicoplast-targeted GFP has also provided an unprecedented opportunity to observe this structure throughout the lifecycle of *P. falciparum* with observations of many hundreds of live cells. These studies reveal a continuous, intimate association of the apicoplast with the mitochondrion (Figure 6A) (Waller *et al.*, 2000). This union has also been observed at the electron microscope level (Figure 3B'-C') (Hopkins *et al.*, 1999), and is even more pronounced in some of the non-human malaria parasites where the apicoplast sits within a deep invagination of the mitochondrion (Aikawa, 1966; Aikawa and Jordan, 1968; Aikawa, 1971; Hopkins *et al.*, 1999). The significance of this association is unclear.

At the beginning of the asexual lifecycle of *P. falciparum*, when it has newly infected an erythrocyte, the apicoplast is rod shaped and slightly curved, perhaps due to being appressed against the plasma membrane in these tiny cells (Figure 5A). As this early ring stage develops the apicoplast rounds up into a small spherical structure (Figure 5B). This spherical conformation persists for most of the asexual life cycle, with only a gradual increase in apicoplast size taking place as the trophozoite grows (Figure 5C). The mitochondrion, on the other hand, elongates and elaborates early on in parasite development (Divo *et al.*, 1985), although it continues to remain in contact with the apicoplast (Figures 3B'-C', and 6A). When the parasite has enlarged to fill most of the erythrocyte the sequence of cell division must commence. It is now that we see a dramatic change in apicoplast morphology.

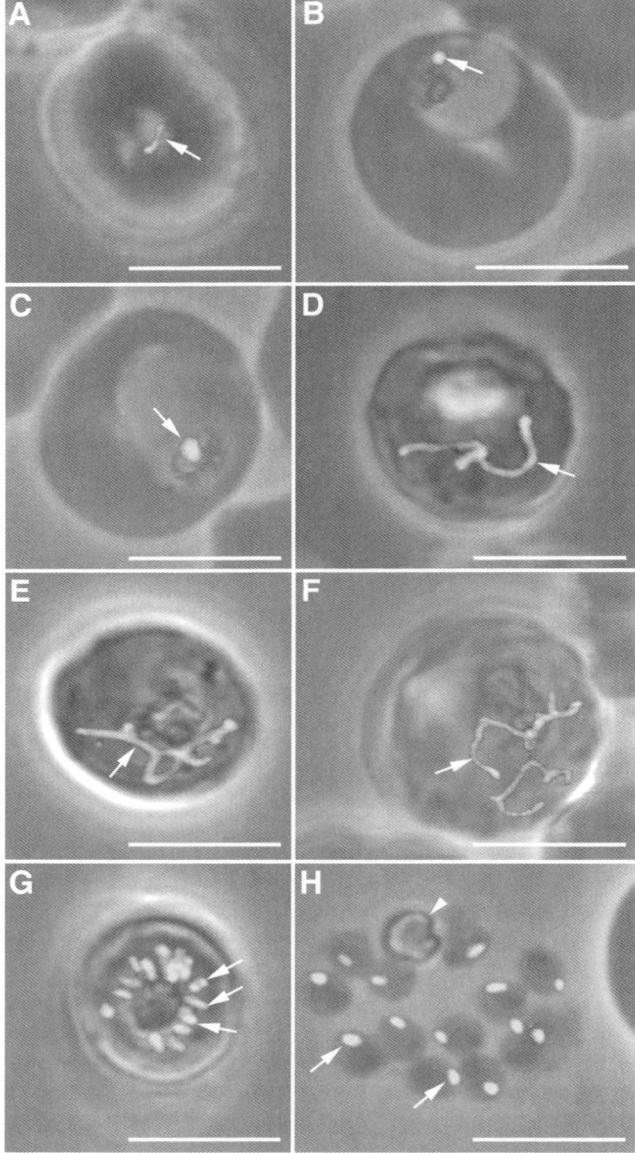

**Figure 5**. *Plasmodium falciparum* apicoplast morphology in live cells throughout the asexual lifecycle of blood stage parasites expressing apicoplast-targeted GFP (arrows). Stages represent ring form (A), small trophozoite (B), large trophozoite (C), three stages of early schizonts (D-F), late schizont (G) and recently released merozoites (remnant food vacuole is seen amongst the merozoites, arrowhead) (H). Single apicoplasts enlarge slowly, elongate to branched forms during schizogony and then divide and segregate prior to merozoite egression. These stages were seen for all transfectants with apicoplast-targeted GFP. Panels A-E and G-H represent ACP$_{presequence}$-GFP expressing parasites and panel F represents a FabH$_{presequence}$-GFP expressing parasite. Scale bars = 5μm.

## 3.3.1. Apicoplast Division

During the late trophozoite/early schizont stage the apicoplast elongates and develops into a multi-branched reticulum (Figures 4, 5D-F). This elongate, branched form persists through successive rounds of nuclear division (Figure 7A) and well into late stage schizonts where organelle segregation has commenced (Figures 4B-C). In GFP expressing live cells the next observed stage is of multiple, squat rod-shaped apicoplasts in mature schizonts (Figures 5G and 7F). This implies that the multiple division of the apicoplast is either simultaneous or occurs as an extremely short chain of events. Upon schizont rupture each merozoite inherits only one apicoplast, and there is no left over organelle with the remaining cell debris (Figure 5H).

This intriguing morphology and timing of apicoplast division has likely co-evolved with the unusual pattern of cell division (schizogony) seen in *Plasmodium* spp. and many other apicomplexans. The process of schizogony is unusual in that multiple nuclei are produced prior to organelle segregation or cytokinesis (see Figure 7). In *P. falciparum* the number of daughter cells (merozoites) produced by schizonts is variable, ranging anywhere from 8 to 24 (Kudo, 1971). This number must be tied to the final number of nuclei produced. DNA replication and nuclear division in *P. falciparum* is an apparently disordered affair where synchrony of mitosis is quickly lost after the first couple of mitotic events (Read *et al.*, 1993). This more flexible regime is probably the key to the variability in daughter cell number, and may allow schizogony to be better tuned to cell resources and fitness. A disorderly conclusion to nuclear replication (Read *et al.*, 1993) (rather than a co-ordinated final round of mitosis (Bannister *et al.*, 2000)) determines the final number of nuclei, which may be an odd number. For the apicoplast to segregate such that each daughter cell receives one apicoplast, apicoplast division needs to receive some cue as to the number of nuclei and divide accordingly. Delaying apicoplast division until the completion of nuclear division may thus be necessary.

How then does the apicoplast know how many nuclei have been produced? This is most likely through a close association with the centrosomes. Electron microscopist have noted in coccidia that the apicoplast is often seen associated with centrosomes (Vivier and Desportes, 1990; Speer and Dubey, 1999), and a recent elegant study in *Toxoplasma gondii* (Striepen *et al.*, 2000) confirmed this association by immuno-fluorescence throughout the cell cycle. Moreover, dividing apicoplasts are seen to be drawn apart by centrosomes at each end, with an apicoplast DNA nucleoid just beneath the limiting membranes at these points of centrosome association (Striepen *et al.*, 2000). The apicoplast is then apparently cleaved in two with the development of the pellicle of the daughter cells (Striepen *et al.*, 2000). This model presents an attractive explanation for the complex branched apicoplasts seen in *P. falciparum* schizonts. An association with

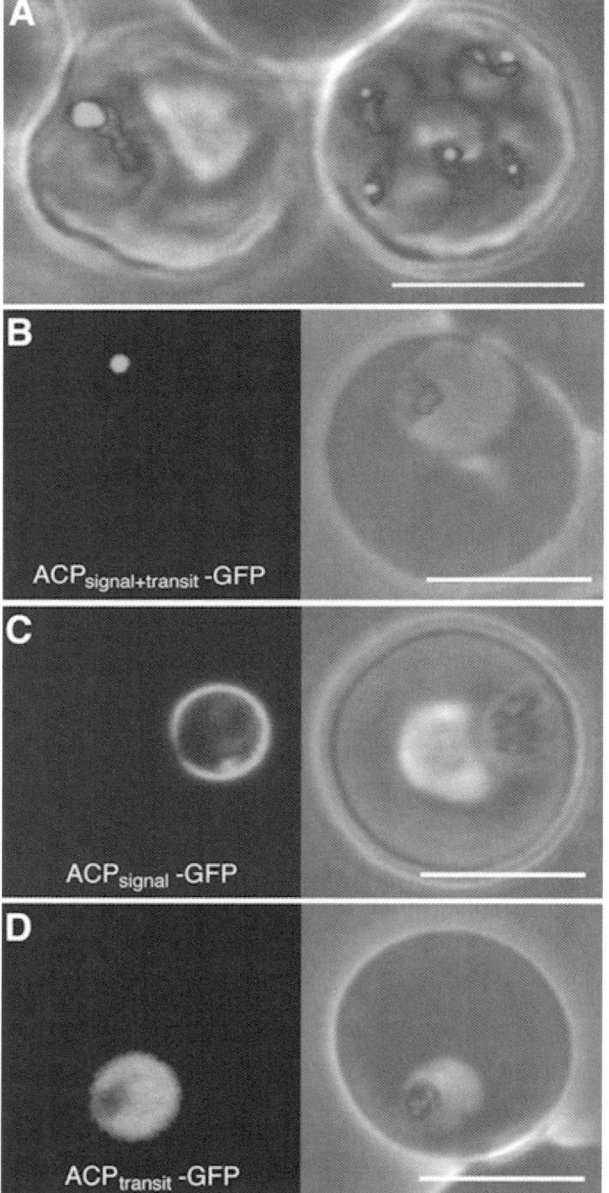

**Figure 6**. (A) Live intra-erythrocytic *Plasmodium falciparum* parasites expressing the apicoplast-targeted GFP fusion protein ACP$_{presequence}$-GFP and co-stained for mitochondria (red) [right-hand erythrocyte contains multiple (5) infections]. (B) Parasite expressing ACP$_{presequence}$-GFP (presequence containing the signal peptide plus the transit peptide) with GFP accumulating in the apicoplast. (C) Parasite expressing ACP$_{signal}$-GFP with GFP accumulating in the parasitophorous vacuole (some GFP re-enters the parasites into the food vacuole). (D). Parasite expressing ACP$_{signal}$-GFP with GFP accumulating in the parasitophorous vacuole (some GFP re-enters the parasites into the food vacuole). (C) Parasite expressing ACP$_{transit}$-GFP with GFP accumulating in the cytosol. Scale bar = 5µm. **See Colour Plate at the back of the book.**

centrosomes replicating during nuclear replication would draw the apicoplast into an elaborate conformation as nucleus number increases. By analogy, in *T. gondii* if apicoplast division is blocked by prolonged microtubule inhibition with dinitroaniline herbicides, apicoplasts with several centrosomes develop and these apicoplasts are elongate and multiply branched (Striepen *et al.*, 2000). In *Plasmodium*, delaying apicoplast division until the final stages of segregation while maintaining this centrosome association would be necessary to ensure that each nucleus, and hence each daughter cell, receives an equal apicoplast inheritance (including copies of the apicoplast genome). The close association of the mitochondrion with the apicoplast may be a mechanism to ensure that this organelle is appropriately partitioned also. We await confirmation of a link between *Plasmodium* apicoplasts and centrosomes to further substantiate this model.

### 3.3.2. Division Molecules

It is a nice testimony to the prokaryotic nature of plastids that many have retained the molecular apparatus for organelle division that bacteria use for cell division. In bacteria the tubulin homologue FtsZ forms a division ring at the center of the cell coordinated by the a family of Min proteins. In most plant and algal groups surveyed gene homologues for FtsZ and some Min proteins have been found, and these have been shown to be linked to plastid division (Gilson and Beech, 2001; Osteryoung, 2001). Interestingly, FtsZs have been implicated in mitochondrial division also, although many eukaryotes (including animals, fungi and plants) have apparently evolved alternate division mechanisms independent of FtsZ (Osteryoung, 2001). The heterokont algae, which are likely relatives of apicomplexa (see previous section), have been shown to use FtsZ molecules in the division of both their plastids and mitochondria (Beech *et al.*, 2000; Gilson and Beech, 2001). It therefore seemed probable that FtsZs would be found in apicomplexa. To date, however, no FtsZs have been found despite extensive searching of the *Plasmodium falciparum* genome and PCR screening of *Toxoplasma gondii* (Striepen *et al.*, 2000). Though such molecules may be discovered, it is possible that apicomplexa have found alternate division strategies. These may involve dynamin-like proteins as have apparently been adopted by yeast for mitochondrion division. Matsuzki *et al.*, (2001) report conspicuous darkly staining patches and even rings at the constriction between dividing apicoplasts in *T. gondii* that might be evidence of some form of division apparatus. Streipen and colleagues (Striepen *et al.*, 2000), on the other hand, have been unable to observe any compelling division structures, and suggest that apicoplast division may be achieved by a slicing force generated by the mitotic spindle pulling the apicoplast against the extending daughter cell pellicle. Presently, this is a topic with more questions than answers.

# 4. Protein Targeting to Apicoplasts

While no plastid is known to have completely lost its prokaryote-type genome, all plastids have undergone extensive transfer of genes to the host cell nucleus (Martin and Herrmann, 1998). This gene transfer has left these organelles heavily dependent upon proteins targeted to the organelle from the host cytosol. In cells with primary plastids (see earlier section) a system of plastid protein import has evolved where a peptide extension, referred to as the transit peptide, occurs at the N-terminus of the protein to be targeted. This transit peptide interacts specifically with the plastid membrane lipids and protein complexes Toc and Tic (Translocon at the outer/inner envelope membrane of chloroplasts) which facilitate protein import into the plastid (Figure 1A) (Bruce, 2001; Jarvis and Soll, 2001). This system is analogous to protein import into mitochondria, and plant cells must discriminate between proteins destined for mitochondria and those destined for plastids. Cells that have acquired a secondary plastid (see earlier section) most likely inherited this Toc/Tic system for protein targeting to plastids. However, upon gene transfer to the new host cell nucleus, cytosolically translated plastid proteins are faced with additional membranes that must be passed on route to the plastid (Figure 1B). In all secondary plastids studied to date, an additional targeting motif has been appended to the transit peptide in response to the extra obstacles of these multi-membrane bound plastids (van Dooren *et al.*, 2001). This motif is a short hydrophobic peptide that apparently acts as a signal peptide for entry into the endomembrane system. This appears to be a logical solution given that the plastid now resides within a derived food vacuole, which is part of the host's endomembrane system (Figure 1B).

Apicoplasts are secondary plastids and are therefore faced with the same extra membrane obstacles for protein import as their algal counter-parts. When the first nucleus-encoded genes whose protein products target to the apicoplast were described, from *Plasmodium falciparum* and *Toxoplasma gondii*, it was abundantly clear that apicoplasts employ a similar targeting mechanism as algae with secondary plastids. From these genes the predicted proteins (fatty acid biosynthetic proteins ACP, FabH and FabZ, and ribosomal proteins S9 and L28) all had a substantial N-terminal presequence that by bioinformatic analysis consisted of two parts (Waller *et al.*, 1998). The most N-terminal portions were identified as classic von Heijne-type signal peptides, and these were followed by domains similar to plant-type transit peptides. *P. falciparum* and *T. gondii* are unique amongst secondary plastid containing organisms in that they are amenable to genetic manipulation and this has allowed in apicomplexa a thorough exploration of protein trafficking to secondary plastids.

## 4.1. Targeting Presequences Comprise a Signal Peptide and Transit Peptide

Using GFP as a reporter molecule several studies have demonstrated that the N-terminal bipartite presequences, of apicoplast-targeted proteins, are necessary and sufficient for apicoplast import in *Plasmodium falciparum* (Figures 3, 4, 5, 6A-B) (Waller *et al.*, 2000) and *Toxoplasma gondii* (Waller *et al.*, 1998; Yung and Lang-Unnasch, 1999; DeRocher *et al.*, 2000; Jelenska *et al.*, 2001; Yung *et al.*, 2001). It has also been shown that *P. falciparum* presequences successfully targets GFP to *T. gondii* apicoplasts (Jomaa *et al.*, 1999) and *vice versa* (Waller *et al.*, 2000). This exchangeability of targeting presequences demonstrates the high level of conservation of trafficking route to apicoplasts between disparate apicomplexa.

The bipartite nature of the presequences has also been examined by deleting either the signal peptide domain or transit peptide domain. If the transit peptide domain of ACP (acyl carrier protein) is removed, leaving only the signal peptide domain, GFP is secreted outside of the cell into the parasitophorous vacuolar space for both *Plasmodium falciparum* (Figure 6B) (Waller *et al.*, 2000) and *Toxoplasma gondii* (Roos *et al.*, 1999). This result is also obtained with the S9 presequence (DeRocher *et al.*, 2000; Yung *et al.*, 2001). This validates the bioinformatic identification of a signal peptide in the bipartite presequences, and indicates that the first trafficking step of proteins destined for the apicoplast is entry into the endomembrane system. Moreover, it suggests that once within the endomembrane system, the transit peptide domain is required to sort the protein to the apicoplast. In the absence of this sorting signal GFP follows a default route of secretion out of the cell. If the ACP signal peptide is removed from the GFP construct, leaving only the transit peptide domain, GFP accumulates in the cytosol (Figure 6D) (Roos *et al.*, 1999; Waller *et al.*, 2000). This suggests that the apicoplast sorting machinery that recognizes the transit peptide is within the endomembrane system and cannot facilitate apicoplast sorting from the cytosol.

Protein targeting to apicoplasts is therefore at least a two step process, with co-translational import into the endomembrane system the first step, and subsequent sorting to the apicoplast the second step. A signal peptide followed by a transit peptide is sufficient for these two trafficking events. These bipartite presequences apparently lack any further cryptic sorting signals as is demonstrated by some elegant domain swapping in *Toxoplasma gondii*. Either the signal peptide was replaced with one from a secreted protein (P30, *T. gondii*), or the transit peptide replaced with a plant transit peptide (FtsZ, *Arabidopsis thaliana*) without effect upon apicoplast targeting (Roos *et al.*, 1999). Furthermore, DeRocher *et al.* (2000) show that the *T. gondii* S9 transit peptide is recognized by, and facilitates, protein import into isolated pea chloroplasts. It is curious, though not surprising, that in *T. gondii* the S9 transit peptide (in the absence of the signal peptide) can facilitate protein

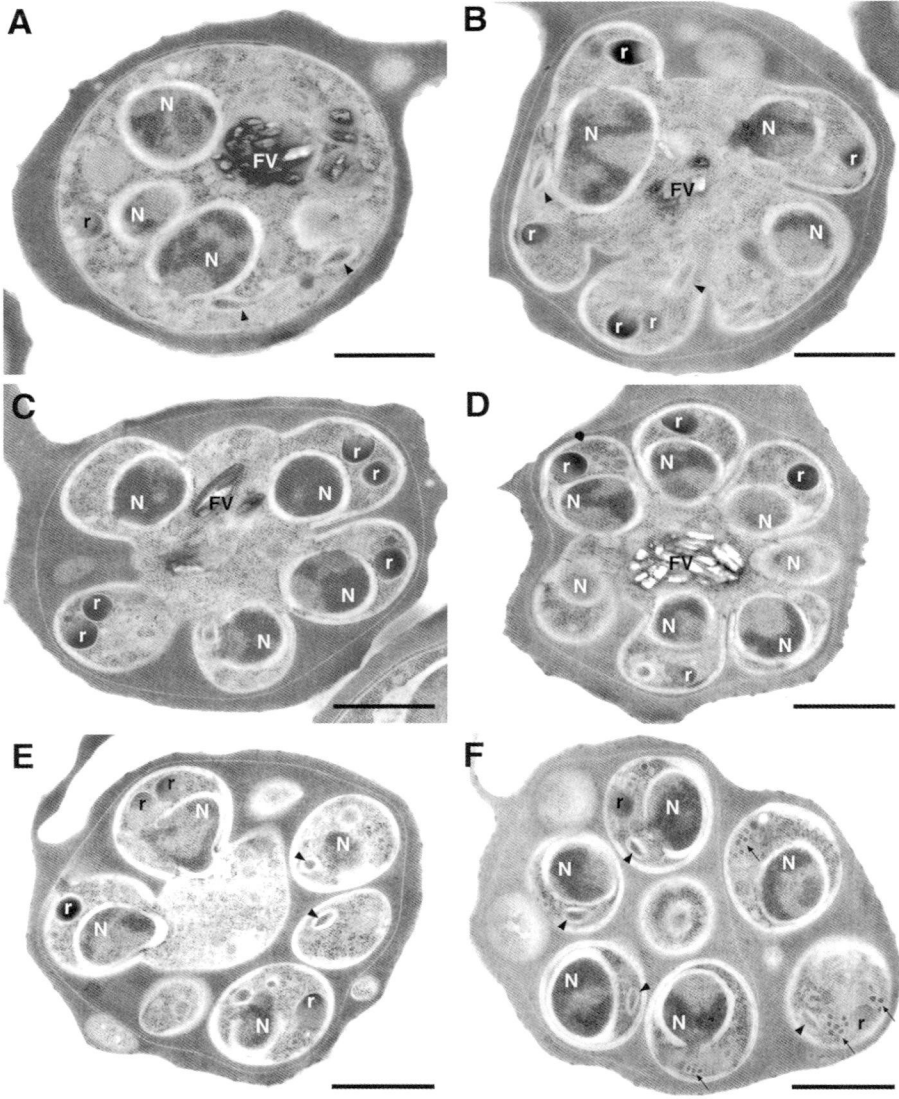

**Figure 7.** Ultrastructure of intra-erythrocytic *Plasmodium falciparum* during schizogony. Cryopreserved cells represent pre-segregation stage with multiple nuclei (A), stages during organelle segregation into new apical cups and cytokinesis (B – E), and mature schizonts consisting of separate daughter cells escaping from the parasitophorous vacuole (F). Whilst thin sections cannot provide full detail of apicoplast morphology in these stages, occasional profiles reveal the elongate nature of the apicoplast (A). Arrowhead, apicoplast; N, nucleus; r, rhoptry; FV, food vacuole; arrows, micronemes. Scale bars = 1µm.

import into the parasite's mitochondrion (DeRocher *et al.*, 2000; Yung *et al.*, 2001). Dual targeting of proteins to plant chloroplasts and mitochondria is seen for a number of proteins (Peeters and Small, 2001) and this result only further emphasizes the relatively generic nature of the apicoplast transit peptide.

## 4.2. Apicoplast Transit Peptides

While signal peptides are relatively simple, and their mode of action extensively studied from a variety of systems, plastid-targeting transit peptides remain enigmatic (see Bruce, 2001 for review). They vary greatly in size, have no primary sequence conservation, and when analyzed in aqueous conditions apparently lack any stable secondary structure (this may change upon interaction with plastid membrane lipids, see below). Their main defining features are considered to be an enrichment in hydroxylated residues (particularly serine and threonine) and a net positive charge. These criteria have held up quite well across algal and plant transit peptides studied to date (Cline and Henry, 1996; Liaud *et al.*, 1997; Sulli *et al.*, 1999; Deane *et al.*, 2000). How then, do apicomplexan transit peptides compare?

The transit peptides of *Plasmodium falciparum* and *Toxoplasma gondii* both show a net positive charge though this bias is predominantly at the N-terminal region of the transit peptide (Waller *et al.*, 1998). Longer transit peptides may become more neutral downstream of this region. A net positive charge in the first 17 residues is such a consistent feature of *P. falciparum* transit peptides that it has been adopted as a useful defining character for predictive software for *P. falciparum* transit peptides (discussed below). An enrichment for serine and threonine residues is also observed in *T. gondii* transit peptides, but this is not the case for *P. falciparum* (Waller *et al.*, 1998; Zuegge *et al.*, 2001). In fact, despite the apparent importance of these residues in plant transit peptides, the mutational removal of all remaining serine and threonines from the transit peptide of FabH apparently had no effect upon apicoplast targeting (Waller *et al.*, 2000). *P. falciparum* does, on the other hand, show a strong preference for lysine and asparagine residues in its transit peptides (Waller *et al.*, 1998; Zuegge *et al.*, 2001). While the significance of this is unclear, it may be a consequence of the *P. falciparum* AT bias (82%) given that both residues can be encoded without a C or G. The lysines also lend much of the positive charge to the transit peptide. Given that *P. falciparum* transit peptides have been demonstrated to work in *T. gondii* and *vice versa* (Jomaa *et al.*, 1999; Waller *et al.*, 2000), it seems more likely that the differences between them reflect genome pressures rather that modified trafficking mechanisms.

Transit peptide length has also been studied by a number of groups. From a set of 35 putative *Plasmodium falciparum* apicoplast-targeted proteins, Zuegge *et al.*, (2001) record a median length of 78 residues, with approximately 20 and 580 the extremities of this range. In *Toxoplasma gondii* two groups have performed serial deletion experiments with transit peptides and show that there is apparently enormous redundancy in some of these longer transit peptides (DeRocher *et al.*, 2000; Yung *et al.*, 2001). From the approximately 150 residue presequence of ribosomal protein S9, all but 55 could be deleted from the C-terminal end without loss of apicoplast targeting. Upon the deletion of a further six residues, transit peptide function was lost and the reporter protein was secreted from the cell (Yung *et al.*, 2001). Whilst this suggested that one or several of these six residues is essential for function, a further construct that lacked only these six out of the 150 residue presequence trafficked normally to the apicoplast (Yung *et al.*, 2001). Therefore, perhaps a minimal length is more important than specific residues. Similar redundancy of transit peptide length has also been shown for *T. gondii* ACP (Roos *et al.*, 1999).

If we return our attention to plants it is worth noting that several transit peptide features are emerging that may offer insights into general transit peptide function (for review see Bruce, 2001). Four distinct plant transit peptides features are; (1) phosphorylation and dephosphorylation of specific motifs in transit peptides prior to import (this may explain in part the need for hydroxylated residues that can be phosphorylated) (Waegemann and Soll, 1996; Jarvis and Soll, 2001), (2) the presence of Hsp70 recognition domains that mediate binding of these and other cytosolic proteins on route to the chloroplasts (Ivey *et al.*, 2000), (3) specific interactions of transit peptides with chloroplast-specific membrane lipids, and (4) one or more transit peptide domains forming $\alpha$-helical structures upon contact with the chloroplast membrane lipids. It has recently been noted that *Plasmodium* transit peptides also typically contain predicted Hsp70 binding domains, and that transit peptide function is sensitive to alterations at these sites (Foth *et al.*, 2003). It remains to be tested whether any more of these features play a part in apicoplast transit peptide function.

### 4.3. Tocs And Tics – What Recognizes the Transit Peptide?

The presence of transit peptides in the bipartite leaders of apicoplast-targeted proteins strongly implies that the mechanism of protein import into apicoplasts, at least for the second stage of transport across the inner membranes, has been retained from the primary plastid-containing progenitor. This means that the protein import complexes Toc and Tic are likely to occur still in apicoplasts. Indeed this appears to be the case. ClpC, also known as Hsp93, is a stromal chaperone of plastids that forms part of the Tic complex (Jackson-Constan *et al.*, 2001). The gene for ClpC is encoded on the apicoplast of *Plasmodium falciparum* and *Toxoplasma gondii* (Wilson *et al.*, 1996b; Köhler *et al.*, 1997, http://www.sas.upenn.edu/~jkissing/toxomap.html). Furthermore, a

*P. falciparum* homologue of Tic22 has been found and it has a bipartite leader that targets GFP to the apicoplast suggesting it is a component of apicoplasts (C. J. Tonkin and G. I. McFadden unpublished). A putative homologue of the Toc complex, Toc34, has also been identified from *P. falciparum* (C. J. Tonkin and G. I. McFadden unpublished). Further homologues will most likely also present themselves from the *P. falciparum* genome data. Additionally, we are now at the stage where "pull down" experiments of these entire complexes can be considered for a more powerful analysis of these protein import machines.

### 4.4. Bipartite Presequences are Sequentially Removed During Protein Targeting

During protein trafficking signal peptides and transit peptides are both removed after proteins reach their destination. Western blot analyses of both *Plasmodium falciparum* and *Toxoplasma gondii* confirm that their bipartite leaders are removed and that this removal occurs sequentially. The signal peptide is presumed to be removed during co-translational import into the endoplasmic reticulum (ER). Consistent with this, full length proteins that include the signal peptide are not detected by Western blot analysis (Waller *et al.*, 1998; 2000). However, protein bands that correspond in size to the mature protein plus the transit peptide are seen in Western blots (Waller *et al.*, 1998; 2000; Vollmer *et al.*, 2001; Yung *et al.*, 2001). These species are thought to represent either proteins that haven't reached their final destination in the apicoplast lumen, or those that await final processing therein. Microscopy studies (either by fluorescence or electron microscopy) do not detect protein trafficking intermediates in the ER or other parts of the cells (Waller *et al.*, 1998; 2000), suggesting that the later explanation is the correct one, although some protein may still be on route through the apicoplast membranes. The smallest apicoplast protein bands detected by Western analysis are the fully processed forms that lack the entire bipartite presequence, and these are typically the predominant bands.

In plants, the removal of the transit peptide from chloroplast-targeted proteins is achieved by the Stromal Processing Peptidase (SPP) once the proteins arrives in the chloroplast (Richter and Lamppa, 1999). A nucleus-encoded homologue of plant SPP has recently been identified from *Plasmodium falciparum*. The *P. falciparum* SPP homologue bears a bipartite presequence for apicoplast-targeting suggesting it is apicoplast located (van Dooren *et al.*, 2002). This presents a very likely candidate for the role of transit peptide processing in the apicoplast. It is of note that for all apicoplast-targeted proteins studied to date (both for *P. falciparum* and *Toxoplasma gondii*), substantial amounts of the unprocessed form, with transit peptide still attached, are observed on Western blots (Waller *et al.*, 1998; 2000; He *et al.*, 2001; Vollmer *et al.*, 2001; Yung *et al.*, 2001). This suggests that processing is either delayed or relatively inefficient in apicoplasts in comparison to plant and algal plastids where only the mature, processed proteins are generally detectable. To characterize further the kinetics of protein targeting to apicoplasts and processing in *P. falciparum,* van Dooren *et al.,* (2002) performed $^{35}$S-Met/Cys pulse-chase labeling and immuno-precipitation of an apicoplast-targeted ACP-GFP fusion protein. In these experiments the larger band is the first to appear followed by the smaller band, consistent with this representing transit peptide processing. Within 45 minutes of label addition (the shortest period that gave sufficient labeling to observe), some processed form of this targeted protein had already appeared, and it took up to four hours for all labeled protein to be processed. Hence, it takes less than 45 minutes for at least some newly synthesized protein to reach the apicoplast and be processed. DeRocher *et al.* (2001) have examined the kinetics of apicoplast targeting in *Toxoplasma gondii* by a novel method. These authors fused the presequence of ribosomal protein S9 to a conditional aggregation domain (CAD) motif (DeRocher *et al.,* 2001). In the absence of a small molecule (AP21998), this fusion protein self associated in the ER of *T. gondii* parasites, presumably shortly after protein synthesis. In this form it was unable to traffic to the apicoplast. Upon the addition of AP21998 the fusion protein can dissociate, and fusion protein reached the apicoplast within 2.5 to 5 minutes. This indicates remarkably fast trafficking in these parasites, and implies that unprocessed stocks of protein seen in Western analyses reveals slow processing.

Purification of two *Plasmodium falciparum* proteins targeted to the apicoplast has enabled N-terminal protein sequencing of the processed forms of these proteins. Hence for FabI (Jomaa *et al.,* 1999) and ACP (van Dooren *et al.,* 2002) the cleavage site, most likely recognized by SPP, is known. For plants and algae, cleavage consensus motifs have been characterized that are useful for further predictions of transit peptide boundaries (Emanuelsson *et al.,* 1999). However, the *P. falciparum* cleavage sites so far conform to none of these "rules" (van Dooren *et al.,* 2002) and offer no clear features that might be useful for further predictions. More data is clearly required to understand better this cleavage event.

### 4.5. From the ER to the Apicoplast – The Events in Between

Our current model for protein targeting to the apicoplast can account for protein entry into the ER (via the signal peptide) and the subsequent passage into the apicoplast across the inner membranes most likely spanned by Toc and Tic import complexes (via the transit peptide). Yet there remain unresolved questions about the transport events in between. (1) How do proteins get from the ER to the apicoplast for Toc/Tic mediated final transport? (2) If the inner two membranes of apicoplasts are derived from the primary plastid and retain their Toc and Tic complexes, how is the third membrane (counting from the inside) of four membrane-bound apicoplasts crossed by proteins?

### 4.5.1. From the ER to the Apicoplast

To address the first question there are several possibilities that we can consider. The first and simplest has a precedent in several algal groups. This is that proteins are co-translationally inserted directly into the outer membrane of the apicoplast. Secondary plastids in many algal groups have ribosomes on the outermost membrane, which is contiguous with the rough ER, and co-translational insertion into the endomembrane lumen containing the plastid is thought to occur (van Dooren *et al.*, 2001). No such ribosomes are observed on the apicoplast limiting membrane so an alternative delivery mechanism must occur. Furthermore, DeRocher's CAD fusion protein accumulated in the ER, not in an outer shell of the apicoplast (see above) (DeRocher *et al.*, 2001). Another possibility is that the apicoplast outer membrane is continuous with the ER, and therefore ER proteins have direct access to the apicoplast. Such connections have not been observed by microscopy, although transient tubular connections such as occur between plastids in plants (Kohler *et al.*, 1997) may be labile and difficult to observe.

If no continuity occurs between the ER and the apicoplast then vesicular traffic must be considered and recent data suggests that NSF (N-ethylmaleimide-sensitive factor)-mediated vesicle docking does occur at the outer apicoplast membrane (Hayashi *et al.*, 2001). What then directs vesicles to the apicoplast? The simplest solution is nothing, simply that the apicoplast sits in the default secretory pathway. This implies that in the absence of any further signal, all proteins from the ER will wash past the apicoplast, allowing for proteins baring transit peptides to be selected from this mix. In this scenario the apicoplast could be positioned anywhere post ER, including before or after (or theoretically even within) the Golgi apparatus. The alternative possibility is that directed vesicle traffic to the apicoplast does occur following a sorting event of apicoplast proteins from other proteins in the endomembrane system. Given that a plant transit peptide can mediate apicoplast transport from the ER (Roos *et al.*, 1999) we know that no novel signals have been added to apicomplexan transit peptides. Could then protein sorting utilize the transit peptide itself? In plants, a class of cytosolic, targeting factors (including Hsp70s and 14-3-3 proteins) have recently been shown to bind the transit peptides of chloroplast-targeted proteins, prior to their arrival at the chloroplast membrane, and deliver them to chloroplasts in a heightened import-competent state (Bruce, 2001; Jackson-Constan *et al.*, 2001; Jarvis and Soll, 2001). This shows that recognition of transit peptides can occur remote from plastids. It is therefore possible that a receptor-mediated transport system in apicomplexa could have evolved from such a plant system, although the system would need to located within the lumen of the endomembrane system for apicomplexa.

Until recently there has been little data to support any of the above scenarios, but attention is now being drawn towards these outstanding questions. Parasite ER retention signals (HDEL) have been appended to apicoplast targeted fusions in *Toxoplasma gondii* and show no effect on apicoplast targeting (D. S. Roos, pers. comm.; DeRocher *et al.*, 2001). This suggest that apicoplast-sorting may occur before the cis Golgi where HDEL resorting to the ER occurs (Van Wye *et al.*, 1996). Conversely, apicoplast targeting has recently been shown to be sensitive to brefeldin A in *Plasmodium falciparum* (Cheresh *et al.*, 2002). Brefeldin A blocks protein movement beyond the Golgi in *P. falciparum* (Crary and Haldar, 1992), therefore this data argues that protein targeting to the apicoplast occurs via the Golgi. Cheresh *et al.* (2002) also provide data for a rather surprising hypothesis. Using stage-regulated expression of a transgene during young ring stage *P. falciparum* parasites these authors report protein accumulation in the parasitophorous vacuole prior to protein sorting to the apicoplast. This routing was not evident by our previous studies of the same fusion protein under direction of a different promoter (Waller *et al.*, 2000). However, the use of a differently timed promoter system may account for these differences. No evidence of such a routing has been reported for *T. gondii*. Cheresh *et al.*'s hypothesis presents a somewhat unanticipated route for apicoplast targeting. Nevertheless, it needs to be remembered that *P. falciparum* possesses a highly developed secretory system that traffics proteins specifically to several destinations beyond the parasite in the erythrocyte cytosol and plasma membrane (Foley and Tilley, 1998). Recent studies suggest that proteins destined for these sites also travel via the parasitophorous vacuole (Wickham *et al.*, 2001). Could it be that *Plasmodium* has redefined this "external" compartment as a sorting station for secretory proteins? If so, we may find that other internal organelles, such as micronemes and rhoptries, also use this pathway for protein trafficking via the parasitophorous vacuole.

## 4.5.2. One Membrane To Go!

The second unresolved question concerns how the third apicoplast membrane is crossed by apicoplast-targeted proteins. This is a question that has vexed all who have considered protein trafficking to four membrane-bound plastids (van Dooren *et al.*, 2001). A simple solution is presented where pores or non specific protein transporters allow proteins to simply diffuse across the third membrane, as if it wasn't there (Kroth and Strotmann, 1999). The origin of such pores is unstated. Perhaps a more attractive option, though equally untested, is that the Toc/Tic machinery has come to span all three inner membranes. This might occur by simply having Toc complexes in both the second and third membranes. Proteins would then pass through two Toc complexes before being passed through a Tic complex. Both of these models suggest that the third membrane is redundant, and cannot explain why it has not been lost, as is the case in some algal groups. It is possible that the third membrane retains some useful role in apicoplast trafficking, but at present, such a role eludes us.

## 4.6. Genome Screening for Apicoplast-Targeted Sequences

Except for a select band of secondary-plastid-protein-targeting aficionados, the greatest interest in protein trafficking routes to the apicoplast is due to their ability to tell us more about the protein content of apicoplasts. In *Arabidopsis thaliana* 3574 proteins are thought to be targeted to chloroplasts (The Arabidopsis Genome Initiative, 2000). Even though the apicoplast lacks proteins associated with photosynthesis and, in the case of *Plasmodium falciparum* at least, the shikimate pathway for aromatic amino acid synthesis (Keeling *et al.*, 1998), its protein count could exceed one thousand. When one considers that the apicoplast genome encodes a mere 64 genes, and many of these are rRNAs and tRNAs, it is obvious that the major portion of the apicoplast proteome is coded for in the nucleus. The distinctive bipartite presequences necessary for apicoplast targeting thus offer a useful handle for identifying putative apicoplast proteins from genomics data.

In response to the tide of *Plasmodium falciparum* genome data, efforts have been made to develop predictive software for screening the genome data. Signal peptide predictors such as neural network-based SignalP, and "knowledge-based" PSORT have proved generally reliable for identifying the signal peptides on the bipartite presequences (Waller *et al.*, 1998; Zuegge *et al.*, 2001). However, the neural network-based transit peptides predictor ChloroP is generally not very reliable for non-plant transit peptides (Bruce, 2001). Indeed, ChoroP performs poorly in predicting *Plasmodium falciparum* transit peptides, probably because it was trained with plant data sets (Zuegge *et al.*, 2001). Therefore, Zuegge *et al.* (2001) set about training a neural network specifically for *P. falciparum* targeting presequences. Using a training set of 84 sequences most probably targeted to the apicoplast, and a negative set of 102 non-apicoplast proteins, Zuegge *et al.* developed a predictive system called PATS (predict apicoplast-targeted sequences) that upon reclassification of the training data was 97% accurate. When used in conjunction with SignalP for signal peptide assignments this power was increased. When PATS was applied to the *P. falciparum* chromosome 2 and 3 data it predicted that 69 genes (15% of total genes on these chromosomes) encoded apicoplast-targeted proteins (Zuegge *et al.*, 2001). Previously only a few of the genes had been annotated as encoding apicoplast targeted proteins (Gardner *et al.*, 1998).

An alternate "knowledge-based" strategy has also been developed where observed features of *Plasmodium falciparum* presequences have been used to model a set of criteria for discriminating apicoplast-targeting presequences (Foth *et al.*, 2003). Foth *et al.* have defined four key criteria in their predictor (called PlasmoAP and accessible online at http://www.PlasmoDB.org). These criteria are as follows. (1) SignalP identifies a likely signal peptide (which is removed at the predicted cleavage site for the rest of the analysis which focuses on the transit peptide component). (2) Within the first 22 residues of the transit peptide the ratio of acid to basic residue must not exceed 0.7. (3) Within the first 80 amino acids a stretch of 40 residues with a total of at least 9 asparagines and /or lysines occurs. (4) The above 40 residue window must have a ratio of acid to basic residues of no greater than 0.9. Using these criteria, this predictor enabled good recovery of the training set used by Zuegge *et al.* (2001) and when applied to chromosome 2 and 3 data recovered a very similar set of genes to PATS (Foth *et al.*, 2003). Moreover, Foth *et al.* (2003) have set about testing the parameters of their predictor *in vivo* through transit peptide mutagenesis experiments. These experiments validate the importance of positively charged residues at the N-terminal extremity of the transit peptide, but also point to possible refinements of the criteria to enhance the power of this predictor.

These custom made predictors of *Plasmodium falciparum* apicoplast-targeting bipartite presequences have provided a valuable primary screen of the *P. falciparum* genome. Together they identify over 500 genes that likely encode plastid-targeted proteins (Gardner *et al.*, 2002; Foth *et al.*, 2003). This represents approximately 10% of the gene content of the *P. falciparum* genome. However, this likely represents a minimal figure given that genes with introns that interrupt the presequence area (of which there are many already known) will be overlooked by these methods. Some genes can even share a single bipartite presequence coding region that is alternatively spliced onto one of two unrelated gene mRNAs. van Dooren *et al.* (2002) have recently identified such a case with *P. falciparum* SPP and ALAD (delta-aminolevulinic acid dehydratase). Such problems, however, are common to the annotation of any eukaryotic genomic sequences and demonstrate that bioinformatics tools represent only the first step in the analysis of genome data. Nevertheless these methods have offered us our first overview of the apicoplast protein content.

# 5. Apicoplast Metabolism – A Target for Drugs

Why apicomplexan parasites should possess a plastid was a confronting question to greet malariologists who first took an interest in this unexpected discovery. What could "plant biology" have to offer such a superb group of intracellular animal parasites? Indeed many questioned the utility of the apicoplast, and considered that it may simply be useless evolutionary baggage.

However, it is now clear that the apicoplast is indispensable to parasites. This was first demonstrated in *Toxoplasma gondii* by selectively blocking apicoplast DNA replication leading to eventual loss of apicoplast DNA and cell death (Fichera and Roos, 1997). Similarly, in *Plasmodium falciparum* agents that reduce apicoplast transcription kill parasites (McConkey *et al.*, 1997). More recently and elegantly, He *et al.* (2001) engineered a *T. gondii* apicoplast segregation mutant that resulted in parasites completely lacking an apicoplast. Interestingly, these parasites were able to invade host cells and divide once or twice, however, growth then stopped and the parasites died. In addition there is a great deal of indirect drug inhibition data that supports apicoplast metabolism being vital to parasites (see Table 1). But why do apicomplexa still require this plastid, despite not using it for photosynthesis? To address this question fully, it is useful to consider the processes of endosymbiotic partnership.

The establishment of an autotrophic endosymbiont offers a radical change of lifestyle for any heterotroph. A cell formerly dependant on prey capture for reduced carbon as an energy source, and potentially numerous other macromolecular building blocks (such as the many "vitamins", certain amino acids and fatty acids, etc), now has the chance to exploit the endosymbiont as a supplier of these items. Indeed, should this cell now become exclusively autotrophic, it will be completely dependant on this "in-house" production, just as the autotrophic endosymbiont would have been before. This partnership, however, also initially creates many duplications of equivalent, though sometimes very divergent (due to the prokaryotic nature of plastids), metabolic pathways. In evolution, redundancy is generally countered by streamlining, and it may be either the host or endosymbiont version of a given pathway that is lost. Cases of the endosymbiont pathway being selected over that of the host include the chloroplast pathways for fatty acid and heme synthesis in most plants (Brown *et al.*, 1990; Harwood, 1996). "Mix and match" combinations of both host and endosymbiont components of a single pathway can also occur, such as fatty acid synthesis in grasses (the first enzyme, ACC, is host-derived, the subsequent ones, FAS, are plastid-derived) (Harwood, 1996). Thus, the endosymbiont may come to fulfill further roles beyond those that drove the initial symbiosis. Should this cell revert to a heterotrophic lifestyle, all or part of the initial utility of the endosymbiont may be lost. However, endosymbiont functions that have displaced host ones (and are not satisfied by heterotrophy) will remain necessary. If these functions have not, or cannot, be relocated from the endosymbiont, then the endosymbiont must be retained. In other words, the endosymbiont may have integrated beyond the point of no return.

Apicomplexans have apparently followed the course of this scenario. An ancestral heterotrophic cell acquired the apicoplast progenitor which was most probably photosynthetic (see first section). Perhaps after a long period of autotrophy, apicomplexans diverged from their algal sisters with

**Table 1.** Drugs proposed to target apicoplast metabolism

| Metabolic Activity | Putative Target | Drug | IC50[a] | Comments | Refs |
|---|---|---|---|---|---|
| **DNA Replication** | Apicoplast DNA gyrase | Ciprofloxacin | Pf 8-38μM Tg 30μM | Confirmed to block apicoplast DNA replication, | 1, 2 3 |
| | | Trovofloxacin | Tg 0.77-0.98μg/ml | causes delayed death in Tg | 4 |
| **RNA Transcription** | Apicoplast RNA polymerase β-subunit | Rifampicin | Pf 3μM Tg 3μM[b] | Confirmed by Northern analysis in Pf | 5, 6 |
| | | Rifabutin | Tg 26.5μg/ml | | 7, 8 |
| **Protein Translation** | Apicoplast 23S rRNA | Clindamycin | Pf 20nM[b] Tg 10nM[b] | Causes delayed death in Tg | 9, 10, 11, 12 |
| | | Azithromycin | Pf 2μM[b] Tg 2μM | Causes delayed death in Tg | 9, 10, 12, 13 |
| | | Spiramycin | Tg 40ng/ml | Causes delayed death in Tg | 9, 10 |
| | | Thiostrepton | Pf 2μM Tg NS[b] | Drug-target interaction cofirmed | 14, 15, 16,17 |
| | | Micrococcin | Pf 35nM | | 18 |
| | | Chloramphenicol | Pf 10μM[b] Tg 5μM[b] | Causes delayed death in Tg | 19 10, 12, 13 |
| | Apicoplast 16S rRNA | Doxycycline | Pf 11.3μM | May also target mitochondrion | 19, 20 |

| Target | Drug | Concentration | Notes | Ref |
|---|---|---|---|---|
| | Tetracycline | Pf 10µM<br>Tg 20µM | No delayed death in Tg, may also target Mitochondrion | 13, 19<br>12 |
| Apicoplast elongation factor TufA | Amythiamicin | Pf 10nM | Inferred by polysome distruption in Pf | 21 |
| **Fatty acid biosynthesis** | | | | |
| Apicoplast-targeted β-ketoacyl-ACP synthase II and III (FabF and FabH) | Thiolactomycin | Pf 50µM<br>Tg 100µM[b] | | 22 |
| | Thiolactomycin analogues | Pf ≥8µM | | 23 |
| Apicoplast-targeted enoyl-ACP reductase (FabI) | Triclosan | Pf 1µM | FabI inhibition confirmed | 24, 25 |
| Apicoplast-targeted β-ketoacyl-ACP synthase II (FabF) | Cerulenin | 11µM | | 23 |
| Apicoplast-targeted Acetyl-CoA carboxylase (ACC) | Clodinafop | Tg 10 µM | ACC inhibition confirmed | 26, 27 |
| | Quizlofop | Tg 100µM | ACC inhibition confirmed | 26 |
| | Haloxypop | Tg 100µM | ACC inhibition confirmed | 26, 27 |
| | Fenoxaprop | Pf 144µM | | 23 |

Table 1, continued

| Metabolic Activity | Putative Target | Drug | IC50[a] | Comments | Refs |
|---|---|---|---|---|---|
| | | Tralkoxydim | Pf 181 μM | | 23 |
| | | Dicolfop | Pf 210 μM | | 23 |
| **Isoprenoid biosynthesis** | Apicoplast-targeted DOXP reductoisomerase | Fosmidomycin | Pf 290-370nM | DOXP reductoisomerase inhibition confirmed | 28 |
| | | FR-900098 | Pf 90-170nM | DOXP reductoisomerase inhibition confirmed | 28 |

[a] IC50s are recorded from a diversity of assays (including growth inhibition, death and enzyme assays) and are not directly comparable

[b] D. S. Roos unpublished results

Abbreviations: ACP, acyl carrier protein;NS, not sensitive;

References: (1) Divo et al., 1988, (2) Weissig et al., 1997, (3) Fichera and Roos, 1997, (4) Khan et al., 1996, (5) Strath et al., 1993, (6)Pukrittayakamee et al., 1994, (7) Araujo et al., 1994, (8) Olliaro et al., 1994, (9) Pfefferkorn and Borotz, 1994, 10) Fichera et al., 1995, (11) Fichera and Roos, 1997, (12) Woods et al., 1996, (13) Beckers et al., 1995, (14) Clough et al., 1997, (15) McConkey et al., 1997, (16) Rogers et al., 1997, (17) Sullivan et al., 2000, (18) Rogers et al., 1998, (19) Budimulja et al., 1997, (20) Pradines et al., 2000 (21) Clough et al., 1999, (22) Waller et al., 1998, (23) Waller et al., 2003, (24) Surolia and Surolia, 2001, (25) McLeod et al., 2001, (26) Zuther et al., 1999, (27) Zagnitko et al., 2001, (28) Jomaa et al., 1999.

a high degree of specialization as parasites. Clearly photosynthesis became redundant and was lost, but the apicoplast has persisted. Given that the apicoplast is indispensable, either some unique metabolism(s) acquired with the endosymbiont continues to drive its utility in the cell, or else it has come to fulfill one or more previous functions of the hosts and therefore cannot be lost. Of course it might be that both of these are true. Whatever the case, all of the additional machinery for apicoplast maintenance is also indispensable.

A thorough survey of apicoplast metabolism is therefore required in order to address the question of its significance in parasites. Separating those functions that directly benefit the host cell, from those that are simply necessary for apicoplast maintenance is a greater challenge. Nevertheless, the parasite's dependence on this organelle, and its plastid-type prokaryotic nature, make it an excellent target for drugs. Presently the majority of our knowledge of apicoplast metabolism has been gleaned from genomic date. This includes genes from the apicoplast genome, but more significantly genes from the nuclear genome through the identification of apicoplast-targeting presequences. The completion of the *Plasmodium falciparum* genome and automated screening for apicoplast genes (see last section) has provided us with an excellent starting point for examining apicoplast metabolism. The following provides a summary of our current understanding of this metabolism and, where known, drugs that interrupt these pathways in parasites.

## 5.1. DNA Replication

A circular genome has served as the apicoplasts defining feature and its replication can be considered the first metabolic process of the apicoplast. Interestingly, Williamson *et al.* (2001) have recently reported that half of the copies of the *Toxoplasma gondii* apicoplast genome actually occur as precise tandem linear arrays. From this they invoke a rolling circle hypothesis for DNA replication in this parasite with the origin of replication at the centre of the inverted repeat of rRNA genes. It is noted that other apicomplexa also display an element of linearity with their apicoplast genomes (*Eimeria tenella* (Dunn *et al.*, 1998) and *Neospora caninum* (Gleeson, 2000)) although in *Plasmodium falciparum* at least 90% of its apicoplast DNA is said to be in the circular conformation (Williamson *et al.*, 2001). It remains to be seen what mode of DNA replication occurs in these.

### 5.1.1. Drugs That Target DNA Replication

The fluoroquinolone ciprofloxacin is a selective-inhibitor of prokaryote-type DNA gyrases (type II topoisomerases in the eukaryotic nomenclature) that results in stabilized nicked DNA complexes. Coding sequences for two putative apicoplast-targeted proteins (GyrA and GyrB) have been identified from *Plasmodium falciparum* genome data (Ralph *et al.*, 2001) and suggest that the apicoplast utilizes these molecules for its DNA replication. Ciprofloxacin has been shown to inhibit apicoplast DNA replication in *P. falciparum* and *Toxoplasma gondii* without affecting parasite nuclear DNA replication (Divo *et al.*, 1988; Fichera and Roos, 1997; Weissig *et al.*, 1997). Parasite death results from this drug treatment. Several derivatives of quinalones and fluoroquinolones have also been shown to have parasiticidal activity against *Toxoplasma gondii*, further supporting apicoplast DNA replication as a selective drug target in apicomplexan parasites (Khan *et al.*, 1996; Gozalbes *et al.*, 2000).

## 5.2. Transcription

Northern blot analyses of *Plasmodium falciparum* report transcription of several genes of the apicoplast genome, often as large polycistronic transcripts (Gardner *et al.*, 1991; Feagin and Drew, 1995; Wilson *et al.*, 1996b). Transcript abundances of some genes (*ssu* and *lsu* rRNA genes, *rpo*B and *rpo*C) have been shown to change throughout the cell cycle and provide evidence of regulated transcription of apicoplast genes (Feagin and Drew, 1995). Plastid transcription utilizes a eubacterial ($\alpha_2\beta\beta'$) system of DNA-dependant RNA polymerases that was inherited with the cyanobacterial endosymbiont (Gray and Lang, 1998). The apicoplast genome encodes part of this system ($\beta$ subunit by *rpo*B, and $\beta'$ subunit by *rpo*C1 and *rpo*C2) while the remainder ($\alpha_2$ subunit, as well as sigma factor) are apparently encoded in the nucleus and targeted back to the apicoplast. Despite the eubacterial origin of mitochondria, these endosymbiotic organelles have apparently adopted a single protein phage-like RNA polymerase (one unusual protist, *Reclinomonas americana*, is the only exception to this rule (Gray and Lang, 1998)), and there is evidence for this form of transcription in the *P. falciparum* mitochondrion (Li *et al.*, 2001). Recently, plant chloroplasts have been found also to employ a copy of this phage-like RNA polymerase, which is used in addition to the eubacterial system. It is currently unknown if *P. falciparum* uses both systems for plastid transcription, or only the eubacterial form.

### 5.2.1. Drugs That Target Transcription

The eubacterial form of transcription is highly sensitive to the antibiotic rifampicin and this drug has anti-malarial activity both *in vitro* and *in vivo* (Strath *et al.*, 1993; Pukrittayakamee *et al.*, 1994). Furthermore, rifampicin is shown to selectively diminish transcripts of apicoplast-encoded genes and, hence, the apicoplast is most probably the site of drug action (Wilson *et al.*, 1996b; McConkey *et al.*, 1997). A derivative of rifampicin, rifabutin, also shows activity against *Toxoplasma gondii in vitro* and *in vivo* (Araujo *et al.*, 1994; Olliaro *et al.*, 1994).

## 5.3. Protein Translation

The case for protein translation in the apicoplast is currently built on indirect evidence, yet is nevertheless extremely strong. Most of the apicoplast genome is dedicated to the machinery of translation, including a complete set of tRNAs, rRNA genes, numerous ribosomal protein genes, and the translation elongation factor TufA (Wilson *et al.*, 1996b). Although apicoplast-encoded genes alone are insufficient for an entire translational machinery, further components have been shown to be nucleus-encoded (for example ribosomal proteins S9 and L28) and targeted to the apicoplast as proteins (Waller *et al.*, 1998) as is the case for all plastids. Prokaryote 70S ribosomal particles are observed in the apicoplasts by electron microscopy suggesting proper ribosomal assembly (McFadden *et al.*, 1996; Hopkins *et al.*, 1999), and polysomes carrying apicoplast-specific mRNAs and rRNAs can be purified from blood-stage parasites supporting their functionality (Roy *et al.*, 1999).

### 5.3.1. Drugs That Target Protein Translation

Numerous anti-bacterial agents work by targeting protein synthesis. Even prior to the discovery of the apicoplast several of these were known to be parasiticidal (clindamycin, chloramphenicol, doxycycline and tetracycline) and some are used clinically for the treatment of malaria and toxoplasmosis (including doxycycline, clindamycin and spiramycin). Understanding their modes of action in parasites is vital for their proper application. Lincosamides (clindamycin) and macrolides (azithromycin) inhibit protein synthesis by binding to the peptidyl transferase domain of the prokaryote-type 23S rRNA. Two thiopeptides (thiostrepton and micrococcin) also bind to 23S rRNAs, specifically the GTPase domain where they interact with key nucleotides represented in eubacteria and some plastids. *Plasmodium falciparum* shares this genotype, which is shown to bind to thiostrepton, and parasites are highly sensitive to thiostrepton (both *in vitro* and *in vivo*) and micrococcin (Clough *et al.*, 1997; McConkey *et al.*, 1997; Rogers *et al.*, 1997; Sullivan *et al.*, 2000). *Toxoplasma gondii*, on the other hand, does not share the complete eubacterial genotype and is insensitive to thiostrepton (Clough *et al.*, 1997; D. S. Roos, pers. comm.). Other inhibitors of prokaryote-type translation are parasiticidal and may inhibit apicoplast translation (Table 1), but their precise mode of action is undefined.

One interesting feature of some of these inhibitors is a "delayed death" phenotype seen in *Toxoplasma gondii* (see Table 1) (Pfefferkorn and Borotz, 1994; Fichera *et al.*, 1995; Fichera and Roos, 1997). This effect is identical to the growth pattern of apicoplast segregation mutants lacking an apicoplast (He *et al.*, 2001). This suggest that the delayed death relates to apicoplast function (one theory suggests a role in parasitophorous vacuole function) and not simply to drug accessibility to the parasites. In *Plasmodium falciparum* a case for delayed death is less clear. Chloramphenicol does not have a delayed effect (Budimulja *et al.*, 1997) (compared to *T. gondii*), and while thiostrepton is reported as slow acting, a delay in drug action has not been demonstrated (Sullivan *et al.*, 2000). Inhibitors of fatty acid synthesis (see below) on the other hand do have very rapid inhibitory effects on *P. falciparum*. Differences in apicoplast-drug response kinetics between *P. falciparum* and *T. gondii* may point to some differences in the roles that these apicoplasts fulfill in parasites.

## 5.4. Other Apicoplast-Encoded Genes

Aside from genes clearly coding for transcription and translation functions on the apicoplast genome, there are 9 remaining open reading frames (ORFs) that could contribute to some further apicoplast metabolism. Within these final ORFs must lie the reason(s) for retaining the apicoplast genome and all of the associated metabolism necessary for its maintenance and expression. Some of these ORFs, however, have been suggested to represent further ribosomal and associated proteins, based on their size and conserved positions, but are too divergent for confident assignment (Blanchard and Hicks, 1999).

*clp*C, on the other hand, is distinguished by being one apicoplast gene that is clearly not involved in either transcription or translation. ClpC (also known as Hsp93) forms part of the Tic complex for protein import in to plastids (Jackson-Constan *et al.*, 2001). ClpC's role, therefore, is to help supplement apicoplast translation, for all apicoplast proteins that are not encoded on the apicoplast genome.

One apicoplast ORF, known as *orf470* or *ycf24* and more recently re-assigned *sufB* shares a high level of similarity with a gene encoded in plastids in some algae and as part of an operon conserved in most bacteria (Wilson *et al.*, 1996b; Blanchard and Hicks, 1999; Ellis *et al.*, 2001). Knockout experiments in the cyanobacterium *Synechocystis* sp. PCC6803 suggest that this gene is essential and it was tentatively assigned a gene component of an ABC transporter (Law *et al.*, 2000). Ellis *et al.* (2001) revise this view and now consider this gene a likely member of the *suf* operon of *E. coli* that is involved in iron homeostasis and [Fe-S] cluster formation. Plastid metalloproteins with [Fe-S] clusters include ferredoxin which is known to target to apicoplasts (see below) (Vollmer *et al.*, 2001). SufB could therefore be essential for post-translational modification of imported protein and provides a rationale for persistence of the apicoplast genome that now awaits further confirmation.

## 5.5. Fatty Acid Biosynthesis

Fatty acids play a critical role in cells as metabolic precursors for biological membranes and energy stores. Their synthesis occurs as iterative elongations of acyl chains utilizing the 2-carbon donor malonyl-CoA. Fatty acid synthase (FAS) is the principal enzymatic unit of this process and in bacterial systems separate proteins constitute the several enzyme activities of FAS. This system is known as the type II or dissociated pathway and is believed to represent the ancestral state (Smith, 1994). In animals, gene fusion events have resulted in FAS evolving into a single large multi-functional protein (Smith, 1994) and this cytosolic pathway is known as the type I or associated pathway. Plants utilize a plastid-based pathway and this is of the bacterial type II system, which reflects the prokaryotic origin of plastids (Harwood, 1996).

*Plasmodium* spp. have long been believed unable to make their own fatty acids, depending instead on scavenged fatty acids from the host erythrocytes and serum (Holz, 1977; Vial and Ancelin, 1992; Fish, 1995). This notion is now being revised, with strong evidence for a type II pathway for *de novo* fatty acid synthesis occurring in apicoplasts. An indication of fatty acid synthesis in apicoplasts first occurred with the discovery of nucleus-encoded FAS genes whose products are targeted to the apicoplast. These include acyl carrier protein (ACP) and β-ketoacyl-ACP synthases III (FabH) in *Plasmodium falciparum* (Waller *et al.*, 2000), and ACP in *Toxoplasma gondii* (Waller *et al.*, 1998). Since then, genes corresponding to all enzymes of FAS have been identified, except for a thioesterase required for acyl chain termination (McLeod *et al.*, 2001; Surolia and Surolia, 2001; Gardner *et al.*, 2002). Additionally, acetyl-CoA carboxylase (ACC), which generates the carbon donor malonyl-CoA for FAS, is also represented in *P. falciparum* genome data (Zuther *et al.*, 1999) and shown to be apicoplast-targeted in *T. gondii* (Jelenska *et al.*, 2001). Biochemical support for fatty acid synthesis in *P. falciparum* has followed with functional characterization of several enzymes of FAS (Surolia and Surolia, 2001; Waters *et al.*, 2002; Prigge *et al.*, 2003). Moreover, [14]C-labeled precursors (acetate and malonyl-CoA) are shown in *P. falciparum* to incorporate into fatty acids (predominantly C10 to C14) in both *in vivo* and *in vitro* systems (Surolia and Surolia, 2001). This incorporation is sensitive to triclosan, a known inhibitor of type II FAS, that is shown to bind specifically to *P. falciparum* FabI (Surolia and Surolia, 2001; Perozzo *et al.*, 2002). In concert, these data strongly implicates that apicoplast as the site of this fatty acid biosynthetic activity.

No evidence of a type I FAS has been found in the *Plasmodium falciparum* genome data. A type II pathway in the apicoplast is likely therefore to fulfill all of the parasite's need for *de novo* synthesized fatty acids. In contrast, *Cryptosporidium parvum* apparently lacks genes for a type II pathway, yet does have a large coding sequence for a type I FAS (note, it remains formally possible that this is a polyketide synthase) (Zhu *et al.*, 2000b). Interestingly, no evidence has yet been found for an apicoplast in *C. parvum* (Zhu *et al.*, 2000a) which suggests that these parasites have maintained a cytosolic type I system in the absence (most likely following apicoplast loss) of an apicoplast-based type II system.

### 5.5.1. Drugs That Target Fatty Acid Biosynthesis

The presence of a distinct, prokaryote pathway for the biosynthesis of fatty acids in parasites offers tremendous potential as a focus for parasite-selective drugs (Table 1). Thiolactomycin is one such drug that inhibits type II FAS in plants and bacteria and by enzyme kinetics and crystal structure studies is known to bind specifically to β-ketoacyl-ACP synthases (most strongly to synthases I and II). Thiolactomycin kills parasites *in vitro* (Table 1) (Waller *et al.*, 1998) and several analogues show up to five-fold greater efficacy (Waller *et al.*, 2003). Triclosan is another drug selective for the prokaryote pathway that shows even greater efficacy against parasites and is also effective as an anti-malarial in mouse models (McLeod *et al.*, 2001; Surolia and Surolia, 2001; Perozzo *et al.*, 2002). This drug has a long record of safety for human use (as an anti-microbial in toothpastes and other products) and triclosan analogues are being explored for anti-malarial use (Perozzo *et al.*, 2002). Both thiolactomycin and triclosan show rapid inhibition of *Plasmodium falciparum* with greatest parasite sensitivity seen in ring-stage parasites (Waller *et al.*, 2003). Cerulenin also targets β-ketoacyl-ACP synthases I and II (although type I FAS is also sensitive) and inhibits *P. falciparum* (Table 1) (Waller *et al.*, 2003).

Acetyl-CoA carboxylase (ACC) is also being explored as a target for drugs in parasites. As in grasses, apicomplexans have adopted the eukaryotic ACC that has become plastid-targeted and supports the type II FAS (Harwood, 1996; Zuther *et al.*, 1999; Jelenska *et al.*, 2001). This chloroplast-localized eukaryotic ACC is sensitive to aryloxyphenoxypropionate herbicides (Zagnitko *et al.*, 2001). The *Toxoplasma gondii* apicoplast ACC enzyme shares this sensitivity to these compounds which also inhibit parasite growth (Table 1) (Zuther *et al.*, 1999). *Plasmodium falciparum* is also sensitive to these compounds (Waller *et al.*, 2002).

### 5.6. Isopreniod Biosynthesis

Isoprenoids form an extremely diverse class of compounds including sterols, ubiquinones, dolichols and prenylation moieties (Sacchettini and Poulter, 1997). All isoprenoid compounds depend on the precursor isomers isopentenyl diphosphate (IPP) and dimethylallyl diphosphate (DMAPP). There are two alternate pathways for IPP/DMAPP synthesis. Animals and fungi rely upon the mevalonate pathway which was the first described. More recently, a mevalonate-independent pathway has been described from eubacteria and plastids proceeding via 1-deoxy-D-xylulose 5-phosphate (DOXP) instead. Plants and some algae possess both pathways, the mevalonate pathway in the cytosol and the DOXP pathway in the plastids (Lichtenthaler, 1999). Three genes for the DOXP pathway, DOXP synthase, DOXP reductoisomerase and MECDP synthase (IspF or YgbB), have recently been identified from *Plasmodium falciparum* and all possess bipartite presequences (Gardner *et al.*, 1998; Jomaa *et al.*, 1999; Gardner *et al.*, 2002). Moreover, DOXP reductoisomerase is shown to target to apicoplasts, and the recombinant expression of this enzyme verified DOXP reductoisomerase activity which has since also been measured from *P. falciparum* extracts (Jomaa *et al.*, 1999; Wiesner *et al.*, 2000). This data presents a strong case for a DOXP pathway for isoprenoid synthesis in the apicoplast of *P. falciparum*. There is no evidence of genes for the mevalonate pathway in *P. falciparum* nor substantial incorporation of labeled mevalonate into isoprenoids (Mbaya *et al.*, 1990). An apicoplast-based pathway may be the sole provider of *de novo* synthesized IPP/DMAPP in parasites.

### 5.6.1. Drugs That Target DOXP Isoprenoid Biosynthesis

DOXP reductoisomerase is the target of the antibiotic fosmidomycin, and both fosmidomycin and a derivative, FR-900098, show effective inhibition of *Plasmodium falciparum* in the nanomolar range (Table 1). Inhibition of recombinantly expressed enzyme is verified, and both compounds are very effective at parasite clearance in mouse models (Jomaa *et al.*, 1999). Recently, modified forms of FR-900098 showed even greater potency (Reichenberg *et al.*, 2001) making these drugs excellent candidates for further development for clinical use.

## 5.7. Ferredoxin – A Source of Reducing Power

A tantalizing insight into apicoplast metabolism is offered my the identification of genes for the plastid redox pair [2Fe-2S] ferredoxin (Fd) and ferredoxin-NADP+ reductase (FNR) from *Plasmodium falciparum* and *Toxoplasma gondii* respectively (Vollmer *et al.*, 2001). FNR catalyses the reversible reaction $2Fd_{red} + NADP^+ + H^+ \rightarrow 2Fd_{ox} + NADPH$. In photosynthetic systems Fd accepts an electron from photosystem I and transfers it to $NADP^+$ driving the above reaction forward and producing NADPH required for Calvin cycle reactions. In the non-photosynthetic plastids of roots, this reaction is driven in the reverse direction by a structurally distinct isoform of FNR. The result is a source of reduced Fd, that provides essential reductant for plastid enzymes such as glutamate synthase, fatty acid desaturases, nitrite reductase, and sulfite reductase (Shanklin and Cahoon, 1998; Neuaus and Emes, 2000). In apicomplexans both genes encode bipartite presequences for apicoplast targeting that are processed off to form mature proteins (Vollmer *et al.*, 2001). Recombinant expression of *P. falciparum* Fd confirms that this protein binds iron, and analysis of the *T. gondii* FNR identifies strong similarities to the root form of FNR (Vollmer *et al.*, 2001). These proteins thus provide good circumstantial evidence for a supply of Fd-reductant in apicoplasts. While no candidate enzymes that utilize this reductant have yet been identified in the apicoplast, the presence of a pathway for fatty acid biosynthesis indicates that fatty acid desaturases are one likely group.

## 5.8. Heme Biosynthesis

Heme is an iron bound tetrapyrrole that serves as an electron carrying prosthetic group in parasite cytochromes and also appears to play a role in protein synthesis in *Plasomdium falciparum* (Surolia and Padmanaban, 1991). In plants, chlorophyll pigments are derived from a magnesium bound tetrapyrrole, and hence both molecules share a common biosynthetic pathway. The first committed step of tetrapyrrole synthesis is the formation of δ-aminolaevulinic acid (ALA), but there are two pathways to ALA. In animals and fungi ALA synthase (ALAS) uses glycine and succinyl-CoA to create ALA in the mitochondrion – the so-called Shemin pathway. An alternate pathway is found in plastids and most bacteria. This $C_5$ or glutamate pathway generates ALA from glutamate, and is believed to be the only source of ALA in plants (Beale and Castelfranco, 1974). Euglenoid algae are unusual in that they possess both pathways, the Shemin in the mitochondrion and the $C_5$ in the plastid (Weinstein and Beale, 1983). From ALA to tetrapyrroles the pathways are equivalent for both plants and animals and require a further seven enzymes. Curiously, in euglenoids there may only be a plastid set of these downstream enzymes (Shashidhara and Smith, 1991), suggesting that mitochondrial derived ALA may be imported to plastids for further processing.

*Plasmodium falciparum* is apparently unable to utilize the heme derived from erythrocyte hemoglobin and instead synthesizes its own (Surolia and Pasmanaban, 1992; Wilson *et al.*, 1996a). Glycine is incorporated into ALA, implicating a Shemin pathway for ALA synthesis in the mitochondrion (Surolia and Pasmanaban, 1992) and indeed ALAS is found in the genome (Wilson *et al.*, 1996a; Varadharajan *et al.*, 2002; Gardner *et al.*, 2002). However, from here the story becomes more complex. Remarkably the next enzyme in this pathway, ALA dehydrase (ALAD), is apparently imported into *Plasmodium* from the host erythrocyte cytoplasm (Bonday *et al.*, 1997; Bonday *et al.*, 2000; Padmanaban and Rangarajan, 2000). Yet *Plasmodium* spp. also encode their own ALAD. This version shares features with plant ALAD including a preference for $Mg^{2+}$ cofactors rather than $Zn^{2+}$ used by the mitochondrial ALAD, and is phylogenetically closer to plant ALADs (Sato *et al.*, 2000). Moreover, the gene encodes a bipartite presequence for apicoplast targeting (van Dooren *et al.*, 2002). This apparent duplicity is difficult to rationalize, however at least two of the next three enzymes are represented in the genome (porpobilinogen deaminase and urophophyrinogen III decarboxylase), and these also have plastid-targeting presequences (Gardner *et al.*, 2002). This, therefore, provides a firm case for this portion of the heme biosynthetic pathway being apicoplast localized. A final twist in this story is that the remaining three steps of heme synthesis (catalyzed by coproporphorinogen oxidase, protoporphorinogen oxidase and ferrochelatase) most likely occur in the mitochondrion (Gardner *et al.*, 2002).

This somewhat convoluted pathway, alternating from mitochodrion to apicoplast to mitochondrion, can in fact be rationalized quite simply. An apicoplast-based segment of heme biosynthesis has taken over the part of this pathway that in animals (and most probably the apicoplast ancestor) occurs in the cytosol. This is apparently a further case where duplicated pathways have been rationalized, this time utilising elements of both. The mechanism of exchange of heme intermediates between these two organelles remains to be elucidated, however such exchange also occurs in plants and apparently euglenoid algae. A role for the imported host-derived ALAD remains to be substantiated.

# 6. Conclusions

The metabolism of the apicoplast is an unfolding story and the completion of the *Plasmodium falciparum* genome brings us to an extremely exciting point in this process. However, the challenge now must begin to swing towards defining the end products of apicoplast metabolic pathways if we are to understanding the full significance of this organelle and any drug targets that it may hold. Presently we have good evidence for fatty acid synthesis and isoprenoid synthesis. However, it is yet to be demonstrated that any of their products are exported from the apicoplast to the rest of the cell. Fatty acids could be expected to incorporate into phospholipids, glycosylphosphati-dylinositols and other acyl moieties such as on proteins. Apicoplast-derived isoprenoids may well find their way into the mitochondrion as ubiquinones, or as dolicols or isoprenyl groups on proteins and other molecules. A further interesting question is, are all apicoplasts equal? Genome data from other *Plasmodium* species, and other apicomplexans will be able to tell us whether all apicoplasts contain a similar suite of metabolic pathways, or whether some have been culled or refined to suit each parasite's niche. Finally, with a near complete set of apicoplast genes in our hands, the regulation of these pathways throughout the parasites life cycle can be addressed (through micro array technologies for instance) in order to shed further light on the significance of apicoplast function.

# 7. References

Aikawa, M. 1966. The fine structure of the erythrocytic stage of three avian malarial parasites, *Plasmodium fallax*, *P. lophyrae*, and *P. cathemerium*. Amer. J. Trop. Med Hyg. 15: 449-471.

Aikawa, M. 1971. *Plasmodium*: the fine structure of malarial parasites. Exp. Parasitol. 30: 284-320.

Aikawa, M. and Jordan, H. 1968. Fine structure of reptilian malarial parasite. J. Parasitol. 54: 1023-1033.

Araujo, F.G., Slifer, T. and Remington, J.S. 1994. Rifabutin is active in murine models of toxoplasmosis. Antimicrob. Agents Chemother. 38: 570-575.

Baldauf, S., Roger, A., Wenk-Siefert, I. and Doolittle, W. 2000. A kingdom-level phylogeny of eukaryotes based on combined protein data. Science 290: 972-977.

Bannister, L.H., Hopkins, J.M., Fowler, R.E., Krishna, S., Mitchell, G.H. 2000. A brief illustrated guide to the ultrastructure of *Plasmodium falciparum* asexual blood stages. Parasitol Today. 16(10): 427-433.

Beale, S. and Castelfranco, P. 1974. The biosynthesis of delta-aminolevulinic acid in higher plants. II. Formation of $^{14}$C-delta-aminolevulinic acid from labeled precursors in greening plant tissues. Plant Physiol. 53: 297-303.

Beckers, C.J.M., Roos, D.S., Donald, R.G.K., Luft, B.J., Schwab, J.C., Cao, Y. and Joiner, K.A. 1995. Inhibition of cytoplasmic and organellar protein synthesis in *Toxoplasma gondii* - implications for the target of macrolide antibiotics. J. Clin. Invest. 95: 367-376.

Beech, P.L., Nheu, T., Schultz, T., Herbert, S., Lithgow, T., Gilson, P.R. and McFadden, G.I. 2000. Mitochondrial FtsZ in a chromophyte alga. Science 287: 1276-1279.

Blanchard, J. and Hicks, J.S. 1999. The non-photosynthetic plastid in malarial parasites and other apicomplexans is derived from outside the green plastid lineage. J. Euk. Microbiol. 46: 367-375.

Bonday, Z.Q., Dhanasekaran, S., Rangarajan, P.N. and Padmanaban, G. 2000. Import of host delta-aminolevulinate dehydratase into the malarial parasite: identification of a new drug target. Nat. Med. 6: 898-903.

Bonday, Z.Q., Taketani, S., Gupta, P.D. and Padmanaban, G. 1997. Heme biosynthesis by the malarial parasite. Import of delta- aminolevulinate dehydrase from the host red cell. J. Biol. Chem. 272: 21839-21846.

Borst, P., Overdlve, J.P., Weijers, P.J., Fase-Fowler, F. and Berg, M.V.D. 1984. DNA circles with cruciforms from *Isospora* (*Toxoplasma*) *gondii*. Biochim. Biophys. Acta 781: 100-111.

Brown, S.B., Houghton, J.D. and Vernon, D.I. 1990. Biosynthesis of phycobilins. Formation of the chromophore of phytochrome, phycocyanin and phycoerythrin. J. Photochem. Photobiol. B 5: 3-23.

Bruce, B.D. 2001. The paradox of plastid transit peptides: conservation of function despite divergence in primary structure. Biochim. Biophys. Acta 1541: 2-21.

Budimulja, A.S., Syafruddin, Tapchaisri, P., Wilairat, P. and Marzuki, S. 1997. The sensitivity of *Plasmodium* protein synthesis to prokaryotic ribosomal inhibitors. Mol. Biochem. Parasitol. 84: 137-141.

Cavalier-Smith, T. 1982. The origins of plastids. Biol. J. Linn. Soc. 17: 289-306.

Cheresh, P., Harrison, T., Fujioka, H. and Haldar, K. 2002. Targeting the malarial plastid via the parasitophorous vacuole. J. Biol. Chem. 277: 16265-16277.

Cline, K. and Henry, R. 1996. Import and routing of nucleus-encoded chloroplast proteins. Ann. Rev. Cell Dev. Biol. 12: 1-26.

Clough, B., Rangachari, K., Strath, M., Preiser, P.R. and Wilson, R. 1999. Antibiotic inhibitors of organellar protein synthesis in *Plasmodium falciparum*. Protist 150: 189-195.

Clough, B., Strath, M., Preiser, P., Denny, P. and Wilson, R. 1997. Thiostrepton binds to malarial plastid rRNA. FEBS Lett. 406: 123-125.

Crary, J.L. and Haldar, K. 1992. Brefeldin A inhibits protein secretion and parasite maturation in the ring stage of *Plasmodium falciparum*. Mol Biochem Parasitol 53: 185-192.

Creasey, A., Mendis, K., Carlton, J., Williamson, D., Wilson, I. and Carter, R. 1994. Maternal inheritance of extrachromosomal DNA in malaria parasites. Mol. Biochem. Parasitol. 65: 95-98.

Deane, J.A., Fraunholz, M., Su, V., Maier, U.G., Martin, W., Durnford, D.G. and McFadden, G.I. 2000. Evidence for nucleomorph to host nucleus gene transfer: light-harvesting complex proteins from cryptomonads and chlorarachniophytes. Protist 151: 239-252.

Delwiche, C. 1999. Tracing the tread of plastid diversity through the tapestry of life. Am. Nat. 154: S164-S177.

Delwiche, C.F. and Palmer, J.D. 1997. The origin of plastids and their spread via secondary endosymbiosis. Pl. Syst. Evol. [Suppl] 11: 51-86.

Denny, P., Preisser, P., Williamson, D. and Wilson, I. 1998. Evidence for a single origin of the 35kb plastid DNA in apicomplexans. Protist 149: 51-59.

DeRocher, A., Feagin, J. and Parsons, M. 2001. ER to apicoplast targeting in *Toxoplasma gondii*. Mol. Biol. Cell 12: 382a.

DeRocher, A., Hagen, C.B., Froehlich, J.E., Feagin, J.E. and Parsons, M. 2000. Analysis of targeting sequences demonstrates that trafficking to the *Toxoplasma gondii* plastid branches off the secretory system. J. Cell Sci. 113: 3969-3977.

Diniz, J., Silva, E., Lainson, R. and Souza, W.d. 2000. The fine structure of *Garnia gonadati* and its association with the host cell. Parasitol. Res. 86: 971-977.

Divo, A., Sartorelli, A., Patton, C. and Bia, F. 1988. Activity of fluoroqinolone antibiotics against *Plasmodium falciparum in vitro*. Antimicrob. Agents Chemother. 32: 1182-1186.

Divo, A.A., Geary, T.G., Jensen, J.B. and Ginsburg, H. 1985. The mitochondrion of *Plasmodium falciparum* visualized by rhodamine 123 fluorescence. J. Protozool. 32: 442-446.

Dore, E., Frontali, C., Forte, T. and Fratarcangeli, S. 1983. Further studies and electron microscopic characterization of *Plasmodium berghei* DNA. Mol. Biochem. Parasitol. 8: 339-352.

Dubremetz, J.F. 1995. *Toxoplasma gondii*: cell biology update. In: Molecular Approaches to Parasitology. Wiley-Liss Inc. p. 345-358.

Dunn, P.P.J., Stephens, P.J. and Shirley, M.W. 1998. *Eimeria tenella* - two species of extrachromosomal DNA revealed by pulsed-field gel electrophoresis. Parasitol. Res. 84: 272-275.

Egea, N. and Lang-Unnasch, N. 1995. Phylogeny of the large extrachromosomal DNA of organisms in the phylum Apicomplexa. J. Euk. Microbiol. 42: 679-684.

Ellis, K.E., Clough, B., Saldanha, J.W. and Wilson, R.J. 2001. Nifs and Sufs in malaria. Mol. Microbiol. 41: 973-981.

Emanuelsson, O., Nielsen, H. and von Heijne, G. 1999. ChloroP, a neural network-based method for predicting chloroplast transit peptides and their cleavage sites. Protein Sci. 8: 978-984.

Fagan, T., Hastings, J. and Morse, D. 1998. Glyceraldehyde-3-phosphate dehydrogenase phylogeny supports and independent origin of the dinoflagellate chloroplast. Endocyt. Cell Res. 13 (supplement): 33.

Fast, N.M., Kissinger, J.C., Roos, D.S. and Keeling, P.J. 2001. Nuclear-encoded, plastid-targeted genes suggest a single common origin for apicomplexan and dinoflagellate plastids. Mol. Biol. Evol. 18: 418-426.

Feagin, J.E. and Drew, M.E. 1995. *Plasmodium falciparum*: alterations in organelle transcript abundance during the erythrocytic cycle. Exp. Parasitol. 80: 430-440.

Fichera, M.E., Bhopale, M.K. and Roos, D.S. 1995. *In vitro* assays elucidate peculiar kinetics of clindamycin action against *Toxoplasma gondii*. Antimicrob. Agents Chemother. 39: 1530-1537.

Fichera, M.E. and Roos, D.S. 1997. A plastid organelle as a drug target in apicomplexan parasites. Nature 390: 407-409.

Fish, W.R. 1995. Lipid and membrane metabolism of the malaria parasite and the African trypanosome. In: Biochemistry and Molecular Biology of Parasites. Marr, J.J. and Müller, M. eds. Academic Press, London. p. 133-145.

Foley, M. and Tilley, L. 1998. Protein trafficking in malaria-infected erythrocytes. Int. J. Parasitol. 28: 1671-1680.

Foth, B.J., Ralph, S.A., Tonkin, C.J., Struck, N.S., Fraunholz, M., Roos, D.S. Cowman, A.F. and McFadden, G.I. (2003) Dissection apicoplast targeting in the malaria parasite *Plasmodium falciparum*. Science 299: 705-708.

Gardner, M.J., Feagin, J.E., Moore, D.J., Spencer, D.F., Gray, M.W., Williamson, D.H. and Wilson, R.J. 1991. Organization and expression of small subunit ribosomal RNA genes encoded by a 35-kilobase circular DNA in *Plasmodium falciparum*. Mol. Biochem. Parasitol. 48: 77-88.

Gardner, M.J., Goldman, N., Barnett, P., Moore, P.W., Rangachari, K., Strath, M., Whyte, A., Williamson, D.H. and Wilson, R.J. 1994. Phylogenetic analysis of the *rpo*B gene from the plastid-like DNA of *Plasmodium falciparum*. Mol. Biochem. Parasitol. 66: 221-231.

Gardner, M.J., Tettelin, H., Carucci, D.J., Cummings, L.M., Aravind, L., Koonin, E.V., Shallom, S., Mason, T., Yu, K., Fujii, C., Pederson, J., Shen, K., Jing, J.P., Aston, C., Lai, Z.W., Schwartz, D.C., Pertea, M., Salzberg, S., Zhou, L.X., Sutton, G.G., Clayton, R., White, O., Smith, H.O., Fraser, C.M., Adams, M.D., Hoffman, S.L. 1998. Chromosome 2 sequence of the human malaria parasite *Plasmodium falciparum*. Science 282: 1126-1132.

Gardner, M.J., Hall, N., Fung, E., White, O., Berriman, M., Hyman, R.W., Carlton, J.M., Pain, A., Nelson, K.E., Bowman, S., Paulsen, I.T., James, K., Eisen, J.A., Rutherford, K., Salzberg, S.L., Craig, A., Kyes, S., Chan, M.S., Nene, V., Shallom, S.J., Suh, B., Peterson, J., Angiuoli, S., Pertea, M., Allen, J., Selengut, J., Haft, D., Mather, M.W., Vaidya, A.B., Martin, D.M., Fairlamb, A.H., Fraunholz, M.J., Roos, D.S., Ralph, S.A., McFadden, G.I., Cummings, L.M., Subramanian, G.M., Mungall, C., Venter, J.C., Carucci, D.J., Hoffman, S.L., Newbold, C., Davis, R.W., Fraser, C.M. and Barrell, B. 2002. Genome sequence of the human malaria parasite *Plasmodium falciparum*. Nature 2002 419: 498-511

Gilson, P.R. and Beech, P.L. 2001. Cell division protein FtsZ: running rings around bacteria, chloroplasts and mitochondria. Res. Microbiol. 152: 3-10.

Gleeson, M.T. 2000. The plastid in Apicomplexa: what use is it? Int. J. Parasitol. 30: 1053-1070.

Gozalbes, R., Brun-Pascaud, M., Garcia-Domenech, R., Galvez, J., Girard, P.M., Doucet, J.P. and Derouin, F. 2000. Anti-Toxoplasma activities of 24 quinolones and fluoroquinolones in vitro: Prediction of activity by molecular topology and virtual computational techniques. Antimicrob. Agents Chemother. 44: 2771-2776.

Gray, M.W. and Lang, B.F. 1998. Transcription in chloroplasts and mitochondria: a tale of two polymerases. Trends Microbiol. 6: 1-3.

Harwood, J. 1996. Recent advances in the biosynthesis of plant fatty-acids. Biochim. Biophys. Acta. 1301: 7-56.

Hayashi, M., Taniguchi, S., Ishizuka, Y., Kim, H.S., Wataya, Y., Yamamoto, A. and Moriyama, Y. 2001. A homologue of N-ethylmaleimide-sensitive factor in the malaria parasite *Plasmodium falciparum* is exported and localized in vesicular structures in the cytoplasm of infected erythrocytes in the brefeldin A-sensitive pathway. J. Biol. Chem. 276: 15249-15255.

He, C.Y., Shaw, M.K., Pletcher, C.H., Striepen, B., Tilney, L.G. and Roos, D.S. 2001. A plastid segregation defect in the protozoan parasite *Toxoplasma gondii*. EMBO J. 20: 330-339.

Holz, G.G. 1977. Lipids and the malarial parasite. Bull. WHO 55: 237-248.

Hopkins, J., Fowler, R., Krishna, S., Wilson, I., Mitchell, G. and Bannister, L. 1999. The plastid in *Plasmodium falciparum* asexual blood stages: a three-dimensional ultrastructural analysis. Protist 150: 283-295.

Howe, C.J. 1992. Plastid origin of an extrachromosomal DNA molecule from *Plasmodium*, the causative agent of malaria. J. Theor. Biol. 158: 199-205.

Ivey, R.A., 3rd, Subramanian, C. and Bruce, B.D. 2000. Identification of a Hsp70 recognition domain within the rubisco small subunit transit peptide. Plant Physiol. 122: 1289-1299.

Jackson-Constan, D., Akita, M. and Keegstra, K. 2001. Molecular chaperones involved in chloroplast protein import. Biochim. Biophys. Acta 1541: 102-113.

Jarvis, P. and Soll, J. 2001. Toc, Tic, and chloroplast protein import. Biochim. Biophys. Acta 1541: 64-79.

Jeffries, A.C. and Johnson, A.M. 1996. The growing importance of the plastid-like DNAs of the Apicomplexa. Int. J. Parasitol. 26: 1139-1150.

Jelenska, J., Crawford, M.J., Harb, O.S., Zuther, E., Haselkorn, R., Roos, D.S. and Gornicki, P. 2001. Subcellular localization of acetyl-CoA carboxylase in the apicomplexan parasite *Toxoplasma gondii*. Proc. Natl. Acad. Sci. USA 98: 2723-2728.

Jomaa, H., Wiesner, J., Sanderbrand, S., Altincicek, B., Weidemeyer, C., Hintz, M., Turbachova, I., Eberl, M., Zeidler, J., Lichtenthaler, H.K., Soldati, D. and Beck, E. 1999. Inhibitors of the nonmevalonate pathway of isoprenoid biosynthesis as antimalarial drugs. Science 285: 1573-1576.

Keeling, P.J., Palmer, J.D., Donald, R.G.K., Roos, D.S., Waller, R.F. and McFadden, G.I. 1998. Shikimate pathway in apicomplexan parasites. Nature 397: 219-220.

Khan, A.A., Slifer, T., Araujo F.G. and Remington, J.S. 1996. Trovafloxacin is active against *Toxoplasma gondii*. Antimicrob. Agents Chemother. 40: 1855-1859.

Kilejian, A. 1975. Circular mitochondrial DNA from the avian malarial parasite *Plasmodium lophurae*. Biochim. Biophys. Acta 390: 276-284.

Kohler, R., Cao, J., Zipfel, W., Webb, W. and Hanson, M. 1997. Exchange of protein molecules through connections between higher plant plastids. Science 276: 2039-2042.

Köhler, S., Delwiche, C.F., Denny, P.W., Tilney, L.G., Webster, P., Wilson, R.J.M., Palmer, J.D. and Roos, D.S. 1997. A plastid of probable green algal origin in apicomplexan parasites. Science 275: 1485-1488.

Kroth, P. and Strotmann, H. 1999. Diatom plastids: Secondary endocytobiosis, plastid genome and protein import. Physiologia Plantarum 107: 136-141.

Kudo, R. 1971. Protozoology. Charles C Thomas, Springfield, Illinois.

Lang-Unnasch, N., Reith, M., Munholland, J. and Barta, J. 1998. Plastids are widespread and ancient in parasites of the phylum Apicomplexa. J. Int. Parasitol. 28: 1743-1754.

Law, A.E., Mullineaux, C.W., Hirst, E.M., Saldanha, J. and Wilson, R.J. 2000. Bacterial orthologues indicate the malarial plastid gene *ycf24* is essential. Protist 151: 317-327.

Li, J.N., Maga, J.A., Cermakian, N., Cedergren, R. and Feagin, J.E. 2001. Identification and characterization of a *Plasmodium falciparum* RNA polymerase gene with similarity to mitochondrial RNA polymerases. Mol. Biochem. Parasitol. 113: 261-269.

Liaud, M.-F., Brandt, U., Scherzinger, M. and Cerff, R. 1997. Evolutionary origin of cryptomonad microalgae: two novel chloroplast/cytosol-specific GAPDH genes as potential markers of ancestral endosymbiont and host cell components. J. Mol. Evol. 44 Suppl. 1: S28-37.

Lichtenthaler, H.K. 1999. The 1-deoxy-D-xylulose-5-phosphate pathway of isoprenoid biosynthesis in plants [Review]. Ann. Rev. Plant Physiol. Plant Mol. Biol. 50: 47-65.

Martin, W. and Herrmann, R.G. 1998. Gene transfer from organelles to the nucleus: how much, what happens, and why? Plant Physiol. 118: 9-17.

Matsuzaki, M., Kikuchi, T., Kita, K., Kojima, S. and Kuroiwa, T. 2001. Large amounts of apicoplast nucleoid DNA and its segregation in *Toxoplasma gondii*. Protoplasma 218: 180-191.

Mbaya, B., Rigomier, D., Edorh, G.G., Karst, F. and Schrevel, J. 1990. Isoprenoid metabolism in *Plasmodium falciparum* during the intraerythrocytic phase of malaria. Biochem. Biophys. Res. Commun. 173: 849-854.

McConkey, G.A., Rogers, M.J. and McCutchan, T.F. 1997. Inhibition of *Plasmodium falciparum* protein synthesis: targeting the plastid-like organelle with thiostrepton. J. Biol. Chem. 272: 2046-2049.

McFadden, G.I., Reith, M., Munholland, J. and Lang-Unnasch, N. 1996. Plastid in human parasites. Nature 381: 482.

McFadden, G.I. and Roos, D.S. 1999. Apicomplexan plastids as drug targets. Trends Microbiol. 6: 328-333.

McFadden, G.I. and Waller, R.F. 1997. Plastids in parasites of humans. BioEssays 19: 1033-1040.

McFadden, G.I., Waller, R.F., Reith, M., Munholland, J. and Lang-Unnasch, N. 1997. Plastids in apicomplexan parasites. Pl. Syst. Evol. [Suppl.] 11: 261-287.

McLeod, R., Muench, S.P., Rafferty, J.B., Kyle, D.E., Mui, E.J., Kirisits, M.J., Mack, D.G., Roberts, C.W., Samuel, B.U., Lyons, R.E., Dorris, M., Milhous, W.K. and Rice, D.W. 2001. Triclosan inhibits the growth of *Plasmodium falciparum* and *Toxoplasma gondii* by inhibition of apicomplexan FabI. Int. J. Parasitol. 31: 109-113.

Neuhaus, H.E., and Emes, M.J. 2000. Nonphotosynthetic metabolism in plastids. Annu. Rev. Plant Physiol. and Plant Mol. Biol. 51: 111-140.

Olliaro, P., Gorini, G., Jabes, D., Regazzetti, A., Rossi, R., Marchetti, A., Tinelli, C. and Della Bruna, C. 1994. *In-vitro* and *in-vivo* activity of rifabutin against *Toxoplasma gondii*. J. Antimicrob. Chemother. 34: 649-657.

Osteryoung, K.W. 2001. Organelle fission in eukaryotes. Curr. Opin. Microbiol. 4: 639-646.

Padmanaban, G. and Rangarajan, P.N. 2000. Heme metabolism of *Plasmodium* is a major antimalarial target. Biochem. Biophys. Res. Commun. 268: 665-668.

Peeters, N. and Small, I. 2001. Dual targeting to mitochondria and chloroplasts. Biochim. Biophys. Acta. 1541: 54-63.

Perozzo, R., Kuo, M., bir Singh Sidhu, A., Valiyaveettil, J.T., Bittman, R., Jacobs, W.R., Jr., Fidock, D.A. and Sacchettini, J.C. 2002. Structural elucidation of the specificity of the antibacterial agent triclosan for malarial enoyl ACP reductase. J. Biol. Chem. 277: 13106-13114.

Pfefferkorn, E.R. and Borotz, S.E. 1994. Comparison of mutants of *Toxoplasma gondii* selected for resistance to azithromycin, spiramycin, or clindamycin. Antimicrob. Agents Chemother. 338: 31-37.

Pradines, B., Spiegel, A., Regier, C., Tall, A., Mosnier, J., Fusai, T., Trape, J.F. and Parzy, D. 2000. Antibiotics for prophylaxis of *Plasmodium falciparum* infections: in vitro activity of doxycycline against Senegalese isolates. Am. J. Trop. Med. Hyg. 62: 82-85.

Prigge, S.T., He, X., Gerena, L., Waters, N.C. and Reynolds, K.A. 2003. The initiating steps of a type II fatty acid synthase in *Plasmodium falciparum* are catalyzed by pfacp, pfmcat, and pfKASIII. Biochemistry 42: 1160-1169.

Pukrittayakamee, S., Viravan. C., Charoenlarp, P., Yeamput, C., Wilson, R.J. and White, N.J. 1994. Antimalarial effects of rifampin in *Plasmodium vivax* malaria. Antimicrob. Agents Chemother. 38: 511-514.

Ralph, S.A., D'Ombrain, M.C. and McFadden, G.I. 2001. The apicoplast as an antimalarial drug target. Drug Resist. Updat. 4: 145-151.

Read, M., Sherwin, T., Holloway, S., Gull, K. and Hyde, J. 1993. Microtubular organization visualized by immunofluorescence microscopy during erythrocytic schizogony in *Plasmodium falciparum* and investigation of post-translational modifications of parasite tubulin. Parasitology 106: 223-232.

Reichenberg, A., Wiesner, J., Weidemeyer, C., Dreiseidler, E., Sanderbrand, S., Altincicek, B., Beck, E., Schlitzer, M. and Jomaa, H. 2001. Diaryl ester prodrugs of FR900098 with improved *in vivo* antimalarial activity. Bioorg. Med. Chem. Lett. 11: 833-835.

Richter, S. and Lamppa, G.K. 1999. Stromal processing peptidase binds transit peptides and initiates their ATP-dependent turnover in chloroplasts. J. Cell Biol. 147: 33-43.

Rogers, M.J., Burkham, Y.V., McCutchan, T.F. and Draper, D.E. 1997. Interaction of thiostrepton with an RNA fragment derived from the plastid-encoded ribosomal RNA of the malaria parasite. RNA 3: 815-820.

Rogers, M.J., Cundliffe, E. and McCutchan, T.F. 1998. The antibiotic micrococcin is a potent inhibitor of growth and protein synthesis in the malaria parasite. Antimicrob. Agents Chemoth. 42: 715-716.

Roos, D.S., Crawford, M.J., Donald, R.G.K., Kissinger, J.C., Klimczak, L.J. and Striepen, B. 1999. Origin, targeting, and function of the apicomplexan plastid. Curr. Opin. Microbiol. 2: 426-432.

Roy, A., Cox, R.A., Williamson, D.H. and Wilson, R.J. 1999. Protein synthesis in the plastid of *Plasmodium falciparum*. Protist 150: 183-188.

Sacchettini, J.C. and Poulter, C.D. 1997. Creating isoprenoid diversity. Science 277: 1788-1789.

Sato, S., Tews, I. and Wilson, R.J.M. 2000. Impact of a plastid-bearing endocytobiont on apicomplexan genomes. Int. J. Parasitol. 30: 427-439.

Schrevel, J. 1971. Contribution a l'étude des *Selenidiidae* parasites d'annélides polychétes II. ultrastructure de quelques trophozoïtes. Protistologica 7: 101-130.

Shanklin, J., and Cahoon, E.B. 1998. Desaturation and related modifications of fatty acids. Annu. Rev. Plant Physiol. Plant Mol. Biol. 49: 611-641

Shashidhara, L.S. and Smith, A.G. 1991. Expression and subcellular location of the tetrapyrrole synthesis enzyme porphobilinogen deaminase in light-grown *Euglena gracilis* and three nonchlorophyllous cell lines. Proc. Natl. Acad. Sci. USA 88: 63-67.

Smith, S. 1994. The animal fatty acid synthase: one gene, one polypeptide, seven enzymes. FASEB J. 8: 1248-1259.

Speer, C.A. and Dubey, J.P. 1999. Ultrastructure of schizonts and merozoites of *Sarcocystis falcatula* in the lungs of budgerigars (*Melopsittacus undulatus*). J. Parasitol. 85: 630-637.

Strath, M., Scott, F.T., Gardner, M., Williamson, D. and Wilson, I. 1993. Antimalarial activity of rifampicin in vitro and in rodent models. Trans. R. Soc. Trop. Med. Hyg. 87: 211-216.

Striepen, B., Crawford, M.J., Shaw, M.K., Tilney, L.G., Seeber, F. and Roos, D.S. 2000. The plastid of *Toxoplasma gondii* is divided by association with the centrosomes. J. Cell Biol. 151: 1423-1434.

Sulli, C., Fang, Z.W., Muchal, U. and Schwartzbach, S.D. 1999. Topology of *Euglena* chloroplast protein precursors within the endoplasmic reticulum to Golgi to chloroplast transport vesicles. J. Biol. Chem. 274: 457-463.

Sullivan, M., Li, J., Kumar, S., Rogers, M.J. and McCutchan, T.F. 2000. Effects of interruption of apicoplast function on malaria infection, development, and transmission. Mol. Biochem. Parasitol. 109: 17-23.

Surolia, N. and Padmanaban, G. 1991. Chloroquine inhibits heme-dependent protein synthesis in *Plasmodium falciparum*. Proc. Natl. Acad. Sci. USA 88: 4786-4790.

Surolia, N. and Pasmanaban, G. 1992. *De novo* biosynthesis of heme offers a new chemotherapeutic target in the human malarial parasite. Biochem. Biophys. Res. Com. 187: 744-750.

Surolia, N. and Surolia, A. 2001. Triclosan offers protection against blood stages of malaria by inhibiting enoyl-ACP reductase of *Plasmodium falciparum*. Nat. Med. 7: 167-173.

Takishita, K. and Uchida, A. 1999. Molecular cloning and nucleotide sequence analysis of *psbA* from dinoflagellates: origin of the dinoflagellate plastid. Phycol. Res. 47: 207-216.

The Arabidopsis Genome Initiative. 2000. Analysis of the genome sequence of the flowering plant *Arabidopsis thaliana*. Nature 408: 796-815.

Varadharajan S., Dhanasekaran S., Bonday Z.Q., Rangarajan P.N., Padmanaban G. 2002. Involvement of delta-aminolaevulinate synthase encoded by the parasite gene in de novo haem synthesis by *Plasmodium falciparum*. Biochem J. 367: 321-327.

Van de Peer, Y., Baldauf, S., Doolittle, W. and Meyer, A. 2000. An updated and comprehensive rRNA phylogeny of (crown) eukaryotes based on rate-calibrated evolutionary distances. J. Mol. Evol. 51: 565-576.

van Dooren, G.G., Schwartzbach, S.D., Osafune, T. and McFadden, G.I. 2001. Translocation of proteins across the multiple membranes of complex plastids. Biochim. Biophys. Acta 1541: 34-53.

van Dooren, G.G., Su, V., D'Ombrain, M.C. and McFadden, G.I. 2002. Processing of an apicoplast leader sequence in *Plasmodium falciparum* and the identification of a putative leader cleavage enzyme. J. Biol. Chem. 277: 23612-23619.

Van Wye, J., Ghori, N., Webster, P., Mitschler, R.R., Elmendorf, H.G. and Haldar, K. 1996. Identification and localization of rab6, separation of rab6 from ERD2 and implications for an 'unstacked' Golgi, in *Plasmodium falciparum*. Mol. Biochem. Parasitol. 83: 107-120.

Vial, G.J. and Ancelin, M.L. 1992. Malarial lipids. An overview. Subcell. Biochem. 18: 259-306.

Vivier, E. and Desportes, I. 1990. Apicomplexa. In: Handbook of Protoctista. Margulis, L., Corliss, J.O., Melkonian, M. and Chapman, D.J. eds., Jones and Bartlett Publishers, Boston. p. 549-573.

Vivier, E., Petitprez, A. and Landau, I. 1972. Observations ultrastructurales sur la sporoblastogenése de l'hémogregarine, *Hepatozoon domerguei*, Coccide Adeleidea. Protistologica 8: 315-334.

Viviere, E. and Hennere, E. 1965. Ultrastrcture des stades végétatifs de la coccidie *Coelotropha durchoni*. Prctistologica 1: 89-104.

Vollmer, M., Thomsen, N., Wiek, S. and Seeber, F. 2001. Apicomplexan Parasites Possess Distinct Nuclear-encoded, but Apicoplast-localized, Plant-type Ferredoxin-NADP+ Reductase and Ferredoxin. J. Biol. Chem. 276: 5483-5490.

Waegemann, K. and Soll, J. 1996. Phosphorylation of the transit sequence of chloroplast precursor proteins. J. Biol. Chem. 271: 6545-6554.

Waller, R.F., Keeling, P.J., Donald, R.G.K., Striepen, B., Handman, E., Lang-Unnasch, N., Cowman, A.F., Besra, G.S., Roos, D.S. and McFadden, G.I. 1998. Nuclear-encoded proteins target to the plastid in *Toxoplasma gondii* and *Plasmodium falciparum*. Proc. Natl. Acad. Sci. U.S.A. 95: 12352-12357.

Waller, R.F., Reed, M.B., Cowman, A.F. and McFadden, G.I. 2000. Protein trafficking to the plastid of *Plasmodium falciparum* is via the secretory pathway. EMBO J. 19: 1794-1802.

Waller, R.F., Ralph, S.A., Reed, M.B., Su, V., Douglas, J.D., Minnikin, D.E., Cowman, A.F., Besra, G.S. and McFadden G.I. 2003. A type II pathway for fatty acid biosynthesis presents drug targets in *Plasmodium falciparum*. Atntimicrob. Agents Chemother. 47: 297-301.

Waters, N.C., Kopydlowski, K.M., Guszcznski, T., Wei, L., Sellers, P., Ferlan, J.T., Lee, P.J., Li, Z., Woodard, C.L., Shallom, S., Gardner, M.J. and Prigge, S.T. 2002. Functional characterization of the acyl carrier protein (PfACP) and beta-ketoacyl ACP synthase II (PfKASIII) from *Plasmodium falciparum*. Mol. Biochem. Parasitol. 123: 85-94.

Weinstein, J.D. and Beale, S.I. 1983. Separate physiological roles and subcellular compartments for two tetrapyrrole biosynthetic pathways in *Euglena gracilis*. J. Biol. Chem. 258: 6799-6807.

Weissig, V., Vetro-Widenhouse, T. and Rowe, T. 1997. Topoisomerase II inhibitors induce cleavage of nuclear and 35-kb plastid DNAs in the malarial parasite *Plasmodium falciparum*. DNA Cell Biol. 16: 1483-1492.

Wickham, M.E., Rug, M., Ralph, S.A., Klonis, N., McFadden, G.I., Tilley, L. and Cowman, A.F. 2001. Trafficking and assembly of the cytoadherence complex in *Plasmodium falciparum*-infected human erythrocytes. EMBO J. 20: 5636-5649.

Wiesner, J., Hintz, M., Altincicek, B., Sanderbrand, S., Weidemeyer, C., Beck, E. and Jomaa, H. 2000. *Plasmodium falciparum*: Detection of the Deoxyxylulose 5-Phosphate Reductoisomerase Activity. Exp. Parasitol. 96: 182-186.

Williamson, D.H., Denny, P.W., Moore, P.W., Sato, S., McCready, S. and Wilson, R.J. 2001. The *in vivo* conformation of the plastid DNA of *Toxoplasma gondii*: implications for replication. J. Mol. Biol. 306: 159-168.

Wilson, C.M., Smith, A.B. and Baylon, R.V. 1996a. Characterization of the delta-aminolevulinate synthase gene homologue in *P. falciparum*. Mol. Biochem. Parasitol. 75: 271-276.

Wilson, R.J., Fry, M., Gardner, M.J., Feagin, J.E. and Williamson, D.H. 1992. Subcellular fractionation of the two organelle DNAs of malaria parasites. Curr. Genet. 21: 405-408.

Wilson, R.J.M., Denny, P.W., Preiser, P.R., Rangachari, K., Roberts, K., Roy, A., Whyte, A., Strath, M., Moore, D.J., Moore, P.W. and Williamson, D.H. 1996b. Complete gene map of the plastid-like DNA of the malaria parasite *Plasmodium falciparum*. J. Mol. Biol. 261: 155-172.

Wilson, R. J. M., Gardner, M. J., Feagin, J. E. and Williamson, D. H. 1991. Have malaria parasites three genomes? Parasitol. Today 7: 134-136.

Wilson, R.J.M. and Williamson, D.H. 1997. Extrachromosomal DNA in the Apicomplexa. Microbiol. Mol. Biol. Rev. 61: 1-16.

Woods, K.M., Nesterenko, M.V. and Upton, S.J. 1996. Efficacy of 101 antimicrobials and other agents on the development of *Cryptosporidium parvum in vitro*. Ann. Trop. Med. Parasitol. 90: 603-615.

Yung, S. and Lang-Unnasch, N. 1999. Targeting of a nuclear encoded protein to the apicoplast of *Toxoplasma gondii*. J. Euk. Microbiol. 46: 79S-80S.

Yung, S., Unnasch, T.R. and Lang-Unnasch, N. 2001. Analysis of apicoplast targeting and transit peptide processing in *Toxoplasma gondii* by deletional and insertional mutagenesis. Mol. Biochem. Parasitol. 118: 11-21.

Zagnitko, O., Jelenska, J., Tevzadze, G., Haselkorn, R. and Gornicki, P. 2001. An isoleucine/leucine residue in the carboxyltransferase domain of acetyl-CoA carboxylase is critical for interaction with aryloxyphenoxypropionate and cyclohexanedione inhibitors. Proc. Natl. Acad. Sci. USA 98: 6617-6622.

Zhang, Z., Cavalier-Smith, T. and Green, B. 2001. A family of selfish minicircular chromosomes with jumbled chloroplast gene fragments from a dinoflagellate. Mol. Biol. Evol. 18: 1558-1565.

Zhang, Z., Green, B. and Cavalier-Smith, T. 1999. Single gene circles in dinoflagellate chloroplast genomes. Nature 400: 155-159.

Zhu, G., Marchewka, M.J. and Keithly, J.S. 2000a. *Cryptosporidium parvum* appears to lack a plastid genome. Microbiology-Uk 146: 315-321.

Zhu, G., Marchewka, M.J., Woods, K.M., Upton, S.J. and Keithly, J.S. 2000b. Molecular analysis of a Type I fatty acid synthase in *Cryptosporidium parvum*. Mol. Biochem. Parasitol. 105: 253-260.

Zuegge, J., Ralph, S., Schmuker, M., McFadden, G.I. and Schneider, G. 2001. Deciphering apicoplast targeting signals - feature extraction from nuclear-encoded precursors of *Plasmodium falciparum* apicoplast proteins. Gene 280: 19-26.

Zuther, E., Johnson, J.J., Haselkorn, R., McLeod, R. and Gornicki, P. 1999. Growth of *Toxoplasma gondii* is inhibited by aryloxyphenoxypropionate herbicides targeting acetyl-CoA carboxylase. Proc. Natl. Acad. Sci. USA 96: 13387-13392.

From: Malaria Parasites: Genomes and Molecular Biology
Edited by: A.P. Waters and C.J. Janse

# Chapter 12

# The Surface of the *Plasmodium falciparum*-infected Erythrocyte

# Joseph D. Smith and Alister G. Craig

## Abstract

In order to navigate its complex lifecycle, the malaria parasites must interact with a range of host cells. Examples of this are the invasion of hepatocytes by sporozoites and erythrocyte invasion by merozoites. This requirement for cell recognition brings with it the need to display cognate ligands on the parasite surface, and therefore the capacity of the host to develop defences against the infection. Even at a stage where the intracellular nature of erythrocyte development would appear to offer an opportunity for the parasite to be immunologically "silent", parasite-derived proteins are found on the surface of the infected erythrocyte. This review will discuss the proteins found on or associated with the surface of the infected erythrocyte and the resulting phenotypes.

## 1. Modifications Occurring to the Surface of Infected Erythrocytes

*Plasmodium* parasites have a complex life cycle that includes multiple stages of development both within a vertebrate and an anopheles mosquito host. Although *Plasmodium* are capable of establishing chronic infections in humans that can last as long as a year, survival of the parasite species is ultimately conditioned upon effective transmission between hosts. Indeed, the cycle of transmission requires a sexual replication cycle that is initiated in the vertebrate host but can only be completed in the mosquito. Thus, while malaria parasites replicate asexually in vertebrates to increase numbers and presumably enhance transmission success, they have evolved sophisticated mechanisms to ensure their transmission back to the mosquito. This review focuses on alterations occurring at the surface of infected erythrocytes with a special emphasis on *P. falciparum*, the most important *Plasmodium* species that infects humans. Reference is made to parallel mechanisms operating throughout the *Plasmodium* genus.

The initial period of parasite development in humans begins in the liver and last for approximately 7 to 10 days. This stage is not associated with disease but is an important period of growth and amplification that allows the parasite to overcome a transmission bottleneck due to the fact that few parasites are inoculated when a mosquito bites. While only a single parasite may infect a hepatocyte, 20-40,000 parasites are released at the end of the liver stage to infect erythrocytes.

The erythrocytic stage of parasite development produces a chronic infection that can last for over a year and during which disease can occur. In erythrocytes, *P. falciparum* grows, differentiates, and divides in a compartment within the erythrocyte cytoplasm called the parasitophorous vacuole. Asexual division of *P. falciparum* requires approximately 48 hrs. For each parasite that infects an erythrocyte approximately 10 to 24 are released to infect new red blood cells leading to amplification of the infection. Importantly, the number of infected erythrocytes (IE) is a significant risk factor for disease so that an important component of host immunity is directed at limiting parasite growth. Also within erythrocytes some parasites differentiate to sexual forms in a process that is still not completely understood. These gametocyte-infected erythrocytes are infective for mosquitoes and are responsible for completing the cycle of transmission the next time a mosquito feeds.

*P. falciparum*-infected erythrocytes display several dramatic morphological changes that affect membrane rigidity, surface antigenic character, and permeability. These changes are intimately connected to *Plasmodium* biology and involved in nutrient acquisition, the establishment of chronic infections, and the evolution of new adhesive properties displayed by some *Plasmodium* species. Parasite-induced modifications occur both to the erythrocyte cytoskeleton and the extracellular face of the membrane. Although sub-cellular modifications are critical to new adhesive properties exhibited by *P. falciparum*-infected erythrocytes these will only be briefly described here (for a review see (Cooke *et al.*, 2001)). Rather, the major focus of this review will be parasite proteins demonstrated or proposed to be surface-exposed on asexually-parasitised erythrocytes. Several excellent reviews on parasite proteins exported to the erythrocyte cytoskeleton have recently been written.

Until relatively recently the infected erythrocyte surface was regarded as having very few parasite-derived proteins on it. The earliest candidates were a modification of an existing erythrocyte protein, band 3 (Winograd and Sherman, 1989), and a biochemically defined, surface-labellable variant protein known as *P. falciparum* erythrocyte membrane protein 1 (PfEMP1) (Leech *et al.*, 1984). A combination of further biochemical characterisation and the advent of the genome sequence has provided a range of new IE surface candidates, turning a rather sparse molecular "landscape" into a potentially complex interface between parasite and host.

Before describing these different proteins and the evidence placing them at the erythrocyte surface it is constructive to briefly review our understanding of the natural immune response to *P. falciparum* infection and resulting expectations for surface-exposed parasite proteins. Protective immunity to *P. falciparum* is acquired slowly and only after repeated infections. The immunity that develops does not appear to ever provide complete protection from infection but does protect against disease and is still imprecisely understood. However, there is increasing evidence that natural malaria immunity is comprised of different elements including anti-disease and anti-parasite components that evolve with characteristic and distinct kinetics. In addition, there is evidence for clonal antigenic variation of antigens at the surface of infected erythrocytes that may allow parasites to establish chronic infections (for a review see (Bull and Marsh, 2002)). The variant antigens are highly immunodominant and antibodies that develop to them are typically strain-specific but appear to have a protective role against infection. Given the slow evolution of malaria immunity the parasite is highly successful at evading immunity. Strategies that a parasite might use to avoid immunity to surface-exposed proteins are to have these proteins belonging to large, diverse and varying protein families so that it takes an individual a long time to learn the different variants. Alternatively, parasite surface proteins may be exposed for only limited times or relatively inaccessible to antibody. Using these criteria we will review the list of potential surface-exposed parasite proteins beginning with the best-characterised example, PfEMP1.

## 1.1. PfEMP1

Evidence for the presence of neo-antigens on the IE surface was first suggested by Brown and Brown (Brown and Brown, 1965) in their seminal paper on antigenic variation in the primate malaria *Plasmodium knowlesi*. In the mid-1980's more direct molecular evidence was produced in two laboratories using radio-iodination of infected erythrocytes from primate (*P. knowlesi*; Howard *et al.*, 1983) and human (*P. falciparum*; Leech *et al.*, 1984) malarias. The latter identified a protein of variable molecular weight (200-350kDa) between different parasite lines that was Triton X-100 insoluble and sensitive to protease digestion of intact IE. Subsequent studies have demonstrated that this protease sensitivity is not universal (Chaiyaroj *et al.*, 1994; Gardner *et al.*, 1996) but this initial result supported the presence of this protein on the erythrocyte surface. This protein, termed *Plasmodium falciparum* erythrocyte membrane protein 1 (PfEMP1), became the target for intense study over the next ten years but proved difficult to work with due to relatively low abundance and the paucity of specific immunological reagents. As often seen in science, the breakthrough, when it came, derived from the work of several groups in identifying the gene family encoding PfEMP1, namely the *var* genes (Baruch *et al.*, 1995; Smith *et al.*, 1995; Su *et al.*, 1995). The basis for this discovery came from several sources but mainly the production of a monoclonal antibody specific for a single PfEMP1, the identification of a candidate gene from a sequencing study (looking for the chloroquine-resitance gene!), the production of a phenotypically characterised parasite clone tree, and the development of antibodies to recombinant protein fragments of the genes.

The structure of the *var* gene matched the scientific expectations very well, with it being a multi-gene family with 50-60 copies per haploid genome and having a high degree of sequence divergence between different family members. This could not only explain the ability of parasites to switch antigenically, through differential expression of *var* genes, but also why sera from people infected by *P. falciparum* only agglutinated IEs from samples taken from earlier waves of parasitemia and not those from later peaks ((Brown and Brown, 1965; Hommel *et al.*, 1983)). By having a repertoire of surface-expressed PfEMP1s with little similarity and a mechanism of mutually exclusive gene expression, the parasite would be able to switch from one antigenic type to another in the face of host immune pressure. This form of immune evasion is believed to be an important factor in the establishment of chronic infection and presumably enhances transmission. Moreover, not only is the *var* repertoire of a single parasite genotype highly diverse but there is extensive diversity of *var* genes between different parasite genotypes. Estimates of *var* diversity based upon degenerate primers have demonstrated that there is little similarity in repertoires between different parasite genotypes (Fowler *et al.*, 2002; Kyes *et al.*, 1997). Thus, at the population level there is an incredible variety of sequence that may only be limited by the functional requirement of the protein to cytoadhere (see below). Strain-specific variation between *var* repertoires might explain why convalescent sera from children infected by *P. falciparum* agglutinate homologous parasites but frequently show little or no reactivity with heterologous parasites (Bull *et al.*, 1998).

Unlike many clonally variant antigens that appear to act only as immunological "smokescreens" or for which other functions have not yet been defined, PfEMP1 also encode binding properties. Around sixteen hours after merozoite invasion IE are able to adhere to a number of host receptors. In people suffering from malaria this can be seen as a sequestering of the parasites from the peripheral circulation into a number of microvascular sites around the body. Sequestration of infected erythrocytes has been recognised as a characteristic trait of *P. falciparum* infection for over 100 years and is a major pathogenic feature of disease. Perhaps the most famous example of how sequestration causes disease is cerebral malaria, which is associated with parasite adhesion to brain microvasculature and carries with it a high case-fatality (MacPherson *et al.*, 1985). However, sequestration appears to have an equally important role in pathogenesis that occurs in pregnant women (Fried and Duffy, 1996) and might impact malaria pathogenesis in multiple ways through other parasite-host cellular interactions discussed in the following sections.

PfEMP1 are key mediators in binding interactions between infected erythrocytes and host cells. While infected erythrocytes display a range of different binding properties (for reviews see (Cooke *et al.*, 2001; Craig and Scherf, 2001; Kyes *et al.*, 2001), individual parasites differ in their receptor specificity depending upon the expressed PfEMP1. An important question that is being addressed is how parasite receptor specificity influences parasite tropism for different tissues and cells and the impact on disease. To begin to understand PfEMP1 function binding assays have been developed to examine recombinant proteins from these genes and sequence analyses have been performed (Smith *et al.*, 2000a; 2001; 2000b).

*var* genes are encoded in two exons. The first exon codes for the variable extracellular binding region and a transmembrane domain, while the second exon encodes a more conserved cytoplasmic tail. The PfEMP1 binding region is comprised of four different domains. These domains are the N-terminal segment (NTS), Duffy-binding-like (DBL) domain, the cysteine-rich interdomain region (CIDR), and the C2 domain. Both the DBL and CIDR domains have been demonstrated to possess adhesive properties and a system of adhesive domain classification has been developed to define their sequence relatedness. From a study of 20 different PfEMP1(Smith *et al.*, 2001), DBL domains grouped into five different sequence types: α, β, γ, δ, and ε. In contrast, CIDR domains grouped into three types: α, β, and γ. Within each adhesive domain classification there are a variety of different sequences, however, domains of a type share characteristic and distinctive amino acid features. As the number, location, and type of DBL and CIDR domains vary between PfEMP1 proteins, a nomenclature has been introduced that describes both the numeric position and sequence type of the domain (Figure 1).

Although PfEMP1 proteins are variably sized, an important concept that has emerged from sequence analysis is that the binding region is not created through a random assortment of domains. Rather adhesive domain sequence types tend to occupy characteristic positions in the protein and associate in favoured tandems. For instance, one tandem association is the DBLα

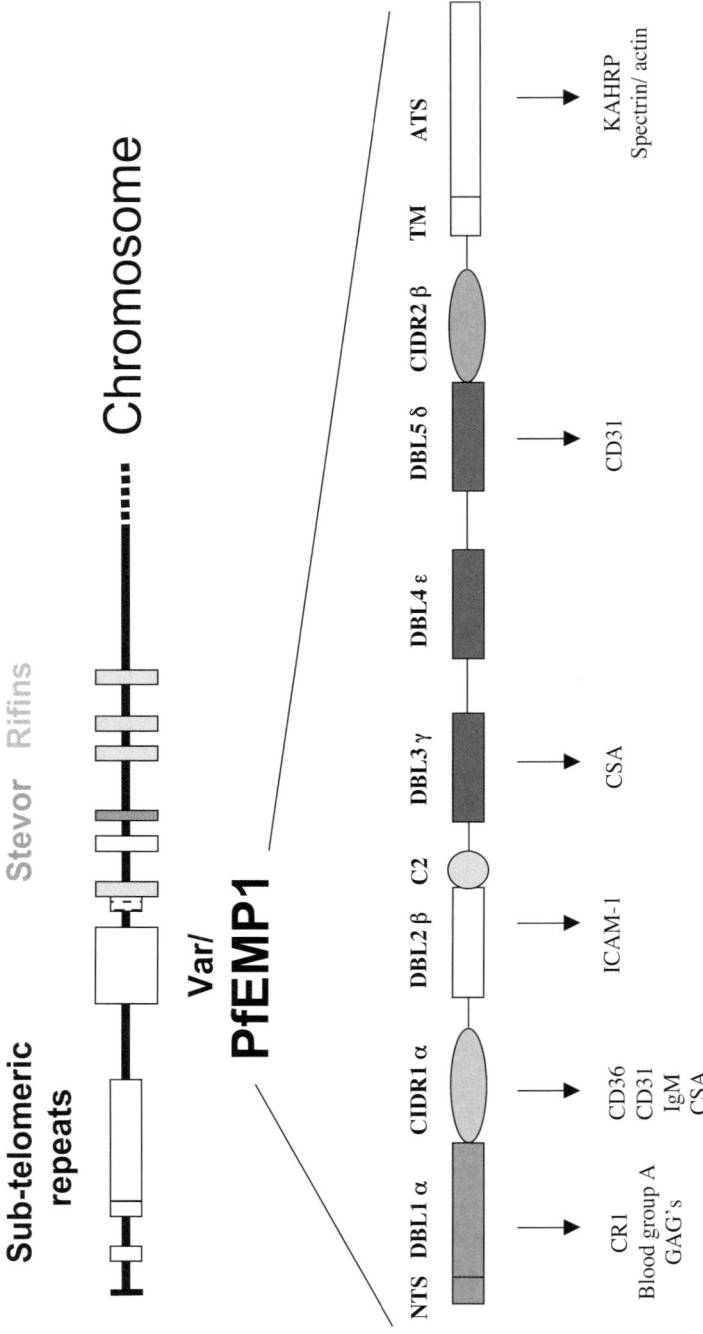

**Figure 1.** Schematic showing the general organisation of the sub-telomeric region of *P.falciparum* chromosomes (Bowman *et al.*, 1998; Gardner *et al.*, 1999) and structure of PfEMP1. At the top, left-hand side of the figure chromosomal elements are indicated for a telomeric repeat sequence followed by sub-telomeric repetitive elements (R-CG7, rep11, and rep20, respectively). Repetitive elements are followed by a *var* gene, R-FA3 repeat sequence and *rif*/ *stevor* genes. Regions of PfEMP1 are indicated in the text except the transmembrane region (TM) and acidic terminal segment (ATS). The nomenclature for motifs within PfEMP1 indicates the position in the molecule (1-5) and the sequence type (α-ε). Note that the sequence types can be in different orders, except for DBL1α, which is always the N-terminal domain. The receptors for the various regions of PfEMP1 are indicated below the schematic.

and CIDR domains that form the semi-conserved head structure of almost all PfEMP1 (Chen *et al.*, 2000; Su *et al.*, 1995)(Figure 1). Other tandem combinations, DBLβ-C2 and DBLδ-CIDR are found internal to the head structure. Thus, adhesive domain sequence classification provides insight into PfEMP1 protein architecture and may have implications for binding function. At present, only limited numbers of PfEMP1 adhesive domains have been functionally characterised but already there are multiple examples in which different parasites adhesion traits have been shown to map to the same adhesive domain type. For instance, two different CSA-binding parasites used a DBLγ type domain to bind CSA even though the DBLγ sequences were distinct. One hypothesis that is being tested is the extent to which different PfEMP1 proteins that bind the same receptor use the same adhesive domain sequence type to bind. If there were functional constraints that caused like-binding domains to have a related structure and antigenicity this could have implications for vaccine design (Duffy *et al.*, 2001). Investigations have already begun to test for cross-reactive epitopes in different PfEMP1 domains (Gamain *et al.*, 2001). Our understanding of malaria pathogenesis will also likely improve with further investigation of PfEMP1 function. For instance, small and large PfEMP1 have quite distinctive protein architectures and are comprised of different adhesive domain types. Large PfEMP1 may possess special adhesive characteristics responsible for a distinct spectrum of sequestration and disease outcomes.

From the beginning studies that established a surface localisation for PfEMP1 proteins using iodination and trypsin sensitivity to the development of serological reagents and functional assays, there are now multiple lines of evidence confirming PfEMP1s surface expression.

## 1.2. Rifins

The Rifins are comprised of two related but slightly distinctive multiple gene families, the *rifs* and the *stevors*, discussed separately below.

### 1.2.1. Rifs

Rifs were originally identified as a repetitive gene sequence (*rif*), which could be used to characterise different parasite lines due to the highly variable banding pattern seen on southern blots of genomic DNA from different lines of *P. falciparum* when probed with this sequence (Weber, 1988). The possibility that *rifs* encoded a set of surface proteins that were initially linked with rosetting (termed rosettins (Helmby *et al.*, 1993)), and subsequently with binding to CD31 (Fernandez *et al.*, 1999) was not recognised until the advent of the *P. falciparum* genome project (Gardner *et al.*, 1998). Sequences with similarity to the original *rif* gene were found as tandem arrays within the sub-telomeric regions. Initially it was not clear that they were expressed as they appeared to lack a proper 5′ region, but careful inspection of the DNA sequence revealed a small 5′ exon. It now appears that *rif* genes have a two-exon structure, with the first exon coding for a predicted signal peptide and the second for a protein that is highly variable but contain stretches of relative amino acid conservation and conserved cysteine residues. The actual orientation of Rif proteins within the membrane is still debated because the second exon codes for more than one block of hydrophobic residues that could potentially act as a transmembrane domain. Thus, it is not known whether the bulk of the protein is exposed or whether Rif proteins make multiple passes through the membrane. However, most models predict that Rif proteins have at least one transmembrane domain near the end of the protein that is followed by a short, highly conserved cytoplasmic tail. Members of the *rif* gene constitute the largest gene family in *P. falciparum* discovered so far, with in excess of two hundred copies per genome.

The physical proximity of rif genes to *var* genes raised some early speculation that both gene families might be regulated in a co-ordinated fashion, but evidence to support this scenario remains to be established. A transcriptional analysis of *var* and *rif* expression indicated that *var* messages were expressed relatively early after erythrocyte invasion with a peak around 12h post-invasion while *rif* messages were limited to late ring and early trophozoite stages (Kyes *et al.*, 2000). Despite transcriptional differences, evidence has been presented that Rifs are co-expressed on the infected erythrocyte surface with PfEMP1 (Fernandez *et al.*, 1999; Kyes *et al.*, 1999). This evidence consists of similar criteria that were originally developed to establish the surface expression of PfEMP1. Thus, it has been demonstrated that Rif proteins are iodinatable and trypsin-sensitive albeit with a much higher trypsin requirement than PfEMP1. Unlike PfEMP1, there is no published evidence that antibodies raised to Rif recombinant proteins react with the infected erythrocyte surface although hyper-immune sera from adults in malaria endemic regions are able to immunoprecipitate the proteins indicating their natural immunogenecity (Fernandez *et al.*, 1999). It is curious that Rif proteins are predicted to act as clonal variant antigens, yet it has been difficult to establish their surface expression with antibodies. Equally elusive is the function of Rif proteins despite attempts to link their expression with specific adhesive phenotypes. However, the investment by *P. falciparum* in maintaining a large repertoire of these genes and the differential expression of *rifs* by parasite lines would indicate that they play an important role in the interaction between parasite and human host.

### 1.2.2. Stevor

*Stevor* genes are also found in close proximity to *var* and *rif* genes. Like *rif*, *Stevor* began as a multicopy probe for distinguishing *P. falciparum* isolates (Limpaiboon *et al.*, 1991) but was later recognized to have a two-exon gene structure in which the first exon is predicted to code for a signal peptide and the second for the protein (Cheng *et al.*, 1998). Within the protein, Rif and Stevor share some conserved cysteines and align with each other by Blast analysis but the two protein families have characteristic and distinct features including different numbers of conserved cysteines (Cheng *et al.*, 1998). Although the cytoplasmic domain of Stevors is relatively well conserved between different proteins it is distinct from that of Rif proteins. Like *rif*, *stevor* probes generate multiple bands on *P. falciparum* genomic DNA. However, unlike rif there is no published data indicating expression on the IE surface, although unpublished data suggest that *stevor* are transcribed in sexual and asexual stages and localised in the Maurer's clefts in the latter. Since no clear functions have been described for either *rif* or *stevor* gene products it is uncertain whether these protein families have similar or distinct activities.

### 1.3. KAHRP and CLAG9

Upon adaptation of *P. falciparum* to in vitro cultivation, many parasite cultures gradually lose the capacity to cytoadhere over a period of weeks (Udeinya *et al.*, 1983). Chromosomal deletions are frequently observed in *P. falciparum*, particularly during *in vitro* propagation. These genetic deletions are often associated with chromosome ends and can extend for as much as 500kb. While the deletion events appear to offer an *in vitro* growth advantage to the parasite outside of the context of a human host, non-cytoadherent parasite strains have proven a valuable resource to investigate the molecular requirements for infected erythrocyte cytoadherence.

The first chromosomal deletion to be characterised in molecular detail occurred in a telomere end of chromosome 2 and resulted in the loss of a protein called the knob-associated histidine-rich protein (KAHRP) (Pologe and Ravetch, 1986). KAHRP is transported by the parasite out of the parsitophorous vacuole and forms an association with the erythrocyte membrane cytoskeleton (Kilejian *et al.*, 1991; Waller *et al.*, 1999). This association leads to the production of knob-like membrane protuberances at the erythrocyte surface that are the point of contact between infected erythrocytes and endothelium. Although KAHRP is completely intracellular and not a parasite adhesion receptor it has been shown to be essential for knob formation and the binding of infected erythrocytes under conditions of flow (Crabb *et al.*, 1997).

Another common deletion in laboratory-adapted parasite isolates occurs on chromosome 9 (Barnes *et al.*, 1994; Day *et al.*, 1993; Shirley *et al.*, 1990). This deletion has been linked with the loss of cytoadherence to C32 melanoma cells (Barnes *et al.*, 1994), but not all cytoadherence (Chaiyaroj *et al.*, 1994). Initially it was thought that a gene encoding

PfEMP1 might map within this region but subsequent detailed analysis of the locus showed that parasites with smaller deletions but still showing the loss of adherence retained the *var* gene at the end of chromosome 9 (Holt *et al.*, 1998). Sequencing of the deleted region revealed a complex structure consisting of nine exons, producing a mature transcript of around 7kb. This gene was called cytoadherence-linked asexual gene (*clag9*) (Trenholme *et al.*, 2000) and encodes a protein of about 220kDa that is associated with the IE membrane, but has not been proven to be on the surface. A number of transfection-based experiments have indicated that this gene is essential for adhesion to C32 via CD36, such that *clag9* knockout or antisense parasites lose the ability to bind to these cells (Gardiner *et al.*, 2000; Trenholme *et al.*, 2000). However it is not clear at what stage in the adhesion process clag9 acts and it is possible that the gene product is involved in the translocation of PfEMP1 from the cytoplasm to the IE surface, rather than being directly involved in adhesion.

It is perhaps surprising given the loss of phenotype seen in the *clag9* knockout that there are several *clag* genes in the *P. falciparum* genome. The genome-sequencing project has revealed at least five related sequences (Holt *et al.*, 1999), located close to the *var/rif/stevor* clusters at the end of the chromosomes. One of these, on chromosome 3, has been identified recently as encoding RhopH1, one of three proteins that make up the RhopH complex in the rhopteries of the malarial merozoite (Kaneko *et al.*, 2001). Further evidence for the involvement of *clag* paralogues in merozoite function has come from the discovery of two *clag*-related sequences in the rodent malaria *P.yoelii* (*pyrhoph1a* and *pyrhoph1a-p*). It appears that this family of sequences, now termed *rhoph1/clag*, mediate broader cell-cell interactions that was previously thought and it will be interesting to see at a structural level how *Plasmodium* has used this gene framework to address cell adhesion in multiple contexts.

## 1.4. Others

A number of other potential parasite ligands on the IE surface have been described as well as permeation pathway involving a voltage-dependent channel which allows the parasite to acquire the metabolites it needs to develop in what is essentially a metabolically inactive cell.

### 1.4.1. Band 3

The best characterised of this group of molecules is in fact a host protein that is modified by the parasite. Band 3 is a highly abundant erythrocyte surface protein that acts as an anion transporter. Modification of this protein is involved in the clearance of senescent erythrocytes, through the exposure of cryptic epitopes. Invasion by *P. falciparum* also causes modification to Band 3, which can be recognised by specific monoclonal antibodies (Crandall and Sherman, 1991). The basis of the modification is unknown but early experiments suggested that modified Band 3 was involved in adhesion to a range of cells via CD36. However, more recent data have implicated thrombospondin as the receptor (Eda *et al.*, 1999).

### 1.4.2. Sequestrin

Another potential parasite adhesion receptor is called Sequestrin (Ockenhouse *et al.*, 1991). Sequestrin was identified using anti-idiotype antibodies to an anti-CD36 monoclonal antibody that could block infected erythrocyte cytoadherence. The sera precipitated an approximately 270 kDa surface-iodinatable protein. Unpublished observations have described a gene for the Sequestrin protein but some doubts on the function of this protein are raised by a lack of an effect on cytoadherence in genetic knockout experiments (Trenholme *et al.*, 2000).

### 1.4.3. RSP1 and RSP2

One recent surprising result was the description of a new form of cytoadherence involving immature ring-stage forms of infected erythrocytes. It has been a long accepted dogma that immature forms of parasites circulate in the blood until approximately 16 hrs post-invasion when sequestration begins for more mature parasite forms. The new work investigating the stage-specific regulation of cytoadherence focused on mature parasites that had been selected to bind CSA but were simultaneously selected to express distinct adhesion ligands at the immature parasite stage. The fascinating observation was the ring-infected erythrocytes of parasites selected to bind CSA were able to adhere to some types of cultured endothelium and placenta tissue sections, but not via CSA (Pouvelle *et al.*, 2000). The binding appears to be mediated by two parasite-encoded proteins, RSP1 and 2, that are expressed on the erythrocyte surface shortly after parasite invasion. Adhesion of immature parasite forms appears to be mechanistically distinct from mature parasites since it occurs independently of knobs. In addition, there are still unanswered questions about the relationship of this binding phenotype to particular parasite adhesion traits. However, these results suggest that there may be non-circulating (cryptic) ring-infected erythrocyte subpopulations in malaria patients. Interestingly, a recent study by (Silamut *et al.*, 1999) was the first to describe a cryptic population of ring-infected erythrocytes in the brain microvasculature of individuals who died from cerebral malaria.

*1.4.4. "New Permeation Pathway" (NPP)*

Other surface changes reflect nutritional requirements of the growing intraerythrocytic parasite. Erythrocytes are an interesting cell to parasitise because they are nearly metabolically inert and have been commonly referred to as "bags of haemoglobin". Despite their metabolic limitations, another protozoan parasite, *Babesia*, also invades and develops within erythrocytes during a period of its life cycle. During their development, *Plasmodium* parasites sample the contents of the erythrocyte cytoplasm and use haemoglobin as an important nutritional source. Although many of the amino acids required for parasite growth are supplied from haemoglobin metabolism, parasitised erythrocytes also acquire enhanced permeability characteristics that allow the uptake of factors from blood serum. This so-called "New Permeation Pathway" (NPP) (Desai *et al.*, 2000) is indicated by the increased uptake of many solutes including anions, sugars, purines, amino acids and organic cations and has been measured experimentally using a whole-cell voltage-clamp method on infected erythrocytes. Identifying the voltage-gated channel is a subject of intense molecular investigation, but it is not yet demonstrated whether it is caused by the insertion of parasite protein(s) in the erythrocyte membrane or modification of host red cell proteins.

# 2. Phenotypes Associated With the IE Surface

From the description above it can be seen that the major roles for the proteins on the surface of the IE are antigenic variation and adhesion. While all *Plasmodium* species that have been studied to date possess a capacity to vary surface antigenic character, it is not clear why an ability to cytoadhere has only developed for *P. falciparum* among the human malarias. However, this property appears a key determinant in the enhanced virulence of this species.

## 2.1. Clonal Antigenic Variation

At first glance it may appear curious that a parasite living within a erythrocyte, a cell that is relatively inaccessible to cellular immunity because it does not present peptide antigens through either the class I or class II antigen-presenting systems, would expose itself to antibody by exporting proteins to the erythrocyte surface. However, it has been pointed out that clonal antigenic variation is a common theme of different pathogens that rely on insect vectors to complete their life cycle (Kyes *et al.*, 2001). Clonal antigenic variation is probably an important factor that allows parasites to establish chronic infections and ensure transmission during periods when host to host contact is sporadic (e.g. during the dry season for malaria parasites). The importance of clonal antigenic variation to *Plasmodium* biology is demonstrated by the fact that it exists throughout the genus. Recently, several different large and diverse protein families expressed during the erythrocytic period of development have been described from several *Plasmodium* species. Although direct evidence for surface variant exposure of these proteins is still being established for all but *P. falciparum*, it is interesting that only some of these protein families are common while some species like *P. falciparum* appear to have evolved distinct and unique protein families. To date, no sequences sharing significant similarity to *var* or *rifin* genes have been reported from other *Plasmodium* species even though some, like *P. falciparum*, cytoadhere. It will be interesting to test for *var* and *rifin* genes in *P. falciparum*'s nearest evolutionary relatives. However, a gene coding for the variant surface antigen of *P. knowlesi* has been cloned and differs both in sequence and organisation from its *P. falciparum* counterpart (al-Khedery *et al.*, 1999). In addition a major gene family has been discovered in *P. vivax* called the *vir* genes which is also located within the sub-telomeric regions (del Portillo *et al.*, 2001). Subsequently genes related to *vir* were identified in the rodent malarias *P. chabaudi*, *P. yoelii*, and *P. berghei* (Carlton and Carucci, 2002; Janssen *et al.*, 2002). The telomeric location of variant gene families may have functional significance as recent work in *P. falciparum* has shown that telomere clusters are formed during mitosis, facilitating intergenic transfer and thereby the generation of genetic variation (Freitas-Junior *et al.*, 2000). The accumulating evidence that different *Plasmodium* species have independently evolved and developed distinct variant protein families to alter the surface antigenic profile of infected erythrocytes supports an important function for this adaptation.

## 2.2. Adhesion to Endothelium (Cytoadherence)

Erythrocytes infected with *P. falciparum* "disappear" from the circulation during their development in a process called sequestration. One of the major mechanisms associated with this event is adhesion to the endothelial cells lining small blood vessels, where the shear flow forces are sufficiently reduced to allow the relatively low avidity IE/ endothelial cell (EC) interactions to operate. A large number of endothelial receptors have been identified (see Table 1 for details) but not all of these are commonly used in patient isolates. A number of studies have attempted to correlate specific receptor usage with disease severity but these experiments are complicated by a wide range of confounders, not the least that the parasites available on admission to hospital (prior to treatment) may not accurately represent the sequestered mass.

An important and reproducible finding that has emerged from these investigations is that the host cell receptor CD36 is a major endothelial receptor

**Table 1.** Host receptors for *P.falciparum*-infected erythrocytes

| Host Receptor | Cellular Target(s) | Parasite Ligand |
|---|---|---|
| Thrombospondin (Roberts *et al.*, 1985) | Endothelium | Modified Band 3 (Eda *et al.*, 1999) PfEMP1 |
| CD36 (Barnwell *et al.*, 1989; Ockenhouse *et al.*, 1989b; Oquendo *et al.*, 1989) | Endothelium, Dendritic Cells (Urban *et al.*, 2001), Uninfected Erythrocytes (Handunnetti *et al.*, 1992), Platelet-bridged Infected Erythrocytes (Pain *et al.*, 2001) | CIDRα (Baruch *et al.*, 1996; Baruch *et al.*, 1997) |
| ICAM-1 (Berendt *et al.*, 1989) | Endothelium | DBLβC2 (Baruch *et al.*, 1996; Smith *et al.*, 2000a) |
| VCAM-1, E-selectin (Ockenhouse *et al.*, 1992), α$_v$β$_3$ (Siano *et al.*, 1998) | Endothelium | |
| Chondroitin-4-sulfate (CSA) (Fried & Duffy, 1996; Rogerson *et al.*, 1995) | Placenta, Endothelium | DBLγ (Buffet *et al.*, 1999; Reeder *et al.*, 1999) |
| P-Selectin (Udomsangpetch *et al.*, 1997) | Endothelium | PfEMP1 (Senczuk *et al.*, 2001) |
| PECAM-1 (CD31) (Newbold *et al.*, 1997; Treutiger *et al.*, 1997) | Endothelium | CIDRα/ DBLδ (Chen *et al.*, 2000) |
| Hyaluronic acid (Beeson *et al.*, 2000), non-immune IgG (Flick *et al.*, 2001) | Placenta | |
| CR1 (Rowe *et al.*, 1997), HS-like GAG (Chen *et al.*, 1998), Blood Group A antigen (Carlson & Wahlgren, 1992) | Uninfected Erythrocytes | DBLα (Chen *et al.*, 1998; Rowe *et al.*, 1997) |
| IgM (Scholander *et al.*, 1996) | Uninfected Erythrocytes | CIDRα (Chen *et al.*, 2000) |

for parasite cytoadherence both in terms of how frequently the receptor is utilised by different parasite isolates and the strength of binding. Indeed, most infections are characterised by parasites that bind CD36 to some extent (Newbold *et al.*, 1997), with the important exception of parasites sequestered in the placenta (below). In addition, infected erythrocytes adhere very avidly to CD36 such that they can stably bind under conditions of flow that mimic those in microvasculature while many other parasite-receptor interactions are only able to support rolling adhesion (Cooke *et al.*, 1994). Thus, the available evidence indicates that *P. falciparum* has made a significant investment into CD36 binding with the widespread cellular distribution of this receptor offering the parasite numerous different opportunities for host interactions. Roles for the parasite-CD36 interaction beyond sequestration are discussed below. By comparison to CD36, other receptor adhesion traits are less common or else support weaker binding. However, there is evidence that multiple receptor interactions can act cooperatively to anchor the infected erythrocyte (McCormick *et al.*, 1997).

In terms of parasite adhesion traits that may be involved in organ-specific sequestration this has generally been more difficult to investigate mainly because tissues that malarial researchers would like to study, such as brain, are not readily accessible. However, there is some evidence from sampling circulating blood and allowing parasites to mature *in vitro* that infected erythrocytes binding ICAM-1 are slightly increased in patients with cerebral malaria (Newbold *et al.*, 1997). In addition, cerebral sequestered parasites specifically co-localise to endothelium expressing ICAM-1 in post-mortem histological investigations (Turner *et al.*, 1994). An important area of future research is to further define receptors involved in cerebral sequestration in order to understand the pathological basis of this disease.

Recently, it was reported that isolates from patients with severe malaria were more likely to bind multiple receptors (Heddini *et al.*, 2001). Many of the receptor binding events described from severe isolates also participated in rosette formation between infected and uninfected erythrocytes (below), a property that had previously associated with severe disease. Thus, a combination of binding events acting synergistically may bring about severe disease.

## 2.3. Adhesion to Uninfected Erythrocytes (Rosetting)

Another form of adhesion is rosetting, the binding of two of more uninfected erythrocytes to an infected cell (Udomsangpetch *et al.*, 1989). While not all parasite isolates form rosettes and the degree of rosette formation can vary dramatically between isolates, higher rosetting rates have generally been associated with more severe disease (Carlson *et al.*, 1990; Rowe *et al.*, 1995) with some exceptions (al-Yaman *et al.*, 1995).

In a *P. falciparum* clonal lineage, rosetting was shown to be a clonally variant property of parasites that was eventually demonstrated to be mediated by specific PfEMP1 proteins (Chen *et al.*, 1998; Rowe *et al.*, 1997). The rosettins, now recognised as *rif* genes, were also initially believed to participate in rosette formation but direct evidence for this role is lacking. There are a number of receptors on the red cell surface that can participate in rosette formation including complement receptor 1 (CR1) (Rowe *et al.*, 1997), Heparan-sulfate (Chen *et al.*, 1998), and the ABO blood group (particularly group A) (Carlson and Wahlgren, 1992). In addition, rouleaux-forming serum proteins are also involved in rosetting (Treutiger *et al.*, 1999). Interestingly, CR1 polymorphisms are common in Africans suggesting the possibility that they may have a protective role against severe malaria (Moulds *et al.*, 2001). Rosetting parasites differ in their receptor specificity. For instance, not all rosetting parasites bind CR1, parasites differ in their susceptibility to rosette disruption with heparin (Rogerson *et al.*, 1994), and parasites can form small or large rosettes. This variability highlights a limitation of correlative studies that tend to lump multifactorial binding phenotypes together or do not account for the possibility that binding properties may exist in different adhesive contexts on different PfEMP1. Thus, it may be a simplification to assume that all rosetting or all ICAM-1 binding parasites possess the same potential for disease.

## 2.4. Adhesion to Other Infected Erythrocytes (Autoagglutination)

Besides binding uninfected erythrocytes, a subset of parasites form autoagglutinates with other infected erythrocytes. Autoagglutination has also been shown to be a clonally variant property (Roberts *et al.*, 1992). Recently, it was demonstrated that autoagglutinate clumps were bridged by platelet cells (Pain *et al.*, 2001). One of the important receptors supporting platelet-mediated clumping of infected erythrocytes is CD36. Since most parasites bind CD36 but do not form autoagglutinates, differences in parasite affinity or specificity for CD36 may be important for clump formation. Alternatively, there may be additional platelet-specific receptors required in combination to CD36 for autoagglutinates to form. Autoagglutination is observed in field isolates and associated with more severe disease (Roberts *et al.*, 2000).

## 2.5. Adhesion to Monocytes and Dendritic Cells

Infected erythrocytes also bind other cells in the blood circulation including monocytes and dendritic cells (Ockenhouse *et al.*, 1989a; Urban *et al.*, 1999). Again, CD36 has been demonstrated to be an important receptor bridging these interactions (Urban *et al.*, 2001). Infected erythrocyte adhesion to monocytes has been demonstrated to induce an oxidative burst (Ockenhouse *et al.*, 1989a). In contrast, infected erythrocyte adhesion to dendritic cells *in vitro* has been shown to down-modulate their antigen-presenting activity and their capacity to stimulate T cells (Urban *et al.*, 1999). Whether this same effect occurs during natural malaria infections in the cellular environment in which T cells are activated is unknown but because dendritic cells are crucial for inducing immune responses this effect could contribute to the immunosuppression typical of malaria infection. One of the functions of CD36 is to act as a receptor in the uptake of apoptotic cells. It has been hypothesised that infected erythrocytes binding to CD36 on dendritic cells may subvert an immodulatory pathway that evolved to prevent autoimmune disease (Urban *et al.*, 2001).

## 2.6. Adhesion to Placenta

The best understood system of organ-specific sequestration is the placenta due to the fact that sequestered parasites can be directly studied from this tissue after delivery. For a long time it had been recognised that women in Africa, even those with significant pre-existing malaria immunity, became susceptible to malaria infection during pregnancy. Malaria during pregnancy is associated with a massive accumulation of infected erythrocytes in the placenta, which can cause severe anaemia in mothers but also appears to be responsible for the development of low-birth weight babies at increased risk of death (Brabin, 1983; Walter *et al.*, 1982). The severity of malaria during pregnancy typically diminishes with successive pregnancies and is correlated with less sequestered parasites in the placentas of multi-gravid women and the acquisition of antibodies to placental-binding parasites (Fried *et al.*, 1998; Staalsoe *et al.*, 2001).

Fried and Duffy performed the seminal experiments investigating the binding properties of placental-sequestered parasites. The remarkable observation that they made was that unlike most infections, placenta sequestered parasites did not bind CD36 but rather these parasites had special affinity for chondroitin-sulfate A (CSA) (Fried and Duffy, 1996; Rogerson *et al.*, 1995). Interestingly, CSA-adherent parasites are rare in non-pregnant individuals. The model that emerged from these studies is that the placenta enriches for parasite binding variants that do not adhere well to microvasculature but have a special affinity for the placenta. Consequently, pregnant mothers become infected with parasite binding variants to which to which they have not previously developed immunity (for review see Beeson *et al.* (2001), Scherf *et al.* (2001)).

Although CSA-containing proteoglycans are widely distributed throughout the vascular endothelium, placental CSA has a unique chemistry with an especially low sulfate content to which infected erythrocytes adhere more avidly (Achur *et al.*, 2000; Alkhalil *et al.*, 2000). Indeed, the placental intervillous spaces contain a web-like matrix of the low sulfate proteoglycans which might "capture" infected erythrocytes and explain the mystery of why many placental sequestered infected erythocytes are not closely associated to cells. In addition, placental syncytiotrophoblasts are also known to bind infected erythrocytes. Thrombomodulin, a CSA-containing proteoglycan expressed by syncytiotrophoblast cells, has been shown to support parasite adhesion in vitro (Gysin *et al.*, 1997) and may be one of the sources for two cell-associated proteoglycans that were also characterised from placenta (Achur *et al.*, 2000).

Since the first binding descriptions it has been reported that placental-sequestered parasites can also bind hyaluronic acid (Beeson *et al.*, 2000) and non-immune immunoglobulins (Flick *et al.*, 2001), although there is still some controversy over whether hyaluronic acid is a receptor (Valiyaveettil *et al.*, 2001). Current efforts are directed at defining the relative importance of these different receptors for placental sequestration and other possible undefined placental sequestration receptors. Another major avenue of research is to understand a protective immune response that develops in pregnant mothers and prevents parasite adhesion to CSA. Remarkably, the anti-adhesion antibodies are pan-reactive and recognise placental sequestered parasites from all over the world (Fried *et al.*, 1998). This contrasts with typical anti-adhesion antibodies detected in children and adults that are typically highly strain-specific (Bull *et al.*, 1998). The evolution of broadly protective anti-adhesion antibodies in pregnant mothers is an encouraging sign and suggests that it may be possible to develop vaccines directed at parasite-encoded, variant erythrocyte surface proteins. Understanding the molecular basis and specificity of these antibodies is an important area of research (Duffy *et al.*, 2001).

# 3. Conclusions

This review summarises the evidence for molecules at the surface of infected erythrocytes. Our increased understanding of these proteins is an important goal because of their central importance to many essential features of the parasite biology including nutrient acquisition, establishment of chronic infection with greater opportunities for transmission, and pathogenesis. In addition to the molecules discussed there is biochemical evidence for additional surface proteins (Fernandez *et al.*, 1999; Howard, 1988). The Malaria Genome Sequencing Effort and post-genomic approaches may provide new opportunities to identify these molecules and thereby develop new strategies to combat this disease.

# 4. References

Achur, R.N., Valiyaveettil, M., Alkhalil, A., Ockenhouse, C.F. and Gowda, D.C., 2000. Characterization of proteoglycans of human placenta and identification of unique chondroitin sulfate proteoglycans of the intervillous spaces that mediate the adherence of *Plasmodium falciparum*-infected erythrocytes to the placenta. J. Biol. Chem. 275(51): 40344-40356.

Alkhalil, A., Achur, R.N., Valiyaveettil, M., Ockenhouse, C.F. and Gowda, D.C., 2000. Structural requirements for the adherence of *Plasmodium falciparum*-infected erythrocytes to chondroitin sulfate proteoglycans of human placenta. J. Biol. Chem. 275(51): 40357-40364.

al-Khedery, B., Barnwell, J.W. and Galinski, M.R., 1999. Antigenic variation in malaria: a 3' genomic alteration associated with the expression of a *P. knowlesi* variant antigen. J. Biol. Chem. Cell. 3(2): 131-141.

al-Yaman, F., Genton, B., Mokela, D., Raiko, A., Kati, S., Rogerson, S., Reeder, J. and Alpers, M. 1995. Human cerebral malaria: lack of significant association between erythrocyte rosetting and disease severity. Trans. R. Soc. Trop. Med. Hyg. 89(1): 55-58.

Barnes, D.A., Thompson, J., Triglia, T., Day, K. and Kemp, D.J., 1994. Mapping the genetic locus implicated in cytoadherence of *Plasmodium falciparum* to melanoma cells. Mol. Biochem. Parasitol. 66(1): 21-29.

Barnwell, J.W., Asch, A.S., Nachman, R.L., Yamaya, M., Aikawa, M. and Ingravallo, P. 1989. A human 88-kD membrane glycoprotein (CD36) functions *in vitro* as a receptor for a cytoadherence ligand on *Plasmodium falciparum*-infected erythrocytes. J. Clin. Invest. 84(3): 765-772.

Baruch, D.I., Gormely, J.A. Ma, C., Howard, R.J. and Pasloske, B.L., 1996. *Plasmodium falciparum* erythrocyte membrane protein 1 is a parasitized erythrocyte receptor for adherence to CD36, thrombospondin, and intercellular adhesion molecule 1. Proc. Natl. Acad. Sci. USA. 93(8): 3497-3502.

Baruch, D.I., Gormely, J.A., Ma, C., Howard, R.J. and Pasloske, B.L. 1997. Identification of a region of PfEMP1 that mediates adherence of *Plasmodium falciparum* infected erythrocytes to CD36: conserved function with variant sequence. Blood. 90(9): 3766-3775.

Baruch, D.I., Ma, X.C., Singh, H.B., Bi, X., Pasloske, B.L. and Howard, R.J. 1995. Cloning the P. falciparum gene encoding PfEMP1, a malarial variant antigen and adherence receptor on the surface of parasitized human erythrocytes. Cell. 82(1): 77-87.

Beeson, J.G., Reeder, J.C., Rogerson, S.J. and Brown, G.V., 2001. Parasite adhesion and immune evasion in placental malaria. Trends Parasitol. 17(7): 331-337.

Beeson, J.G., Rogerson, S.J., Cooke, B.M., Reeder, J.C., Chai, W., Lawson, A.M., Molyneux, M.E. and Brown, G.V. 2000. Adhesion of *Plasmodium falciparum*-infected erythrocytes to hyaluronic acid in placental malaria. Nat. Med. 6(1): 86-90.

Berendt, A.R., Simmons, D.L., Tansey, J., Newbold, C.I. and Marsh, K., 1989. Intercellular adhesion molecule-1 is an endothelial cell adhesion receptor for *Plasmodium falciparum*. Nature. 341(6237): 57-59.

Bowman, S., Lawson, D., Basham, D., Brown, D., Chillingworth, T., Churcher, C.M., Craig, A., Davies, R.M., Devlin, K., Feltwell, T., Gentles, S., Gwilliam, R., Hamlin, N., Harris, D., Holroyd, S., Hornsby, T., Horrocks, P., Jagels, K., Jassal, B., Kyes, S., McLean, J. Moule, S., Mungall, K., Murphy, L., Barrell, B.G. *et al.* 1999. The complete nucleotide sequence of chromosome 3 of *Plasmodium falciparum.* Nature. 400(6744): 532-538.

Brabin, B.J., 1983. An analysis of malaria in pregnancy in Africa. Bull World Health Organ. 61(6): 1005-1016.

Brown, K.N. and Brown, I.N., 1965. Immunity to malaria antigenic variation in chronic infections in P.knowlesi. Nature. 208: 1286-1288.

Buffet, P.A., Gamain, B., Scheidig, C., Baruch, D., Smith, J.D., Hernandez-Rivas, R., Pouvelle, B., Oishi, S., Fujii, N., Fusai, T., Parzy, D., Miller, L.H., Gysin, J. and Scherf, A. 1999. *Plasmodium falciparum* domain mediating adhesion to chondroitin sulfate A: A receptor for human placental infection. Proc. Natl. Acad. Sci. USA. 96(22): 12743-12748.

Bull, P.C., Lowe, B.S., Kortok, M., Molyneux, C.S., Newbold, C.I. and Marsh, K. 1998. Parasite antigens on the infected red cell surface are targets for naturally acquired immunity to malaria. Nat Med. 4(3): 358-360.

Bull, P.C. and Marsh, K., 2002. The role of antibodies to *Plasmodium falciparum-*infected-erythrocyte surface antigens in naturally acquired immunity to malaria. Trends Microbiol. 10(2): 55-58.

Carlson, J., Helmby, H., Hill, A.V., Brewster, D., Greenwood, B.M. and Wahlgren, M. 1990. Human cerebral malaria: association with erythrocyte rosetting and lack of anti-rosetting antibodies. Lancet. 336(8729): 1457-1460.

Carlson, J. and Wahlgren, M., 1992. *Plasmodium falciparum* erythrocyte rosetting is mediated by promiscuous lectin-like interactions. J. Exp. Med. 176(5): 1311-1317.

Carlton, J.M. and Carucci, D.J., 2002. Rodent models of malaria in the genomics era. Trends Parasitol. 18(3): 100-102.

Chaiyaroj, S.C., Coppel, R.L., Magowan, C. and Brown, G.V., 1994. A *Plasmodium falciparum* isolate with a chromosome 9 deletion expresses a trypsin-resistant cytoadherence molecule. Mol. Biochem. Parasitol. 67(1): 21-30.

Chen, Q., Barragan, A., Fernandez, V., Sundstrom, A., Schlichtherle, M., Sahlen, A., Carlson, J., Datta, S. and Wahlgren, M. 1998. Identification of *Plasmodium falciparum* erythrocyte membrane protein 1 (PfEMP1) as the rosetting ligand of the malaria parasite *P. falciparum.* J. Exp. Med. 187(1): 15-23.

Chen, Q., Heddini, A., Barragan, A., Fernandez, V., Pearce, S.F. and Wahlgren, M., 2000. The semiconserved head structure of *Plasmodium falciparum* erythrocyte membrane protein 1 mediates binding to multiple independent host receptors. J. Exp. Med. 192(1): 1-10.

Cheng, Q., Cloonan, N., Fischer, K., Thompson, J., Waine, G., Lanzer, M. and Saul, A. 1998. *stevor* and *rif* are *Plasmodium falciparum* multicopy gene families which potentially encode variant antigens. Mol. Biochem. Parasitol. 97(1-2): 161-176.

Cooke, B.M., Berendt, A.R., Craig, A.G., MacGregor, J., Newbold, C.I. and Nash, G.B. 1994. Rolling and stationary cytoadhesion of red blood cells parasitized by *Plasmodium falciparum*: separate roles for ICAM-1, CD36 and thrombospondin. Br. J. Haematol. 87(1): 162-170.

Cooke, B.M., Mohandas, N. and Coppel, R.L., 2001. The malaria-infected red blood cell: structural and functional changes. Adv. Parasitol. 50: 1-86.

Crabb, B.S., Cooke, B.M., Reeder, J.C., Waller, R.F., Caruana, S.R., Davern, K.M., Wickham, M.E., Brown, G.V., Coppel, R.L. and Cowman, A.F. 1997. Targeted gene disruption shows that knobs enable malaria-infected red cells to cytoadhere under physiological shear stress. Cell. 89(2): 287-296.

Craig, A. and Scherf, A. 2001. Molecules on the surface of the *Plasmodium falciparum* infected erythrocyte and their role in malaria pathogenesis and immune evasion. Mol. Biochem. Parasitol. 115(2): 129-143.

Crandall, I. and Sherman, I.W., 1991. *Plasmodium falciparum* (human malaria)-induced modifications in human erythrocyte band 3 protein. Parasitology. 102 Pt 3: 335-340.

Day, K.P., Karamalis, F., Thompson, J., Barnes, D.A., Peterson, C., Brown, H., Brown, G.V. and Kemp, D.J. 1993. Genes necessary for expression of a virulence determinant and for transmission of *Plasmodium falciparum* are located on a 0.3-megabase region of chromosome 9. Proc. Natl. Acad. Sci. USA. 90(17): 8292-8296.

del Portillo, H.A., Fernandez-Becerra, C., Bowman, S., Oliver, K., Preuss, M., Sanchez, C.P., Schneider, N.K., Villalobos, J.M., Rajandream, M.A., Harris, D., Pereira da Silva, L.H., Barrell, B. and Lanzer, M. 2001. A superfamily of variant genes encoded in the subtelomeric region of *Plasmodium vivax*. Nature. 410(6830): 839-842.

Desai, S.A., Bezrukov, S.M. and Zimmerberg, J., 2000. A voltage-dependent channel involved in nutrient uptake by red blood cells infected with the malaria parasite. Nature. 406(6799): 1001-1005.

Duffy, P.E., Craig, A.G. and Baruch, D.I., 2001. Variant proteins on the surface of malaria-infected erythrocytes--developing vaccines. Trends Parasitol. 17(8): 354-356.

Eda, S., Lawler, J. and Sherman, I.W., 1999. *Plasmodium falciparum*-infected erythrocyte adhesion to the type 3 repeat domain of thrombospondin-1 is mediated by a modified band 3 protein. Mol. Biochem. Parasitol. 100(2): 195-205.

Fernandez, V., Hommel, M., Chen, Q., Hagblom, P. and Wahlgren, M., 1999. small, clonally variant antigens expressed on the surface of the *Plasmodium falciparum*-infected erythrocyte are encoded by the *rif* gene family and are the target of human immune responses. J. Exp. Med. 190(10): 1393-1404.

Flick, K., Scholander, C., Chen, Q., Fernandez, V., Pouvelle, B., Gysin, J. and Wahlgren, M. 2001. Role of nonimmune IgG bound to PfEMP1 in placental malaria. Science. 293(5537): 2098-2100.

Fowler, E.V., Peters, J.M., Gatton, M.L., Chen, N. and Cheng, Q., 2002. Genetic diversity of the DBLalpha region in *Plasmodium falciparum var* genes among Asia-Pacific isolates. Mol Biochem. Parasitol. 120(1): 117-126.

Freitas-Junior, L.H., Bottius, E., Pirrit, L.A., Deitsch, K.W., Scheidig, C., Guinet, F., Nehrbass, U., Wellems, T.E. and Scherf, A. 2000. Frequent ectopic recombination of virulence factor genes in telomeric chromosome clusters of *P.falciparum*. Nature. 407: 1018-1022.

Fried, M. and Duffy, P.E., 1996. Adherence of *Plasmodium falciparum* to chondroitin sulfate A in the human placenta. Science. 272(5267): 1502-1504.

Fried, M., Nosten, F., Brockman, A., Brabin, B.J. and Duffy, P.E., 1998. Maternal antibodies block malaria. Nature. 395(6705): 851-852.

Gamain, B., Miller, L.H. and Baruch, D.I., 2001. The surface variant antigens of *Plasmodium falciparum* contain cross-reactive epitopes. Proc. Natl. Acad. Sci. USA. 98(5): 2664-2669.

Gardiner, D.L., Holt, D.C., Thomas, E.A., Kemp, D.J. and Trenholme, K.R., 2000. Inhibition of *Plasmodium falciparum* clag9 gene function by antisense RNA. Mol. Biochem. Parasitol. 110(1): 33-41.

Gardner, J.P., Pinches, R.A., Roberts, D.J. and Newbold, C.I., 1996. Variant antigens and endothelial receptor adhesion in *Plasmodium falciparum*. Proc. Natl. Acad. Sci. USA. 93(8): 3503-3508.

Gardner, M.J., Tettelin, H., Carucci, D.J., Cummings, L.M., Aravind, L., Koonin, E.V., Shallom, S., Mason, T., Yu, K., Fujii, C., Pederson, J., Shen, K., Jing, J., Aston, C., Lai, Z., Schwartz, D.C., Pertea, M., Salzberg, S., Zhou, L., Sutton, G.G., Clayton, R., White, O., Smith, H.O., Fraser, C.M., Hoffman, S.L. *et al.* 1998. Chromosome 2 sequence of the human malaria parasite *Plasmodium falciparum*. Science. 282(5391): 1126-1132.

Gysin, J., Pouvelle, B., Le Tonqueze, M., Edelman, L. and Boffa, M.C., 1997. Chondroitin sulfate of thrombomodulin is an adhesion receptor for *Plasmodium falciparum*-infected erythrocytes. Mol. Biochem. Parasitol. 88(1-2): 267-271.

Handunnetti, S.M., van Schravendijk, M.R., Hasler, T., Barnwell, J.W., Greenwalt, D.E. and Howard, R.J. 1992. Involvement of CD36 on erythrocytes as a rosetting receptor for *Plasmodium falciparum*-infected erythrocytes. Blood. 80(8): 2097-2104.

Heddini, A., Pettersson, F., Kai, O., Shafi, J., Obiero, J., Chen, Q., Barragan, A., Wahlgren, M. and Marsh, K. 2001. Fresh isolates from children with severe *Plasmodium falciparum* malaria bind to multiple receptors. Infect. Immun. 69(9): 5849-5856.

Helmby, H., Cavelier, L., Pettersson, U. and Wahlgren, M., 1993. Rosetting *Plasmodium falciparum*-infected erythrocytes express unique strain-specific antigens on their surface. Infect Immun. 61(1): 284-288.

Holt, D.C., Bourke, P.F., Mayo, M. and Kemp, D.J., 1998. A high resolution map of chromosome 9 of *Plasmodium falciparum*. Mol. Biochem. Parasitol. 97(1-2): 229-233.

Holt, D.C., Gardiner, D.L., Thomas, E.A., Mayo, M., Bourke, P.F., Sutherland, C.J., Carter, R., Myers, G., Kemp, D.J. and Trenholme, K.R. 1999. The cytoadherence linked asexual gene family of *Plasmodium falciparum*: are there roles other than cytoadherence? Int. J. Parasitol. 29(6): 939-944.

Hommel, M., David, P.H. and Oligino, L.D., 1983. Surface alterations of erythrocytes in *Plasmodium falciparum* malaria. Antigenic variation, antigenic diversity, and the role of the spleen. J. Exp. Med. 157(4): 1137-1148.

Howard, R.J., 1988. Malarial proteins at the membrane of *Plasmodium falciparum*-infected erythrocytes and their involvement in cytoadherence to endothelial cells. Prog. Allergy. 41: 98-147.

Howard, R.J., Barnwell, J.W. and Kao, V., 1983. Antigenic variation of Plasmodium knowlesi malaria: identification of the variant antigen on infected erythrocytes. Proc. Natl. Acad. Sci. USA. 80(13): 4129-4133.

Janssen, C.S., Barrett, M.P., Turner, C.M. and Phillips, R.S., 2002. A large gene family for putative variant antigens shared by human and rodent malaria parasites. Proc. R. Soc. Lond. B. Biol. Sci. 269(1489): 431-436.

Kaneko, O., Tsuboi, T., Ling, I.T., Howell, S., Shirano, M., Tachibana, M., Cao, Y.M., Holder, A.A. and Torii, M. 2001. The high molecular mass rhoptry protein, RhopH1, is encoded by members of the *clag* multigene family in *Plasmodium falciparum* and *Plasmodium yoelii*. Mol. Biochem. Parasitol. 118(2): 223-231.

Kilejian, A., Rashid, M.A., Aikawa, M., Aji, T. and Yang, Y.F., 1991. Selective association of a fragment of the knob protein with spectrin, actin and the red cell membrane. Mol. Biochem. Parasitol. 44(2): 175-181.

Kyes, S., Horrocks, P. and Newbold, C., 2001. Antigenic variation at the infected red cell surface in malaria. Annu. Rev. Microbiol. 55: 673-707.

Kyes, S., Pinches, R. and Newbold, C., 2000. A simple RNA analysis method shows *var* and *rif* multigene family expression patterns in *Plasmodium falciparum*. Mol. Biochem. Parasitol. 105(2): 311-315.

Kyes, S., Taylor, H., Craig, A., Marsh, K. and Newbold, C., 1997. Genomic representation of var gene sequences in *Plasmodium falciparum* field isolates from different geographic regions. Mol. Biochem. Parasitol. 87(2): 235-238.

Kyes, S.A., Rowe, J.A., Kriek, N. and Newbold, C.I., 1999. Rifins: a second family of clonally variant proteins expressed on the surface of red cells infected with *Plasmodium falciparum*. Proc. Natl. Acad. Sci. USA. 96(16): 9333-9338.

Leech, J.H., Barnwell, J.W., Miller, L.H. and Howard, R.J., 1984. Identification of a strain-specific malarial antigen exposed on the surface of *Plasmodium falciparum*-infected erythrocytes. J. Exp. Med. 159(6): 1567-1575.

Limpaiboon, T., Shirley, M.W., Kemp, D.J. and Saul, A., 1991. 7H8/6, a multicopy DNA probe for distinguishing isolates of *Plasmodium falciparum*. Mol. Biochem. Parasitol. 47(2): 197-206.

MacPherson, G.G., Warrell, M.J., White, N.J., Looareesuwan, S. and Warrell, D.A., 1985. Human cerebral malaria. A quantitative ultrastructural analysis of parasitized erythrocyte sequestration. Am. J. Pathol. 119(3): 385-401.

McCormick, C.J., Craig, A., Roberts, D., Newbold, C.I. and Berendt, A.R., 1997. Intercellular adhesion molecule-1 and CD36 synergize to mediate adherence of *Plasmodium falciparum*-infected erythrocytes to cultured human microvascular endothelial cells. J. Clin. Invest. 100(10): 2521-2529.

Moulds, J.M., Zimmerman, P.A., Doumbo, O.K., Kassambara, L., Sagara, I., Diallo, D.A., Atkinson, J.P., Krych-Goldberg, M., Hauhart, R.E., Hourcade, D.E., McNamara, D.T., Birmingham, D.J., Rowe, J.A., Moulds, J.J. and Miller, L.H. 2001. Molecular identification of Knops blood group polymorphisms found in long homologous region D of complement receptor 1. Blood. 97(9): 2879-2885.

Newbold, C., Warn, P., Black, G., Berendt, A., Craig, A., Snow, B., Msobo, M., Peshu, N. and Marsh, K.1997. Receptor-specific adhesion and clinical disease in *Plasmodium falciparum*. Am. J. Trop. Med. Hyg. 57(4): 389-398.

Ockenhouse, C.F., Klotz, F.W. Tandon, N.N. and Jamieson, G.A., 1991. Sequestrin, a CD36 recognition protein on *Plasmodium falciparum* malaria- infected erythrocytes identified by anti-idiotype antibodies. Proc. Natl. Acad. Sci. USA. 88(8): 3175-3179.

Ockenhouse, C.F., Magowan, C. and Chulay, J.D., 1989a. Activation of monocytes and platelets by monoclonal antibodies or malaria-infected erythrocytes binding to the CD36 surface receptor in vitro. J. Clin. Invest. 84(2): 468-475.

Ockenhouse, C.F., Tandon, N.N., Magowan, C., Jamieson, G.A. and Chulay, J.D., 1989b. Identification of a platelet membrane glycoprotein as a falciparum malaria sequestration receptor. Science. 243(4897): 1469-1471.

Ockenhouse, C.F., Tegoshi, T., Maeno, Y., Benjamin, C., Ho, M., Kan, K.E., Thway, Y., Win, K., Aikawa, M. and Lobb, R.R. 1992. Human vascular endothelial cell adhesion receptors for *Plasmodium falciparum*-infected erythrocytes: roles for endothelial leukocyte adhesion molecule 1 and vascular cell adhesion molecule 1. J. Exp. Med. 176(4): 1183-1189.

Oquendo, P., Hundt, E., Lawler, J. and Seed, B., 1989. CD36 directly mediates cytoadherence of *Plasmodium falciparum* parasitized erythrocytes. Cell. 58(1): 95-101.

Pain, A., Ferguson, D.J., Kai, O., Urban, B.C., Lowe, B., Marsh, K. and Roberts, D.J. 2001. Platelet-mediated clumping of *Plasmodium falciparum*-infected erythrocytes is a common adhesive phenotype and is associated with severe malaria. Proc. Natl. Acad. Sci. USA. 98(4): 1805-1810.

Pologe, L.G. and Ravetch, J.V., 1986. A chromosomal rearrangement in a *P. falciparum* histidine-rich protein gene is associated with the knobless phenotype. Nature. 322(6078): 474-477.

Pouvelle, B., Buffet, P.A., Lepolard, C., Scherf, A. and Gysin, J., 2000. Cytoadhesion of *Plasmodium falciparum* ring-stage-infected erythrocytes. Nature Med. 6(11): 1264-1268.

Reeder, J.C., Cowman, A.F., Davern, K.M., Beeson, J.G., Thompson, J.K., Rogerson, S.J. and Brown, G.V. 1999. The adhesion of *Plasmodium falciparum*-infected erythrocytes to chondroitin sulfate A is mediated by *P. falciparum* erythrocyte membrane protein 1. Proc. Natl. Acad. Sci. USA. 96(9): 5198-5202.

Roberts, D.D., Sherwood, J.A., Spitalnik, S.L., Panton, L.J., Howard, R.J., Dixit, V.M., Frazier, W.A., Miller, L.H. and Ginsburg, V.1985. Thrombospondin binds falciparum malaria parasitized erythrocytes and may mediate cytoadherence. Nature. 318(6041): 64-66.

Roberts, D.J., Craig, A.G., Berendt, A.R., Pinches, R., Nash, G., Marsh, K. and Newbold, C.I. 1992. Rapid switching to multiple antigenic and adhesive phenotypes in malaria. Nature. 357(6380): 689-692.

Roberts, D.J., Pain, A., Kai, O., Kortok, M. and Marsh, K., 2000. Autoagglutination of malaria-infected red blood cells and malaria severity. Lancet. 355(9213): 1427-1428.

Rogerson, S.J., Chaiyaroj, S.C., Ng, K., Reeder, J.C. and Brown, G.V., 1995. Chondroitin sulfate A is a cell surface receptor for *Plasmodium falciparum*-infected erythrocytes. J. Exp. Med. 182(1): 15-20.

Rogerson, S.J., Reeder, J.C., al-Yaman, F. and Brown, G.V., 1994. Sulfated glycoconjugates as disrupters of *Plasmodium falciparum* erythrocyte rosettes. Am. J. Trop. Med. Hyg. 51(2): 198-203.

Rowe, A., Obeiro, J., Newbold, C.I. and Marsh, K., 1995. *Plasmodium falciparum* rosetting is associated with malaria severity in Kenya. Infect. Immun. 63(6): 2323-2326.

Rowe, J.A., Moulds, J.M., Newbold, C.I. and Miller, L.H., 1997. *P. falciparum* rosetting mediated by a parasite-variant erythrocyte membrane protein and complement-receptor 1. Nature. 388(6639): 292-295.

Scherf, A., Pouvelle, B., Buffet, P.A. and Gysin, J., 2001. Molecular mechanisms of *Plasmodium falciparum* placental adhesion. Cell Microbiol. 3(3): 125-131.

Scholander, C., Treutiger, C.J., Hultenby, K. and Wahlgren, M., 1996. Novel fibrillar structure confers adhesive property to malaria-infected erythrocytes. Nat. Med. 2(2): 204-208.

Senczuk, A.M., Reeder, J.C., Kosmala, M.M. and Ho, M., 2001. *Plasmodium falciparum* erythrocyte membrane protein 1 functions as a ligand for P-selectin. Blood. 98(10): 3132-3135.

Shirley, M.W., Biggs, B.A., Forsyth, K.P., Brown, H.J., Thompson, J.K., Brown, G.V. and Kemp, D.J. 1990. Chromosome 9 from independent clones and isolates of *Plasmodium falciparum* undergoes subtelomeric deletions with similar breakpoints *in vitro*. Mol. Biochem. Parasitol. 40(1): 137-145.

Siano, J.P., Grady, K.K., Millet, P. and Wick, T.M., 1998. Short report: *Plasmodium falciparum*: cytoadherence to alpha(v)beta3 on human microvascular endothelial cells. Am. J. Trop. Med. Hyg. 59(1): 77-79.

Silamut, K., Phu, N.H., Whitty, C., Turner, G.D., Louwrier, K., Mai, N.T., Simpson, J.A., Hien, T.T. and White, N.J. 1999. A quantitative analysis of the microvascular sequestration of malaria parasites in the human brain. Am. J. Pathol. 155(2): 395-410.

Smith, J.D., Chitnis, C.E., Craig, A.G., Roberts, D.J., Hudson-Taylor, D.E., Peterson, D.S., Pinches, R., Newbold, C.I. and Miller, L.H. 1995. Switches in expression of *Plasmodium falciparum* var genes correlate with changes in antigenic and cytoadherent phenotypes of infected erythrocytes. Cell. 82(1): 101-110.

Smith, J.D., Craig, A.G., Kriek, N., Hudson-Taylor, D., Kyes, S., Fagen, T., Pinches, R., Baruch, D.I., Newbold, C.I. and Miller, L.H. 2000a. Identification of a *Plasmodium falciparum* intercellular adhesion molecule-1 binding domain: a parasite adhesion trait implicated in cerebral malaria. Proc. Natl. Acad. Sci. USA. 97(4): 1766-1771.

Smith, J.D., Gamain, B., Baruch, D.I. and Kyes, S., 2001. Decoding the language of var genes and *Plasmodium falciparum* sequestration. Trends Parasitol. 17(11): 538-545.

Smith, J.D., Subramanian, G., Gamain, B., Baruch, D.I. and Miller, L.H., 2000b. Classification of adhesive domains in the *Plasmodium falciparum* erythrocyte membrane protein 1 family. Mol. Biochem. Parasitol. 110: 293-310.

Staalsoe, T. *et al.*, 2001. Acquisition and decay of antibodies to pregnancy-associated variant antigens on the surface of *Plasmodium falciparum*-infected erythrocytes that protect against placental parasitemia. J. Infect. Dis. 184(5): 618-626.

Su, X.Z., Heatwole, V.M., Wertheimer, S.P., Guinet, F., Herrfeldt, J.A., Peterson, D.S., Ravetch, J.A. and Wellems, T.E. 1995. The large diverse gene family var encodes proteins involved in cytoadherence and antigenic variation of *Plasmodium falciparum*-infected erythrocytes. Cell. 82(1): 89-100.

Trenholme, K.R., Gardiner, D.L., Holt, D.C., Thomas, E.A., Cowman, A.F. and Kemp, D.J. 2000. *clag9*: A cytoadherence gene in *Plasmodium falciparum* essential for binding of parasitized erythrocytes to CD36. Proc. Natl. Acad. Sci. USA. 97(8): 4029-4033.

Treutiger, C.J., Heddini, A., Fernandez, V., Muller, W.A. and Wahlgren, M., 1997. PECAM-1/CD31, an endothelial receptor for binding *Plasmodium falciparum*-infected erythrocytes. Nat. Med. 3(12): 1405-1408.

Treutiger, C.J., Scholander, C., Carlson, J., McAdam, K.P., Raynes, J.G., Falksveden, L. and Wahlgren, M. 1999. Rouleaux-forming serum proteins are involved in the rosetting of *Plasmodium falciparum*-infected erythrocytes. Exp. Parasitol. 93(4): 215-224.

Turner, G.D., Morrison, H., Jones, M., Davis, T.M., Looareesuwan, S., Buley, I.D., Gatter, K.C., Newbold, C.I., Pukritayakamee, S., Nagachinta, B. *et al.* 1994. An immunohistochemical study of the pathology of fatal malaria. Evidence for widespread endothelial activation and a potential role for intercellular adhesion molecule-1 in cerebral sequestration. Am. J. Pathol. 145(5): 1057-1069.

Udeinya, I.J., Graves, P.M., Carter, R., Aikawa, M. and Miller, L.H., 1983. *Plasmodium falciparum*: effect of time in continuous culture on binding to human endothelial cells and amelanotic melanoma cells. Exp. Parasitol. 56(2): 207-214.

Udomsangpetch, R., Reinhardt, P.H., Schollaardt, T., Elliott, J.F., Kubes, P. and Ho, M. 1997. Promiscuity of clinical *Plasmodium falciparum* isolates for multiple adhesion molecules under flow conditions. J. Immunol. 158(9): 4358-4364.

Udomsangpetch, R., Wahlin, B., Carlson, J., Berzins, K., Torii, M., Aikawa, M., Perlmann, P. and Wahlgren, M. 1989. *Plasmodium falciparum*-infected erythrocytes form spontaneous erythrocyte rosettes. J. Exp. Med. 169(5): 1835-1840.

Urban, B.C., Ferguson, D.J., Pain, A., Willcox, N., Plebanski, M., Austyn, J.M. and Roberts, D.J. 1999. *Plasmodium falciparum*-infected erythrocytes modulate the maturation of dendritic cells. Nature. 400(6739): 73-77.

Urban, B.C., Willcox, N. and Roberts, D.J., 2001. A role for CD36 in the regulation of dendritic cell function. Proc. Natl. Acad. Sci. USA. 98(15): 8750-8755.

Valiyaveettil, M., Achur, R.N., Alkhalil, A., Ockenhouse, C.F. and Gowda, D.C., 2001. *Plasmodium falciparum* cytoadherence to human placenta: evaluation of hyaluronic acid and chondroitin 4-sulfate for binding of infected erythrocytes. Exp. Parasitol. 99(2): 57-65.

Waller, K.L., Cooke, B.M., Nunomura, W., Mohandas, N. and Coppel, R.L., 1999. Mapping the binding domains involved in the interaction between the *Plasmodium falciparum* knob-associated histidine-rich protein (KAHRP) and the cytoadherence ligand *P. falciparum* erythrocyte membrane protein 1 (PfEMP1). J. Biol. Chem. 274(34): 23808-23813.

Walter, P.R., Garin, Y. and Blot, P., 1982. Placental pathologic changes in malaria. A histologic and ultrastructural study. Am. J. Pathol. 109(3): 330-342.

Weber, J.L., 1988. Interspersed repetitive DNA from *Plasmodium falciparum*. Mol. Biochem. Parasitol. 29(2-3): 117-124.

Winograd, E. and Sherman, I.W., 1989. Characterization of a modified red cell membrane protein expressed on erythrocytes infected with the human malaria parasite *Plasmodium falciparum*: possible role as a cytoadherent mediating protein. J. Cell. Biol. 108(1): 23-30.

From: Malaria Parasites: Genomes and Molecular Biology
Edited by: A.P. Waters and C.J. Janse

# Chapter 13

## Merozoite Cell Biology

# Agnieszka E. Topolska, Lina Wang, Casilda G. Black and Ross L. Coppel

## Abstract

The merozoite of the asexual stage parasite is adapted for the invasion of the red blood cell. Specific structures of the merozoite such as the filamentous surface coat and the secretory organelles of the rhoptries, micronemes and dense granules participate in this process. The repertoire of proteins in these locations participate in a specific set of receptor-ligand interactions, some still poorly understood, that trigger the secretion of products from specialized organelles and the activation of a molecular motor. There is redundancy in the parasite ligands involved and genomic analysis is revealing a large set of gene products that are potentially involved in invasion.

## 1. Introduction

Within the vertebrate host, only two developmental stages of the parasite are found outside host cells, the sporozoite and the merozoite. Both are highly specialised forms of the parasite, that are present only transiently within the blood stream before entry into a host cell which in the case of the merozoite, is the circulating red blood cell (RBC). The requirement to enter a non-nucleated cell with no endocytic machinery necessitates that the merozoite utilise an active method of entry, while the lack of intermediary metabolism within the red cell to preserve the homeostatic state or replace degraded haemoglobin results in fairly rapid deterioration of the host cell, requiring the parasite to move on to a new host cell. The merozoite is extremely well adapted to red cell invasion and has evolved an invasion apparatus that is highly efficient and capable of invading both normal and modified red cells. Analysis of the *P. falciparum* genome is revealing a degree of gene duplication that provides the basis for an invasion process that is highly redundant.

Merozoites are released from hepatic cells at the end of the pre-erythrocytic cycle and from red cells at the end of the asexual blood stage cycle. It is estimated that hepatic schizonts give rise to approximately 10,000 merozoites per cell whereas blood-stage schizonts typically give rise to 16-24 merozoites. This may simply reflect the more robust nature of the hepatic cell such that the process of mitotic division is allowed to proceed for longer or there may be major differences in control of cell division between the two types of schizonts. The difficulty of obtaining large numbers of hepatic-derived merozoites has made detailed study of this form difficult and the rest of this review will confine itself to discussion of the blood-stage merozoite. The little that can be said is that studies by immunofluorescence of hepatic stages identified a number of typical blood-stage merozoite proteins within the hepatic schizont (Szarfman *et al.*, 1988a). For example, a panel of monoclonal antibodies (Mabs) to variant forms of MSP1 showed that both hepatic and blood-stage merozoites expressed the same epitopes (Szarfman *et al.*, 1988b). Use of methods such as laser capture microdissection or sorting of cells containing fluorescent markers coupled with sensitive proteomics analysis may enable researchers to develop a better understanding of the protein repertoire of this stage.

The recent development of molecular genetic approaches that can modify malaria asexual stages and the availability of the malaria genome will enable new experimental approaches to understanding protein function in the merozoite. Nevertheless there remain substantial technical difficulties such as the imperfect nature of currently available genetic systems, the difficulty of subcellular fractionation and the relative fragility of merozoites, particularly those of *P. falciparum* still provide formidable obstacles to full understanding of merozoite cell biology.

## 2. Merozoite Cell Structure

The merozoite is variously described as oval, lemon or tear-shaped with a flattened prominence at one end (the apex). In this prominence are found a number of membranous, secretory vesicles including the rhoptries, the micronemes and the dense granules. Nearby are the apicoplast, the mitochondrion with the nucleus taking up most of the basal area of the merozoite (Figure 1). Merozoites are quite small and those of *P. falciparum* are among the smallest with dimensions of 1.6 μm in length by 1.0 μm in width. This small size has complicated protein localization studies,

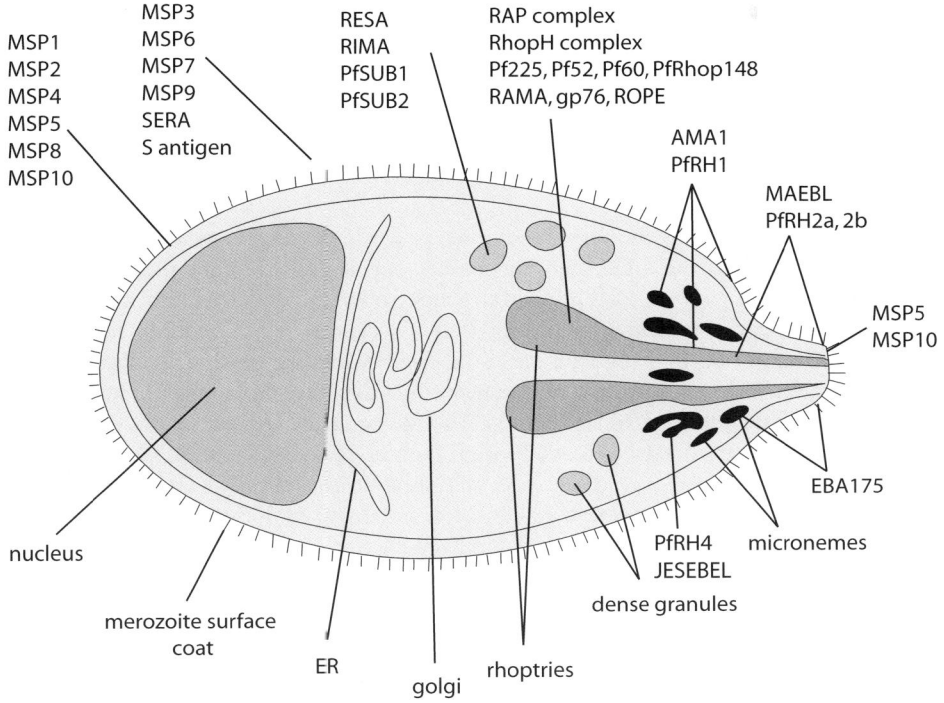

MSP1
MSP2
MSP4
MSP5
MSP8
MSP10

MSP3
MSP6
MSP7
MSP9
SERA
S antigen

RESA
RIMA
PfSUB1
PfSUB2

RAP complex
RhopH complex
Pf225, Pf52, Pf60, PfRhop148
RAMA, gp76, ROPE

AMA1
PfRH1

MAEBL
PfRH2a, 2b

MSP5
MSP10

EBA175

nucleus

merozoite surface
coat

ER

golgi

rhoptries

PfRH4
JESEBEL

dense granules

micronemes

**Figure 1.** Organisation of a *Plasmodium falciparum* merozoite. The filamentous surface coat and the various secretory organelles of the merozoite (the paired rhoptries, micronemes and dense granules) are indicated. The nucleus and components of the secretory pathway (ER and golgi) are also shown. The experimentally determined locations of various proteins discussed in the text are indicated.

particularly those involving co-localization or assignment of proteins to organellar locations, with many examples of proteins being re-assigned to new locations as further data becomes available. Technical limitations also bedevil current studies so that it still remains difficult to successfully label and visualize merozoite surface proteins by immuno-electron microscopy.

Our current state of knowledge of merozoite morphology and ultrastructure has been summarized in a number of reviews including a particularly elegant report by Bannister and co-workers who have used serial electron microscopic sections to reconstruct three-dimensional views of blood stage parasites (Aikawa *et al.*, 1980; Bannister *et al.*, 2000a).

Although the merozoite is designed for cell invasion, the rapidity of the process is such that the internal structural changes that accompany cell entry are difficult to visualize. Thus biochemical evidence strongly implicates the action of a molecular motor during this process, yet the cytoskeleton of the merozoite is surprisingly sparse. At the apex of the merozoite there are three dense polar rings that anchor a set of two or three microtubules that run longitudinally towards the base of the parasite. Drugs that depolymerise microtubules arrest invasion supporting an important role for these structures (Fowler *et al.*, 1998). Other aspects of a putative actin-myosin motor have yet to be visualized although both proteins have been localized to the merozoite periphery (Pinder *et al.*, 1998; Tardieux *et al.*, 1998a).

Surrounding the free merozoite is a filamentous coat that differs in thickness among different Plasmodial species. In *P. falciparum*, it is a somewhat gauzy, poorly defined structure, whereas in *P. knowlesi* for example it is thicker and better demarcated (Aikawa *et al.*, 1980). The filaments are of the order of 2-3 nm and are anchored directly into the plasma membrane. As the most external portion of the merozoite, they are the point of first contact with RBCs that are to be invaded. The filaments are lost during the invasion process presumably as a result of proteolytic cleavage. The substantiality of the coat seems to correlate with the duration of viability of the merozoite as measured by retention of the capacity to invade. Thus the viability of *P. falciparum* merozoites is generally reckoned to be a handful of minutes, whereas *P. knowlesi* can retain the capacity to invade red cells after up to 2 hours of extracellular incubation (Dvorak *et al.*, 1975). Species differences in the coat are presumably related to differences in the identity of constituent proteins, and differences in sequence and abundance of common proteins. Without full genome information for several malaria species, the extent of the differences is still unclear, but at least one surface protein, merozoite surface protein 2 (MSP2) is found only in *P. falciparum* and *P. reichenowi* and not in other species such as *P. vivax*, *P. knowlesi*, *P. chabaudi*, *P. berghei* or *P. yoelii* (Dubbeld *et al.*, 1998; Black *et al.*, 1999; Kedzierski *et al.*, 2000a; Black *et al.*, 2002).

At the apex of the merozoite there are three types of secretory organelles: rhoptries (2 per cell), micronemes (few) and dense granules (numerous). Rhoptries and micronemes are part of the apical complex and are connected via a common duct that extends to the apical membrane, whereas dense granules are round vesicles located more deeply within the merozoite cytoplasm (Blackman *et al.*, 2001). Secretory organelles discharge their contents during the merozoites' invasion of RBCs in a regulated and orderly fashion, with micronemes releasing their material first, followed by rhoptries and (after parasite internalisation) dense granules. The material released from secretory organelles is involved in the recognition, attachment and entry into the host cells and in the formation and maintenance of the parasitophorous vacuole (PV) (vide infra). With each intraerythrocytic cycle secretory organelles have to be formed *de novo*. The elucidation of the steps of organelle biogenesis of malaria parasites has been hindered by the lack of early organelle markers. Most of our knowledge is based on microscopic examinations, which have concentrated on rhoptry development. However, it is widely believed that the formation of all three types of merozoite secretory organelles follows the same general pattern, and that micronemes and dense granules are formed shortly after rhoptries (Ward *et al.*, 1997).

Rhoptries, the largest among the merozoite secretory organelles, are club-shaped with an easily distinguishable bulbous body and a narrow neck, which extends into the common duct leading to the merozoite membrane. Rhoptry contents do not appear to be uniformly distributed within the organelle, and dense, granular as well as lamellar sections can be seen. Rhoptry organelles have been partially purified from *P. falciparum*, as well as from the rodent malaria species: *P. berghei*, *P. yoelii* and *P. chabaudi* and shown to contain a large number of proteins and peptides with sizes ranging from less than 18kDa to more than 200kDa (Etzion *et al.*, 1991). Some cross-reactive epitopes have been identified, indicating that rhoptries of different malaria species contain related proteins (Sam-Yellowe *et al.*, 1998). The lamellar sections of rhoptries in particular are thought to contain lipid material and some experimental observations support this notion. Fluorescent lipid probes localized to rhoptries and were discharged during RBC invasion and incorporated into the PV (Mikkelsen *et al.*, 1988). Further, a known rhoptry protein was observed to be associated with lipid structures, an association that could be abolished by phospholipase treatment (Etzion *et al.*, 1991). Finally, a number of studies have shown the presence of membrane-like whorls located in rhoptries or being discharged from the organelles during RBC invasion (Bannister *et al.*, 1977; Bannister *et al.*, 1986a; Stewart *et al.*, 1986; Bannister *et al.*, 1989; Dluzewski *et al.*, 1989).

Micronemes are much smaller than rhoptries, being fusiform in shape. Several micronemes are present within each merozoite, and they are grouped in a "crown" at the apical tip. Micronemes are in contact with rhoptries and the apical membrane via the common duct. Under microscopic examination micronemes are 120 nm in length with dense granular contents. Micronemes contain a number of proteins that have been implicated in the process of merozoite invasion of RBCs. Discharge of microneme contents precedes that of rhoptries, and micronemal proteins play a role in both recognition and attachment of the merozoite during the invasion process (Langreth *et al.*, 1978; Bannister *et al.*, 1989; Wilson, 1990).

Dense granules are numerous, small, spherical organelles located within the cytoplasm of the merozoite. Under microscopic examination they are 80 nm in diameter, have a dense granular interior and an electron-lucent periphery. Dense granules are the last set of secretory organelles of malaria parasites to release their contents during the invasion process. In fact, exocytosis from dense granules occurs when the parasite has already entered the RBC, after internalisation and the closure of the PV (Bannister *et al.*, 1975a; Bannister *et al.*, 1989). The few dense granule proteins that have been identified and studied seem to play a role in the establishment of the PV within the host cell. It is unclear whether micronemes or dense granules are homogeneous populations or whether there may be subpopulations. The availability of a range of specific organellar markers will allow co-localization studies that will elucidate the situation.

## 3. Biogenesis of Merozoite Structures

The merozoite surface is a combination of integral membrane proteins inserted into the lipid bilayer of the cell membrane and a number of peripheral membrane proteins that are presumably stabilized at the surface by protein-protein interactions. These interactions have not yet been mapped in detail, neither is the identity of the proteins responsible for the filaments determined. Thus for example, there is evidence that merozoite surface protein 3 (MSP3) and acidic basic repeat antigen (ABRA) associate or at least are co-transported (McColl *et al.*, 1994; Mills *et al.*, 2002) and MSP3 has been shown to have domains that are strongly helical and may have the capacity to form filaments (Mulhern *et al.*, 1995). Merozoite surface protein 1 (MSP1) may also be a component as it appears to be the most abundant of the surface proteins and most of the molecule is lost during invasion, a fate that mirrors that of the filaments. Transport of proteins to the surface seems to occur by conventional means with an endoplasmic reticulum (ER) and a Golgi apparatus being identified in the parasite, and the various surface proteins containing conventional signal sequences and in the case of integral membrane proteins, classical glycosylphosphatidyl-inositol (GPI)-attachment or membrane spanning sequences (Van Wye *et al.*, 1996; Noe *et al.*, 2000). One interesting and as yet unexplained observation is that all integral membrane proteins that are exported to the surface and are resident there at the time of merozoite release are attached by GPI-anchors, whereas proteins that make their way to the surface following release from secretory organelles have conventional membrane spanning sequences and cytoplasmic tails. Examples of the former class of proteins are MSP1, MSP2 and merozoite surface protein 4 (MSP4) and of the latter class, apical membrane antigen 1 (AMA1) and the 175 kDa erythrocyte binding antigen (EBA175) (Peterson *et al.*, 1989; Sim *et al.*, 1990; Coppel *et al.*, 1994; Marshall *et al.*, 1997b). This may reflect the requirement that surface proteins be capable of being mobile within the merozoite membrane, to allow concentration into microdomains, a fate common among GPI-anchored proteins (Brown *et al.*, 1998). Thus proteins with weak affinity for the red cell surface, may cluster to allow an interaction between merozoite and red cell to be of higher avidity.

Secretory organelles are formed in a stage-specific manner, during schizont stages. The onset of schizogony coincides with re-organisation of the secretory pathway. In immature stages of the parasite, the ER forms a tubular reticular network, while the Golgi apparatus is dispersed and unstacked and no cisterna can be observed. The "classical" ER and Golgi are re-assembled in schizont stages, probably to support protein trafficking to nascent organelles and to the merozoite surface (Van Wye *et al.*, 1996; Noe *et al.*, 2000). It is now widely accepted that secretory organelle formation follows the secretory pathway route.

*Plasmodium* rhoptries are first observed in the vicinity of the post-Golgi cluster of coated vesicles. These early rhoptries are small, round vesicles of about 150 nm in diameter, containing loosely packed granular material. Smaller vesicles are constantly docking and fusing with the pre-rhoptries, and the latter grow in diameter and become more densely packed (Bannister *et al.*, 2000b). Jaikaria and colleagues showed that the addition of new material to rhoptries is sequential and coincides with the timed onset of expression of different rhoptry proteins (Jaikaria *et al.*, 1993). When the young rhoptry grows to approximately 350nm in diameter, an electron-lucent area appears at one end, a conical projection is formed on the surface of this region and extends progressively, with differential fusion of Golgi-derived coated vesicles, to form the rhoptry neck and duct. The growing rhoptry assumes its mature, pear shape (Bannister *et al.*, 2000b). The exact molecular events that govern the shaping of the rhoptry organelle are not yet known. It is possible that, similarly to zymogen granules in mammalian pancreatic cells, an underlying submembranous matrix may be present, formed by a GPI-anchored membrane protein(s) and proteoglycans (Scheele *et al.*, 1994), that stabilises the shape of the rhoptry body while allowing for the extension of the rhoptry neck. In the Golgi apparatus, clustering of rhoptry-destined GPI-anchored proteins into lipid rafts may aid in the formation of small vesicles that pinch out and traffick to the forming rhoptries (Brown *et al.*, 1992). The presence of lipid rafts in the rhoptry membrane may ensure proper distribution of contents and facilitate differential vesicle docking and shaping of the rhoptry organelle. Plasmodial rhoptry organelles are enclosed by a 7 nm-thick membrane. Numerous intramembranous particles (IMPs) are observed on both faces of the membrane in freeze-fractured and freeze-etched preparations. The role of IMPs in the rhoptry membrane remains to be elucidated, however possible functions include pore formation, and ionic and water regulation. At later stages of organelle maturation, rhoptry contents become progressively more dense, suggesting aggregation of the rhoptry material and removal of water from the organelle (Bannister *et al.*, 2000b). Similar observation have been made for zymogen granules in pancreatic cells (Goncz *et al.*, 1995), and it has been shown that a drop in pH causes an aggregation of zymogen material (Dartsch *et al.*, 1998). Coincidentally, forming and mature rhoptries of *Toxoplasma gondii* have been reported to be acidic (Shaw *et al.*, 1998). The acidification was stronger and more uniform in young organelles, while in more mature rhoptries the pH distribution was more heterogeneous and the more acidic material was seen in distal, expanding parts, suggesting continuous addition and aggregation of fresh material. Examination of rhoptries in *T. gondii* also suggested that cysteine and serine proteases play a role during rhoptry development, since addition of specific protease inhibitors resulted in marked alterations of rhoptry assembly process and of rhoptry structure (Shaw *et al.*, 2002).

Development of micronemes and dense granules follows that of rhoptries and also occurs by vesicular budding and fusion in the post-Golgi compartments (Ward *et al.*, 1997). Ordered organelle formation coinciding with timed protein expression probably ensures correct targeting of proteins to destined organelles, together with some form of default destination in the absence of contrary signals. Thus, the examination of protein targeting in *T. gondii* has shown that dense granules are the default destination of soluble proteins (Karsten *et al.*, 1998). The importance of timing of expression for correct organelle localization has been demonstrated for AMA1 in *P. falciparum* where expression under a heterologous promoter led to aberrant accumulation in the cytoplasm rather than the rhoptry. (Kocken *et al.*, 1998). There is also a requirement for specific targeting motifs, at least in *T. gondii,* where a tyrosine-containing tetra-peptide motif in the cytoplasmic domain of transmembrane rhoptry proteins has been shown to play a role in proper trafficking (Hoppe *et al.*, 2000).

# 4. Invasion of Red Blood Cells

The major function of the merozoite is the invasion of new RBCs (Figure 2). The process of invasion is quite rapid and has been studied by a combination of video-microscopy (Dvorak *et al.*, 1975; Hermentin *et al.*, 1984) and transmission electron microscopy (Mitchell *et al.*, 1988). The studies have predominantly focused on *P. knowlesi*, because of the ease of obtaining relatively pure preparations of viable merozoites, but it seems likely that the invasion process of all species is similar. These studies have led to a general model of invasion being proposed, consisting of phases of recognition and attachment, reorientation and entry. We will describe the invasion process and discuss in detail our molecular knowledge of merozoite proteins that may be involved in this process. Some species such as *P. falciparum*, are able to invade RBCs of all stages, whereas others, such as *P. vivax* preferentially invade reticulocytes.

Merozoites released into the bloodstream following schizont rupture come into contact with RBCs moving through the microcirculation. There may be a preference for nearby RBCs immobilized in rosettes surrounding the rupturing schizont, although the evidence for this is not yet clear-cut. The initial contact, accidental in nature in the flowing blood stream, may occur via any part of the merozoite surface, is mediated by merozoite coat filaments, and is not strong, specific or irreversible. Merozoites are frequently observed to break off contact and abut fresh RBCs. If the invasion process continues the merozoite glides, a process promoted by movement of actin microtubules

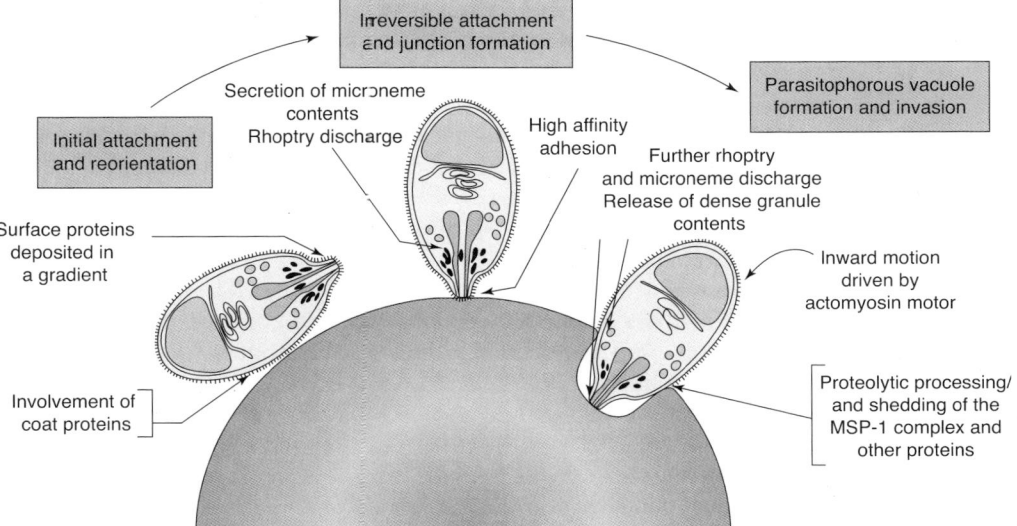

**Figure 2.** Schematic depicting the early stages of red blood cell (RBC) invasion by the malaria merozoite, and the putative roles of the various organelles and some proteins. The initial, low-affinity interaction with the host RBC may involve adhesive ligands stably resident on the merozoite surface such as MSP1 and other components of the associated protein complex. Alternatively, it may involve AMA1, which is secreted onto the parasite surface in a truncated form from a primary location in the neck of the rhoptries, or MAEBL. Reorientation of the bound parasite may be favoured by the presence of higher-avidity ligands clustered around the apical prominence, such as PvRBP1 and 2 in *P. vivax* (or functionally equivalent molecules in *P. falciparum* and *P. yoelii*). Upon reorientation, tight attachment and junction formation may be initiated by discharge of micronemal components such as EBA-175 and other DBL-EBPs. The effects of anterior to posterior trafficking and/or proteolytic shedding of these and other RBC-binding proteins, linked to the action of a sub-pellicular actomyosin motor, may then aid in driving the parasite into the nascent parasitophorous vacuole. *Reprinted in modified form from Parasitology Today, Vol. 16, C. E. Chitnis and M. J. Blackman, Host Cell Invasion by Malaria Parasites, pp.411-415, 2000, with permission from Elsevier Science.*

and myosin microfilaments. to alter its orientation and bring the apical end of the merozoite in contact with the RBC. It is not clear what molecular event triggers the merozoite re-orientation but once it commences, the interaction between merozoite and RBC becomes much stronger until it is functionally irreversible. Actual entry of merozoite into the red cell has only been observed with the merozoite in this apically-attached position. Between the merozoite apex and the RBC surface, in the area of molecular interactions between the merozoite adhesins and the RBC membrane receptors, a so-called junction zone is formed, and a dense undercoating of the RBC membrane can be observed in this region (Bannister *et al.*, 1986b).

Apical attachment and the formation of the junction zone triggers discharge from the rhoptres and micronemes, located at the apical end of merozoite. The exact molecular bases of this trigger remain to be elucidated, however increased cytosolic concentration of calcium ions $Ca^{2+}$, activation of a phospholipid-dependent protein kinase C, activation of phospholipase C (PLC), and subsequent release of inositol triphosphate ($IP_3$) and other secondary messengers have been implicated in *Toxoplasma* (Lovett *et al.*, 2002). Rhoptries discharge their contents, which appear to be of both, protein and lipid nature, directly onto the RBC membrane at the point of merozoite attachment. The discharged material seems to participate in the initiation of structural changes in the RBC membrane and cytoskeleton that lead to membrane invagination and formation of the PV (Wilson, 1990). Protease and phospholipase activities have been reported in the released material (Braun-Breton *et al.*, 1988a; Braun-Breton *et al.*, 1988b), and the degradation of band 3 protein by the gp76 serine protease has been shown (Roggwiller *et al.*, 1996). Phosphorylation of RBC cytoskeleton components and a decrease in the RBC membrane rigidity have also been suggested to facilitate merozoite entry (Rangachari *et al.*, 1986; Wilson, 1990). The detachment of the RBC membrane from the cytoskeleton, and structural changes in the membrane layer allow for the incorporation of some of the discharged rhoptry material, and the growing invagination of the RBC membrane. Merozoites enter the developing vacuole by a gliding motion, using the actin-myosin motor (Dobrowolski *et al.*, 1996; 1997b). The internalisation of the merozoite involves lateral distribution of the junction zone and encapsulation of the parasite within the PV. The final step in the invasion process, once the merozoite is encircled, is the discharge of contents from the dense granules of the merozoite (Bannister *et al.*, 1975b). The parasite then resides in the PV that will accommodate it for the next 48-72 hours while it grows and multiplies.

## 5. The Parasitophorous Vacuole

In passing it is worth discussing the formation of the PV as secreted products of the merozoite play an important role in this process. We still do not have a complete understanding of how the vacuole is formed, however it seems that the PV and the parasitophorous vacuolar membrane (PVM) are derived from both, host cell and the invading parasite, and they consist of lipid and protein components acquired from each. The formation of the PV is initiated by junction formation between the merozoite and the RBC membrane, and discharge of apical organelles. Ultrastrucural examination of apically-attached and invading merozoites showed marked changes in the composition of RBC membrane at and beyond the point of the moving junction. Small satellite vesicles appear within the host cell at the point of parasite entry as a result of rhoptry discharge. These vesicles contain rhoptry proteins later found in the PVM, and are seen to fuse with the parasite-containing vacuole bringing more material to the forming PV (Miller *et al.*, 1979; Aikawa *et al.*, 1981; Bannister *et al.*, 1986a; Dluzewski *et al.*, 1989; Hakansson *et al.*, 2001). Accumulation of intramembranous particles at the point of junction and their exclusion from the invaginating region, reflect the dissociation and exclusion of RBC cytoskeleton components such as ankyrin, spectrin and band 3 (Aikawa *et al.*, 1981; Atkinson *et al.*, 1988; Dluzewski *et al.*, 1989). Discharge from rhoptry organelles and deposition of their contents onto the RBC membrane has been observed (Aikawa *et al.*, 1981; Bannister *et al.*, 1989). Part of the released rhoptry material has a membranous appearance and is often referred to as lamellar whorls. These whorls assemble inside the rhoptry organelles around the time of invasion and are often observed in association with the RBC membrane or in the growing vacuolar space (Bannister *et al.*, 1986a; Bannister *et al.*, 1989), and are presumed to be composed of both protein and lipid material. Indeed, incorporation of lipid labels and lipid staining has confirmed the presence of lipid species in rhoptries of malaria parasites (Stewart *et al.*, 1986; Mikkelsen *et al.*, 1988). Rhoptry-derived lipids contribute to the formation of the PV, since they have been shown to be incroporated into the PVM during invasion, while some RBC-derived lipid species, such phosphatidylethanolamine (PE) appeared to be excluded (Mikkelsen *et al.*, 1988; Dluzewski *et al.*, 1992). Rhoptry proteins have been found to be incorporated into the PVM as well, such as rhoptry high molecular weight complex proteins 1 (RhopH1), 2 (RhopH2) and 3 (RhopH 3) (Narum *et al.*, 1994; Perkins *et al.*, 1994). This again indicates that protein components of PVM are partially derived from the rhoptry-discharged material. Lauer and colleagues have shown that lipid rafts are involved in the formation of the PVM, and that GPI-anchored as well as transmembrane and soluble host cell proteins are recruited into the PVM by virtue of their association with lipid rafts. Furthermore, the raft-associated lipids, cholesterol and sphingomyelin appear to be important in the establishment and maintenance of the PV (Lauer *et al.*, 2000).

In *Plasmodium* parasites, which reside in RBCs, cells that do not contain intracellular organelles, the PVM was proposed to associate with the so-called tubovesicular network (TVN) (Elmendorf *et al.*, 1993). The TVN is a network of tubular and vesicular membrane structures that extends into the RBC cytosol and is proposed to mediate molecular transport in the infected RBC, a mechanism that also seems to be lipid raft-dependent (reviewed in Haldar *et al.*, 2001). The final contributions to the PV are made by dense granules. These organelles release their contents after the internalisation of the parasite, and promote further maturation of the PV. Dense granule proteins are found to be incorporated into the PVM or cross it to interact with host cell components. For example, the *P. falciparum* ring-infected RBC surface antigen (RESA) has been shown to interact with the RBC cytoskeleton (Foley *et al.*, 1991).

**Table 1: Merozoite proteins**

**Merozoite Surface Proteins**

| Protein | Full Name and Alternative Nomenclature | Species[a] | MW (kDa) | Location | Antigenic Polymorphism | Comments | References |
|---|---|---|---|---|---|---|---|
| MSP1 | Merozoite surface protein 1; MSA-1, PMMSA, MSPP, p190, gp185, gp195, PSA | Pf, Pk, Pv, Pcy, Py | 185-220 | Merozoite surface | +++ | Dimorphic forms undergo recombination to generate recombinant variants | (Holder et al., 1984; Holder, 1988) |
| MSP2 | Merozoite surface protein 2; MSA2: QF122; GYMMSA: gp56 | Pf, Pr | 45-52 | Merozoite surface | +++ | GPI anchored protein Extensive repetitive sequences | (Smythe et al., 1988; Smythe et al., 1990; Thomas et al., 1990; Smythe et al., 1991) |
| MSP3 | Merozoite surface protein 3; SPAM | Pf, Pk, Pv | 45-76 | Parasitophorous vacuole and non-covalently attached to merozoite surface | ++ | Extensive heptad based alpha helical structure | (McColl et al., 1994; Oeuvray et al., 1994b) |
| MSP4 | Merozoite surface protein 4 | Pf, Pv, Py, Pb, Pch, Pk | 40 | Merozoite surface | +/- | GPI anchored protein Single EGF-like domain | (Marshall et al., 1997b; Marshall et al., 1998; Wang et al., 1999; Black et al., 2002) |
| MSP5 | Merozoite surface protein 5 | Pf, Pv, Py, Pb, Pch, Pk | 40 | Merozoite surface and apical region | - | Membrane-associated protein Single EGF-like domain | (Marshall et al., 1998; Wu et al., 1999; Black et al., 2002) |
| MSP6 | Merozoite surface protein 6 | Pf | 36 | Non-covalently attached to merozoite surface | - | Exist in MSP1 complex | (Trucco et al., 2001) |
| MSP7 | Merozoite surface protein 7 | Pf, Py | 22 | Non-covalently attached to merozoite surface | + | Exist in MSP1 complex | (Pachebat et al., 2001) |
| MSP8 | Merozoite surface protein 8 | Pf | 98 | Merozoite surface | +/- | Membrane-associated protein Two EGF-like domains Evidence of processing | (Black et al., 2001) |
| MSP9 | Merozoite surface protein 9; ABRA, p110 | Pf, Pk, Pv, Pcy, Py | 110 | Parasitophorous vacuole and non-covalently attached to merozoite surface | - | Extensive repetitive sequences | (Stahl et al., 1986; Vargas-Serrato et al., 2002) |

| | | | | | +/- | Membrane-associated protein Two EGF-like domains Evidence of processing | (Black et al., 2003) |
|---|---|---|---|---|---|---|---|
| MSP10 | Merozoite surface protein 10 | Pf, Pk, Py | 80 | Merozoite surface and apical region | | | |
| S antigen | | Pf | 45-220 | parasitophorous vacuole of schizonts | +++ | Extensive repetitive sequences. | (Coppel et al., 1983) |
| SERA | Serine rich antigen; SERP, p113, p126, Pf 140 | Pf, Pk, Pv, Py, Pch | 113-140 | parasitophorous vacuole of schizonts | +/- | Binds RBC membrane phospholipids | (Bzik et al., 1988; Perkins et al., 1994; Miller et al., 2002) |
| **Proteins in Apical Organelles** | | | | | | | |
| RAP1 | Rhoptry associated protein 1 | Pf, Pr, Pk, Pv | 80/65 | Rhoptries | - | Targets RAP2 to rhoptries | (Ridley et al., 1990; Baldi et al., 2000) |
| RAP2 | Rhoptry associated protein 2 | Pf, Pk, Pv, Py, Pb | 43/40 | Rhoptries | - | Forms a tri-molecular complex with RAP1 and RAP3 | (Saul et al., 1992) |
| RAP3 | Rhoptry associated protein 3 | Pf, Pk, Pv, Py, Pb | 40/37 | Rhoptries | - | | (Baldi et al., 2002) |
| RhopH1 | High molecular weight component 1 | Pf, Py, Pk, Pb | 145-155 | Rhoptries | unknown | Member of the clag multigene family | (Kaneko et al., 2001) |
| RhopH2 | High molecular weight component 2 | Pf, Py, Pk, Pb | 135-140 | Rhoptries | unknown | Forms a tri-molecular complex with RhopH1 and RhopH3 | (Sam-Yellowe et al., 1995) |
| RhopH3 | High molecular weight component 3 | Pf, Py, Pk, Pv | 103-110 | Rhoptry body | - | Binds to RBC surface during invasion | (Brown et al., 1991; Perkins et al., 1994) |
| AMA1 | Apical membrane antigen 1; Pf83 | Pf, Pk, Pv, Pcy, Pr, Pfr, Py, Pb, Pch, Tg | 82/66 | Apical region; micronemes, neck of rhoptry and merozoite surface | + | Transmembrane, translocates and is processed during invasion, associates with the PV | (Peterson et al., 1989; Narum et al., 1994; Howell et al., 2001) |
| MAEBL | | Pf, Py, Pb, Pk, Pv, Pcy | 47/39/30 | Apical region; rhoptries, merozoite surface | unknown | Transmembrane | (Blair et al., 2002) |
| Pf225 | | Pf | 225 | Neck of rhoptry | unknown | | (Roger et al., 1988) |
| Pf52 | | Pf | 52 | Rhoptries | unknown | | (Storey, 1992) |
| Pf60 | | Pf, Bsp | 60 | Rhoptries | +++ | Multigene family | (Carcy et al., 1994; Grellier et al., 1994) |
| RAMA | Rhoptry associated membrane antigen; Ag512 | Pf, Pk, Pv, Py, Pb, Pch | 170/60 | Rhoptries | - | GPI anchored protein | (Smythe et al., 1988) |

*Table 1, continued*

| Protein | Full Name and Alternative Nomenclature | Species[a] | MW (kDa) | Location | Antigenic Polymorphism | Comments | References |
|---|---|---|---|---|---|---|---|
| gp76 | | Pf, Pch | 76 | Rhoptries | unknown | GPI anchored protein, PI-PLC cleaved to activate protease activity, cleaves band 3 and glycophorin A | (Braun-Breton et al., 1988a; Braun-Breton et al., 1988b; Roggwiller et al., 1996) |
| PfRH1 | Rhoptry protein homologue 1; PfNBP1 | Pf | >300/195 | Apical region; micronemes, rhoptries, merozoite surface | unknown | Transmembrane, binds receptor "Y" during invasion | (Rayner et al., 2001; Taylor et al., 2002) |
| PfRH2a | Rhoptry protein homologue 2a; PfRBP2Ha | Pf | >300/>200 | Apical region; rhoptries, mezoroite surface | unknown | Transmembrane | (Rayner et al., 2000; Triglia et al., 2001) |
| PfRH2b | Rhoptry protein homologue 2b; PfRBP2Hb | Pf | >300/>200 | Apical region; rhoptries, mezoroite surface | unknown | Transmembrane, binds receptor "Z" during invasion | (Rayner et al., 2000; Triglia et al., 2001; Duraisingh et al., 2003) |
| PfRH4 | Rhoptry protein homologue 4 | Pf | 220 | Apical region; micronemes and merozoite surface | unknown | Transmembrane | (Kaneko et al., 2002) |
| PfRhop148 | | Pf | 148 | Rhoptries | unknown | Transmembrane, asparagine-rich | (Lobo et al., 2003a) |
| Py235 | | Py | 235 | Rhoptries | +++ | Multigene family; bind RBC membrane | (Borre et al., 1995; Ogun et al., 1996) |
| ROPE | Repetitive organellar protein | Pch, Py, Pb, Pk | 240/225 | Rhoptries | unknown | Spectrin-like structure | (Werner et al., 1998) |
| Py140 | | Py | 140 | Rhoptries | unknown | Co-precipitates with PyAMA1 | (Narum et al., 2000) |
| EBA175 | Erythrocyte binding antigen 175; SABP | Pf | 175 | Apical region; micronemes and merozoite surface | - | Transmembrane, then cleaved for release, binds glycophorin A during invasion | (Camus et al., 1985; Sim et al., 1990; Sim et al., 1994b) |
| BAEBL | BA erythrocyte binding ligand; EBA140 | Pf | 135 | Micronemes | + | Binds glycophorin C, D and other receptors; alternative invasion pathways | (Mayer et al., 2001; Thompson et al., 2001; Mayer et al., 2002) |
| JESEBL | EBA181 | Pf | 190/170 | Micronemes | unknown | Binds receptor "E" during invasion | (Gilberger et al., 2003) |
| DBP | Duffy binding protein; EBP | Pk, Pv, Pcy, Pr, Py, Pb, Pch | 135-140 | Micronemes | unknown | Binds Duffy antigen | (Adams et al., 1990; Adams et al., 1992; Prasad et al., 2003) |

| | | | | | | | |
|---|---|---|---|---|---|---|---|
| PvRBP1 | Reticulocyte binding protein 1 | Pv, Pcy | 325 | Micronemes | unknown | Binds reticulocyte membrane | (Galinski *et al.*, 1992; Galinski *et al.*, 2000) |
| PvRBP2 | Reticulocyte binding protein 2 | Pv, Pcy | 330 | Micronemes | unknown | Binds reticulocyte membrane | (Galinski *et al.*, 1992; Galinski *et al.*, 2000) |
| RESA | Ring-infected erythrocyte surface antigen; Pf155 | Pf, *Pk* | 155 | Dense granules and RBC internal surface in ring-infected RBC | +/- | Binds to RBC spectrin. Domain homologous to chaperonins. Extensive repetitive sequences. | (Favaloro *et al.*, 1986; Foley *et al.*, 1991) |
| RIMA | Ring membrane antigen | Pf | 14 | Dense granules and membrane of new ring | unknown | | (Trager *et al.*, 1992) |
| PfSUB1 | Subtilisin-like protease 1 | Pf, Tg, *Pk, Py* | 82/47 | Dense granules (?) | unknown | Subtilisin-like serine protease | (Blackman *et al.*, 1998) |
| PfSUB 2 | Subtilisin-like protease 2 | Pf, *Pk, Pv, Py, Pb* | 160/65 | Dense granules (?) | unknown | Subtilisin-like serine protease | (Barale *et al.*, 1999; Hackett *et al.*, 1999) |

[a] Listed are species in which the presence of the protein has been reported (normal font) or determined by BLAST searches in PlasmoDB or found in the GenBank (italics).

Abbreviations: Pf – *P. falciparum*, Pk – *P. knowlesi*, Pv – *P. vivax*, Pr – *P. reichenowi*, Pcy – *P. cynomolgi*, Pfr – *P. fragile*, Py – *P. yoelii*, Pb – *P. berghei*, Pch – *P. chabaudi*, Tg – *Toxoplasma gondii*, Bsp – *Babesia spp.*,

# 6. Merozoite Surface Proteins

The initial event in invasion commences once the merozoite surface comes into contact with the RBC surface. This occurs as a result of protein-protein interactions between parasite molecules in the filamentous merozoite coat and molecules on the RBC. These are likely to be protein in nature, but it has not been formally demonstrated that all the important players are proteins (Table 1). A number of *P. falciparum* merozoite surface proteins (MSPs) have been reported, and as discussed these may be either integral membrane or peripheral membrane proteins. MSP1 (Holder *et al.*, 1984), MSP2, MSP4 (Marshall *et al.*, 1997b), MSP5 (Marshall *et al.*, 1998), MSP8 (Black *et al.*, 2001) and MSP10 (Black *et al.*, 2003) are anchored to the plasma membrane by GPI moieties, whereas MSP3 (McColl *et al.*, 1994; Oeuvray *et al.*, 1994b), MSP6 (Trucco *et al.*, 2001), MSP7 (Pachebat *et al.*, 2001) and ABRA/MSP9 (Weber *et al.*, 1988; Vargas-Serrato *et al.*, 2002), serine rich antigen (SERA) (Bzik *et al.*, 1988) and S antigen (Coppel *et al.*, 1983) are soluble proteins that are in part associated with the merozoite surface by protein-protein interactions. The availability of a number of partial genome sequences from malaria species other than *P. falciparum*, offers the opportunity of identifying homologues of these proteins. We present some information on the presence of homologues in Table1, however as these assignments are based on preliminary sequence data, they must in many cases be viewed as provisional until the publication of reports confirming other aspects of the proteins such as shared localisation, stage-specific expression and function.

## 6.1. MSP1

MSP1 is probably the best-characterised and most abundant merozoite surface protein. Apparently found in all species of *Plasmodium,* it has been most extensively studied in *P. falciparum* because of the availability of a relatively simple culture system that facilitates biochemical analysis. We will discuss findings for MSP1 of this species, but the general principles are similar for MSP1 in other species with differences being confined to details of molecular weight, timing of synthesis and details of protelytic processing. MSP1 is synthesised by the intracellular schizont as a high-molecular-mass precursor (185-220 kDa) and undergoes two steps of proteolytic processing during the maturation of merozoite (Holder *et al.*, 1984; Blackman *et al.*, 1990; 1991a; 1991b; 1992; 1993; 1994). Just prior to or at the point of merozoite release from the mature schizont, the MSP1 precursor is cleaved into four fragments with approximate molecular masses of 83, 28-30, 38 and 42 kDa (Holder *et al.*, 1984; Holder, 1988). These fragments, referred to as $MSP1_{83}$, $MSP1_{30}$, $MSP1_{38}$ and $MSP1_{42}$, exist as a non-covalently associated complex on the free merozoite surface. At some point following merozoite release, a second processing event takes place, in which the C-terminal $MSP1_{42}$ is cleaved into two fragments of approximately 33 and 19 kDa (refereed as $MSP1_{33}$ and $MSP1_{19}$) (Blackman *et al.*, 1992). $MSP1_{19}$, which contains two epidermal growth factor (EGF)-like domains, remains attached to the merozoite surface and is carried into the newly invaded RBCs, whereas $MSP1_{33}$ is shed in a soluble form with other components of the merozoite surface complex (Blackman *et al.*, 1990; Blackman *et al.*, 1991a). Proteolytic processing has been described for *P. knowlesi* MSP1 (Blackman *et al.*, 1996) and is believed to occur for MSP1 of *P. vivax* (Longacre *et al.*, 1994) and *P. cynomolgi* (Longacre, 1995).

A recent report describing the *in vitro* reconstitution of the MSP1 complex of *P. falciparum* from its heterologously produced subunits has shown that only one subunit ($MSP1_{30}$) interacts with all other partners. In contrast, $MSP1_{83}$ interacts exclusively with $MSP1_{30}$ whereas $MSP1_{38}$ and $MSP1_{42}$ bind each other and $MSP1_{30}$. Based on these results, the authors proposed the first structural model for the MSP1 complex (Kauth *et al.*, 2003).

MSP1 is encoded by a single-copy gene on chromosome 9 in *P. falciparum*. Comparison of the sequences from a number of *P. falciparum* isolates indicate that the gene can be divided into 17 blocks of conserved and variable regions (Tanabe *et al.*, 1987; Miller *et al.*, 1993). With the exception of the highly polymorphic block 2, the sequences appear to fall essentially into two allelic families (MAD20-type and Wellcome-type) with additional variants being produced by limited intragenic recombination between the families (Tanabe *et al.*, 1987; Miller *et al.*, 1993). Block 17 at the extreme C-terminus, which corresponds to $MSP1_{19}$ and contains the two EGF-like domains, is highly conserved with polymorphism reproducibly observed at only five of the 96 amino acid residues (Miller *et al.*, 1993; Kang *et al.*, 1995; Tolle *et al.*, 1995).

$MSP1_{19}$ has been shown to be a target of protective immunity (reviewed in (Holder, 1996)) and it is a leading vaccine candidate currently under development. Some $MSP1_{19}$-specifc antibodies that inhibit merozoite invasion also inhibit the secondary processing of MSP1 (Blackman *et al.*, 1994) and this is proposed to be the basis of their protective mechanism. However, the binding of these inhibitory antibodies can be blocked by another group of antibodies, which recognise adjacent or overlapping epitopes, but themselves have no effect on either MSP1 processing or merozoite invasion (Guevara Patino *et al.*, 1997). These blocking antibodies have the effect of reducing the protective efficacy of an anti-MSP-$1_{19}$ antibody response and are therefore deleterious to the host. Making use of the available structural information (Morgan *et al.*, 1999), the residues that are crucial to recognition by both inhibitory and blocking antibodies have been identified and can be modified to engineer mutant forms of recombinant MSP119, which are no longer recognised by known blocking antibodies but retain the structures required for recognition by and induction of processing inhibitory antibodies (Uthaipibull *et al.*, 2001).

Despite the immunological evidence of the importance of MSP1 as a target for protective antibodies, its function in merozoite invasion is not fully understood. Given its location and large size, MSP1 could correspond to the long fibrils that tether merozoites to RBCs at distances of 40 nm or greater, or comprise the 20 nm fibrillar bundles (Bannister *et al.*, 1986b). It remains controversial as to whether MSP1 has any capacity to bind structures on the RBC surface, although the balance of reports favour some binding role for MSP1. Burghaus and colleagues expressed MSP1 on the surface of mammalian cells, but failed to observe binding of RBCs (Burghaus *et al.*, 1999). The protein was surface exposed and reactive with at least two Mabs directed to MSP1. In contrast, a set of experiments demonstrated binding of biosynthetically labelled MSP1 to RBCs. Thus, using *in vitro* RBC-binding assays, *P. falciparum* MSP1 was shown to bind to human and primate RBCs (Camus *et al.*, 1985; Perkins *et al.*, 1988b; Su *et al.*, 1993). Similarly, the *P. vivax* MSP1 binds to human reticulocytes, which *P. vivax* merozoites preferentially invade (Rodriguez *et al.*, 2002). The binding capacity of MSP1 was said to be either dependent on terminal sialic acid residues present on RBC membrane glycoproteins, such as glycophorin A (Perkins *et al.*, 1988b) or independent of glycophorin A (Nikodem *et al.*, 2000), suggesting that there were at least two different protein-protein interactions between MSP1 and RBCs. Only the intact, unprocessed form of *P. falciparum* MSP1 was shown to partake in glycophorin-dependent RBC binding (Perkins *et al.*, 1988b), whereas the binding domain in the glycophorin A-independent interaction was mapped to a 115 amino acid region in the processed fragment $MSP_{38}$ (Nikodem *et al.*, 2000). The target of this glycophorin A-independent interaction of MSP1 may be the band 3 protein (Goel *et al.* 2003). Goel and colleagues have demonstrated binding between residues 720-761 and 807-826 of band 3 and $MSP1_{38}$ and $MSP1_{19.}$ The interaction was shown to be chymotrypsin-sensitive and sialic acid-independent and the affinity of interaction varied between 2.1nM and 67 $\mu$M. It is uncertain at present, how accessible these regions of MSP1 are to proteins on the red cell surface. A first structure of MSP1 has been proposed, and in this $MSP1_{38}$ is reasonably accessible, whereas $MSP1_{19}$ appears to be buried (Kauth *et al.*, 2003). This structure does not include the MSP6 and MSP7 polypeptides, which probably mask some portions of MSP1. Thus, band 3 interactions may only occur once the MSP1 complex has been partly perturbed. A plausible, though still speculative, sequence of events is that the initial interaction is between intact MSP1 and glycophorin A, and that this interaction perturbs the MSP1 complex to expose regions that take part in a secondary interaction with band 3.

Attempts to disrupt the $MSP1_{19}$ coding sequence by gene targeting did not result in recovery of viable parasites suggesting this protein is essential (O'Donnell *et al.*, 2001). However, the function of *P. falciparum* $MSP1_{19}$ can be successfully complemented with the corresponding sequence from *P. chabaudi*, suggesting that the role of this region in RBC invasion is conserved across distantly related *Plasmodium* species (O'Donnell *et al.*, 2001). The capacity to tolerate such large-scale changes is somewhat odd given the relatively strong sequence conservation of $MSP1_{19}$ in different isolates (Miller *et al.*, 1993; Kang *et al.*, 1995; Tolle *et al.*, 1995). It may be that in the wild, particular mutations may require simultaneous compensatory mutations, and such an occurrence is quite unlikely.

MSP1 has been reported to take part in at least four other protein-protein interactions. Two of these are with MSP6 and MSP7 to form a multiprotein complex (Stafford *et al.*, 1994; Stafford *et al.*, 1996). The others are with spectrin (Herrera *et al.*, 1993) and S antigen (Perkins *et al.*, 1990). The interaction with S antigen is presumably responsible for anchoring some S antigen at the merozoite surface, although the bulk of this protein is found free in the circulation. The binding site for spectrin has been mapped to a 30-residue, linear sequence in a conserved block of MSP1, but the importance of this interaction is unclear. MSP1 is only likely to come in contact with spectrin during RBC lysis during which merozoites are released from RBCs. The rationale for an interaction that would tend to trap the merozoite within the shell of the exploded RBC is not obvious.

## 6.2. MSP2

Merozoite surface protein 2 (MSP2) of *P. falciparum*, encoded by a single copy gene on chromosome 2, is a 45-52 kDa integral membrane glycoprotein anchored on the merozoite surface by a GPI moiety. The MSP2 protein is comprised of highly conserved N- and C-termini flanking a central variable region. The central variable region consists of centrally located repeats, which are flanked by non-repetitive sequences. There are several hundred variant forms of this protein described which differ predominantly in repeat unit number and sequence, making it a useful strain marker in epidemiological studies. MSP2 sequences are assigned to one of two families, FC27 and IC-1/3D7, on the basis of the non-repetitive sequences (Smythe *et al.*, 1988; Smythe *et al.*, 1990; Thomas *et al.*, 1990; Fenton *et al.*, 1991; Marshall *et al.*, 1991; Smythe *et al.*, 1991; Snewin *et al.*, 1991; Marshall *et al.*, 1992; Felger *et al.*, 1994; Marshall *et al.*, 1994; Felger *et al.*, 1997; Irion *et al.*, 1997).

In *P. falciparum*, the *MSP2* gene is located between the genes encoding the enzyme adenylosuccinate lysase (ASL) and MSP5 on chromosome 2 (Marshall *et al.*, 1997a; 1998). The homologue of *MSP2* is not present at the corresponding position in the malaria species *P. chabaudi*, *P. berghei*, *P. yoelii* (Black *et al.*, 1999; Kedzierski *et al.*, 2000a), *P. vivax* (Black *et al.*, 2002) or *P. knowlesi* (C.G. Black, unpublished). However, the possibility that *MSP2* is situated at a different chromosomal locus in these species cannot be eliminated until the entire genome sequences are available. The only definitive determination of an MSP2 homologue in a malaria parasite species other than *P. falciparum* is the closely related primate malaria species *P. reichenowi* (Dubbeld *et al.*, 1998).

There is some evidence to implicate MSP2 as a target of host protective immune responses and by implication suggest that it has a role in RBC invasion. For example, there is a specific monoclonal antibody to MSP2 that has been found to inhibit parasite growth (Epping *et al.*, 1988). Antibodies to MSP2 are frequently detected in sera from individuals living in areas of endemicity (Thomas *et al.*, 1990; Taylor *et al.*, 1995; Ranford-Cartwright *et al.*, 1996; Weisman *et al.*, 2001) and the presence of IgG3 antibodies to the 3D7 family MSP2 protein was negatively associated with the risk of clinical malaria in the Gambia and Papua New Guinea (al-Yaman *et al.*, 1995; Taylor *et al.*, 1998). The extent of antibody reactivity to MSP2 is sequence dependent, such that antibodies that are inhibitory to parasites expressing a particular form of MSP2 do not inhibit parasites expressing a different form (Saul *et al.*, 1989; Ranford-Cartwright *et al.*, 1996). The results of several studies have indicated that infection induces a type of strain-specific immune response against the MSP2 antigen and biases against reinfection by parasites expressing identical forms of MSP2 (Eisen *et al.*, 1998; Weisman *et al.*, 2001). Immunization with a vaccine formulation that contains the 3D7 form of MSP2 led to a preponderance of breakthrough infections that contained the FC27 form of the parasite (Genton *et al.*, 2002). These results taken together with the enormous strain variation exhibited by this protein suggests that effective immune responses to MSP2 are deleterious to the parasite and by implication the protein plays an important role in *P. falciparum* invasion.

### 6.3. MSP3

MSP3 (Oeuvray *et al.*, 1994b), also referred to as secreted polymorphic antigen associated with merozoites (SPAM) (McColl *et al.*, 1994), is a polymorphic antigen (45-76 kDa) that is synthesised by mature stage parasites and secreted into the PV where it undergoes proteolytic cleavage (McColl *et al.*, 1994). The predicted polypeptide sequence of MSP3 has a typical N-terminal secretion signal but lacks the structural features of an integral membrane protein. There is no other hydrophobic region which could provide a membrane spanning domain nor is there a signal for the attachment of a GPI moity (McColl *et al.*, 1994). However, indirect immunofluorescence studies have suggested the merozoite surface location of this molecule. Therefore, MSP3 is not an integral membrane protein but associated with the merozoite surface, although the mechanism and significance of this association is not clear. The deduced amino acid sequence of MSP3 has three contiguous regions consisting of four heptad repeats, with the hydrophobic amino acid alanine in the first and fourth positions of each heptad (McColl *et al.*, 1994). The solution structure of the heptad repeats has been solved, and each block appears to form an α-helix, which may interact to form a somewhat unstable coiled-coil (Mulhern *et al.*, 1995). Despite some sequence diversity of the MSP3 gene between *P. falciparum* isolates, the alanine residues within the heptad repeat regions and the C-terminal half of the protein, which includes a glutamic acid rich region and a putative leucine zipper motif, are strongly conserved (McColl *et al.*, 1994; 1997; Escalante *et al.*, 1998).

Several MSP3 homologues have been reported in other *Plasmodium* species. In *P. vivax*, three MSP3-like proteins designated as MSP3α, MSP3β and MSP3γ have a similar structure with blocks of heptad repeats predicted to form a coiled coil to MSP3 (Galinski *et al.*, 1999; 2001). However, the homology to *P. falciparum* is not particularly strong and they could also be related to MSP6 or MSP7 for example. The availability of the full genome sequence of *P. vivax* will allow a more complete determination of the relationship between these proteins. A previously identified 140 kDa protein in *P. knowlesi* is also characterised as a member of the MSP3 family (Hudson *et al.*, 1983). Members of the MSP3 family have neither a classical hydrophobic transmembrane domain nor a signal sequence for the attachment of a GPI moity (McColl *et al.*, 1994; Galinski *et al.*, 2001). They could be non-covalently harnessed to the surface of the merozoite by interaction with other membrane-bound MSPs. Since proteins with coiled-coil α-helices have the potential to form homotypic or heterotypic multimeric protein bundles (Lupas, 1996), it is possible that the coiled-coil α-helices of MSP3 become entwined and contribute to the plush appearance of the organised fibrillar bundled clusters that have been observed at the surface of merozoite (Bannister *et al.*, 1986b).

MSP3 was first detected using human hyperimmune serum (McColl *et al.*, 1994) and also with antibodies that inhibit *P. falciparum* growth *in vitro* by co-operation with blood monocytes in an antibody-dependent cellular inhibition assay (Oeuvray *et al.*, 1994a). In an immunocompromised mouse model infected with *P. falciparum*, MSP3-specific human antibodies are able to suppress *P. falciparum* growth in the presence of human monocytes (Badell *et al.*, 2000), and in a primate model vaccination with MSP3 can protect monkeys from a challenge infection (Hisaeda *et al.*, 2002). Truncation of MSP3 by gene targeting to remove the leucine zipper but to retain the glutamic acid-rich region and the heptad repeats, led to interference with normal targeting of MSP3 to the PV (Mills *et al.*, 2002). In addition, ABRA, another secreted protein, also failed to reach its normal location. Transgenic parasites that lacked both these proteins in their normal locations were less efficient at invading RBCs. Thus MSP3 would appear not to be absolutely essential for blood stage growth in an *in vitro* culture system. This finding is somewhat at odds with its suggested importance based on immunological studies. It may be that relatively modest changes in invasion efficiency in an *in vitro* system, translate into significant effects *in vivo*. The leucine zipper region appears to be important for trafficking of both MSP3 and ABRA to the PV.

## 6.4. MSP 4 and MSP5

The *MSP4* and *MSP5* genes are found adjacent to each other on chromosome 2, immediately upstream of *MSP2* (Marshall *et al.*, 1998), and probably arose through a gene duplication event. The integral membrane proteins encoded by MSP4 and MSP5 are each 272 amino acids in length and have an apparent molecular mass of 40 kDa (Marshall *et al.*, 1997b; 1998). Both proteins have an N-terminal signal peptide, a C-terminal GPI anchor sequence, a single EGF-like domain and have been localised to the surface of merozoites (Marshall *et al.*, 1997b; Wu *et al.*, 1999). Human antibodies to the EGF-like domains of MSP4 and MSP5 are highly conformational and lose reactivity if the target proteins are reduced and alkylated ((Wang *et al.*, 1999) and T. Wu, unpublished). Furthermore, the entire protein sequences including the EGF-like domains of MSP4 and MSP5 are conserved among several malaria species examined (Wu *et al.*, 1999; Wang *et al.*, 2002). Antibodies raised to recombinant MSP4 and MSP5 can inhibit parasite growth *in vitro* (T. Wu, unpublished). In other *Plasmodium* species, the tandem head-to-tail arrangement of the *MSP4*, *MSP5* and *MSP2* genes is replaced in the syntenic region either by a single gene (designated MSP4/5) showing homology to both *MSP4* and *MSP5* (Black *et al.*, 1999; Kedzierski *et al.*, 2000a) or by two *MSP4* and *MSP5*-like genes (Black *et al.*, 2002; C.G. Black, unpublished). The existence of a single form in rodent malaria species supports the concept of gene duplication as a mechanism of generating MSP4 and MSP5 in *P. falciparum* and *P. vivax*. Immunisation with recombinant *P. yoelii* MSP4/5 is capable of conferring significant protection against lethal challenge in mice (Kedzierski *et al.*, 2000b). More recently, we have shown that immunisation with a combination of PyMSP4/5 and PyMSP1$_{19}$ enhances protection against lethal challenge when compared to each antigen alone (Kedzierski *et al.*, 2002). To date, it has not been possible to disrupt the *P. falciparum MSP4*, *MSP5* or *MSP2* genes by targeted homologous recombination, suggesting that these genes may play an essential role in the erythrocytic cycle (Cowman *et al.*, 2002). The function of these proteins is unknown, however EGF-like domains are frequently involved in binding to other proteins or to chelating divalent cations (Campbell *et al.*, 1993). In *T. gondii*, a family of EGF-like domain containing proteins act as escorters of other proteins to the micronemes (Meissner *et al.*, 2002a). It may be that MSP4 and MSP5 act as chaperones in this way or they may be involved in stabilization of the merozoite coat.

## 6.5. MSP6 and MSP7

The MSP1 complex shed as a result of secondary processing also contains three polypeptides that are not encoded by the MSP1 gene (Stafford *et al.*, 1994; 1996). The 36 kDa polypeptide is derived from a precursor named MSP6 (Trucco *et al.*, 2001), and the 22 and 19 kDa polypeptides are both from a precursor named MSP7 (Pachebat *et al.*, 2001). Similar to MSP3, both MSP6$_{36}$ and MSP7$_{22}$ are partially associated with the parasite surface and partially released as soluble proteins at the time of merozoite release (Pachebat *et al.*, 2001; Trucco *et al.*, 2001). MSP7$_{22}$ is further cleaved into a 19 kDa fragment in the shed MSP1 complex (Stafford *et al.*, 1996; Pachebat *et al.*, 2001). The sequence of MSP6 has similarity with the C-terminal region of MSP3, which contains a unique sequence motif, ILGWEFGGG- (AV)-P, that may be an important structure related to the function of these molecules (Trucco *et al.*, 2001). There are two clusters of hydrophobic amino acids at the C-terminus of MSP6$_{36}$, which are suggested to form an intramolecular anti-parallel coiled-coil structure and be involved in MSP6$_{36}$ binding to MSP1 or other molecules (Trucco *et al.*, 2001). In contrast to the polymorphism of MSP1, both MSP6 and MSP7 genes are conserved between parasite lines, which may suggest that they are not under immune pressure. However, sera from individuals naturally exposed to malaria contain antibodies reacting with MSP6 and MSP7 (L. Wang *et al.* unpublished data).

### 6.6. MSP8

Examination of the *P. falciparum* genome for additional EGF-like domain-containing molecules resulted in the identification of novel proteins present at the merozoite surface. One such protein was MSP8, which shares sequence features with MSP1, including a signal sequence at the N-terminus, a GPI attachment motif at the C-terminus and two EGF-like domains near the C-terminus (Black *et al.*, 2001). The *MSP8* gene is located on chromosome 5 and has a single exon structure with the potential to encode a polypeptide of 597 residues. The protein was found to be membrane-associated by Triton X-114 fractionation and is localised to the surface of trophozoites, schizonts and free merozoites. MSP8 appears to undergo post-translational processing in a similar way to MSP1. There is very limited sequence diversity in the *MSP8* gene sequences from several *P. falciparum* laboratory isolates. It is not known whether MSP8 is absolutely essential for growth, but it appears to play a role of some importance. Evidence for this comes from the work of Burns and colleagues who identified the *P. yoelii* homologue of MSP8 which they named pypAg-2 (Burns *et al.*, 2000). They demonstrated that immunisation with pypAg-2 induced a degree of protection in mice against lethal *P. yoelii* challenge (Burns *et al.*, 2000). Combined with the findings obtained in *P. falciparum*, this suggests that MSP8/pypAg-2 may play an important role in the process of red cell invasion, perhaps as part of an alternative invasion pathway.

### 6.7. MSP9/ ABRA

Four members of the MSP9 family have been characterised to date from *P. falciparum, P. vivax, P. knowlesi* and *P. cynomolgi* (Vargas-Serrato *et al.*, 2002). The MSP9 in *P. falciparum* was originally named the acid basic repeat antigen

(ABRA) on the basis of repeating dimers of lysine and glutamic acid residues encoded in 3' end of the coding region (Stahl *et al.*, 1986). It is regrettable that already well-characterised proteins such as ABRA are renamed in the literature as this causes confusion. In an attempt at consistency, we will use the MSP9 designation in this review to summarise papers published using the name ABRA. MSP9 is found on the surface of the merozoite, although there is neither a hydrophobic transmembrane domain nor a consensus motif for modification with a GPI anchor in the sequences (Stahl *et al.*, 1986; Weber *et al.*, 1988; Vargas-Serrato *et al.*, 2002). As is suggested for MSP3, MSP9 may be complexed to other MSPs at the merozoite surface (Vargas-Serrato *et al.*, 2002). MSP3 and MSP9 are apparently trafficked to the PV together (Mills *et al.*, 2002), where they may form a complex that also contains other proteins, in a manner similar to the MSP1 complex. A number of genes have recently been identified in the *P. falciparum* genome that have structural similarities to MSP1 and the proteins encoded by them are located on the merozoite surface (Black *et al.*, 2001; 2003). These proteins are candidates to act as binding partners for MSP3 and MSP9 in a complex on the merozoite surface.

MSP9 is found in immune clusters of merozoites formed at the time of mature schizont rupture by the action of immune serum (Lyon *et al.*, 1986b; Chulay *et al.*, 1987). Some synthetic MSP9 peptides are able to elicit specific antibodies that inhibit merozoite invasion of human erythrocytes *in vitro* by up to 90% (Sharma *et al.*, 1998). It has been suggested that MSP9 could be involved in protease-mediated processes prior to or during merozoite invasion, since MSP9 shows chymotrypsin-like enzyme activity localised in its N-terminal region (Nwagwu *et al.*, 1992; Kushwaha *et al.*, 2000). Two independent studies have shown that MSP9 interacts with human erythrocytes in a highly specific manner through its N-terminus (Curtidor *et al.*, 2001; Kushwaha *et al.*, 2002). It binds band 3 protein, a major membrane-spanning protein of RBCs and a possible receptor during invasion (Okoye *et al.*, 1985). Proteolytic cleavage of band 3 as a requirement for RBC invasion has been described (Breton *et al.*, 1992; Roggwiller *et al.*, 1996), however the protease responsible for this has been suggested to be a GPI-anchored serine protease (Braun-Breton *et al.*, 1988b) and not MSP9. The binding domain in MSP9 is fully conserved in different field isolates (Kushwaha *et al.*, 2002), emphasising the potentially significant role of this region in the biology of the parasite.

## 6.8. MSP10

More recently, we have identified another gene encoding a double EGF-like domain containing membrane protein with similarities to MSP1 and MSP8 (Black *et al.*, 2003). The novel protein, designated MSP10, is 524 amino acids in length and is encoded by a single exon on chromosome 6. The protein partitions in the Triton X-114 detergent phase, indicative of a membrane association. Our results suggest that MSP10 may be processed in a similar way to MSP1 and MSP8, but with fewer cleavage events. Unlike MSP1 and MSP8, the MSP10 protein is found at or near the merozoite surface in an apical location. This finding is analogous to MSP5 of *P. falciparum* and *P. vivax* which has a complex pattern of distribution being found as both small dots at the apical end or distributed over the surface of merozoites (Wu *et al.*, 1999; Black *et al.*, 2002).

Recombinant MSP10 fusion proteins are recognised by human immune sera and as found for MSP8, very little sequence variation has been detected in a region encompassing the two EGF-like domains. Co-precipitation of MSP10 with other known merozoite surface proteins (MSP1, MSP4, MSP5 and MSP8) has not been detected, providing no evidence to suggest that these proteins form a complex in *P. falciparum* (L. Wang, unpublished). Again the presence of EGF-like domains in this protein suggest a role in the formation of protein complexes. An MSP10 homologue has been identified in the available *P. yoelii* genomic sequence (C.G. Black, unpublished), offering the possibility of ascertaining whether this novel protein can induce host protective responses in an *in vivo* model.

## 6.9. SERA

The serine-rich antigen (SERA) is a polypeptide of about 110-120 kDa that is synthesized late in schizogony and is found in the PV, apparently bathing the developing schizont/merozoite surface (Delplace *et al.*, 1987). SERA is processed at about the time of schizont rupture to two fragments, one of 73 kDa which is in turn composed of two disulphide bound components of 47 and 18 kDa, and one of 50 kDa. The protein is acidic and characterized by a high number of serine residues, including a stretch of between 12 and 35 serine residues, depending on the isolate, located in the N-terminal half of the protein (Bzik *et al.*, 1988). The gene encodes an N-terminal signal sequence but no hydrophobic stretches compatible with a membrane anchor. A limited number of gene sequences have been reported for this protein and regions of sequence variation have been reported (Coppel *et al.*, 1988). These may account for the molecular weight differences noted above but it is currently not known how extensive the sequence diversity may be. Based on its gene structure, which shares similarities with the catalytic sites of papain-like proteases, SERA is suggested to be a protease. Since both the processing of SERA and the rupture of mature schizont-infected erythrocytes are inhibited by leupeptin, it has been suggested that SERA might be involved in merozoite release (Delplace *et al.*, 1988). However, there is no evidence that SERA homologues exhibit any protease activity. Perkins and Ziefer have shown that SERA is able to interact with RBC membrane components and that it binds an inner leaflet phospholipid, phosphatidylserine (Perkins *et al.*, 1994). This suggests a role for SERA during the invasion process and formation of the PV.

Monoclonal antibodies (mAbs) directed to the amino terminus of SERA inhibit parasite growth *in vitro* (Banyal *et al.*, 1985; Horii *et al.*, 1988). Antibodies to SERA can be eluted from agglutinated merozoite clusters suggesting that SERA may be loosely bound to the merozoite surface (Lyon *et al.*, 1986a; Chulay *et al.*, 1987). The purified protein has been used in primate and rodent challenge trials and gave good protection overall (Perrin *et al.*, 1984; Inselburg *et al.*, 1991; 1993). Collectively these studies suggest that the protein may have a role in RBC invasion.

Further work identified two SERA-related proteins in *P. falciparum* (Knapp *et al.*, 1991; Fox *et al.*, 1994), and when the genome sequence of chromosome 2 was published it became apparent that there were in fact eight SERA-like genes located in a head to tail configuration on this chromosome of which SERA was the fifth gene (Gardner *et al.*, 1998). An additional SERA gene was located on chromosome 9 by further genome searching (Miller *et al.*, 2002). Of these various SERA homologues, it appears that SERA 4-6 are the most abundantly expressed, although SERA2 and SERA3 may also be expressed in some parasite strains. Gene disruption studies showed that genes at the periphery of the cluster could be interrupted without compromising viability of parasites. However SERA4, SERA5 and SERA6 appeared to be essential as no viable parasites could be obtained when attempts were made to disrupt these genes (Miller *et al.*, 2002).

### 6.10. S Antigen

S antigen remains the Cinderella protein of the asexual stage. It is a protein that exhibits an extraordinary degree of antigenic polymorphism suggesting extreme selective pressure by the immune system and accordingly an important role in parasite biology (Anders *et al.*, 1983). Nevertheless it has been only sparsely studied and much basic information is still lacking. S antigen is encoded by a single exon in which a central block of tandem repeats is flanked by non-repeat sequences containing regions rich in charged amino acids that show considerable homology between different parasite isolates. The central tandem repeats are highly variable between isolates but quite strongly conserved within a particular sequence. Thus, the S antigen of the *P.falciparum* isolate FC27 contains about 100 repeats of the 11-mer PAKASQGGLED, whereas that of NF7 contains about 40 repeats of an octamer with two variants, ARKSDEAE and ALKSDEAE. The FC27 and NF7 repeats are so different that they do not cross hybridize (Coppel *et al.*, 1983; Cowman *et al.*, 1985) and natural antibody responses directed against epitopes encoded by the repeats do not cross react immunologically. This results in an enormous degree of serological diversity of S antigens based on the large repertoire of different repeats that may vary in sequence, size, number or reading frame (Brown *et al.*, 1987; Saint *et al.*, 1987; Nicholls *et al.*, 1988; Bickle *et al.*, 1992; 1993; Kyes *et al.*, 1993; Black *et al.*, 2002).

S antigen is secreted into the PV and released upon schizont rupture. A fraction of the released protein associates with the merozoite surface where it is complexed to MSP1 (Perkins *et al.*, 1990). mAbs to the S antigen of the FC27 isolated do inhibit parasite growth, but as might be expected inhibition is isolate-specific (Saul *et al.*, 1984). Most of the S antigen produced is in fact disseminated throughout the bloodstream where it may act as circulating sink for induced inhibitory antibodies. The functional part of the molecule is likely to be the relatively conserved N- and C- termini, although even these vary somewhat, belonging to at least 4 distinct families (Bickle *et al.*, 1992; 1993). The repetitive nature of the protein makes it difficult to clone and this makes gene disruption studies technically difficult.

# 7. Rhoptry Proteins

An increasing number of rhoptry proteins are being identified (Table 1). General features shared by these proteins are the presence of an N-terminal signal sequence and strong sequence conservation of the genes between different parasite isolates. Proteomic studies on isolated rhoptries should provide a more complete inventory of these proteins in the near future.

## 7.1. RAP1

The low molecular weight complex of *P. falciparum* rhoptries consist of three, non-covalently associated proteins, called rhoptry associated protein 1, 2, and 3 (RAP1, RAP2 and RAP3). RAP1 is a 782 residue polypeptide encoded by a single intronless open reading frame (ORF) on chromosome 14. Apart from a classical signal peptide of 22 residues, there is a serine-rich region containing five degenerate repeats of a KSSSPS motif near the N-terminus. The significance of this region is as yet unknown (Ridley *et al.*, 1990). RAP1 is a soluble protein expressed in a stage-specific manner during schizogony, first at 38h post infection (Jaikaria *et al.*, 1993). The primary translation product of RAP1 is an 84 kDa molecule, which is modified to produce an 86 kDa form by an unknown process. The 86kDa protein is proteolytically processed at the amino-terminus in a two-step process. The p86 form is quickly cleaved to produce an 82 kDa protein, a process that can be blocked by incubation at low temperatures or by brefeldin A, in agreement with the presence of the protein within compartments of the secretory pathway (Howard *et al.*, 1995). After some delay, the p82 protein is again cleaved to produce a 67 kDa RAP1 product, an event that seems to occur in nascent rhoptries. The second cleavage site has been identified and it has been determined to fall between residues $A_{190}$ and $D_{191}$ (Ridley *et al.*, 1991). Both forms of RAP1, the p82 and the p67 are present in rhoptry organelles in schizont and segmenter stages, and are also present in rhoptries of free extracellular merozoites (Howard *et al.*, 1998b), and both are discharged from the rhoptry organelles. Interestingly, only the p82 protein is later detected in ring-stage parasites, with some diffuse fluorescence staining observed in the infected RBC (Howard *et al.*, 1984).

Early studies have shown growth inhibition of *P. falciparum* in *in vitro* cultures by mAbs raised against RAP1 or against parasite material and recognising RAP1 (Perrin *et al.*, 1981; Schofield *et al.*, 1986). A later report has shown that sera from mice and rabbits immunised with RAP1 recognise the N-terminal one third of the protein and partially inhibit RBC invasion *in vitro* (Stowers *et al.*, 1995). Finally, immunoglobulins purified from anti-RAP1 mAbs inhibit parasite growth *in vitro*. The epitope recognised by these inhibitory antibodies has been mapped to a peptide of 10-20 residues after the p82 processing site (position $A_{190}$) (Harnyuttanakorn *et al.*, 1992). A second inhibitory B-cell epitope of RAP1 has also been identified. Both epitopes are located within the amino-terminal part of RAP1, found to be the immunodominant portion of the protein in owl monkeys immunised with *P. falciparum* (Howard *et al.*, 1998a). In line with these observations, 11 out of 13 mAbs raised against recombinant RAP1 protein comprising residues 23 to 711 have been found to react with an N-terminal region between residues 23 and 294. Five of these 11 mAbs recognised a region spanning residues 225 to 250. The sequences of epitopes recognised by these antibodies have been determined. Interestingly, none of the raised mAbs reacted with the short repeat region (residues 123 to 164), a result similar to observations based on examination of natural humoral responses. Although the mAbs recognizing the N-terminal epitopes have not inhibited *P. falciparum* growth on their own, they did exhibit enhancement of growth-inhibitory activity of a noninhibitory antibody directed against the $Y_{218}KYSL_{222}$ sequence (Moreno *et al.*, 2001). This effect suggests that the efficacy of RAP1-conferred immunity may be dependent on a combination of protective epitopes. Collectively these studies suggest that RAP1 has an important role in invasion. Nevertheless, targeted disruption of the gene encoding RAP1 -to produce a truncated form of the parasite still led to viable merozoites capable of RBC invasion and growth at similar rates to parasites with intact RAP1 (Baldi *et al.*, 2000). The parasite was still able to produce the N-terminal region of RAP1 implicated as important by mAb inhibition studies, as truncation occurred at residue 345. Truncated RAP1 was present in rhoptries suggesting that rhoptry targeting signals were still present, but it was not detected in newly invaded rings, in contrast to full length RAP1. Neither RAP2 nor RAP3 were associated with RAP1 in mutant parasites with RAP2 being found in the ER. This suggests the C-terminal region of RAP1 is important for RAP2 binding and appears essential for correct trafficking of RAP2 to rhoptries (Baldi *et al.*, 2000). Further, since the truncated RAP1 was not carried into ring-stage parasites, the C-terminal domain appears to play a role in this process as well (Baldi *et al.*, 2000).

## 7.2. RAP2

The second member of the low molecular weight complex is RAP2. RAP2, endoded by a single intronless ORF on chromosome 5, is a 398 residue protein that shows little sequence variation between *P. falciparum* strains. RAP2 contains a classical signal peptide of 21 residues, and 5 conserved cysteine residues, but no repeat regions (Saul *et al.*, 1992). RAP2 has an apparent molecular mass of 42 kDa (also reported as 39 kDa), is a soluble protein and its expression coincides with RAP1 (Jaikaria *et al.*, 1993). In contrast to RAP1, the RAP2 protein does not appear to be proteolytically processed. Following RBC invasion by merozoites, RAP2 can be detected in ring-stage parasites along with RAP1 (Baldi *et al.*, 2000). Rhoptry localisation of RAP2 is not essential for red cell invasion as the RAP1 mutant parasites discussed above were still invasion competent despite the marooning of RAP2 in the ER. This study clearly indicated that RAP2 lacks its own rhoptry-targeting signals, which were presumably provided by RAP1 or some protein that interacts with RAP1. It is likely that mistrafficked RAP2 does not partake in the invasion process, but a formal proof that it is not required for invasion awaits the report of a RAP2 null parasite.

Purified IgG fractions taken from mice immunized with recombinant RAP2 recognize an octapeptide $E_{25}$TEFSKLY$_{32}$ present at the N-terminus of the protein. Unexpectedly, sera from mice immunised with this peptide have reacted not only with RAP2 but also with RAP1. The cross-reactive epitope of RAP1 has been identified and found to be $L_{202}$TPLEELY$_{209}$, which falls within the inhibitory epitope between residues 200 and 211. Not surprisingly, the purified IgG fractions exhibited growth inhibition in *in vitro* culture (Stowers *et al.*, 1996).

## 7.3. RAP3

RAP3 has not been extensively studied and the *RAP3* gene sequence has been reported relatively recently. The *RAP3* ORF is located on chromosome 5 adjacent to *RAP2*, is intronless, and encodes a protein of 400 residues. There is 68% similarity and 44% identity between the two proteins, and the 5 cysteine residues present in RAP2 are conserved in RAP3 (Baldi *et al.*, 2002). On SDS-PAGE gels, RAP3 appears as a protein with an apparent molecular mass of 40 kDa (also reported as 37 kDa) that can be easily co-precipitated with the two other RAP proteins. RAP3 seems to be expressed around the same time as RAP1 and RAP2 during schizogony, and also seems to be carried into ring stages after the RBC invasion (Howard *et al.*, 1984; Howard *et al.*, 1990). RAP3 does not appear to be essential for parasite survival, as evidenced by a targeted disruption experiment. The loss of the RAP3 protein had no effect on parasite growth rate or invasion, nor did it affect the trafficking of RAP1 and RAP2 proteins to rhoptry organelles. It is possible that RAP2 can complement the loss of RAP3 since the two proteins share a significant level of similarity (Baldi *et al.*, 2002).

Homologues of RAP2 and RAP3 have been identified recently in *P. yoelii*, *P. berghei*, *P. vivax* and *P. knowlesi*. Interestingly, in these species the homologues are encoded by single ORFs, which give chimeric protein products, RAP2/3 (Baldi *et al.*, 2002).This is akin to the MSP4 and MSP5 of *P. falciparum*, which in rodent malaria species are represented by single MSP4/5 proteins. The levels of identity and similarity of PyRAP2/3 and PbRAP2/3 to PfRAP2 and PfRAP3 range at 30-40% and 49-63%, respectively. The signal peptide cleavage site and 3 of 5 cysteine residues are conserved between the species

## 7.4. RhopH1

The high molecular weight complex of rhoptry proteins contains three non-covalently associated proteins: RhopH1, RhopH2 and RhopH3. The RhopH1 protein is encoded by a member of the *clag* multigene family. The *rhoph1* ORF is located on chromosome 3 and has a complex structure consisting of nine exons. Apart from the signal peptide, there are 15 cysteine residues within the RhopH1 sequence, 10 of which are conserved between the parasite species, while 5 are conserved between the members of the clag family. The primogeniture of this family is the clag9 protein, which has been implicated to play a role in cytoadherence (Trenholme *et al.*, 2000). Surprisingly there are no differences in general protein structure that might explain why one member of this family targets to the RBC surface and the other to the rhoptries. There are several further members of this family and it is possible that these may also be rhoptry proteins. Perhaps the common feature is that both rhoptry and cytoadherence proteins are involved in interactions with RBCs (Kaneko *et al.*, 2001).

RhopH1 seems to be expressed around the same time as RhopH3 (see below) and can be localised to rhoptry organelles. The apparent molecular mass of RhopH1 is 140 kDa (Jaikaria *et al.*, 1993), although some differences are present between individual reports. The entire high molecular weight complex has been shown to bind to phospholipids of the inner leaflet of the RBC, but it is not clear whether RhopH1 and RhopH2 are capable of binding or are present by virtue of their association with RhopH3 (see below). These proteins may have a role in formation of the PV (Perkins *et al.*, 1994; Ndengele *et al.*, 1995).

A homologue of RhopH1 has been identified in *P. yoelii* parasites by using monoclonal antibodies reactive to PfRhopH1. Similarly to *P. falciparum*, PyRhopH1 seems to belong to a family of proteins, and a second member has been identified in the same study. The complex ORF structure of *rhoph1*, the signal peptide and 10 cysteine residues are conserved in the rodent species. Immunoelectron microscopic studies localised PyRhopH1 to the body of rhoptry organelles (Kaneko *et al.*, 2001).

## 7.5. RhopH2

RhopH2 is encoded by a single-copy ORF on chromosome 9. The *rhoph2* gene comprises ten exons, including three mini-exons located at the ORF termini. The predicted protein product of 1378 residues contains a classical signal peptide and 18 cysteine residues, 17 of which are conserved across the species. RhopH2 has an apparent molecular mass of 130 kDa, although there are differences between individual reports. The protein is expressed during schizogony at the time of expression of RhopH1 and RhopH3, and can be localised to rhoptry organelles (Jaikaria *et al.*, 1993; Ling *et al.*, 2003).

A homologue of RhopH2 has been identified in *P. yoelii*. PyRhopH2 is encoded by a single-copy gene made up of ten exons. The gene structure is conserved across the species, and the intron boundries in *pyrhoph2* ORF correspond to those in *pfrhoph2*. Similarly to PfRhopH2, the predicted protein sequence contains 18 cysteines, 17 in the same position as in PfRhopH2. The predicted protein product is 1362 residues in length and PyRhopH2 has an apparent molecular mass of 140 kDa (Ling *et al.*, 2003).

## 7.6. RhopH3

RhopH3 is encoded by a single ORF with a complex structure, containing six introns and seven exons, two of which are "miniexons". The complexity of the ORF structure seems to be a characteristic feature of the RhopH proteins. The *rhoph3* gene encodes a predicted protein product of 895 residues, which lacks the antigenic diversity of many malarial proteins. RhopH3 is a soluble protein that contains a 24 residue-long signal peptide and 13 cysteine residues (Brown *et al.*, 1991). RhopH3 is expressed in a stage-specific manner during schizogony, first at 30 h post infection (Jaikaria *et al.*, 1993). The primary translation product of RhopH3 was reported to have an apparent molecular mass of 103 kDa, be modified to produce a 105 kDa molecule in schizont stages and in free merozoites. In ring-stage parasites RhopH3 seemed to undergo yet another modification, resulting in a molecular mass shift to 110 kDa (Lustigman *et al.*, 1988). In later reports RhopH3 is observed as a 110 kDa or 105 kDa molecule in mature parasite stages, and Doury and colleagues (Doury *et al.*, 1997) have demonstrated that the 105 kDa molecule is derived from the 110 kDa protein by proteolytic processing which removes 45-50 residues from the C-terminus. However the small size differences noted and differences in standards employed by various laboratories make the significance of these size differences hard to assess. By immunofluorescence RhopH3 showed a characteristic punctate pattern of rhoptry localisation in schizonts and free merozoites, and diffuse staining within ring-infected RBCs. Immunoelectron microscopy indicated that RhopH3 is located in the rhoptry body (Coppel *et al.*, 1987). During examination of isolated rhoptry organelles it has been observed that RhopH3 is partially associated with a membrane-bound fraction, from which it can be released by treatment with phospholipases (Etzion *et al.*, 1991). This suggests that RhopH3 is either associated with lipid species and possibly the membranous whorls in the rhoptry body, or that it is indirectly linked to the rhoptry membrane via a GPI-anchored protein. RhopH3 has been reported not only to be discharged from rhoptries during RBC invasion and then localise within newly-invaded erythrocytes, but also to be able to interact with RBC membrane phospholipids (Ndengele *et al.*, 1995). Perkins and Ziefer have been able to identify phosphatidylserine (PS) and phosphatidylinositol (PI) as the preferential lipid binding ligands for RhopH3. These species of phosphoplipids are associated with the inner leaflet of the RBC membrane and the authors proposed that during the invasion process RhopH3 initiates changes on the inner surface of the RBC membrane that lead to the invagination and formation of the PV (Perkins *et al.*, 1994). An important function for RhopH3 is suggested by reports demonstrating growth inhibitory properties of Mabs reacting with this protein (Doury *et al.*, 1994) and unsuccesful attempts to disrupt the RhopH3 gene (Cowman *et al.*, 2000).

A RhopH3 homologue has been detected in *P. yoelii* (Anthony *et al.*, 2000). The *PyRhopH3* gene exists in a single copy in the *P. yoelii* genome and has a similar complex gene structure to that found for *PfRhopH3*. The coding sequence spans seven exons, two of which are "miniexons", and gives a predicted protein product of 882 residues. Based on the similarity to PfRhopH3, three types of blocks: conserved, semi-conserved and variable have been identified within PyRhopH3. The conserved blocks seem to cluster within the amino-terminal part of the protein, whereas the carboxy-terminal region is variable. Twelve of the 13 cysteine residues are conserved between PyRhopH3 and PfRhopH3. Additionally, a serine-rich repeat region of four degenerate octapeptide repeats is present within the PyRhopH3 sequence. The apparent molecular mass of PyRhopH3 is 100 kDa, and in schizont stages the protein shows a punctate pattern of rhoptry localisation (Shirano *et al.*, 2001).

## 7.7. AMA1

Apical membrane antigen 1 (AMA1) is encoded by a single intronless ORF on chromosome 11 giving a protein product of 622 residues. In addition to a classical signal peptide, there is a hydrophobic region of 21 residues located 55 residues from the C-terminus, which is a membrane-spanning domain. The AMA1 sequence does not contain any repeat regions, but there are 16 conserved cysteine residues located in the extracellular part of the protein (Peterson *et al.*, 1989). The cysteines are involved in the formation of 8 disulphide bonds, which divide the extracellular domain of AMA1 into 3 ectodomains. A considerable degree of sequence diversity characterises AMA1, which suggests that selective pressure is exerted on the protein because of an important role in parasite biology and anti-malarial immunity. AMA1 is expressed in a stage specific manner, with the onset of expression around 41h post infection (Jaikaria *et al.*, 1993; Narum *et al.*, 1994). The primary translation product of *AMA1* has an apparent molecular mass of 83 kDa. This protein is rapidly processed by proteolytic cleavage at the N-terminus, to produce a 66 kDa molecule, an event once thought to occur in rhoptry organelles (Crewther *et al.*, 1990; Narum *et al.*, 1994) but recently shown to take place in micronemes (Healer *et al.*, 2002). The processing site has been identified and shown to fall between $S_{96}$ and $I_{97}$ (Howell *et al.*, 2001). The 66-kDa processed form of AMA1 is translocated to rhoptries where it resides within the organelle neck (Crewther *et al.*, 1990). Just prior to schizont rupture AMA1 translocates again to the membrane of the apical tip of free merozoites (Narum *et al.*, 1994). After the release of free merozoites the 66-kDa AMA1 is alternatively processed by proteolytic cleavage resulting in the shedding of 48 kDa and 44 kDa AMA1 fragments. $AMA1_{48}$ is the major shed form, generated by the cleavage between residues $T_{517}$ and $S_{518}$, apparently mediated by the same serine protease as that involved in secondary processing of MSP1. Approximately one third of the shed AMA1 molecules are also cleaved at the internal site between residues $N_{464}$ and $D_{465}$, generating the $AMA1_{44}$ form. During RBC invasion by merozoites the remaining membrane-bound portion of AMA1 is carried into the invaded host cell, and in early ring stages can be detected associated with the PV (Narum *et al.*, 1994; Howell *et al.*, 2001; 2003). Interestingly, examinations of binding activity of AMA1-derived peptides indicated that the RBC binding regions are located within ectodomains I and II of AMA1 (Urquiza *et al.*, 2000; Fraser *et al.*, 2001). A study of AMA1 expression by Kocken and colleagues gave some insight into the processes that govern trafficking of proteins to apical organelles in *Plasmodium* parasites. It was determined that, at least in the case of AMA1, the timing of protein expression is an important factor. When expressed under its native promoter, and therefore at the appropriate time with respect to parasite stage of late in schizogony, AMA1 was properly routed to rhoptry organelles and localised in the rhoptry neck. However, when expressed under the control of the *dhfr-ts* promoter AMA1 showed diffuse, cytoplasmic localisation, as well as circumferential staining in merozoites. Some rhoptry localisation was also detected in late schizonts, however AMA1 was observed in the rhoptry body instead of the rhoptry neck (Kocken *et al.*, 1998). This study clearly indicated that proper localisation in rhoptry organelles depends on other factors than a mere presence of specific targeting signals and motifs within the protein sequence.

AMA1 homologues have been identified in a number of *Plasmodium* species, including *P. knowlesi*, *P. vivax*, *P. fragile*, *P. cynomolgi*, *P. reichenowi*, *P. berghei*, *P. yoelii* and *P. chabaudi*. The homologues lack an amino-terminal region present in PfAMA1 and have an apparent molecular mass of 66 kDa for the precursor form and 44 kDa for the processed form. However, the general characteristics – onset of expression late in schizogony, proteolytic processing and subsequent differential localisation after the schizont rupture are shared among the species (Marshall *et al.*, 1989; Peterson *et al.*, 1990; Waters *et al.*, 1990; Cheng *et al.*, 1994; Dutta *et al.*, 1995; Kappe *et al.*, 1996; Barnwell *et al.*, 1998; Kocken *et al.*, 2000). AMA1 appears to have an important role in invasion as immune responses to it interfere with parasite replication. Most of the supporting data comes from studies of AMA1 homologues from primate or rodent malaria species, however PfAMA1 has been examined as well. Early studies have shown that rat monoclonal antibodies directed against PkAMA1 inhibit growth of *P. knowlesi in vitro*. Use of Fab fragments indicated that the observed effect was due to the blocking of merozoite attachment to RBCs and subsequent invasion of host cells (Deans *et al.*, 1982; Thomas *et al.*, 1984). A study of human malaria parasite AMA1 has shown that a rat mAbs raised against recombinant PfAMA1 can block invasion of RBCs *in vitro* (Kocken *et al.*, 1998). Interestingly, this antibody and a second that shared the same inhibitory characteristics, both inhibited *P. reichenowi* growth *in vitro* (Kocken *et al.*, 2000). This result indicates that anti-AMA1 antibodies can act in a heterologous system. This may be due to the fact that, despite the sequence differences, PfAMA1 and PrAMA1 are very closely related and are the only homologues that share the most N-terminal region. In a recent study, IgG fractions of sera from rabbits immunised with conformationally-correct PfAMA1 inhibited *P. falciparum* invasion of RBCs (Kocken *et al.*, 2002). Immunization of mice with *P. chabaudi* AMA1 induced high levels of immunity against homologous parasites but not those expressing a variant form of the parasite (Crewther *et al.*, 1996). This suggests that the AMA1 sequence can tolerate many changes and still perform its function and that immune responses are frequently induced to these variable regions. Supporting an important role for AMA1 in cell invasion are genetic studies in which attempts were made to disrupt the AMA1 gene (Triglia *et al.*, 2000). It was not possible to produce viable parasites, even in parasites in which the *P. chabaudi* AMA1 gene was also being expressed (Triglia *et al.*, 2000). Interestingly, the function of PfAMA1 could be partially complemented by this PcAMA1 transgene and these parasites showed an improved capacity to invade mouse RBCs. The parasites were also able to invade human RBCs to about one-third the normal level in the presence of inhibitory antibodies to PfAMA1 that normally completely prevent invasion. Consistent with previous studies, correct AMA1 trafficking required expression at the right stage of the parasite life cycle as the PcAMA1 transgene expressed under the PfAMA1 promoter reached the rhoptries whereas expression under the calmodulin promoter led to aberrant localization at the parasite plasma membrane (Triglia *et al.*, 2000).

## 7.8. MAEBL

Chimeric rhoptry proteins called MAEBL have been identified in the rodent malaria species *P. yoelii* and *P. berghei*. In both species the proteins are encoded by a single ORF with a complex structure, consisting of 6 exons and 5 introns, of which one intron is cryptic. The sequences of *P. yoelii* and *P. berghei* MAEBL are highly similar, and the gene and protein structures are identical. The MAEBL proteins have a chimeric nature, being homologous with AMA1 at their N-termini, and with Duffy-binding-like erythrocyte-binding ligands (DBL-EBPs) (see below) at their C-termini. Following the classical signal peptide, there are two cysteine-rich domains, M1 and M2, each corresponding to the first two AMA1 subdomains. These are followed by 13-residue tandem repeats, and by the C-terminal cysteine-rich region (C-cys) homologous to the *P. vivax* and *P. knowlesi* Duffy-binding proteins (DBPs) and the *P. falciparum* EBA175. A transmembrane domain and a putative cytoplasmic domain follow the conserved cysteine-rich region. It appears that in *P. yoelii* *maebl* is alternatively spliced, with two transcripts being detected in Northern blots. A full-length transcript of approximately 8 kb can be observed in addition to a shorter transcript of 5.6 kb, spanning only the 5' region of *maebl* and including the M1 domain and the cryptic intron, which can give rise to a shorter, soluble protein product (Kappe *et al.*, 1998). PyMAEBL is observed as a protein doublet of 128/120 kDa, and is expressed in a stage-specific manner during schizogony. PyMAEBL is detected at the two-nucleus stage, with onset of expression preceding PyAMA1. In schizont stages PyMAEBL shows a punctate pattern of rhoptry localisation, while in mature and rupturing segmenters labelling is also observed on the merozoite surface. The protein is not detected in culture supernatants, indicating that it may be involved in interactions with the RBC membrane and the formation of the parasitophorous vacuole (Noe *et al.*, 1998). Indeed, the cysteine-rich domains M1 and M2 have been shown to bind mouse erythrocytes (Kappe *et al.*, 1998).

MAEBL has also been identified in *P. berghei* parasites and, surprisingly, the pattern of expression and localisation of PbMAEBL is quite different to that of PyMAEBL (Kariu *et al.*, 2002). PbMAEBL was observed as a 200 kDa molecule present in midgut sporozoites but not detected in intraerythrocytic stages including merozoites. Interestingly, PbMAEBL was found in micronemes (not rhoptries) of mature sporozoites in oocysts. Targeted gene disruption has indicated that PbMAEBL is necessary for sporozoite entry into the salivary gland, and it has been proposed that PbMAEBL may be involved in the attachment to the gland surface by means of binding to salivary gland receptors. Clearly further work is required to define the stage-specificity of this protein. There are precedents for a single membrane protein being present in both liver stages and asexual blood stages, but as yet none for a membrane protein that is common between sporozoites and asexual stages.

*P. falciparum* MAEBL is encoded by a single copy ORF on chromosome 11 consisting of five exons. One transcript is generated, coding for a product of 2056 residues. The chimeric protein structure is conserved, and apart from the signal peptide PfMAEBL contains the two N-terminal cysteine-rich domains M1 and M2, the repeat region, the C-terminal cysteine-rich domain, the transmembrane domain, and the cytoplasmic tail. The primary translation product with a predicted molecular mass of 243 kDa is not observed on Western blots. It seems that PfMAEBL is rapidly proteolytically processed to give protein products of apparent molecular masses of 47, 39 and 30 kDa. The latter has a potential precursor form of 58 kDa. PfMAEBL is expressed in a similar fashion to its homologue in *P. yoelii*, and localises in rhoptry organelles in schizonts and merozoites and on the merozoite surface after schizont rupture (Blair *et al.*, 2002). No function for this protein is currently known, although a separate report by Ghai and collegues (Ghai *et al.*, 2002) suggested that MAEBL may be involved in the invasion process by virtue of binding to the RBC surface. The putative receptor for MAEBL has been determined to be of protein nature, cleavable by papain and disrupted by trypsin treatment. Furthermore, Ghai and colleagues observed MAEBL as a 154 kDa protein present in rhoptries of late schizonts and merozoites but also on the surface of salivary gland sporozoites. The latter finding underlines the importance of further characterisation of MAEBL and addressing the issue of its differential localisation in different *Plasmodium* species.

## 7.9. The 225kDa Rhoptry Protein

MAbs raised against *P. falciparum* culture supernatant identified a 225 kDa rhoptry protein but no sequence data is available for this protein at present. The protein appears to be expressed late during schizogony, with the onset of expression falling around 30 h post infection. The primary translation product has an apparent molecular mass of 240 kDa. This protein is proteolytically processed approximately 30 min after its expression to produce the 225 kDa form. In schizont stages of the parasite the protein shows a characteristic punctate pattern of rhoptry localisation. Immunoelectron-microscopy revealed that the 225 kDa protein is located laterally in young rhoptries, while in mature organelles the protein localises to the rhoptry neck (Roger *et al.*, 1988).

## 7.10. The 52kDa Rhoptry Protein

Immunochemical studies using mAbs raised against *P. falciparum* parasites have identified another rhoptry protein with an apparent molecular weight of 52 kDa. No gene sequence data has been reported. The onset of expression of this protein appears to fall between 12 and 20 h post infection. No proteolytic processing has been observed for the 52 kDa protein product and it does not appear to be membrane-bound. In schizont stages, the protein is found in rhoptry organelles, but following RBC invasion by merozoites no labelling is observed, indicating that the 52 kDa protein is not carried into ring stages (Storey, 1992). The protein appears to have a role in invasion as mouse polyclonal antibodies directed against the 52 kDa rhoptry protein of *P. falciparum* inhibited parasite invasion *in vitro* (Storey, 1992). However studies involving mouse antisera need to be treated with caution as such reagents can non-specifically inhibit parasite growth. The 52 kDa protein is apparently distinct from the 55 kDa protein of Smythe and co-workers (Smythe *et al.*, 1988), now known as rhoptry associated membrane antigen (RAMA) (A. E. Topolska *et al.*, unpublished).

## 7.11. The 60 kDa Rhoptry Protein Superfamily

A multigene family of 60 kDa rhoptry proteins has been identified using sera raised against a *Babesia divergens* exoantigen. Proteins are encoded by a family of genes present in about 140 copies within the *P. falciparum* genome, although it is not clear how many of these are transcribed. The predicted protein products share some sequence similarity with the RAP1 family of *Babesia spp.*, including a highly conserved stretch of 14 residues. The proteins are expressed in a stage-specific manner during late schizont stages, first at 42 h post infection. The apparent molecular mass of the protein varies slightly among *P. falciparum* strains, and is detected as 58-60 kDa polypeptides. No proteolytic processing has been observed for this protein, and the 60 kDa form appears to be the primary translation product. The protein is not membrane-associated. A punctate pattern of rhoptry localisation can be observed for this protein in schizont stages and in free merozoites. However, no labelling was detected in invading merozoites, nor in early ring stages, indicating that the 60 kDa rhoptry protein is not carried into rings during the RBC invasion process. Moreover, the protein can be detected in culture supernatant after the invasion, suggesting that the protein is discharged from rhoptries but does not participate in the formation of the PV (Carcy *et al.*, 1994; Grellier *et al.*, 1994).

## 7.12. RAMA

In 1988, Smythe and co-workers described a 55 kDa rhoptry protein that had been identified by screening cDNA expression libraries with immune human sera that was affinity-purified on the Triton-X114 fraction of parasites (Smythe *et al.*, 1988). The protein was present in the Triton X-114 detergent-enriched phase and could be labelled with [3H] glucosamine and [3H] myristate, indicating membrane association by means of a GPI anchor. Neither sequence data, nor timing of expression was reported at that stage. The protein gave a punctate pattern of rhoptry localisation in late schizont stages. Solubility studies indicated that the 55-kDa protein was completely detergent soluble in late schizonts and in free merozoites, but in ring and trophozoite stages the protein was also detected in the Triton X-114 insoluble pellet (Smythe *et al.*, 1988). We have performed further studies on this protein, now renamed as the rhoptry associated membrane antigen (RAMA) and the results show that the 55kDa protein, which we observe as a 60kDa molecule, is in fact a product of proteolytic cleavage of a full-length primary translation product of 170 kDa (A. E. Topolska *et al.*, unpublished).

### 7.13. Gp76

The gp76 rhoptry protein is a serine protease, expressed in late trophozoite and schizont stages. The primary translation product has an apparent molecular mass of 83 kDa, and is rapidly processed by proteolytic cleavage at the N-terminus to give a product of 76 kDa. The gp76 protein is membrane-associated by means of a GPI moiety. Following the merozoite's attachment to the RBC, a phosphatidylinositol-specific phospholipase C (PI-PLC), also located at the merozoite apex, cleaves the GPI anchor of gp76. This event not only releases gp76 and allows for its secretion from the merozoite, but also activates its proteolytic property (Braun-Breton *et al.*, 1988a; 1988b). The targets of gp76 have been identified as RBC proteins, band 3 and glycophorin A. It has been proposed that gp76 plays an important role during invasion of RBCs by merozoites, and that the degradation of these target proteins facilitates internalisation of the merozoite, incorporation of phospholipids into the RBC membrane, and the formation of the PV (Roggwiller *et al.*, 1996). The importance of the gp76 protein is further supoorted by the fact that the apparent homologue in *P. chabaudi*, gp68 is necessary for merozoite invasion of RBCs. Invasion of merozoites treated with the serine protease inhibitors diisopropyl fluorophosphate or Pefabloc SC was greatly inhibited. This effect, however, could be reversed by incubation of RBCs with purified gp68. Similarly to gp76, the gp68 protein is a membrane-bound serine protease activated by PI-PLC cleavage of its GPI anchor, and its target in the RBC membrane is band 3 protein (Breton *et al.*, 1992).

### 7.14. The PfRH Protein Family

The PfRH (*P. falciparum* rhoptry protein homologue) family has been identified on the basis of similarity to adhesion proteins of *P. vivax* and *P. yoelii*. The first described members of the PfRH family (PfRH2a and PfRH2b) share a significant level of similarity to the *P. vivax* reticulocyte binding protein 2, PvRBP-2, hence their alternative names are *P. falciparum* RBP-2 homologue a and b (PfRBP2-Ha and –Hb). They also show similarity to the 235-kDa protein superfamily of *P. yoelii* (Rayner *et al.*, 2000; Triglia *et al.*, 2001a). Further searching of the genome revealed a third member of the PfRH family

(PfRH3) that was found to be a transcribed pseudogene (Taylor *et al.*, 2001), and a fourth member PfRH4 that is expressed (Kaneko *et al.*, 2002). Another member of the PfRH family, the homologue of *P. vivax* reticulocyte binding protein 1 named PfRH1 (Taylor *et al.*, 2002) is identical to the *P. falciparum* normocyte binding protein 1 (PfNBP1) identified by Rayner and colleagues (Rayner *et al.*, 2001). The PfRH proteins are encoded by two-exon genes, with the short first exon coding mainly for the signal peptide. The conserved features of predicted protein products include a transmembrane domain, a large extracellular domain containing shared regions of high similarity to PvRBP-2 and the *P. yoelii* 235 kDa rhoptry proteins (Py235), and a short cytoplasmic domain. The PfRH2a and b proteins also contain unique regions close to the transmembrane domain, as well as distinctive regions of degenerate repeats within their extracellular domains. Repeats are also present in extracellular and cytoplasmic domains of PfRH4. The PfRH1, 2a and 2b proteins are expressed late during schizogony. The observed molecular masses differ between individual reports, however the proteins appear to be expressed as precursor molecules of over 300 kDa which seem to be proteolytically processed at the N-terminus to give products of over 200 kDa (PfRH2a and b) or 195 kDa (PfRH1). Preliminary analysis of PfRH4 indicates a protein of 220 kDa, however it is not clear whether it is the primary translation product or the mature protein. In immunofluorescence, PfRH1 was localised to micronemes, rhoptries, and also to the membrane of the apical tip, whereas the PfRH2 proteins have been localised to rhoptries and the apical membrane. The localisation of PfRH4 is different from other PfRH proteins and appears to be in the micronemes (Rayner *et al.*, 2000; 2001; Triglia *et al.*, 2001a; Kaneko *et al.*, 2002; Taylor *et al.*, 2002). It is likely that the PfRH proteins are translocated from secretory organelles to the apical membrane in a similar fashion to proteins such as AMA1 or MAEBL (Blair *et al.*, 2002; Healer *et al.*, 2002). Interestingly, the PfRH1 and PfRH2 proteins seem to be expressed alternatively in different *P. falciparum* strains (Taylor *et al.*, 2002). This is akin to the expression of different variants of the Py235 superfamily (Preiser *et al.*, 1998; 1999). As putative RBC-binding ligands, the possibility of PfRH1 and 2 being present at differing levels may provide means for alternative invasion pathways. Previously, the PvRBP proteins have been shown to bind reticulocytes, and the Py235 proteins appear to participate in binding of *P. yoelii* parasites to RBCs (Galinski *et al.*, 1992; Ogun *et al.*, 1996; Galinski *et al.*, 2000). In line with this, the PfRH1 protein (PfNBP1) has been shown to bind RBCs, and its cognate receptor has been determined to be distinct from those identified so far (Rayner *et al.*, 2001). Furthermore, involvement of the PfRH2b protein in a novel invasion pathway, independent of PfRH1 or PfRH2a has been demonstrated. The RBC receptor (termed receptor Z) is distinct from the known ligands for parasite adhesion molecules providing yet another alternative invasion pathway (Duraisingh *et al.*, 2003). Duraisingh and colleagues have also demonstrated that there exists a phenotypic variation in expression of PfRH proteins in different parasite strains. This variation alters the pattern of RBC receptor usage and the invasion pathway, thus providing the parasite with a strategy to overcome the RBC receptor polymorphism and host immune responses.

## 7.15. PfRhop148

PfRhop148, is encoded by a single-exon gene and has an apparent molecular mass of 148 kDa (Lobo *et al.*, 2003a). The predicted protein product has 1262 residues and contains a putative transmembrane domain near its C-terminus, but surprisingly it seems to lack a signal peptide. All other known rhoptry proteins move to the rhoptry via the secretory pathway, using a signal sequence to traverse the ER, suggesting that PfRhop148 may need to be chaperoned to this destination. An interesting characteristic of PfRhop148 is the high content (22%) of asparagine, found in clusters and in tandem repeats. The localisation of PfRhop148 in the rhoptry organelles was evidenced by both, double-staining immunofluorescence and by electron microscopy. PfRhop148 is first expressed between 21 and 24 h post infection, which is also unusual for a rhoptry protein. In trophozoite stages PfRhop148 is observed in a diffuse pattern, reminiscent of the secretory pathway network. PfRhop148 awaits further, more detailed characterisation.

## 7.16. Py235

A multigene family of rhoptry proteins has been identified in *P. yoelii* and *P. berghei* parasites. The family includes up to 50 separate genes within the parasite genome, located at chromosomal ends (Borre *et al.*, 1995; Owen *et al.*, 1999). The ORFs appear to be intronless, and encode for proteins of 2294 residues. The apparent molecular mass of the proteins is about 235 kDa, and these appear to be processing products produced by a proteolytic cleavage at the N-terminus of the precursor proteins, an event that occurs late within the secretory pathway (Ogun *et al.*, 1994). Apart from a classical signal peptide of 19 residues, the proteins contain a transmembrane domain of 16 residues, followed by a short 46 residue-long cytoplasmic domain. A 500-residue region within the carboxy-terminal half of the 235 kDa protein shares some similarity with the *P. vivax* RBP2 protein. There is also a short region of tripeptide repeats located close to the transmembrane domain, within the ectodomain (Keen *et al.*, 1994). The 235kDa protein was localised to rhoptry organelles by immunoelectron microscopy (Oka *et al.*, 1984). The onset of expression of the 235kDa protein seems to occur late in schizogony, similarly to AMA1 (Noe *et al.*, 1998). In agreement with the presence of the PvRBP-2-like region within its sequence, the 235 kDa protein has been shown to be released from a membrane and bind mouse erythrocytes, possibly via glycophorin A, Duffy or band 3 molecules (Ogun *et al.*, 1996; 2000). Interestingly, some differences in the expression of individual variants of the 235kDa protein have been observed between the lethal and non-lethal

*P. yoelii* strains (Preiser *et al.*, 1998) and between single parasites within the same strain (Preiser *et al.*, 1999). This suggests that the 235kDa proteins may be involved in a clonal phenotypic variation utilised by the malaria parasite to maximise its survival.

## 7.17. ROPE

A rhoptry protein named ROPE, an acronym for repetitive organellar protein, has been identified in *P. chabaudi*. The protein is encoded by a single intronless ORF and is 1939 residues in length. Sequence analysis of ROPE revealed the presence of three blocks of eight 11-mer repeats divided by two 6-mer repeats, that form a predicted leucine-histidine zipper. The repeats are flanked by predicted coil-coil regions, suggesting that the protein may form dimers. Overall, ROPE seems to have a spectrin-like structure. The deduced molecular mass of ROPE is 229 kDa, and is in agreement with the observed protein bands of 240 kDa and 225 kDa. It is not clear at this stage whether the 240kDa form is a precursor protein that is proteolytically processed to produce the 225kDa protein. In immunofluorescent microscopy, ROPE showed a punctate pattern of rhoptry localisation in late schizonts, as well as in free merozoites. No staining was observed in earlier schizonts suggesting onset of expression late during schizogony. It has been proposed that ROPE is discharged from rhoptries during RBC invasion by merozoites and that it interacts with the RBC cytoskeleton by means of its structural similarity to spectrin. Additionally, the acidic environment of rhoptry organelles would keep the leucine-histidine zipper disasembled and ensure that ROPE is in an inactive form until it is deposited onto the RBC membrane (Werner *et al.*, 1998).

## 7.18. The 140kDa Protein

A 140kDa protein has been identified while investigating protective efficacy of AMA1 in *P. yoelii*. The protein co-precipitated with PyAMA1 suggesting protein-protein interaction between the two. Mabs raised against this protein have precipitated a protein doublet of approximately 140 kDa and have shown a typical rhoptry staining in immunofluorescence. The protein, and its potential interaction with AMA1 awaits further investigation (Narum *et al.*, 2000).

# 8. Microneme Proteins

The majority of well-studied proteins found in micronemes of malaria parasites contain conserved domains. The most important is the DBL domain, which is characterised by the presence of 12 conserved cysteine residues within tryptophan and tyrosine-rich sequences (Adams *et al.*, 1992). These domains are involved in binding to the RBC surface (Chitnis *et al.*, 1994) and are important in determining the particular RBC receptor utilised by the parasite and the type of RBC invaded.

### 8.1. DBPs and EBPs

The DBL-EBP family members have been first identified in *P. vivax* and *P. knowlesi*. These were the PvDBP and PkDBP, and PkEBP(β) and PkEBP(γ). The genes encoding these molecules have a partly conserved intron-exon structure, with a predominant second exon coding for the extracellular portion of the protein. DBPs and EBPs have conserved features including the signal peptide, the single DBL domain, often referred to as the amino cysteine-rich domain, the C-terminal cysteine-rich (C-cys) domain, the transmembrane domain and the cytoplasmic tail. Based on sequence comparison, seven distinct regions have been identified within Duffy binding proteins, with the DBL domain placed in region II, the C-cys domain in region VI and the transmembrane and cytoplasmic domains in region VII (Adams *et al.*, 1992). DBL domains of DBPs and EBPs have been shown to be the ligand domains, involved in the RBC binding, and recognition and interaction of Duffy antigens (Chitnis *et al.*, 1994). Mild proteolysis of DBL domains yields two subdomains of which the one containing cysteines 1 to 4 shows no detectable binding capacity, whereas the subdomain formed by cysteines 5 to 12 retains the capacity to bind the Duffy antigen. The binding sequence appears to be contained in the region between cysteines 5 and 8 (Singh *et al.*, 2003). It is not known if DBL-based binding regions in other proteins show a similar localisation of binding subdomains. PkDBP and PkEBPs have been localised to micronemes of *P. knowlesi* parasites (Adams *et al.*, 1990). In support of functional importance of the DBL domain, a seroepidemiological study has shown that the naturally-acquired antibodies were directed against the DBL domain, and effectively inhibited *in vitro* binding of this ligand to human RBCs (Michon *et al.*, 2000).

### 8.2. EBA175

The erythrocyte binding antigen-175 (EBA175), also known as the sialic acid binding protein (SABP), is an adhesion molecule of *P. falciparum* merozoites, and a member of the DBL-EBP family. EBA175 is encoded by a single-copy gene located on chromosome 7. The *EBA175* ORF has a conserved four-exon structure, with the first exon comprising most of the coding sequence. The predicted protein product is 1475 residues and contains the seven characteristic regions and the features conserved within the DBL-EBP family (Camus *et al.*, 1985; Sim *et al.*, 1990; Adams *et al.*, 1992). The DBL domains are the ligand domains, and have been shown to mediate RBC binding (Sim *et al.*, 1994b). The role of the C-cys domain is as yet unknown, although conservation of this region among members of this family suggests it has an important function. The cytoplasmic domain is thought to be involved in the signalling of recognition and attachment. The RBC ligand for EBA175 has been identified as glycophorin A, and the binding was found to be sialic acid-dependent. Both carbohydrate and peptide sequences in glycophorin A contribute to EBA175 binding (Sim *et al.*, 1994b). EBA175 is expressed as a 175kDa molecule in *P. falciparum* and has been shown to localise in micronemes (Sim *et al.*, 1992). The biological role of EBA175 as a glycophorin A binding ligand was confirmed by genetic studies in which the EBA175 gene was truncated, generating parasites that expressed regions I-V. The resultant parasites expressed approximately normal amounts of the shorter polypeptide in micronemes, suggesting the loss of regions VI and VII (C-cys, transmembrane and cytoplasmic domains) had no effect on proper trafficking of EBA175. The truncated protein was less efficient at binding to the RBC surface. This is somewhat surprising as the mapped glycophorin A binding site is found in region II containing the DBL domains, which is still present in truncated EBA175. Importantly, the mutant parasites showed a switch in invasion phenotype now entering RBCs by a sialic acid-independent pathway, presumably via glycophorin B (Kaneko *et al.*, 2000; Reed *et al.*, 2000). In support of EBA175 having a role in RBC invasion, antibodies directed against a peptide derived from the EBA175 ectodomain efficiently block merozoite invasion of RBCs *in vitro*. The peptide target is a 42 residue putative adhesion ligand (EBA-peptide 4) and growth inhibition seemed to be mediated by blocking the binding of native EBA175 to the RBC membrane (Sim *et al.*, 1990). Further epitope mapping identified a 19 residue subfragment that is the target of the growth inhibitory antibodies (Sim *et al.*, 1994a).

### 8.3. BAEBL

BAEBL is an adhesion molecule of *P. falciparum* and a member of the DBL-EBP family, identified on the basis of sequence similarity to EBA175. BAEBL (also known as EBA140) is encoded by a single-copy gene on chromosome 13. The *baebl* ORF shows the conserved four-exon structure of the DBL-EBP family. The predicted protein product of 1210 residues contains all the conserved features, including the two DBL domains and the C-cys domain. BAEBL is expressed as a 135kDa molecule and has been localised to micronemes by immunofluorescence. Similarly to EBA175, and as may be expected from the presence of a DBL domain, BAEBL has been shown to bind RBCs. The binding is sialic acid-dependent and trypsin-sensitive, however the receptor specificity of EBA175 and BAEBL are different, and it seems that glycophorins C and/or D may be involved (Mayer *et al.*, 2001). A recent study using targeted disruption of BAEBL, enzymatic treatment of RBCs, and assays with mutant phenotype RBCs has indicated that glycophorin C is the main BAEBL receptor on the RBC surface (Maier *et al.*, 2003). This finding was confirmed by the study of Lobo and colleagues who determined that the binding region on glycophorin C lies within exon 2, between residues 14 and 22 (Lobo *et al.*, 2003b). Interestingly, sequence polymorphism of the first DBL domain of BAEBL has been shown to result in changes of RBC ligand specificity resulting in differing invasion patterns of a panel of *P. falciparum* strains (Mayer *et al.*, 2002).

## 8.4. JESEBL

JESEBL (also known as EBA181) is a member of the DBL-EBP family identified by *P. falciparum* genome searches. The JESEBL gene exists in a single copy on chromosome 1 and has a typical four-exon structure. The predicted protein product of 1567 residues contains the characteristic features of the DBL-EBP family (Adams et al., 2001). The apparent molecular mass of JESEBL is 190 kDa although a second band of 170 kDa, which might represent a processing product, is also observed. JESEBL colocalises with EBA175 in micronemes and is expressed around the same time during the parasite life cycle. Interestingly, JESEBL appears to be expressed at different levels in different parasite cell lines. In the W2mef strain JESEBL is expressed at lower levels and seems to be redundant, since targeted disruption of the JESEBL gene had no effect on parasite's invasion of RBCs. In the 3D7 strain, however, JESEBL is expressed at significantly higher levels and appears to play an important role during the invasion process, since the disruption of JESEBL proved to be lethal to parasites. JESEBL appears to be a functional analogue of EBA175 and BAEBL, since it was shown to bind RBC surface via a novel sialic-acid dependent receptor, termed receptor E (Gilberger *et al.*, 2003).

## 8.5. Proteins with Unconfirmed Microneme Location

### 8.5.1. RBPs

Two adhesive proteins of *P. vivax* have been identified on the basis of their binding to reticulocytes, and named the reticulocyte binding proteins 1 and 2 (PvRBP1 and PvRBP2). PvRBPs are encoded by single-copy genes with a conserved two-exon structure, and are 2869 residues (PvRBP1) and 2867 residues (PvRBP2) in length. Although the overall similarity between the two proteins is about 45%, they do share a similar structure, including conserved signal peptides, the transmembrane domains, and the short cytoplasmic tails. PvRBP1 contains a number of cysteine residues that form disulphide bonds, whereas PvRBP2 has two repeat regions located close to the transmembrane domain, within the extracellular portion. In addition, PvRBP-2 has an approximately 500 residue region showing similarity to a member of the Py235 superfamily, which also has a similar overall ORF and protein structure. The PvRBP proteins are localised at the apical pole of *P. vivax* merozoites and have been shown to bind specifically to reticulocytes, which are the cells preferentially invaded by *P. vivax* parasites. That the RBPs mediate specific recognition of reticulocytes seem to be confirmed by the fact that homologous genes have been identified in the genome of *P. cynomolgi*, which also preferentially invades reticulocytes, but not in *P. knowlesi*, which invades all types of RBCs (Galinski *et al.*, 1992; 2000). Interestingly, the most similar proteins to the RBPs in *P. falciparum* are found in the rhoptries, but the RBPs certainly do not appear to be in these organelles unless they are present in a circumscribed subdomain near the merozoite apex.

### 8.5.2. EBL-1 and PEBL

EBL1 is a member of the DBL-EBP family identified on the basis of sequence similarity to EBA175. EBL1 is encoded by a single-copy gene on chromosome 13 of the *P. falciparum* genome. The *ebl1* ORF is transcribed in late schizont stages, and encodes a predicted protein of 2647 residues (Peterson *et al.*, 2000). No information on expression, localisation and potential RBC binding properties is available as yet.

P. falciparum genome searches identified an additional member of the family, PEBL (also known as EBA165). The *PEBL* gene exists in a single copy on chromosome 4 and encodes a protein of 1431 residues. The *PEBL* ORF has the conserved four-exon structure and the predicted protein product contains the characteristic features of the DBL-EBP family (Adams *et al.*, 2001). However, PEBL appears to be a pseudogene that is transcribed but not translated in *P. falciparum* parasites, and targeted disruption of the *PEBL* gene had no effect on parasite's ability to invade RBCs (Triglia *et al.*, 2001b).

### 8.5.3. PcyEBP and PrEBP

Members of the DBL-EBP family have been identified in *P. cynomolgi* (PcyEBP) and *P. reichenowi* (PrEBP) on the basis of their sequence similarity to the known DBL-EBPs in *P. falciparum*, *P. vivax* and *P. knowlesi*. The PcyEBP and PrEBP proteins are coded by genes with the conserved intron-exon structure. At least two homologous *EBP* genes have been found in the *P. cynomolgi* genome, one of which has been sequenced, whereas only a single *PrEBP* ORF has been identified. The predicted protein products of 1045 residues (PcyEBP) and 1454 residues (PrEBP) contain the characteristic features: the signal peptide, the DBL domains (one in PcyEBP, and two in PrEBP), the C-cys domain, the transmembrane domain and the cytoplasmic tail (Okenu *et al.*, 1997; Ozwara *et al.*, 2001). Both proteins await localisation and functional examination.

*8.5.4. PyEBP*

The first rodent malaria member of the DBL-EBP family has been identified recently on the basis of sequence similarity to the PvDBP. The predicted protein product of 839 residues is encoded by a five-exon gene and has the features characteristic of the family including the signal peptide, a single DBL domain, the C-cys domain, the transmembrane domain and the cytoplasmic tail. The DBL domain of PyEBP was shown to bind mouse erythrocytes specifically in a reaction that was trypsin and neuraminidase resistant but chymotrypsin sensitive, indicating that the ligand molecule was the Dyffy antigen (Prasad *et al.*, 2003).

# 9. Dense Granule Proteins

Contents of the dense granules are released after entry of the merozoite to the RBC is complete. It is unlikely therefore that any of these proteins would be essential for invasion, but may have a role in the early stages of intra-erythrocytic development of the parasite.

## 9.1. RESA

Ring infected erythrocyte surface antigen (RESA) of *P. falciparum* is a 755 residue protein encoded by a single-copy gene. The protein lacks a classical signal peptide, but instead has a buried hydrophobic region typical of malaria proteins exported to the membrane skeleton. The predicted protein product contains two distinct blocks of repetitive sequence (Favaloro *et al.*, 1986). Initially located within dense granules, RESA is discharged from these organelles shortly after merozoite internalisation into the RBC. After its release, RESA has been localised in discrete regions the PV, as well as within the host cell at the membrane skeleton of the RBC (Aikawa *et al.*, 1990; Culvenor *et al.*, 1991), and has been shown to interact with RBC spectrin (Foley *et al.*, 1991). The spectrin-binding domain of RESA has been identified and is a 48-residue sequence located between the repeat blocks (Foley *et al.*, 1994). The exact function of RESA and the role of its interaction with spectrin remain to be elucidated, however it has been suggested to have a role in maintaining the stability of the RBC membrane (Da Silva *et al.*, 1994). A recent study in which mutant parasite strains were examined for their cytoadherence ability has shown that the loss of RESA results in increased adherence of parasitised RBCs to the CD36 molecule. It has been proposed that the absence of RESA might result in decreased RBC membrane rigidity which could promote cytoadherence events (Cooke *et al.*, 2002).

## 9.2. RIMA

Ring membrane antigen (RIMA) of *P. falciparum* is a small, 14kDa protein. RIMA has been localised to dense granules and, after its release, to the membrane of the early ring stages (Trager *et al.*, 1992). No further information is available at present.

## 9.3. PfSUB1 and PfSUB2

PfSUB1 and PfSUB2 are putative *P. falciparum* proteases. PfSUB1 is encoded by a single-copy, intronless gene and is 690 residues in length. PfSUB1 is a soluble protein expressed during schizogony as a primary translation product of 82 kDa. This protein is proteolytically processed in a multistep fashion, first in the ER to give a 54kDa (p54) product (via 60/ 61kDa intermediates), and then within the secretory pathway to produce a 47kDa (p47) peptide. This protein is further processed to a 43kDa form, a product which can be detected in culture media following schizont rupture, but is not carried into rings. Despite the sequence similarity and the presence of conserved subtilase residues, attempts to show protease activity of an external substrate have failed (Blackman *et al.*, 1998). Sajid and colleagues have detected proteolytic activity in which PfSUB1 itself is the substrate, leading to processing of the 54kDa form to the 47kDa. The processing site was identified to fall between $Asp_{251}$ and $Ala_{252}$, and the autoproteolytic activity was found to be calcium-dependent. The current view is that PfSUB1 is a subtilisin-like protease with an unusual target specificity, activated by an autocatalytic proteolytic cleavage most likely taking place in the target organelle (Sajid *et al.*, 2000) Jean and colleagues have shown that the the N-terminus of the PfSUB-1 propeptide, p31 is non-covalently bound to both p54 and p47. The C-terminus of p31 binds the active site of the enzyme and acts as a high affinity inhibitor of protease activity (Jean *et al.* 2003). A location for PfSUB1 in dense granules is in our view still uncertain, as the electron micrographs presented in the report do not provide unambiguous evidence. Micronemes and dense granules are often difficult to distinguish during microscopic examination. That PfSUB1 may be a microneme protein is supported by its apically-restricted localisation determined by both immunofluorescence and immunoelectron-microscopy, and by the fact that it is discharged from merozoites during early stages of the RBC invasion process. Additionally, a similar soluble subtilisin-like protease of *T. gondii*, TgSUB1 has been localised to microneme organelles and shown to be a part of the micronemal secretion (Miller *et al.*, 2001). Clearly, double-staining immunofluorescence microscopy with known organelle markers is needed to ascertain the correct PfSUB1 localisation.

PfSUB2 is encoded by a single-copy gene on chromosome 11. The *PfSUB2* ORF contains two exons and encodes a predicted protein product of 1337 residues. Interestingly, in PfSUB2 the intron separates the predicted subtilisin active site. Apart from a signal peptide, PfSUB2 also contains a hydrophobic membrane-spanning region. PfSUB-2 is expressed during schizogony as an integral membrane protein with an apparent molecular mass of 160 kDa. This protein is proteolytically processed at the N-terminus to produce a 65 kDa form present in free merozoites (Barale *et al.*, 1999). In a separate report, Hackett and colleagues showed that PfSUB2 matures via a two-step process, from a 160kDa molecule to a 74kDa intermediate, to a 72kDa mature protein present in free merozoites. Again this is thought to be activation of a zymogen into an active enzyme (Hackett *et al.*, 1999). It has been suggested that PfSUB2 might be the MSP1 maturase that catalyses the processing of $MSP1_{42}$ to $MSP1_{19}$. However, despite the presence of the conserved subtilase domain, no enzymatic activity has been shown for PfSUB2 (Barale *et al.*, 1999; Hackett *et al.*, 1999). Similarly to PfSUB1, the reported localisation of PfSUB2 in a subset of apically-located dense granules is questionable for exactly the same reasons as for PfSUB1.

## 10. RBC Receptors of Merozoite Adhesion Molecules Involved in Invasion

The redundancy and molecular diversity of merozoite molecules involved in recognition and attachment to the RBC surface is an evolutionarily conserved mechanism that malaria parasites have developed to enhance the chances of survival. Confronted with RBC surface molecule polymorphism and immune responses of mammalian hosts, *Plasmodium* species, particularly *P. falciparum*, have created a wide array of adhesion molecules present either in the secretory organelles or on the apical membrane (see above) that allow the parasite to utilise alternative invasion pathways through multiple RBC receptors (Hadley *et al.*, 1987; Perkins *et al.*, 1988a; Mayer *et al.*, 2002). This varying adhesin repertoire is thought to explain the observation that malaria parasites of different species differ in their efficiency of invasion of RBCs. For example *P. vivax* and *P. cynomolgi* invade reticulocytes, whereas *P. knowlesi* and *P. falciparum* are capable of invading RBC of all ages. Furthermore, experiments using enzyme treatment of RBCs or genetic modifications of adhesion molecules have demonstrated that *P. falciparum* parasites can switch the invasion pathway they utilise (Dolan *et al.*, 1990; Reed *et al.*, 2000).

The earliest identified receptor molecules for merozoite invasion on the RBC membrane are members of the Duffy blood-group surface antigens, now known to be chemokine receptors and renamed the Duffy antigen/receptor for chemokines (DARC) (Horuk *et al.*, 1993). DARC has been shown to be the receptor for *P. knowlesi* and *P. vivax* adhesion proteins (Miller *et al.*, 1976; Haynes *et al.*, 1988), and these adhesion proteins have been later shown to belong to a superfamily of proteins containing the characteristic

DBL domain (see above). Studies of *P. falciparum* that followed led to identification of a micronemal adhesion molecule EBA175 (see above) and the major RBC receptors for *P. falciparum*, glycophorin A and B. The EBA175-glycophorin A interaction was found to be mediated by both protein and sialic acid components of the receptor molecule (Dolan *et al.*, 1994; Sim *et al.*, 1994b). Identification of the BAEBL adhesion protein was accompanied by recognition of glycophorins C and D as yet other parasite-binding proteins (Mayer *et al.*, 2001). However, glycophorins may not be the only RBC receptors for parasite entry and target molecules are still being defined on the basis of altered invasion efficiency of mutant or enzymatically-treated RBCs. More recently DARC has also been shown to be a receptor for *P. yoelii* (Swardson-Olver *et al.*, 2002), with the most likely candidate for the parasite ligand(s) being the newly identified *P. yoelii* EBP (Prasad *et al.*, 2003). Swardson-Olver and colleagues also showed the presence of a DARC-independent invasion pathway and demonstrated that alternative pathways exist for invasion of reticulocytes and normocytes by *P. yoelii*.

Changes in the invasion pathways used by parasites can be caused experimentally such as by truncation of EBA175 (Reed *et al.*, 2000) or may occur naturally as a result of polymorphism in binding domains such as the DBL domain of BAEBL (Mayer *et al.*, 2002). In addition, different expression levels of adhesion proteins such as JESEBL may influence the invasion pattern of different parasite strains (Gilberger *et al.*, 2003). Similarly, recent identification and examination of members of the PfRH family (see above) have demonstrated that the phenotypic variation in PfRH proteins in different *P. falciparum* strains is associated with altered RBC receptor usage (Duraisingh *et al.*, 2003). To date the presence of at least four glycophorin-independent, as yet unidentified RBC receptors, named X, Y, Z and E has been proposed (Dolan *et al*, 1994; Rayner *et al.*, 2001; Duraisingh *et al.*, 2003; Gilberger *et al.*, 2003). There are undoubtedly further complications in this story still to be elucidated, and it is likely that field isolates may manifest even greater variability than that found in the comparatively small number of laboratory isolates in which these pathways have been defined (Okoyeh *et al.*, 1999).

# 11. Other Merozoite Proteins Molecules Implicated in Invasion

## 11.1. Components of the Molecular Motor

The entry of merozoites into the RBC is an active process that requires motility and expenditure of energy. Since the RBC itself is not capable of either phagocytosis or receptor-mediated endocytosis, most of the events that occur during invasion must be driven by the parasite. The RBC is not a completely passive container however as it has been shown that RBC-derived ATP and cytoskeletal phosphorylation by an RBC kinase are also required for invasion (Rangachari *et al.*, 1986; Rangachari *et al.*, 1989).

## 11.2. Actins

*P. falciparum* and other apicomplexan parasites have been shown to invade host cells using an actin-myosin motor (King, 1988; Bannister *et al.*, 1995; Dobrowolski *et al.*, 1997b; Pinder *et al.*, 1998). This parasite-mediated driving force underlying the junctional movement has been investigated in studies using cytochalasins, actin-capping and depolymerising agents. Pre-treatment of merozoites with cytochalasin B has been shown to prevent invasion of erythrocytes. This finding was first demonstrated for *P. knowlesi* (Miller *et al.*, 1979) and later confirmed in *P. falciparum* (Field *et al.*, 1993). Merozoites treated in this manner remain tightly attached to the RBC surface and secrete material from the apical organelles causing local vacuolation of the RBC membrane, however the subsequent steps in the invasion process do not occur (Miller *et al.*, 1979; Dluzewski *et al.*, 1989). The cytochalasin studies strongly suggest that actin plays a role. Actin microfilaments have never been detected in merozoites by electron microscopy, however high concentrations of the protein have been found in *P. falciparum* merozoites, predominantly in the filamentous form (F-actin) (Field *et al.*, 1993). F-actin was shown to be localised at the poles and periphery of merozoites (Webb *et al.*, 1996). Myosin, which is known to play a role in gliding motility and invasion of *T. gondii* tachyzoites (Dobrowolski *et al.*, 1997a) has also been shown to have a similar distribution in merozoites of that parasite (Webb *et al.*, 1996).

Two actin genes have been identified in *P. falciparum*. *pf-actinI* is expressed throughout the life cycle, whereas *pf-actinII* is expressed only in sexual stages (Wesseling *et al.*, 1989). A family of actin-capping and -uncapping proteins have been identified in *P. knowlesi* merozoites, including a protein homologous to heat-shock protein 70 of *P. falciparum* (Tardieux *et al.*, 1998a). It was found that the HSC70/32/34 complex exhibited capping activity by inhibiting the polymerisation of rabbit skeletal muscle actin *in vitro* and preventing shortening the average length of actin filaments. The authors suggested that capping or uncapping of actin filaments during invasion could provide a mechanism for localised actin filament growth and movement of the merozoite into the host RBC (Tardieux *et al.*, 1998a). Tardieux and colleagues (Tardieux *et al.*, 1998b) have also identified a *P. falciparum* homologue of an actin-associated protein called coronin. A homologue of the widely conserved actin-depolymerising factor (ADF) has been identified in *T. gondii* (Allen *et al.*, 1997) but not in *Plasmodium* species. Not all the components of a conventional actin-myosin motor have yet been identified. To date, a homologue of the actin monomer binding protein profilin has not been identified in apicomplexan parasites. However, there is a substitute protein in *Toxoplasma*, called toxofilin, which acts by inhibiting actin polymerisation and slowing microfilament disassembly (Poupel *et al.*, 2000).

## 11.3. Myosins

Myosins are ATPases that contain head domains that bind to and are activated by actin, and tails that bind to other proteins. When an ATP molecule is hydrolysed, there is a conformational change in the myosin head that generates force and movement, with the actin and the other partner moving relative to each other (Pinder *et al.*, 2000).

Recently, three myosin genes (*Pfmyo-A*, *Pfmyo-B* and *Pfmyo-C*) have been identified in *P. falciparum* (Pinder *et al.*, 1998; Hettmann *et al.*, 2000). In the asexual blood stage of the malaria parasite, Pfmyo-A is only expressed in mature merozoites (Pinder *et al.*, 1998) and is peripherally located with a concentration towards the apex. Electron microscopy indicates that it is located between the plasma membrane and the cisternal membranes (Pinder *et al.*, 2000). As a recent study has shown that Tgmyo-A is essential for gliding motility and host cell invasion of tachyzoites (Meissner *et al.*, 2002b), it seems reasonable to propose that Pfmyo-A forms part of the actomyosin motor that drives merozoite invasion. In support of this, the reversible myosin inhibitor 2,3-butane-dione monoxime (BDM) blocks invasion (Dobrowolski *et al.*, 1997b; Pinder *et al.*, 1998).

In *T. gondii* myo-A (Tgmyo-A), the tail domain of this class XIV myosin is required for targeting the protein to the tachyzoite periphery and a dibasic motif in the tail region is an essential determinant of plasma membrane localisation (Hettmann *et al.*, 2000). Similar motifs may be important in Pfmyo-A localization, although this motif might anchor myosin at the membrane via another protein. A novel protein that localises to the inner membrane complex and interacts with the 15 residue C-terminal tail region of Pfmyo-A has recently been identified using the yeast two-hybrid system (Bergman *et al.*, 2003). This protein, dubbed MTIP (Myosin A tail domain interacting protein), apparently binds to this dibasic motif in myosin. A homolog of MTIP (TgMLC1) has been identified and shown to associate with Tgmyo-A (Herm-Gotz *et al.*, 2002) and this probably localises Tgmyo-A to the tachyzoite periphery (Bergman *et al.*, 2003). The results of Bergman and colleagues (Bergman *et al.*, 2003) appear to favour the model for an actomyosin motor configuration in which the MTIP/myo-A complex would remain stationary, tethered to the outer membrane of the IMC by an unidentified protein. This model predicts that actin, somehow linked (possibly via aldolase) to the cytoplasmic domain of a transmembrane protein (now known to be a member of the thrombospondin-related anonymous protein

(TRAP) family (Sultan *et al.*, 1997; Kappe *et al.*, 1999; Matuschewski *et al.*, 2002)), is then moved along this complex. A similar model may also apply to merozoite invasion of erythrocytes (Chitnis *et al.*, 2000; Pinder *et al.*, 2000), although TRAP itself is not expressed in this stage and another protein to serve this role must be sought.

## 11.4. Microtubules

Depolymerisation of microtubules has been shown to inhibit invasion by merozoites (Bejon *et al.*, 1997; Fowler *et al.*, 1998). The *P. falciparum* merozoite assemblage of subpellicular microtubules (*f*-MAST) is a narrow band of two or three microtubules running in parallel along one side of the merozoite from the third polar ring to the posterior, just beneath the cisternal membranes (Read *et al.*, 1993; Bannister *et al.*, 1995; Fowler *et al.*, 1998). Antimicrotubule drugs have been used to investigate the role of the *f*-MAST in invasion. One such study showed that treatment with colchicine at concentrations that did not affect merozoite development moderately inhibited invasion (Bejon *et al.*, 1997). In another study, it was found that depolymerisation of microtubules using dinitrianilines reduced invasion rates whereas stabilising microtubules with Taxol had no effect on invasion (Fowler *et al.*, 1998) Recently, two microtubule-associated motor proteins, kinesin and dynein, have been characterised in *P. falciparum* merozoites (Fowler *et al.*, 2001).

## 11.5. Junction Proteins

Adhesins from the apical organelles are translocated along the length of the parasite and are shed at the site of the moving junction. The moving junction is a highly specialised interface of the parasite with the RBC that would seem to employ cytoskeletal proteins, signalling molecules, and receptors (see Pinder *et al.*, (2000) for a review). One merozoite submembranous protein that has been described at this junction is merozoite capping protein 1 (MCP1). This protein follows the distribution of the moving junction during merozoite invasion of erythrocytes (Klotz *et al.*, 1989). MCP1 lacks a signal sequence and transmembrane domains and is located in the parasite cytosol. The N-terminal domain includes a region that is conserved in a large family of bacterial and eukaryotic oxidoreductases (Hudson-Taylor *et al.*, 1995). A group of microneme proteins have been identified in *T. gondii* that also localise to the moving junction and the exposed region during invasion (Fourmaux *et al.*, 1996; Carruthers *et al.*, 1997; 1999; Garcia-Reguet *et al.*, 2000).

There are still a number of details missing within this model including how events at the surface such as red cell binding, or rhoptry discharge lead to the activation of the motor. Alternatively it may be possible that the motor is continually active as soon as lysis occurs and it is only when the merozoite is bound to the RBC surface that action of the motor leads to entry. Prior to this it might be involved in random motion within the bloodstream and ensuring that the released merozoite comes into contact with target RBCs.

## 12. Proteases and Merozoite Invasion

A number of studies have shown that invasion by *P. falciparum* merozoites can be blocked by certain serine and cysteine protease inhibitors, such as chymostatin and leupeptin (Dluzewski *et al.*, 1986; Braun-Breton *et al.*, 1988b; Olaya *et al.*, 1991; Braun-Breton *et al.*, 1992), suggesting that proteolytic events are essential for invasion. More than one distinct protease appears to be involved and these proteases may function at different stages of the invasion pathway (Hadley *et al.*, 1983; Dluzewski *et al.*, 1986). For example, inhibition of *P. falciparum* merozoite invasion by the serine and cysteine protease inhibitors chymostatin, but not leupeptin, can be reversed by pretreatment of target erythrocytes with chymotrypsin, suggesting that chymostatin and leupeptin may be inhibiting different activities (Dluzewski *et al.*, 1986). Studies with isolated, invasive *P. knowlesi* merozoites have shown that whereas N-alpha-p-tosyl-L-lysine chloromethyl ketone (TLCK; trypsin-like protease inhibitor) and L-1-tosylamide-2-phenylethylchloromethyl ketone (TPCK; chymotrypsin-like protease inhibitor) both prevent attachment of the merozoite to the RBC, chymostatin blocks a later stage in the invasion (Hadley *et al.*, 1983).

Parasite proteases are believed to play a crucial role in at least two distinct aspects of RBC invasion: modification of parasite proteins involved in host cell recognition and entry; and restructuring of the RBC surface and cytoskeleton during and following invasion. They also appear to have a role in parasite release from the RBC (Blackman, 2000; Salmon *et al.*, 2001). A number of merozoite surface proteins and organellar components of the merozoite have been shown to undergo proteolytic modification around the point of RBC invasion (Adams *et al.*, 1990; Blackman *et al.*, 1992; Narum *et al.*, 1994; Ogun *et al.*, 1996; Howard *et al.*, 1998b; Pachebat *et al.*, 2001; Trucco *et al.*, 2001). The best-characterised example is the secondary processing of MSP1, which cleaves $MSP1_{42}$ into $MSP1_{19}$ and $MSP1_{33}$ by a membrane-associated serine protease (Blackman *et al.*, 1992; 1993). Although the function of this processing step is unclear, it appears to be important for invasion as monoclonal antibodies that inhibit the protease-mediated processing also inhibit merozoite invasion of the RBC and vice versa (Blackman *et al.*, 1994). It is proposed that processing allows the EGF-like domains, which are known in other proteins to participate in molecular adhesion, to interact with host or parasite molecules (Holder *et al.*, 1994). Proteolytic processing of AMA1 has also been studied in some detail as described above and the same serine protease that cleaves MSP1 has been implicated (Howell *et al.*, 2003).

Proteolytic modification of the host RBC is also a necessary step for merozoite invasion (Braun-Breton *et al.*, 1992; McPherson *et al.*, 1993; Roggwiller *et al.*, 1996). The first evidence for this derived from the study of Dluzewski and colleagues (Dluzewski *et al.*, 1986), indicating an obligatory role for a chymostatin-sensitive protease at invasion, which acts on an RBC surface substrate. Subsequent studies showed that chymotrypsin-treatment of RBCs results in degradation of band 3 and induces a localised disruption of the RBC membrane (McPherson *et al.*, 1993). This destabilised region of membrane may represent the site for the insertion of parasite-derived phospholipids and/or proteins allowing the formation of the PVM. Peptides corresponding to the sequence of the predicted band 3 cleavage sites are able to inhibit merozoite invasion (Braun-Breton *et al.*, 1992; Roggwiller *et al.*, 1996). These results have suggested that proteolysis of band 3 is a necessary process in invasion, and this activity facilitates RBC membrane invagination and the PV formation by restructuring the red cell cytoskeleton.

A substantial body of evidences indicates that merozoite release from the mature schizont requires protease activity. Although the identity of proteases involved in the degradation of the parasitophorous vacuole membrane and the infected RBC membrane of mature schizonts is still being uncovered, treatment of schizonts with some protease inhibitors has been shown to have a substantial or a partial inhibitory effect on merozoite release (Hadley *et al.*, 1983; Dluzewski *et al.*, 1986; Salmon *et al.*, 2001). Delplace and colleagues showed clearly that release of *P. falciparum* merozoites *in vitro* could be prevented by treatment with leupeptin, which appeared to block RBC membrane rupture (Delplace *et al.*, 1988). Lyon and colleagues (Lyon *et al.*, 1986b) found a combination of leupeptin, chymostatin, pepstatin and antipain could dramatically affect schizont release, resulting in the accumulation of aggregates of merozoite similar to those obtained by culturing schizonts in the presence of immune sera (Chulay *et al.*, 1987). A similar observation has been made in a different system, in which addition of both leupeptin and chymostatin to maturing *P. knowlesi* schizonts could allow parasite maturation to the segment stage but prevent normal rupture of the schizonts (David *et al.*, 1984). More recently, RBCs incubated in the presence of the cysteine protease inhibitor, 1-transepoxy-succinyl-leucylamido-(4-guanidino)butane (E64) resulted in accumulation of extraerythrocytic merozoites enclosed within the PVM, suggesting that proteolysis is required for this phase of the release process (Salmon *et al.*, 2001). To date several proteases in mature stages and merozoites have been described. At least two of these are implicated in merozoite release from infected RBCs and invasion, gp76 (see above) and falcipain-2. *P. falciparum* cysteine protease falcipain-2 has been shown to cleave ankyrin and protein 4.1, which it is suggested will destabilise the RBC membrane cytoskeleton and facilitate merozoite release (Raphael *et al.*, 2000; Dua *et al.*, 2001). Further studies identified the cleavage sites on ankyrin and protein 4.1, and showed that recombinant peptides containing those sites could inhibit falcipain-2 activity and parasite growth (Hanspal *et al.*, 2002; Dhawan *et al.*, 2003). The role of other proteases such as PfSUB1 and PfSUB2 are still being defined (see above).

## 13. The Genome Project and Merozoite Cell Biology Studies

Clearly the malaria genome has the capacity to rapidly advance our knowledge of merozoite cell biology. Early searches revealed the presence of a number of genes sharing homology to genes already identified to have a role in invasion. Thus the SERA gene is one of nine such genes in the *P. falciparum* 3D7 genome whose expression appears to vary in different isolates. (Gardner *et al.*, 1998; Miller *et al.*, 2002). Homologues of the reticulocyte binding protein in *P. vivax* have an important role in defining which RBC receptors the parasite recognizes and presumably confer differing efficiencies of invasion on different isolates in the wild (Rayner *et al.*, 2001; Duraisingh *et al.*, 2003). Proteins that share important functional domains were also found (Michon *et al.*, 2002). Thus the two EGF-like domains of MSP1 are also found in MSP8 and MSP10, proteins with distinct expression patterns and locations in the merozoite (Black *et al.*, 2001; 2003). With the completion of at least a first draft of the genome, an iterative search to define all related sequences and sequence families provides a number of new groups to examine. The clag family for example appears to have a number of members, at least one of which appears to be a secreted rhoptry product. There are plans to partially sequence a number of other rodent and primate malarias to partial coverage. This information will allow identification of conserved and variable regions within proteins of interest that may offer information about functional domains. It will also assist the construction of gene models for multi-exon genes, particularly those with small exons, which may currently be missed or incorrectly predicted. Finally, synteny between species might allow identification of poorly conserved homologues of proteins of importance. Thus in pre-genome project times, the MSP4/5 homologue in rodent species was identified by cross-species hybridisation of the highly conserved adenylosuccinate lyase gene, which is located nearby (Black *et al*,. 1999). Such laborious laboratory-based approaches are now replaced by a few keystrokes at the computer.

It would also be of interest to try and define the complete set of possible merozoite surface and organellar proteins. Thus, a systematic search of the data for encoded proteins containing signal sequences should give us a set of proteins among which will be found the secreted and anchored proteins of the merozoite surface, rhoptries and micronemes. At present this search is still difficult to do as the gene prediction models for multi-exon genes are worst at predicting first and last exons, particularly when these exons are short. These are precisely the positions of sequences that encode signal and anchor sequences, and it is not uncommon for first exons of malaria genes to be quite short. Further the programs often include an upstream region of non-coding sequence when they miss the exon boundary. This non-coding sequence frequently contain AT dinucleotide repeats, which are translated as isoleucine-tyrosine repeats which give a false patina of hydrophobicity and lead to artifactual prediction of signal sequences. Collectively these factors lead to a high number of false positive and false negative genes in the list of potentially exported proteins. Further work on proteins transported to organelles may define specific transport motifs associated with micronemal or rhoptry importation, in the same way as has been noted for *Toxoplasma* rhoptry proteins. Back searching of these against the genome could identify a set of candidate organellar proteins. Such a process has been successfully used to identify the set of proteins likely to be present in the apicoplast, by identification of the hybrid signal and import peptides (Foth *et al.*, 2003). Proteomic data from detergent extracts of the merozoite or purified organelles would be a valuable additional data source, notwithstanding the difficulties of making pure preparations from these sources.

Beyond the identification of one or several genes transported to particular locations is the identification of the full set of proteins involved in particular cellular processes. Here, information from other well-studied systems such as yeast, *Toxoplasma* and the mammalian cell is used to prepare a list of proteins whose homologues are sought in the malaria genome. Processes of interest to merozoite cell biology include protein transport, the molecular motor and the components of DNA replication. Parasite proteins that differ significantly from homologues in higher eukaryotic cells may be potential targets for specific drug development. The total set of proteins involved in signalling processes including phosphorylation and dephosphorylation would provide a starting set of candidates that might be involved in the regulation of the invasion process. Such studies on the human kinase repertoire have already identified interesting candidates for particular processes (Manning *et al.*, 2002). It seems reasonable to suppose that the cytoplasmic tail of proteins such as EBA175, AMA1 and the PfRH proteins may undergo post-translational modification upon binding of the red cells and trigger later steps in the invasion process. Kinases are common in these sequential cascades although they do not mediate the only possible modifications that might take place.

# 14. Conclusions

The merozoite is difficult to study because of its fastidious requirements to maintain function, its transient existence as an independent cell, the fragile nature of many cellular components and its general refractoriness to genetic manipulation. Notwithstanding these limitations significant progress has been made in recent years to identify the set of proteins involved in invasion and define their function. Useful advances would include: the development of fractionation procedures that produce purified cellular fractions of organelles and membranes that would facilitate proteomic studies; improvements in genetic technology that increase the efficiency of transformation, better vectors with a wider selection of positive and negative selectable markers and the development of an inducible promoter that tightly controls the expression of genes. The latter development particularly could allow production of mutant lines with modified, dominant negative or anti-sense genes under tight control. Once the lines were generated and synchronized, specific genes could be turned on or off and the resulting change in invasion behaviour directly observed. However even with current tools, the genome will allow the design of experiments that will significantly advance our understanding of how these cells function.

## 15. Acknowledgments

This work was supported by grants from the Australian National Health and Medical Research Council, the United States National Institute of Health DK-32094, the Burroughs Wellcome Fund, the Howard Hughes Medical Institute International Scholars in Infectious Diseases and Parasitology Program and the UNDP/World Bank/WHO Special Programme for Research and Training in Tropical Diseases.

## 16. References

Adams, J.H., Blair, P.L., Kaneko, O., and Peterson, D.S. 2001. An expanding ebl family of *Plasmodium falciparum*. Trends Parasitol. 17: 297-299.

Adams, J.H., Hudson, D.E., Torii, M., Ward, G.E., Wellems, T.E., Aikawa, M., and Miller, L.H. 1990. The Duffy receptor family of *Plasmodium knowlesi* is located within the micronemes of invasive malaria merozoites. Cell. 63: 141-153.

Adams, J.H., Sim, B.K., Dolan, S.A., Fang, X., Kaslow, D.C., and Miller, L.H. 1992. A family of erythrocyte binding proteins of malaria parasites. Proc. Natl. Acad. Sci. USA. 89: 7085-7089.

Aikawa, M., Miller, L.H., Rabbege, J.R., and Epstein, N. 1981. Freeze-fracture study on the erythrocyte membrane during malarial parasite invasion. J. Cell. Biol. 91: 55-62.

Aikawa, M., and Seed, T.M. 1980. Morphology of Plasmodia. In: Malaria. J.P. Kreier, ed. Academic Press. p. 285-344.

Aikawa, M., Torii, M., Sjolander, A., Berzins, K., Perlmann, P., and Miller, L.H. 1990. Pf155/RESA antigen is localized in dense granules of *Plasmodium falciparum* merozoites. Exp. Parasitol. 71: 326-329.

Allen, M.L., Dobrowolski, J.M., Muller, H., Sibley, L.D., and Mansour, T.E. 1997. Cloning and characterization of actin depolymerizing factor from *Toxoplasma gondii*. Mol. Biochem. Parasitol. 88: 43-52.

al-Yaman, F., Genton, B., Anders, R., Taraika, J., Ginny, M., Mellor, S., and Alpers, M.P. 1995. Assessment of the role of the humoral response to *Plasmodium falciparum* MSP2 compared to RESA and SPf66 in protecting Papua New Guinean children from clinical malaria. Parasite Immunol. 17: 493-501.

Anders, R.F., Brown, G.V., and Edwards, A. 1983. Characterization of an S antigen synthesized by several isolates of *Plasmodium falciparum*. Proc. Natl. Acad. Sci. USA. 80: 6652-6656.

Anthony, R.N., Yang, J., Krall, J.A., and Sam-Yellowe, T.Y. 2000. Sequence analysis of the Rhop-3 gene of *Plasmodium yoelii*. J. Eukaryot. Microbiol. 47: 319-322.

Atkinson, C.T., Aikawa, M., Perry, G., Fujino, T., Bennett, V., Davidson, E.A., and Howard, R.J. 1988. Ultrastructural localization of erythrocyte cytoskeletal and integral membrane proteins in *Plasmodium falciparum*-infected erythrocytes. Eur. J. Cell. Biol. 45: 192-199.

Badell, E., Oeuvray, C., Moreno, A., Soe, S., van Rooijen, N., Bouzidi, A., and Druilhe, P. 2000. Human malaria in immunocompromised mice: an *in vivo* model to study defense mechanisms against *Plasmodium falciparum*. J. Exp. Med. 192: 1653-1660.

Baldi, D.L., Andrews, K.T., Waller, R.F., Roos, D.S., Howard, R.F., Crabb, B.S., and Cowman, A.F. 2000. RAP1 controls rhoptry targeting of RAP2 in the malaria parasite *Plasmodium falciparum*. EMBO J. 19: 2435-2443.

Baldi, D.L., Good, R., Duraisingh, M.T., Crabb, B.S., and Cowman, A.F. 2002. Identification and disruption of the gene encoding the third member of the low-molecular-mass rhoptry complex in *Plasmodium falciparum*. Infect. Immun. 70: 5236-5245.

Bannister, L.H., Butcher, G.A., Dennis, E.D., and Mitchell, G.H. 1975a. Structure and invasive behaviour of *Plasmodium knowlesi* merozoites *in vitro*. Parasitology. 71: 483-491.

Bannister, L.H., Butcher, G.A., Dennis, E.D., and Mitchell, G.H. 1975b. Studies on the structure and invasive behaviour of merozoites of *Plasmodium knowlesi*. Trans. R. Soc. Trop. Med. Hyg. 69: 5.

Bannister, L.H., Butcher, G.A., and Mitchell, G.H. 1977. Recent advances in understanding the invasion of erythrocytes by merozoites of *Plasmodium knowlesi*. Bull. World Health Organ. 55: 163-169.

Bannister, L.H., Hopkins, J.M., Fowler, R.E., Krishna, S., and Mitchell, G.H. 2000a. A brief illustrated guide to the ultrastructure of *Plasmodium falciparum* asexual blood stages. Parasitol. Today. 16: 427-433.

Bannister, L.H., Hopkins, J.M., Fowler, R.E., Krishna, S., and Mitchell, G.H. 2000b. Ultrastructure of rhoptry development in *Plasmodium falciparum* erythrocytic schizonts [In Process Citation]. Parasitology. 121: 273-287.

Bannister, L.H., and Mitchell, G.H. 1989. The fine structure of secretion by *Plasmodium knowlesi* merozoites during red cell invasion. J. Protozool. 36: 362-367.

Bannister, L.H., and Mitchell, G.H. 1995. The role of the cytoskeleton in *Plasmodium falciparum* merozoite biology: an electron-microscopic view. Ann. Trop. Med. Parasitol. 89: 105-111.

Bannister, L.H., Mitchell, G.H., Butcher, G.A., and Dennis, E.D. 1986a. Lamellar membranes associated with rhoptries in erythrocytic merozoites of *Plasmodium knowlesi*: a clue to the mechanism of invasion. Parasitology. 92: 291-303.

Bannister, L.H., Mitchell, G.H., Butcher, G.A., Dennis, E.D., and Cohen, S. 1986b. Structure and development of the surface coat of erythrocytic merozoites of *Plasmodium knowlesi*. Cell Tissue Res. 245: 281-290.

Banyal, H.S., and Inselburg, J. 1985. Isolation and characterization of parasite-inhibitory *Plasmodium falciparum* monoclonal antibodies. Am. J. Trop. Med. Hyg. 34: 1055-1064.

Barale, J.C., Blisnick, T., Fujioka, H., Alzari, P.M., Aikawa, M., Braun-Breton, C., and Langsley, G. 1999. *Plasmodium falciparum* subtilisin-like protease 2, a merozoite candidate for the merozoite surface protein 1-42 maturase. Proc. Natl. Acad. Sci. USA. 96: 6445-6450.

Barnwell, J.W., and Galinski, M.R. 1998. Invasion of vertebrate cells: erythrocytes. In: Malaria - parasite biology, pathogenesis, and protection. I.W. Sherman, ed. ASM Press. p. 93-120.

Bejon, P.A., Bannister, L.H., Fowler, R.E., Fookes, R.E., Webb, S.E., Wright, A., and Mitchell, G.H. 1997. A role for microtubules in *Plasmodium falciparum* merozoite invasion. Parasitology. 114: 1-6.

Bergman, L.W., Kaiser, K., Fujioka, H., Coppens, I., Daly, T.M., Fox, S., Matuschewski, K., Nussenzweig, V., and Kappe, S.H. 2003. Myosin A tail domain interacting protein (MTIP) localizes to the inner membrane complex of Plasmodium sporozoites. J. Cell. Sci. 116: 39-49.

Bickle, Q., Anders, R.F., Day, K., and Coppel, R.L. 1993. The S-antigen of *Plasmodium falciparum*: Repertoire and origin of diversity. Mol. Biochem. Parasitol. 61: 189-196.

Bickle, Q., and Coppel, R.L. 1992. A fourth family of the *Plasmodium falciparum* S-antigen. Mol. Biochem. Parasitol. 56: 141-150.

Black, C.G., Barnwell, J.W., Huber, C.S., Galinski, M.R., and Coppel, R.L. 2002. The *Plasmodium vivax* homologues of merozoite surface proteins 4 and 5 from *Plasmodium falciparum* are expressed at different locations in the merozoite. Mol. Biochem. Parasitol. 120: 215-224.

Black, C.G., Wang, L., Hibbs. A.R., Werner, E., and Coppel, R.L. 1999. Identification of the *Plasmodium chabaudi* homologue of merozoite surface proteins 4 and 5 of *Plasmodium falciparum*. Infect. Immun. 67: 2075-2081.

Black, C.G., Wang, L., Wu, T., and Coppel, R.L. 2003. Apical location of a novel EGF-like domain-containing protein of *Plasmodium falciparum*. Mol. Biochem. Parasitol. 127: 59-68.

Black, C.G., Wu, T., Wang, L., Hibbs, A.R., and Coppel, R.L. 2001. Merozoite surface protein 8 of *Plasmodium falciparum* contains two epidermal growth factor-like domains. Mol. Biochem. Parasitol. 114: 217-226.

Blackman, M.J. 2000. Proteases involved in erythrocyte invasion by the malaria parasite: function and potential as chemotherapeutic targets. Curr. Drug. Targets. 1: 59-83.

Blackman, M.J., and Bannister, L.H. 2001. Apical organelles of Apicomplexa: biology and isolation by subcellular fractionation. Mol. Biochem. Parasitol. 117: 11-25.

Blackman, M.J., Chappel, J.A., Shai, S., and Holder, A.A. 1993. A conserved parasite serine protease processes the *Plasmodium falciparum* merozoite surface protein-1. Mol. Biochem. Parasitol. 62: 103-114.

Blackman, M.J., Dennis, E.D., Hirst, E.M., Kocken, C.H., Scott-Finnigan, T.J., and Thomas, A.W. 1996. *Plasmodium knowlesi*: secondary processing of the malaria merozoite surface protein-1. Exp. Parasitol. 83: 229-239.

Blackman, M.J., Fujioka, H., Stafford, W.H., Sajid, M., Clough, B., Fleck, S.L., Aikawa, M., Grainger, M., and Hackett, F. 1998. A subtilisin-like protein in secretory organelles of *Plasmodium falciparum* merozoites. J. Biol. Chem. 273: 23398-23409.

Blackman, M.J., Heidrich, H.G., Donachie, S., McBride, J.S., and Holder, A.A. 1990. A single fragment of a malaria merozoite surface protein remains on the parasite during red cell invasion and is the target of invasion-inhibiting antibodies. J. Exp. Med. 172: 379-382.

Blackman, M.J., and Holder, A.A. 1992. Secondary processing of the *Plasmodium falciparum* merozoite surface protein-1 (MSP1) by a calcium-dependent membrane-bound serine protease: shedding of MSP133 as a noncovalently associated complex with other fragments of the MSP1. Mol. Biochem. Parasitol. 50: 307-315.

Blackman, M.J., Ling, I.T., Nicholls, S.C., and Holder, A.A. 1991a. Proteolytic processing of the *Plasmodium falciparum* merozoite surface protein-1 produces a membrane-bound fragment containing two epidermal growth factor-like domains. Mol. Biochem. Parasitol. 49: 29-33.

Blackman, M.J., Scott-Finnigan, T.J., Shai, S., and Holder, A.A. 1994. Antibodies inhibit the protease-mediated processing of a malaria merozoite surface protein. J. Exp. Med. 180: 389-393.

Blackman, M.J., Whittle, H., and Holder, A.A. 1991b. Processing of the *Plasmodium falciparum* major merozoite surface protein-1: identification of a 33-kilodalton secondary processing product which is shed prior to erythrocyte invasion. Mol. Biochem. Parasitol. 49: 35-44.

Blair, P.L., Kappe, S.H., Maciel, J.E., Balu, B., and Adams, J.H. 2002. *Plasmodium falciparum* MAEBL is a unique member of the ebl family. Mol. Biochem. Parasitol. 122: 35-44.

Borre, M.B., Owen, C.A., Keen, J.K., Sinha, K.A., and Holder, A.A. 1995. Multiple genes code for high-molecular-mass rhoptry proteins of *Plasmodium yoelii*. Mol. Biochem. Parasitol. 70: 149-155.

Braun-Breton, C., Blisnick, T., Barbot, P., Bulow, R., Pereira da Silva, L., and Langsley, G. 1992. *Plasmodium falciparum* and Plasmodium chabaudi: characterization of glycosylphosphatidylinositol-degrading activities. Exp. Parasitol. 74: 452-462.

Braun-Breton, C., and Pereira da Silva, L. 1988a. Activation of a *Plasmodium falciparum* protease correlated with merozoite maturation and erythrocyte invasion. Biol Cell. 64: 223-231.

Braun-Breton, C., Rosenberry, T.L., and da Silva, L.P. 1988b. Induction of the proteolytic activity of a membrane protein in *Plasmodium falciparum* by phosphatidyl inositol-specific phospholipase C. Nature. 332: 457-459.

Breton, C.B., Blisnick, T., Jouin, H., Barale, J.C., Rabilloud, T., Langsley, G., and Pereira da Silva, L.H. 1992. *Plasmodium chabaudi* p68 serine protease activity required for merozoite entry into mouse erythrocytes. Proc. Natl. Acad. Sci. USA. 89: 9647-9651.

Brown, D.A., and London, E. 1998. Functions of lipid rafts in biological membranes. Annu. Rev. Cell. Dev. Biol. 14: 111-136.

Brown, D.A., and Rose, J.K. 1992. Sorting of GPI-anchored proteins to glycolipid-enriched membrane subdomains during transport to the apical cell surface. Cell. 68: 533-544.

Brown, H., Kemp, D.J., Barzaga, N., Brown, G.V., Anders, R.F., and Coppel, R.L. 1987. Sequence variation in S-antigen genes of *Plasmodium falciparum*. Mol. Biol. Med. 4: 365-376.

Brown, H.J., and Coppel, R.L. 1991. Primary structure of a *Plasmodium falciparum* rhoptry antigen. Mol. Biochem. Parasitol. 49: 99-110.

Burghaus, P.A., Gerold, P., Pan, W., Schwarz, R.T., Lingelbach, K., and Bujard, H. 1999. Analysis of recombinant merozoite surface protein-1 of *Plasmodium falciparum* expressed in mammalian cells. Mol. Biochem. Parasitol. 104: 171-183.

Burns, J.M. Jr., Belk, C.C., and Dunn, P.D. 2000. A protective glycosylphosphatid ylinositol-anchored membrane protein of *Plasmodium yoelii* trophozoites and merozoites contains two epidermal growth factor-like domains. Infect. Immun. 68: 6189-6195.

Bzik, D.J., Li, W.B., Horii, T., and Inselburg, J. 1988. Amino acid sequence of the serine-repeat antigen (SERA) of *Plasmodium falciparum* determined from cloned cDNA. Mol. Biochem. Parasitol. 30: 279-288.

Campbell, I.D., and Bork, P. 1993. Epidermal growth factor-like modules. Curr. Opin. Struct. Biol. 3: 385-392.

Camus, D., and Hadley, T.J. 1985. A *Plasmodium falciparum* antigen that binds to host erythrocytes and merozoites. Science. 230: 553-556.

Carcy, B., Bonnefoy, S., Guillotte, M., Le Scanf, C., Grellier, P., Schrevel, J., Fandeur, T., and Mercereau-Puijalon, O. 1994. A large multigene family expressed during the erythrocytic schizogony of *Plasmodium falciparum*. Mol. Biochem. Parasitol. 68: 221-233.

Carruthers, V.B., Giddings, O.K., and Sibley, L.D. 1999. Secretion of micronemal proteins is associated with toxoplasma invasion of host cells. Cell. Microbiol. 1: 225-235.

Carruthers, V.B., and Sibley, L.D. 1997. Sequential protein secretion from three distinct organelles of *Toxoplasma gondii* accompanies invasion of human fibroblasts. Eur. J. Cell. Biol. 73: 114-123.

Cheng, Q., and Saul, A. 1994. Sequence analysis of the apical membrane antigen 1 (AMA-1) of *Plasmodium vivax*. Mol. Biochem. Parasitol. 65: 183-187.

Chitnis, C.E., and Blackman, M.J. 2000. Host cell invasion by malaria parasites. Parasitol. Today. 16: 411-415.

Chitnis, C.E., and Miller, L.H. 1994. Identification of the erythrocyte binding domains of *Plasmodium vivax* and *Plasmodium knowlesi* proteins involved in erythrocyte invasion. J. Exp. Med. 180: 497-506.

Chulay, J.D., Lyon, J.A., Haynes, J.D., Meierovics, A.I., Atkinson, C.T., and Aikawa, M. 1987. Monoclonal antibody characterization of *Plasmodium falciparum* antigens in immune complexes formed when schizonts rupture in the presence of immune serum. J. Immunol. 139: 2768-2774.

Cooke, B.M., Glenister, F.K., Mohandas, N., and Coppel, R.L. 2002. Assignment of functional roles to parasite proteins in malaria-infected red blood cells by competitive flow-based adhesion assay. Br. J. Haematol. 117: 203-211.

Coppel, R.L., Bianco, A.E., Culvenor, J.G., Crewther, P.E., Brown, G.V., Anders, R.F., and Kemp, D.J. 1987. A cDNA clone expressing a rhoptry protein of *Plasmodium falciparum*. Mol. Biochem. Parasitol. 25: 73-81.

Coppel, R.L., Cowman, A.F., Lingelbach, K.R., Brown, G.V., Saint, R.B., Kemp, D.J., and Anders, R.F. 1983. Isolate-specific S-antigen of *Plasmodium falciparum* contains a repeated sequence of eleven amino acids. Nature. 306: 751-756.

Coppel, R.L., Crewther, P.E., Culvenor, J.G., Perrin, L.H., Brown, G.V., Kemp, D.J., and Anders, R.F. 1988. Variation in p126, a parasitophorous vacuole antigen of *Plasmodium falciparum*. Mol. Biol. Med. 5: 155-166.

Coppel, R.L., Davern, K.M., and McConville, M.J. 1994. Immunochemistry of Parasite Antigens. In: Immunochemistry. C.J. van Oss, and M.H.V. van Regenmortel, ed. Marcel Dekker Inc. p. 475-532.

Cowman, A.F., Baldi, D.L., Duraisingh, M., Healer, J., Mills, K.E., O'Donnell, R.A., Thompson, J., Triglia, T., Wickham, M.E., and Crabb, B.S. 2002. Functional analysis of *Plasmodium falciparum* merozoite antigens: implications for erythrocyte invasion and vaccine development. Philos. Trans. R. Soc. Lond. B. Biol. Sci. 357: 25-33.

Cowman, A.F., Baldi, D.L., Healer, J., Mills, K.E., O'Donnell, R.A., Reed, M.B., Triglia, T., Wickham, M.E., and Crabb, B.S. 2000. Functional analysis of proteins involved in *Plasmodium falciparum* merozoite invasion of red blood cells. FEBS Lett. 476: 84-88.

Cowman, A.F., Saint, R.B., Coppel, R.L., Brown, G.V., Anders, R.F., and Kemp, D.J. 1985. Conserved sequences flank variable tandem repeats in two S-antigen genes of *Plasmodium falciparum*. Cell. 40: 775-783.

Crewther, P.E., Culvenor, J.G., Silva, A., Cooper, J.A., and Anders, R.F. 1990. *Plasmodium falciparum*: two antigens of similar size are located in different compartments of the rhoptry. Exp. Parasitol. 70: 193-206.

Crewther, P.E., Matthew, M., Flegg, R.H., and Anders, R.F. 1996. Protective Immune responses to apical membrane antigen 1 of *Plasmodium chabaudi* involve recognition of strain-specific epitopes. Infect. Immun. 64: 3310-3317.

Culvenor, J.G., Day, K.P., and Anders, R.F. 1991. *Plasmodium falciparum* ring-infected erythrocyte surface antigen is released from merozoite dense granules after erythrocyte invasion. Infect. Immun. 59: 1183-1187.

Curtidor, H., Urquiza, M., Suarez, J.E., Rodriguez, L.E., Ocampo, M., Puentes, A., Garcia, J.E., Vera, R., Lopez, R., Ramirez, L.E., Pinzon, M., and Patarroyo, M.E. 2001. *Plasmodium falciparum* acid basic repeat antigen (ABRA) peptides: erythrocyte binding and biological activity. Vaccine. 19: 4496-4504.

Da Silva, E., Foley, M., Dluzewski, A.R., Murray, L.J., Anders, R.F., and Tilley, L. 1994. The *Plasmodium falciparum* protein RESA interacts with the erythrocyte cytoskeleton and modifies erythrocyte thermal stability. Mol. Biochem. Parasitol. 66: 59-69.

Dartsch, H., Kleene, R., and Kern, H.F. 1998. *In vitro* condensation-sorting of enzyme proteins isolated from rat pancreatic acinar cells. Eur. J. Cell. Biol. 75: 211-222.

David, P.H., Hadley, T.J., Aikawa, M., and Miller, L.H. 1984. Processing of a major parasite surface glycoprotein during the ultimate stages of differentiation in *Plasmodium knowlesi*. Mol. Biochem. Parasitol. 11: 267-282.

Deans, J.A., Alderson, T., Thomas, A.W., Mitchell, G.H., Lennox, E.S., and Cohen, S. 1982. Rat monoclonal antibodies which inhibit the *in vitro* multiplication of *Plasmodium knowlesi*. Clin. Exp. Immunol. 49: 297-309.

Delplace, P., Bhatia, A., Cagnard, M., Camus, D., Colombet, G., Debrabant, A., Dubremetz, J.F., Dubreuil, N., Prensier, G., Fortier, B., *et al.* 1988. Protein p126: a parasitophorous vacuole antigen associated with the release of *Plasmodium falciparum* merozoites. Biol. Cell. 64: 215-221.

Delplace, P., Fortier, B., Tronchin, G., Dubremetz, J.F., and Vernes, A. 1987. Localization, biosynthesis, processing and isolation of a major 126 kDa antigen of the parasitophorous vacuole of *Plasmodium falciparum*. Mol. Biochem. Parasitol. 23: 193-201.

Dhawan, S., Dua, M., Chishti, A.H., and Hanspal, M. 2003. Ankyrin peptide blocks falcipain-2-mediated malaria parasite release from red blood cells. J. Biol. Chem. 278: 30180-30186.

.Dluzewski, A.R., Fryer, P.R., Griffiths, S., Wilson, R.J., and Gratzer, W.B. 1989. Red cell membrane protein distribution during malarial invasion. J. Cell. Sci. 92: 691-699.

Dluzewski, A.R., Mitchell, G.H., Fryer, P.R., Griffiths, S., Wilson, R.J., and Gratzer, W.B. 1992. Origins of the parasitophorous vacuole membrane of the malaria parasite, *Plasmodium falciparum*, in human red blood cells. J. Cell. Sci. 102: 527-532.

Dluzewski, A.R., Rangachari, K., Wilson, R.J., and Gratzer, W.B. 1986. *Plasmodium falciparum*: protease inhibitors and inhibition of erythrocyte invasion. Exp. Parasitol. 62: 416-422.

Dobrowolski, J., and Sibley, L.D. 1997a. The role of the cytoskeleton in host cell invasion by *Toxoplasma gondii*. Behring. Inst. Mitt. 90-96.

Dobrowolski, J.M., Carruthers, V.B., and Sibley, L.D. 1997b. Participation of myosin in gliding motility and host cell invasion by *Toxoplasma gondii*. Mol. Microbiol. 26: 163-173.

Dobrowolski, J.M., and Sibley, L.D. 1996. Toxoplasma invasion of mammalian cells is powered by the actin cytoskeleton of the parasite. Cell. 84: 933-939.

Dolan, S.A., Miller, L.H., and Wellems, T.E. 1990. Evidence for a switching mechanism in the invasion of erythrocytes by *Plasmodium falciparum*. J. Clin. Invest. 86: 618-624.

Dolan, S.A., Proctor, J.L., Alling, D.W., Okubo, Y., Wellems, T.E., and Miller, L.H. 1994. Glycophorin B as an EBA-175 independent *Plasmodium falciparum* receptor of human erythrocytes. Mol. Biochem. Parasitol. 64: 55-63.

Doury, J.C., Bonnefoy, S., Roger, N., Dubremetz, J.F., and Mercereau-Puijalon, O. 1994. Analysis of the high molecular weight rhoptry complex of *Plasmodium falciparum* using monoclonal antibodies. Parasitology. 108 ( Pt 3): 269-280.

Doury, J.C., Goasdoue, J.L., Tolou, H., Martelloni, M., Bonnefoy, S., and Mercereau-Puijalon, O. 1997. Characterisation of the binding sites of monoclonal antibodies reacting with the *Plasmodium falciparum* rhoptry protein RhopH3. Mol. Biochem. Parasitol. 85: 149-159.

Dua, M., Raphael, P., Sijwali, P.S., Rosenthal, P.J., and Hanspal, M. 2001. Recombinant falcipain-2 cleaves erythrocyte membrane ankyrin and protein 4.1. Mol. Biochem. Parasitol. 116: 95-99.

Dubbeld, M.A., Kocken, C.H., and Thomas, A.W. 1998. Merozoite surface protein 2 of *Plasmodium reichenowi* is a unique mosaic of *Plasmodium falciparum* allelic forms and species-specific elements. Mol. Biochem. Parasitol. 92: 187-192.

Duraisingh, M.T., Triglia, T., Ralph, S.A., Rayner, J.C., Barnwell, J.W., McFadden, G.I., and Cowman, A.F. 2003. Phenotypic variation of *Plasmodium falciparum* merozoite proteins directs receptor targeting for invasion of human erythrocytes. EMBO J. 22: 1047-1057.

Dutta, S., Malhotra, P., and Chauhan, V.S. 1995. Sequnece analysis of apical membrane antigen 1 (AMA-1) of *Plasmodium cynomolgi bastianelli*. Mol. Biochem. Parasitol. 73: 267-270.

Dvorak, J.A., Miller, L.H., Whitehouse, W.C., and Shiroishi, T. 1975. Invasion of erythrocytes by malaria merozoites. Science. 187: 748-750.

Eisen, D., Billman-Jacobe, H., Marshall, V.F., Fryauff, D., and Coppel, R.L. 1998. Temporal variation of the merozoite surface protein-2 gene of *Plasmodium falciparum*. Infect. Immun. 66: 239-246.

Elmendorf, H.G., and Haldar, K. 1993. Secretory transport in Plasmodium. Parasitol. Today. 9: 98-102.

Epping, R.J., Goldstone, S.D., Ingram, L.T., Upcroft, J.A., Ramasamy, R., Cooper, J.A., Bushell, G.R., and Geysen, H.M. 1988. An epitope recognised by inhibitory monoclonal antibodies that react with a 51 kilodalton merozoite surface antigen in *Plasmodium falciparum*. Mol. Biochem. Parasitol. 28: 1-10.

Escalante, A.A., Lal, A.A., and Ayala, F.J. 1998. Genetic polymorphism and natural selection in the malaria parasite *Plasmodium falciparum*. Genetics. 149: 189-202.

Etzion, Z., Murray, M.C., and Perkins, M.E. 1991. Isolation and characterization of rhoptries of *Plasmodium falciparum*. Mol. Biochem. Parasitol. 47: 51-61.

Favaloro, J.M., Coppel, R.L., Corcoran, L.M., Foote, S.J., Brown, G.V., Anders, R.F., and Kemp, D.J. 1986. Structure of the RESA gene of *Plasmodium falciparum*. Nucleic Acids Res. 14: 8265-8277.

Felger, I., Marshal, V.M., Reeder, J.C., Hunt, J.A., Mgone, C.S., and Beck, H.P. 1997. Sequence diversity and molecular evolution of the merozoite surface antigen 2 of *Plasmodium falciparum*. J. Mol. Evol. 45: 154-160.

Felger, I., Tavul, L., Kabintik, S., Marshall, V., Genton, B., Alpers, M., and Beck, H.P. 1994. *Plasmodium falciparum*: extensive polymorphism in merozoite surface antigen 2 alleles in an area with endemic malaria in Papua New Guinea. Exp. Parasitol. 79: 106-116.

Fenton, B., Clark, J.T., Khan, C.M., Robinson, J.V., Walliker, D., Ridley, R., Scaife, J.G., and McBride, J.S. 1991. Structural and antigenic polymorphism of the 35- to 48-kilodalton merozoite surface antigen (MSA-2) of the malaria parasite *Plasmodium falciparum*. Mol. Cell. Biol. 11: 963-974.

Field, S.J., Pinder, J.C., Clough, B., Dluzewski, A.R., Wilson, R.J., and Gratzer, W.B. 1993. Actin in the merozoite of the malaria parasite, *Plasmodium falciparum*. Cell. Motil. Cytoskeleton. 25: 43-48.

Foley, M., Corcoran, L., Tilley, L., and Anders, R. 1994. *Plasmodium falciparum*: mapping the membrane-binding domain in the ring-infected erythrocyte surface antigen. Exp. Parasitol. 79: 340-350.

Foley, M., Tilley, L., W.H., S., and Anders, R.F. 1991. The ring-infected erythrocyte surface antigen of *Plasmodium falciparum* associates with spectrin in the erythrocyte membrane. Mol. Biochem. Parasitol. 46: 137-148.

Foth, B.J., Ralph, S.A., Tonkin, C.J., Struck, N.S., Fraunholz, M., Roos, D.S., Cowman, A.F., and McFadden, G.I. 2003. Dissecting apicoplast targeting in the malaria parasite *Plasmodium falciparum*. Science. 299: 705-708.

Fourmaux, M.N., Achbarou, A., Mercereau-Puijalon, O., Biderre, C., Briche, I., Loyens, A., Odberg-Ferragut, C., Camus, D., and Dubremetz, J.F. 1996. The MIC1 microneme protein of *Toxoplasma gondii* contains a duplicated receptor-like domain and binds to host cell surface. Mol. Biochem. Parasitol. 83: 201-210.

Fowler, R.E., Fookes, R.E., Lavin, F., Bannister, L.H., and Mitchell, G.H. 1998. Microtubules in *Plasmodium falciparum* merozoites and their importance for invasion of erythrocytes. Parasitology. 117 ( Pt 5): 425-433.

Fowler, R.E., Smith, A.M., Whitehorn, J., Williams, I.T., Bannister, L.H., and Mitchell, G.H. 2001. Microtubule associated motor proteins of *Plasmodium falciparum* merozoites. Mol. Biochem. Parasitol. 117: 187-200.

Fox, B.A., and Bzik, D.J. 1994. Analysis of stage-specific transcripts of the *Plasmodium falciparum* serine repeat antigen (SERA) gene and transcription from the SERA locus. Mol. Biochem. Parasitol. 68: 133-144.

Fraser, T.S., Kappe, S.H., Narum, D.L., VanBuskirk, K.M., and Adams, J.H. 2001. Erythrocyte-binding activity of Plasmodium yoelii apical membrane antigen-1 expressed on the surface of transfected COS-7 cells. Mol. Biochem. Parasitol. 117: 49-59.

Galinski, M.R., Corredor-Medina, C., Povoa, M., Crosby, J., Ingravallo, P., and Barnwell, J.W. 1999. *Plasmodium vivax* merozoite surface protein-3 contains coiled-coil motifs in an alanine-rich central domain. Mol. Biochem. Parasitol. 101: 131-147.

Galinski, M.R., Ingravallo, P., Corredor-Medina, C., Al-Khedery, B., Povoa, M., and Barnwell, J.W. 2001. *Plasmodium vivax* merozoite surface proteins-3beta and-3gamma share structural similarities with *P. vivax* merozoite surface protein-3alpha and define a new gene family. Mol. Biochem. Parasitol. 115: 41-53.

Galinski, M.R., Medina, C.C., Ingravallo, P., and Barnwell, J.W. 1992. A reticulocyte-binding protein complex of *Plasmodium vivax* merozoites. Cell. 69: 1213-1226.

Galinski, M.R., Xu, M., and Barnwell, J.W. 2000. *Plasmodium vivax* reticulocyte binding protein-2 (PvRBP-2) shares structural features with PvRBP-1 and the *Plasmodium yoelii* 235 kDa rhoptry protein family. Mol. Biochem. Parasitol. 108: 257-262.

Garcia-Reguet, N., Lebrun, M., Fourmaux, M.N., Mercereau-Puijalon, O., Mann, T., Beckers, C.J., Samyn, B., Van Beeumen, J., Bout, D., and Dubremetz, J.F. 2000. The microneme protein MIC3 of *Toxoplasma gondii* is a secretory adhesin that binds to both the surface of the host cells and the surface of the parasite. Cell .Microbiol. 2: 353-364.

Gardner, M.J., Tettelin, H., Carucci, D.J., Cummings, L.M., Aravind, L., Koonin, E.V., Shallom, S., Mason, T., Yu, K., Fujii, C., Pederson, J., Shen, K., Jing, J.P., Aston, C., Lai, Z.W., Schwartz, D.C., Pertea, M., Salzberg, S., Zhou, L.X., Sutton, G.G., Clayton, R., White, O., Smith, H.O., Fraser, C.M., Adams, M.D., Hoffman, S.L., *et al.* 1998. Chromosome 2 sequence of the human malaria parasite *Plasmodium falciparum*. Science. 282 1126-1132.

Genton, B., Betuela, I., Felger, I., Al-Yaman, F., Anders, R.F., Saul, A., Rare, L., Baisor, M., Lorry, K., Brown, G.V., Pye, D., Irving, D.O., Smith, T.A., Beck, H.P., and Alpers, M.P. 2002. A recombinant blood-stage malaria vaccine reduces *Plasmodium falciparum* density and exerts selective pressure on parasite populations in a phase 1-2b trial in Papua New Guinea. J. Infect. Dis. 185: 820-827.

Ghai, M., Dutta, S., Hall, T., Freilich, D., and Ockenhouse, C.F. 2002. Identification, expression, and functional characterization of MAEBL, a sporozoite and asexual blood stage chimeric erythrocyte-binding protein of *Plasmodium falciparum*. Mol. Biochem. Parasitol. 123: 35-45.

Gilberger, T.W., Thompson, J.K., Triglia, T., Good, R.T., Duraisingh, M.T., and Cowman, A.F. 2003. A novel erythrocyte binding antigen-175 paralogue from *Plasmodium falciparum* defines a new trypsin-resistant receptor on human erythrocytes. J. Biol. Chem. 278: 14480-14486.

Goel, V.K., Li, X., Chen, H., Liu, S.C., Chishti, A.H., and Oh, S.S. 2003. Band 3 is a host receptor binding merozoite surface protein 1 during the *Plasmodium falciparum* invasion of erythrocytes. Proc. Natl. Acad. Sci. USA. 100: 5164-5169.

Goncz, K.K., Behrsing, R., and Rothman, S.S. 1995. The protein content and morphogenesis of zymogen granules. Cell Tissue Res. 280: 519-530.

Grellier, P., Precigout, E., Valentin, A., Carcy, B., and Schrevel, J. 1994. Characterization of a new 60 kDa apical protein of *Plasmodium falciparum* merozoite expressed in late schizogony. Biol Cell. 82: 129-138.

Guevara Patino, J.A., Holder, A.A., McBride, J.S., and Blackman, M.J. 1997. Antibodies that inhibit malaria merozoite surface protein-1 processing and erythrocyte invasion are blocked by naturally acquired human antibodies. J. Exp. Med. 186: 1689-1699.

Hackett, F., Sajid, M., Withers-Martinez, C., Grainger, M., and Blackman, M.J. 1999. PfSUB-2: a second subtilisin-like protein in *Plasmodium falciparum* merozoites. Mol. Biochem. Parasitol. 103: 183-195.

Hadley, T., Aikawa, M., and Miller, L.H. 1983. *Plasmodium knowlesi*: studies on invasion of rhesus erythrocytes by merozoites in the presence of protease inhibitors. Exp. Parasitol. 55: 306-311.

Hadley, T.J., Klotz, F.W., Pasvol, G., Haynes, J.D., McGinniss, M.H., Okubo, Y., and Miller, L.H. 1987. Falciparum malaria parasites invade erythrocytes that lack glycophorin A and B (MkMk). Strain differences indicate receptor heterogeneity and two pathways for invasion. J. Clin. Invest. 80: 1190-1193.

Hakansson, S., Charron, A.J., and Sibley, L.D. 2001. Toxoplasma evacuoles: a two-step process of secretion and fusion forms the parasitophorous vacuole. EMBO J. 20: 3132-3144.

Haldar, K., Samuel, B.U., Mohandas, N., Harrison, T., and Hiller, N.L. 2001. Transport mechanisms in Plasmodium-infected erythrocytes: lipid rafts and a tubovesicular network. Int. J. Parasitol. 31: 1393-1401.

Hanspal, M., Dua, M., Takakuwa, Y., Chishti, A.H., and Mizuno, A. 2002. *Plasmodium falciparum* cysteine protease falcipain-2 cleaves erythrocyte membrane skeletal proteins at late stages of parasite development. Blood. 100: 1048-1054.

Harnyuttanakorn, P., McBride, J.S., Donachie, S., Heidrich, H.G., and Ridley, R.G. 1992. Inhibitory monoclonal antibodies recognise epitopes adjacent to a proteolytic cleavage site on the RAP-1 protein of *Plasmodium falciparum*. Mol. Biochem. Parasitol. 55: 177-186.

Haynes, J.D., Dalton, J.P., Klotz, F.W., McGinniss, M.H., Hadley, T.J., Hudson, D.E., and Miller, L.H. 1988. Receptor-like specificity of a *Plasmodium knowlesi* malarial protein that binds to Duffy antigen ligands on erythrocytes. J. Exp. Med. 167: 1873-1881.

Healer, J., Crawford, S., Ralph, S., McFadden, G., and Cowman, A.F. 2002. Independent translocation of two micronemal proteins in developing *Plasmodium falciparum* merozoites. Infect. Immun. 70: 5751-5758.

Hermentin, P., and Enders, B. 1984. Erythrocyte invasion by malaria (*Plasmodium falciparum*) merozoites: recent advances in the evaluation of receptor sites. Behring. Inst. Mitt. 121-141.

Herm-Gotz, A., Weiss, S., Stratmann, R., Fujita-Becker, S., Ruff, C., Meyhofer, E., Soldati, T., Manstein, D.J., Geeves, M.A., and Soldati, D. 2002. *Toxoplasma gondii* myosin A and its light chain: a fast, single-headed, plus-end-directed motor. EMBO J. 21: 2149-2158.

Herrera, S., Rudin, W., Herrera, M., Clavijo, P., Mancilla, L., De, P.C., Matile, H., and Certa, U. 1993. A conserved region of the MSP-1 surface protein of *Plasmodium falciparum* contains a recognition sequence for erythrocyte spectrin. EMBO J. 12: 1607-1614.

Hettmann, C., Herm, A., Geiter, A., Frank, B., Schwarz, E., Soldati, T., and Soldati, D. 2000. A dibasic motif in the tail of a class XIV apicomplexan myosin is an essential determinant of plasma membrane localization. Mol. Biol. Cell. 11: 1385-1400.

Hisaeda, H., Saul, A., Reece, J.J., Kennedy, M.C., Long, C.A., Miller, L.H., and Stowers, A.W. 2002. Merozoite surface protein 3 and protection against malaria in Aotus nancymai monkeys. J. Infect. Dis. 185: 657-664.

Holder, A.A. 1988. The precursor to major merozoite surface antigens: structure and role in immunity. Prog Allergy. 41: 72-97.

Holder, A.A. 1996. Preventing merozoite invasion of erythrocytes. In: Malria Vaccine Development: A Multi-Immune Response Approach. S.L. Hoffman, ed. ASM press. p. 77-104.

Holder, A.A., and Blackman, M.J. 1994. What is the function of MSP-1 on the malaria merozoite? Parasitol. Today. 10: 182-184.

Holder, A.A., and Freeman, R.R. 1984. The three major antigens on the surface of *Plasmodium falciparum* merozoites are derived from a single high molecular weight precursor. J. Exp. Med. 160: 624-629.

Hoppe, H.C., Ngo, H.M., Yang, M., and Joiner, K.A. 2000. Targeting to rhoptry organelles of *Toxoplasma gondii* involves evolutionarily conserved mechanisms. Nat. Cell Biol. 2: 449-456.

Horii, T., Bzik, D.J., and Inselburg, J. 1988. Characterization of antigen-expressing *Plasmodium falciparum* cDNA clones that are reactive with parasite inhibitory antibodies. Mol. Biochem. Parasitol. 30: 9-18.

Horuk, R., Chitnis, C.E., Darbonne, W.C., Colby, T.J., Rybicki, A., Hadley, T.J., and Miller, L.H. 1993. A receptor for the malarial parasite *Plasmodium vivax*: the erythrocyte chemokine receptor. Science. 261: 1182-1184.

Howard, R.F., Jacobson, K.C., Rickel, E., and Thurman, J. 1998a. Analysis of inhibitory epitopes in the *Plasmodium falciparum* rhoptry protein RAP-1 including identification of a second inhibitory epitope. Infect. Immun. 66: 380-386.

Howard, R.F., Narum, D.L., Blackman, M., and Thurman, J. 1998b. Analysis of the processing of *Plasmodium falciparum* rhoptry-associated protein 1 and localization of Pr86 to schizont rhoptries and p67 to free merozoites. Mol. Biochem. Parasitol. 92: 111-122.

Howard, R.F., and Reese, R.T. 1990. *Plasmodium falciparum*: hetero-oligomeric complexes of rhoptry polypeptides. Exp. Parasitol. 71: 330-342.

Howard, R.F., and Schmidt, C.M. 1995. The secretory pathway of *Plasmodium falciparum* regulates transport of p82/RAP1 to the rhoptries. Mol. Biochem. Parasitol. 74: 43-54.

Howard, R.F., Stanley, H.A., Campbell, G.H., and Reese, R.T. 1984. Proteins responsible for a punctate fluorescence pattern in *Plasmodium falciparum* merozoites. Am. J. Trop. Med. Hyg. 33: 1055-1059.

Howell, S.A., Withers-Martinez, C., Kocken, C.H., Thomas, A.W., and Blackman, M.J. 2001. Proteolytic processing and primary structure of *Plasmodium falciparum* apical membrane antigen-1. J. Biol. Chem. 276: 31311-31320.

Howell, S.A., Well, I., Fleck, S.L., Kettleborough, C., Collins, C., and Blackman, M.J. 2003. A single malaria merozoite serine protease mediates shedding of multiple surface proteins by juxtamembrane cleavage. J. Biol. Chem. 278: 23890-23898.

Hudson, D.E., Miller, L.H., Richards, R.L., David, P.H., Alving, C.R., and Gitler, C. 1983. The malaria merozoite surface: a 140,000 m.w. protein antigenically unrelated to other surface components on *Plasmodium knowlesi* merozoites. J. Immunol. 130: 2886-2890.

Hudson-Taylor, D.E., Dolan, S.A., Klotz, F.W., Fujioka, H., Aikawa, M., Koonin, E.V., and Miller, L.H. 1995. *Plasmodium falciparum* protein associated with the invasion junction contains a conserved oxidoreductase domain. Mol. Microbiol. 15: 463-471.

Inselburg, J., Bathurst, I.C., Kansopon, J., Barchfeld, G.L., Barr, P.J., and Rossan, R.N. 1993. Protective immunity induced in Aotus monkeys by a recombinant SERA protein of *Plasmodium falciparum*: Adjuvant effects on induction of protective immunity. Infect. Immun. 61: 2041-2047.

Inselburg, J., Bzik, D.J., Li, W.B., Green, K.M., Kansopon, J., Hahm, B.K., Bathurst, I.C., Barr, P.J., and Rossan, R.N. 1991. Protective immunity induced in Aotus monkeys by recombinant SERA proteins of *Plasmodium falciparum*. Infect. Immun. 59: 1247-1250.

Irion, A., Beck, H.P., and Felger, I. 1997. New repeat unit and hot spot of recombination in FC27-type alleles of the gene coding for *Plasmodium falciparum* merozoite surface protein 2. Mol. Biochem. Parasitol. 90: 367-370.

Jaikaria, N.S., Rozario, C., Ridley, R.G., and Perkins, M.E. 1993. Biogenesis of rhoptry organelles in *Plasmodium falciparum*. Mol. Biochem. Parasitol. 57: 269-279.

Jean, L., Hackett, F., Martin, S.R., and Blackman, M.J. 2003. Functional characterisation of the propeptide of *Plasmodium falciparum* subtilisin-like protease-1. J. Biol. Chem. 278: 28572-28579.

Kaneko, O., Fidock, D.A., Schwartz, O.M., and Miller, L.H. 2000. Disruption of the C-terminal region of EBA-175 in the Dd2/Nm clone of *Plasmodium falciparum* does not affect erythrocyte invasion. Mol. Biochem. Parasitol. 110: 135-146.

Kaneko, O., Mu, J., Tsuboi, T., Su, X., and Torii, M. 2002. Gene structure and expression of a *Plasmodium falciparum* 220-kDa protein homologous to the *Plasmodium vivax* reticulocyte binding proteins. Mol. Biochem. Parasitol. 121: 275-278.

Kaneko, O., Tsuboi, T., Ling, I.T., Howell, S., Shirano, M., Tachibana, M., Cao, Y.M., Holder, A.A., and Torii, M. 2001. The high molecular mass rhoptry protein, RhopH1, is encoded by members of the *clag* multigene family in *Plasmodium falciparum* and *Plasmodium yoelii*. Mol. Biochem. Parasitol. 118: 223-231.

Kang, Y., and Long, C.A. 1995. Sequence heterogeneity of the C-terminal, Cys-rich region of the merozoite surface protein-1 (MSP-1) in field samples of *Plasmodium falciparum*. Mol. Biochem. Parasitol. 73: 103-110.

Kappe, S., Bruderer, T., Gantt, S., Fujioka, H., Nussenzweig, V., and Menard, R. 1999. Conservation of a gliding motility and cell invasion machinery in Apicomplexan parasites. J. Cell. Biol. 147: 937-944.

Kappe, S.H.I., and Adams, J.H. 1996. Sequence analysis of the apical membrane antigen-1 genes (AMA-1) of *Plasmodium yoelii yoelii* and *Plasmodium berghei*. Mol. Biochem. Parasitol. 78: 279-283.

Kappe, S.H.I., Noe, A.R., Fraser, T.S., Blair, P.L., and Adams, J.H. 1998. A family of chimeric erythrocyte binding proteins of malaria parasites. Proc. Natl. Acad. Sci. USA. 95: 1230-1235.

Kariu, T., Yuda, M., Yano, K., and Chinzei, Y. 2002. MAEBL is essential for malarial sporozoite infection of the mosquito salivary gland. J. Exp. Med. 195: 1317-1323.

Karsten, V., Qi, H., Beckers, C.J., Reddy, A., Dubremetz, J.F., Webster, P., and Joiner, K.A. 1998. The protozoan parasite *Toxoplasma gondii* targets proteins to dense granules and the vacuolar space using both conserved and unusual mechanisms. J. Cell. Biol. 141: 1323-1333.

Kauth, C.W., Epp, C., Bujard, H., and Lutz, R. 2003. The merozoite surface protein 1 complex of human malaria parasite *Plasmodium falciparum*: interactions and arrangements of subunits. J. Biol. Chem. 278: 22257-22264.

Kedzierski, L., Black, C.G., and Coppel, R.L. 2000a. Characterization of the merozoite surface protein 4/5 gene of *Plasmodium berghei* and *Plasmodium yoelii*. Mol. Biochem. Parasitol. 105: 137-147.

Kedzierski, L., Black, C.G., and Coppel, R.L. 2000b. Immunization with recombinant *Plasmodium yoelii* merozoite surface protein 4/5 protects mice against lethal challenge. Infect. Immun. 68: 6034-6037.

Kedzierski, L., Black, C.G., Goschnick, M.W., Stowers, A.W., and Coppel, R.L. 2002. Immunization with a combination of merozoite surface proteins 4/5 and 1 enhances protection against lethal challenge with *Plasmodium yoelii*. Infect. Immun. 70: 6606-6613.

Keen, J.K., Sinha, K.A., Brown, K.N., and Holder, A.A. 1994. A gene coding for a high-molecular mass rhoptry protein of *Plasmodium yoelii*. Mol. Biochem. Parasitol. 65: 171-177.

King, C. 1988. Cell motility of sporozoan protozoa. Parasitol. Today. 4: 315-319.

Klotz, F.W., Hadley, T.J., Aikawa, M., Leech, J., Howard, R.J., and Miller, L.H. 1989. A 60-kDa *Plasmodium falciparum* protein at the moving junction formed between merozoite and erythrocyte during invasion. Mol. Biochem. Parasitol. 36: 177-185.

Knapp, B., Nau, U., Hundt, E., and Küpper, H.A. 1991. A new blood stage antigen of *Plasmodium falciparum* highly homologous to the serine-stretch protein SERP. Mol. Biochem. Parasitol. 44: 1-14.

Kocken, C.H., Narum, D.L., Massougbodji, A., Ayivi, B., Dubbeld, M.A., van der Wel, A., Conway, D.J., Sanni, A., and Thomas, A.W. 2000. Molecular characterisation of Plasmodium reichenowi apical membrane antigen-1 (AMA-1), comparison with *P. falciparum* AMA-1, and antibody- mediated inhibition of red cell invasion. Mol. Biochem. Parasitol. 109: 147-156.

Kocken, C.H., van der Wel, A.M., Dubbeld, M.A., Narum, D.L., van de Rijke, F.M., van Gemert, G.J., van der Linde, X., Bannister, L.H., Janse, C., Waters, A.P., and Thomas, A.W. 1998. Precise timing of expression of a *Plasmodium falciparum*-derived transgene in *Plasmodium berghei* is a critical determinant of subsequent subcellular localization. J. Biol. Chem. 273: 15119-15124.

Kocken, C.H., Withers-Martinez, C., Dubbeld, M.A., van der Wel, A., Hackett, F., Blackman, M.J., and Thomas, A.W. 2002. High-level expression of the malaria blood-stage vaccine candidate *Plasmodium falciparum* apical membrane antigen 1 and induction of antibodies that inhibit erythrocyte invasion. Infect. Immun. 70: 4471-4476.

Kushwaha, A., Perween, A., Mukund, S., Majumdar, S., Bhardwaj, D., Chowdhury, N.R., and Chauhan, V.S 2002. Amino terminus of *Plasmodium falciparum* acidic basic repeat antigen interacts with the erythrocyte membrane through band 3 protein. Mol. Biochem. Parasitol. 122: 45-54.

Kushwaha, A., Rao, P.P., Duttu, V.S., Malhotra, P., and Chauhan, V.S. 2000. Expression and characterisation of *Plasmodium falciparum* acidic basic repeat antigen expressed in *Escherichia coli*. Mol. Biochem. Parasitol. 106: 213-224.

Kyes, S., Craig, A.G., Marsh, K., and Newbold, C.I. 1993. *Plasmodium falciparum*: a method for the amplification of S antigens and its application to laboratory and field samples. Exp. Parasitol. 77: 473-483.

Langreth, S.G., Jensen, J.B., Reese, R.T., and Trager, W. 1978. Fine structure of human malaria *in vitro*. J. Protozool. 25: 443-452.

Lauer, S., VanWye, J., Harrison, T., McManus, H., Samuel, B.U., Hiller, N.L., Mohandas, N., and Haldar, K. 2000. Vacuolar uptake of host components, and a role for cholesterol and sphingomyelin in malarial infection. EMBO J. 19: 3556-3564.

Ling, I.T., Kaneko, O., Narum, D.L., Tsuboi, T., Howell, S., Taylor, H.M., Scott-Finnigan, T..J., Torii, M., and Holder, A.A. 2003. Characterisation of the rhoph2 gene of *Plasmodium falciparum* and *Plasmodium yoelii*. Mol. Biochem. Parasitol. 127: 47-57.

Lobo, C.A., Rodriguez, M., Hou, G., Perkins, M., Oskov, Y., and Lustigman, S. 2003a. Characterization of PfRhop148, a novel rhoptry protein of *Plasmodium falciparum*. Mol. Biochem. Parasitol. 128: 59-65.

Lobo, C.A., Rodriguez, M., Reid, M., and Lustigman, S. 2003b. Glycophorin C is the receptor for the *Plasmodium falciparum* erythrocyte binding ligand PfEBP-2 (baebl). Blood. 101: 4628-4631.

Longacre, S. 1995. The Plasmodium cynomolgi merozoite surface protein 1 C-terminal sequence and its homologies with other Plasmodium species. Mol. Biochem. Parasitol. 74: 105-111.

Longacre, S., Mendis, K.N., and David, P.H. 1994. *Plasmodium vivax* merozoite surface protein 1 C-terminal recombinant proteins in baculovirus. Mol. Biochem. Parasitol. 64: 191-205.

Lovett, J.L., Marchesini, N., Moreno, S.N., and Sibley, L.D. 2002. *Toxoplasma gondii* microneme secretion involves intracellular Ca(2+) release from inositol 1,4,5-triphosphate (IP(3))/ryanodine-sensitive stores. J. Biol. Chem. 277: 25870-25876.

Lupas, A. 1996. Coiled coils: new structures and new functions. Trends Biochem. Sci. 21: 375-382.

Lustigman, S., Anders, R.F., Brown, G.V., and Coppel, R.L. 1988. A component of an antigenic rhoptry complex of *Plasmodium falciparum* is modified after merozoite invasion. Mol. Biochem. Parasitol. 30: 217-224.

Lyon, J.A., and Haynes, J.D. 1986a. *Plasmodium falciparum* antigens synthesized by schizonts and stabilized at the merozoite surface when schizonts mature in the presence of protease inhibitors. J. Immunol. 136: 2245-2251.

Lyon, J.A., Haynes, J.D., Diggs, C.L., Chulay, J.D., and Pratt-Rossiter, J.M. 1986b. *Plasmodium falciparum* antigens synthesized by schizonts and stabilized at the merozoite surface by antibodies when schizonts mature in the presence of growth inhibitory immune serum. J. Immunol. 136: 2252-2258.

Maier, A.G., Duraisingh, M.T., Reeder, J.C., Patel, S.S., Kazura, J.W., Zimmerman, P.A., and Cowman, A.F. 2003. *Plasmodium falciparum* erythrocyte invasion through glycophorin C and selection for Gerbich negativity in human populations. Nat. Med. 9: 87-92.

Manning, G., Whyte, D.B., Martinez, R., Hunter, T., and Sudarsanam, S. 2002. The protein kinase complement of the human genome. Science. 298: 1912-1934.

Marshall, V.M., Anthony, R.L., Bangs, M.J., Purnomo, Anders, R.F., and Coppel, R.L. 1994. Allelic variants of the *Plasmodium falciparum* merozoite surface antigen 2 (MSA-2) in a geographically restricted area of Irian Jaya. Mol. Biochem. Parasitol. 63: 13-21.

Marshall, V.M., and Coppel, R.L. 1997a. Characterisation of the gene encoding adenylosuccinate lyase of *Plasmodium falciparum*. Mol. Biochem. Parasitol. 88: 237-241.

Marshall, V.M., Coppel, R.L., Anders, R.F., and Kemp, D.J. 1992. Two novel alleles within subfamilies of the merozoite surface antigen 2 (MSA-2) of *Plasmodium falciparum*. Mol. Biochem. Parasitol. 50: 181-184.

Marshall, V.M., Coppel, R.L., Martin, R.K., Oduola, A.M., Anders, R.F., and Kemp, D.J. 1991. A *Plasmodium falciparum MSA-2* gene apparently generated by intragenic recombination between the two allelic families. Mol. Biochem. Parasitol. 45: 349-351.

Marshall, V.M., Peterson, M.G., Lew, A.M., and Kemp, D.J. 1989. Structure of the apical membrane antigen I (AMA-1) of *Plasmodium chabaudi*. Mol. Biochem. Parasitol. 37: 281-283.

Marshall, V.M., Silva, A., Foley, M., Cranmer, S., Wang, L., McColl, D.J., Kemp, D.J., and Coppel, R.L. 1997b. A second merozoite surface protein (MSP-4) of *Plasmodium falciparum* that contains an epidermal growth factor-like domain. Infect. Immun. 65: 4460-4467.

Marshall, V.M., Tieqiao, W., and Coppel, R.L. 1998. Close linkage of three merozoite surface protein genes on chromosome 2 of *Plasmodium falciparum*. Mol. Biochem. Parasitol. 94: 13-25.

Matuschewski, K., Nunes, A.C., Nussenzweig, V., and Menard, R. 2002. Plasmodium sporozoite invasion into insect and mammalian cells is directed by the same dual binding system. EMBO J. 21: 1597-1606.

Mayer, D.C., Kaneko, O., Hudson-Taylor, D.E., Reid, M.E., and Miller, L.H. 2001. Characterization of a *Plasmodium falciparum* erythrocyte-binding protein paralogous to EBA-175. Proc. Natl. Acad. Sci. USA. 98: 5222-5227.

Mayer, D.C., Mu, J.B., Feng, X., Su, X.Z., and Miller, L.H. 2002. Polymorphism in a *Plasmodium falciparum* erythrocyte-binding ligand changes its receptor specificity. J. Exp. Med. 196: 1523-1528.

McColl, D.J., and Anders, R.F. 1997. Conservation of structural motifs and antigenic diversity in the *Plasmodium falciparum* merozoite surface protein-3 (MSP-3). Mol. Biochem. Parasitol. 90: 21-31.

McColl, D.J., Silva, A., Foley, M., Kun, J.F., Favaloro, J.M., Thompson, J.K., Marshall, V.M., Coppel, R.L., Kemp. D.J., and Anders, R.F. 1994. Molecular variation in a novel polymorphic antigen associated with *Plasmodium falciparum* merozoites. Mol. Biochem. Parasitol. 68: 53-67.

McPherson, R.A., Donald, D.R., Sawyer, W.H., and Tilley, L. 1993. Proteolytic digestion of band 3 at an external site alters the erythrocyte membrane organisation and may facilitate malarial invasion. Mol. Biochem. Parasitol. 62: 233-242.

Meissner, M., Reiss, M., Viebig, N., Carruthers, V.B., Toursel, C., Tomavo, S., Ajioka, J.W., and Soldati, D. 2002a. A family of transmembrane microneme proteins of *Toxoplasma gondii* contain EGF-like domains and function as escorters. J. Cell. Sci. 115: 563-574.

Meissner, M., Schluter, D., and Soldati, D. 2002b. Role of *Toxoplasma gondii* myosin A in powering parasite gliding and host cell invasion. Science. 298: 837-840.

Michon, P., Fraser, T., and Adams, J.H. 2000. Naturally acquired and vaccine-elicited antibodies block erythrocyte cytoadherence of the *Plasmodium vivax* Duffy binding protein. Infect. Immun. 68: 3164-3171.

Michon, P., Stevens, J.R., Kaneko, O., and Adams, J.H. 2002. Evolutionary relationships of conserved cysteine-rich motifs in adhesive molecules of malaria parasites. Mol. Biol. Evol. 19: 1128-1142.

Mikkelsen, R.B., Kamber, M., Wadwa, K.S., Lin, P.S., and Schmidt-Ullrich, R. 1988. The role of lipids in *Plasmodium falciparum* invasion of erythrocytes: a coordinated biochemical and microscopic analysis. Proc. Natl. Acad. Sci. USA. 85: 5956-5960.

Miller, L.H., Aikawa, M., Johnson, J.G., and Shiroishi, T. 1979. Interaction between cytochalasin B-treated malarial parasites and erythrocytes. Attachment and junction formation. J. Exp. Med. 149: 172-184.

Miller, L.H., Mason, S.J., Clyde, D.F., and McGinniss, M.H. 1976. The resistance factor to *Plasmodium vivax* in blacks. The Duffy-blood- group genotype, FyFy. N. Engl. J. Med. 295: 302-304.

Miller, L.H., Roberts, T., Shahabuddin, M., and McCutchan, T.F. 1993. Analysis of sequence diversity in the *Plasmodium falciparum* merozoite surface protein-1 (MSP-1). Mol. Biochem. Parasitol. 59: 1-14.

Miller, S.A., Binder, E.M., Blackman, M.J., Carruthers, V.B., and Kim, K. 2001. A conserved subtilisin-like protein TgSUB1 in microneme organelles of *Toxoplasma gondii*. J. Biol. Chem. 276: 45341-45348.

Miller, S.K., Good, R.T., Drew, D.R., Delorenzi, M., Sanders, P.R., Hodder, A.N., Speed, T.P., Cowman, A.F., de Koning-Ward, T.F., and Crabb, B.S. 2002. A subset of *Plasmodium falciparum SERA* genes are expressed and appear to play an important role in the erythrocytic cycle. J. Biol. Chem. 277: 47524-47532.

Mills, K.E., Pearce, J.A., Crabb, B.S., and Cowman, A.F. 2002. Truncation of merozoite surface protein 3 disrupts its trafficking and that of acidic-basic repeat protein to the surface of *Plasmodium falciparum* merozoites. Mol. Microbiol. 43: 1401-1411.

Mitchell, G.H., and Bannister, L.H. 1988. Malaria parasite invasion: interactions with the red cell membrane. Crit. Rev. Oncol. Hematol. 8: 225-310.

Mordue, D.G., Desai, N., Dustin, M., and Sibley, L.D. 1999. Invasion by *Toxoplasma gondii* establishes a moving junction that selectively excludes host cell plasma membrane proteins on the basis of their membrane anchoring. J. Exp. Med. 190: 1783-1792.

Moreno, R., Poltl-Frank, F., Stuber, D., Matile, H., Mutz, M., Weiss, N.A., and Pluschke, G. 2001. Rhoptry-associated protein 1-binding monoclonal antibody raised against a heterologous peptide sequence inhibits *Plasmodium falciparum* growth *in vitro*. Infect. Immun. 69: 2558-2568.

Morgan, W.D., Birdsall, B., Frenkiel, T.A., Gradwell, M.G., Burghaus, P.A., Syed, S.E., Uthaipibull, C., Holder, A.A., and Feeney, J. 1999. Solution structure of an EGF module pair from the *Plasmodium falciparum* merozoite surface protein 1. J. Mol. Biol. 289: 113-122.

Mulhern, T.D., Howlett, G.J., Reid, G.E., Simpson, R.J., McColl, D.J., Anders, R.F., and Norton, R.S. 1995. Solution structure of a polypeptide containing four heptad repeat units from a merozoite surface antigen of *Plasmodium falciparum*. Biochemistry. 34: 3479-3491.

Narum, D.L., Ogun, S.A., Thomas, A.W., and Holder, A.A. 2000. Immunization with parasite-derived apical membrane antigen 1 or passive immunization with a specific monoclonal antibody protects BALB/c mice against lethal *Plasmodium yoelii yoelii* YM blood-stage infection. Infect. Immun. 68: 2899-2906.

Narum, D.L., and Thomas, A.W. 1994. Differential localization of full-length and processed forms of PF83/AMA-1 an apical membrane antigen of *Plasmodium falciparum* merozoites. Mol. Biochem. Parasitol. 67: 59-68.

Ndengele, M.M., Messineo, D.G., Sam-Yellowe, T., and Harwalkar, J.A. 1995. *Plasmodium falciparum*: effects of membrane modulating agents on direct binding of rhoptry proteins to human erythrocytes. Exp. Parasitol. 81: 191-201.

Nicholls, S.C., Hillman, Y., Lockyer, M.J., Odink, K.G., and Holder, A.A. 1988. An S antigen gene from *Plasmodium falciparum* contains a novel repetitive sequence. Mol. Biochem. Parasitol. 28: 11-19.

Nikodem, D., and Davidson, E. 2000. Identification of a novel antigenic domain of *Plasmodium falciparum* merozoite surface protein-1 that specifically binds to human erythrocytes and inhibits parasite invasion, *in vitro*. Mol. Biochem. Parasitol. 108: 79-91.

Noe, A.R., and Adams, J.H. 1998. *Plasmodium yoelii* YM MAEBL protein is coexpressed and colocalizes with rhoptry proteins. Mol. Biochem. Parasitol. 96: 27-35.

Noe, A.R., Fishkind, D.J., and Adams, J.H. 2000. Spatial and temporal dynamics of the secretory pathway during differentiation of the *Plasmodium yoelii* schizont. Mol. Biochem. Parasitol. 108: 169-185.

Nwagwu, M., Haynes, J.D., Orlandi, P.A., and Chulay, J.D. 1992. *Plasmodium falciparum*: chymotryptic-like proteolysis associated with a 101-kDa acidic-basic repeat antigen. Exp. Parasitol. 75: 399-414.

O'Donnell, R.A., de Koning-Ward, T.F., Burt, R.A., Bockarie, M., Reeder, J.C., Cowman, A.F., and Crabb, B.S. 2001. Antibodies against merozoite surface protein (MSP)-1(19) are a major component of the invasion-inhibitory response in individuals immune to malaria. J. Exp. Med. 193: 1403-1412.

Oeuvray, C., Bouharoun-Tayoun, H., Gras-Masse, H., Bottius, E., Kaidoh, T., Aikawa, M., Filgueira, M.C., Tartar, A., and Druilhe, P. 1994a. Merozoite surface protein-3: a malaria protein inducing antibodies that promote *Plasmodium falciparum* killing by cooperation with blood monocytes. Blood. 84: 1594-1602.

Oeuvray, C., Bouharoun-Tayoun, H., Grass-Masse, H., Lepers, J.P., Ralamboranto, L., Tartar, A., and Druilhe, P. 1994b. A novel merozoite surface antigen of *Plasmodium falciparum* (MSP-3) identified by cellular-antibody cooperative mechanism antigenicity and biological activity of antibodies. Mem. Inst. Oswaldo Cruz. 89 Suppl 2: 77-80.

Ogun, S.A., and Holder, A.A. 1994. *Plasmodium yoelii*: Brefeldin A-sensitive processing of proteins targeted to the rhoptries. Exp. Parasitol. 79: 270-278.

Ogun, S.A., and Holder, A.A. 1996. A high molecular mass *Plasmodium yoelii* rhoptry protein binds to erythrocytes. Mol. Biochem. Parasitol. 76: 321-324.

Ogun, S.A., Scott-Finnigan, T.J., Narum, D.L., and Holder, A.A. 2000. *Plasmodium yoelii*: effects of red blood cell modification and antibodies on the binding characteristics of the 235-kDa rhoptry protein. Exp. Parasitol. 95: 187-195.

Oka, M., Aikawa, M., Freeman, R.R., Holder, A.A., and Fine, E. 1984. Ultrastructural localization of protective antigens of *Plasmodium yoelii* merozoites by the use of monoclonal antibodies and ultrathin cryomicrotomy. Am. J. Trop. Med. Hyg. 33: 342-346.

Okenu, D.M., Malhotra, P., Lalitha, P.V., Chitnis, C.E., and Chauhan, V.S. 1997. Cloning and sequence analysis of a gene encoding an erythrocyte binding protein from *Plasmodium cynomolgi*. Mol. Biochem. Parasitol. 89: 301-306.

Okoye, V.C., and Bennett, V. 1985. *Plasmodium falciparum* malaria: band 3 as a possible receptor during invasion of human erythrocytes. Science. 227: 169-171.

Okoyeh, J.N., Pillai, C.R., and Chitnis, C.E. 1999. *Plasmodium falciparum* field isolates commonly use erythrocyte invasion pathways that are independent of sialic acid residues of glycophorin A. Infect. Immun. 67: 5784-5791.

Olaya, P., and Wasserman, M. 1991. Effect of calpain inhibitors on the invasion of human erythrocytes by the parasite *Plasmodium falciparum*. Biochim Biophys Acta. 1096: 217-221.

Owen, C.A., Sinha, K.A., Keen, J.K., Ogun, S.A., and Holder, A.A. 1999. Chromosomal organisation of a gene family encoding rhoptry proteins in *Plasmodium yoelii*. Mol. Biochem. Parasitol. 99: 183-192.

Ozwara, H., Kocken, C.H., Conway, D.J., Mwenda, J.M., and Thomas, A.W. 2001. Comparative analysis of *Plasmodium reichenowi* and *P. falciparum* erythrocyte-binding proteins reveals selection to maintain polymorphism in the erythrocyte-binding region of EBA-175. Mol. Biochem. Parasitol. 116: 81-84.

Pachebat, J.A., Ling, I.T., Grainger, M., Trucco, C., Howell, S., Fernandez-Reyes, D., Gunaratne, R., and Holder, A.A. 2001. The 22 kDa component of the protein complex on the surface of *Plasmodium falciparum* merozoites is derived from a larger precursor, merozoite surface protein 7. Mol. Biochem. Parasitol. 117: 83-89.

Perkins, M.E., and Holt, E.H. 1988a. Erythrocyte receptor recognition varies in *Plasmodium falciparum* isolates. Mol. Biochem. Parasitol. 27: 23-34.

Perkins, M.E., and Rocco, L.J. 1988b. Sialic acid-dependent binding of *Plasmodium falciparum* merozoite surface antigen, Pf200, to human erythrocytes. J. Immunol. 141: 3190-3196.

Perkins, M.E., and Rocco, L.J. 1990. Chemical cross-linking of *Plasmodium falciparum* glycoprotein, Pf200 (190-205 kDa), to the S-antigen at the merozoite surface. Exp. Parasitol. 70: 207-216.

Perkins, M.E., and Ziefer, A. 1994. Preferential binding of *Plasmodium falciparum* SERA and rhoptry proteins to erythrocyte membrane inner leaflet phospholipids. Infect. Immun. 62: 1207-1212.

Perrin, L.H., Merkli, B., Loche, M., Chizzolini, C., Smart, J., and Richle. 1984. Antimalarial immunity in Saimiri monkeys. Immunization with surface components of asexual blood stages. J. Exp. Med. 160: 441-451.

Perrin, L.H., Ramirez, E., Lambert, P.H., and Miescher, P.A. 1981. Inhibition of *P. falciparum* growth in human erythrocytes by monoclonal antibodies. Nature. 289: 301-303.

Peterson, D.S., and Wellems, T.E. 2000. EBL-1, a putative erythrocyte binding protein of *Plasmodium falciparum*, maps within a favored linkage group in two genetic crosses. Mol. Biochem. Parasitol. 105: 105-113.

Peterson, M.G., Marshall, V.M., Smythe, J.A., Crewther, P.E., Lew, A., Silva, A., Anders, R.F., and Kemp, D.J. 1989. Integral membrane protein located in the apical complex of *Plasmodium falciparum*. Mol. Cell. Biol. 9: 3151-3154.

Peterson, M.G., Nguyen-Dinh, P., Marshall, V.M., Elliott, J.F., Collins, W.E., Anders, R.F., and Kemp, D.J. 1990. Apical membrane antigen of *Plasmodium fragile*. Mol. Biochem. Parasitol. 39: 279-283.

Pinder, J., Fowler, R., Bannister, L., Dluzewski, A., and Mitchell, G.H. 2000. Motile systems in malaria merozoites: how is the red blood cell invaded? Parasitol. Today. 16: 240-245.

Pinder, J.C., Fowler, R.E., Dluzewski, A.R., Bannister, L.H., Lavin, F.M., Mitchell, G.H., Wilson, R.J., and Gratzer, W.B. 1998. Actomyosin motor in the merozoite of the malaria parasite, *Plasmodium falciparum*: implications for red cell invasion. J. Cell. Sci. 111: 1831-1839.

Poupel, O., Boleti, H., Axisa, S., Couture-Tosi, E., and Tardieux, I. 2000. Toxofilin, a novel actin-binding protein from *Toxoplasma gondii*, sequesters actin monomers and caps actin filaments. Mol. Biol. Cell. 11: 355-368.

Prasad, C.D., Prasad Singh, A., Chitnis, C.E., and Sharma, A. 2003. A *Plasmodium yoelii yoelii* erythrocyte binding protein that uses Duffy binding-like domain for invasion: a rodent model for studying erythrocyte invasion. Mol. Biochem. Parasitol. 128: 101-105.

Preiser, P.R., and Jarra, W. 1998. P*lasmodium yoelii*: differences in the transcription of the 235-kDa rhoptry protein multigene family in lethal and nonlethal lines. Exp. Parasitol. 89: 50-57.

Preiser, P.R., Jarra, W., Capiod, T., and Snounou, G. 1999. A rhoptry-protein-associated mechanism of clonal phenotypic variation in rodent malaria [see comments]. Nature. 398: 618-622.

Ranford-Cartwright, L.C., Taylor, R.R., Asgari-Jirhandeh, N., Smith, D.B., Roberts, P.E., Robinson, V.I., Babiker, H.A., Riley, E.M., Walliker, D., and McBride, J.S. 1996. Differential antibody recognition of FC27-like *Plasmodium falciparum* merozoite surface protein MSP2 antigens which lack 12 amino acid repeats. Parasite Immunol. 18: 411-420.

Rangachari, K., Beaven, G.H., Nash, G.B., Clough, B., Dluzewski, A.R., Myint, O., Wilson, R.J., and Gratzer, W.B. 1989. A study of red cell membrane properties in relation to malarial invasion. Mol. Biochem. Parasitol. 34: 63-74.

Rangachari, K., Dluzewski, A., Wilson, R.J., and Gratzer, W.B. 1986. Control of malarial invasion by phosphorylation of the host cell membrane cytoskeleton. Nature. 324: 364-365.

Raphael, P., Takakuwa, Y., Manno, S., Liu, S.C., Chishti, A.H., and Hanspal, M. 2000. A cysteine protease activity from *Plasmodium falciparum* cleaves human erythrocyte ankyrin. Mol. Biochem. Parasitol. 110: 259-272.

Rayner, J.C., Galinski, M.R., Ingravallo, P., and Barnwell, J.W. 2000. Two *Plasmodium falciparum* genes express merozoite proteins that are related to *Plasmodium vivax* and *Plasmodium yoelii* adhesive proteins involved in host cell selection and invasion. Proc. Natl. Acad. Sci. USA. 97: 9648-9653.

Rayner, J.C., Vargas-Serrato, E., Huber, C.S., Galinski, M.R., and Barnwell, J.W. 2001. A *Plasmodium falciparum* homologue of *Plasmodium vivax* reticulocyte binding protein (PvRBP1) defines a trypsin-resistant erythrocyte invasion pathway. J. Exp. Med. 194: 1571-1581.

Read, M., Sherwin, T., Holloway, S.P., Gull, K., and Hyde, J.E. 1993. Microtubular organization visualized by immunofluorescence microscopy during erythrocytic schizogony in *Plasmodium falciparum* and investigation of post-translational modifications of parasite tubulin. Parasitology. 106 ( Pt 3): 223-232.

Reed, M.B., Caruana, S.R., Batchelor, A.H., Thompson, J.K., Crabb, B.S., and Cowman, A.F. 2000. Targeted disruption of an erythrocyte binding antigen in *Plasmodium falciparum* is associated with a switch toward a sialic acid-independent pathway of invasion. Proc. Natl. Acad. Sci. USA. 97: 7509-7514.

Ridley, R.G., Lahm, H.W., Takacs, B., and Scaife, J.G. 1991. Genetic and structural relationships between components of a protective rhoptry antigen complex from *Plasmodium falciparum*. Mol. Biochem. Parasitol. 47: 245-246.

Ridley, R.G., Takacs, B., Lahm, H.W., Delves, C.J., Goman, M., Certa, U., Matile, H., Woollett, G.R., and Scaife, J.G. 1990. Characterisation and sequence of a protective rhoptry antigen from *Plasmodium falciparum*. Mol. Biochem. Parasitol. 41: 125-134.

Rodriguez, L.E., Urquiza, M., Ocampo, M., Curtidor, H., Suarez, J., Garcia, J., Vera, R., Puentes, A., Lopez, R., Pinto, M., Rivera, Z., and Patarroyo, M.E. 2002. *Plasmodium vivax* MSP-1 peptides have high specific binding activity to human reticulocytes. Vaccine. 20: 1331-1339.

Roger, N., Dubremetz, J.F., Delplace, P., Fortier, B., Tronchin, G., and Vernes, A. 1988. Characterization of a 225 kilodalton rhoptry protein of *Plasmodium falciparum*. Mol. Biochem. Parasitol. 27: 135-141.

Roggwiller, E., Betoulle, M.E., Blisnick, T., and Braun Breton, C. 1996. A role for erythrocyte band 3 degradation by the parasite gp76 serine protease in the formation of the parasitophorous vacuole during invasion of erythrocytes by *Plasmodium falciparum*. Mol. Biochem. Parasitol. 82: 13-24.

Saint, R.B., Coppel, R.L., Cowman, A.F., Brown, G.V., Shi, P.T., Barzaga, N., Kemp, D.J., and Anders, R.F. 1987. Changes in repeat number, sequence, and reading frame in S-antigen genes of *Plasmodium falciparum*. Mol. Cell. Biol. 7: 2968-2973.

Sajid, M., Withers-Martinez, C., and Blackman, M.J. 2000. Maturation and specificity of *Plasmodium falciparum* subtilisin-like protease-1, a malaria merozoite subtilisin-like serine protease. J. Biol. Chem. 275: 631-641.

Salmon, B.L., Oksman, A., and Goldberg, D.E. 2001. Malaria parasite exit from the host erythrocyte: a two-step process requiring extraerythrocytic proteolysis. Proc. Natl. Acad. Sci. USA. 98: 271-276.

Sam-Yellowe, T.Y., Del Rio, R.A., Fujioka, H., Aikawa, M., Yang, J.C., and Yakubu, Z. 1998. Isolation of merozoite rhoptries, identification of novel rhoptry-associated proteins from *Plasmodium yoelii*, *P. chabaudi*, *P. berghei*, and conserved interspecies reactivity of organelles and proteins with *P. falciparum* rhoptry-specific antibodies. Exp. Parasitol. 89: 271-284.

Sam-Yellowe, T.Y., Fujioka, H., Aikawa, M., and Messineo, D.G. 1995. *Plasmodium falciparum* rhoptry proteins of 140/130/110 kd (Rhop-H) are located in an electron lucent compartment in the neck of the rhoptries. J. Eukaryot. Microbiol. 42: 224-231.

Saul, A., Cooper, J., Hauquitz, D., Irving, D., Cheng, Q., Stowers, A., and Limpaiboon, T. 1992. The 42-kilodalton rhoptry-associated protein of *Plasmodium falciparum*. Mol. Biochem. Parasitol. 50: 139-149.

Saul, A., Lord, R., Jones, G., Geysen, H.M., Gale, J., and Mollard, R. 1989. Cross-reactivity of antibody against an epitope of the *Plasmodium falciparum* second merozoite surface antigen. Parasite Immunol. 11: 593-601.

Saul, A., Myler, P., Schofield, L., and Kidson, C. 1984. A high molecular weight antigen in *Plasmodium falciparum* recognized by inhibitory monoclonal antibodies. Parasite Immunol. 6: 39-50.

Scheele, G.A., Fukuoka, S., and Freedman, S.D. 1994. Role of the GP2/THP family of GPI-anchored proteins in membrane trafficking during regulated exocrine secretion. Pancreas. 9: 139-149.

Schofield, L., Bushell, G.R., Cooper, J.A., Saul, A.J., Upcroft, J.A., and Kidson, C. 1986. A rhoptry antigen of *Plasmodium falciparum* contains conserved and variable epitopes recognized by inhibitory monoclonal antibodies. Mol. Biochem. Parasitol. 18: 183-195.

Sharma, P., Kumar, A., Singh, B., Bharadwaj, A., Sailaja, V.N., Adak, T., Kushwaha, A., Malhotra, P., and Chauhan, V.S. 1998. Characterization of protective epitopes in a highly conserved *Plasmodium falciparum* antigenic protein containing repeats of acidic and basic residues. Infect. Immun. 66: 2895-2904.

Shaw, M.K., Roos, D.S., and Tilney, L.G. 1998. Acidic compartments and rhoptry formation in *Toxoplasma gondii*. Parasitology. 117: 435-443.

Shaw, M.K., Roos, D.S., and Tilney, L.G. 2002. Cysteine and serine protease inhibitors block intracellular development and disrupt the secretory pathway of *Toxoplasma gondii*. Microbes Infect. 4: 119-132.

Shirano, M., Tsuboi, T., Kaneko, O., Tachibana, M., Adams, J.H., and Torii, M. 2001. Conserved regions of the *Plasmodium yoelii rhoptry* protein RhopH3 revealed by comparison with the *P. falciparum* homologue. Mol. Biochem. Parasitol. 112: 297-299.

Sim, B., Orlandi, P.A., Haynes, J.D., Klotz, F.W., Carter, J.M., Camus, D., Zegans, M.E., and Chulay, J.D. 1990. Primary structure of the 175K *Plasmodium falciparum* erythrocyte binding antigen and identification of a peptide which elicits antibodies that inhibit malaria merozoite invasion. J. Cell. Biol. 111: 1877-1884.

Sim, B.K., Carter, J.M., Deal, C.D., Holland, C., Haynes, J.D., and Gross, M. 1994a. *Plasmodium falciparum*: further characterization of a functionally active region of the merozoite invasion ligand EBA-175. Exp. Parasitol. 78: 259-268.

Sim, B.K., Chitnis, C.E., Wasniowska, K., Hadley, T.J., and Miller, L.H. 1994b. Receptor and ligand domains for invasion of erythrocytes by *Plasmodium falciparum*. Science. 264: 1941-1944.

Sim, B.K., Toyoshima, T., Haynes, J.D., and Aikawa, M. 1992. Localization of the 175-kilodalton erythrocyte binding antigen in micronemes of *Plasmodium falciparum* merozoites. Mol. Biochem. Parasitol. 51: 157-159.

Sigh, S.K., Sinhg, A.P., Pandey, S., Yazdani, S.S., Chitnis, C.E., and Sharma, A. 2003. Definition of structural elements in *Plasmodium vivax* and *Plasmodium knowlesi* Duffy binding domains necessary for erythrocyte invasion. Biochem. J. 374: 193-198.

Smythe, J.A., Coppel, R.L., Brown, G.V., Ramasamy, R., Kemp, D.J., and Anders, R.F. 1988. Identification of two integral membrane proteins of *Plasmodium falciparum*. Proc. Natl. Acad. Sci. USA. 85: 5195-5199.

Smythe, J.A., Coppel, R.L., Day, K.P., Martin, R.K., Oduola, A.M., Kemp, D.J., and Anders, R.F. 1991. Structural diversity in the *Plasmodium falciparum* merozoite surface antigen 2. Proc. Natl. Acad. Sci. USA. 88: 1751-1755.

Smythe, J.A., Peterson, M.G., Coppel, R.L., Saul, A.J., Kemp, D.J., and Anders, R.F. 1990. Structural diversity in the 45-kilodalton merozoite surface antigen of *Plasmodium falciparum*. Mol. Biochem. Parasitol. 39: 227-234.

Snewin, V.A., Herrera, M., Sanchez, G., Scherf, A., Langsley, G., and Herrera, S. 1991. Polymorphism of the alleles of the merozoite surface antigens MSA1 and MSA2 in *Plasmodium falciparum* wild isolates from Colombia. Mol. Biochem. Parasitol. 49: 265-275.

Stafford, W.H., Blackman, M.J., Harris, A., Shai, S., Grainger, M., and Holder, A.A. 1994. N-terminal amino acid sequence of the *Plasmodium falciparum* merozoite surface protein-1 polypeptides. Mol. Biochem. Parasitol. 66: 157-160.

Stafford, W.H., Gunder, B., Harris, A., Heidrich, H.G., Holder, A.A., and Blackman, M.J. 1996. A 22 kDa protein associated with the *Plasmodium falciparum* merozoite surface protein-1 complex. Mol. Biochem. Parasitol. 80: 159-169.

Stahl, H.D., Bianco, A.E., Crewther, P.E., Anders, R.F., Kyne, A.P., Coppel, R.L., Mitchell, G.F., Kemp, D.J., and Brown, G.V. 1986. Sorting large numbers of clones expressing *Plasmodium falciparum* antigens in *Escherichia coli* by differential antibody screening. Mol. Biol. Med. 3: 351-368.

Stewart, M.J., Schulman, S., and Vanderberg, J.P. 1986. Rhoptry secretion of membranous whorls by *Plasmodium falciparum* merozoites. Am. J. Trop. Med. Hyg. 35: 37-44.

Storey, E. 1992. A polyclonal but not a monoclonal antibody to an M(r) 52-kD protein responsible for a punctate fluorescence pattern in *Plasmodium falciparum* merozoites inhibits invasion *in vitro*. Am. J. Trop. Med. Hyg. 47: 663-674.

Stowers, A., Prescott, N., Cooper, J., Takacs, B., Stueber, D., Kennedy, P., and Saul, A. 1995. Immunogenicity of recombinant *Plasmodium falciparum* rhoptry associated proteins 1 and 2. Parasite Immunol. 17: 631-642.

Stowers, A.W., Cooper, J.A., Ehrhardt, T., and Saul, A. 1996. A peptide derived from a B cell epitope of *Plasmodium falciparum* rhoptry associated protein 2 specifically raises antibodies to rhoptry associated protein 1. Mol. Biochem. Parasitol. 82: 167-180.

Su, S., Sanadi, A.R., Ifon, E., and Davidson, E.A. 1993. A monoclonal antibody capable of blocking the binding of Pf200 (MSA-1) to human erythrocytes and inhibiting the invasion of *Plasmodium falciparum* merozoites into human erythrocytes. J. Immunol. 151: 2309-2317.

Sultan, A.A., Thathy, V., Frevert, U., Robson, K.J., Crisanti, A., Nussenzweig, V., Nussenzweig, R.S., and Menard, R. 1997. TRAP is necessary for gliding motility and infectivity of plasmodium sporozoites. Cell. 90: 511-522.

Swardson-Olver, C.J., Dawson, T.C., Burnett, R.C., Peiper, S.C., Maeda, N., and Avery, A.C. 2002. *Plasmodium yoelii* uses the murine Duffy antigen receptor for chemokines as a receptor for normocyte invasion and an alternative receptor for reticulocyte invasion. Blood. 99: 2677-2684.

Szarfman, A., Lyon, J.A., Walliker, D., Quakyi, I., Howard, R.J., Sun, S., Ballou, W.R., Esser, K., London, W.T., Wirtz, R.A., and Carter, R. 1988a. Mature liver stages of cloned *Plasmodium falciparum* share epitopes with proteins from sporozoites and asexual blood stages. Parasite Immunol. 10: 339-351.

Szarfman, A., Walliker, D., McBride, J.S., Lyon, J.A., Quakyi, I.A., and Carter, R. 1988b. Allelic forms of gp195, a major blood-stage antigen of *Plasmodium falciparum*, are expressed in liver stages. J. Exp. Med. 167: 231-236.

Tanabe, K., Mackay, M., Goman, M., and Scaife, J.G. 1987. Allelic dimorphism in a surface antigen gene of the malaria parasite *Plasmodium falciparum*. J. Mol. Biol. 195: 273-287.

Tardieux, I., Baines, I., Mossakowska, M., and Ward, G.E. 1998a. Actin-binding proteins of invasive malaria parasites and the regulation of actin polymerization by a complex of 32/34-kDa proteins associated with heat shock protein 70kDa. Mol. Biochem. Parasitol. 93: 295-308.

Tardieux, I., Liu, X., Poupel, O., Parzy, D., Dehoux, P., and Langsley, G. 1998b. A *Plasmodium falciparum* novel gene encoding a coronin-like protein which associates with actin filaments. FEBS Lett. 441: 251-256.

Taylor, H.M., Grainger, M., and Holder, A.A. 2002. Variation in the expression of a *Plasmodium falciparum* protein family implicated in erythrocyte invasion. Infect. Immun. 70: 5779-5789.

Taylor, H.M., Triglia, T., Thompson, J., Sajid, M., Fowler, R., Wickham, M.E., Cowman, A.F., and Holder, A.A. 2001. *Plasmodium falciparum* homologue of the genes for *Plasmodium vivax* and P*lasmodium yoelii* adhesive proteins, which is transcribed but not translated. Infect. Immun. 69: 3635-3645.

Taylor, R.R., Allen, S.J., Greenwood, B.M., and Riley, E.M. 1998. IgG3 antibodies to *Plasmodium falciparum* merozoite surface protein 2 (MSP2): increasing prevalence with age and association with clinical immunity to malaria. Am. J. Trop. Med. Hyg. 58: 406-413.

Taylor, R.R., Smith, D.B., Robinson, V.J., McBride, J.S., and Riley, E.M. 1995. Human antibody response to *Plasmodium falciparum* merozoite surface protein 2 is serogroup specific and predominantly of the immunoglobulin G3 subclass. Infect. Immun. 63: 4382-4388.

Thomas, A.W., Carr, D.A., Carter, J.M., and Lyon, J.A. 1990. Sequence comparison of allelic forms of the *Plasmodium falciparum* merozoite surface antigen MSA2. Mol. Biochem. Parasitol. 43: 211-220.

Thomas, A.W., Deans, J.A., Mitchell, G.H., Alderson, T., and Cohen, S. 1984. The Fab fragments of monoclonal IgG to a merozoite surface antigen inhibit *Plasmodium knowlesi* invasion of erythrocytes. Mol. Biochem. Parasitol. 13: 187-199.

Thompson, J.K., Triglia, T., Reed, M.B., and Cowman, A.F. 2001. A novel ligand from *Plasmodium falciparum* that binds to a sialic acid-containing receptor on the surface of human erythrocytes. Mol. Microbiol. 41: 47-58.

Tolle, R., Bujard, H., and Cooper, J.A. 1995. *Plasmodium falciparum*: variations within the C-terminal region of merozoite surface antigen-1. Exp. Parasitol. 81: 47-54.

Trager, W., Rozario, C., Shio, H., Williams, J., and Perkins, M.E. 1992. Transfer of a dense granule protein of *Plasmodium falciparum* to the membrane of ring stages and isolation of dense granules. Infect. Immun. 60: 4656-4661.

Trenholme, K.R., Gardiner, D.L., Holt, D.C., Thomas, E.A., Cowman, A.F., and Kemp, D.J. 2000. clag9: A cytoadherence gene in *Plasmodium falciparum* essential for binding of parasitized erythrocytes to CD36. Proc. Natl. Acad. Sci. USA. 97: 4029-4033.

Triglia, T., Healer, J., Caruana, S.R., Hodder, A.N., Anders, R.F., Crabb, B.S., and Cowman, A.F. 2000. Apical membrane antigen 1 plays a central role in erythrocyte invasion by Plasmodium species. Mol. Microbiol. 38: 706-718.

Triglia, T., Thompson, J., Caruana, S.R., Delorenzi, M., Speed, T., and Cowman, A.F. 2001a. Identification of proteins from *Plasmodium falciparum* that are homologous to reticulocyte binding proteins in *Plasmodium vivax*. Infect. Immun. 69: 1084-1092.

Triglia, T., Thompson, J.K., and Cowman, A.F. 2001b. An EBA175 homologue which is transcribed but not translated in erythrocytic stages of *Plasmodium falciparum*. Mol. Biochem. Parasitol. 116: 55-63.

Trucco, C., Fernandez-Reyes, D., Howell, S., Stafford, W.H., Scott-Finnigan, T.J., Grainger, M., Ogun, S.A., Taylor, W.R., and Holder, A.A. 2001. The merozoite surface protein 6 gene codes for a 36 kDa protein associated with the *Plasmodium falciparum* merozoite surface protein-1 complex. Mol. Biochem. Parasitol. 112: 91-101.

Urquiza, M., Suarez, J.E., Cardenas, C., Lopez, R., Puentes, A., Chavez, F., Calvo, J.C., and Patarroyo, M.E. 2000. *Plasmodium falciparum* AMA-1 erythrocyte binding peptides implicate AMA- 1 as erythrocyte binding protein. Vaccine. 19: 508-513.

Uthaipibull, C., Aufiero, B., Syed, S.E., Hansen, B., Guevara Patino, J.A., Angov, E., Ling, I.T., Fegeding, K., Morgan, W.D., Ockenhouse, C., Birdsall, B., Feeney, J., Lyon, J.A., and Holder, A.A. 2001. Inhibitory and blocking monoclonal antibody epitopes on merozoite surface protein 1 of the malaria parasite *Plasmodium falciparum*. J. Mol. Biol. 307: 1381-1394.

Van Wye, J., Ghori, N., Webster, P., Mitschler, R.R., Elmendorf, H.G., and Haldar, K. 1996. Identification and localization of rab6, separation of rab6 from ERD2 and implications for an 'unstacked' Golgi, in *Plasmodium falciparum*. Mol. Biochem. Parasitol. 83: 107-120.

Vargas-Serrato, E., Barnwell, J.W., Ingravallo, P., Perler, F.B., and Galinski, M.R. 2002. Merozoite surface protein-9 of *Plasmodium vivax* and related simian malaria parasites is orthologous to p101/ABRA of *P. falciparum*. Mol. Biochem. Parasitol. 120: 41-52.

Wang, L., Black, C.G., Marshall, V.M., and Coppel, R.L. 1999. Structural and antigenic properties of merozoite surface protein 4 of *Plasmodium falciparum*. Infect. Immun. 67: 2193-2200.

Wang, L., Marshall, V.M., and Coppel, R.L. 2002. Limited polymorphism of the vaccine candidate merozoite surface protein 4 of *Plasmodium falciparum*. Mol. Biochem. Parasitol. 120: 301-303.

Ward, G.E., Tilney, L.G., and Langsley, G. 1997. Rab GTPases and the unusual secretory pathway of Plasmodium. Parasitol. Today. 13: 57-62.

Waters, A.P., Thomas, A.W., Deans, J.A., Mitchell, G.H., Hudson, D.E., Miller, L.H., McCutchan, T.F., and Cohen, S. 1990. A merozoite receptor protein from *Plasmodium knowlesi* is highly conserved and distributed throughout Plasmodium. J. Biol. Chem. 265: 17974-17979.

Webb, S.E., Fowler, R.E., O'Shaughnessy, C., Pinder, J.C., Dluzewski, A.R., Gratzer, W.B., Bannister, L.H., and Mitchell, G.H. 1996. Contractile protein system in the asexual stages of the malaria parasite *Plasmodium falciparum*. Parasitology. 112 ( Pt 5): 451-457.

Weber, J.L., Lyon, J.A., Wolff, R.H., Hall, T., Lowell, G.H., and Chulay, J.D. 1988. Primary structure of a *Plasmodium falciparum* malaria antigen located at the merozoite surface and within the parasitophorous vacuole. J. Biol. Chem. 263: 11421-11425.

Weisman, S., Wang, L., Billman-Jacobe, H., Nhan, D.H., Richie, T.L., and Coppel, R.L. 2001. Antibody responses to infections with strains of *Plasmodium falciparum* expressing diverse forms of merozoite surface protein 2. Infect. Immun. 69: 959-967.

Werner, E.B., Taylor, W.R., and Holder, A.A. 1998. A *Plasmodium chabaudi* protein contains a repetitive region with a predicted spectrin-like structure. Mol. Biochem. Parasitol. 94: 185-196.

Wesseling, J.G., Snijders, P.J., van Someren, P., Jansen, J., Smits, M.A., and Schoenmakers, J.G. 1989. Stage-specific expression and genomic organization of the actin genes of the malaria parasite *Plasmodium falciparum*. Mol. Biochem. Parasitol. 35: 167-176.

Wilson, R.J. 1990. Biochemistry of red cell invasion. Blood Cells. 16: 237-252.

Wu, T., Black, C.G., Wang, L., Hibbs, A.R., and Coppel, R.L. 1999. Lack of sequence diversity in the gene encoding merozoite surface protein 5 of *Plasmodium falciparum*. Mol. Biochem. Parasitol. 103: 243-250.

From: Malaria Parasites: Genomes and Molecular Biology
Edited by: A.P. Waters and C.J. Janse

# Chapter 14

# Sexual Development of Malaria Parasites

## C.J. Janse and A.P. Waters

## Abstract

Malaria parasites switch from asexual proliferation to a sexual cycle of development in the blood of a vertebrate host in order to transmit between the vertebrate and mosquito host. The central role of sexual development in the life cycle and transmission of *Plasmodium* parasites make the sexual stages, such as the gametes and zygotes, attractive targets for interruption strategies to prevent transmission of the disease (for example transmission blocking vaccines). Here we review the knowledge on the cellular and molecular processes involved in sexual differentiation, i.e. the development of gametocytes and gametes, fertilization and zygote formation. In contrast to rapidly expanding molecular knowledge of asexual multiplication of malaria parasites, the molecular mechanisms of sexual differentiation are still largely unknown. However, with the availability of the genome sequence of different *Plasmodium* species combined with post-genomic technologies this situation will change. Insight into molecular processes underlying sexual development and the proteins involved will maximise numbers of candidate proteins available for consideration as transmission blocking vaccine candidates and will allow researchers and policy makers to make rational, informed decisions about the proteins considered for components of a multi-stage vaccine.

# 1. Introduction

Like birds and bees, malaria parasites do it; sex is an obligate feature of the life cycle of malaria parasites. These unicellular parasites must switch from asexual proliferation to a sexual cycle of development in the blood of a vertebrate host in order to transmit between the vertebrate and mosquito host. The sexual cycle is initiated by the switch of haploid asexual forms that multiply by syncitial, mitotic division into haploid, sexually differentiated, male and female cells. These cells, the so-called gametocytes, are the precursor forms of the male and female gametes. Gametocytes circulate in the blood of the vertebrate host in a state of developmental arrest until triggered to transform into gametes by environmental cues associated with the mosquito midgut. The gametes fertilize in the midgut of the mosquito and the resulting zygote develops into a motile form, the ookinete. The ookinete traverses the midgut wall, forming an oocyst on the hemolymph side of the midgut to establish the parasitic infection of the mosquito vector.

The central role of sexual development in the life cycle and transmission of *Plasmodium* parasites make the sexual stages, such as the gametes and zygotes, attractive targets for interruption strategies to prevent transmission of the disease. For example, an actively pursued approach to block transmission of malaria parasites is immunization of the human host with surface proteins of gametes or zygotes. Immune responses, principally through the action of antibodies, have successfully targeted events during sexual development, resulting in reduction of transmission capacity (Kaslow, 1996; Carter *et al.*, 2000). In addition to the central role of the sexual cells in transmission, the obligatory nature of sex in the lifecycle of *Plasmodium* is an evolved trait and must offer other evolutionary advantages to pure asexual proliferation as is the case for other sexually reproducing organisms. However, presumably this does not include as a defence mechanism to parasitism (Howard and Lively, 1994). Moreover, it is notable that several other protozoal organisms that are mosquito borne do not engage in sexual reproduction in the vector. Most theories about the evolutionary advantage of sex emphasize the benefits of meiotic recombination in removing harmful mutations and allowing new combinations of genes, providing more opportunities for improved fitness and offering the flexibility to new environments (Barton and Charlesworth, 1998). In *Plasmodium* there is ample evidence that genetic recombination occurs during the sexual cycle but the rate of recombination and its effect on generation of genetic diversity and adaptive capacity of the different *Plasmodium* species is largely unknown (Hey, 1999; Rich *et al.*, 2000; Awadalla *et al.* 2001).

In this review we will not address questions concerning the evolutionary aspects of sexual development in *Plasmodium* and the effects of recombination on genetic diversity in natural populations. Here we will focus on the knowledge on the cellular and molecular processes involved in sexual differentiation, i.e. the development of gametocytes and gametes, fertilization and zygote formation. In contrast to rapidly expanding molecular knowledge of asexual multiplication of malaria parasites, the molecular mechanisms of sexual differentiation are still largely unknown. Therefore, reviews on the molecular basis of cellular processes in sexual cells are necessarily based more on speculation then experimental data. It is expected that the availability of the genome sequence of different *Plasmodium* species combined with post-genomic technologies will radically change this situation. It is expected that increased knowledge of sexual differentiation will be valuable for further development of transmission blocking strategies and in addition will provide more insight into processes underlying the generation of genetic diversity in *Plasmodium*.

# 2. Gametocytes: The Switch From Asexual to Sexual Development

## 2.1. Fundamental Features of Gametocyte Development Differ Between Plasmodium Species

The switch from asexual multiplication to sexual development in *Plasmodium* parasites occurs in the haploid blood stages. A relatively small percentage of the blood stages stop multiplying by mitotic (asexual) division and differentiate into sexual cells, gametocytes, that are the precursor cells of the gametes. Two types of gametocytes are formed, female (macro) and male (micro) gametocytes. The molecular basis of the switch to sexual development is unknown and *Plasmodium* lacks characteristic sex chromosomes yet differentiation is a clearly regulated developmental pathway characterised by a specific patttern of stage (and presumably sex) specific transcription. In this section we will describe several of the distinctive (morphological) characteristics of gametocytes of the human parasite *P. falciparum* and the rodent parasite, *P. berghei*. The developing (young) gametocytes of *P. falciparum* show a number of features that are different from developing gametocytes of most other human and mammalian malaria parasites of which *P. berghei* is typical (Garnham 1966, Table 1). These unique features of *P. falciparum* gametocyte development are particularly striking in view of the similarities in later sexual development between all mammalian malaria parasites such as the morphology and development of the mature gametocytes, gametes and zygotes. A clear understanding of differences and similarities of the cellular biology of gametocyte and gamete development is crucial for a meaningful interpretation and comparison of genome, microarray and proteomic data from the different *Plasmodium* species.

The development of *P. falciparum* gametocytes has been divided into five stages (stage I-V) covering about 7-8 days from merozoite invasion to the mature gametocyte (Figure 1) (Hawking *et al.*, 1971; Ponnudurai *et*

**Table 1**. Differences and similarities between the gametocytes and gametes of *P. falciparum* and *P. berghei*

| Differences | *P. falciparum*[1] | *P. berghei*[2] |
|---|---|---|
| - sexual/asexual commitment | in 'parental'schizont | in young trophozoite? |
| - shape of (mature) gametocytes | crescent (banana-shaped) | round to oval shape |
| - developmental time | long (7-8 days) | short (26-30h, only a few hours longer then the asexual cycle) |
| - sequestration of young stages | in spleen and bone marrow | no sequestration? |
| - subpellicular cytoskeleton | complete | incomplete or absent third membrane, few or none microtubules |
| - mitochondria | cristate | few or no cristae |
| - nucleolus | yes | No |
| - expression of the sex-specific proteins | expression of pf27 and α-tubulin-II | not yet described, but two α-tubulin genes are present in the genome |

| Similarities between gametocytes | *P. falciparum and P. berghei* |
|---|---|
| - mitochondria | increase in number in female gametocytes |
| - osmiophilic bodies | higher numbers present in female then male gametocytes |
| - ribosomes/ER in females | High amounts of (single) ribosomes and (extended) ER |
| - storage of untranslated RNA in females | RNA coding for proteins expressed in zygotes, such as P25/P28 accumulates in the female gametocytes |
| - ribosomes/ER in males | Low amounts of ribosomes (polysomes) and ER in mature males |
| - DNA replication in males | rapid genome replication during the 10 minutes after induction of gamete formation. Sensitive to the DNA polymerase-α inhibitor aphidicolin |
| - expression of sex specific surface proteins | members of the P48/45 protein family, P48/45 and P230, are expressed in male/female gametocytes |
| - presence/expression of other sex-specific genes | In the *P. berghei* genome homologous genes have already been located for the gametocytes specific *P. falciparum* genes pf16, pf77, pfg377. In the *P. falciparum* genome homologous genes are present for the gametocytes specific *P. berghei* genes 230p, P36, P38, P47 |

| Similarities between Gametes | |
|---|---|
| - Formation of gametes (gametogenesis) | Gametocytes are induced to form gametes by drop in temperature and increase of pH; formation of gametes takes 10-12 minutes |
| - morphology of female gametes | female gametes are round to oval shaped cells, no differences reported in ultra-structural morphology |
| - morphology of male gametes | the male gametocyte produces 8 flagellar, motile gametes, no differences reported in ultra-structural morphology |
| - expression of proteins | expression of the EGF-containing proteins, P25 and P28 starts in (unfertilized) female gametes |
| - function of gamete specific proteins | P48/45 surface protein of gametes function in the process of fertilization |

[1] Many of the characteristics of *P. falciparum* gametocytes are shared with gametocytes of a subgroup of the avian malaria parasites.
[2] Many of the characteristics of *P. berghei* gametocytes are shared with gametocytes of most mammalian parasites belonging to the subgenus *Plasmodium* (including the human parasites *P. vivax, P. ovale, P. malariae*) and *Vinckeria* (rodent parasites).

*al.*, 1986). Blood stage merozoites of *P. falciparum* are already committed to either differentiate into gametocytes or to again develop into a schizont (see below). However, sexual parasites cannot be distinguished from asexual parasites up to the young trophozoite stage, 24-30 hours after invasion. The immature stages of gametocytes (stage I-IV) sequester in the capillaries of bone marrow and spleen (Thomson and Robertson, 1935; Smalley *et al.*, 1980). Knobs are present at the surface of erythrocytes infected with stage I/II gametocytes but not at the later stages (Day *et al.*, 1998). These knobs, electron-dense protuberances at the surface of infected erythrocytes are unique to *P. falciparum* and have been associated with adhesion and sequestration of the asexual trophozoite/schizont stage. The adhesion phenotype of stage I/II gametocytes has been reported to be similar to that of asexual trophozoites, based on their binding to several host cell receptors and trypsin sensitivity of binding. Receptors that mediate the binding of later stages (Stage II-IV) has not been defined yet but seem to be different from those binding young gametocytes (Day *et al.*, 1998). Male gametocytes can be distinguished from females only at stage III-V. The development of the mature stage V gametocytes, as with all mature gametocytes of *Plasmodium* that can form gametes, is arrested. They circulate in the peripheral blood and are only activated to continue development forming gametes once taken up by a mosquito in their midgut. It is in the later stages (III and IV) of gametocyte development that the distinctive scythe-like shape of *P. falciparum* gametocytes is (temporarily) formed, involving the formation of distinctive structures such as the subpellicular complex (see Figure 1) that are not overt in the other mammalian-infective species of *Plasmodium*.

**A**

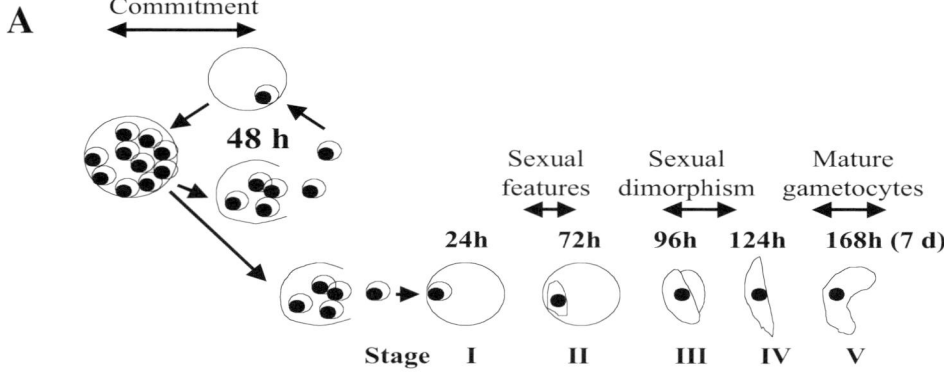

**B**

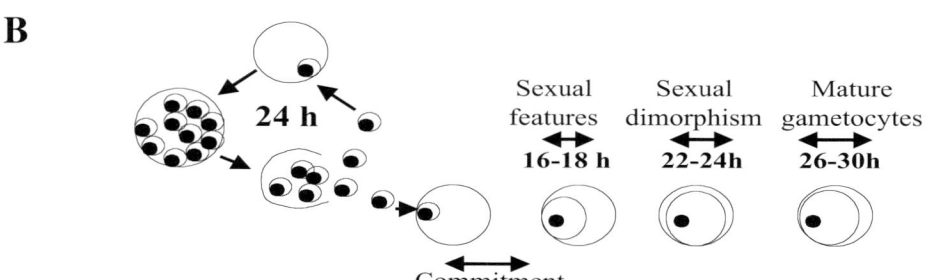

**C**

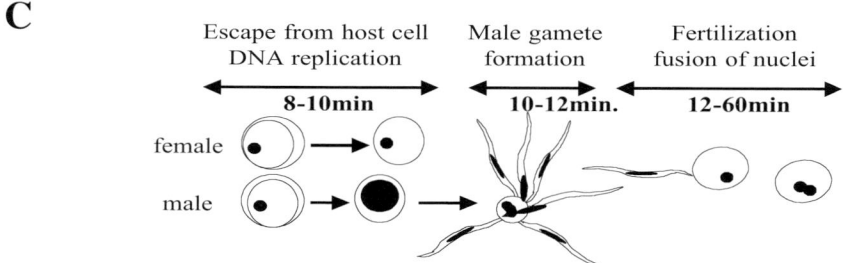

**Figure 1**. Development of gametocytes of *P. falciparum* and *P. berghei*. **A:** In *P. falciparum* meroziotes are already committed to either sexual or asexual development. The sexual merozoites cannot be distinguished from asexual merozoites up to the young trophozoite stage 24-30 hours after invasion. Sex specific features become clear during the development of Stage II parasites. A new subpellicular cytoskeleton is laid down, resulting in the presence of three membranes (parasitophorous, plasmalemma and 'third' double membrane, surrounding the gametocyte) and a layer of subpellicular microtubules; this is what gives the gametocytes their characteristic elongated shape and pointed ends. At stage III male and female forms can be distinguished. Stage I-III synthesize numerous ribosomes and endoplasmatic reticulum, correlating with the active synthesis of RNA and protein in these stages. Stage IV gametocytes are symmetrical, correlating with the complete encirclement of the parasites by the subpellicular cytoskeleton. From stage IV onwards small ovoid to elongated electron-dense or osmiophilic bodies become apparent. Sexual dimorphism is more clear at this stage, with a decrease in the density of ribosomes,

endoplasmatic reticulum, Golgi vessels and mitochondria in the male gametocyte compared to the female gametocyte. Only few osmiophilic bodies are present in stage IV and V male gametocytes whereas in the stage V female osmiophilic bodies are numerous and are localized beneath the double membrane. They are supposed to play a role in the escape from the (female) gametocytes from their host cell when activated to produce gametes. Female gametocytes have several mitochondria (Krunkrai 2000) that show a greater density of the cristae than that of asexual stage parasites. Transformation into the morphologically mature stage V gametocyte is accompanied by a brake down of the subpellicular complex followed by collapse of the pointed, spindle shape into a crescent shape cell. Mature gametocytes have a survival time of several days and a half-life of about 2 days have been reported.

**B:** In *P. berghei* evidence has been found that sexual commitment may take place in the young trophozoite stage. The development of gametocytes of *P .berghei* takes only about 30 hours from merozoite invasion to mature gametocyte. Up to sixteen hours after invasion of the merozoites, parasites that develop into gametocytes cannot be distinguished from asexual trophozoites at the ultra-structural and light-microscopic level. From 16-18h gametocyte specific morphological features become visible, such as eccentric located nucleus, randomly distributed hematin and osmiophilic bodies, while at the same time asexual trophozoites start nuclear division. Between 22-24 h male and female specific features develop. The females show a compact eccentric nucleus, numerous ribosomes, partially cristate mitochondria and extended endoplasmatic reticulum (ER), while males develop an enlarged nucleus and have a decreased density of ribosomes and limited ER. In contrast to *P.falciparum*, a third membrane under the plasma lemma is incomplete (the inner membrane layers of the pellicle are far less pronounced than those of *P. falciparum*) and seems to be only present in males. Membrane associated microtubules are absent (or occasionally found scattered beneath the pellicle (Sinden, 1978). Gametocytes are mature, i.e. capable of producing fertile gametes, between 26-30h after invasion of the merozoite and they have a survival time of about 24-30h in the blood of the host.

**C.** Upon transfer of gametocytes from the blood of the host to the midgut of a mosquito, these gametocytes are triggered to form gametes. The formation of the female gamete involves at the structural level little more than escape from the host cell. Upon activation, the gametocyte increases in size and the plasmalemma of the host erythrocyte becomes less distinct. At the ultra-structural level, the osmiophilic bodies can bee seen to be attached to the parasite plasmalemma by small ducts, supposedly to disorganize the host cell membrane and favouring the escape of the female gamete. The process of male gamete formation is referred to as exflagellation. At the light microscope level, an increase in the size of the nucleus is observed within 2-10 minutes after activation and the nucleus is migrating to the center of the cell, displacing the cytoplasm to the edge of the cell. This size increase coincides with three rounds of DNA replication and genome segregation. After 10-12 minutes exflagellating starts and eight highly active and motile gametes emerge. At the ultrastructural level many, rapid changes occur, both in the nucleus and the cytoplasm of the activated male gametocyte (for a detailed description see Sinden, 1978). Within 15 sec the cytoplasmic kinetosomes are formed in the cytoplasm upon which the axonemes start to polymerize whereas in the nucleus spindles are formed with associated kinetochores. In a period of less than 10 minutes three rounds of genome replication and segregation take place within a persistent nuclear envelope without condensation of the individual chromosomes. The processes of genome replication and /segregation in the nucleus and the formation of the axonemes in the cytoplasm are linked. These axonemes are formed by assembly of microtubeles on the kinetosomes. During the process of exflagellation the eight flagellar axonemes become highly active and they extend through the gametocyte plasmalemma. Each kinetosome then drags a single haploid genome into each of the eight gametes after which the gametes are torn away from the plasmalemma by the activity of the flagellar axoneme. The normal male gametocyte contains a single axoneme, a cigar-shaped nucleus, no mitochondrion and no significant cytoplasm.

The development of *P. berghei* gametocytes is much more rapid than that of *P. falciparum* and more typical of the genus taking only about 30 hours from merozoite invasion to mature gametocyte (Figure 1)(Mons *et al.*, 1985). In *P. berghei* there is evidence that the blood stage merozoites/ring forms are not already committed to sexual development as is the case in *P. falciparum* (see below). Up to 16 hours after invasion of the merozoites, parasites that develop into gametocytes cannot be distinguished from asexual trophozoites at the ultrastructural and light-microscopic level (Mons, 1986). From 16-19 hours onwards transcription of sex specific genes is switched on (Vervenne *et al.*, 1994; Thompson and Sinden, 1994), which is the first described feature of sexual development in *P. berghei*. From 18h gametocyte specific morphological features become visible and between 22-24h male and female specific features develop (Figure 1). A distinct subpellicular complex is lacking in *P. berghei* gametocytes. At 26-30h the ovoid gametocytes are mature, i.e. capable of producing fertile gametes. In comparative studies on synchronous development of gametocytes *in vivo* and *in vitro* no evidence was found for sequestration of either young or mature gametocytes (Mons, 1985; Janse *et al.*, 1985). For other rodent parasites it has been suggested that infectious gametocytes sequester in capillaries in the skin that the mosquito will feed from (Landau and Chabaud, 1994).

The mature male gametocyte in *Plasmodium* is a 'terminal' cell in which all essential processes of transcription and translation are reduced to the bare minimum for survival in the blood as shown by the low amount of ribosomes and endoplasmatic reticulum (ER). This cell is prepared for the formation of the eight flagellar gametes within a very short period of 10-12 minutes when activated in the mosquito midgut. The female gametocyte, although temporarily inactivated, is clearly prepared for renewed synthetic activity as the zygote after fertilization as shown by the high amount of ribosomes and ER and storage of mRNA pools (see below).

## 2.2. Metabolism and Drug-sensitivity Differences Between Gametocytes and Asexual Parasites

Although the metabolism of gametocytes has not been studied in detail no significant differences in metabolic processes have yet been reported between young gametocytes and asexual blood stages. Reported changes in mature gametocytes that might result in different metabolic capabilities might be related to the different metabolic requirements during further development of the zygotes and oocysts within the mosquito midgut (Lang-Unnasch and Murphy, 1998). For example, the changes in the number and structure of mitochondria in female gametocytes (Sinden, 1982), the switch between expression of different ribosomal-RNA genes (Waters *et al.*, 1989), and increased levels and activity of phosphenol pyruvate kinase (Hayward, 2000). The latter observation, obtained from the first DNA microarray experiments, indicate the gametocyte elevates gluconeogenesis probably in preparation for the glucose-poor environment in the mosquito midgut and is supported by observations on the gametocyte proteome where phosphofructokinase and malate dehydrogenase are abundant (Lasonder *et al.*, 2002). In view of the comparable metabolic requirements of asexual parasites and young gametocytes, such as digestion of haemoglobin, synthesis of RNA, proteins, plasma membranes and preparation for genome replication, it is not surprising that most anti-malarial drugs that kill asexual parasites also inhibit the young sexual stages (Lang-Unnasch and Murphy, 1998). Some clear examples are the antimalarials pyrimethamine, atovaquone, artemisinin and chloroquine. In contrast, the mature stage gametocyte is often insensitive to these drugs as a result of reduction of their metabolic activities to a minimum level. Some exceptions are riboflavin (Akompong *et al.*, 2000) and primaquine (Bhasin and Trager, 1987) that also kill mature gametocytes. No chemotherapeutics have been reported yet that selectively kill sexual parasites. One of the promises of post-genome technologies is that they will reveal unique features of gametocyte metabolism that might allow the development of gametocidal drugs.

### 2.3. Innate and Environmental Factors Determine Sexual Commitment

In *P. falciparum* the switch to sexual development (i.e. the sexual commitment of parasites) takes place in the 'parental' schizont. The young merozoite/ring form parasite is thus already committed to differentiate either into a gametocyte or develop into another schizont. Parasites that are derived from a single 'parental' schizont are always either gametocytes or asexual parasites (Bruce *et al.*, 1990; Smith *et al.*, 2000; Silvestrini *et al.*, 2000). Whether this timing of commitment is true for all mammalian *Plasmodium* species is unknown. For example, the rodent parasite *P. berghei* is thought to be still uncommitted at early ring stage retaining the potential to either differentiate into a gametocyte or remain asexual (Mons, 1986).

Cloned lines of *P. falciparum* and *P. berghei* show stable, innate differences in the rate of gametocyte production (Burkot *et al.*, 1984; Alano *et al.*, 1995; Janse *et al.*, 1992). A well known phenomenon in *Plasmodium* is the (irreversible) loss of the capacity to produce gametocytes of clones and lines during prolonged periods of asexual multiplication. In both *P. berghei* and *P. falciparum* the loss of gametocyte production has been correlated with large scale chromosomal rearrangements (Janse *et al.*, 1992; Alano *et al.*, 1995, Scherf *et al.*, 1992), indicating that these rearrangements might affect genes that are essential for sexual development. Mutant parasite lines that are defective in sexual development will be a useful source of probes for DNA microarrays, for example, the *P. falciparum* mutant that is defective in its production of mature male gametocytes (Guinet *et al.*, 1996). However, the limited number of such mutants that are currently available means that meaningful associations between genome alterations and failure to produce gametocytes might be difficult to obtain. The percentage of parasites that switch from asexual to sexual differentiation is not 'genetically fixed'. From *in vitro* studies on cloned *P. falciparum* parasites ample evidence is available that environmental stimuli influence sexual commitment (Carter and Miller, 1979; Bruce *et al.*, 1990). For example, stimuli associated with less favourable culture conditions enhance sexual commitment (Graves *et al.*, 1984). Numerous stimuli have been implicated in sexual commitment (reviewed in Lobo and Kumar (1998), Dyer and Day (2000b)) and a quantitative model for sexual commitment based on the effects of the various stimuli has been proposed by Dyer and Day (2000b). Central to this model is that individual parasites from a clone vary in their sensitivity to the environmental stimuli, that many different stimuli modulate the rate of commitment and that specific signal transduction pathways, for example those involving trimeric G proteins (Dyer and Day, 2000a) link the stimuli to the developmental switch. Environmental modulation of sexual commitment has also been described in the rodent parasites *P. chabaudi* and *P. berghei*. For example, induction of reticulocytosis in the hosts appears to enhance sexual commitment in both *P. berghei* (Mons, 1986) and *P. chabaudi* (Gautret *et al.*, 1996) and also influences sex ratio in *P. vinckei* (see below). It is unknown whether serum-factors or the cellular environment (reticulocytes) stimulate sexual commitment in the rodent parasites. All these studies on specific stimuli have not yet resulted in a demonstration of the molecular pathways involved in sexual commitment and have not provided defined stimuli for reproducible induction of gametocyte production. Post genome methodologies including the use of genetically manipulated parasite lines will be central to further research on this aspect of *Plasmodium* biology.

## 2.4. Environmental Factors Influence Sex Ratio

The molecular mechanisms underlying sex determination (male versus female) are unknown. The timing of sex determination is not known in *Plasmodium* but in *P. falciparum* the sex is already determined in the 'parental' schizont (Silvestrini *et al.*, 2000). Sexual merozoites/ringforms are thus already committed to either sex. In general more female gametocytes are produced then male gametocytes but there appears to be considerable variation in the gametocyte sex ratio (see references in Paul *et al.* (2002a)). Both innate and environmental factors appear to influence variation in sex ratio (Burkot *et al.*, 1984; for review see Paul *et al.*, 2002c). Direct evidence for environmental modulation of sex ratio has recently been obtained by the natural and artificial induction of erythropoiesis in the vertebrate host, that provoked a shift toward male parasite production (Paul *et al.*, 2000a ). This shift could be experimentally triggered by the hormone erythropoietin but the molecular mechanisms whereby this hormone modulates sex ratio remain to be elucidated and may well involve signal transduction pathways (kinase pathways) within the parasite similar to those found in other eukaryotes (Paul *et al.*, 2000b).

# 3. Conserved Processes of Gamete Formation and Fertilization

In contrast to the species specific features of early development of *Plasmodium* gametocytes notably of *P. falciparum*, the formation of gametes, their fertilisation and subsequent development of the zygote into an ookinete are remarkably conserved between all species. Environmental changes associated with transfer of the mature gametocytes from vertebrate host into the mosquito midgut trigger the initiation of gamete formation in *Plasmodium*. Within seconds the processes of gamete formation are triggered and fertilization is largely complete within 15-20 minutes. Induction of gamete formation in different species can be achieved routinely *in vitro* by simultaneous exposure of the gametocytes to two or three stimuli: a drop in temperature to at least 5°C below 37°C and either a rise in pH from 7.4 to 7.8-8.0 or addition of certain gametocyte activating factors (GAFs) derived from the mosquito or vertebrate host (Arai *et al.*, 2001). Recently it has been demonstrated that a mosquito-derived GAF is xanthurenic acid (Billker *et al.*, 1998). The parasite receptors for GAFs and the induced secondary pathways have not been yet defined, but activation of the parasite cGMP signalling pathway might be involved (Carucci *et al.*, 2000; Muhia *et al.*, 2001). Induction of gamete formation involves the simultaneous activation of several cellular events such as disruption of the host erythrocyte of the gametocytes, the reorganization of the male cytoplasmic structures to form the eight gametes and replication of the haploid male genome to provide the haploid genomes of the individual male gametes. The formation of the female gamete involves at the structural level little more than escape from the host cell. In the male gametocyte an increase in the size of the nucleus occurs within 2-10 minutes, coinciding with three rounds of genome replication and segregation (Janse *et al.*, 1986a,b; Janse *et al.*, 1988). After 10-12 minutes the process of 'exflagellation' starts and eight highly active and motile gametes emerge (for a more detailed description see Figure 1 and Sinden, 1983). Male gametes are initially highly and constantly active, an activity that declines over a period of 40 minutes. No directed movement of the male gamete towards the female gamete has been described (see Sinden, 1983) for a description of the motility of male gametes) and microgametes avidly adhere to female gametes as well as to neighboring erythrocytes. The receptor protein(s) involved in recognition/adhesion of the female gametes have not been characterized yet although the male gametes of a *P. berghei* line that cannot express P48/45 are infertile and fail to attach to fertile female gametes (van Dijk *et al.*, 2001). Sialic acid and one or more glycophorins of erythrocytes have been proposed to play a role in the adhesion of male gametes to erythrocytes (Templeton *et al.*, 1998). Following contact between male and female, the male slides back and forth over the surface of the female gamete. The plasmalemmas of both gametes then fuse and with a characteristic wave of moving activity the axoneme and the attached nucleus of the male move into the cytoplasm of the female cell. The male nucleus dissociates from the axoneme and fuses with the female nucleus. In the single, fused nucleus of the zygote DNA synthesis occurs up to the tetraploid value (Janse *et al.*, 1986a) coinciding with meiotic division, as shown by ultrastructural changes in the nucleus of the zygote (Sinden and Hartley, 1985). The developing zygote has a single, large nucleus containing the different haploid genomes and division of this nucleus start only in the oocyst stage. Clearly many of the classical processes associated with fertilisation and initial zygote development will be conserved in comparison with other organisms and performed by equally conserved proteins that will be pinpointed in the genome sequence largely by bioinformatics. Equally, other processes such as gamete formation, motility and gamete/gamete recognition might be expected to be genus (or even species) specific. Whilst some of the proteins involved have been recognised, it is one of the challenges of the post-genome technologies to identify the players. It is reassuring that these categories of proteins were readily identified in the emerging proteomes of gametocytes and gametes (Lasonder *et al.*, 2002; Florens *et al.*, 2002).

# 4. Specific Features of DNA and RNA in Relation to Sexual Differentiation

### 4.1. Clustering of Sex-specific Genes in the Genome

Although general synteny between the genomes of different species of *Plasmodium* is considered elsewhere in this book (see review by Carlton *et al*), it is relevant to consider that the need to undergo sexual differentiation may have helped shape the parasite genome and contributed to the observed patterns of synteny in the absence of dedicated sex chromosomes. Early studies on gene content and genome organisation had indicated a significant number of conserved genes expressed in sexual and post-fertilisation stages of *Plasmodium* were located on chromosome 5 of *P. berghei* (Janse *et al.*, 1995; van Lin *et al.*, 1998). A comprehensive study of a specific 13.6kb locus of the same chromosome revealed a complex arrangement of 6 genes, 3 of which are transcribed in gametocytes, 4 of which overlap, two in their coding regions. The degree of overlap is significant as the 13.6kb locus produces 15.5kb of mRNA and three of the genes generate multiple mRNA species through alternative splicing (van Lin *et al.*, 2001). The organisation of the orthologous locus of *P. falciparum* (on chromosome 10) was shown to be absolutely conserved. An intricate organisation of the genome, as shown for this locus, might have a positive influence on the maintenance of synteny in *Plasmodium*; spontaneous large-scale chromosomal rearrangements like translocations are not prone to be neutral, but likely to have a significant effect on gene expression of at least one gene. Thus, complex genome organisation may serve to keep clusters of genes linked in the genome for different evolutionary strategies, one being co-ordinated expression during a specific developmental phase such as gametocyte development. Certainly, preliminary data shows that syntenic regions can be large, for example, synteny of *P. berghei* chromosome 5 and *P. falciparum* chromosome 10 extends to ~650kb (Carlton *et al.*, 2002). Assembly of the complete genomes of *Plasmodium spp.* supported by detailed transcriptome analyses will reveal the full extent of co-ordinately expressed gene clusters a topic which has only recently begun to be considered (Florens *et al.*, 2002).

## 4.2. Regulation of Transcription and Storage of Untranslated mRNA in Female Gametocytes

Transcriptional differences are the earliest indicators of the switch to gametocyte development (Alano *et al.*, 1991; Bruce *et al.*, 1994), yet although the transcription of many sexual stage and sex-specific genes will be up or down-regulated during sexual development, only a few of these genes have been characterised. The regulation of gene expression in *Plasmodium* is poorly understood and in general there is scant knowledge of the DNA elements that direct transcriptional activity in this parasite (reviewed in Horrocks *et al.*, 1998). It has been shown that gene structure appears to be typically eukaryotic and therefore promoter activity generally lies upstream of transcriptional start sites and relies upon certain, ill-defined elements that show little intra- or interspecies conservation. A thus far unique transcription factor binding site has been mapped in detail upstream of the sex-specific gene *pfs25* of *P. falciparum* that is expressed in female gametocytes, gametes and zygotes (Dechering *et al.*, 1999). By analysis of the upstream UTR sequence of the sex-specific gene *pbs21* of *P. berghei* that is homologous to *pfs25,* it was demonstrated that upstream elements are able to control stage specific expression. Transcriptional control of this gene was lost in transgenic parasites in which *pbs21* had been introduced with only 200 bp of the upstream UTR region present and in these parasites, *pbs21* was transcribed in both asexual and sexual blood stages (Margos *et al.*, 1998). For several (housekeeping) genes it has been found that they produce different sized transcripts in asexual and sexual stages (Ridley *et al.*, 1991, Delves *et al.*, 1990), that might imply the activation of different promoters or other regulatory elements for the same gene in different stages [the recent discovery of extensive antisense transcripts notwithstanding (Pantakar *et al.*, 2001)]. The production of different sized transcripts of the single copy gene B7 have been analysed in detail (Pace *et al.*, 1998). Two mechanisms explain size differences in B7-mRNA between asexual and sexual stages: the use of alternative transcription initiation sites that implies the presence of separate, stage-specific promoters and the splicing of a gametocyte-specific intron from the upstream UTR region. As a result, expression of this gene can be modulated either at the transcriptional level or the post-transcriptional level. Mature female gametocytes are able to store untranslated RNA in large amounts that is only derepressed during further development of gametes and zygotes. This has been shown in *P. berghei* for mRNA of the P25 gene (Paton *et al.*, 1993), which has orthologues and paralogues in all species of *Plasmodium* (Tomas *et al.*, 2001). The mechanism of this kind of post-transcriptional control is also unknown although the UTR region of all P25 mRNA species exhibit oligo(U) runs typical of RNA elements that are known to interact with RNA binding proteins. This suggests that translational control (activation and/or repression) might be protein-mediated. Recently proteins associated with repression of translation have beeen described in the *P. falciparum* genome that contain *puf* domains, highly conserved repeat elements (typically 8 domains of 36aa) that are known to bind directly to a target mRNA and assemble a complex of proteins typically in the 3′UTR of the mRNA that mediates translational repression (Cui *et al.*, 2002). Post-translational control of gene expression provides a compelling reason why post-genome approaches to understanding *Plasmodium* sexual development should combine analyses of transcription (transcriptome) and protein content (proteome – see below). Such analyses will only prove fruitful if the relevant forms can be isolated with sufficient quantity as well as purity and with their physiological status intact (*e.g.* pure, mature non-activated gametocytes) and the model malarias may well present the best opportunity to generate such forms (see Table 2).

## 4.3. Rapid Genome Replication in Male Gametocytes

The three rounds of genome replication for the production of the eight haploid gametes takes place within 10 minutes after triggering the gametocyte to produce the eight gametes. It has been calculated that the entire haploid genome is replicated in about 3 min during this period (Janse *et al.*, 1986a, b). The genome replication in male gametocytes can be reversibly inhibited by the compound aphidicolin, which is a specific inhibitor for eukaryotic polymerase-$\alpha$ (Ohashi *et al.*, 1978). Since aphidicolin specifically inhibits male gamete formation while the female gametes are unaffected, this inhibitor can be used in cross-fertilization studies for obtaining specific cross-fertilization between different parasite clones (van Dijk *et al.*, 2001). It has been reported that during gametocyte development of *P. falciparum* and *P. berghei* limited DNA synthesis occurs, up to 1.3-2 times the amount of haploid DNA (Janse *et al.*, 1986a, 1986b; 1988). This has been demonstrated by quantitative determination of the DNA content of nuclei after staining with DNA-specific fluorescent dyes, but has not yet been confirmed by independent methods for analysis of the nuclear DNA (own observations).

**Table 2.** Available experimental tools to study sexual development

| | *P. falciparum* | *P. berghei* |
|---|---|---|
| - *In vitro* culture of gametocytes | yes | Yes |
| - defined stimuli for reproducible induction of gametocyte production | no, but certain culture conditions enhance gametocyte production | no, but phenylhydrazine treatment of hosts enhance gametocyte production |
| - synchronous development of gametocytes | yes, in cultures | yes, in cultures and in hosts |
| - purification of different gametocyte stages | yes, stage I to stage V gametocytes | yes, only young (24-26h) and mature (30-35h) gametocytes |
| - separation of male and female gametocytes/gametes | no | no |
| - *in vitro* gamete formation | yes | yes |
| - *in vitro* fertilization and zygote development | very inefficient | yes |
| - purification of zygotes | not in large numbers | yes |
| - fertilization and zygote development in the mosquito | yes, via membrane feeding of gametocytes to mosquitoes | yes, via host feeding and membrane feeding of gametocytes or in vitro formed zygotes |
| - experimentally cross-fertilization between parasite clones | yes, via membrane feeding of mixtures of gametocytes | yes, also *in vitro* cross-fertilization assays |
| - Mabs/antiserum against sexual stage proteins | yes, but to less then 15 proteins | yes, but to less then 5 proteins |
| - studies on Mabs inhibition of fertilization and zygote formation | yes, via membrane feeding of mosquitoes | yes, via membrane feeding of mosquitoes |
| - Mabs/antiserum that block fertilization or zygote development | yes, Mabs against P48/45, P230, P25, P28 | yes, Mabs/antiserum against P25, P28 |
| - studies of function of sexual genes by reverse genetics | yes, gene knock-out and transgenic possibilities | yes, gene knock-out and transgenic possibilities |
| - natural or transgenic mutants with known defects in sexual development | Only a few | Only a few |
| - drugs that specifically interfere with sexual development that can be used experimentally | | aphidicolin blocks male gamete formation; (erythropoeitin enhances sex ratio in *P. vinckei*) |

**Table 3.** Genes and proteins that are exclusively (predominantly) transcribed or expressed in gametocytes and gametes*

| Gene/protein | Transcription | Expression | Location | Function | Reference |
|---|---|---|---|---|---|
| **Genes of the P48/45 gene family transcribed in gametocytes** | | | | | |
| P48/45 | gametocyte | gametocyte and gamete | surface of gametocyte gamete | role in fertilisation | Kocken et al. 1993; van Dijk et al., 2001 |
| P230 | gametocyte | gametocyte and gamete | surface of gametocyte and gamete | Mabs inhibit zygote formation | Williamson et al., 1993; |
| P230p, P47, P38, P36 | gametocyte | ? | predicted surface | ? | Thompson et al., 2001 |
| **Genes/proteins specifically transcribed and expressed in gametocytes/gametes** | | | | | |
| α-Tubulin II | gametocyte | male gametocyte and gamete | axoneme | formation of the microtubules | Rawlings et al., 1992 |
| pf77 | female gametocyte | ? | ? | ? | Baker et al., 1995 |
| pfg377 | gametocyte | gametocyte | osmiophilic bodies | maturation gametocyte, escape from host cell of the gamete | Alano et al., 1995; Severini et al., 1999 |
| Pfg27/25* | gametocyte (stage II, III) | gametocyte (stage II-?) | ? | Knock-out parasites did not produce gametocytes | Alano et al., 1991; Lobo et al., 1999 |
| Pf16 | gametocyte (stage I-?) | gametocyte (stage II-V) | parasitophorous vacuole membrane | | Moelans et al., 1991; Bruce et al., 1994 |
| Pfl1-1* | gametocyte | gametocyte | cytoplasm | escape from host cell of the gamete? | Scherf et al., 1992 |

**Genes/proteins transcribed/expressed in gametocytes/gametes but with function in later stages**

| Gene/protein | Transcription | Stage | Location | Function | References |
|---|---|---|---|---|---|
| P25, P28 | gametocytes, gametes, zygotes | gametes and zygotes and later stages | surface of gametes, zygotes and later stages | interactions of the zygote/ookinete with the mosquito | Kaslow *et al.*, 1988; Duffy and kaslow, 1997; Tomas *et al.*, 2001 |
| pfS-type rRNA | gametocytes (only mRNA precursors) | Mature forms in zygote | cytoplasm | mature rRNA molecules formed in ookinetes/oocysts for ribosome formation | Waters *et al.*, 1989 |
| **protein kinases, cyclases, ATPases and phosphatases, significantly upregulated in gametocytes** | | | | | |
| pfmrk, pfmap-2, pfcrk-1, pflammer, pfcdpk, pfcdk3, pfkin | predominantly in gametocytes | ? | ? | homology with different eukaryotic protein kinases | Li *et al.*, 1997; Dorin *et al.*, (1999); Doerig *et al.*, (1995); Li *et al.*, (2001); Bracchi *et al.*, 1996); Kappes *et al.*, (1999) |
| pfgcalpha, Pfgcbeta | predominantly in gametocytes | gametocytes | (pfgcalpha) parasite and parasitophorous membrane | homology with eukaryotic guanyl cyclases | Carucci *et al.*, 2000 |
| pfpp-beta, Pfpp-alpha | predominantly in gametocytes | predominantly in gametocytes | | homology with eukaryotic protein phosphatases | Li and Baker., 1997 |
| P-type ATPase | predominantly gametocytes | ? | cytoplasm | Homology with eukaryotic P-type ATPases | Krishna *et al.*, (1993); Trottein and Cowman (1995) |

*Indicates genes that are NOT conserved between *P. falciparum* and other species of *Plasmodium* according to BLAST searches performed (30.4.2).

## 4.4. The Extrachromosomal DNA's of Plasmodium Parasites are Inherited Maternally

*Plasmodium* has two extrachromosomal genomes, the mitochondrial 6-kb DNA element and the 35-kb circular DNA. Only the female gamete and not the male contain the extrachromosomal genomes, resulting in maternal transfer of these DNA elements during sexual development and fertilisation (Vaidya *et al.*, 1993; Creasey *et al.*, 1994).

# 5. Characterized Genes Expressed During Sexual Development

## 5.1. Only a Small Number of Sex-specific Genes Have Been Characterized

The fact that sexual development involves a gross reprogramming of gene transcription has been emphasised by the ranking of a known gametocyte-specific transcriptional event (transcription of Pfs48/45) at number 479 of 3648 clones assessed using DNA microarrays indicating that at least 14% of the genes are upregulated during bloodstage sexual development (P. Rathod, personal communication; Hayward *et al.*, 1999). Clearly the further characterisation of this pool of clones will be of great interest. Currently, however, a remarkably limited number of genes are identified that are either transcribed exclusively or significantly upregulated during sexual differentiation (Table 3) and this is merely a reflection of the relative lack of research interest in these stages. These identified gametocyte implicated genes mainly fall into two major classes, genes encoding either (*Plasmodium* specific) surface proteins or regulatory proteins such as phosphatases (Li and Baker, 1997) kinases and nucleotide cyclases (see Kappes *et al.* (1999) for an overview of *Plasmodium* protein kinases). Of these identified genes that are transcriptionally upregulated in gametocytes all but two predicted cytoplasmic proteins (pf11-1 and pfg25/27) have been shown to be conserved between *Plasmodium* species (Table 3). The function of only two of these proteins is in any way clear: the male-specific role in gamete fertilisation of the conserved P48/45 protein (See below; van Dijk *et al.*, 2001), and the essential nature of pfg25/27 for early development of gametocytes (Lobo *et al.*, 1999). However, gametocytes are an ideal target for the currently available technologies for genetic manipulation of *Plasmodium* and it should be possible to begin to a comprehensive study of the remainder (see chapter by Carvalho and Menard) It can be anticipated that signalling pathways will play a central role in translating the change in environment that a gametocyte experiences upon bloodmeal ingestion into the observed cellular responses. Many of the gametocyte genes already described such as kinases and phosphatases could be key in these processes.

## 5.2. A Gene Family, Containing the Gametocyte Specific Genes P48/45 and P230, and the Organisation of Duplicated Genes

Genomics has already contributed to the definition of a family of 8 additional proteins similar to the gametocyte/gamete surface proteins Pfs48/45 and pfs230 (P48/45 family; Templeton and Kaslow, 1999; Thompson *et al.*, 2001). The P48/45 family appears to be distributed and conserved throughout the different *Plasmodium* species. The function of P48/45 has recently been investigated by genetic manipulation. Targeted disruption of the P48/45 gene resulted in a strong inhibition of fertilisation by affecting the male's capacity to bind to and penetrate female gametes (van Dijk *et al.*, 2001). Transcription of several of the other members of the gene family are now known to be also up-regulated during gametocyte development. Although the exact function of these other members is still unclear (Table 3), these proteins may also play a role in fertilisation and, therefore, may be considered as possible candidate antigens for a transmission blocking vaccine like P48/45 (see below). From analysis of the P48/45 gene family it became clear that several genes forming closest paralogue pairs are closely linked to one another in the genome. For example, P47 is linked to P48/45, its closest paralogue, and lies only 1.5 kb immediately upstream of P48/45 (van Dijk *et al.*, 2001). Tight linkage of closely related genes that are specifically transcribed in gametocytes and encode surface proteins appears to be a recurrent theme in *Plasmodium*. For example of the 8 of the 10 genes encoding P48/45 and its homologues are arranged in this manner (Thompson *et al.*, 2001; J.T. unpublished data; Figure 2) as well as the genes encoding P28 and its orthologue P25, that are surface proteins of ookinetes first transcribed in gametocytes (Thompson *et al.*, 2001). Gene duplication resulting in paralogous pairs appears, therefore, to be an ancient and coordinated event in *Plasmodium* and the complete genome may well reveal more examples.

# 6. Transmission Blocking Vaccines: Only a Few Candidate Antigens for the Sexual Stages

The premise behind a vaccine that would block malaria parasite transmission is simple. Antibodies to determinants (to date exclusively protein) block gamete development, fertilisation or zygote development and thereby transmission of the parasite given the central nature of sexual development. Modeling studies and practical evidence demonstrates that reduction of the rate transmission by transmission blocking (TB) vaccines would reduce incidence of disease (for a reviews on TB vaccines see Kaslow (1996), Carter *et al.* (2000), and Stowers and Carter (2001)). A perceived advantage is that antigens expressed by the sexual stages, such as gametes and zygotes, are not normally exposed to immune pressure and therefore are less likely to exhibit (extensive) variation (Niederwieser *et al.* 2001). Furthermore, a transmission

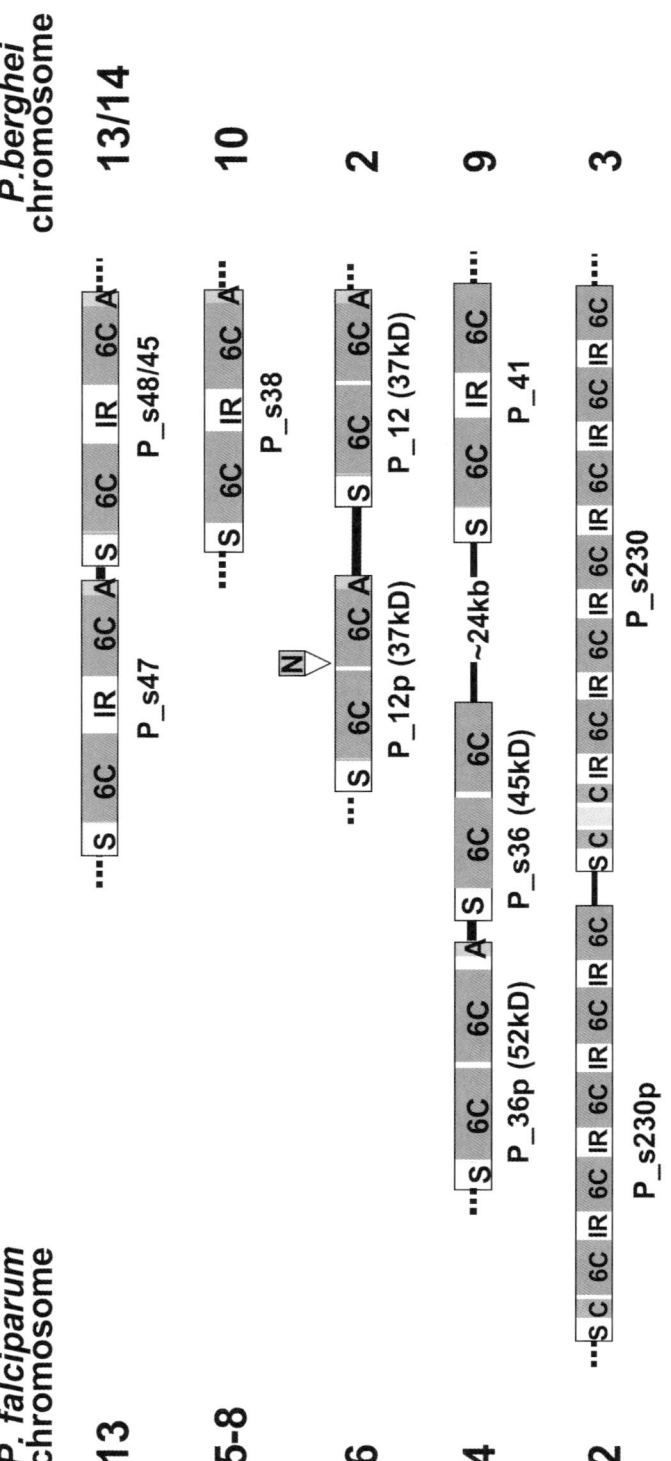

**Figure 2. Conserved organisation of the P48/45 family in two species of *Plasmodium*.** Cartoon depiction of the modular structure of the different members of the P48/45 gene superfamily. The 6-cys modular regions are indicated by 6C; the intermodular regions by IR; the predicted gpi anchor regions by A; the Cysteine rich regions by C and the signal sequence by S. An asparagine rich region of P12p that is specific to *P. falciparum* and appears to be a low complexity insertion typical of many proteins of the species is indicated by the box labelled N above the box and its approximate position of insertion is indicated. The chromosome number that the genes lie on are given to the left (*P. falciparum*) and to the right (*P. berghei*) of the cartoons. Each pair of genes are separated by approximately 2kb of intergenic sequence. Transcription is always left to right.

blocking vaccine in combination with other types of malaria vaccine or antimalarial drugs, could be effective in preventing the escape and spread of mutants resistant to those vaccines or drugs. Only a few of the current prime antigen candidates are sexual stage associated, the gametocyte/gamete surface proteins Pfs48/45, Pfs230 and the zygote/ookinete surface proteins P25/P28. The development of the different molecules as a transmission blocking vaccine highlights current difficulties with the proposed formulations with available candidates. For example, for the gamete surface protein Pfs48/45 evidence has been reported that it might vary in a regional manner (ref) that may, therefore, be associated with the gamete recognition and compatibility (incipient speciation)(Conway *et al.*, 2001; Escalante *et al.*, 2002). Given that the members of the P48/45 family, such as P48/45 itself and P230 contain modular structure determined by disulphide bonds it has proven to be difficult to generate large amounts of authentically folded material suitable for vaccination (Milek *et al.*, 2000). Moreover, genetic redundancy of sexual stage antigens such as P48/45 and the zygote surface proteins, P25 and p28 (van Dijk *et al.*, 2001, Tomas *et al.*, 2001), cautions against reliance upon a single antigen as a vaccine immunogen against transmission stages, even in a multistage cocktail vaccine. For these reasons the search for other sexual stage antigens must continue that might also be used in tandem in vaccine formulations. Obviously genome research is a vital tool to assist this search and is already bearing fruit (see below).

### 6.1. Genome Project, Post Genomics, Comparative Genomics and Sexual Development

The recently completed and annotated genome sequence of *P. falciparum* (Gardner *et al.*, 2002) complemented by the significant additional genome sequence data available for *P. yoelii* (Carlton *et al.*, 2002), *P. berghei*, *P. chabaudi* and *P. knowlesi* (all by Sanger) has allowed a more comprehensive picture of the conserved and species specific gene content of these parasites to be compiled. Exploitation of this data awaits detailed post-genomic analyses that assess transcription and protein content of the different life cycle forms. The combined datasets will then provide a complete picture of the protein repertoire of the gametocytes, gametes and zygotes. Independent analyses are already available for the proteome of *P. falciparum* gametocytes (Florens *et al.*, 2002) and one that is probably best considered as a combined proteome for both gametocytes and gametes (Lasonder *et al.*, 2002). These data are starting to confirm that it is correct to anticipate a high level of conservation of content and structure of both gametocytes and gametes of all *Plasmodium* species and the genetic basis for the more varied nature of the gametocytes (*P. falciparum* c.f. the other mammalian parasites) remains unclear. The proteome informs all of the aspects of gametocyte biology that have been considered throughout this review. Most but not all of the known gametocyte/gamete specific proteins were detected, including the known surface proteins, pfs48/45, pfs230 as well as the male specific α-tubulin II (Rawlings *et al.,* 1992). Numerous proteins that are predicted to be expressed on the surface are apparent including a conserved family of 5 proteins that are exclusive to gametocytes and contain multiple domains including lectin binding domains and 4 contain a highly conserved LCCL domain that is typically found in extracellular proteins in association with other modular domains and is thought to be involved in protein-protein interactions (Lasonder *et al.,* 2002). Indeed, one LCCL domain protein has subsequently been shown to have a role in oocyst development (Claudianos *et al.,* 2002). Furthermore, the proteome analysis also demonstrated upregulated expression of a large number of proteins associated with motility including a repertoire of dyneins; RNA binding proteins that might be involved in translational repression; protein kinases in signal transduction, chromatin remodeling factors and meiosis. One must remember that fully 50% of the predicted proteome of *P. falciparum* gametes and gametocytes consists of hypothetical proteins with unknown function lacking homologies to other proteins so far described. Comprehensive programmes of structure-function and genetics-based analysis are needed to ascertain if any of these "unknown" proteins are likely to be vaccine and drug targets. However, the insights already apparent make one optimistic that additional medically important proteins will be characterised.

Sexual development is a series of complex developmental programmes carried out by *Plasmodium* parasites in two radically different environments. The blood stage parasite commits to sexual development and produces, from an initially undifferentiated but committed cell, both male and female gameotcytes that are themselves developmentally arrested. Gamete formation and fertilisation and subsequent development of the zygote to an infectious ookinete occurs in sequence in the mosquito midgut within 24 hours after ingestion of gametocytes as part of the bloodmeal. We lack insight into the molecular complexity of the processes outlined above which have different degrees of potential to generate therapeutic strategies. For example, sexual commitment and sex determination are intriguing biological processes that might reveal understanding of molecular processes of gene regulation and differentiation in malaria that might have direct application for strategies to interrupt the life cycle. However, (defining) stimuli that significantly increase gametocyte production or influence sex ratio would be useful tools for improving production and purification of the different sexual stages benefitting the investigation of surface proteins of gametes/zygotes with relevance for vaccine development and identifying the proteins and genes involved is only the beginning. Although an understanding of the molecular mechanisms of translation inhibition in gametocytes and storage of untranslated mRNA pools for later translation may have no direct therapeutic relevance, nevertheless similar mechanisms of translational control probably exist in asexual stages. Therefore, this kind of research might lead to discovery and development of new drugs that interfere with those mechanisms and thus block sexual and possibly asexual development. Annotated, comparitive post-genome data will relate genome organisation to stage specific expression, determine the potential metabolic status of sexual cells and catalogue the proteins that are likely to be involved in host-gametocyte interactions and mediate gamete fertilisation. Although a few proteins and members of at least one family have already been shown to play a role in fertilisation, the genome and post-genome activities will maximise the numbers of different candidate proteins available for consideration as transmission blocking vaccine candidates and herein lies the medical value of studying sexual development. However beyond this simplistic level of identification, application of functional assays through genetic modification and the availability of batteries of appropriate monoclonal antibodies will allow researchers and policy makers to make rational, informed decisions about which and how many proteins should be considered for vaccine development. These arguments can, of course, be applied to all stages of the parasite to which we seek to develop components of a multi-stage vaccine.

# 7. References

Akompong, T., Eksi, S., Williamson, K., and Haldar, K. 2000. Gametocytocidal activity and synergistic interactions of riboflavin with standard antimalarial drugs against growth of *Plasmodium falciparum in vitro*. Antimicrob. Agents Chemother. 44: 3107-3111.

Alano, P., Premawansa, S., Bruce, M.C., and Carter, R. 1991. A stage specific gene expressed at the onset of gametocytogenesis in *Plasmodium falciparum*. Mol. Biochem. Parasitol. 46: 81-88.

Alano, P. and Carter, R. 1990. Sexual differentiation in malaria parasites. Annu. Rev. Microbiol. 44: 429-449.

Alano, P., Roca, L., Smith, D., Read, D., Carter, R., and Day, K. 1995. *Plasmodium falciparum*: parasites defective in early stages of gametocytogenesis. Exp. Parasitol. 81: 227-235.

Alano, P., Read, D., Bruce, M., Aikawa, M., Kaido, T., Tegoshi, T., Bhatti, S., Smith, D.K., Luo, C., Hansra, S., Carter, R., and Elliott, J.F. 1995. COS cell expression cloning of Pfg377, a *Plasmodium falciparum* gametocyte antigen associated with osmiophilic bodies. Mol. Biochem. Parasitol. 74: 143-156.

Arai, M., Billker, O., Morris, H.R., Panico, M., Delcroix, M., Dixon, D., Ley, S.V., and Sinden, R.E. 2001. Both mosquito-derived xanthurenic acid and a host blood-derived factor regulate gametogenesis of *Plasmodium* in the midgut of the mosquito. Mol. Biochem. Parasitol. 116: 17-24.

Awadalla, P., Walliker, D., Babiker, H., and Mackinnon, M. 2001. The question of *Plasmodium falciparum* population structure. Trends Parasitol. 17: 351-353.

Baker, D.A., Thomson, J., Daramola, O.O., Carlton, J.M., and Tagett, G.A. 1995. Sexual-stage-specific RNA expression of a new *Plasmodium falciparum* gene detected by *in situ* hybridisation. Mol. Biochem. Parasitol. 72: 193-201.

Barton, N.H. and Charlesworth, B. 1998. Why sex and recombination? Science. 281: 1986-1989.

Bhasin, V.K. and Trager, W. 1987. Gametocidal effects *in vitro* of primaquine and related compounds on *Plasmodium falciparum*. In: Primaquine: Phamacokinetics, Metabolism, Toxicity And Activity. P.I. Triggg, ed. Wiley and Sons, New York. p145-153.

Billker, O., Lindo, V., Panico, M., Etienne, A.E., Paxton, T., Dell, A., Rogers, M., Sinden, R.E., and Morris, H.R. 1998. Identification of xanthurenic acid as the putative inducer of malaria development in the mosquito. Nature. 392: 289-292.

Bracchi, V., Langsley, G., Thelu, J., Eling, W., and Ambroise-Thomas, P. 1996. PfKIN, an SNF1 type protein kinase of *Plasmodium falciparum* predominantly expressed in gametocytes. Mol. Biochem. Parasitol. 76: 299-303

Bruce, M.C., Alano, P., Duthie, S. and Carter, R. 1990. Commitment of the malaria parasite *Plasmodium falciparum* to sexual and asexual development. Parasitology. 100: 191-200.

Bruce, M.C., Carter, R.N., Nakamura, K., Aikawa, M., and Carter, R. 1994. Cellular location and temporal expression of the *Plasmodium falciparum* sexual stage antigen Pfs16. Mol. Biochem. Parasitol. 65: 11-22.

Burkot, T.R., Williams, J.L. and Schneider, I. 1984 Infectivity to mosquitoes of *Plasmodium falciparum* clones grown *in vitro* from the same isolate. Trans. R. Soc. Trop. Med. Hyg. 78: 339-341.

Carlton, J. M.-R. Angiuoli S.V., Suh B.B., Kooij T.W., Pertea M., Silva J.C., Ermolaeva M.D., Allen J.E., Selengut J.D., Koo H.L., Peterson J.D., Pop, M., Kosack, D.S., Shumway, M.F., Bidwell, S.L., Shallom, S.J., van Aken, S,E,, Riedmuller, S.B., Feldblyum, T.V., Cho, J.K., Quackenbush, J., Sedegah, M., Shoaibi, A., Cummings, L.M., Florens, L., Yates, J.R., Raine, J.D., Sinden, R.E., Harris, M.A., Cunningham, D.A., Preiser, P.R., Bergman, L.W., Vaidya, A.B., van Lin, L.H., Janse, C.J., Waters, A.P., Smith, H.O., White, O.R., Salzberg, S.L., Venter, J.C., Fraser, C.M., Hoffman, S.L., Gardner, M.J., and Carucci, D.J. 2002. Genome sequence and comparative analysis of the model rodent malaria parasite *Plasmodium yoelii yoelii*. Nature. 419: 512-519.

Carter, R. and Miller, L.H. 1979 Evidence for environmental modulation of gametocytogenesis in *Plasmodium falciparum* in continous culture. Bull. WHO. 57: 37-52

Carter, R., Mendis, K.N., Miller, L.H., Molineaux, L., and Saul, A. 2000. Malaria transmission-blocking vaccines--how can their development be supported? Nat. Med. 6: 241-244.

Carucci, D.J., Witney, A.A., Muhia, D.K., Warhurst, D.C., Schaap, P., Meima, M., Li, J.L., Taylor, M.C., Kelly, J.M., and Baker, D.A. 2000. Guanylyl cyclase activity associated with putative bifunctional integral membrane proteins in *Plasmodium falciparum*. J. Biol. Chem. 275: 22147-22156.

Claudianos, C., Dessens, J.T., Trueman, H.E., Arai, M., Mendoza, J., Butcher, G.A., Crompton, T., and Sinden, R.E. 2002. A malaria scavenger receptor-like protein essential for parasite development. Mol. Microbiol. 45: 1473-1484.

Conway, D.J., Machado, R.L., Singh, B., Dessert, P., Mikes, Z.S., Povoa, M.M., Oduola, A.M., and Roper, C. 2001. Extreme geographical fixation of variation in the *Plasmodium falciparum* gamete surface protein gene Pfs48/45 compared with microsatellite loci. Mol. Biochem. Parasitol. 115: 145-156.

Creasy, A., Mendis, K., Carlton, J., Willimason, D., Wilson, I and Carter, R. 1994. Maternal inheritance of extrachromosomal DNA in malaria parasites. Mol. Biochem. Parasitol. 65: 95-98.

Cui, L., Fan, Q., and Li, J. 2002. The malaria parasite *Plasmodium falciparum* encodes members of the Puf RNA-binding protein family with conserved RNA binding activity. Nucleic Acids Res. 30: 4607-17.

Day, K.P., Hayward, R.E., Smith, D. and Culvenor, J.G. 1998. CD36-dependent adhesion and knob expression of the transmission stages of *Plasmodium falciparum* is stage-specific. Mol. Biochem. Parasitol. 93: 167-177.

Dechering, K.J., Kaan, A.M., Mbacham, W., Wirth, D.F., Eling, W., Konings, R.N., and Stunnenberg, H.G. 1999. Isolation and functional characterization of two distinct sexual-stage-specific promoters of the human malaria parasite *Plasmodium falciparum*. Mol. Cell. Biol. 19: 967-978.

Delves, C.J., Alano, P., Ridley, R.G., Goman, M., Holloway, S.P., Hyde, J.E., and Scaife, J.G. 1990. Expression of alpha and beta tubulin genes during the asexual and sexual blood stages of *Plasmodium falciparum*. Mol. Biochem. Parasitol. 43: 271-278.

Dijk van, M.R., Janse, C.J., Thompson, J., Waters, A.P., Braks, J.A., Dodemont, H.J., Stunnenberg, H.G., van Gemert, G.J., Sauerwein, R.W., and Eling, W. 2001. A central role for P48/45 in malaria parasite male gamete fertility. Cell. 104: 153-164.

Doerig, C., Doerig, C., Horrocks, P., Coyle, J., Carlton, J., Sultan, A., Arnot, D., and Carter, R. 1995. Pfcrk-1, a developmentally regulated cdc2-related protein kinase of *Plasmodium falciparum*. Mol. Biochem. Parasitol. 70: 167-174.

Dorin, D., Alano, P., Boccaccio, I., Ciceron, L., Doerig, C., Sulpice, R., Parzy, D., and Doerig, C. 1999. An atypical mitogen-activated protein kinase (MAPK) homologue expressed in gametocytes of the human malaria parasite *Plasmodium falciparum*. Identification of a MAPK signature. J. Biol. Chem. 274: 29912-29920.

Dyer, M. and Day, K. 2000.Expression of *Plasmodium falciparum* trimeric G proteins and their involvement in switching to sexual development. Mol. Biochem. Parasitol. 108: 67-78.

Dyer M. and Day, K.P. 2000. Commitment to gametogenesis in *Plasmodium falciparum*. Parasitol. Today. 16: 102-107.

Duffy, P.E., and Kaslow, D.C. 1997. A novel malaria protein, Pfs28, and Pfs25 are genetically linked and synergistic as falciparum malaria transmission-blocking vaccines. Infect. Immun. 65: 1109-1113.

Escalante, A.A., Grebert, H.M., Chaiyaroj, S.C., Riggione, F., Biswas, S., Nahlen, B.L., and Lal, A.A. 2002. Polymorphism in the gene encoding the Pfs48/45 antigen of *Plasmodium falciparum*. XI. Asembo Bay Cohort Project. Mol. Biochem. Parasitol. 119: 17-22.

Florens, L., Washburn, M.P., Raine, J.D., Anthony, R.M., Grainger, M., Haynes, J.D., Moch, J.K., Muster, N., Sacci, J.B., Tabb, D.L., Witney, A.A., Wolters, D., Wu, Y., Gardner, M.J., Holder, A.A., Sinden, R.E., Yates, J.R., and Carucci, D.J. 2002. A proteomic view of the *Plasmodium falciparum* life cycle. Nature. 419: 520-526.

Gardner, M.J., Hall, N., Fung, E., White, O., Berriman, M., Hyman, R.W., Carlton, J.M., Pain, A., Nelson, K.E., Bowman, S., Paulsen, I.T., James, K., Eisen, J.A., Rutherford, K., Salzberg, S.L., Craig, A., Kyes, S., Chan, M.S., Nene, V., Shallom, S.J., Suh, B., Peterson, J., Angiuoli, S., Pertea, M., Allen, J., Selengut, J., Haft, D., Mather, M.W., Vaidya, A.B., Martin, D.M., Fairlamb, A.H., Fraunholz, M.J., Roos, D.S., Ralph, S.A., McFadden, G.I., Cummings, L.M., Subramanian, G.M., Mungall, C., Venter. J.C., Carucci, D.J., Hoffman, S.L., Newbold, C., Davis, R.W., Fraser, C.M., and Barrell, B. 2002. Genome sequence of the human malaria parasite *Plasmodium falciparum*. Nature. 419: 498-511.

Garnham, P.C.C. 1966. Malaria Parasites and Other Haemosporidia. Blackwell, Oxford.

Gautret, P., Miltgen, F., Gantier, J.C., Chabaud, A.G., and Landau I. 1996. Enhanced gametocyte formation by P*lasmodium chabaudi* in immature erythrocytes: pattern of production, sequestration, and infectivity to mosquitoes. J. Parasitol. 82: 900-906.

Graves, P.M., Carter, R., and McNeill, K.M. 1984 Gametocyte production in cloned lines of *Plasmodium falciparum*. Am. J. Trop. Med. Hyg. 33: 1045-1050

Guinet, F., Dvorak, J.A., Fujioka, H., Keister, D.B., Muratova, O., Kaslow, D.C., Aikawa, M., Vaidya, A.B. and Wellems, T.E. 1996. A developmental effect in *Plasmodium falciparum* male gametogenesis. J. Cell. Biol. 135: 269-278.

Hawking, F., Wilson, M.E. and Gammage, K. 1971 Evidence for cyclic development and short lived maturity in the gametocytes of *Plasmodium falciparum*. Trans. R. Soc. Trop. Med. Hyg. 65: 549-559

Hayward, R.E. 2000. *Plasmodium falciparum* phosphoenolpyruvate carboxykinase is developmentally regulated in gametocytes. Mol. Biochem. Parasitol. 107: 227-240.

Hayward, R.E., Derisi, J.L., Alfadhli, S., Kaslow, D.C., Brown, P.O., and Rathod, P.K. 2000. Shotgun DNA microarrays and stage-specific gene expression in *Plasmodium falciparum* malaria. Mol. Microbiol. 35: 6-14

Hey, J. 1999. Parasite populations: the puzzle of *Plasmodium*. Curr. Biol. 12: R565-R567

Horrocks, P., Dechering, K., and Lanzer, M. 1998. Control of gene expression in *Plasmodium falciparum*. Mol. Biochem. Parasitol. 95: 171-181.

Howard, R.S. and Lively, C.M. 1994. Parasitism, mutation accumulation and the maintenance of sex. Nature. 367: 554-557

Janse, C.J., Carlton, J.M-R., Walliker, D. and Waters, A.P. 1994. Conserved location of genes on polymorphic chromosomes of four species of malaria parasites. Mol. Biochem. Parasitol. 68: 285-296.

Janse, C.J., Mons, B., Rouwenhorst, R.J., Van der Klooster, P.F., Overdulve, J.P., and Van der Kaay, H.J. 1985. *In vitro* formation of ookinetes and functional maturity of *Plasmodium berghei* gametocytes. Parasitology. 91: 19-29

Janse, C.J., Van der Klooster, P.F., Van der Kaay, H.J., Van der Ploeg, M., and Overdulve, J.P. 1986a. Rapid repeated DNA replication during microgametogenesis and DNA synthesis in young zygotes of *Plasmodium berghei*. Trans. R. Soc. Trop. Med. Hyg. 80: 154-157.

Janse, C.J., van der Klooster, P.F., van der Kaay, H.J., van der Ploeg, M., and Overdulve, J.P. 1986b. DNA synthesis in *Plasmodium berghei* during asexual and sexual development. Mol. Biochem. Parasitol. 20: 173-182.

Janse, C.J., Ponnudurai, T., Lensen, A.H., Meuwissen, J.H., Ramesar, J., Van der Ploeg, M., and Overdulve, J.P. 1988. DNA synthesis in gametocytes of *Plasmodium falciparum*. Parasitology. 96: 1-7.

Janse, C.J., Ramesar, J., van den Berg, F.M., and Mons, B. 1992. *Plasmodium berghei*: *in vivo* generation and selection of karyotype mutants and non-gametocyte producer mutants. Exp. Parasitol. 74: 1-10.

Kappes, B., Doerig, C.D. and Greaser, R. 1999. An overview of *Plasmodium* protein kinases. Parastol. Today 15: 449-453.

Kaslow, D.C., Quakyi, I.A., Syin, C., Raum, M.G., Keister, D.B., Coligan, J.E., McCutchan, T.F., and Miller, L.H. 1988. A vaccine candidate from the sexual stage of human malaria that contains EGF-like domains. Nature. 333: 74-6.

Kaslow, D.C. 1996. Transmission blocking vaccines. In: Malaria Vaccine Development. S.L. Hoffman, ed. ASM Press, Washington, D.C. p. 181-228.

Kocken, C.H., Jansen, J., Kaan, A.M., Beckers, P.J., Ponnudurai, T., Kaslow, D.C., Konings, R.N., and Schoenmakers, J.G. 1993. Cloning and expression of the gene coding for the transmission blocking target antigen Pfs48/45 of *Plasmodium falciparum*. Mol. Biochem. Parasitol. 61: 59-68.

Krishna, S., Cowan, G., Meade, J.C., Wells, R.A., Stringer, J.R., and Robson, K.J. 1993. A family of cation ATPase-like molecules from *Plasmodium falciparum*. J. Cell. Biol. 120: 385-398.

Landau, I., and Chabaud, A. 1994. *Plasmodium* species infecting *Thamnomys rutilans*: a zoological study. Adv. Parasitol. 33: 49-90.

Lang-Unnasch, N. and Murphy, A.D. 1998. Metabolic changes of the malaria parasite during the transition from the human to the mosquito host. Annu. Rev. Microbiol. 52: 561-590.

Lasonder, E., Ishihama, Y., Andersen, J.S., Vermunt, A., Pain, A., Sauerwein, R.W., Eling,W., Hall, N., Waters, A.P., Stunnenberg, H.G., and Mann, M. 2002. Analysis of the *Plasmodium falciparum* proteome by high accuracy mass spectrometry. Nature. 419: 537-542.

Li, J.L., and Baker, D.A. 1997. Protein phosphatase beta, a putative type-2A protein phosphatase from the human malaria parasite *Plasmodium falciparum*. Eur. J. Biochem. 249: 98-106.

Li, J.L., Robson, K.J., Chen, J.L., Targett, G.A., and Baker, D.A. 1997. Pfmrk, a MO15-related protein kinase from *Plasmodium falciparum*. Gene cloning, sequence, stage-specific expression and chromosome localization. Eur. J. Biochem. 24: 805-813.

Li, J.L., Targett, G.A., and Baker, D.A. 2001. Primary structure and sexual stage-specific expression of a LAMMER protein kinase of P*lasmodium falciparum*. Int. J. Parasitol. 31: 387-392.

Lobo, C.A. and Kumar, N. 1998. Sexual differentiation and development in the malaria parasite. Parasitology Today. 14: 146-150.

Lobo, C.A., Fujioka, H., Aikawa, M., and Kumar, N. 1999. Disruption of the Pfg27 locus by homologous recombination leads to loss of the sexual phenotype in *P. falciparum*. Mol. Cell. 3: 793-798.

Margos, G., van Dijk, M.R., Ramesar,J., Janse, C.J., Waters, A.P. and Sinden, R.E. 1998. Transgenic expression of a mosquito stage malarial protein, Pbs21, in blood stages of transformed *Plasmodium berghei* and the induction of an immune response upon infection. Infect. Immun. 66: 3884-3891.

Milek, R.L., Stunnenberg, H.G., and Konings, R.N. 2000. Assembly and expression of a synthetic gene encoding the antigen Pfs48/45 of the human malaria parasite *Plasmodium falciparum* in yeast. Vaccine. 18: 1402-1411.

Moelans, I.I., Klaassen, C.H., Kaslow, D.C., Konings, R.N., and Schoenmakers, J.G. 1991. Minimal variation in Pfs16, a novel protein located in the membrane of gametes and sporozoites of *Plasmodium falciparum*. Mol. Biochem. Parasitol.46: 311-313.

Mons, B., Janse, C.J., Boorsma, E.G., and Van der Kaay, H.J. 1985. Synchronized erythrocytic schizogony and gametocytogenesis of *Plasmodium berghei in vivo* and *in vitro*. Parasitology. 91: 423-430.

Mons, B. 1986 Intra-erythrocytic differentiation of *Plasmodium berghei*. Acta Leiden. 54: 1-124.

Muhia, D.K., Swales, C.A., Deng, W., Kelly, J.M., and Baker, D.A. 2001. The gametocyte-activating factor xanthurenic acid stimulates an increase in membrane-associated guanylyl cyclase activity in the human malaria parasite *Plasmodium falciparum*. Mol. Microbiol. 42: 553-560.

Niederwieser, I., Felger, I., and Beck, H.P. 2001. Limited polymorphism in *Plasmodium falciparum* sexual-stage antigens. Am. J. Trop. Med. Hyg. 64: 9-11.

Ohashi, M., Taguchi, T., Iand kegami, S. 1978 Aphidicolin: a specific inhibitor of DNA polymerases in the cytosol of rat liver. Biochem. Biophys Res. Commun. 82: 1084-1090.

Pace, T., Birago, C., Janse, C.J., Picci, L., and Ponzi, M. 1998. Developmental regulation of a *Plasmodium* gene involves the generation of stage-specific 5' untranslated sequences. Mol. Biochem. Parasitol. 97: 45-53.

Patankar, S., Munasinghe, A., Shoaibi, A., Cummings, L.M., and Wirth, D.F. 2001. Serial analysis of gene expression in *Plasmodium falciparum* reveals the global expression profile of erythrocytic stages and the presence of anti-sense transcripts in the malarial parasite. Mol. Biol Cell. 12: 3114-3125.

Paton, M.G., Barker, G.C., Matsuoka, H., Ramesar, J., Janse, C.J., Waters, A.P., and Sinden, R.E. 1993. Structure and expression of a post-transcriptionally regulated malaria gene encoding a surface protein from the sexual stages of *Plasmodium berghei*. Mol. Biochem. Parasitol. 59: 263-275.

Paul, R.E.L., Coulson, T.N., Raibaud, A. and Brey, P.T. 2000a. Sex determination in malaria parasites. Science. 287: 128-131.

Paul, R.E., Doerig, C. and Brey, P.T. 2000b. Erythropoiesis and molecular mechanisms for sexual determination in malaria parasites. IUBMB Life 49L: 245-248.

Paul, R.E.L., Brey, P.T. and Robert, V. 2002c. *Plasmodium* sex determination and transmission to mosquitoes. Trends in Parasitology. 18: 32-38.

Ponnudurai, T., Lensen, A.H., Meis, J.F. and Meuwissen, J.H. 1986 Synchronization of *Plasmodium falciparum* gametocytes using an automated suspension culture system. Parasitoloy. 93: 263-274.

Rawlings, D.J., Fujioka, H., Fried, M., Keister, D.B., Aikawa, M. and Kaslow, D. 1992. Alpha-tubulin II is a male-specific protein in *Plasmodium falciparum*. Mol. Biochem. Parasitol. 56: 239-250.

Rich, S.M. Ferreiar, M.U. and Ayala, F.J. 2000. The origin of antigenic diversity in *Plasmodium falciparum*. Parasitol Today. 16: 390-396.

Ridley, R.G., Lahm, H.W., Takacs, B., and Scaife, J.G. 1991. Genetic and structural relationships between components of a protective rhoptry antigen complex from *Plasmodium falciparum*. Mol. Biochem. Parasitol. 47: 245-246.

Scherf, A., Carter, R., Petersen, C., Alano, P., Nelson, R., Aikawa, M., Mattei, D., Pereira da Silva, L., and Leech, J. 1992. Gene inactivation of Pf11-1 of *Plasmodium falciparum* by chromosome breakage and healing: identification of a gametocyte-specific protein with a potential role in gametogenesis. EMBO J. 11: 2293-2301.

Severini, C., Silvestrini, F., Sannella, A., Barca, S., Gradoni, L., and Alano, P. 1999. The production of the osmiophilic body protein Pfg377 is associated with stage of maturation and sex in *Plasmodium falciparum* gametocytes. Mol. Biochem. Parasitol. 100: 247-252.

Silvestrini, F., Alano, P. and Williams, J.L. 2000. Commitment to the production of male and female gametocytes in the human malaria parasite *Plasmodium falciparum*. Parasitology. 121: 465-471.

Sinden, R.E. 1978. cell Biology. In: Rodent Malaria. R. Killick-Kendrick and W. Peters, eds. Academic Press, London. p.85-168.

Sinden, R.E. 1982. Gametocytogenesis of *Plasmodium falciparum in vitro:* an electron microscopic study. Parasitology. 84: 1-11.

Sinden, R.E. 1983. Sexual development of malarial parasites. Adv. Parasitol. 22: 153-216.

Sinden, R.E. and Hartley, R.H. 1985. Identification of the meiotic division of malarial parasites. J. Protozool. 32: 742-744.

Smalley, M.E., Abdalla, S. and Brown, J. 1980. The distribution of *Plasmodium falciparum* in the peripheral blood and bone marrow of Gambian children. Trans. R. Soc. Trop. Med. Hyg. 75: 103-105.

Smith, T.G., Lourenco, P., Carter, R., Walliker, D., and Ranford-Cartwright, L.C. 2000. Commitment to sexual differentiation in the human malaria parasite, *Plasmodium falciparum*. Parasitology. 121: 127-133.

Stowers, A. and Carter, R. 2001. Current developments in malaria transmission-blocking vaccines. Expert Opin. Biol. Ther. 1: 619-628.

Templeton, T.J., Keister, D.B., Muratova, O., Procter, J.L., and Kaslow, D.C. 1998. Adherence of erythrocytes during exflagellation of *Plasmodium falciparum* microgametes is dependent on erythrocyte surface sialic acid and glycophorins. J. Exp. Med. 187: 1599-609.

Templeton, T.J., and Kaslow, D.C.. 1999. Identification of additional members define a *Plasmodium falciparum* gene superfamily which includes Pfs48/45 and Pfs230. Mol. Biochem. Parasitol. 101: 223-227.

Thompson, J., Janse, C.J., and Waters, A.P. 2001. Comparative genomics in *Plasmodium*: a tool for the identification of genes and functional analysis. Mol. Biochem. Parasitol. 118: 147-154.

Thompson, J., and Sinden, R.E. 1994. *In situ* detection of Pbs21 mRNA during sexual development of *Plasmodium berghei*. Mol. Biochem. Parasitol. 68: 189-196.

Thomson, J.G. and Robertson, A. 1935 The structure and development of *Plasmodium falciparum* gametocytes in the internal organs and peripheral circulation. Trans. R. Soc. Trop. Med. Hyg. 29: 31-40.

Tomas, A.M., Margos, G., Dimopoulos, G., van Lin, L.H., de Koning-Ward, T.F., Sinha, R., Lupetti, P., Beetsma, A.L., Rodriguez, M.C., Karras, M., Hager, A., Mendoza, J., Butcher, G.A., Kafatos, F., Janse, C.J., Waters, A.P., and Sinden, R.E. 2001. P25 and P28 proteins of the malaria ookinete surface have multiple and partially redundant functions. EMBO J. 20: 3975-3983.

Trottein, F., and Cowman, A.F. 1995. Molecular cloning and sequence of two novel P-type adenosinetriphosphatases from *Plasmodium falciparum*. Eur. J. Biochem. 227: 214-225.

Vaidya, A.B., Morrisey, J., Plowe, C.V., Kaslow, D.C., and Wellems, T.E. 1993. Unidirectional dominance of cytoplasmic inheritance in two genetic crosses of *Plasmodium falciparum*. Mol. Cell. Biol. 13: 7349-7357.

van Lin, L.H., Pace, T., Janse, C.J., Birago, C., Ramesar, J., Picci, L., Ponzi, M., and Waters, A.P. 2001. Interspecies conservation of gene order and intron-exon structure in a genomic locus of high gene density and complexity in *Plasmodium*. Nucleic Acids Res. 29: 2059-2068.

van Lin, L.H., Pace, T., Janse, C.J., Scotti, R., and Ponzi, M. 1997. A long range restriction map of chromosome 5 of *Plasmodium berghei* demonstrates a chromosome specific symmetrical subtelomeric organisation. Mol. Biochem. Parasitol. 86: 111-115.

Vervenne, R.A., Dirks, R.W., Ramesar, J., Waters, A.P., and Janse, C.J. 1994. Differential expression in blood stages of the gene coding for the 21-kilodalton surface protein of ookinetes of *Plasmodium berghei* as detected by RNA *in situ* hybridisation. Mol. Biochem. Parasitol. 68: 259-266.

Waters, A.P., Syin, C., and McCutchan, T.F. 1989. Developmental regulation of stage-specific ribosome populations in *Plasmodium*. Nature. 342: 438-440.

Williamson, K.C., Criscio, M.D., and Kaslow, D.C. 1993. Cloning and expression of the gene for *Plasmodium falciparum* transmission-blocking target antigen, Pfs230. Mol. Biochem. Parasitol. 58: 355-358.

From: Malaria Parasites: Genomes and Molecular Biology
Edited by: A.P. Waters and C.J. Janse

# Chapter 15

## Ookinete Cell Biology

R.E. Sinden, Y.I.H. Alavi, G.A.Butcher,
J.T.Dessens, J.D.Raine,
and H.E. Trueman

### Abstract

Recent advances in our understanding of the molecular and morphological development of the malarial zygote into a mature invasive ookinete are described. The similarities and more particularly the differences of this development to sporozoite and merozoite differentiation are emphasised. Notable differences include the absences rhoptries, the formation of just one apical complex during meiosis, and the formation of a pore-containing inner pellicular complex. The impact of both vertebrate and mosquito components of the bloodmeal on ookinete development are discussed. The destructive events of invasion of the midgut wall and oocyst differentiation are described and the potential interactions between the parasite and the mosquito immune system outlined. The very successful efforts to develop vaccines targeted to the ookinete are summarised. The contrasting lack of drugs that block transmission through the vector is highlighted.

### 1. Introduction

Following escape of the gametocytes from the erythrocyte, and the formation of the gametes in the mosquito bloodmeal, fertilization is often completed within an hour of the female mosquito taking a blood feed. In the resulting diploid zygote an array of 'fertilization-dependent' proteins are expressed.

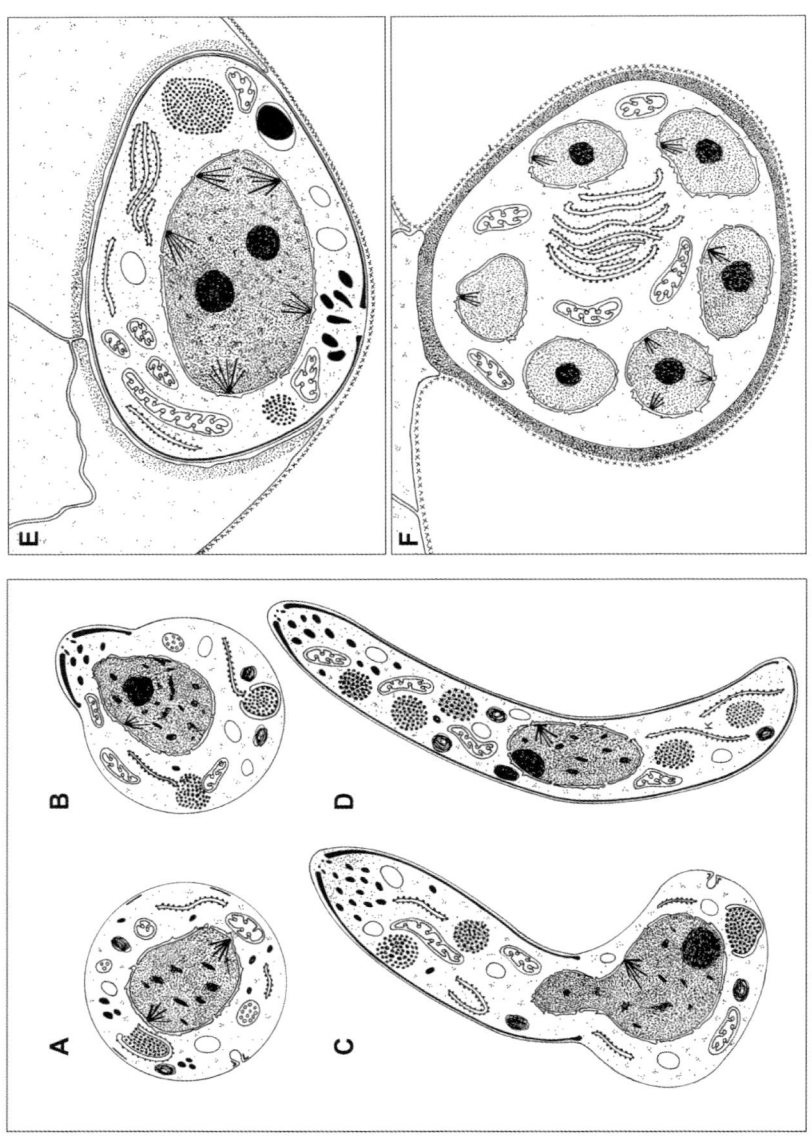

**Figure 1.** A–F) Diagram illustrating zygote-oocyst development in *P. gallinaceum* (*taken from Mehlhorn et al.*, 1980). A) zygote; B) early retort; C) late retort; D) mature ookinete; E) ookinete-oocyst transformation; F) young oocyst.

The activities of these molecules underpin the morphological transformation of the spherical zygote into the banana-shaped motile and invasive ookinete. The ookinete in turn then develops into a spherical oocyst. The ookinete maturation period varies between parasite species, and within a species is temperature dependent. *P. berghei* matures in *Anopheles stephensi* in 18-24 hours (19-21°C), *P. yoelii* and *P. falciparum* which both develop at 26°C in *An. stephensi* take 8-30hrs, and 24-36 hours respectively. The underlying reasons for these inter-specific differences are unknown.

## 1.1. Zygote-ookinete Transformation

Differentiation of each zygote into a single ookinete embraces three key events (Figure 1). First the parasite changes its cell surface from that of a fertilization-receptive macrogamete, into an invasive cell capable both of interacting with mosquito tissues, and of resisting attack by the hostile environment of the midgut. Second the cell builds the cytoskeleton and apical complex that contain the molecular motor and secretory apparatus required to escape the bloodmeal and invade the gut wall. Third the zygote undergoes the critical process of meiosis (Figure 2A).

At the molecular level we understand little of the precise regulation of zygote development. Some proteins critical to the survival and function of the zygote e.g. P25 and 28 are translated following gametogenesis from stable mRNA's retained in the mature gametocytes (Paton *et al.*, 1993; Rodriguez *et al.*, 2000). Other proteins that will also be secreted onto the parasite surface (but via the micronemes) are encoded by genes that are only transcribed following gametocyte activation e.g. CTRP (circumsporozoite protein and thrombospondin-related adhesive protein (TRAP)-related protein), WARP (von Willebrand factor A domain-related protein) and SOAP (secreted ookinete adhesive protein) (Dessens *et al.*, 1999; Yuda *et al.*, 2001; Dessens, unpublished).

At the cellular level we first recognise the differentiation of the zygote by the formation of a flattened, disc-shaped vacuole with subtending microtubules immediately beneath the plasma-membrane adjacent to the nucleus (Figure 1B), these become the cytoskeleton and apical complex (Figure 1C-D). By comparison with the analogous formation of both merozoites and sporozoites it is interesting that despite the presence of two (at meiosis I) or four (at meiosis II) spindle poles, only one apical complex is formed. Why, at merozoite and sporozoite formation is an apical complex formed at every spindle pole during mitosis whereas at meiosis only one? Other events of cell division including the formation of cytoskeletal elements of the apical complex, the secretory organelles (micronemes) and the migration of the essential extra-nuclear genomes of the mitochondrion and apicoplast are comparable in the analogous life-processes (schizogony and sporogony). Briefly, the overall shape of the ookinete as it emerges

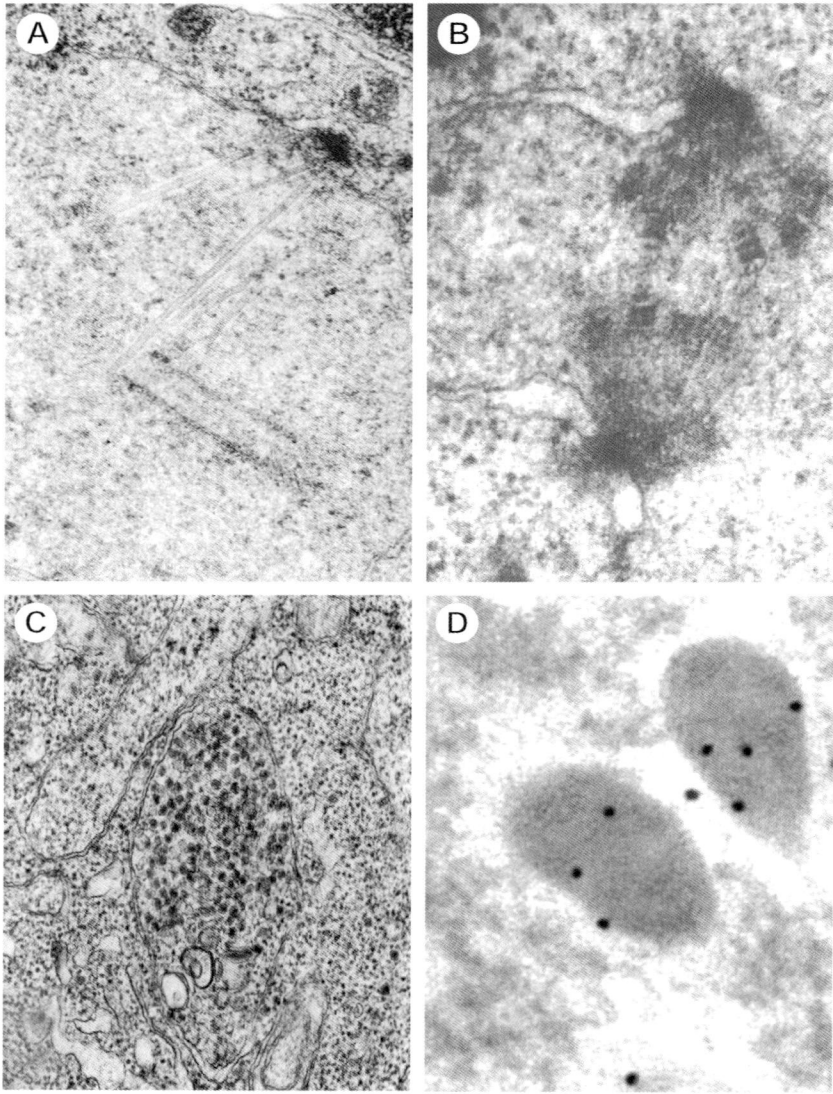

**Figure 2.** A) Transmission electron micrograph illustrating the synaptonemal complex between adjacent chromatids in meiotic division within persistent nuclear envelope (*P. yoelii nigeriensis*, from Sinden *et al.*, 1985). B) Transmission electron micrograph of mitotic spindle in a lobe of the oocyst nucleus, note chromosomes are not condensed on the kinetochores (*Plasmodium* does not express histone-1 which is normally essential for chromosomal condensation) (from Canning and Sinden, 1973). C) Transmission electron micrograph of vesicular network that gives rise to the micronemes and crystalloid during differentiation of the zygote into an ookinete (*P. yoelii nigeriensis*). D) Transmission electron micrograph illustrating the localization of CTRP to the micronemes by the immunogold technique (*P. berghei*).

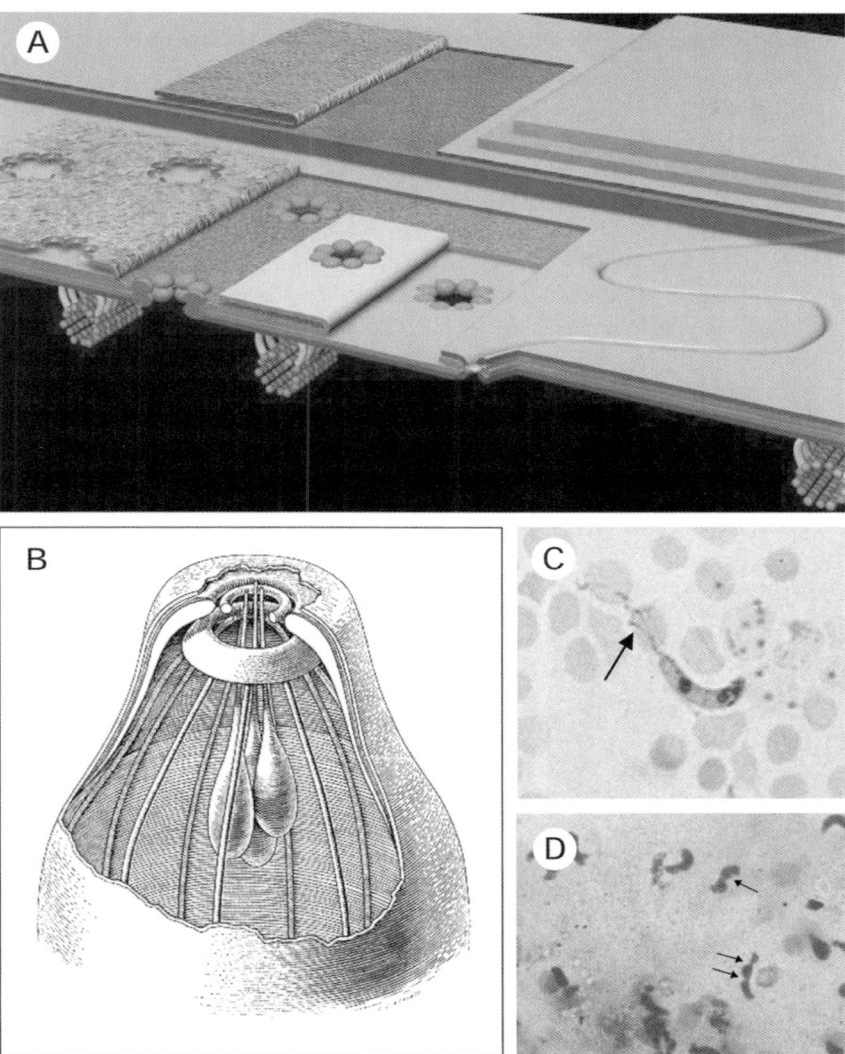

**Figure 3.** A) Reconstruction from transmission eslectron micrograph of freeze-fractured ookinete illustrating distribution of large pores in inner pellicle membranes of the apical complex of *P. gallinaceum* (from Raibaud *et al.,* 2001). B) 3D reconstruction from transmission electron micrographs of the ookinete apical complex (from Canning and Sinden, 1973). C) Immunogold silver enhanced micrograph of *P. berghei* ookinetes showing the shedding of Pbs21 by the migrating ookinete (arrow) (from Sinden *et al.,* 1987). D) Giemsa stained preparation of *An. stephensi* peritrophic matrix illustrating the constrictions in *P. berghei* ookinetes (arrows) as they pass through this tissue. **See Colour Plate at the back of the book.**

from the zygote is dictated by the expansion of the cytoskeletal vacuole and microtubules (Figure 1B,C). The inner membrane complex and associated subpellicular microtubules of the ookinete lay down the polarity of the cytoskeleton, the molecular motor and the cell. This large vacuole under the ookinete plasma membrane contains organised rows of large (43nm diameter) pores in the membranes (Raibaud *et al.*, 2001) (Figure 3A). Whilst these pores have not been seen in the merozoite or sporozoite it must be recognised the techniques used to reveal their presence in the ookinete have not yet been applied to other parasite stages. These pores offer unique opportunities for molecular exchange between the synthetic apparatus in the cytosol and the peri-alveolar space (analogous to the nuclear pore).

The ookinete apical complex is more elaborate than that in either the merozoite or sporozoite in that it has an extra, rigid collar (Figure 3B) that forms the apical, and apparently retractable, protuberance. Chitinase has been localized in this collar, though the secretory pathway of the enzyme is far from clear. Micronemes (which contain CTRP, SOAP, WARP and chitinase Figure 4), and possibly the crystalloid (not a viral inclusion as previously described) form from an extensive array of vesicles between the nucleus and the apical complex (Figure 2C). Unlike other invasive stages, mature ookinetes do not possess rhoptries. Thus the presence in the ookinete proteome of molecules nominally called 'rhoptry-associated' e.g. RAP1 (Raine *et al.*, unpublished) may suggest that the original descriptions of their cellular localization may be insecure, and that a micronemal location is appropriate.

Movement of the ookinete is by gliding motility. It migrates through the bloodmeal and invades the midgut wall, driven by the directed capping of surface ligands (see Soldati *et al.*, page 135-166). Gene knock-out studies on CTRP (Dessens *et al.*,1999) suggest this trans-membranous, adhesin-like molecule, like the TRAP homologue in the sporozoite, is secreted from the micronemes onto the parasite surface following $Ca^{2+}$ signalling (Figure 2D). The extracellular adhesive domains (thrombospondin-related motifs (TRM) and A-domains) typically recognise high-density sulphated glycosaminoglycans (GAGs) on the surface of the target structure. The surface proteins P25 and P28 are concomitantly shed from the parasite surface as the ookinete moves forward (Figure 3C).

An intriguing, and as yet unexplained aspect of ookinete development is the switch in expression from asexual (A) to sexual (S) and oocyst (O) forms of the ribosomal RNA genes. Current evidence suggests this change in the rRNA scaffolds is not accompanied by changes in the expression of ribosomal proteins (Raine *et al.*, unpublished). Recognising that mRNA encoding sexual-stage proteins can be expressed on A form ribosomes, and house-keeping proteins can be expressed in all forms of ribosome – the rationale for this 'expensive' duplication of rRNA genes of such fundamental importance remains difficult to explain – clearly we have a lot to learn.

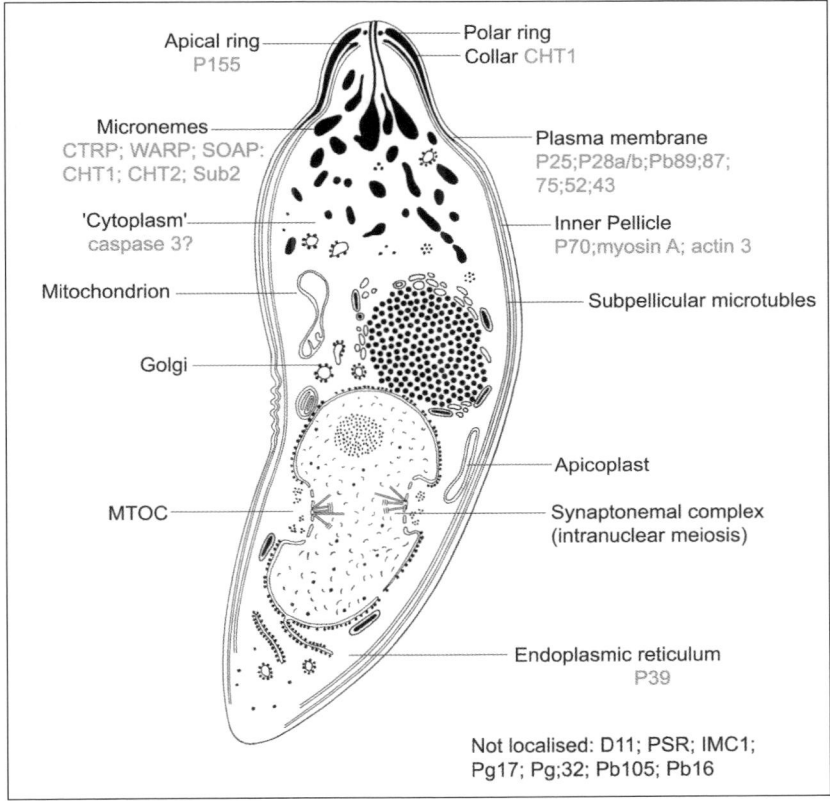

Apical ring — P155
Polar ring
Collar CHT1
Micronemes — CTRP; WARP; SOAP; CHT1; CHT2; Sub2
Plasma membrane P25;P28a/b;Pb89;87; 75;52;43
'Cytoplasm' — caspase 3?
Inner Pellicle P70;myosin A; actin 3
Mitochondrion
Subpellicular microtubles
Golgi
Apicoplast
MTOC
Synaptonemal complex (intranuclear meiosis)
Endoplasmic reticulum P39
Not localised: D11; PSR; IMC1; Pg17; Pg;32; Pb105; Pb16

**Figure 4**. Diagram summarizing the structure and location of known proteins in the *Plasmodium* ookinete.

A major switch in surface protein repertoire occurs about 5-7 hours after ingestion of the bloodmeal, all known pre-existing proteins >55kDa are shed and replaced by the ookinete molecules required for defence and host interaction (Kaushal *et al.*, 1983). The surface of the ookinete is characterised by a 'new' repertoire of proteins the dominant two of which, P25 and P28, are GPI anchored to the membrane. How these two proteins are transported to the parasite surface is unknown, but it is not via the micronemes.

Survival of the ookinete in the bloodmeal, its traversal of the peritrophic matrix, the microvillar network, and the midgut epithelium represents one of the major hurdles to malaria development in the mosquito vector. Even in the most 'compatible' of parasite-vector combinations losses at this stage may be as high as 85%, and in other species-pairings the reduction can be absolute. Losses are incurred for a variety of reasons. Within the bloodmeal the zygote is sensitive to complement ingested in the vertebrate blood (Grotendorst *et al.*,1986; Grotendorst and Carter, 1987; Margos *et al.*,2001).

The mature ookinete despite being more sensitive, is less at risk because many complement factors (Factor B, D and $C_3$) will have been destroyed by the proteolytic enzymes secreted by the mosquito. Phagocytes ingested with the bloodmeal also attack the gametes/zygotes/ookinetes, as do antibodies directed against the briefly retained antigens inherited from the blood stage gametocytes e.g. P230, P48/45. The mature ookinete is sensitive to attack by mosquito proteases (Gass, 1977; Gass and Yeates, 1979) and is dangerously exposed to the highest concentrations of digestive enzymes if it becomes entangled in the peritrophic matrix, or delayed in the ecto-peritrophic space. Phenotypic analysis of P25 and P28 single- and double-knockout *P. berghei* ookinetes suggest these proteins are necessary for ookinete maturation and have mutually redundant functions including protection against proteolysis. P25/28 double knockout lines are more sensitive to trypsin type 1 than both the P25 or P28 single knockout lines and the wild type. Yeast two-hybrid assays indicate that Pbs25 will both homodimerise and heterodimerise with Pbs21 (Siden-Kiamos *et al.*, 2000) suggesting, they may mediate the parasite clumping which is believed to enhance ookinete survival in the hostile environment of the midgut. Up to 50% of ookinetes formed in the midgut undergo apoptosis (Al-Olayan *et al.*, 2002b).

Invasion of the peritrophic matrix by the *P. gallinaceum* ookinete has been described in detail by Huber *et al.* (1991). Prochitinase secreted by the *P. gallinaceum* ookinete is cleaved and activated by a midgut protease, resulting in the disorganisation of the chitinous layer. Inactivation of the cleaved enzyme by allosamidin, a chitin analogue, or by the addition of an anti-trypsin antibody to the mosquito bloodmeal both result in reduced oocyst intensities compared to their respective controls (Shahabuddin and Kaslow, 1999). Thereafter the ookinete meets the microvillar network (Zieler and Dvorak, 2000), and finally binds to the midgut epithelial cell. Potential carbohydrate-like receptor molecules identified on the surface of the midgut epithelial cells include *N*-acetyl-D-glucosamine (*Ae. aegypti*) *N*-acetyl-D-galactosamine (*An. stephensi*) (Rudin and Hecker 1989; Zieler *et al.*, 1999), and a mucin (*An. gambiae*, AgMuc1. Shen *et al.,*1999). However, we know nothing of significance about the molecular interactions of the ookinete with any of these insect tissues, other than that sialic acid-like moieties are involved in the binding of the ookinete to isolated midgut epithelia (Zieler and Dvorak, 2000), and that these interactions are somehow inhibited by the expression of the peptide SM1 (PCQRAIFQSICN) (Ito *et al.,*2002), or the pre-treatment of the guts with the snake venom PLA2 (Zieler *et al.,*2001).

### 1.1.1. Ookinete Invasion of the Midgut

Having crossed the peritrophic matrix and microvillar network the ookinete attaches to and then actively invades the epithelial cell employing the gliding motility described above. CTRP knockout parasites are non-motile and unable to invade the gut wall, implicating CTRP, like the homologous sporozoite surface molecule, TRAP, is a crucial molecule for motility and invasion (Dessens *et al.,*1999). The opposing cell surfaces of the midgut epithelium differ significantly. The apical/lumenal side is dominated by microvilli beneath which lie numerous secretory vesicles containing digestive enzymes. The basal/haemocoele surface is bounded by a basal lamina rich in laminin, collagen IV and fibronectin (Beier and Vanderberg, 1998).

*In vitro* studies indicate that ookinetes can undergo extensive gliding for up to 30 mins along the surface and between the crevices of midgut epithelial cells. *In vitro* adhesion assays show ookinetes (but not zygotes and macrogametes) specifically adhere to the lumen side the epithelium (Zieler and Dvorak, 2000). Several studies have suggested that midgut epithelial cells invaded by the ookinete differ from non-invaded cells (Cociancich *et al.,* 1999; Han *et al.,* 2000; Shahabuddin and Pimenta, 1998; Zieler and Dvorak, 2000). Studies in the mosquito *Aedes aegypti* have produced the hypothesis that the *P. gallinaceum* ookinete specifically invades 'Ross' cells that express high levels of vesicular ATPase (vATPase), have few microvilli, contain minimal endoplasmic reticulum and lack secretory granules (Shahabuddin and Pimenta, 1998). In contrast studies with the rodent malaria parasite *P. berghei* and the mosquito *An. stephensi* indicate that the ookinete invades cells of the midgut epithelium at random (Han *et al.,* 2000). The second observation was supported by studies on both *P. gallinaceum*/*Ae. aegypti* and *P. berghei*/*An. stephensi* (Zieler and Dvorak, 2000). Invasion of the midgut epithelium results in extensive cell death amongst the invaded cells. These cells, and even cells which have simply come into contact with ookinetes gliding along their surface, undergo nuclear swelling, microvillar breakdown and cell surface blebbing (Zieler and Dvorak, 2000) and substantial DNA fragmentation (Han *et al.,* 2000). Activation of a caspase-3 like protease and DNA fragmentation suggests that cell death is apoptotic (Zieler and Dvorak, 2000). Cells damaged by invasion protrude from the midgut epithelium and by means of an actin purse-string mechanism 'bud off' into the midgut lumen (Han *et al.,* 2000). The molecular interactions inducing cell death are unknown but members of the GPI-anchored surface protein P25/28 gene family are shed into the invaded cell. High levels of Pbs21 have been found in areas where ookinetes could not be found (Han *et al.,* 2000), possibly representing a site of intra-cellular ookinete lysis. Secretion of Pfs25 (Meis and Ponnudurai, 1987) and the subtilisin-like protease PbSub2 (Han *et al.,* 2000) has also been reported. PbSub2 reportedly forms protein aggregates frequently seen to associate with the actin cytoskeleton (Han *et al.,* 2000). A putative role for PbSub2 in the modification of the host's cell cytoskeleton is therefore feasible. Evidence that the ookinete invades neighbouring cells is convincing. Pbs21 (P28) trails extend through a number of adjacent cells (Han *et al.,* 2000); contiguous midgut cells undergo apoptosis (Han *et al.,* 2000; Zieler and Dvorak, 2000) and individual parasites have been observed moving between cells *ex vivo* (Zieler and Dvorak, 2000). Baso-lateral

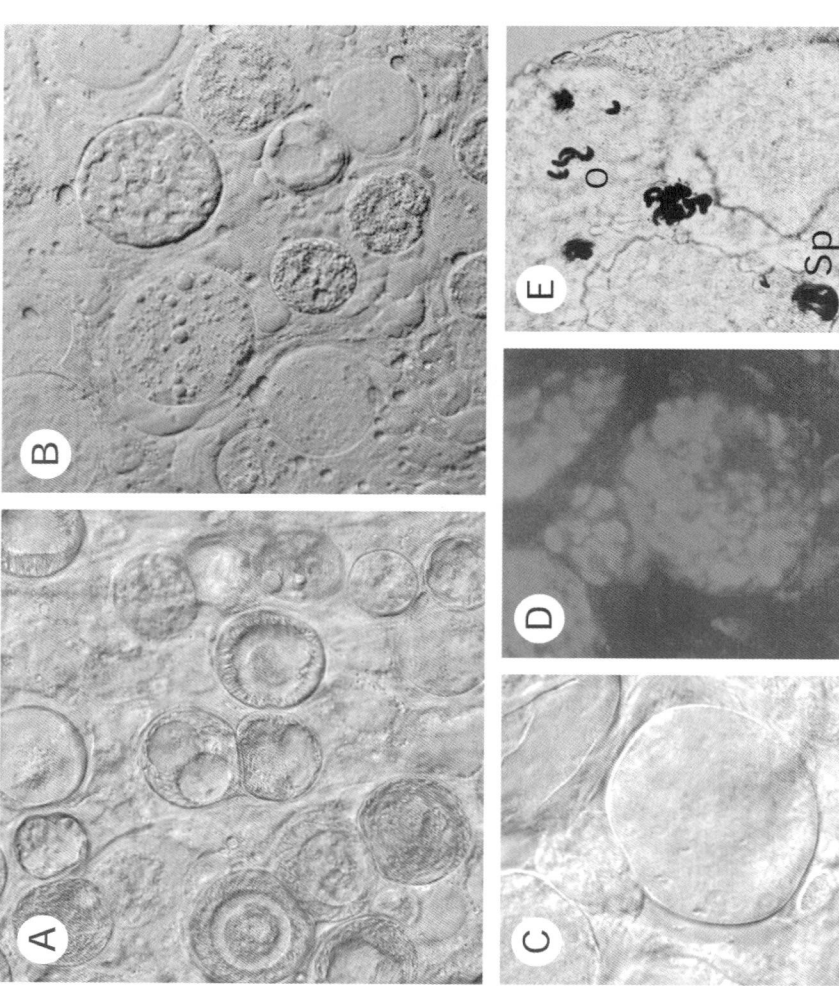

**Figure 5.** A-B) Interference contrast micrographs of the wt+ *P. berghei* (A) and PbSRko (B) oocysts showing the failure of the PbSRko to undergo sporoblast and sporozoite differentiation (from Claudianos *et al.*, 2002). C-D) Confocal laser scanning micrographs illustrating the nuclear organisation of the *P. berghei* SRko malarial oocyst as revealed by interference contrast (C) and DAPI (D) staining. E) Brightfield light micrograph illustrating simultaneously the melanization of both ookinetes (O) and sporozoite-containing oocyst (Sp) of *P. berghei* in *An. stephensi*.

movements of the ookinete between cells could represent a parasite strategy to escape from high levels of toxic substances (see below). Ookinetes are astonishingly plastic and have been observed to undergo constrictions as they migrate through the peritrophic matrix (Figure 3D); laterally between adjacent midgut cells (Freyvogel 1967; Vernick *et al.*, 1999) and as they emerge on the basal side of the midgut (Han *et al.*, 2000).

Ookinetes penetrate the basement membrane but do not continue through the basal lamina into the haemocoele (Sinden, 1984). Is ookinete motility arrested simply because it is unable to cross the basal lamina or are there environmental cues that are responsible for the ookinete becoming sessile and rounding up to form an oocyst? Evidence that basal lamina proteins, e.g. laminin and collagen IV (Paulsson, 1992), interact with parasite surface proteins and trigger parasite differentiation is substantial (Weathersby, 1952; Warburg and Miller, 1992; Adini and Warburg, 1999; Schneider and Shahabuddin, 2000; Arrighi and Hurd, 2002; Vlachou *et al.*, 2001).

## 1.2. Ookinete-to-Oocyst Differentiation

Ookinetes surviving the onslaught of the midgut reach the basal lamina. In 'immunologically-stressed' mosquitoes some, or all, emerging ookinetes can here be subject to melanization by the insects' immune system (Figure 5E). In vector mosquitoes, contact of the ookinete with the basal lamina triggers developmental transition, via the oocyst, into sporozoites. Whilst this 10-20 day development is intrinsically programmed to 'run to completion', nutritional, immunological and other extraneous pressures from the surrounding environment can suppress the efficiency with which infectious sporozoites are formed.

Initially the ookinete rounds up and loses the enigmatic crystalloid, the micronemes and apical/pellicular organelles associated with invasion and locomotion (Figure 1E). Thereafter the small (4 μm diameter) cell undergoes massive cytoplasmic expansion, notably of the endoplasmic reticulum, Golgi apparatus and secretory vesicles (Figure 1F) (Sinden 1984). The single ookinete nucleus arrested with its unusual complement of 4 haploid (potentially recombinant), genomes undergoes 11 synchronised endomitotic divisions (Figure 2B) to yield 2000-8000 haploid nuclei (Figure 5C,D). The location of the spindle poles of the final mitotic divisions dictates the positions at which the daughter sporozoites will emerge (Figure 6A,B). It remains unclear whether the nucleus is a syncitial structure or whether karyokinesis is completed at the end of every mitotic division. Each oocyst is the product of a single mating event (Ranford-Cartwright *et al.*, 1993), thus the unusual poly-genomic nature of the oocyst cytoplasm offers unique opportunities to study inter-genic/allelic relationships in *Plasmodium*.

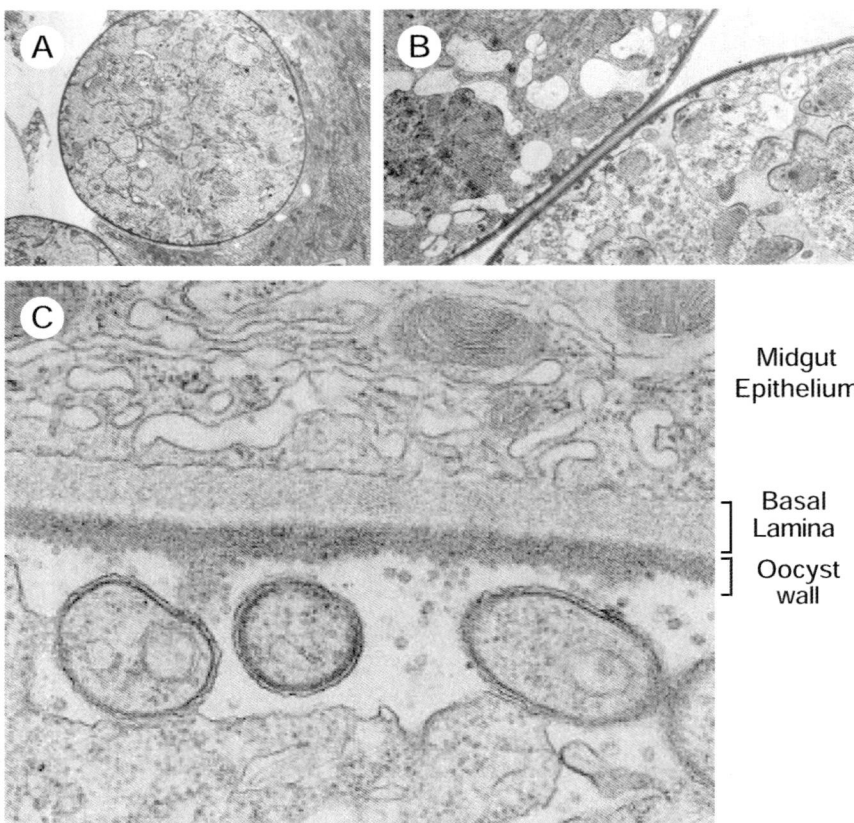

**Figure 6.** A) Transmission electron micrograph illustrating growing oocyst (*P. falciparum*: from Sinden and Strong, 1978). B) Transmission electron micrograph illustrating sporozoite formation adjacent to the dividing nuclei of the final mitotic division in the oocyst (*P. falciparum*: from Sinden and Strong, 1978). C) Transmission electron micrograph illustrating the hybrid structure of the oocyst capsule. The amorphous, parasite derived oocyst wall, and the structured basal lamina of the midgut walL (*P. falciparum*: from Sinden and Strong, 1978).

The major biological role of the oocyst is to amplify parasite numbers in the vector, i.e. the number of invasive sporozoites that will eventually be injected into the vertebrate host in the mosquito's saliva. The oocyst grows rapidly in size to reach approximately 40-60μm in diameter (Figure 6A). The plasma membrane lies beneath a newly formed electron dense oocyst wall/capsule that is in direct contact with the mosquito basal lamina (Figure 6B). Electron microscopy shows clearly that within the oocyst wall, as seen in the light microscope, the parasite tissue merges almost seamlessly with the basal lamina of the midgut (Figure 6C), thus explaining the recent description of lamina proteins in the oocyst wall (Adini and Warburg, 1999). Parasite transglutaminase is now believed to contribute to oocyst capsule formation by cross-linking mosquito-derived basement membrane components with parasite proteins (Adini et al., 2001). It has not been shown whether large proteins can pass through the oocyst capsule, but amino acids and nucleotides do so freely (Vanderberg et. al, 1967), as can the small immune peptide defensin (Shahabuddin *et al.*, 1998). After 6-8 days of growth the solid cytoplasm of the multi-nucleated oocyst becomes segregated by vacuolisation and cleft formation into a number of cytoplasmic islands (sporoblasts) within the oocyst capsule (Figure 6B). Within the sporoblasts the nuclei polarize along the periphery of the cytoplasm and then cytokinesis occurs dividing into uninucleate sporozoites.

The oocyst is one of the least understood life stages, at the molecular level. Two molecules with an identified role in the growth and cell differentiation are a scavenger receptor-like protein (PxSR/PSLAP), which is required for sporoblast development (Delrieu *et al.*, 2002; Claudianos *et al.*, 2002) and the circumsporozoite protein (CSP), which is involved in differentiation of the sporozoite (see Ménard *et al.*, 1997). PxSR is a predicted secreted, soluble protein that has a mosaic eight-domain structure involved in immune recognition/activation and protein/lipid interactions at the cell surface, namely: (i) scavenger receptor cysteine rich (SRCR); (ii) pentraxin (PTX); (iii) polycystine-1, lipoxygenase, alpha toxin (LH2/PLAT); (iv) Limulus clotting factor C, Coch-5b2 and Lgl1 (LCCL).. PbSR gene disruption in *P. berghei* shows a clear phenotype in oocyst infections in *An. stephensi* (Claudianos *et al.*, 2002). Oocysts are formed with the same efficiency as the wild type parasite and nuclear division is normal (Figure 5A-D) however sporoblast formation (cytokinesis) does not occur (Figure 5B). Moreover, oocysts of *PbSR*-disrupted parasites reach a significantly greater size than wt *P. berghei* oocysts.

## 1.3. Mosquito Immune Response and Ookinete Defence

Following invasion of the midgut epithelium significant parasite losses result from a variety of defence mechanisms mounted by the insect. Occasionally the mosquito immune response completely blocks parasite development, rendering the insect refractory. Two classes of refractory mechanism have been described in the human malaria vector *An. gambiae*. Melanization (Figure 5E) causes the melanotic encapsulation of ookinetes, oocysts and sporozoites on the basal side of the midgut epithelium by activation of a phenol oxidase reaction cascade (Collins *et al.*, 1986). The second refractory mechanism results in the lysis of ookinetes during their invasion of the midgut epithelium (Vernick *et al.*, 1995).

In malaria-infected mosquitoes a temporal and spatial correlation exists between the activation of immune-responsive genes and the critical stages of midgut and salivary gland invasion by ookinetes and sporozoites, respectively (Dimopoulos *et al.,* 1998). While the induction of many such genes occurs locally in the invaded tissues, their expression can also be detected in the fat body (Dimopoulos *et al.,* 1998). Mosquito immune responses against the malarial ookinete and other microbial infections have received increasing scientific interest in recent years, and the number of inducible immune-responsive genes identified has been growing rapidly (Dimopoulos *et al.,* 1996; 1997; 2000; 2002; Oduol *et al.,* 2000; Richman *et al.,* 1997). Reflecting the complexity of these immune responses, genes fall into different classes including recognition and binding of non-self; signal amplification and regulation and the neutralisation of the invading organism such as melanization, lysis, coagulation or phagocytosis.

Pattern-recognition receptors are soluble or surface-bound proteins that recognise and bind invading microorganisms. They activate the reaction cascades of specific defence mechanisms. Putative pattern-recognition molecules identified in *An. gambiae* include various Gram-negative bacteria binding proteins that are thought to bind lipopolysaccharides; ICHIT, a protein containing both chitin-binding and mucin domains; lectins; proteins with fibrinogen-like domains and a *Drosophila* dSR-C1 homologue, containing scavenger receptor domains (Dimopoulos *et al.,* 1997; 1998; 2000, 2002). Some proteins may act as opsonins for phagocytosis by haemocytes e.g. the recently reported haemocyte-specific aTEP-1 protein, binds to the surface of Gram-negative bacteria via a complement-like thioester bond (Levashina *et al.,* 2001). ATEPs have been identified that have different target specificities, including the malaria ookinete in the midgut epithelium (Dimopoulos *et al.,* 2002).

A large number of immune-response genes encoding serine proteases have been identified in *Anopheles* (Dimopoulos *et al.,* 1996; 1997; 2000; Danielli *et al.,* 2000; Gorman *et al.,* 2000ab; Paskewitz *et al.,* 1999; Oduol *et al.,* 2000). These proteases amplify specific defence signalling-pathways, e.g. for melanization and coagulation. This class of serine proteases often contain additional domains implicated in extracellular matrix interactions, e.g. SP22D, a haemocyte-expressed serine protease that contains chitin binding, mucin-like and scavenger receptor cysteine-rich domains (Danielli *et al.,* 2000; Gorman *et al.,* 2000a). Additionally several serine protease inhibitors (serpins) have been identified, which are thought to control and regulate the protease-activated reaction cascades (Dimopoulos *et al.,* 2000; Kanost, 1999; Oduol *et al.,* 2000).

Inducible antimicrobial peptides usually kill invading microorganisms by disrupting the target-cell membrane (Cociancich *et al.,* 1994; Hoffmann *et al.,* 1996). Three classes of antimicrobial peptide have been described in *An. gambiae*: one defensin, two cecropins (A and B), and gambicin (Muta and Iwanaga, 1996; Hoffmann and Reichart 1997; Gillespie *et al.,* 1997; Cociancich *et al.,* 1994; Dimopoulos *et al.,* 1997, Vizioli *et al.,* 2000, 2001).

Following parasite infection, peptide expression predominantly occurs in the anterior midgut, thorax and abdomen. These peptides are active against a broad spectrum of microorganisms, including Gram-positive and Gram-negative bacteria and fungi (Vizioli *et al.*, 2001). Two insect defensins have been shown to suppress development of *P. gallinaceum* oocysts *in vivo* and to be highly toxic to *P. gallinaceum* sporozoites *in vitro*, however there were no adverse effects on zygotes and ookinetes (Shahabuddin *et al.*, 1998). Two artificial peptides SHIVA-1 and LDEV-3 inhibit early parasite development (Rodriguez *et al.*, 1995; Arrighi *et al.*, 2002).

*Plasmodium* is killed by nitric oxide (Mellouk *et al.*, 1994; Naotunne *et al.*, 1993). Ookinete invasion of *An. stephensi* midgut elevates expression of nitric oxide synthase (NOS), resulting in the production of highly diffusible NO (Richman *et al.*, 1997; Dimopoulos *et al.*, 1997; Luckhart *et al.*, 1998; Han *et al.*, 2000). Inhibition of NOS activity significantly increased oocyst numbers in *An. stephensi* (Luckhart *et al.*, 1998). NOS expression is also upregulated in the abdomen of malaria-infected *Anopheles* at later time points that coincide with sporozoite release from the oocyst and salivary gland invasion (Dimopoulos *et al.*, 1998; Luckhart *et al.*, 1998). The action of NOS may thus provide a mechanism that adversely affects ookinete, oocyst and sporozoite development.

Given the co-evolution of parasite and mosquito, it is likely that the parasite has developed mechanisms to limit the damage inflicted by the mosquito's immune responses. One mechanism is based on the 'strength in numbers' approach; each ookinete that successfully transforms into an oocyst may produce up to 8,000 sporozoites, some of which are likely to evade immune attack. At the molecular level also, there are likely to be mechanisms that block or modulate the immune response. . The ookinete surface protein P25, as well as a secreted ookinete protein PbSOAP, a secreted ookinete protein are important for ookinete-to-oocyst development *in vivo*, but not *in vitro* (Tomas *et al.*, 2001; Siden-Kiamos *et al.*, 2001; Dessens, unpublished) and may play roles in the survival of the ookinete in the mosquito. Arguably a more obvious candidate for potential involvement in immune modulation is the recently described scavenger receptor-like protein PxSR (Claudianos *et al.*, 2002). Recognising the structural features of PxSR, it is conceivable that it could scavenge or compete with the molecular ligands of receptors that are required for immune activation in the mosquito.

## 2. Ookinete-Based Transmission-Blocking Vaccines

Stemming from the pioneering work of Huff (1957), who showed that formalin-treated parasites induced transmission-blocking immunity in birds, interest in the possibility of developing a transmission-blocking vaccine (TBV) has grown in recent years with Carter and Gwadz (1980) initiating the modern era of research aimed at developing a vaccine that will induce transmission-blocking immunity in humans. Interestingly, in a recent paper modelling the potential of parasites to evolve mechanisms to defeat vaccines it was concluded that infection-blocking vaccines (as opposed to growth-inhibiting and anti-toxic vaccines) should have high priority as they may select for lower parasite virulence (Gandon *et al.*, 2002).

In a number of recent reviews covering the TBV field (e.g. Carter, 2001; Stowers and Carter 2001; Richie and Saul 2002) the authors describe antibodies that arise naturally to proteins found on the surface of gametes and/or gametocytes (Pfs/Pvs48/45 and Pfs/Pvs230) and these may reduce the infectivity of humans with *P. falciparum* or *P. vivax* to mosquitoes. The respective antigens are thus current contenders for inclusion in a potential TBV against these parasites. However, antibodies generated experimentally to zygotes and ookinetes may also totally block parasite development in the mosquito, and immunity aimed at these stages has particular advantages. The high efficacy levels achieved for TBVs based on the ookinete may correlate with the fact that ookinetes are exposed for many hours to effective antibody levels, consequently there is often a good correlation between the presence of transmission-blocking antibody to these stages and infectivity in membrane-feeding systems (Ranawaka *et al.*, 1994). The search for potential vaccine candidates using immune antibody is thus made somewhat easier. This is in contrast to other types of vaccine, where there are no generally agreed *in vitro* tests that correlate with protective immunity.

Immunization regimes have generated antibodies that damage the ookinete surface and/or inhibit penetration of the mosquito midgut wall. Such antibody can be remarkably effective: for example, in the mouse-*P. berghei* model in which infectivity is invariably reliable and high numbers of oocysts per midgut are not uncommon, >99% blockade of infectivity was observed with a monoclonal antibody to Pbs21/P28. The same antibody given after the ookinete has invaded the midgut epithelium is still effective (Ranawaka *et al.*, 1994). Immune antibody is also rendered more effective in the presence of complement (Healer *et al.*, 1997) at least for a limited time (Margos *et al.*, 2001) and also by blood leukocytes (Rutledge *et al.*, 1969; Sinden and Smalley, 1976; Ranawaka *et al.*, 1993, 1994; Lensen *et al.*, 1997). Experiments with knockout lines of *P. berghei* reveal that the presence of only one of the two paralogues Pbs21/P28 and 25 is sufficient to ensure transmission, but in the absence of both proteins transmission is effectively blocked (Tomas *et al.*, 2001). It is interesting in this regard that antibodies to Pfs 25 and 28 of *P. falciparum* act synergistically (Duffy and Kaslow, 1997). Thus for a vaccine based on these antigens to be effective in the long term, both must be included.

Ookinete proteins P25 and 28 are further advanced in terms of vaccine development than the gametocyte/fertilization antigens (Pfs/Pvs48/45 and Pfs/Pvs230) in that pre-clinical grade material is, or soon will be, available

(WHO). As post-fertilization antigens (i.e. not exposed to the vertebrate immune system *in vivo*) (Kaslow *et al.*, 1989) they are highly conserved in *P. falciparum*, though less so in *P. vivax*, whilst *P. ovale* currently shows the greatest diversity. *P. ovale* is also unusual in that it has one P25 gene but two for P28 (Tachibana *et al.*, 2002). Ookinete antigens have the disadvantage, however, that not being expressed in the vertebrate host immunity cannot therefore be boosted by natural infections.

Matsuoka *et al.*, (1996) generated usable quantities of recombinant that P28 protein was recognised by transmission-blocking antibodies in a baculovirus-silkworm expression system. Hisaeda *et al.*, (2001) have produced clinical grade Pvs25/28, and Gozar *et al.*, (1998) produced Pfs25/ 28 in yeasts. In both P28 (Pbs21) and P25 (Pfs25) EGF domains contain critical TB epitopes. In Pbs21 it lies between aa95-98 on the first EGF domain (Spano *et al.*, 1996) whereas it is the second of the four EGF-like domains of Pfs25 (Stowers *et al.*, 2000).

The search for new prospective vaccine components is made easier by the continued use of animal malarias, of which *P. berghei* and *P. gallinaceum* are the most convenient and most researched. Two new potential vaccine candidates have been identified recently: ookinete chitinase(s), and CTRP. Chitinase(s) aid parasite penetration of the chitin-containing peritrophic matrix surrounding the blood meal. *P. gallinaceum* has two chitinase enzymes (PgCHT1, PgCHT2), whereas in other species only one has been described to date (PfCHT1 and PbCHT1). It appears that *P berghei* and *P. falciparum* could have each acquired a different chitinase from their common (avian-type) ancestor. The *berghei* chitinase is insensitive to allosamidin, a chitinase inhibitor, unlike the avian and human parasites. A knockout line of this parasite exhibited a reduction in infectivity of 90% in *An. stephensi* mosquitoes so its importance in sporogonic development marks it as a possible vaccine component.

CTRP (circumsporozoite-TRAP related protein) is the first ookinete microneme protein to be identified (Dessens *et al.*, 1999; Yuda *et al.*, 1999) and belongs to the TRAP family. Parasites with a disrupted gene fail to invade the midgut epithelium. This molecule therefore deserves consideration as a potential vaccine component - transmission-blocking antibodies have been raised to PfCTRP (Templeton and Vinetz, pers. comm.).

Earlier work demonstrated that the *P. berghei* ookinete has 10-15 surface proteins (Sinden *et al.*, 1996), but current technology is likely to reveal far more, at least some of which may be appropriate for consideration in future vaccines. Using multidimensional protein identification technology (MudPIT), a novel non-gel based proteomic technique that couples on-line high-resolution liquid chromatography with tandem mass spectrometry and database searching (Washburn *et al.*,2001), almost 2000 parasite proteins were identified in *P. berghei.* Of these 1230 were in ookinetes, with over 470 being unique to this stage compared to mixed asexuals, gametocytes and oocysts (Raine *et al.,* unpublished).

# 3. Drugs, Ookinetes and Transmission

The interactions of antimalarials with the sporogonic phase of parasite development have been grossly neglected, despite the obvious fact that for drug-resistant parasites to spread in the wild they need to be mosquito transmitted. It is now recognised, however, that new combination drugs that inhibit transmission as well as those targeting different biochemical pathways in the asexual stages are needed (Nosten and Brasseur, 2002).

Much of the older data on malaria transmission and drug action has been reviewed previously (Butcher, 1997). Relatively little information exists on the details of the interactions of antimalarials on the sporogonic stages of *Plasmodium.* Sulpha-drugs, particularly sulfadoxine (as pyrimethamine/ sulfadoxine) cause increased gametocyte generation (Butcher, 1997) and infectivity (Targett *et al.,* 2001; von Seidlein *et al.,* 2001). Pyronaridine targets gametocytes *in vitro* through action on DNA topoisomerase (Chavalitshewinkoon-Petmitr *et al.,* 2000) while thiostrepton, interacts with the apicoplast and this has now been reported to reduce transmission, indicating a possible role for this organelle in sporogonic development. Another recently discovered compound, G25 that inhibits phosphpatidylcholine biosynthesis might also affect early gametocyte development (Wengelnik *et al.,*2002).

To our knowledge, the most detailed study of drug interaction with mosquito stages of parasites concerns atovaquone, combined with proguanil as Malarone™. Atovaquone targets cytochrome b, thus affecting mitochondrial respiration (Fivelman *et al.,* 2002); *in vivo* it inhibits oocyst formation and *in vitro* prevents ookinete formation of *P. berghei* (Fowler *et al.,* 1995). Sera taken from human volunteers treated with proguanil/atovaquone or atovaquone given separately, also inhibited *P. berghei* ookinete maturation and asexual stage development (Butcher *et al.,* 2000). Surprisingly, sera taken up to 7 weeks after atovaquone treatment alone retained this inhibitory activity *in vitro*, despite published data that atovaquone is excreted within a week of treatment (Hudson, 1993; Beerahee, 1999). This work was extended to *P. falciparum* (Butcher and Sinden, 2003) and again the inhibition of transmission and asexual stage development also occurred with sera taken 7 weeks after treatment. In contrast to the results with atovaquone, sera from proguanil-treated subjects inhibited transmission for less than one week after treatment. The transmission-blocking activity and the inhibition of asexual stages *in vitro* was completely lost by the atovaquone sera tested against an atovaquone-resistant isolate of *P. falciparum*. The resistant parasite had a mutation in cytochrome b, tyr268Asn, not seen previously (Fivelman *et al.*, 2002), although resistance to atovaquone alone is known to be relatively easy to generate experimentally. The persistence of low levels of atovaquone, probably bound to serum, may facilitate the generation of resistant strains of *P. falciparum* and thus shorten the useful life of Malarone™.

The unpredictability of the effects of antimalarials on transmission is illustrated by the observation that sera from chloroquine-treated subjects taken three to four weeks after treatment enhanced individual gametocyte infectivity, as judged by oocyst numbers, with *P. falciparum* and *P. berghei*, (Høgh *et al.*, 1998). Consistent with the rise in oocysts of *P. berghei* was an increase in ookinete numbers in mosquitoes fed gametocytes plus sera from rats, also taken three weeks after chloroquine treatment (Butcher, unpublished data). No enhancement was observed by sera from mice treated with chloroquine and may explain the failure to find a late effect of chloroquine on infectivity of *P. chabaudi* by others (Buckling and Read, 1999). Chloroquine thus has quite different effects on the sporogonic development of *Plasmodium* in comparison with its activity against the asexual erythrocytic stages, emphasising the need to investigate drug interactions in the mosquito as well as in the other phases of the life cycle of malaria parasites.

# 4. References

Adini, A., Krugliak, M., Ginsburg, H., Li, L., Lavie, L., and Warburg, A. 2001. Transglutaminase in *Plasmodium* parasites: activity and putative role in oocysts and blood stages. Mol. Biochem. Parasitol. 117: 161-168.

Adini, A., and Warburg, A. 1999. Interaction of *Plasmodium gallinaceum* ookinetes and oocysts with extracellular matrix proteins. Parasitol. 199: 331-336.

Al-Olayan, E. M., Beetsma, A. L., Butcher, G. A., Sinden, R. E., and Hurd, H. 2002a. Complete development of mosquito phases of the malaria parasite *in vitro*. Science 295: 677-679.

Al-Olayan, E.M., Williams, G.T., and Hurd, H. 2002b. Apoptosis in the malaria protozoan, *Plasmodium berghei*: a possible mechanism for limiting intensity of infection in the mosquito. Int. J. Parasitol. 32: 1133-1143.

Arrighi, R.B., Nakamura, C., Miyake, J., Hurd, H., and Burgess, J.G. 2002. Design and activity of antimicrobial peptides against sporogonic-stage parasites causing murine malarias. Antimicrob. Agents Chemother. 46: 2104-2110.

Arrighi, R. B. G., and Hurd, H. 2002. The role of *Plasmodium berghei* ookinete proteins in binding to basal lamina components and transformation into oocysts. Int. J. Parasitol. 32: 91-98.

Beerahee, M. 1999. Clinical pharmacology of atovaquone and proguanil hydrochloride. J. Travel Med. 6: S13-S17.

Beier, J. C., and Vanderberg, J. P. 1998. Sporogonic development in the mosquito. In: Malara: Parasite Biology, Pathogenesis, and Protection, I. W. Sherman, ed. Washington, D.C., ASM Press. p. 49-61.

Buckling, A.G.J., and Read, A.F. 1999. The effect of chloroquine treatment on the infectivity of *Plasmodium chabaudi* gametocytes. Int. J. Parasitol. 29: 619-625.

Butcher, G.A. 1997. Antimalarial drugs and the mosquito transmission of *Plasmodium*. Int. J. Parasitol. 27: 975-987.

Butcher, G.A., Mendoza, J., and Sinden, R.E. 2000. Inhibition of the mosquito transmission of *Plasmodium berghei* by Malarone (atovaquone-proguanil). Ann. Trop. Med. Parasitol. 94, 429-436.

Butcher, G.A., and Sinden, R.E. 2003. Persistence of atovaquone in human sera following treatment:inhibition of *P. falciparum* development *in vivo* and *in vitro*. Am. J. Trop. Medi. Hyg. 68: 111-114.

Canning, E. U., and Sinden, R. E. 1973. The organization of the ookinete and observations on nuclear division in oocysts of *Plasmodium berghei*. Parasitol. 67: 29-40.

Carter, R. 2001. Transmission blocking malaria vaccines. Vaccine. 19: 2309-2314.

Carter, R. and Gwadz, R. W. 1980. Infectiousness and gamete immunization in malaria. In: Malaria, Volume 3: Immunology and Immunization. Kreier, J. P., ed. New York: Academic Press p. 263-298.

Chavalitshewinkoon-Petmitr P., Pongvilairat G., Auparakkitanon S., and Wilairat P. 2000. Gametocytocidal activity of pyronaridine and DNA topoisomerase II inhibitors against multidrug-resistant *Plasmodium falciparum in vitro*. Parasitol. Int. 48: 275-80.

Claudianos, C., Dessens J.T., Trueman, H.E., Arai M., Mendoza, J., Butcher, G.A., Crompton, T., and Sinden R.E. 2002. A malaria scavenger receptor-like protein essential for parasite development. Mol. Microbiol. 45: 1473-1484.

Cociancich, S. O., Bulet, P., and Hoffman, J.A. 1994. The inducible antibacterial peptides of insects. Parasitol. Today 10: 132-139.

Cociancich, S.O., Park, S.S., Fidock, D.A., and Shahabuddin, M. 1999. Vesicular ATPase-overexpressing cells determine the distribution of malaria parasite oocysts on the midguts of mosquitoes. J. Biol. Chem. 274: 12650-12655.

Collins, F.H., Sakai, R.K., Vernick, K.D., Paskewitz, S., Seeley, D.C., Miller, L.H., Collins, W.E., Campbell, C.C., and Gwadz, R.W. 1986. Genetic selection of a Plasmodium-refractory strain of the malaria vector *Anopheles gambiae*. Science. 234: 607-610.

Danielli, A., Loukeris, T. G., Lagueux, M., Muller, H. M., Richman, A., and Kafatos, F. C. 2000. A modular chitin-binding protease associated with hemocytes and hemolymph in the mosquito *Anopheles gambiae*. Proc. Natl. Acad. Sci. USA. 97: 7136-7141.

Delrieu, I., Waller, C., Mota, M., Grainger, M., Langhorne, J., and Holder, A. 2002. *P*SLAP, a protein with multiple adhesive motifs, is expressed in *Plasmodium falciparum* gametocytes. Mol. Biochem. Parasitol. 121: 11-20.

Dessens, J. T., Beetsma, A., Dimopoulos, G., Wengelnik, K., Crisanti, A., Kafatos, F., and Sinden, R. E. 1999. CTRP is essential for mosquito infection by malaria ookinetes. EMBO J. 18: 6221-6227.

Dessens, J.T., Mendoza, J. Claudianos, C., Vinetz, J.M., Khater, E., Hassard, S., Ranawaka, G.R. and Sinden R.E. 2001 Knockout of the rodent malaria parasite chitinase pbCHT1 reduces infectivity to mosqitoes. Infect. Imm. 69: 4041-4047.

Dimopoulos, G., Seeley,D., Wolf, A., Kafatos, F.C. 1998. Malaria infection of the mosquito *Anopheles gambiae* activates immune-responsive genes during critical transition stages of the parasite life cycle. EMBO J. 17: 6115-6123.

Dimopoulos, G., Christophides, G.K., Meister, S., Schultz, J., White, K.P., Barillas-Mury, C. and Kafatos, F.C. 2002. Genome expression analysis of *Anopheles gambiae*: responses to injury, bacterial challenge, and malaria infection. Proc. Natl. Acad. Sci USA. 99: 8814-8819.

Dimopoulos, G., Richman, A., Dellatorre, A., Kafatos, F. C., and Louis, C. 1996. Identification and characterization of differentially expressed cDNAs of the vector mosquito, *Anopheles gambiae*. Proc. Natl. Acad. Sci. USA. 93: 13066-13071.

Dimopoulos, G., Richman, A., Muller, H. M., and Kafatos, F. C. 1997. Molecular immune responses of the mosquito *Anopheles gambiae* to bacteria and malaria parasites. Proc. Natl. Acad. Sci. USA. 94: 11508-11513.

Dimopoulos, G., Seeley, D., Wolf, A., and Kafatos, F. C. 1998. Malaria infection of the mosquito *Anopheles gambiae* activates immune-responsive genes during critical transition stages of the parasite life cycle. EMBO J. 17: 6115-6123.

Duffy, P. E. and Kaslow, D. C. 1997. A novel malaria protein, Pfs28, and Pfs25 are genetically linked and synergistic as falciparum malaria transmission- blocking vaccines. Infect. Immun. 65: 1109-1113.

Fivelman Q.L., Butcher G.A., Adagu I.S., Warhurst D.C., and Pasvol G. 2002. Malarone treatment failure and *in vitro* confirmation of resistance of *Plasmodium falciparum* isolate from Lagos, Nigeria. Malaria J. 1: 1-4.

Fowler, R. E., Sinden, R. E., and Pudney, M. 1995. Inhibitory activity of the anti-malarial atovaquone (566C80) against ookinete, oocysts and sporozoites of *Plasmodium berghei*. J. Parasitol. 8: 454-458.

Freyvogel, T. A. 1966. Shape, movement *in situ* and locomotion of plasmodial ookinetes. Acta Tropica. 23: 201-221.

Gandon S., Mackinon M. J., Nee S., and Read A.F. 2002. Imperfect vacines and the evolution of parasite virulence. Nature. 414: 751-755.

Gass, R. F. 1977. Influences of blood digestion on the development of *Plasmodium gallinaceum* in the midgut of *Aedes aegypti*. Acta Tropica. 34: 127-140.

Gass, R. F. and Yeates, R. A. 1979. *In vitro* damage of cultured ookinetes of *Plasmodium gallinaceum* by digestive proteinases from susceptible *Aedes aegypti*. Acta Tropica. 36: 243-252.

Gillespie, J.P., Kanost, M.R., and Trenczek, T. 1997. Biological mediators of insect immunity. Ann. Rev. Entomol. 42: 611-643.

Gorman, M.J., Andreeva, O.V., and Paskewitz, S.M. 2000a. SP22D: a multidomain serine protease with a putative role in insect immunity. Gene. 251: 9-17.

Gorman, M.J., Andreeva, O.V., and Paskewitz, S.M. 2000b. Molecular characterization of five serine protease genes cloned from *Anopheles gambiae* hemolymph. Insect Biochem. Mol. Biol. 30: 35-46.

Gozar, M. M. G., Price, V. L., and Kaslow, D. C. 1998. Saccharomyces cerevisiae-secreted fusion proteins Pfs25 and Pfs28 elicit potent *Plasmodium falciparum* transmission- blocking antibodies in mice. Infect. Immun. 66: 59-64.

Grotendorst, C. A., and Carter, R. 1987. Complement effects on the infectivity of *Plasmodium gallinaceum* to *Aedes aegypti* mosquitoes. 2. Changes in sensitivity to complement-like factors during zygote development. J. Parasitol. 73: 980-984.

Grotendorst, C. A., Carter, R., Rosenberg, R., and Koontz, L. C. 1986. Complement effects on the infectivity of *Plasmodium gallinaceum* to *Aedes aegypti* mosquitoes. 1. Resistance of zygotes to the alternative pathway of complement. J. Immunol. 136: 4270-4274.

Han, Y. S., Thompson, J., Kafatos, F. C., and Barillas-Mury, C. 2000. Molecular interactions between *Anopheles stephensi* midgut cells and *Plasmodium berghei*: the time bomb theory of ookinete invasion of mosquitoes. EMBO J. 19: 6030-6040.

Healer, J., Mcguinness, D., Hopcroft, P., Haley, S., Carter, R., and Riley, E. 1997. Complement-mediated lysis of *Plasmodium falciparum* gametes by malaria-immune human sera is associated with antibodies to the gamete surface antigen Pfs230. Infect. Immun. 65: 3017-3023.

Hisaeda, H., Collins, W.E., Saul, A., and Stowers, A.W. 2002. Antibodies to *Plasmodium vivax* transmission-blocking vaccine candidate antigens Pvs25 and Pvs28 do not show synergism. Vaccine. 20: 763-770.

Hoffman, J.A., Reichhard, J-M., and Hetru, C. 1996. Innate immunity in higher insects. Curr. Opin. Immunol. 8: 8-13.

Hoffman, J.A. and Reichhard, J-M. 1997. *Drosophila* immunity. Trends Cell Biol. 7: 309-316.

Høgh, B., Gamage-Mendis, A., Butcher, G.A., Thompson, R., Begtrup, K., Mendis, C., Enosse, S.M., Dgedge, M., Barreto, J., Eling, W., and Sinden, R.E. 1998. The differing impact of chloroquine and pyrimethamine/sulfadoxine upon the infectivity of malaria spp. to the mosquito vector. Am. J. Trop. Med. Hyg. 58: 176-182.

Huber, M., Cabib, E., and Miller, L. H. 1991. Malaria parasite chitinase and penetration of the mosquito peritrophic membrane. Proc. Natl. Acad. Sci. USA. 88: 2807-2810.

Hudson, A. T. 1993. Atovaquone - A novel broad spectrum anti infective drug. Parasitol.Today 9: 66-68.

Huff, C.G., Marchbank, D.F., and Shiroishi,T. 1957. Changes in infectiousness of malarial gametocytes. II. Analysis of the possible causative factors. Exp. Parasitol. 7: 399-417.

Ito, J., Ghosh, A., Moreira, L.A., Wimmer, E.A., and Jacobs-Lorena, M. 2002. Transgenic anopheline mosquitoes impaired in transmission of a malaria parasite. Nature. 417: 452-455.

Kanost, M.R. 1999. Serine proteinase inhibitors in arthropod immunity. Dev. Comp. Immunol. 23: 291-301.

Kaslow, D. C., Quakyi, I. A., and Keister, D. B. 1989. Minimal variation in a candidate from the sexual stage of *Plasmodium falciparum*. Mol. Biochem. Parasitol. 32: 101-104.

Kaushal, D. C., Carter, R., Howard, R. J., and McAuliffe, F. M. 1983. Characterization of antigens on mosquito midgut stages of *Plasmodium gallinaceum* I. zygote surface antigens. Mol. Biochem. Parasitol. 8: 53-69.

Lensen, A. H. W., BolmervandeVegte, M., Vangemert, G. J., Eling, W. M. C., and Sauerwein, R. W. 1997. Leukocytes in a *Plasmodium falciparum*-infected blood meal reduce transmission of malaria to Anopheles mosquitoes. Infect. Immun. 65: 3834-3837.

Levashina, E.A., Moita, L.F., Blandin, S., Vriend, G., Lagueux, M., and Kafatos, F.C. 2001. Conserved role of a complement-like protein in phagocytosis revealed by dsRNA knockout in cultured cells of the mosquito, *Anopheles gambiae*. Cell. 104: 709-718.

Luckhart, S., Vodovotz, Y., Cui, L. W., and Rosenberg, R. 1998. The mosquito *Anopheles stephensi* limits malaria parasite development with inducible synthesis of nitric oxide. Proc. Natl. Acad. Sci. USA. 95: 5700-5705.

Margos, G., Navarrette, S., Butcher, G.A., Davies, A., Willers, C., Sinden. R.E. and Lachmann, P. 2001. Interaction between host complement and the mosquito mid-gut stage of *Plasmodium berghei*. Infect. Immun. 69: 5064-5071.

Matsuoka, H., Kobayashi J., Barker, G.C., Miura, K., Chinzei, Y., Miyajima S., Ishii, A., and Sinden, R.E. 1996. Induction of anti-malarial transmission blocking immunity with a recombinant ookinete surface antigen of *Plasmodium berghei* produced in silkworm larvae using the baculovirus expression system. Vaccine. 14: 120-126.

Mehlhorn, H., Peters, W., and Haberkorn, A. 1980. The formation of kinetes and oocyst in *Plasmodium gallinaceum* (Haemosporidia) and considerations on phylogenetic relationships between Haemosporidia, Piroplasmida and other Coccidia. Protistologica. 16: 135-154.

Meis, J. F. G. M., and Ponnudurai, T. 1987. Ultrastructural studies on the interaction of *Plasmodium falciparum* ookinetes with the midgut epithelium of *Anopheles stephensi* mosquitoes. Parasitol. Res. 73: 500-506.

Mellouk, S., Hoffman, S. L., Liu, Z. Z., Delavega, P., Billiar, T. R., and Nussler, A. K. 1994. Nitric oxide-mediated antiplasmodial activity in human and murine hepatocytes induced by gamma interferon and the parasite itself: enhancement by exogenous tetrahydrobiopterin. Infect. Immun. 62: 4043-4046

Ménard, R., Sultan, A.A., Cortes, C., Altszuler, R., van Dijk, M.R., Janse, C.J., Waters, A.P., Nussenzweig, R.S., and Nussenzweig, V. 1997. Circumsporozoite protein is required for development of malaria sporozoites in mosquitoes. Nature. 385(6614): 336-340.

Muta, T. and Iwanaga, S. 1996. The role of hemolymph coagulation in innate immunity. Curr. Opin. Immunol. 8: 41-47.

Naotunne, T. D., Karunaweera, N. D., Mendis, K. N., and Carter, R. 1993. Cytokine-mediated inactivation of malarial gametocytes is dependent on the presence of white blood cells and involves reactive nitrogen intermediates. Immunol. 78: 555-562.

Nosten, F., and Brasseur, P. 2002. Combination therapy for malaria: the way forward? Drugs 62: 1315-29.

Oduol, F., Xu, J., Niare,O., Natarajan,R., and Vernick, K.D. 2000. Genes identified by an expression screen of the vector mosquito *Anopheles gambiae* display differential molecular immune response to malaria parasites and bacteria. Proc. Natl. Acad. Sci. USA. 97: 11397-11402.

Paskewitz, S.M., Reese-Stardy, S., and Gorman, M.J. 1999. An easter-like serine protease from *Anopheles gambiae* exhibits changes in transcript abundance following immune challenge. Insect Mol. Biol. 8: 329-337.

Paton, M. G., Barker, G. C., Matsuoka, H., Ramesar, J., Janse, C. J., Waters, A. P., and Sinden, R. E. 1993. Structure and expression of a conserved and post-transcriptionally regulated gene encoding a surface protein of the sexual stages from malaria parasite *Plasmodium berghei*. Mol. Biochem. Parasitol. 59: 263-276.

Paulsson, M. 1992. Basement membrane proteins: structure, assembly, and cellular interactions. Crit. Rev. Biochem. Mol. Biol. 27: 93-127.

Peters, W. 1984. Use of drug combinations. In: Antimalarial Drugs. Vol 2. Peters, W., and Richards, W.H.G., eds. London Springer-Verlag. p. 499-509.

Raibaud, A., Lupetti, P., Paul, R.E., Mercati, D., Brey, P.T., Sinden, R.E., Heuser, J.E., and Dallai, R. 2001. Cryofracture electron microscopy of the ookinete pellicle of *Plasmodium gallinaceum* reveals the existence of novel pores in the alveolar membranes. J. Struct. Biol. 135: 47-57.

Ranford-Cartwright, L. C., Balfe, P., Carter, R., and Walliker, D. 1993. Frequency of cross-fertilization in the human malaria parasite *Plasmodium falciparum*. Parasitol. 107: 11-18.

Ranawaka, G. R. R., Alejo-Blanco, A. R., and Sinden, R. E. 1993. The effect of transmission-blocking antibody ingested in primary and secondary bloodfeeds, upon the development of Plasmodium berghei in the mosquito vector. Parasitology. 107 ( Pt 3): 225-231.

Ranawaka, G. R. R., Alejo-Blanco, A. R., and Sinden, R. E. 1994. Characterization of the effector mechanisms of a transmission-blocking antibody upon differentiation of plasmodium berghei gametocytes into ookinetes *in vitro*. Parasitol. 109: 11-17.

Richie T.L., and Saul A. 2002. Progress and challenges for malaria vaccines. Nature. 415: 694-701.

Richman, A. M., Dimopoulos, G., Seeley, D., and Kafatos, F. C. 1997. Plasmodium activates the innate immune response of *Anopheles gambiae* mosquitoes. EMBO J.16: 6114-6119.

Rodriguez, M. D., Zamudio, F., Torres, J. A., Gonzalezceron, L., Possani, L. D., and Rodriguez, M. H. 1995. Effect of a cecropin-like synthetic peptide (Shiva-3) on the sporogonic development of *Plasmodium berghei*. Exp. Parasitol. 80: 596-604.

Rodriguez, M. del C., Gerold, P., Dessens, J., Kurtenbach, K., Schwartz, R. T., Sinden, R. E., and Margos, G. 2000. Characterisation and expression of Pbs25, a sexual and sporogonic stage specific protein of *Plasmodium berghei*. Mol. Biochem. Parasitol. 110: 147-159.

Rudin, W., and Hecker, H. 1989. Lectin-binding sites in the midgut of the mosquitoes *Anopheles stephensi* Liston and *Aedes aegypti* L. (Diptera: Culicidae0. Parasitol. Res. 75: 268-279.

Rutledge, L. C., Gould, D. J., and Tantichareon, B. 1969. Factors affecting the infection of anophelines with human malaria in Thailand. Trans. R. Soc. Trop. Med. Hyg. 63: 613-619.

Schneider, D., and Shahabuddin, M. 2000. Malaria parasite development in a *Drosophila* model. Science 288: 2376-2379.

Von Siedlein L., Jawara, M., Coleman R., Doherty T., Walraven, G., and Targett, G. 2001. Parasitaemia and gametocytaemia after treatment with chloroquine, pyrimethamine/sulfadoxine and pyrimethamine/sulfadoxine with artesunate in young Gambians with uncomplicated malaria. Trop. Med. Int. Hlth.6: 92-98.

Shahabuddin, M., Fields, I., Bulet, P., Hoffmann, J. A., and Miller, L. H. 1998. *Plasmodium gallinaceum*: Differential killing of some mosquito stages of the parasite by insect defensin. Exp. Parasitol. 89: 103-112.

Shahabuddin, M., and Kaslow, D. C. 1994. Plasmodium: parasite chitinase and its role in malaria transmission. Exp. Parasitol. 79: 85-88.

Shahabuddin, M., and Pimenta, P. F. P. 1998. *Plasmodium gallinaceum* preferentially invades vesicular ATPase-expressing cells in *Aedes aegypti* midgut. Proc. Natl. Acad. Sci. USA. 95: 3385-3389.

Shen, Z. C., Dimopoulos, G., Kafatos, F. C., and Jacobs-Lorena, M. 1999. A cell surface mucin specifically expressed in the midgut of the malaria mosquito *Anopheles gambiae*. Proc. Natl. Acad. Sci. USA. 96: 5610-5615.

Siden-Kiamos, I., Vlachou, D., Margos, G., Beetsma, A., Waters, A., Sinden, R., and Louis, C. 2000. Distinct roles for Pbs21 and Pbs25 in the in vitro ookinete to oocyst transformation of *Plasmodium berghei*. J. Cell Sci. 113: 3419-3426.

Sinden, R. E. 1984. The biology of Plasmodium in the mosquito. Experientia. 40: 1330-1343.

Sinden, R.E., Butcher, G.A., Billker, O., and Fleck, S.L. 1996 Regulation of infectivity of *Plasmodium* to the mosquito vector. Ad. Parasitol. 38: 54-117.

Sinden, R.E., Hartley, R., and Winger, L. 1985. The development of *Plasmodium* ookinetes *in vitro*: an ultra-structural study including a description of the meiotic division. Parasitol. 91: 227-244.

Sinden, R. E., and Smalley, M. E. 1976. Gametocytes of *Plasmodium falciparum*: phagocytosis by leucocytes *in vivo* and *in vitro*. Trans. Roy. Soc.Trop. Med. Hyg. 70: 344-345.

Sinden, R. E., and Strong, K. 1978. An ultrastructural study of the sporogonic development of *Plasmodium falciparum* in *Anopheles gambiae*. Trans. Roy. Soc. Trop. Med. Hyg. 27: 477-491.

Sinden, R.E., Winger, L., Carer, E.H., Hartley, R.H., Tirawanchai, N., Davies, C.S. Moore, J., and Sluiters, J G. 1987. Ookinete antigens of *Plasmodium berghei* : a light and electron-microscope immunogold study of expression of the 21 Kda determinant recognised by a transmission-blocking antibody. Proc. Roy. Soc. B. 230: 443-458.

Spano, F., Matsuoka, H., Ozawa, R., Chinzei, Y., Crisanti, A., and Sinden, R. E. 1996. Epitope mapping on the ookinete surface antigen Pbs21 of *Plasmodium berghei*: identification of the site of binding of transmission-blocking monoclonal antibody 13.1. Parassitologia. 38: 559-563.

Stowers, A., and Carter, R. 2001. Current developments in malaria transmission-blocking vaccines. Expert Opin. Biol. Therapeutics 1: 619-628.

Stowers A.W., Keister, D.B. Muratora, O., and Kaslow D.C. 2000. A region of *Plasmodium falciparum* antigen Pfs 25 that is the target of highly potent transmission-blocking antibodies. Infect. Immun. 68: 5530-5538.

Tachibana M., Tsuboi, T., Templeton, T.J., and Keneko O. 2001. Presence of three distinct ookinete surface protein genes, Pos 25, Pos 28-1, and Pos 28-2 in *P. ovale*. Mol. Biochem. Para. 118: 223-231.

Targett, G.A., Drakely, C., Jawara, M., von Seidlein, L., Coleman, R., Dean, J., Pinder, M., Doherty, F., Sutherland, C., Walraven G., and Milligan P. 2001. Artesunate reduces but does not prevent postreatment transmission of *Plasmodium falciparum* to *Anopheles gambiae*. J. Infect. Diseases. 183: 1254-1259.

Tomas, A., Margos, G., Dimcpoulos, G., van Lin L.H.M., de Koning-Ward, T.F., Sinha, R., Lupetti, P., Beetsma, A.L., Rodriguez, M.C., Karras, M., Hager, A., Mendoza, J., Butcher, G.A., Kafatos, F., Janse, C.J., Waters, A., and Sinden R.E. 2001. P25 and P28 proteins of the malaria ookinete surface have multiple and partially redundant functions. EMBO J. 20: 3975-3983.

Vanderberg, J. P., Rhodin, J, and Yoeli, M. 1967. Electron microscopic and histochemical studies of sporozoite formation in *Plasmodium berghei*. J. Protozoo.14: 82-103.

Vernick, K. D., Fujioka, H., Seeley, D. C., Tandler, B., Aikawa, M., and Miller, L. H. 1995. *Plasmodium gallinaceum*: a refractory mechanism of ookinete killing in the mosquito, *Anopheles gambiae*. Exp. Parasitol. 80: 583-595.

Vernick, K. D., Fujioka, H., anc Aikawa, M. 1999. *Plasmodium gallinaceum*: a novel morphology of malaria ookinetes in the midgut of the mosquito vector. Exp. Parasitol. 91: 362-366.

Vizioli, J., Bulet, P., Charlet, M., Lowenberger,C., Blass, C., Muller, H.M., Dimopoulos, G., Hoffman, J., Kafatos, F.C., and Richman, A. 2000. Cloning and analysis of a cercropin gene from the malaria vector mosquito, *Anopheles gambiae*. Insect Mol. Biol. 9: 75-84.

Vizioli, J., Bulet, P., Hoffmann, J.A., Kafatos, F.C., Muller, H-M., and Dimopoulos, G. 2001. Gambicin: A ncvel immune responsive antimicrobial peptide from the malaria vector *Anopheles gambiae*. Proc. Natl. Acad. Sci. USA. 98: 12630-12635.

Vlachou, D., Lycett, G., Siden-Kiamos, I., Blass, C., Sinden, R. E., and Louis, C. 2001. *Anopheles gambiae* laminin interacts with the P25 surface protein of *Plasmodium berghei* ookinetes. Mol. Biochem. Parasitol.112: 229-237.

Warburg, A., and Miller, L. H. 1992. Sporogonic development of a malaria parasite *in vitro*. Science 255: 448-450.

Washburn, M.P., Wolters, D., and Yates, J.R III. 2001. Large-scale analysis of the yeast proteome by multidimensional protein identification technology. Nature Biotech. 19: 242-247

Weathersby, A. B. 1952. The role of the stomach wall in the exogenous development of *Plasmodium gallinaceum* as studied by means of haemocoel injections of susceptible and refractory mosquitoes. J. Infect. Disease. 91:198-205.

Wengelnik, K., Vidal, V., Ancelin, M.L., Cathiard, A.M., Morgat, J.L., Kocken, C.H., Calas, M., Herrera, S., Thomas, A.W., and vial, H.J. 2002. A class of potent antimalarials and their specific accumulation in infected erythrocytes. Science. 15: 1311-1314.

Yuda, M., Sakaida, H., and Chinzei, Y. 1999. Targeted disruption of the *Plasmodium berghei CTRP* gene reveals its essential role in malaria infection of the vector mosquito. J. Exp. Med. 190: 1711-1715.

Yuda, M., Yano, K., Tsuboi, T., Torii, M., and Chinzei, Y. 2001. von Willebrand factor A domain-related protein, a novel microneme protein of the malaria ookinete highly conserved throughout *Plasmodium* parasites. Mol. Biochem. Parasitol.116: 65-72.

Zieler, H., and Dvorak, J. A. 2000. Invasion *in vitro* of mosquito midgut cells by the malaria parasite proceeds by a conserved mechanism and results in death of the invaded midgut cells. Proc. Natl. Acad. Sci. USA. 97: 11516-11521.

Zieler, H., Keister, D. B., Dvorak, J. A., and Ribeiro, J. M. C. 2001. A snake venom phospholipase A(2) blocks malaria parasite development in the mosquito midgut by inhibiting ookinete association with the midgut surface. J. Exp. Biol. 204: 4157-4167.

Zieler, H., Nawrocki, J.P., and Shahabuddin, M. 1999. *Plasmodium gallinaceum* ookinetes adhere specifically to the midgut epithelium of *Aedes aegypti* by interaction with a carbohydrate ligand. J. Exp. Biol. 202: 485

From: Malaria Parasites: Genomes and Molecular Biology
Edited by: A.P. Waters and C.J. Janse

# Chapter 16

# Molecular and Cellular Biology of Chloroquine Resistance in *Plasmodium falciparum*

Karena L. Waller, Sylvia Lee, and David A. Fidock

## Abstract

For decades, the 4-aminoquinoline chloroquine (CQ[1]) was the mainstay of antimalarial chemotherapy, with its characteristic properties of rapid efficacy, safety and affordability. However, resistance to this drug has now become prevalent in the majority of malaria-endemic countries and is severely limiting its use as an effective antimalarial. CQ is believed to act against CQ sensitive (CQS) malaria parasites by accumulating inside the digestive vacuole (DV) of the intra-erythrocytic parasite, where CQ binds to heme (ferriprotoporphyrin IX, or FPIX) that is liberated during hemoglobin degradation. The resulting impairment in heme detoxification results in parasite death. Chloroquine resistance (CQR) in *Plasmodium falciparum* is defined *in vitro* by elevated CQ $IC_{50}$ values, reduced CQ accumulation and the ability of verapamil (VP) to reverse (or "chemosensitize") these two traits. Molecular genetic and epidemiological investigations have recently revealed an association between *P. falciparum* CQR and the presence of multiple amino acid mutations in the DV transmembrane protein PfCRT

---

[1] ADQ, amodiaquine; AO, acridine orange; CQ, chloroquine; CQS, chloroquine sensitive; CQR, chloroquine resistance (or chloroquine resistant); DV, digestive vacuole; FPIX, ferriprotoporphyrin IX; iRBC, infected red blood cells; NHE, $Na^+/H^+$ exchange; Pgh1, P-glycoprotein; RBC, red blood cells; SP, sulfadoxine-pyrimethamine; UTR, untranslated region; VP, verapamil; WHO, World Health Organization.

(*P. falciparum* chloroquine resistance transporter). Allelic exchange experiments have now definitively demonstrated that mutant *pfcrt* alleles present in different malaria-endemic continents can confer CQR to CQS parasites. Investigations into other candidate genetic determinants have demonstrated that mutations in the parasite protein Pgh1 (*P. falciparum* P-glycoprotein homologue 1, also localized to the parasite DV membrane and encoded by the *pfmdr1* gene) can modulate the CQR phenotype of parasites, but are not sufficient to confer CQR *in vitro*. Recent clinical studies demonstrate an increased risk of CQ treatment failure in patients harboring parasites with CQR-associated mutations in *pfcrt* and sometimes *pfmdr1*. Other factors, including host immunity, are also important in determining clinical outcome. *In vitro*, the CQR phenotype may be intimately associated with changes in the physiological processes of the DV, such as altered DV pH. These changes may critically impact the access of CQ to hematin dimers that act as a drug receptor in the DV. Alternatively, mutant PfCRT may contribute to CQR by physically interacting with CQ, leading to increased drug efflux. Ongoing studies, bolstered by recent advances in *Plasmodium* genetics, genomics and proteomics, can be expected to significantly deepen our understanding of the genetic basis of *P. falciparum* CQR, the relationship between CQR and DV physiology, and the effect of CQR on the parasite response to other heme-binding antimalarials. The knowledge gained from these studies will be of greatest benefit when they can be translated into new therapies to treat drug-resistant malaria.

# 1. Introduction

Malaria is the most devastating human parasitic disease. It is currently estimated that each year more than 1 million people die from malaria, a further 300 million have clinical symptoms and 40% of the world's population live in malaria-endemic regions (WHO, 1999). Malaria is caused by infection with any of four *Plasmodium spp.*, and although infections resulting from *P. vivax, P. malariae and P. ovale* can result in serious disease symptoms, these species rarely cause mortality. In contrast, *P. falciparum* infections can cause more severe and sometimes fatal disease complications, including severe anemia, pulmonary edema, metabolic acidosis and cerebral malaria. The effective treatment and control of malaria infections, especially those caused by *P. falciparum*, is proving increasingly difficult due to the emergence and spread of drug resistant parasite populations in most malaria-endemic regions of the world.

Here, we provide a brief review of CQ's activity against *P. falciparum* and the recent advances in our understanding of the molecular basis of *P. falciparum* CQR. Additional information relating to CQ's mode of action, biochemical studies on CQR, and laboratory and clinical evaluation of resistance reversal agents, can be found in a number of earlier reviews (Bray and Ward, 1998; Cowman, 1995; Dorsey *et al.*, 2001a; Fitch, 1998; Ginsburg *et al.*, 1999; Guan *et al.*, 2002; Martiney *et al.*, 1999; O'Neill *et al.*, 1998; Peel, 2001; Ridley, 1998; Tilley *et al.*, 2001; Ursos and Roepe, 2002; Ward and Bray, 2001; Warhurst, 2001b; Wellems and Plowe, 2001).

# 2. History of CQ and CQR

For the past 50 years, CQ has been the gold standard in the prevention and treatment of uncomplicated malaria (Coatney, 1963; Loeb *et al.*, 1946). Defining characteristics of this 4-aminoquinoline compound include its rapid parasiticidal action, low cost, safety and widespread availability (White, 1996). Massive CQ pressure on the parasite population worldwide, exacerbated by problems of non-compliance and sub-therapeutic dosing, led to the first reports of CQR *P. falciparum* malaria in the late 1950s (some 12 years after initial widespread use of this compound (Payne, 1987; Wongsrichanalai *et al.*, 2002)). Two initial foci of resistance were documented, one in Columbia and the other near the Cambodia-Thailand border (Wernsdorfer, 1994). From there, resistance gradually extended during the next 20 years throughout South America and Southeast Asia. Arriving in the late 1970s in East Africa, CQR has since spread across all of sub-Saharan Africa (Peters, 1987). Of particular concern are recent epidemiological reports that the emergence and spread of CQR *P. falciparum* malaria has been associated in Africa with increases in malaria-related mortality and morbidity (Marsh, 1998; Snow *et al.*, 2001; Trape, 2001; Trape *et al.*, 1998; 2002). For example, regions of Senegal that previously had low levels of infant (0 – 4 years) mortality have seen dramatic increases (up to eleven fold) in malaria-associated mortality in this age group since the emergence of CQR (Marsh, 1998; Trape, 2001; Trape *et al.*, 1998). CQ is still widely used throughout Africa, although many countries have officially switched to sulfadoxine-pyrimethamine (SP, also known as Fansidar), the only affordable alternative to CQ therapy. SP is used either alone or in combination with either CQ or amodiaquine (ADQ) (Newton and White, 1999; Winstanley *et al.*, 2002; Wongsrichanalai *et al.*, 2002), but resistance to SP has already been reported in regions of Africa, Southeast Asia and South America ((Wongsrichanalai *et al.*, 2002) and references therein). Other antimalarial drugs such as artesunate, mefloquine, quinine and atovaquone-proguanil have significant drawbacks in terms of cost, compliance, parasite resistance and/or toxicity. The prospects are therefore not encouraging, and only a concerted multinational effort involving the pharmaceutical sector, major funding bodies and the malaria research community can avert a pending disaster (Ridley, 2002).

In addition to *P. falciparum*, CQR has also been reported for *P. vivax,* a less virulent but nonetheless highly prevalent agent of malaria in South America and Southeast Asia (Wellems and Plowe, 2001; Whitby, 1997). Additional information detailing *P. vivax* CQR can be found in (Alecrim *et al.*, 1999; Baird *et al.*, 1991; Carlton *et al.*, 2001b; Dua *et al.*, 1996; Garg *et al.*, 1995; Marlar *et al.*, 1995; Myat *et al.*, 1993; Nomura *et al.*, 2001; Phillips *et al.*, 1996; Rieckmann *et al.*, 1989; Schwartz *et al.*, 1991). Interestingly, CQR in *P. ovale* has not been reported and for *P. malariae*, evidence for CQR has only recently surfaced, in the Indonesian archipelago of Southeast Asia (Maguire *et al.*, 2002; Peters, 1998). Reasons for the reduced frequency of CQR in these species may include their lower global prevalence, combined with the relative rarity in generating mutations that confer CQR.

# 3. CQ Mode of Action

Initial insights into the mechanistic basis of CQ sensitivity included the work of Peters *et al.*, (1970), who observed CQ to be effective only on red blood cell (RBC) forms of the parasite that were actively degrading hemoglobin. This degradation occurs inside the DV, which was noted to swell upon CQ treatment (Aikawa, 1972; Jacobs *et al.*, 1988). Accumulation of this highly soluble, diprotonable (dibasic) compound ($pK_{a1} = 8.1$, $pK_{a2} = 10.2$) in the DV has been postulated to depend, in part, on the presence of a weak-base gradient between the pH-neutral parasite cytosol and the acidic DV (Geary *et al.*, 1990; Ginsburg *et al.*, 1989; Hawley *et al.*, 1996; Krogstad and Schlesinger, 1987; Krogstad *et al.*, 1985; Yayon *et al.*, 1984). Once inside this acidic compartment, CQ should become predominantly diprotonated and membrane-impermeant. In agreement with this postulate, the degree of *P. falciparum* susceptibility to CQ was found to be pH-dependent (Yayon *et al.*, 1985) and CQ action was mitigated by the proton-pump inhibitors bafilomycin A1 and omeprazole (Bray *et al.*, 1992b; Skinner-Adams and Davis, 1999). However, CQ has been reported to accumulate in iRBC at much higher levels than those observed in acidic compartments in eukaryotic cells, implicating more than just pH gradients (Aikawa, 1972; De Duve *et al.*, 1974; Krogstad *et al.*, 1992; Krogstad and Schlesinger, 1987; Macomber *et al.*, 1966; Yayon *et al.*, 1984; 1985). Recent studies now point to CQ:hematin interactions as the major factor leading to CQ retention inside the iRBC of CQS parasites and a core element of CQ's mode of action ((Bray *et al.*, 1998; Chou *et al.*, 1980; Dorn *et al.*, 1995; 1998; Francis *et al.*, 1997; Krogstad and De, 1998; O'Neill *et al.*, 1998); see below). To better discuss this key issue, it is important to describe some fundamental aspects of hemoglobin processing and hematin availability.

## 3.1. Hemoglobin Processing and its Relationship to CQ

Degradation of hemoglobin (present at an intracellular concentration of 5 mM) in the iRBC is believed to provide an essential source of amino acids for parasite growth (see Rosenthal and Meshnick (1998) and Banerjee and Goldberg (2001) for recent comprehensive reviews). This processing apparently also enables the parasite to manipulate the host cell volume and osmolality in order to prevent premature iRBC lysis (Ginsburg, 1990; Lew and Hockaday, 1999). Approximately 60% of RBC hemoglobin is digested by trophozoite stage parasites, with this amount increasing up to almost 80% by schizont stages (Egan *et al.*, 2002; Loria *et al.*, 1999). Hemoglobin degradation also provides iron (present within heme), the bulk of which ends up in the DV, but which can also be incorporated into iron-binding parasite proteins (Egan *et al.*, 2002).

Hemoglobin degradation is facilitated by the action of multiple enzymes, including cysteine and aspartic proteases (Banerjee *et al.*, 2002; Eggleson *et al.*, 1999; Francis *et al.*, 1997; Rosenthal and Meshnick, 1998; Shenai and Rosenthal, 2002; Shenai *et al.*, 2000; Sijwali *et al.*, 2001; Westling *et al.*, 1999). This process is thought to occur primarily in the DV, but may be initiated during vesicle-mediated hemoglobin transport to this vacuole. Degradation releases heme moieties containing iron that can be rapidly oxidized from the relatively innocuous hemoglobin-bound ferrous [Fe(II)] form to the more toxic, liberated ferric [Fe(III)] form (Atamna and Ginsburg, 1993; Foley and Tilley, 1998; Har-El *et al.*, 1993). The mechanism of this oxidation in the parasite remains obscure, but it has been speculated to originate from the reaction of Fe(II)(protoporphyrin-IX) and dioxygen, presumably liberating superoxide ($^{\bullet}O^{2-}$). Free Fe(II)(protoporphyrin-IX) (and perhaps Fe(III)(protoporphyrin-IX, though at a slower rate), may also generate hydroxyl radicals ($^{\bullet}OH$) via the Fenton reaction ($Fe^{2+} + H_2O_2 \rightarrow Fe^{3+} + OH^- + {}^{\bullet}OH$). Intracellular accumulation of $H_2O_2$ and $^{\bullet}O_2^-$, if this occurs at physiologically relevant concentrations as a consequence of hemoglobin degradation (Atamna and Ginsburg, 1993), could be mitigated by the actions of parasite superoxide dismutase, glutathione reductase and perhaps host catalase (Becuwe *et al.*, 1996; Clarebout *et al.*, 1998; Gamain *et al.*, 1996; Sztajer *et al.*, 2001).

Heme toxicity is thought to result from its rapid ability to intercalate into lipid bilayers, interfering with electron transport chains, and ultimately resulting in peroxidative damage to unsaturated lipids and/or membrane-embedded proteins (Fitch *et al.*, 1982; Loria *et al.*, 1999; Tappel, 1953; Zhang *et al.*, 1999). Accumulation of membrane-bound heme above a toxic threshold (possibly combined with a build-up of reactive oxygen radicals) may explain the irreversible nature of CQ action on malaria parasites (Ginsburg and Krugliak, 1992; Tilley *et al.*, 2001). Further experimental validation of this hypothesis is required.

Lacking an effective heme oxygenase pathway, the parasite detoxifies heme, or hematin, by generating crystals of inert hemozoin (Banerjee and Goldberg, 2001; Slater and Cerami, 1992; Slater *et al.*, 1991). Chemical, spectroscopic and crystallographic analyses have shown that hemozoin crystals are composed of β-hematin ((Hempelmann and Egan, 2002;

Pagola *et al.*, 2000) and references therein). β-hematin [formed from heme monomers (Bohle *et al.*, 1997; Francis *et al.*, 1997; Rosenthal and Meshnick, 1998) consists of a unique porphyrin structure, composed of heme dimers that form via reciprocal iron-carboxylate bonds to one of the propionic side chains of each porphyrin. These dimers are linked into chains that can bind to the asymmetric growing faces of each crystal by hydrogen bond linkages (between each pair of carboxylate dioxygens) (Pagola *et al.*, 2000). Initial crystallization of β-hematin, from preformed dimers, may involve parasite proteins, including some histidine-rich proteins (Sullivan *et al.*, 1996a). The extending heme chain can act as a nucleation center for subsequent addition of hematin moieties (Francis *et al.*, 1997; Pagola *et al.*, 2000; Tilley *et al.*, 2001).

### 3.2. Formation of CQ-hematin Complexes and Their Role in Parasite Killing

As alluded to before, many studies have implicated heme (or more precisely, the hematin dimer) as the CQ receptor (Bray *et al.*, 1999a; 1998; Chou *et al.*, 1980; Dorn *et al.*, 1998; Moreau *et al.*, 1982). Among the more definitive are the studies by Bray *et al.* (1999a; 1998), who reported a significant decrease in CQ accumulation in parasites treated with protease inhibitors that reduced hemoglobin processing and formation of soluble heme. Debate continues on whether CQ binds primarily with membrane-associated (or soluble) hematin or instead with the terminal hematin structure present at the actively growing faces of the hemozoin crystal (Dorn *et al.*, 1998; Pagola *et al.*, 2000; Sullivan *et al.*, 1996b; 1998). Presumably, either process would effectively allow the build-up of heme (or CQ-heme complexes) to levels that become irreversibly toxic to the parasite, possibly via heme-mediated DV membrane lipid peroxidation (see below). Interestingly, studies on CQ action have found that drug accumulation follows a biphasic pattern, having a high-affinity, low-capacity component ($K_d = 25 - 150$ nM) plus a higher-capacity non-saturable component that predominates at higher (micromolar) external CQ concentrations (Bray *et al.*, 1996; 1998; Fitch *et al.*, 1974; Tilley *et al.*, 2001). The high-affinity, saturable component is believed to be responsible for CQ's antimalarial activity (Bray *et al.*, 1998).

Whereas most of the current data implicate trapping of protonated CQ within the DV and CQ-heme binding as key to this drug's action, alternative models include CQ inhibition of proteases associated with hemoglobin processing or possibly CQ interference with glutathione-mediated detoxification of reactive oxygen radicals formed during hemoglobin degradation (reviewed in (Ginsburg *et al.*, 1999; Krogstad and De, 1998)).

# 4. The Genetic Basis of CQR

## 4.1. Initial Linkage and Transfection-based Evidence That *pfcrt* is a Critical Genetic Determinant of CQR

The identification of *pfcrt* traces back to the seminal decision of T. Wellems (NIAID/NIH, Bethesda, MD) to search for determinants of *P. falciparum* CQR via the implementation and analysis of a genetic cross (Wellems *et al.*, 1990). This cross was carried out between a CQR (Dd2, from Indochina (Wellems *et al.*, 1990)) and a CQS clone (HB3, from Honduras (Walliker *et al.*, 1987)). Molecular and phenotypic studies of the resulting, individually cloned progeny revealed a 1:1 segregation between VP-reversible CQR and VP-insensitive CQS clones, suggesting that CQR resulted primarily from a single inherited locus (Wellems *et al.*, 1991). Restriction fragment length polymorphism studies subsequently mapped the CQR locus to a 36 kb region containing 8 putative genes (Su *et al.*, 1997). Initial nucleotide sequence analysis of these genes identified *cg2* as a candidate CQR genetic determinant, on the basis of multiple point mutations that segregated mostly with the CQR phenotype (Su *et al.*, 1997). However, *cg2* was later definitively discounted on the basis of results from allelic exchange experiments (Fidock *et al.*, 2000a).

Shortly thereafter, *pfcrt* was identified following more detailed visual and RT-PCR screening of this 36 kb region (Fidock *et al.*, 2000b). This highly interrupted gene, comprised of 13 short exons, was initially overlooked as its genetic arrangement escaped detection by the gene prediction algorithms used at the time. The 424 amino acid PfCRT protein has a predicted molecular mass of 49 kDa and 10 putative transmembrane domains (Fidock *et al.*, 2000b; Nomura *et al.*, 2001). PfCRT has orthologs in other *Plasmodium spp.*, as well as in the slime mold *Dictyostelium discoideum* (Carlton *et al.*, 2001b; Nomura *et al.*, 2001), however these homologies provide precious few clues about the possible function of this protein. Subcellular localization studies of native and epitope-tagged protein as well as parasite organelle fractionation assays have clearly localized PfCRT to the parasite DV membrane (Cooper *et al.*, 2002; Fidock *et al.*, 2000b). The *pfcrt* sequence identified in the genetic cross encoded 8 point mutations (M74I, N75E, K76T, A220S, Q271E, N326S, I356T and R371I) that distinguished CQR from the CQS progeny (Fidock *et al.*, 2000b; Table 1). Interestingly, all of these point mutations cluster within or near the predicted transmembrane domains (Fidock *et al.*, 2000b; Figure 1), which suggests a potential involvement in altered substrate flux through this protein at the level of the DV membrane.

Comparative sequence analyses performed on parasites from several geographical locations throughout Asia and Africa (or "Old World" isolates) indicated a significant degree of association between the CQR phenotype and the presence of seven of the eight PfCRT point mutations identified in

**Table 1.** *pfcrt* and *pfmdr1* haplotypes and chloroquine phenotypes in culture-adapted *P. falciparum* lines

| Origin[a] | *pfcrt* position and encoded amino acid | | | | | | | | | | CQ IC$_{50}$ mean ± SD[b] (nM) | CQ IC$_{50}$ Range (nM) | CQ sensitivity[c] | |
|---|---|---|---|---|---|---|---|---|---|---|---|---|---|---|
| | 72 | 74 | 75 | 76 | 97 | 220 | 271 | 326 | 356 | 371 | | | S | R |
| All Regions (HB3) | C | M | N | K | H | A | Q | N | I | R | 15 ± 7 | 9-32 | 15/15 | 0/15 |
| Africa (106/1) | C | I | E | K | H | S | E | S | I | I | 33 | 33 | 1/1 | 0/1 |
| Asia/Africa (Dd2) | C | I | E | T | H | S | E | S | T | I | 204 ± 26 | 180-246 | 0/6 | 6/6 |
| Asia/Africa | C | I | E | T | H | S | E | S | I | I | 182 ± 70 | 102-320 | 0/12 | 12/12 |
| S. America/Oceania (7G8) | S | M | N | T | H | S | Q | D | L | R | 171 ± 45 | 96-229 | 0/7 | 7/7 |
| S. America | C | M | E | T | Q | S | Q | N | I | T | 153 ± 1 | 152-154 | 0/2 | 2/2 |
| S. America | C | M | N | T | H | S | Q | D | L | R | 127 | 127 | 0/1 | 1/1 |

| Origin[a] | pfmdr1 position and encoded amino acid | | | | | CQ IC$_{50}$ mean ± SD[b] (nM) | CQ IC$_{50}$ Range (nM) | CQ sensitivity[c] | |
|---|---|---|---|---|---|---|---|---|---|
| | 86 | 184 | 1034 | 1042 | 1246 | | | S | R |
| Asia/Africa (3D7) | N | Y | S | N | D | 18±7 | 9-27 | 8/8 | 0/8 |
| Asia/Africa | N | F | S | N | D | 152±174 | 9-620 | 8/12 | 4/12 |
| Asia/Africa | N | S | S | N | D | 12±0 | 12 | 2/2 | 0/2 |
| Asia/Africa/ S.America (K1) | Y | Y | S | N | D | 167±397 | 12-405 | 3/24 | 21/24 |
| Asia/Africa | Y | F | S | N | D | 158±22 | 128-188 | 0/5 | 5/5 |
| Asia/Africa | N | F | S | D | D | 80±67 | 32-127 | 1/2 | 1/2 |
| Asia | N | F | C | D | D | 620±0 | 620 | 0/1 | 1/1 |
| Asia/S.America/ Oceania (7G8) | N | F | C | D | Y | 119±92 | 31-307 | 7/13 | 6/13 |

[a] The brackets following the geographic origin refers to reference lines frequently used to study these genes that have the stated haplotypes.
[b] The mean IC$_{50}$ values (± standard deviations) were calculated from in vitro studies of culture-adapted lines (reported in (Basco et al., 1995; Chen et al., 2001; Duraisingh et al., 2000a; Fidock et al., 2000a; 2000b; Foote et al., 1990; 1989; Maguire et al., 2001; Mehlotra et al., 2001; Sidhu et al., 2002; Su et al., 1997; Wilson et al., 1993; Wootton et al., 2002; T. Wellems, pers. comm). This table does not include data from allelic exchange studies or studies with patient isolates that were not culture-adapted (the latter were therefore tested only once).
[c] The CQ sensitivity value indicates the fraction of the total number of lines reported with the specific haplotype that are either sensitive or resistant.

CQ, chloroquine; R, resistant; SD, standard deviation; S, sensitive.

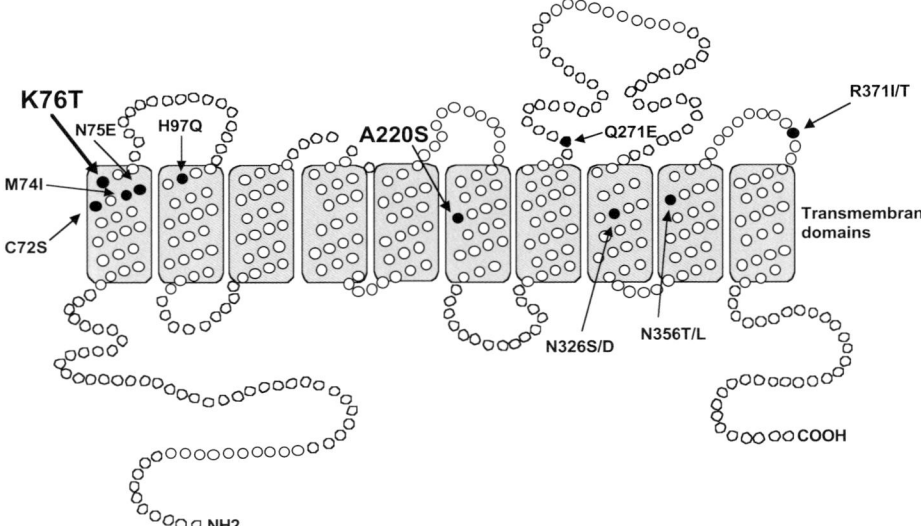

**Figure 1.** Schematic of PfCRT, showing the ten predicted transmembrane domains. Positions of the mutations identified from the analysis of over 50 geographically diverse, culture-adapted isolates are indicated by filled circles. The K (lysine) to T (threonine) change at position 76 is proving to be critical to CQR in *P. falciparum*. Adapted and reproduced from *Current Opinion in Microbiology*, Vol. 4, J. Carlton *et al.*, "Conservation of a novel vacuolar transporter in *Plasmodium* species and its role in chloroquine resistance in falciparum but not vivax malaria", pp. 415-420, Copyright (2001), with permission from Elsevier Science.

Dd2 (Fidock *et al.*, 2000b). The eighth mutation (I356T) was found to be present in approximately half of the "Old World" CQR lines. South American (or "New World") CQR lines contained 3 distinct PfCRT haplotypes that differed from the canonical CQS haplotype of HB3 by a total of 4 - 5 amino acids (Table 1). The heterogeneous PfCRT haplotypes observed in South America reflect multiple independent events that may, in part, account for the rapid dissemination of CQR throughout that region. There remain few, if any, CQS lines in South America (Wellems and Plowe, 2001), attesting to the tremendous selective advantage conferred by the acquisition of CQR. Interestingly, recent studies have found that CQR lines from Papua New Guinea share the same PfCRT haplotype as that of the Brazilian line 7G8, thus indicating their origin as apparently independent mutational events (Chen *et al.*, 2001; Mehlotra *et al.*, 2001; Wootton *et al.*, 2002).

Overall, these linkage data are consistent with historical records, which indicate that CQR arose independently in a handful of sites, including Southeast Asia, two or more sites in South America and an apparently separate site in Papua New Guinea (Payne, 1987; Plowe *et al.*, 2001; Wootton *et al.*, 2002). Interestingly, all CQR lines carrying mutant PfCRT share the K76T and A220S mutations, evoking the possibility that these mutations are critical to conferring CQR.

Characterization of the *pfcrt* sequence from the Sudanese 106/1 clone (Bayoumi *et al.*, 1993; Su *et al.*, 1997) proved particularly informative (Fidock *et al.*, 2000b). The 106/1 CQS line was unique in that it contained all the *pfcrt* point mutations previously associated with the Old World CQR phenotype with the exception of codon 76 (Table 1). This encoded a lysine at this position, as is typical of CQS lines, whereas all CQR lines adapted from field isolates encode a threonine (Fidock *et al.*, 2000b). In independent experiments, the application of CQ pressure to 106/1 parasites resulted in the selection of derivative CQR lines whose *pfcrt* sequence had undergone the single point mutations K76I or K67N (Cooper *et al.*, 2002; Fidock *et al.*, 2000b). Co-expression studies on the K76I variant confirmed the ability of this mutant protein to confer a degree of CQR to CQS parasites (Fidock *et al.*, 2000b). These results suggested that the loss of K76 in PfCRT is a critical component of the CQR phenotype and raised the possibility of this mutation being a molecular epidemiological marker of resistance (see below).

While these linkage studies and drug pressure regimes provided important support for the role of *pfcrt* in CQR, they failed to establish whether *pfcrt* mutations were sufficient to confer CQR to CQS parasites. To address this question, Sidhu *et al.* (2002) implemented a genetic exchange strategy to replace the endogenous, "wild type" *pfcrt* allele in a CQS parasite with mutant *pfcrt* alleles common to CQR isolates from Asia, Africa, South America and Papua New Guinea. Phenotypic analysis of cloned, recombinant lines provided conclusive evidence that mutant *pfcrt* alleles from around the globe could confer CQR. This phenotype was confirmed using the three established markers of resistance, namely elevated CQ $IC_{50}$ values, VP reversibility and reduced drug accumulation. Interestingly, expression of the 7G8 allele, present in South America and Papua New Guinea, led to diminished VP reversibility when compared to the Dd2 allele found in Asia and Africa. This finding mirrored earlier observations of variable VP reversibility in lines of distinct geographic origins (Mehlotra *et al.*, 2001), suggesting that the VP-reversible component of the CQR phenotype may largely be dictated by the tertiary PfCRT structure. These allelic exchange experiments, which necessitated two rounds of genetic modification, were performed with the CQS clone GC03, which is a progeny of the HB3 X Dd2 cross. Notably, this clone carries the *pfmdr1* HB3 allele that encodes the D1042Y mutation, but not the N86Y mutation that is more typically associated with CQR and that is present in Dd2 (see below).

In the GC03 *pfcrt*-modified clones, the degree of CQR (measured as CQ $IC_{50}$) was approximately 70-80% of that observed with the culture-adapted, non-transformed CQR control lines expressing the same *pfcrt* allele (Sidhu *et al.*, 2002). One plausible explanation is that additional genes, possibly including *pfmdr1* (see below), are required to attain high levels of CQR. Another explanation is that the reduced CQ $IC_{50}$ values in the genetically modified lines, relative to CQR control lines, resulted from altered expression of the recombinant *pfcrt* gene. Indeed, although these lines were engineered to maintain the endogenous promoter, thereby maintaining correct timing of expression, almost all of the introns were removed and the 3′ untranslated region (UTR) was changed. Western blot assays confirmed an approximately 40-70% decrease in PfCRT expression levels in the recombinant lines compared to the parental line 7G8 (Sidhu *et al.*, 2002). To address the hypothesis that expression levels of PfCRT could modulate the degree of CQR, a separate allelic exchange experiment was implemented to truncate the *pfcrt* 3′UTR, without altering the coding sequence. Luciferase reporter assays performed with transiently transfected parasites indicated that truncating the *pfcrt* 3′UTR to 148 bp (extending from the stop codon) resulted in a 90% reduction of luciferase expression (Waller *et al.*, 2003). Cloned recombinant 7G8 parasites expressing *pfcrt* with this 148 bp truncated 3′UTR (referred to as *pfcrt* "knockdown" clones) demonstrated a 30-40% decrease in PfCRT expression, as shown by Western blot and immuno-electronmicroscopy. These values closely matched the 30% reduction in CQ $IC_{50}$ values observed in these clones using drug response assays, as compared to the parental 7G8 line (Waller *et al.*, 2003).

Support for *pfcrt* being the key gene responsible for CQR comes from the recent work by Su and colleagues (Wootton *et al.*, 2002), who used 342 highly polymorphic microsatellite markers (Su *et al.*, 1999) to study genomic variation in 87 *P. falciparum* isolates from around the world. This analysis provided evidence of restricted allelic diversity around the *pfcrt* genomic locus in CQR parasites from Asia, Africa, Papua New Guinea and South America, whereas such restricted diversity was not observed in CQS lines (Wootton *et al.*, 2002). These data imply the existence of a major selective advantage for mutant *pfcrt,* which can be attributed to massive CQ pressure, and provide further evidence that the world-wide dissemination of mutant alleles stemmed from a handful of initial mutational events. These findings were also consistent with earlier genetic evidence suggesting that CQR spread from Asia to Africa (Fidock *et al.*, 2000b; Su *et al.*, 1997; Wellems and Plowe, 2001).

It remains a mystery why *pfcrt* mutations leading to CQR did not arise separately in Africa, where *P. falciparum* prevalence and transmission and CQ drug use are of greatest incidence. Perhaps the answer lies in parasite meta-population structures (populations living in patches with a balance between extinction and colonization) that may be unique to Africa (Ariey *et al.*, 2003; Levins, 1969) and that may significantly impact rates of gene flow, levels of genetic linkage disequilibrium, and/or infection-dependent acquired immunity (premunition). Another explanation could be differences in the use and impact of CQ-medicated salt distribution programs. These were conducted to differing extents in parts of South America, Iran, Southeast Asia, Irian Jaya, Indonesian New Guinea and Africa. These programs may well have resulted in widespread under-dosing and, therefore, increased selection for low-level CQR mutants and could have been the major factor contributing to the appearance of CQR in South America and Southeast Asia (Payne, 1987; Verdrager, 1986). Another, particularly interesting possibility may be the existence of hyper-mutator phenotypes in some strains of *P. falciparum*. Evidence for this comes from a seminal study by P. Rathod and colleagues showing that some Southeast Asian drug-resistant strains were able to acquire resistance to unrelated antimalarial compounds at frequencies up to 1,000 fold higher than drug-sensitive strains (Rathod *et al.*, 1997). This phenotype, referred to as "accelerated resistance to multiple drugs (ARMD)", may help to explain anecdotal observations that drug resistance appears to develop more easily in Southeast Asia than in other parts of the world. Possible mechanisms include increases in mutation or recombination rates and/or pre-existing forms of resistance that can adapt to novel drugs. Such a rapid mutator phenotype could potentially result in bursts of localized point mutations in chromosomal segments, which could explain the existence of multiple sequence polymorphisms in genes (including *cg2* and *cg1*) that flank *pfcrt* (Su *et al.*, 1997). These mutations may have occurred during initial hyper-mutational events that resulted in the appearance of point mutations in *pfcrt* that conferred CQR. An alternative model, presented by Hastings *et al.* (Hastings *et al.*, 2002), postulates the sequential accumulation of individual point mutations conferring ever-increasing rates of resistance, culminating in the full set of point mutations. In either scenario, parasites with a full set of *pfcrt* mutations presumably could survive in the face of intense CQ pressure and begin to expand and spread the mutant locus by genetic recombination with other parasites.

## 4.2. Epidemiological and Clinical Evidence Supporting a Key Role for *pfcrt*

Shortly after the discovery of *pfcrt* in late 1998 and the release of the nucleotide sequence to the World Health Organization (WHO), many malaria research groups incorporated screening for the PfCRT K76T mutation into their epidemiological investigations of candidate genetic markers of CQR. The first report, published in 2001 by Djimdé *et al.* (2001), found the PfCRT K76T mutation to be present in all 60 samples from patients with CQR infections that persisted or recurred after treatment. In comparison, this mutation was only found in 47 out of 116 samples tested before treatment (P<0.001). The presence of the PfCRT K76T mutation was more strongly associated with CQ treatment failure (odds ratio, 18.8) than was the presence of the *pfmdr1* N86Y mutation (odds ratio, 3.2) or the presence of both mutations (odds ratio, 9.8). Very recently published studies on parasites from diverse regions of the world, including Southeast Asia, Papua New Guinea, Brazil and several African countries have largely confirmed an extremely tight association between the PfCRT K76T mutation and CQR (Ariey *et al.*, 2002; Babiker *et al.*, 2001; Basco, 2002; Basco and Ringwald, 2001; Chen *et al.*, 2002; 2001; Contreras *et al.*, 2002; Dorsey *et al.*, 2001b; Durand *et al.*, 2001; Labbe *et al.*, 2001; Lim *et al.*, 2003; Lopes *et al.*, 2002; Maguire *et al.*, 2001; Mayor *et al.*, 2001; Pillai *et al.*, 2001; Vieira *et al.*, 2001). Among these, several studies found selection for PfCRT K76T in many or all cases of CQ treatment failure, in populations where *pfcrt* mutations had not attained 100% prevalence (Babiker *et al.*, 2001; Basco *et al.*, 2002; Basco and Ringwald, 2001; Mayor *et al.*, 2001; Sutherland *et al.*, 2003). Other studies noted that *pfcrt* mutations were already present in all the field isolates tested (Dorsey *et al.*, 2001b; Pillai *et al.*, 2001; Vieira *et al.*, 2001), suggesting a significant selective advantage for mutant *pfcrt*.

Significantly, several studies based on the use of the WHO microtiter assay[2], or variations thereof (Desjardins *et al.*, 1979; Le Bras and Deloron, 1983; Moreno *et al.*, 2001; Rieckmann *et al.*, 1978), have reported a decreased association between *in vitro* CQR and the PfCRT K76T mutation (Ariey *et al.*, 2002; Thomas *et al.*, 2002). These may well be valid results that denote the existence of *pfcrt*-independent means of CQR, which would certainly be of interest to discover and further investigate. Micro-test assays however, are well known to be subject to variation and artifact, in part due to the frequent occurrence of mixed infections (that can confound matching of data with PCR results). These assays are more appropriate for estimating levels of *in vitro* resistance in the field than for characterizing individual infections (Carlin *et al.*, 1984; Durand *et al.*, 2001; Wellems and Plowe, 2001). Thus, microtiter results should be interpreted with caution as they provide less confidence than assays with culture-adapted lines for which sequence information and phenotypes can be reproducibly determined.

The evidence is now overwhelming that sustained CQ pressure leads to the selection and maintenance of *pfcrt* mutant alleles in *P. falciparum* populations. Insights into the potential fitness cost of these mutations on the organism come from a recent study involving Malawi. In 1993, Malawi health officials vigorously enforced a ban on CQ use in public and private health sectors, and instead began using SP. Molecular epidemiological and *in vitro* screens recently conducted by C. Plowe and colleagues documented that the prevalence of the *pfcrt* K76T mutation decreased dramatically during this period, from 85% in 1992 to 13% in 2000 and 0% in 2001. In parallel, rates of parasitological resistance fell from over 80% in 1990 to 0% in infected asymptomatic adults in 2001 (Kublin *et al.*, 2003). Similar findings were reported by A. Bjorkman and colleagues (Mita *et al.*, 2003). These results imply a deleterious effect for the parasite of expressing mutant *pfcrt* in the absence of CQ pressure.

Interestingly, G. Targett and colleagues have recently discovered that mutant *pfcrt* may also confer a distinct advantage to gametocyte survival

---

[2] In the classical WHO "Mark III" assay, infected blood is placed in *in vitro* culture conditions for 48 hours in the presence of increasing CQ concentrations and parasite survival is then determined by light microscopy of Geimsa-stained thick blood films.

in CQ-treated individuals. Working in The Gambia, this group discovered that CQ treatment of children led to a significant selection for mutant *pfcrt* and *pfmdr1* in the surviving gametocyte population (Sutherland *et al.*, 2002). Membrane-feeding assays with venous blood collected from these children four to seven days post-treatment revealed that these mutant gametocytes were infectious for Anopheline mosquitoes (Sutherland *et al.*, 2002; Targett *et al.*, 2001). Preliminary molecular data with pre- and post-treatment peripheral blood and midgut oocysts (following membrane feeding) suggest that the *pfcrt* and *pfmdr1* haplotypes in the mosquito gut oocysts reflect those present among the ingested gametocytes (R. Hallett, C. Sutherland and A. Alloueche, pers. comm.). Overall, these data are consistent with favored transmission of CQR haplotypes from treated individuals, into the mosquito population and back to the human host. The most likely explanation is that mutant *pfcrt* confers a selective advantage to the young gametocytes (stages I-III) that need to digest hemoglobin, thereby achieving protection against the drug much the same way as for the asexual stages (see above). The ability of *pfcrt* to not only protect asexual stages against CQ but also enhance the transmissibility of mutant gametocytes may help to explain the dominant role of this gene in spreading CQR around the globe.

Epidemiological studies have also frequently shown that the association between K76T and CQR is typically less pronounced *in vivo* than *in vitro*, indicating that factors other than *pfcrt* contribute to the clinical outcome of CQ treatment. These factors may include differences in host immunity, the presence of concomitant infections, frequency of exposure, nutritional status, and/or the requirement for additional parasite genetic determinants including *pfmdr1* (see below). Critical support for a key role of age-related acquisition of anti-parasitic immunity in clearance of CQR infections was reported by Djimde *et al.*, (2001), in which children over 10 years had a much greater rate of clearance of parasites harboring the PfCRT K76T mutation when compared to younger children ($p<0.001$). Therefore, the overall picture emerging from these studies is that *pfcrt* mutations appear to be necessary for CQR *in vitro* and for CQ treatment failure *in vivo*, but that other factors such as host immunity, and possibly other parasite genes, including *pfmdr1*, are important in determining clinical outcome (Figure 2).

### 4.3. *pfmdr1* and its Association with CQR

Prior to the discovery of *pfcrt*, the majority of research into the genetic basis of *P. falciparum* CQR focused on *pfmdr1*. Research into this gene was stimulated by the finding that *P. falciparum* CQR parasites and multi-drug resistant cancer cell lines shared characteristics of reduced drug accumulation and resistance reversibility by VP. These traits are typically attributed in the cancer cell lines to altered properties of P-glycoproteins (encoded by *mdr* genes), that are believed to actively efflux drugs outside the cells (Krogstad

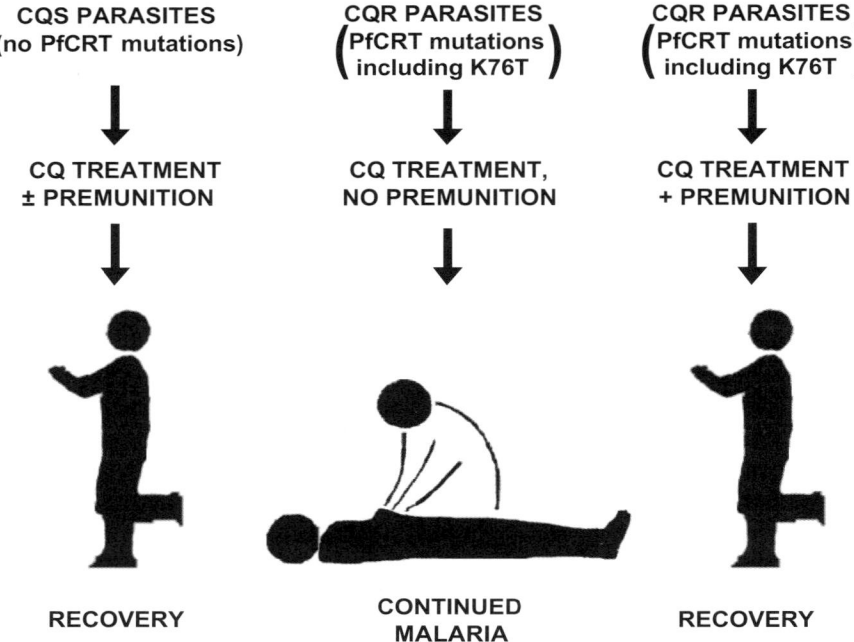

**Figure 2.** Possible outcomes in individuals treated with CQ for *P. falciparum* malaria. Infections with CQR parasites may persist or clear after treatment, depending on the status of pre-existing malaria immunity (premunition). Inter-individual variations in the whole-blood concentrations of CQ and its monodesethyl metabolite also may affect response after treatment. PfCRT and in some cases Pgh1 point mutations are associated with CQ treatment failure. CQ, chloroquine; CQS, CQ sensitive. Reproduced from (Plowe *et al.*, 2001) with permission from the University of Chicago and the authors C. Plowe and T. Wellems. © 2001 by the Infectious Diseases Society of America. All rights reserved.

*et al.*, 1987; Martin *et al.*, 1987). The ensuing search for *mdr*-like genes in *P. falciparum* located *pfmdr1* on chromosome 5 (Cowman and Karcz, 1991; Foote *et al.*, 1989; Peel, 2001; Wellems *et al.*, 1990; Wilson *et al.*, 1989; Zalis *et al.*, 1993). This gene is comprised of a single 4.3 kb exon and encodes a 160 kDa P-glycoprotein (Pgh1) that has been localized to the parasite DV membrane (Cowman *et al.*, 1991; Cowman and Karcz, 1991). This protein belongs to the ABC transporter family and contains two ATP-binding sites and two hydrophobic regions, each containing six predicted transmembrane domains (Foote *et al.*, 1989; Karcz *et al.*, 1993).

Several early investigations into *pfmdr1* revealed variability in gene copy number as well as differences in expression levels of the Pgh1 product between different *P. falciparum* strains and provided evidence for an association between increased expression levels and CQR (Ekong *et al.*, 1993; Foote *et al.*, 1989; Wilson *et al.*, 1989). Some other studies however, have reported no such association (Basco *et al.*, 1995; Price *et al.*, 1999).

Interestingly, Barnes *et al.* (1992) reported that application of CQ pressure to a CQR line that possessed multiple copies of *pfmdr1* resulted in the selection of a highly resistant line that had undergone a de-amplification in *pfmdr1* copy number. This inverse relationship between the level of CQR and *pfmdr1* gene copy number was also found in separate investigations of lines pressured with CQ or mefloquine (Cowman *et al.*, 1994; Peel *et al.*, 1994). A number of other studies, although not all, have also identified a correlation between increased *pfdmr1* copy number and increased resistance to the antimalarial drugs mefloquine, quinine and halofantrine (Barnes *et al.*, 1992; Cowman *et al.*, 1994; Peel *et al.*, 1994; Price *et al.*, 1999; Ritchie *et al.*, 1996).

An initial analysis of *pfmdr1* sequences in parasite lines with known *in vitro* drug resistant or sensitive phenotypes revealed a close association between the CQR phenotype and two distinct *pfmdr1* alleles (Foote *et al.*, 1990). One allele, typified by the K1 strain (Thailand; (Thaithong and Beale, 1981)), encodes a single N86Y mutation and is associated with CQR in Asian and African strains (Table 1). The other allele, typified by 7G8 ((Brazil; (Burkot *et al.*, 1984)), encodes the four point mutations Y184F, S1034C, N1042D and D1246Y, and showed a good association with CQR in lines derived mostly from South America. Many subsequent molecular epidemiological studies of *pfmdr1* point mutations, using field isolates or lab-adapted lines, have yielded conflicting data that show either some or no association between *pfmdr1* point mutations and the falciparum CQR phenotype (Awad-el-Kariem *et al.*, 1992; Basco and Ringwald, 1998; Bhattacharya and Pillai, 1999; Chaiyaroj *et al.*, 1999; Duraisingh *et al.*, 1997; 2000b; Gomez-Saladin *et al.*, 1999; Grobusch *et al.*, 1998; Haruki *et al.*, 1994; Price *et al.*, 1997; von Seidlein *et al.*, 1997; Wilson *et al.*, 1993). Some studies have found evidence of linkage disequillibrium between *pfmdr1* and *pfcrt* (Adagu and Warhust, 2001; Babiker *et al.*, 2001; Mehlotra *et al.*, 2001; Sutherland *et al.*, 2002). These results suggest that mutant *pfmdr1* may be selected either because of a direct contribution to CQR (e.g. by reducing CQ accumulation in the DV) or because mutant *pfmdr1* expression can compensate for physiological perturbations induced by expression of mutant PfCRT. This linkage disequillibrium is however, not universal among CQR parasites, suggesting that *pfmdr1* may not be an important component of CQR in some parasite populations (Flueck *et al.*, 2000; Pillai *et al.*, 2001; Povoa *et al.*, 1998).

Using powerful allelic exchange techniques developed for *P. falciparum* (Crabb and Cowman, 1996; Wu *et al.*, 1995), Reed *et al.*, (2000) recently provided evidence that *pfmdr1* point mutations could modulate the degree of CQR. Working with the CQR 7G8 parasite line, these authors reported that removal of three point mutations (S1034C, N1042D and D1246Y) resulted in a 2-fold decrease of CQ $IC_{50}$ values in the recombinant lines, as compared to the non-transformed control. The opposite experiment, that of introducing the 7G8 CQR-associated mutations into a CQS line (D10, Papua New Guinea) that was wild type for *pfmdr1* (and *pfcrt*) did not alter the CQ susceptibility levels of the recombinant versus control lines. These data, combined with the somewhat conflicting data discussed above, suggest that *pfmdr1* mutations do not themselves confer CQR, however their presence in CQR lines is a reflection of their capacity to modulate the *P. falciparum* CQR phenotype.

Further support for a modulatory role of *pfmdr1* in contributing to CQR for some, but not all parasite lines, comes from several molecular epidemiological and clinical studies that have reported an association between *pfmdr1* point mutations and an increased risk of CQ treatment failure (Babiker *et al.*, 2001; Djimdé *et al.*, 2001; Sutherland *et al.*, 2002). Other studies have not detected such an association (Dorsey *et al.*, 2001b; Pillai *et al.*, 2001). These findings evoke two possibilities: 1) *pfmdr1* point mutations have a significant impact on drug resistance levels by affecting drug accumulation, either directly via protein:drug interactions or indirectly via an effect of these mutations on physiological parameters (e.g. DV pH) that influence drug accumulation; or 2) *pfmdr1* point mutations do not affect drug resistance levels *per se* and are present primarily to compensate for altered physiological function resulting from changes in PfCRT. The strongest argument for the former possibility comes from the work by Reed *et al.* (2000) that documented increased CQ accumulation in CQR parasites modified in their *pfmdr1* haplotype. Further allelic exchange investigations into *pfmdr1* mutations and their effect on multiple genetic backgrounds will be valuable to better understand the overall contribution of *pfmdr1* to CQR.

## 5. Mechanistic Models of CQR – A Brief Overview

A review of biochemical and pharmacological investigations into CQR reveals a complex array of models. The principal ones postulate that CQR may result from: 1) increased efflux of CQ from the parasite DV, possibly as a result of changes in the transport properties of a DV membrane protein (or proteins) that can transport drug (Krogstad *et al.*, 1987); 2) altered ionic flux or CQ uptake as a result of a physiological change at the parasite DV membrane (Martiney *et al.*, 1999; Ursos and Roepe, 2002); 3) reduced drug accumulation via alterations in a CQ "importer"; 4) reduced CQ access to hematin; or 5) accelerated detoxification of CQ-heme complexes by a mechanism involving increased levels of glutathione. In reviewing these models, greatest attention will be paid to those that presently appear the most relevant. For more comprehensive reviews, refer to (Bray and Ward, 1998; Fitch, 1998; Foley and Tilley, 1998; Ginsburg *et al.*, 1999; Tilley *et al.*, 2001; Ursos and Roepe, 2002; Wellems and Plowe, 2001; Wellems *et al.*, 1998).

## 5.1. Increased CQ Efflux

Experimental data in favor of this model, reported by Krogstad *et al.* (1987), showed that CQR parasites released pre-accumulated CQ almost 50 times faster than CQS parasites. Other reports, however, have found similar rates of CQ efflux in resistant and sensitive strains, in addition to reinforcing the finding of reduced CQ accumulation in CQR strains (Bray *et al.*, 1994; 1992a; Geary *et al.*, 1990; Ginsburg and Stein, 1991). If this increased CQ efflux model is correct, then this would suggest that mutant PfCRT and/or Pgh1 can rapidly efflux heme-binding antimalarials out of the DV (see Figure 3).

## 5.2. pH-dependent Physiological Changes at the Parasite DV Membrane

Mechanistic models relating pH-dependent physiological changes to CQR have recently seen intriguing yet contradictory developments. Initial modeling postulated that uptake of diprotonable CQ into the acidic DV (estimated to have a pH in the range of 5.0 to 5.4; (Ginsburg *et al.*, 1989; Krogstad and Schlesinger, 1987; Krogstad *et al.*, 1985; Yayon *et al.*, 1984)) resulted from weak-base ion-trapping as predicted by the Henderson-Hasselbach equation (Homewood *et al.*, 1972; Krogstad *et al.*, 1985; Yayon *et al.*, 1984). *Increased DV pH in CQR parasites*, causing a reduced pH gradient between the DV and the parasite cytosol, would thus conceivably cause reduced CQ accumulation (Bray *et al.*, 1992b; Geary *et al.*, 1990; Ginsburg and Stein, 1991). Recent developments in single cell imaging and photometry of *P. falciparum* parasites, however, have provided evidence consistent with a *decreased* DV pH in CQR parasites (Dzekunov *et al.*, 2000). These experiments, using acridine orange (AO) as a pH-sensitive fluorescent dye, were performed under perfusion with reduced $O_2$ gas conditions routinely used for *P. falciparum in vitro* culture (Trager and Jensen, 1976). DV pH differences between strains were assessed based on several measurements, most critically the linear change in DV fluorescence upon rapid change in the AO concentration exposed to iRBC. The results were consistent with a pH reduction of 0.3 to 0.5 units in the DV of CQR parasites, as compared to CQS parasites (Dzekunov *et al.*, 2000; Fidock *et al.*, 2000b). Additional investigations have suggested that VP might increase the DV pH of CQR parasites to a value more comparable to CQS parasites (Ursos *et al.*, 2000). These results are inconsistent with expectations founded on the Hendersson-Hasselbach equation, which predicts that a more acidic DV pH should increase CQ accumulation. However, CQ accumulation is known to also be driven by the binding of CQ to heme ((Bray *et al.*, 1998; Fitch, 1970); see Section 3), which effectively acts as a sink for CQ retention. One proffered explanation for a more acidic DV pH in CQR came from studies suggesting that pH changes could dramatically affect the conversion of soluble heme to insoluble aggregates (Dorn *et al.*, 1998; Dzekunov *et al.*, 2000). The pH midpoint of this conversion was strikingly close to the DV pH

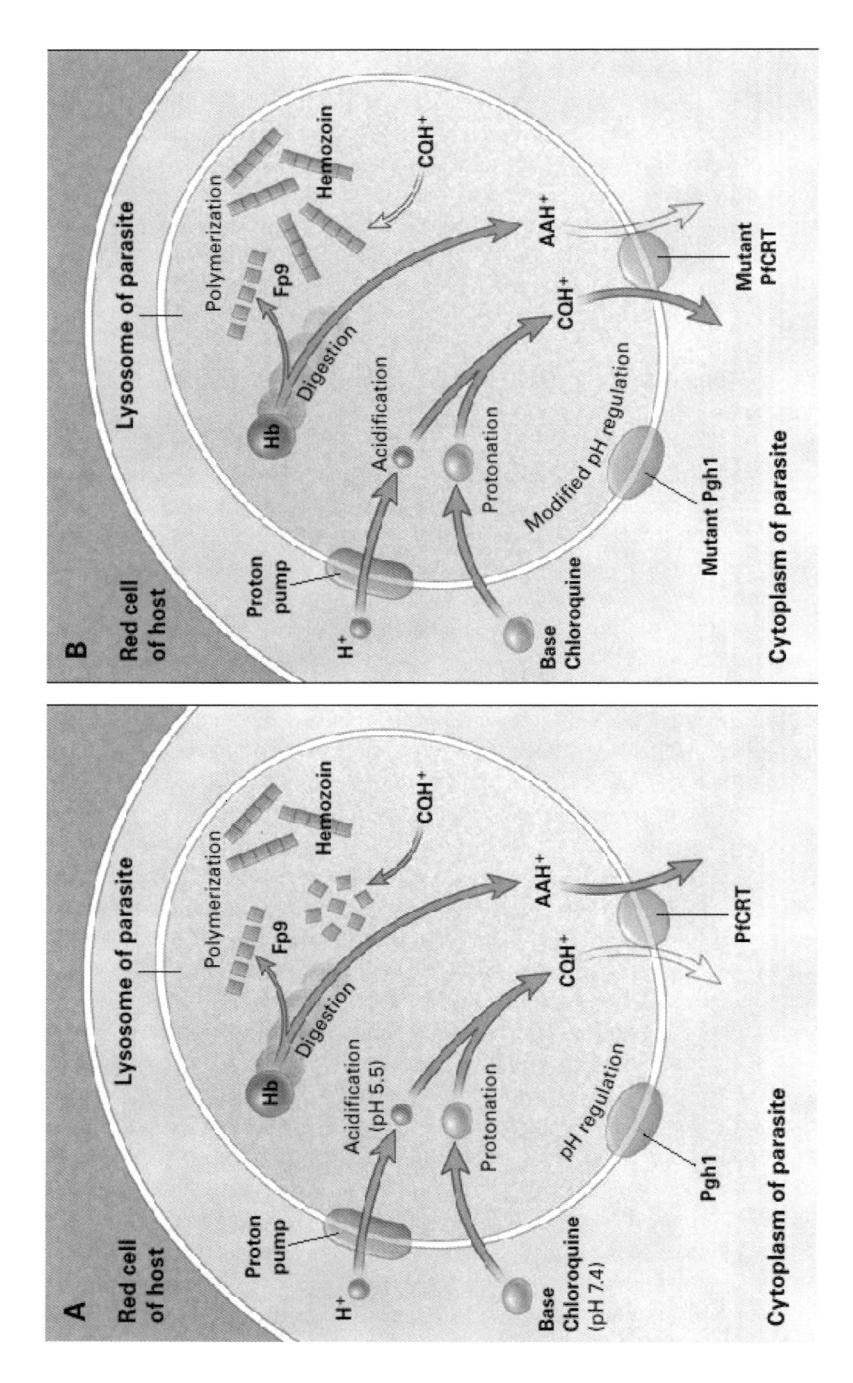

A

Red cell of host

Lysosome of parasite

Polymerization

Hemozoin

Fp9

CQH⁺

Digestion

Hb

Acidification (pH 5.5)

Proton pump

H⁺

Protonation

CQH⁺

AAH⁺

pH regulation

PfCRT

Base Chloroquine (pH 7.4)

Pgh1

Cytoplasm of parasite


B

Red cell of host

Lysosome of parasite

Polymerization

Hemozoin

Fp9

CQH⁺

Digestion

Hb

Acidification

Proton pump

H⁺

Protonation

CQH⁺

AAH⁺

Modified pH regulation

Mutant PfCRT

Base Chloroquine

Mutant Pgh1

Cytoplasm of parasite

**Figure 3.** Model of CQ effect on heme detoxification in the DV (lysosome) of *P. falciparum* CQS (Panel A) or CQR (Panel B) parasites. **(A)** In the DV of a CQS parasite, hydrogen ions enter through a proton pump, acidifying the DV environment (pH 5.5). This process may be regulated by Pgh1, which may release anions into the DV to optimize the difference in the transmembrane charge. During the digestion of hemoglobin (Hb), protonated basic amino acids (AAH+) are released together with toxic ferriprotoporphyrin IX (FPIX). FPIX is detoxified via its incorporation into hemozoin crystals. The weak base CQ, present in the cytoplasm (pH 7.4), crosses the DV membrane and concentrates in the acidic DV, undergoing protonation to a membrane-insoluble form (CQH+). CQH+ binds to FPIX and prevents its detoxification. PfCRT may transport AAH+ out of the DV or have a separate transport function that affects ΔpH. Wild type PfCRT may have a limited affinity for CQH+ and efflux some of this drug. **(B)** The DV of a parasite with CQR-associated mutations in *pfcrt* and *pfmdr1*. Mutant PfCRT may have an increased affinity for CQH+ and be able to export large amounts of the drug, enabling FPIX detoxification and hemozoin formation to proceed normally. Mutant PfCRT may have a reduced capacity for transport of AAH+ (or other charged substrates), which could result in the accumulation of more protons (H+) in the DV. The presence of mutant Pgh1 may partially prevent this accumulation of protons, increasing the fitness of parasites with *pfcrt* and *pfmdr1* mutations. Mutations in *pfmdr1* also increase sensitivity to mefloquine and artemisinin, probably as a result of the partial inactivation of the ability of mutant Pgh1 to export these drugs. Reproduced in an adapted form from (Warhurst, 2001a) with permission from the author, D. Warhurst. © 2001 Massachusetts Medical Society. All rights reserved.

values of CQS parasites, and formation of insoluble heme was much more efficient at the acid DV pH values of CQR parasites (Dzekunov *et al.*, 2000). It was therefore postulated that DV acidification would consequently leave significantly less free heme available for the formation of toxic complexes with CQ, an effect that is consistent with experimental data showing a reduction in high-affinity drug receptor sites (Bray *et al.*, 1998; Fitch, 1970). These studies however, may not accurately reflect parasite heme and hemozoin products (S. Bohle, McGill U. Montreal, pers. comm.). Further, such a pH effect may not adequately explain the reported efficacy of certain CQ side-chain derivatives on CQR parasites, which more readily evokes the notion of structurally-dependent protein:drug interactions (De *et al.*, 1998; 1996; O'Neill *et al.*, 1998; Ridley *et al.*, 1996).

In reviewing these data, it must be noted that doubts have been raised as to whether AO-based single-cell photometric methods are an appropriate methodology for determining the DV pH (Bray *et al.*, 2001; Wissing *et al.*, 2002). Nevertheless, AO studies with congenic *P. falciparum* lines or recombinant *Pichia Pastoris* expressing distinct *pfcrt* sequences (Cooper *et al.*, 2002; Fidock *et al.*, 2000b; Zhang *et al.*, 2002), as well as dextran studies in CHO cells expressing various *pfmdr1* alleles (van Es *et al.*, 1994a; 1994b), favor an impact on intracellular compartmental pH of mutations in both genes. Further work is essential to confidently define the relationship between DV and cytosolic pH, changes in expressed *pfcrt* and *pfmdr1* sequences, and CQR.

## 5.3. Alterations in a CQ "Importer"

Various groups have also postulated the existence of a CQ importer and highlighted a potential role for a $Na^+/H^+$ exchanger (NHE; see below) (Ferrari and Cutler, 1991; Sanchez *et al.*, 1997; Wunsch *et al.*, 1998). The involvement of a parasite NHE in determining CQ activity has since been challenged on experimental grounds (Bray *et al.*, 1999b). While it is possible that a parasite NHE may affect parasite susceptibility to CQ and related drugs, data to support this have not yet been published.

## 5.4. Reduced CQ Access to Hematin

In support of the model that CQR may result from reduced CQ access to hematin, Bray *et al.*, (1998) reported that the antimalarial activity of CQ could be attributed to the saturable component of CQ accumulation and that CQR and CQS parasites could be distinguished by the $K_d$ in saturable CQ binding. These data supported an earlier report (Bray *et al.*, 1996), indicating that the differences in CQ accumulation between CQR and CQS parasites were routinely smaller than the differences in their CQ $IC_{50}$ values, suggesting that CQR involved both alterations in receptor accessibility and drug accumulation.

## 5.5. Glutathione-mediated Detoxification of CQ-heme Complexes

Early experiments using CQR murine malaria parasites indicated that these lines produced less hemozoin and showed higher levels of glutathione relative to CQS lines (Bhatia and Charet, 1984; Dubois *et al.*, 1995; Mahoney and Eaton, 1981; Wood and Eaton, 1993). Comparative studies in *P. falciparum* parasites were able to demonstrate that increasing levels of glutathione led to increased resistance to CQ (Ginsburg *et al.*, 1998). A separate study however, showed no differences in the level of hemozoin accumulation between CQR and CQS parasites (Zhang *et al.*, 1999), providing evidence that the mechanistic basis of CQR in *P. falciparum* was likely to differ from that of other *Plasmodium spp.* (further detail on CQR in rodent malaria parasites can be found in a review by Carlton *et al.*, (2001a)). Potentially, glutathione may participate in the CQR mechanism by its noted ability to bind to heme, resulting in decreased levels of accumulated CQ in the DV, protection of heme-mediated lipid peroxidation and membrane lysis (Sahini *et al.*, 1996; Shviro and Shaklai, 1987).

# 6. CQR and Parasite Susceptibility to Other Antimalarials

Many compounds in addition to CQ are believed to act, at least in part, by binding to heme and inhibiting its detoxification. Chief among these are the quinolinemethanol antimalarials quinine, mefloquine and halofantrine. As mentioned above, several studies have noted an inverse relationship between CQR and parasite response to these quinolinemethanols (Barnes *et al.*, 1992; Cowman *et al.*, 1994; Duraisingh *et al.*, 2000a; 2000b; Peel *et al.*, 1994; Price *et al.*, 1999; Ritchie *et al.*, 1996). Allelic exchange studies indicate that *pfmdr1* point mutations can influence parasite susceptibility to multiple heme-binding antimalarials including those cited above (Reed *et al.*, 2000). Recent investigations also show that *pfcrt* point mutations can contribute to a dichotomous relationship between CQR and the quinolinemethanols (Cooper *et al.*, 2002; Sidhu *et al.*, 2002). Surprisingly, both genes also appear to influence parasite susceptibility to the endoperoxide drug artemisinin, derivatives of which are showing excellent clinical activity against falciparum malaria (Newton and White, 1999). The presence of both *pfcrt* and *pfmdr1* gene products on the DV membrane implicates altered DV physiology and possibly structure-dependent drug interactions as being key to these susceptibility patterns. Interestingly, *pfcrt*-modified lines show only a very small degree of cross-resistance with ADQ, consistent with earlier reports (Geary and Jensen, 1983; Olliaro *et al.*, 1996). These findings are quite promising in view of the fact that ADQ is inexpensive and is currently under serious consideration as a possible replacement for CQ as a component of combination therapy in sub-Saharan Africa (Adjuik *et al.*, 2002; Sowunmi *et al.*, 2001). Nevertheless, the data also suggest that increased clinical use of ADQ might result in selection of resistance-conferring, novel *pfcrt* mutant alleles.

# 7. Final Thoughts and Unresolved Questions on the CQR Mechanism

As is evident from the above discussion, the current understanding of the mode of action of CQ, the basis for its parasitocidal effect and the genetic and physiological basis of CQR remains incomplete. Nonetheless, several findings stand out as being generally agreed upon: 1) CQ appears to bind to hematin, present either in the lipid-rich DV membrane or at the growing faces of the hemozoin crystal; 2) this binding interferes with the heme detoxification pathway and results in parasite death; 3) CQR appears to be related primarily to reduced CQ accumulation, probably as a result of reduced CQ access to its heme receptor; 4) CQR appears to involve biochemical changes that include alterations in DV physiology; 5) PfCRT, and to a lesser extent Pgh1, play a key role in dictating CQR; and 6) pH probably plays a significant role in affecting levels of CQ accumulation, perhaps with the contribution of by altered kinetics of hematin availability and hemozoin formation.

The hypothesis we currently favor is that mutant PfCRT acts principally by enabling more effective efflux of CQ out of the DV (Warhurst *et al.*, 2002), perhaps by stripping the drug out of the DV membrane where presumably it is bound to hematin. If this model is correct, then the effect of PfCRT mutant alleles on parasite susceptibility to other antimalarial heme-binding drugs would be expected to depend, at least in part, on their lipophilicity (i.e. the lipid affinity resulting in membrane insertion). This model would be consistent with the ADQ data, a compound that shares the same quinoline ring as CQ but that is more lipophilic and typically more effective than CQ against CQR parasites, including *pfcrt*-modified lines (Bjorkman and Phillips-Howard, 1990; Geary *et al.*, 1987; Geary and Jensen, 1983; Olliaro *et al.*, 1996; Scott *et al.*, 1987; Sidhu *et al.*, 2002). Kinetic studies have suggested that ADQ may be subject to the same resistance mechanism as CQ, but in an attenuated form, as certain CQR parasites displayed significantly reduced accumulation of ADQ (Bray *et al.*, 1996). Possibly, mutant PfCRT may be less competent at effluxing ADQ out of the DV by virtue of this drug being more tightly embedded in the DV membrane and less accessible to interactions with a PfCRT drug-transporting domain. PfCRT effects on quinine and mefloquine susceptibility may, in this regard, be dependent on the mutant protein's ability to interact with and efflux these drugs out of the DV, which might vary depending on the amino acids present at the polymorphic sites (Table 1). We note that studies in mammalian systems suggest that mutant P-glycoproteins may be able to physically strip drugs out of membranes and directly efflux them through the membrane (Borchers *et al.*, 2002; Romsicki and Sharom, 1999; Safa *et al.*, 1990). This potentially could explain why mutant *pfmdr1* can affect parasite susceptibility to several heme-binding antimalarials, including CQ.

Of particular interest in the question of *pfcrt* and its role in CQR is the loss of the K76 residue and its replacement with threonine, isoleucine or asparagine in CQR lines (Cooper *et al.*, 2002; Fidock *et al.*, 2000a). One explanation could be that replacement of the negatively charged lysine with an uncharged residue leads to an increased affinity of mutant PfCRT for diprotonated CQ inside the DV, leading to increased CQ efflux out of the DV. Reducing the CQ concentration would effectively allow heme detoxification (via biomineralization) to proceed normally, thus ensuring parasite growth in the presence of pharmacological doses of CQ. Other PfCRT mutations (notably the A220S mutation that is always present in CQR lines) may act either to augment or stabilize this protein:drug interaction or to compensate for a loss of o-attenuation of native function associated with the loss of the K76. CQR parasites can be killed with CQ concentrations that are an order of magnitude greater than those required to kill CQS parasites. This may be attributed either to CQ attaining concentrations that overwhelm an efflux mechanism (resulting, therefore, in the buildup of toxic heme or CQ:hematin complexes in the DV cytosol or DV membrane) or via an independent mode of CQ action operating at higher concentrations (for example, via CQ inhibition of parasite proteases or DNA intercalation).

What is the native function of PfCRT? One possibility is transport of protonated amino acids from the DV ((Warhurst, 2001a); Figure 3). Another is ion-dependent transport processes that influence DV pH (Cooper *et al.*, 2002; Fidock *et al.*, 2000b; Mehlotra *et al.*, 2001). In an extension of the pH studies, P. Roepe (Georgetown U., Washington DC) and colleagues now provide intriguing evidence that this function may involve passive movement of $Cl^-$. Using yeast codon-adjusted *pfcrt* sequences, this group expressed HB3 (CQS) or Dd2 (CQR) type PfCRT in plasma membranes of *Saccharomyces cerevisiae* and *Pichia pastoris* (Zhang *et al.*, 2002). Measurements of $H^+$ pumping by inside-out vesicle membranes, in the presence of symmetrical $Cl^-$, suggest that mutant PfCRT can generate a $Cl^-$ dependent, increased $\Delta pH$ across the membrane (as measured using the pH sensor AO). These data were consistent with earlier evidence of increased acidity resulting from mutant PfCRT (Dzekunov *et al.*, 2000), supporting the authors' hypothesis that increased acidity in the plasmodial DV can lead to CQR by preventing CQ: hematin interactions (Howard *et al.*, 2002; Ursos and Roepe, 2002). These authors noted an alternative explanation that mutated PfCRT may mediate $Cl^-$-dependent and VP-inhibitable export of AO, which shares some structural homology with the antimalarial quinacrine (Warhurst *et al.*, 2002; Zhang *et al.*, 2002; see section 5.2). Further studies in this area are clearly needed.

The question of PfCRT's function is one of several key questions, the answers to which will be central in furthering our understanding of CQR:

- Are *pfcrt* point mutations necessary and sufficient to confer VP-reversible CQR *in vitro* in most *P. falciparum* lines or in only a limited subset? Are these mutations sufficient to produce clinical treatment failure in individuals with minimal immunity?
- Do *pfmdr1* point mutations consistently modulate the degree of CQR in lines already carrying mutant *pfcrt*? What other genes can affect CQR?
- What is the physiological reason behind the observed association between *pfcrt* and *pfmdr1* in some CQR lines? Do PfCRT and Pgh1 physically interact and perhaps transport these drugs, or do they work to alter parasite antimalarial susceptibility by generating physiological alterations that change drug access to the heme receptor? What are their physiological functions and natural substrates?
- What other genetic and physiological changes occur in the parasite to compensate for altered function of PfCRT and Pgh1?
- What is the primary mechanism of CQR? Is CQR primarily a result of increased or decreased efflux of CQ, and/or changes in the kinetics of heme: membrane association and hematin or hemozoin formation? What is the role of pH? How does the CQR mechanism affect susceptibility to the other quinoline drugs quinine and mefloquine and the endoperoxide artemisinin?

# 8. Concluding Remarks

The treatment of malaria is reaching a crisis point as CQR continues its inexorable spread across endemic regions of the world. Many countries lack the economic resources and health care infrastructures needed to switch to more expensive combination therapies. The alternative being widely used in Africa, SP, has already encountered resistance and cannot be expected to last much longer. Inhibition of heme detoxification, the major mode of CQ action, is a validated mode of antimalarial action that we believe can still be of tremendous utility for existing antimalarials such as ADQ or for future antimalarials. Continued scientific probing into the molecular and cellular basis of CQR and its impact on parasite susceptibility to other heme-binding antimalarials promises to improve both 1) current means of assessing the threat of CQR in malaria-endemic areas and thus allow appropriate responses at the public health level; and 2) our ability to develop and employ alternative drugs effective against CQR strains. More efforts are urgently required in this area, in order to improve the world's means to combat this disease that tragically afflicts so many of the world's poorest nations.

# 9. Acknowledgments

The authors would like to thank Rebecca Muhle (Einstein), Scott Bohle (McGill U., Montreal), Paul Roepe (Georgetown U., Washington DC), Thierry Fandeur (Institut Pasteur, Cambodia), Frederic Ariey and Milijaona Randrianarivelojosia (Institut Pasteur, Madagascar), David Warhurst, Geoffrey Targett and Colin Sutherland (London School of Tropical Medicine and Hygiene, Pat Bray and Steve Ward (Liverpool School of Tropical Medicine), Mariano Zalis (U. Federal, Rio de Janeiro),  Chris Plowe (U. Maryland) and Thomas Wellems (NIAID/NIH, Bethesda, MD) for their helpful input during the preparation of this manuscript.

# 10. References

Adagu, I.S., and Warhust, D.C. 2001. *Plasmodium falciparum*: linkage disequilibrium between loci in chromosomes 7 and 5 and chloroquine selective pressure in Northern Nigeria. Parasitology. 123: 219-224.

Adjuik, M., Agnamey, P., Babiker, A., Borrmann, S., Brasseur, P., Cisse, M., Cobelens, F., Diallo, S., Faucher, J.F., Garner, P., Gikunda, S., Kremsner, P.G., Krishna, S., Lell, B., Loolpapit, M., Matsiegui, P.B., Missinou, M.A., Mwanza, J., Ntoumi, F., Olliaro, P., Osimbo, P., Rezbach, P., Some, E., and Taylor, W.R. 2002. Amodiaquine-artesunate versus amodiaquine for uncomplicated *Plasmodium falciparum* malaria in African children: a randomised, multicentre trial. Lancet. 359: 1365-1372.

Aikawa, M. 1972. High-resolution autoradiography of malarial prasites treated with $^3$H-chloroquine. Am. J. Pathol. 67: 277-284.

Alecrim, M.G., Alecrim, W., and Macedo, V. 1999. *Plasmodium vivax* resistance to chloroquine (R2) and mefloquine (R3) in Brazilian Amazon region. Rev. Soc. Bras. Med. Trop. 32: 67-68.

Ariey, F., Randrianarivelojosia, M., Duchemin, J.B., Rakotondramarina, D., Ouledi, A., Robert, V., Jambou, R., Jahevitra, M., Andrianantenaina, H., Raharimalala, L., and Mauclere, P. 2002. Mapping of a *Plasmodium falciparum* pfcrt K76T mutation: a useful strategy for controlling chloroquine resistance in Madagascar. J. Infect. Dis. 185: 710-712.

Ariey F, Duchemin JB, Robert V. 2003. Metapopulation concepts applied to falciparum malaria and their impacts on the emergence and spread of chloroquine resistance. Infect. Genet. Evol. 2(3):185-192.

Atamna, H., and Ginsburg, H. 1993. Origin of reactive oxygen species in erythrocytes infected with *Plasmodium falciparum*. Mol. Biochem. Parasitol. 61: 231-241.

Awad-el-Kariem, F.M., Milles, M.A., and Warhust, D.C. 1992. Chloroquine-resistant *Plasmodium falciparum* isolates from the Sudan lack two mutations in the *pfmdr1* gene thought to be associated with chloroquine resistance. Trans. R. Soc. Trop. Med. Hyg. 86: 587-589.

Babiker, H.A., Pringle, S.J., Abdel-Muhsin, A., Mackinnon, M., Hunt, P., and Walliker, D. 2001. High-level chloroquine resistance in Sudanese isolates of *Plasmodium falciparum* is associated with mutations in the chloroquine resistance transporter gene pfcrt and the multidrug resistance gene *pfmdr1*. J. Infect. Dis. 183: 1535-1538.

Baird, J.K., Basri, H., Purnomo, Bangs, M.J., Subianto, B., Patchen, L.C., and Hoffman, S.L. 1991. Resistance to chloroquine by *Plasmodium vivax* in Irian Jaya, Indonesia. Am. J. Trop. Med. Hyg. 44: 547-552.

Banerjee, R., and Goldberg, D.E. 2001. The *Plasmodium* food vacuole. In: Antimalarial Chemotherapy. P.J. Rosenthal, ed. Humana Press, Totowa, NJ. p. 43-63.

Banerjee, R., Liu, J., Beatty, W., Pelosof, L., Klemba, M., and Goldberg, D.E. 2002. Four plasmepsins are active in the *Plasmodium falciparum* food vacuole, including a protease with an active-site histidine. Proc. Natl. Acad. Sci. USA. 99: 990-995.

Barnes, D.A., Foote, S.J., Galatis, D., Kemp, D.J., and Cowman, A.F. 1992. Selection for high-level chloroquine resistance results in deamplification of the *pfmdr1* gene and increased sensitivity to mefloquine in *Plasmodium falciparum*. EMBO J. 11: 3067-3075.

Basco, L.K. 2002. Molecular epidemiology of malaria in Cameroon. XIII. Analysis of *pfcrt* mutations and *in vitro* chloroquine resistance. Am. J. Trop. Med. Hyg. 67: 388-391.

Basco, L.K., Le Bras, J., Rhoades, Z., and Wilson, C.M. 1995. Analysis of *pfmdr1* and drug susceptibility in fresh isolates of *Plasmodium falciparum* from subsaharan Africa. Mol. Biochem. Parasitol. 74: 157-166.

Basco, L.K., Ndounga, M., Ngane, V.F., and Soula, G. 2002. Molecular epidemiology of malaria in Cameroon. XIV. *Plasmodium falciparum* chloroquine resistance transporter (PFCRT) gene sequences of isolates before and after chloroquine treatment. Am. J. Trop. Med. Hyg. 67: 392-395.

Basco, L.K., and Ringwald, P. 1998. Molecular epidemiology of malaria in Yaounde, Cameroon. III. Analysis of chloroquine resistance and point mutations in the multidrug resistance 1 (*pfmdr 1*) gene of *Plasmodium falciparum*. Am. J. Trop. Med. Hyg. 59: 577-581.

Basco, L.K., and Ringwald, P. 2001. Analysis of the key *pfcrt* point mutation and *in vitro* and *in vivo* response to chloroquine in Yaounde, Cameroon. J. Infect. Dis. 183: 1828-1831.

Bayoumi, R.A., Creasey, A.M., Babiker, H.A., Carlton, J.M., Sultan, A.A., Satti, G., Sohal, A.K., Walliker, D., Jensen, J.B., and Arnot, D.E. 1993. Drug response and genetic characterization of *Plasmodium falciparum* clones recently isolated from a Sudanese village. Trans. R. Soc. Trop. Med. Hyg. 87: 454-458.

Becuwe, P., Gratepanche, S., Fourmaux, M.N., Van Beeumen, J., Samyn, B., Mercereau-Puijalon, O., Touzel, J.P., Slomianny, C., Camus, D., and Dive, D. 1996. Characterization of iron-dependent endogenous superoxide dismutase of *Plasmodium falciparum*. Mol. Biochem. Parasitol. 76: 125-134.

Bhatia, A., and Charet, P. 1984. Action of chloroquine on glutathione metabolism in erythroctyes parasitized by *Plasmodium berghei*. Ann. Parasitol. Hum. Comp. 59: 317-320.

Bhattacharya, P.R., and Pillai, C.R. 1999. Strong association, but incomplete correlation, between chloroquine resistance and allelic variation in the *pfmdr-1* gene of *Plasmodium falciparum* isolates from India. Ann. Trop. Med. Parasitol. 93: 679-684.

Bjorkman, A., and Phillips-Howard, P.A. 1990. The epidemiology of drug-resistant malaria. Trans. R. Soc. Trop. Med. Hyg. 84: 177-180.

Bohle, D.S., Dinnebier, R.E., Madsen, S.K., and Stephens, P.W. 1997. Characterization of the products of the heme detoxification pathway in malarial late trophozoites by X-ray diffraction. J. Biol. Chem. 272: 713-716.

Borchers, C., Boer, R., Klemm, K., Figala, V., Denzinger, T., Ulrich, W.R., Haas, S., Ise, W., Gekeler, V., and Przybylski, M. 2002. Characterization of the dexniguldipine binding site in the multidrug resistance-related transport protein P-glycoprotein by photoaffinity labeling and mass spectrometry. Mol. Pharmacol. 61: 1366-1376.

Bray, P.G., Boulter, M.K., Ritchie, G.Y., Howells, R.E., and Ward, S.A. 1994. Relationship of global chloroquine transport and reversal of resistance in *Plasmodium falciparum*. Mol. Biochem. Parasitol. 63: 87-94.

Bray, P.G., Hawley, S.R., and Ward, S.A. 1996. 4-Aminoquinoline resistance of *Plasmodium falciparum*: insights from the study of amodiaquine uptake. Mol. Pharmacol. 50: 1551-1558.

Bray, P.G., Howells, R.E., Ritchie, G.Y., and Ward, S.A. 1992a. Rapid chloroquine efflux phenotype in both chloroquine-sensitive and chloroquine-resistant *Plasmodium falciparum*. A correlation of chloroquine sensitivity with energy-dependent drug accumulation. Biochem. Pharmacol. 44: 1317-1324.

Bray, P.G., Howells, R.E., and Ward, S.A. 1992b. Vacuolar acidification and chloroquine sensitivity in *Plasmodium falciparum*. Biochem. Pharmacol. 43: 1219-1227.

Bray, P.G., Janneh, O., Raynes, K.J., Mungthin, M., Ginsburg, H., and Ward, S.A. 1999a. Cellular uptake of chloroquine is dependent on binding to ferriprotoporphyrin IX and is independent of NHE activity in *Plasmodium falciparum*. J. Cell Biol. 145: 363-376.

Bray, P.G., Mungthin, M., Ridley, R.G., and Ward, S.A. 1998. Access to hematin: the basis of chloroquine resistance. Mol. Pharmacol. 54: 170-179.

Bray, P.G., Saliba, K.J., Davies, J.D., Spiller, D.G., White, M.R.H., Kirk, K., and Ward, S.A. 2001. Distribution of acridine orange fluorescence in *Plasmodium falciparum*-infected erythrocytes and its implications for the evaluation of digestive vacuole pH. Mol. Biochem. Parasitol. 119: 301-304.

Bray, P.G., and Ward, S.A. 1998. A comparison of the phenomenology and genetics of multidrug resistance in cancer cells and quinoline resistance in *Plasmodium falciparum*. Pharmacol. Ther. 77: 1-28.

Bray, P.G., Ward, S.A., and Ginsburg, H. 1999b. Na+/H+ antiporter, chloroquine uptake and drug resistance: inconsistencies in a newly proposed model. Parasitol. Today. 15: 360-363.

Burkot, T.R., Williams, J.L., and Schneider, I. 1984. Infectivity to mosquitoes of *Plasmodium falciparum* clones grown *in vitro* from the same isolate. Trans. R. Soc. Trop. Med. Hyg. 78: 339-341.

Carlin, J.M., Vande Waa, J.A., Jensen, J.B., and Akood, M.A. 1984. African serum interference in the determination of chloroquine sensitivity in *Plasmodium falciparum*. Z Parasitenkd. 70: 589-597.

Carlton, J.M., Hayton, K., Cravo, P.V., and Walliker, D. 2001a. Of mice and malaria mutants: unravelling the genetics of drug resistance using rodent malaria models. Trends Parasitol. 17: 236-242.

Carlton, J.M.-R., Fidock, D.A., Djimde, A., Plowe, C.V., and Wellems, T.E. 2001b. Conservation of a novel vacuolar transporter in Plasmodium species and its role in chloroquine resistance in falciparum but not vivax malaria. Curr. Opinion Microbiol. 4: 415-420.

Chaiyaroj, S.C., Buranakiti, A., Angkasekwinai, P., Looressuwan, S., and Cowman, A.F. 1999. Analysis of mefloquine resistance and amplification of *pfmdr1* in multidrug-resistant *Plasmodium falciparum* isolates from Thailand. Am. J. Trop. Med. Hyg. 61: 780-783.

Chen, N., Russell, B., Fowler, E., Peters, J., and Cheng, Q. 2002. Levels of chloroquine resistance in *Plasmodium falciparum* are determined by loci other than *pfcrt* and *pfmdr1*. J. Infect. Dis. 185: 405-406.

Chen, N., Russell, B., Staley, J., Kotecka, B., Nasveld, P., and Cheng, Q. 2001. Sequence polymorphisms in *pfcrt* are strongly associated with chloroquine resistance in *Plasmodium falciparum*. J. Infect. Dis. 183: 1543-1545.

Chou, A.C., Chevli, R., and Fitch, C.D. 1980. Ferriprotoporphyrin IX fulfills the criteria for identification as the chloroquine receptor of malaria parasites. Biochemistry. 19: 1543-1549.

Clarebout, G., Slomianny, C., Delcourt, P., Leu, B., Masset, A., Camus, D., and Dive, D. 1998. Status of *Plasmodium falciparum* towards catalase. Br J Haematol. 103: 52-59.

Coatney, G.R. 1963. Pitfalls in a discovery: the chronicle of chloroquine. Am. J. Trop. Med. Hyg. 12: 121-128.

Contreras, C.E., Cortese, J.F., Caraballo, A., and Plowe, C.V. 2002. Genetics of drug-resistant *Plasmodium falciparum* malaria in the Venezuelan state of Bolivar. Am. J. Trop. Med. Hyg. 67: 400-405.

Cooper, R.A., Ferdig, M.T., Su, X.Z., Ursos, L.M., Mu, J., Nomura, T., Fujioka, H., Fidock, D.A., Roepe, P.D., and Wellems, T.E. 2002. Alternative mutations at position 76 of the vacuolar transmembrane protein PfCRT are associated with chloroquine resistance and unique stereospecific quinine and quinidine responses in *Plasmodium falciparum*. Mol. Pharmacol. 61: 35-42.

Cowman, A.F. 1995. Mechanisms of drug resistance in malaria. Aust. N. Z. J. Med. 25: 837-844.

Cowman, A.F., Galatis, D., and Thompson, J.K. 1994. Selection for mefloquine resistance in *Plasmodium falciparum* is linked to amplification of the *pfmdr1* gene and cross-resistance to halofantrine and quinine. Proc. Natl. Acad. Sci. USA. 91: 1143-1147.

Cowman, A.F., Karcz, S., Galatis, D., and Culvenor, J.G. 1991. A P-glycoprotein homologue of *Plasmodium falciparum* is localized on the digestive vacuole. J. Cell Biol. 113: 1033-1042.

Cowman, A.F., and Karcz, S.R. 1991. The *pfmdr* gene homologues of *Plasmodium falciparum*. Acta Leiden. 60: 121-129.

Crabb, B.S., and Cowman, A.F. 1996. Characterization of promoters and stable transfection by homologous and nonhomologous recombination in *Plasmodium falciparum*. Proc. Natl. Acad. Sci. USA. 93: 7289-7294.

De, D., Krogstad, F.M., Byers, L.D., and Krogstad, D.J. 1998. Structure-activity relationships for antiplasmodial activity among 7- substituted 4-aminoquinolines. J. Med. Chem. 41: 4918-4926.

De, D., Krogstad, F.M., Cogswell, F.B., and Krogstad, D.J. 1996. Aminoquinolines that circumvent resistance in *Plasmodium falciparum in vitro*. Am. J. Trop. Med. Hyg. 55: 579-583.

De Duve, C., De Barsy, T., Poole, B., Trouet, A., Tulkens, P., and Van Hoof, F. 1974. Lysosomotrophic agents. Biochem. Pharmacol. 23: 2495-2531.

Desjardins, R.E., Canfield, C.J., Haynes, J.D., and Chulay, J.D. 1979. Quantitative assessment of antimalarial activity *in vitro* by a semiautomated microdilution technique. Antimicrob. Agents Chemother. 16: 710-718.

Djimdé, A., Doumbo, M.D., Cortese, J.F., Kayentao, K., Doumbo, S., Diourté, Y., Coulibaly, D., Dicko, A., Su, X.-z., Nomura, T., Fidock, D.A., Wellems, T.E., and Plowe, C.V. 2001. A molecular marker for chloroquine resistant *falciparum* malaria. New Engl. J. Med. 344: 257-263.

Dorn, A., Stoffel, R., Matile, H., Bubendorf, A., and Ridley, R.G. 1995. Malarial haemozoin/beta-haematin supports haem polymerization in the absence of protein. Nature. 374: 269-271.

Dorn, A., Vippagunta, S.R., Matile, H., Jaquet, C., Vennerstrom, J.L., and Ridley, R.G. 1998. An assessment of drug-haematin binding as a mechanism for inhibition of haematin polymerisation by quinoline antimalarials. Biochem. Pharmacol. 55: 727-736.

Dorsey, G., Fidock, D.A., Wellems, T.E., and Rosenthal, P.J. 2001a. Mechanisms of quinoline resistance. In: Antimalarial Chemotherapy. P.J. Rosenthal, ed. Humana Press, Totowa, N.J. p. 153-172.

Dorsey, G., Kamya, M.R., Singh, A., and Rosenthal, P.J. 2001b. Polymorphisms in the *Plasmodium falciparum pfcrt* and *pfmdr-1* genes and clinical response to chloroquine in Kampala, Uganda. J. Infect. Dis. 183: 1417-1420.

Dua, V.K., Kar, P.K., and Sharma, V.P. 1996. Chloroquine resistant *Plasmodium vivax* malaria in India. Trop. Med. Int. Health. 1: 816-819.

Dubois, V.L., Platel, D.F., Pauly, G., and Tribouley-Duret, J. 1995. *Plasmodium berghei*: implication of intracellular glutathione and its related enzyme in chloroquine resistance *in vivo*. Exp Parasitol. 81: 117-124.

Duraisingh, M.T., Drakeley, C.J., Muller, O., Bailey, R., Snounou, G., Targett, G.A., Greenwood, B.M., and Warhurst, D.C. 1997. Evidence for selection for the tyrosine-86 allele of the *pfmdr 1* gene of *Plasmodium falciparum* by chloroquine and amodiaquine. Parasitology. 114: 205-211.

Duraisingh, M.T., Jones, P., Sambou, I., von Seidlein, L., Pinder, M., and Warhurst, D.C. 2000a. The tyrosine-86 allele of the *pfmdr1* gene of *Plasmodium falciparum* is associated with increased sensitivity to the anti-malarials mefloquine and artemisinin. Mol. Biochem. Parasitol. 108: 13-23.

Duraisingh, M.T., Roper, C., Walliker, D., and Warhurst, D.C. 2000b. Increased sensitivity to the antimalarials mefloquine and artemisinin is conferred by mutations in the *pfmdr1* gene of *Plasmodium falciparum*. Mol. Microbiol. 36: 955-961.

Durand, R., Jafari, S., Vauzelle, J., Delabre, J., Jesic, Z., and Le Bras, J. 2001. Analysis of *pfcrt* point mutations and chloroquine susceptibility in isolates of *Plasmodium falciparum*. Mol. Biochem. Parasitol. 114: 95-102.

Dzekunov, S.M., Ursos, L.M.B., and Roepe, P.D. 2000. Digestive vacuolar pH of intact intraerythrocytic *P. falciparum* either sensitive or resistant to chloroquine. Mol. Biochem. Parasitol. 110: 107-124.

Egan, T.J., Combrinck, J.M., Egan, J., Hearne, G.R., Marques, H.M., Ntenteni, S., Sewell, B.T., Smith, P.J., Taylor, D., Van Schalkwyk, D.A., and Walden, J.C. 2002. Fate of haem iron in the malaria parasite *Plasmodium falciparum*. Biochem. J. 365: 343-347.

Eggleson, K.K., Duffin, K.L., and Goldberg, D.E. 1999. Identification and characterization of falcilysin, a metallopeptidase involved in hemoglobin catabolism within the malaria parasite *Plasmodium falciparum*. J. Biol. Chem. 274: 32411-32417.

Ekong, R.M., Robson, K.J., Baker, D.A., and Warhurst, D.C. 1993. Transcripts of the multidrug resistance genes in chloroquine-sensitive and chloroquine-resistant *Plasmodium falciparum*. Parasitology. 106: 107-115.

Ferrari, V., and Cutler, D.J. 1991. Simulation of kinetic data on the influx and efflux of chloroquine by erythrocytes infected with *Plasmodium falciparum*. Evidence for a drug-importer in chloroquine-sensitive strains. Biochem. Pharmacol. 42 Suppl: S167-179.

Fidock, D.A., Nomura, T., Cooper, R.A., Su, X.-Z., Talley, A.K., and Wellems, T.E. 2000a. Allelic modifications of the *cg2* and *cg1* genes do not alter the chloroquine response of drug-resistant *Plasmodium falciparum*. Mol. Biochem. Parasitol. 110: 1-10.

Fidock, D.A., Nomura, T., Talley, A.K., A., C.R., Dzekunov, S.M., Ferdig, M.T., Ursos, L.M., Sidhu, A.B.S., Naude, B., Deitsch, K., Su, X.-z., Wootton, J.C., Roepe, P.D., and Wellems, T.E. 2000b. Mutations in the *P. falciparum* digestive vacuole transmembrane protein PfCRT and evidence for their role in chloroquine resistance. Mol. Cell. 6: 861-871.

Fitch, C.D. 1970. *Plasmodium falciparum* in owl monkeys: drug resistance and chloroquine binding capacity. Science. 169: 289-290.

Fitch, C.D. 1998. Involvement of heme in the antimalarial action of chloroquine. Trans. Am. Clin. Climatol. Assoc. 109: 97-105.

Fitch, C.D., Chevli, R., Banyal, H.S., Phillips, G., Pfaller, M.A., and Krogstad, D.J. 1982. Lysis of *Plasmodium falciparum* by ferriprotoporphyrin IX and a chloroquine-ferriprotoporphyrin IX complex. Antimicrob. Agents Chemother. 21: 819-822.

Fitch, C.D., Yunis, N.G., Chevli, R., and Gonzalez, Y. 1974. High-affinity accumulation of chloroquine by mouse erythrocytes infected with *Plasmodium berghei*. J. Clin. Invest. 54: 24-33.

Flueck, T.P., Jelinek, T., Kilian, A.H., Adagu, I.S., Kabagambe, G., Sonnenburg, F., and Warhurst, D.C. 2000. Correlation of *in vivo*-resistance to chloroquine and allelic polymorphisms in *Plasmodium falciparum* isolates from Uganda. Trop. Med. Int. Health. 5: 174-178.

Foley, M., and Tilley, L. 1998. Quinoline antimalarials: mechanisms of action and resistance and prospects for new agents. Pharmacol Ther. 79: 55-87.

Foote, S.J., Kyle, D.E., Martin, R.K., Oduola, A.M., Forsyth, K., Kemp, D.J., and Cowman, A.F. 1990. Several alleles of the multidrug-resistance gene are closely linked to chloroquine resistance in *Plasmodium falciparum*. Nature. 345: 255-258.

Foote, S.J., Thompson, J.K., Cowman, A.F., and Kemp, D.J. 1989. Amplification of the multidrug resistance gene in some chloroquine-resistant isolates of *P. falciparum*. Cell. 57: 921-930.

Francis, S.E., Sullivan, D.J., Jr., and Goldberg, D.E. 1997. Hemoglobin metabolism in the malaria parasite *Plasmodium falciparum*. Annu. Rev. Microbiol. 51: 97-123.

Gamain, B., Langsley, G., Fourmaux, M.N., Touzel, J.P., Camus, D., Dive, D., and Slomianny, C. 1996. Molecular characterization of the glutathione peroxidase gene of the human malaria parasite *Plasmodium falciparum*. Mol. Biochem. Parasitol. 78: 237-248.

Garg, M., Gopinathan, N., Bodhe, P., and Kshirsagar, N.A. 1995. *Vivax* malaria resistant to chloroquine: case reports from Bombay. Trans. R. Soc. Trop. Med. Hyg. 89: 656-657.

Geary, T.G., Divo, A.A., and Jensen, J.B. 1987. Activity of quinoline-containing antimalarials against chloroquine- sensitive and -resistant strains of *Plasmodium falciparum in vitro*. Trans. R. Soc. Trop. Med. Hyg. 81: 499-503.

Geary, T.G., Divo, A.D., Jensen, J.B., Zangwill, M., and Ginsburg, H. 1990. Kinetic modelling of the response of *Plasmodium falciparum* to chloroquine and its experimental testing *in vitro*. Implications for mechanism of action of and resistance to the drug. Biochem. Pharmacol. 40: 685-691.

Geary, T.G., and Jensen, J.B. 1983. Lack of cross-resistance to 4-aminoquinolines in chloroquine-resistant *Plasmodium falciparum in vitro*. J. Parasitol. 69: 97-105.

Ginsburg, H. 1990. Some reflections concerning host erythrocyte-malarial parasite interrelationships. Blood Cells. 16: 225-235.

Ginsburg, H., Famin, O., Zhang, J., and Krugliak, M. 1998. Inhibition of glutathione-dependent degradation of heme by chloroquine and amodiaquine as a possible basis for their antimalarial mode of action. Biochem. Pharmacol. 56: 1305-1313.

Ginsburg, H., and Krugliak, M. 1992. Quinoline-containing antimalarials--mode of action, drug resistance and its reversal. An update with unresolved puzzles. Biochem. Pharmacol. 43: 63-70.

Ginsburg, H., Nissani, E., and Krugliak, M. 1989. Alkalinization of the food vacuole of malaria parasites by quinoline drugs and alkylamines is not correlated with their antimalarial activity. Biochem. Pharmacol. 38: 2645-2654.

Ginsburg, H., and Stein, W.D. 1991. Kinetic modelling of chloroquine uptake by malaria-infected erythrocytes. Assessment of the factors that may determine drug resistance. Biochem. Pharmacol. 41: 1463-1470.

Ginsburg, H., Ward, S.A., and Bray, P.G. 1999. An integrated model of chloroquine action. Parasitol. Today. 15: 357-360.

Gomez-Saladin, E., Fryauff, D.J., Taylor, W.R., Laksana, B.S., Susanti, A.I., Purnomo, Subianto, B., and Richie, T.L. 1999. *Plasmodium falciparum mdr1* mutations and *in vivo* chloroquine resistance in Indonesia. Am. J. Trop. Med. Hyg. 61: 240-244.

Grobusch, M.P., Adagu, I.S., Kremsner, P.G., and Warhurst, D.C. 1998. *Plasmodium falciparum*: *in vitro* chloroquine susceptibility and allele-specific PCR detection of *Pfmdr1* Asn86Tyr polymorphism in Lambarene, Gabon. Parasitology. 116: 211-217.

Guan, J., Kyle, D.E., Gerena, L., Zhang, Q., Milhous, W.K., and Lin, A.J. 2002. Design, synthesis, and evaluation of new chemosensitizers in multi-drug-resistant *Plasmodium falciparum*. J. Med. Chem. 45: 2741-2748.

Har-El, R., Marva, E., Chevion, M., and Golenser, J. 1993. Is hemin responsible for the susceptibility of *Plasmodia* to oxidant stress? Free Radical Res. Commun. 18: 279-290.

Haruki, K., Bray, P.G., Ward, S.A., Hommel, M., and Ritchie, G.Y. 1994. Chloroquine resistance of *Plasmodium falciparum*: further evidence for a lack of association with mutations of the *pfmdr1* gene. Trans. R. Soc. Trop. Med. Hyg. 88: 694.

Hastings, I.M., Bray, P.G., and Ward, S.A. 2002. Parasitology. A requiem for chloroquine. Science. 298: 74-75.

Hawley, S.R., Bray, P.G., Park, B.K., and Ward, S.A. 1996. Amodiaquine accumulation in *Plasmodium falciparum* as a possible explanation for its superior antimalarial activity over chloroquine. Mol. Biochem. Parasitol. 80: 15-25.

Hempelmann, E., and Egan, T.J. 2002. Pigment biocrystallization in *Plasmodium falciparum*. Trends Parasitol. 18: 11.

Homewood, C.A., Warhurst, D.C., Peters, W., and Baggaley, V.C. 1972. Lysosomes, pH and the anti-malarial action of chloroquine. Nature. 235: 50-52.

Howard, E.M., Zhang, H., and Roepe, P.D. 2002. A novel transporter, *pfcrt*, confers antimalarial drug resistance. J. Membr. Biol. 190: 1-8.

Jacobs, G.H., Oduola, A.M., Kyle, D.E., Milhous, W.K., Marin, S.K., and Aikawa, M. 1988. Ultrasturctural study of the effects of chloroquine and verapmil on *Plasmodium falciparum*. Am. J. Trop. Med. Hyg. 39: 15-20.

Karcz, S.R., Galatis, D., and Cowman, A.F. 1993. Nucleotide binding properties of a P-glycoprotein homologue from *Plasmodium falciparum*. Mol. Biochem. Parasitol. 58: 269-276.

Krogstad, D.J., and De, D. 1998. Chloroquine: modes of action and resistance and the activity of chloroquine analogs. In: Malaria: parasite biology, pathogenesis and protection. I.W. Sherman, ed. ASM Press, Washington, DC. p. 331-340.

Krogstad, D.J., Gluzman, I.Y., Herwaldt, B.L., Schlesinger, P.H., and Wellems, T.E. 1992. Energy dependence of chloroquine accumulation and chloroquine efflux in *Plasmodium falciparum*. Biochem. Pharmacol. 43: 57-62.

Krogstad, D.J., Gluzman, I.Y., Kyle, D.E., Oduola, A.M., Martin, S.K., Milhous, W.K., and Schlesinger, P.H. 1987. Efflux of chloroquine from *Plasmodium falciparum*: mechanism of chloroquine resistance. Science. 238: 1283-1285.

Krogstad, D.J., and Schlesinger, P.H. 1987. The basis of antimalarial action: non-weak base effects of chloroquine on acid vesicle pH. Am. J. Trop. Med. Hyg. 36: 213-220.

Krogstad, D.J., Schlesinger, P.H., and Gluzman, I.Y. 1985. Antimalarials increase vesicle pH in *Plasmodium falciparum*. J. Cell. Biol. 101: 2302-2309.

Kublin, J.G., Cortese, J.F., Njunju, E.M., Mukadam, R.A., Wirima, J.J., Kazembe, P.N., Djimde, A.A., Kouriba, B., Taylor, T.E., and Plowe, C.V. 2003. Reemergence of chloroquine-sensitive *Plasmodium falciparum* malaria after cessation of chloroquine use in Malawi. J. Infect. Dis. 187(12):1870-1875.

Labbe, A.C., Bualombai, P., Pillai, D.R., Zhong, K.J., Vanisaveth, V., Hongvanthong, B., Looareesuwan, S., and Kain, K.C. 2001. Molecular markers for chloroquine-resistant *Plasmodium falciparum* malaria in Thailand and Laos. Ann. Trop. Med. Parasitol. 95: 781-788.

Le Bras, J., and Deloron, P. 1983. *In vitro* study of drug sensitivity of P*lasmodium falciparum*: evaluation of a new semi-micro test. Am. J. Trop. Med. Hyg. 32: 447-451.

Levins, R. 1969. Some demographic and genetic consequences of environmental heterogeneity for biological control. Bull. Entomol. Soc. Am. 15: 237-240.

Lew, V.L., and Hockaday, A.R. 1999. The effects of transport perturbations on the homeostasis of erythrocytes. Novartis Found Symp. 226: 37-50.

Lim, P., Chy, S., Ariey, F., Incardona, S., Chim, P., Sem, R., Denis, M.B., Hewitt, S., Hoyer, S., Socheat, D., Merecreau-Puijalon, O., and Fandeur, T. 2003. pfcrt polymorphism and chloroquine resistance in *Plasmodium falciparum* strains isolated in Cambodia. Antimicrob. Agents Chemother. 47: 87-94.

Loeb, R.F., Clarke, W.M., Coatney, G.R., Coggeshall, L.T., Dieuaide, F.R., Docher, A.R., Hakansson, E.G., Marshall, E.K., Marvel, S.C., McCoy, O.R., Sapero, J.J., H., S.W., Shannon, J.A., and Carden, G., A. 1946. Activity of a new antimalarial agent, chloroquine (SN 7618). J. Am. Med. Assoc. 130: 1069-1070.

Lopes, D., Rungsihirunrat, K., Nogueira, F., Seugorn, A., Gil, J., Do Rosario, V.E., and Cravo, P. 2002. Molecular characterisation of drug-resistant *Plasmodium falciparum* from Thailand. Malar J. 1: 12.

Loria, P., Miller, S., Foley, M., and Tilley, L. 1999. Inhibition of the peroxidative degradation of haem as the basis of action of chloroquine and other quinoline antimalarials. Biochem. J. 339: 363-370.

Macomber, P.B., O'Brien, R.L., and Hahn, F.E. 1966. Chloroquine: physiological basis of drug resistance in *Plasmodium berghei*. Science. 152: 1374-1375.

Maguire, J.D., Sumawinata, I.W., Masbar, S., Laksana, B., Prodjodipuro, P., Susanti, I., Sismadi, P., Mahmud, N., Bangs, M.J., and Baird, J.K. 2002. Chloroquine-resistant *Plasmodium malariae* in south Sumatra, Indonesia. Lancet. 360: 58-60.

Maguire, J.D., Susanti, A.I., Krisin, Sismadi, P., Fryauff, D.J., and Baird, J.K. 2001. The T76 mutation in the *pfcrt* gene of *Plasmodium falciparum* and clinical chloroquine resistance phenotypes in Papua, Indonesia. Ann. Trop. Med. Parasitol. 95: 559-572.

Mahoney, J.R., and Eaton, J.W. 1981. Chloroquine resistant malaria: association with enhanced plasmodial protease activity. Biochem. Biophys. Res. Commun. 100: 1266-1271.

Marlar, T., Myat, P.K., Aye, Y.S., Khiang, K.G., Ma, S., and Myint, O. 1995. Development of resistance to chlorquine by *Plasmodium vivax* in Myanmar. Trans. R. Soc. Trop. Med. Hyg. 89: 307-308.

Marsh, K. 1998. Malaria disaster in Africa. Lancet. 352: 924.

Martin, S.K., Oduola, A.M., and Milhous, W.K. 1987. Reversal of chloroquine resistance in *Plasmodium falciparum* by verapamil. Science. 235: 899-901.

Martiney, J.A., Ferrer, A.S., Cerami, A., Dzekunov, S., and Roepe, P. 1999. Chloroquine uptake, altered partitioning and the basis of drug resistance: evidence for chloride-dependent ionic regulation. In: Tranport and Trafficking in the Malaria-Infected Erythrocyte. Novartis Foundation Symposium 226, Wiley, Chichester. p. 265-280.

Mayor, A.G., Gomez-Olive, X., Aponte, J.J., Casimiro, S., Mabunda, S., Dgedge, M., Barreto, A., and Alonso, P.L. 2001. Prevalence of the K76T mutation in the putative *Plasmodium falciparum* chloroquine resistance transporter (*pfcrt*) gene and its relation to chloroquine resistance in Mozambique. J. Infect. Dis. 183: 1413-1416.

Mehlotra, R.K., Fujioka, H., Roepe, P.D., Janneh, O., Ursos, L.M., Jacobs-Lorena, V., McNamara, D.T., Bockarie, M.J., Kazura, J.W., Kyle, D.E., Fidock, D.A., and Zimmerman, P.A. 2001. Evolution of a unique *Plasmodium falciparum* chloroquine-resistance phenotype in association with *pfcrt* polymorphism in Papua New Guinea and South America. Proc. Natl. Acad. Sci. USA. 98: 12689-12694.

Mita, T., Kaneko, A., Lum, J.K., Bwijo, B., Takechi, M., Zungu, I.L., Tsukahara, T., Tanabe, K., Kobayakawa, T., and Bjorkman, A. 2003. Recovery of chloroquine sensitivity and low prevalence of the *Plasmodium falciparum* chloroquine resistance transporter gene mutation K76T following the discontinuance of chloroquine use in Malawi. Am, J, Trop, Med, Hyg. 2003 68(4): 413-415.

Moreau, S., Perly, B., and Biguet, J. 1982. [Interaction of chloroquine with ferriprotophorphyrin IX. Nuclear magnetic resonance study]. Biochimie. 64: 1015-1025.

Moreno, A., Brasseur, P., Cuzin-Ouattara, N., Blanc, C., and Druilhe, P. 2001. Evaluation under field conditions of the colourimetric DELI-microtest for the assessment of *Plasmodium falciparum* drug resistance. Trans. R. Soc. Trop. Med. Hyg. 95: 100-103.

Myat, P.K., Myint, O., Myint, L., Tahw, Z., Kyin, H.A., and Nwe, N.Y. 1993. Emergence of chloroquine-resistant *Plasmodium vivax* in Myanmar (Burma). Trans. R. Soc. Trop. Med. Hyg. 87: 687.

Newton, P., and White, N. 1999. Malaria: new developments in treatment and prevention. Annu. Rev. Med. 50: 179-192.

Nomura, T., Carlton, J.M.-R., Baird, J.K., Del Portillo, H.A., Fryauff, D.J., Rathore, D., Fidock, D.A., Su, X.-z., Collins, W.E., McCutchan, T.F., Wootton, J.C., and Wellems, T.E. 2001. Evidence for different mechanisms of chloroquine resistance in 2 *Plasmodium* species that cause human malaria. J. Infect. Dis. 183: 1653-1661.

O'Neill, P.M., Bray, P.G., Hawley, S.R., Ward, S.A., and Park, B.K. 1998. 4-Aminoquinolines--past, present, and future: a chemical perspective. Pharmacol. Ther. 77: 29-58.

Olliaro, P., Nevill, C., LeBras, J., Ringwald, P., Mussano, P., Garner, P., and Brasseur, P. 1996. Systematic review of amodiaquine treatment in uncomplicated malaria. Lancet. 348: 1196-1201.

Pagola, S., Stephens, P.W., Bohle, D.S., Kosar, A.D., and Madsen, S.K. 2000. The structure of malaria pigment beta-haematin. Nature. 404: 307-310.

Payne, D. 1987. Spread of chloroquine resistance in *Plasmodium falciparum*. Parasitol. Today. 3: 241-246.

Peel, S.A. 2001. The ABC transporter genes of *Plasmodium falciparum* and drug resistance. Drug Resist. Updat. 4: 66-74.

Peel, S.A., Bright, P., Yount, B., Handy, J., and Baric, R.S. 1994. A strong association between mefloquine and halofantrine resistance and amplification, overexpression, and mutation in the P-glycoprotein gene homolog (*pfmdr*) of *Plasmodium falciparum in vitro*. Am. J. Trop. Med. Hyg. 51: 648-658.

Peters, W. 1970. Chemotherapy and Drug Resistance in Malaria. Academic Press, London.

Peters, W. 1987. Resistance in human malaria IV: 4-aminoquinolines and multiple resistance. In: Chemotherapy and Drug Resistance in Malaria. Vol. 2. Academic Press, London. p. 659-725.

Peters, W. 1998. Drug resistance in malaria parasites of animals and man. Adv. Parasitol. 41: 1-62.

Phillips, E.J., Keystone, J.S., and Kain, K.C. 1996. Failure of combined chloroquine and high-dose primaquine therapy for *Plasmodium vivax* malaria acquired in Guyana, South America. Clin. Infect. Dis. 23: 1171-1173.

Pillai, D.R., Labbe, A.C., Vanisaveth, V., Hongvangthong, B., Pomphida, S., Inkathone, S., Zhong, K., and Kain, K.C. 2001. *Plasmodium falciparum* malaria in Laos: chloroquine treatment outcome and predictive value of molecular markers. J. Infect. Dis. 183: 789-795.

Plowe, C.V., Doumbo, O.K., Djimde, A., Kayentao, K., Diourte, Y., Doumbo, S.N., Coulibaly, D., Thera, M., Wellems, T.E., and Diallo, D.A. 2001. Chloroquine treatment of uncomplicated *Plasmodium falciparum* malaria in Mali: parasitologic resistance versus therapeutic efficacy. Am. J. Trop. Med. Hyg. 64: 242-246.

Povoa, M.M., Adagu, I.S., Oliveira, S.G., Machado, R.L., Miles, M.A., and Warhurst, D.C. 1998. Pfmdr1 Asn1042Asp and Asp1246Tyr polymorphisms, thought to be associated with chloroquine resistance, are present in chloroquine- resistant and -sensitive Brazilian field isolates of *Plasmodium falciparum*. Exp. Parasitol. 88: 64-68.

Price, R., Robinson, G., Brockman, A., Cowman, A., and Krishna, S. 1997. Assessment of *pfmdr 1* gene copy number by tandem competitive polymerase chain reaction. Mol. Biochem. Parasitol. 85: 161-169.

Price, R.N., Cassar, C., Brockman, A., Duraisingh, M., van Vugt, M., White, N.J., Nosten, F., and Krishna, S. 1999. The *pfmdr1* gene is associated with a multidrug-resistant phenotype in *Plasmodium falciparum* from the western border of Thailand. Antimicrob. Agents Chemother. 43: 2943-2949.

Rathod, P.K., McErlean, T., and Lee, P.-C. 1997. Variations in frequencies of drug resistance in *Plasmodium falciparum*. Proc. Natl. Acad. Sci. USA. 94: 9389-9393.

Reed, M.B., Saliba, K.J., Caruana, S.R., Kirk, K., and Cowman, A.F. 2000. Pgh1 modulates sensitivity and resistance to multiple antimalarials in *Plasmodium falciparum*. Nature. 403: 906-909.

Ridley, R.G. 1998. Malaria: dissecting chloroquine resistance. Curr. Biol. 8: R346-349.

Ridley, R.G. 2002. Medical need, scientific opportunity and the drive for antimalarial drugs. Nature. 415: 686-693.

Ridley, R.G., Hofheinz, W., Matile, H., Jaquet, C., Dorn, A., Masciadri, R., Jolidon, S., Richter, W.F., Guenzi, A., Girometta, M.A., Urwyler, H., Huber, W., Thaithong, S., and Peters, W. 1996. 4-aminoquinoline analogs of chloroquine with shortened side chains retain activity against chloroquine-resistant *Plasmodium falciparum*. Antimicrob. Agents Chemother. 40: 1846-1854.

Rieckmann, K.H., Campbell, G.H., Sax, L.J., and Mrema, J.E. 1978. Drug sensitivity of *Plasmodium falciparum*. An *in-vitro* microtechnique. Lancet. 1: 22-23.

Rieckmann, K.H., Davis, D.R., and Hutton, D.C. 1989. *Plasmodium vivax* resistance to chloroquine? Lancet. 2: 1183-1184.

Ritchie, G.Y., Mungthin, M., Green, J.E., Bray, P.G., Hawley, S.R., and Ward, S.A. 1996. *In vitro* selection of halofantrine resistance in *Plasmodium falciparum* is not associated with increased expression of Pgh1. Mol. Biochem. Parasitol. 83: 35-46.

Romsicki, Y., and Sharom, F.J. 1999. The membrane lipid environment modulates drug interactions with the P- glycoprotein multidrug transporter. Biochemistry. 38: 6887-6896.

Rosenthal, P.J., and Meshnick, S.R. 1998. Hemoglobin processing and the metabolism of amino acids, heme and iron. In: Malaria: Parasite Biology, Pathogenesis and Protection. I.W. Sherman, ed. ASM Press, Washington, D. C. p. 145-158.

Safa, A.R., Stern, R.K., Choi, K., Agresti, M., Tamai, I., Mehta, N.D., and Roninson, I.B. 1990. Molecular basis of preferential resistance to colchicine in multidrug-resistant human cells conferred by Gly-185----Val-185 substitution in P-glycoprotein. Proc. Natl. Acad. Sci. USA. 87: 7225-7229.

Sahini, V.E., Dumitrescu, M., Volanschi, E., Birla, L., and Diaconu, C. 1996. Spectral and interferometrical study of the interaction of haemin with glutathione. Biophys Chem. 58: 245-253.

Sanchez, C.P., Wunsch, S., and Lanzer, M. 1997. Identification of a chloroquine importer in *Plasmodium falciparum*. Differences in import kinetics are genetically linked with the chloroquine-resistant phenotype. J. Biol. Chem. 272: 2652-2658.

Schwartz, I.K., Lackritz, E.M., and Patchen, L.C. 1991. Chloroquine-resistant *Plasmodium vivax* from Indonesia. N. Engl. J. Med. 324: 927.

Scott, H.V., Tan, W.L., and Barlin, G.B. 1987. Antimalarial activity of Mannich bases derived from 4-(7'-bromo-1',5'-naphthyridin-4'-ylamino)phenol and 4-(7'-trifluoromethylquinolin-4'-ylamino)phenol against *Plasmodium falciparum in vitro*. Ann. Trop. Med. Parasitol. 81: 85-93.

Shenai, B.R., and Rosenthal, P.J. 2002. Reducing requirements for hemoglobin hydrolysis by *Plasmodium falciparum* cysteine proteases. Mol. Biochem. Parasitol. 122: 99-104.

Shenai, B.R., Sijwali, P.S., Singh, A., and Rosenthal, P.J. 2000. Characterization of native and recombinant falcipain-2, a principal trophozoite cysteine protease and essential hemoglobinase of *Plasmodium falciparum*. J. Biol. Chem. 275: 29000-29010.

Shviro, Y., and Shaklai, N. 1987. Glutathione as a scavenger of free hemin. A mechanism of preventing red cell membrane damage. Biochem. Pharmacol. 36: 3801-3807.

Sidhu, A.B.S., Verdier-Pinard, D., and Fidock, D.A. 2002. Chloroquine resistance in *Plasmodium falciparum* malaria parasites conferred by *pfcrt* mutations. Science. 298: 210-213.

Sijwali, P.S., Shenai, B.R., Gut, J., Singh, A., and Rosenthal, P.J. 2001. Expression and characterization of the *Plasmodium falciparum* haemoglobinase falcipain-3. Biochem. J. 360: 481-489.

Skinner-Adams, T., and Davis, T.M. 1999. Synergistic *in vitro* antimalarial activity of omeprazole and quinine. Antimicrob. Agents Chemother. 43: 1304-1306.

Slater, A.F., and Cerami, A. 1992. Inhibition by chloroquine of a novel haem polymerase enzyme activity in malaria trophozoites. Nature. 355: 167-169.

Slater, A.F., Swiggard, W.J., Orton, B.R., Flitter, W.D., Goldberg, D.E., Cerami, A., and Henderson, G.B. 1991. An iron-carboxylate bond links the heme units of malaria pigment. Proc. Natl. Acad. Sci. USA. 88: 325-329.

Snow, R.W., Trape, J.F., and Marsh, K. 2001. The past, present and future of childhood malaria mortality in Africa. Trends Parasitol. 17: 593-597.

Sowunmi, A., Ayede, A.I., Falade, A.G., Ndikum, V.N., Sowunmi, C.O., Adedeji, A.A., Falade, C.O., Happi, T.C., and Oduola, A.M. 2001. Randomized comparison of chloroquine and amodiaquine in the treatment of acute, uncomplicated, *Plasmodium falciparum* malaria in children. Ann. Trop. Med. Parasitol. 95: 549-558.

Su, X.-z., Ferdig, M.T., Huang, Y., Huynh, C.Q., Liu, A., You, J., Wootton, J.C., and Wellems, T.E. 1999. A genetic map and recombination parameters of the human malaria parasite *P. falciparum*. Science. 286: 1351-1353.

Su, X.-Z., Kirkman, L.S., and Wellems, T.E. 1997. Complex polymorphisms in a ~330 kDa protein are linked to chloroquine-resistant *P. falciparum* in Southeast Asia and Africa. Cell. 91: 593-603.

Sullivan, D.J., Jr., Gluzman, I.Y., and Goldberg, D.E. 1996a. *Plasmodium* hemozoin formation mediated by histidine-rich proteins. Science. 271: 219-222.

Sullivan, D.J., Jr., Gluzman, I.Y., Russell, D.G., and Goldberg, D.E. 1996b. On the molecular mechanism of chloroquine's antimalarial action. Proc. Natl. Acad. Sci. USA. 93: 11865-11870.

Sullivan, D.J., Jr., Matile, H., Ridley, R.G., and Goldberg, D.E. 1998. A common mechanism for blockade of heme polymerization by antimalarial quinolines. J. Biol. Chem. 273: 31103-31107.

Sutherland, C.J., Alloueche, A., Curtis, J., Drakeley, C.J., Ord, R., Duraisingh, M., Greenwood, B.M., Pinder, M., Warhurst, D.C., and Targett, G.A. 2002. Gambian children successfully treated with chloroquine can harbour and transmit *Plasmodium falciparum* gametocytes carrying resistance genes. Am. J. Trop. Med. Hyg. 67: 578-585.

Sztajer, H., Gamain, B., Aumann, K.D., Slomianny, C., Becker, K., Brigelius-Flohe, R., and Flohe, L. 2001. The putative glutathione peroxidase gene of *Plasmodium falciparum* codes for a thioredoxin peroxidase. J. Biol. Chem. 276: 7397-7403.

Tappel, A.L. 1953. The mechanism of oxidation of unsaturated fatty acids by hematin compounds. Arch Biochem Biophys. 44: 378-395.

Targett, G., Drakeley, C., Jawara, M., von Seidlein, L., Coleman, R., Deen, J., Pinder, M., Doherty, T., Sutherland, C., Walraven, G., and Milligan, P. 2001. Artesunate reduces but does not prevent posttreatment transmission of *Plasmodium falciparum* to *Anopheles gambiae*. J. Infect. Dis. 183: 1254-1259.

Thaithong, S., and Beale, G.H. 1981. Resistance of ten Thai isolates of *Plasmodium falciparum* to chloroquine and pyrimethamine by *in vitro* tests. Trans. R. Soc. Trop. Med. Hyg. 75: 271-273.

Thomas, S.M., Ndir, O., Dieng, T., Mboup, S., Wypij, D., Maguire, J.H., and Wirth, D.F. 2002. *In vitro* chloroquine susceptibility and PCR analysis of *pfcrt* and *pfmdr1* polymorphisms in *Plasmodium falciparum* isolates from Senegal. Am. J. Trop. Med. Hyg. 66: 474-480.

Tilley, L., Loria, P., and Foley, M. 2001. Chloroquine and other quinoline antimalarials. In: Antimalarial Chemotherapy: Mechanisms of Action, Resistance and New Directions in Drug Discovery. P.J. Rosenthal, ed. Humana Press Inc., Totowa, NJ. p. 87-121.

Trager, W., and Jensen, J.B. 1976. Human malaria parasites in continuous culture. Science. 193: 673-675.

Trape, J.F. 2001. The public health impact of chloroquine resistance in Africa. Am. J. Trop. Med. Hyg. 64: 12-17.

Trape, J.F., Pison, G., Preziosi, M.P., Enel, C., Desgrees du Lou, A., Delaunay, V., Samb, B., Lagarde, E., Molez, J.F., and Simondon, F. 1998. Impact of chloroquine resistance on malaria mortality. C R Acad Sci III. 321: 689-697.

Trape, J.F., Pison, G., Spiegel, A., Enel, C., and Rogier, C. 2002. Combating malaria in Africa. Trends Parasitol. 18: 224-230.

Ursos, L.M., and Roepe, P.D. 2002. Chloroquine resistance in the malarial parasite, *Plasmodium falciparum*. Med. Res. Reviews. 22: 465-491.

Ursos, L.M.B., Dzekunov, S., and Roepe, P.D. 2000. The effects of chloroquine and verapamil on digestive vacuolar pH of *P. falciparum* either sensitive or resistant to chloroquine. Mol. Biochem. Parasitol. 110: 125-134.

van Es, H.H., Karcz, S., Chu, F., Cowman, A.F., Vidal, S., Gros, P., and Schurr, E. 1994a. Expression of the plasmodial *pfmdr1* gene in mammalian cells is associated with increased susceptibility to chloroquine. Mol. Cell. Biol. 14: 2419-2428.

van Es, H.H., Renkema, H., Aerts, H., and Schurr, E. 1994b. Enhanced lysosomal acidification leads to increased chloroquine accumulation in CHO cells expressing the *pfmdr1* gene. Mol. Biochem. Parasitol. 68: 209-219.

Verdrager, J. 1986. Epidemiology of the emergence and spread of drug-resistant falciparum malaria in South-East Asia and Australasia. J. Trop. Med. Hyg. 89: 277-289.

Vieira, P.P., Alecrim, M., da Silva, L.H., Gonzalez-Jimenez, I., and Zalis, M.G. 2001. Analysis of the PfCRT K76T mutation in *Plasmodium falciparum* isolates from the Amazon region of Brazil. J. Infect. Dis. 183: 1832-1833.

von Seidlein, L., Duraisingh, M.T., Drakeley, C.J., Bailey, R., Greenwood, B.M., and Pinder, M. 1997. Polymorphism of the *Pfmdr1* gene and chloroquine resistance in *Plasmodium falciparum* in The Gambia. Trans. R. Soc. Trop. Med. Hyg. 91: 450-453.

Waller, K.L., Muhle, R.A., Ursos, L.M., Horrocks, P., Verdier-Pinard, D., Sidhu, A.B., Fujioka, H., Roepe, P.D., and Fidock, D.A. 2003. Chloroquine resistance modulated *in vitro* by expression levels of the *Plasmodium falciparum* chloroquine resistance transporter. J. Biol. Chem. 278: 33593-33601.

Walliker, D., Quakyi, I.A., Wellems, T.E., McCutchan, T.F., Szarfman, A., London, W.T., Corcoran, L.M., Burkot, T.R., and Carter, R. 1987. Genetic analysis of the human malaria parasite *Plasmodium falciparum*. Science. 236: 1661-1666.

Ward, S.A., and Bray, P.G. 2001. Is reversal of chloroquine resistance ready for the clinic? Lancet. 357: 904.

Warhurst, D.C. 2001a. A molecular marker for chloroquine-resistant falciparum malaria. N. Engl. J. Med. 344: 299-302.

Warhurst, D.C. 2001b. New developments: chloroquine-resistance in *Plasmodium falciparum*. Drug Resist. Updat. 4: 141-144.

Warhurst, D.C., Craig, J.C., and Adagu, I.S. 2002. Lysosomes and drug resistance in malaria. Lancet. 360: 1527-1529.

Wellems, T.E., Panton, L.J., Gluzman, I.Y., do Rosario, V.E., Gwadz, R.W., Walker-Jonah, A., and Krogstad, D.J. 1990. Chloroquine resistance not linked to *mdr*-like genes in a *Plasmodium falciparum* cross. Nature. 345: 253-255.

Wellems, T.E., and Plowe, C.V. 2001. Chloroquine-resistant malaria. J. Infect. Dis. 184: 770-776.

Wellems, T.E., Walker-Jonah, A., and Panton, L.J. 1991. Genetic mapping of the chloroquine-resistance locus on *Plasmodium falciparum* chromosome 7. Proc. Natl. Acad. Sci. USA. 88: 3382-3386.

Wellems, T.E., Wootton, J.C., Fujioka, H., Su, X.-Z., Cooper, R., Baruch, D., and Fidock, D.A. 1998. *P. falciparum* CG2, linked to chloroquine resistance, does not resemble Na+/H+ exchangers. Cell. 94: 285-286.

Wernsdorfer, W.H. 1994. Epidemiology of drug resistance in malaria. Acta Trop. 56: 143-156.

Westling, J., Cipullo, P., Hung, S.H., Saft, H., Dame, J.B., and Dunn, B.M. 1999. Active site specificity of plasmepsin II. Protein Sci. 8: 2001-2009.

Whitby, M. 1997. Drug resistant Plasmodium vivax malaria. J. Antimicrob. Chemother. 40: 749-752.

White, N.J. 1996. The treatment of malaria. N. Engl. J. Med. 335: 800-806.

WHO. 1999. Malaria, 1982-1997. Wkly Epidemiol Rec. 74: 265-270.

Wilson, C.M., Serrano, A.E., Wasley, A., Bogenschutz, M.P., Shankar, A.H., and Wirth, D.F. 1989. Amplification of a gene related to mammalian *mdr* genes in drug-resistant *Plasmodium falciparum*. Science. 244: 1184-1186.

Wilson, C.M., Volkman, S.K., Thaithong, S., Martin, R.K., Kyle, D.E., Milhous, W.K., and Wirth, D.F. 1993. Amplification of *pfmdr1* associated with mefloquine and halofantrine resistance in *Plasmodium falciparum* from Thailand. Mol. Biochem. Parasitol. 57: 151-160.

Winstanley, P.A., Ward, S.A., and Snow, R.W. 2002. Clinical status and implications of antimalarial drug resistance. Microbes Infect. 4: 157-164.

Wissing, F., Sanchez, C.P., Rohrbach, P., Ricken, S., and Lanzer, M. 2002. Illumination of the malaria parasite *Plasmodium falciparum* alters intracellular pH. Implications for live cell imaging. J. Biol. Chem. 277: 37747-37755.

Wongsrichanalai, C., Pickard, A.L., Wernsdorfer, W.H., and Meshnick, S.R. 2002. Epidemiology of drug-resistant malaria. Lancet Infect. Dis. 2: 209-218.

Wood, P.A., and Eaton, J.W. 1993. Hemoglobin catabolism and host-parasite heme balance in chloroquine-sensitive and chloroquine-resistant *Plasmodium berghei* infections. Am. J. Trop. Med. Hyg. 48: 465-472.

Wootton, J.C., Feng, X., Ferdig, M.T., Cooper, R.A., Mu, J., Baruch, D.I., Magill, A.J., and Su, X.Z. 2002. Genetic diversity and chloroquine selective sweeps in *Plasmodium falciparum*. Nature. 418: 320-323.

Wu, Y., Sifri, C.D., Lei, H.H., Su, X.Z., and Wellems, T.E. 1995. Transfection of *Plasmodium falciparum* within human red blood cells. Proc. Natl. Acad. Sci. USA. 92: 973-977.

Wunsch, S., Sanchez, C.P., Gekle, M., Grosse-Wortmann, L., Wiesner, J., and Lanzer, M. 1998. Differential stimulation of the $Na^+/H^+$ exchanger determines chloroquine uptake in *Plasmodium falciparum*. J. Cell Biol. 140: 335-345.

Yayon, A., Cabantchik, Z.I., and Ginsburg, H. 1984. Identification of the acidic compartment of *Plasmodium falciparum*-infected human erythrocytes as the target of the antimalarial drug chloroquine. EMBO J. 3: 2695-2700.

Yayon, A., Cabantchik, Z.I., and Ginsburg, H. 1985. Susceptibility of human malaria parasites to chloroquine is pH dependent. Proc. Natl. Acad. Sci. USA. 82: 2784-2788.

Zalis, M.G., Wilson, C.M., Zhang, Y., and Wirth, D.F. 1993. Characterization of the *pfmdr2* gene for *Plasmodium falciparum*. Mol. Biochem. Parasitol. 62: 83-92.

Zhang, H., Howard, E.M., and Roepe, P.D. 2002. Analysis of the antimalarial drug resistance protein Pfcrt in yeast. J. Biol. Chem. 277: 49767-49775.

Zhang, J., Krugliak, M., and Ginsburg, H. 1999. The fate of ferriprotorphyrin IX in malaria infected erythrocytes in conjunction with the mode of action of antimalarial drugs. Mol. Biochem. Parasitol. 99: 129-141.

# Index

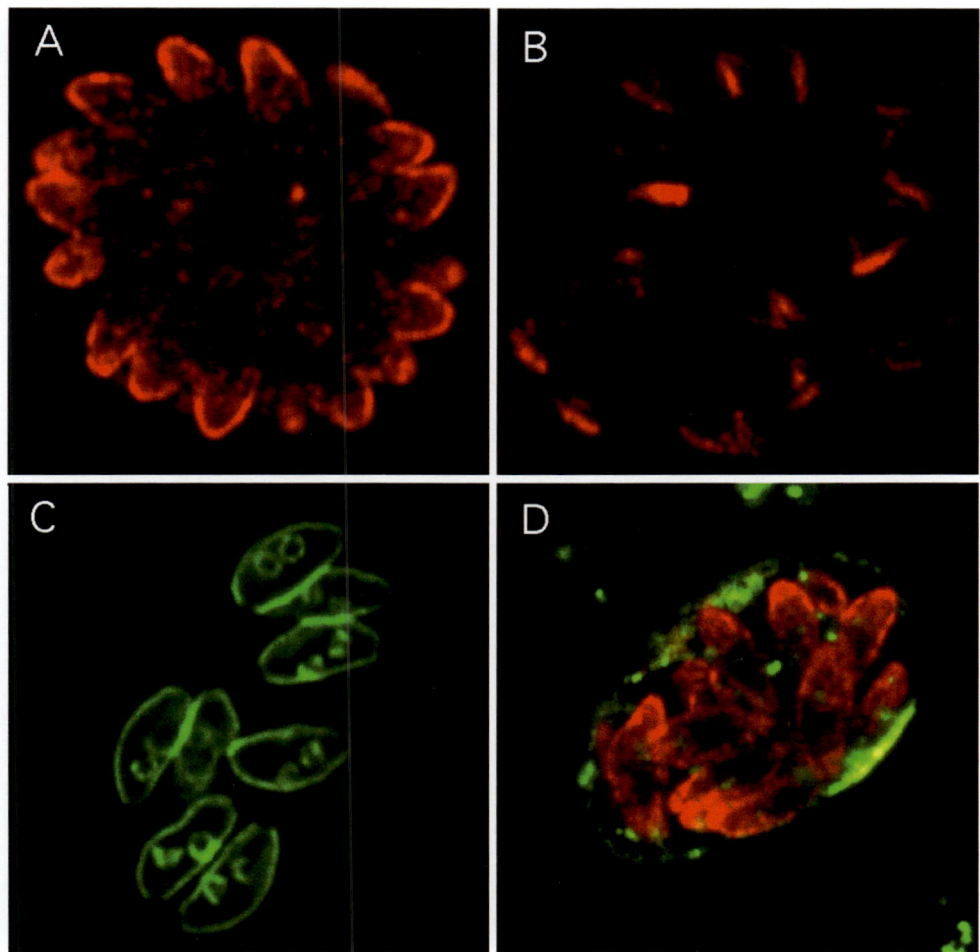

**Chapter 5, Figure 2.** Indirect immunofluorescence confocal microscopy analysis of intracellular *T. gondii* tachyzoites. Panel **A** shows the micronemes stained with anti-MIC6. **B** shows the rhoptires stained with anti-ROP2. **C** shows the inner membrane complex and the nascent daughter cells stained with anti-IMC1. **D** shows the parasitophorous vacuole stained with anti-GRA3 in green and the micronemes are stained in red.

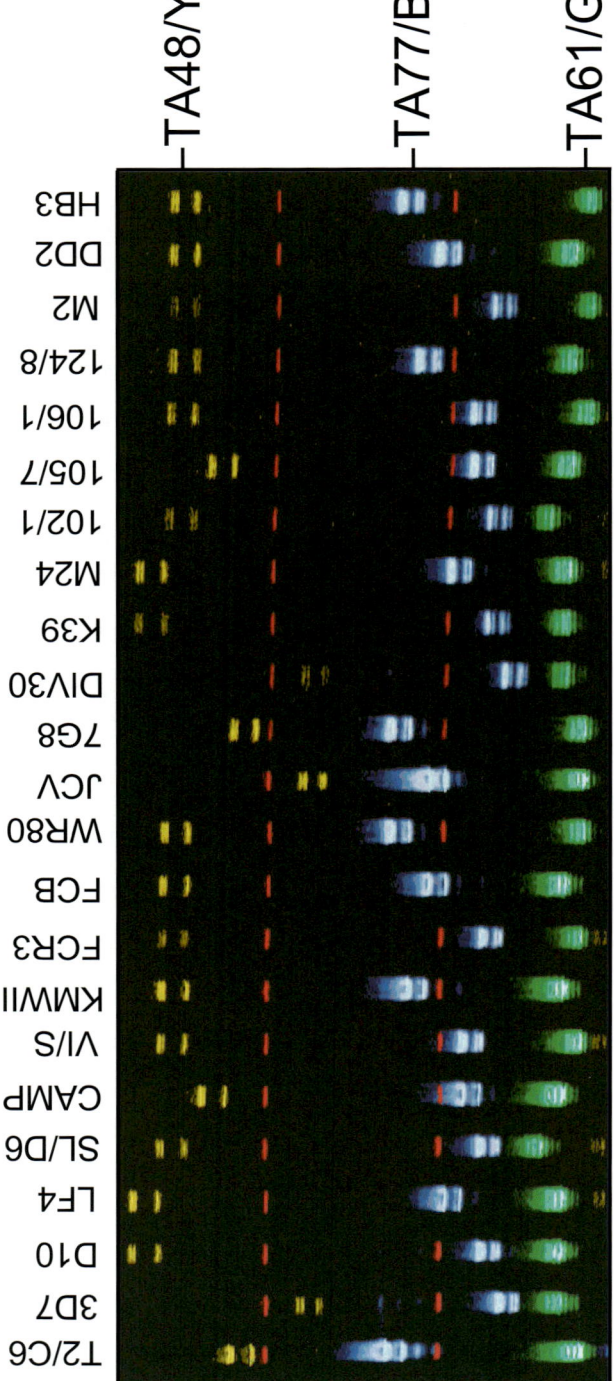

**Chapter 6, Figure 4.** Microsatellite typing of *P.falciparum* isolates. Three microsatellite markers (right) were used to type 23 isolates (top). Most isolates exhibit an unique banding pattern. Red bands indicate size markers. Note: Different chromosomal strands migrate at slightly different rates, giving the appearance of double bands (Blue and yellow markers).

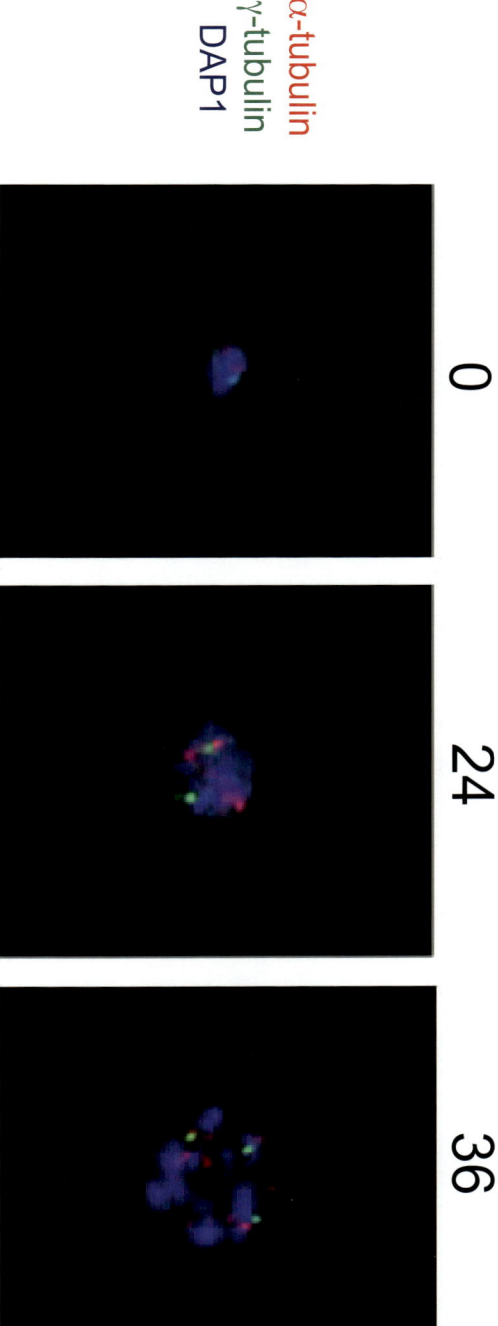

α-tubulin
γ-tubulin
DAP1

0    24    36

**Chapter 10, Figure 1.** DeltaVision deconvolution fluorescence micrograph of cells stained with alpha-tubulin (chicken, red) and gamma-tubulin (human, green) antibodies, illustrating microtubule structure during *Plasmodium falciparum* asexual development. The use of these antibodies to label the *P. falciparum* microtubule-organizing center (MTOC) and microtubules was first reported by Fowler *et al.* (2001). Nuclear stain (blue) is DAPI. Numbers on top indicate the number of hours post synchronisation. (0) Ring. (24), Mature Trophozoite. (36) Segmenter/schizont. In animal cells, gamma-tubulin is associated with the MTOC in the centrosome, from which microtubules (consisting of alpha- and beta-tubulins) spokes radiate. Here, alpha-tubulin can be seen to become organised into spindles as cell division progresses, while gamma-tubulin staining remains concentrated in a small number of punctuations, suggesting that the parasite utilizes the tubulins similarly to higher eukaryotes. As reported in Read *et al.*, 1993, spindles at different stages of development appear to be present in the schizont. This experimental approach does not yield conclusive information on the subcellular location of the MTOC; it would be of interest to determine by immuno-electron microscopy what the relation is between the tubulins and the intranuclear hemicentriolar plaques referred to in the text. For a thorough discussion of microtubules in *Plasmodium* merozoites, see Fowler *et al.*, 2001.

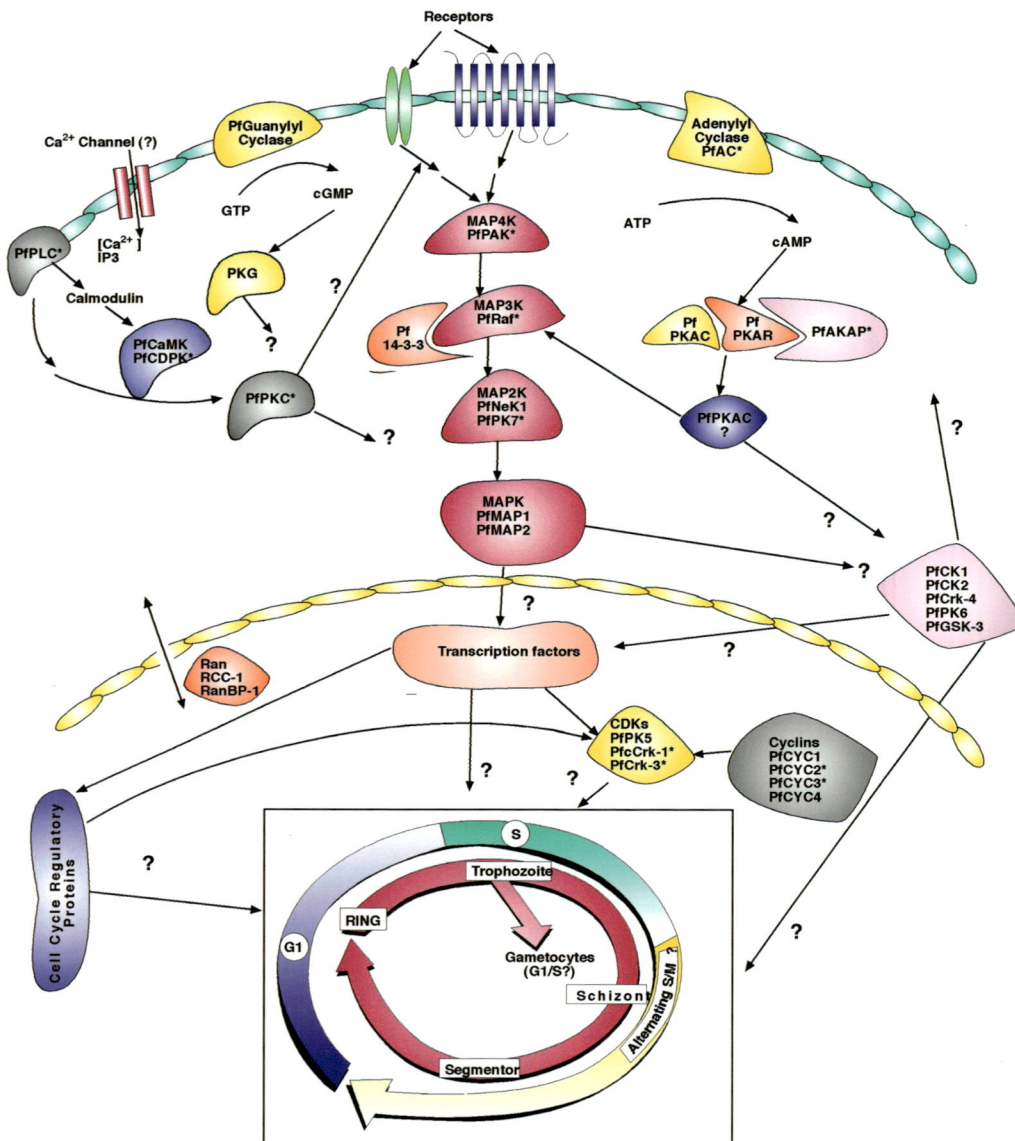

**Chapter 10, Figure 2.** Tentative regulatory map of *P. falciparum* gene products potentially involved in the control of cell proliferation and/or development. All elements showed on the figure have been identified in the *P. falciparum* genome, and are arranged following guidance from other eukaryotic systems in which the pathways are well understood. Those labelled with an asterisk have been tentatively identified on the sole basis of database sequence information. It must be kept in mind that no clear-cut "ultimate" functional data are available for any of the gene products appearing in this figure, which should be taken more as a series of working hypotheses than as a representation of existing pathways (see the text for details). The proteins labelled as "receptors" can be any of the numerous membrane proteins identified at the surface of the parasite. For the sake of clarity, the parasitophorous and erythrocyte membranes have not been included in the picture, but their components may obviously have a potentially important role to play in signal transduction. Note that the depiction of the cell cycle phases at the bottom of the picture is an oversimplification (see text for details).

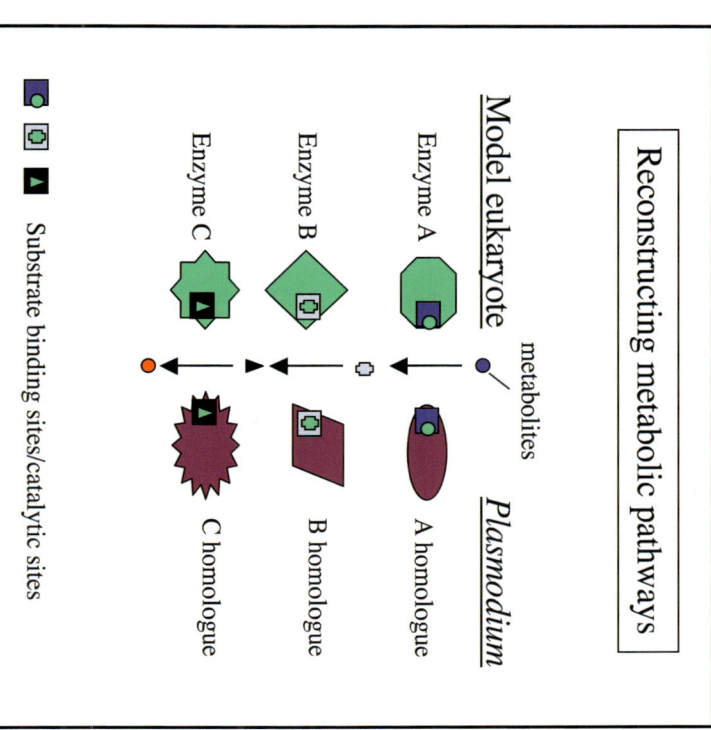

**Reconstructing metabolic pathways**

**Model eukaryote**

Enzyme A

Enzyme B

Enzyme C

*Plasmodium*

A homologue

B homologue

C homologue

metabolites

■ Substrate binding sites/catalytic sites

**Reconstructing regulatory networks**

**Model eukaryote**

*Plasmodium*

Enzyme:
recognisable
(e.g. CDK)

Regulatory element:
not recognisable
(e.g. cyclin or CKI)

► Substrate binding site/catalytic site

**Chapter 10, Figure 3.** A scheme illustrating the difficulty of reconstituting *Plasmodium* regulatory pathways from sequence databases. Left Panel: Reconstruction of metabolic pathways. Presence of substrate binding motifs and catalytic sites allows to position the *Plasmodium* orthologue on the metabolic map, even though the molecule may be overall divergent from those of model organisms (as schematised by different shapes and colors). This is because the metabolite substrates are identical (species-independent) for all members of a given family of orthologues. Right panel: Regulatory elements (shown at the extreme left and right of the panel) have to interact not with invariant substrates, but with species-specific proteins (shown at the center), which are divergent from their orthologues in other systems.

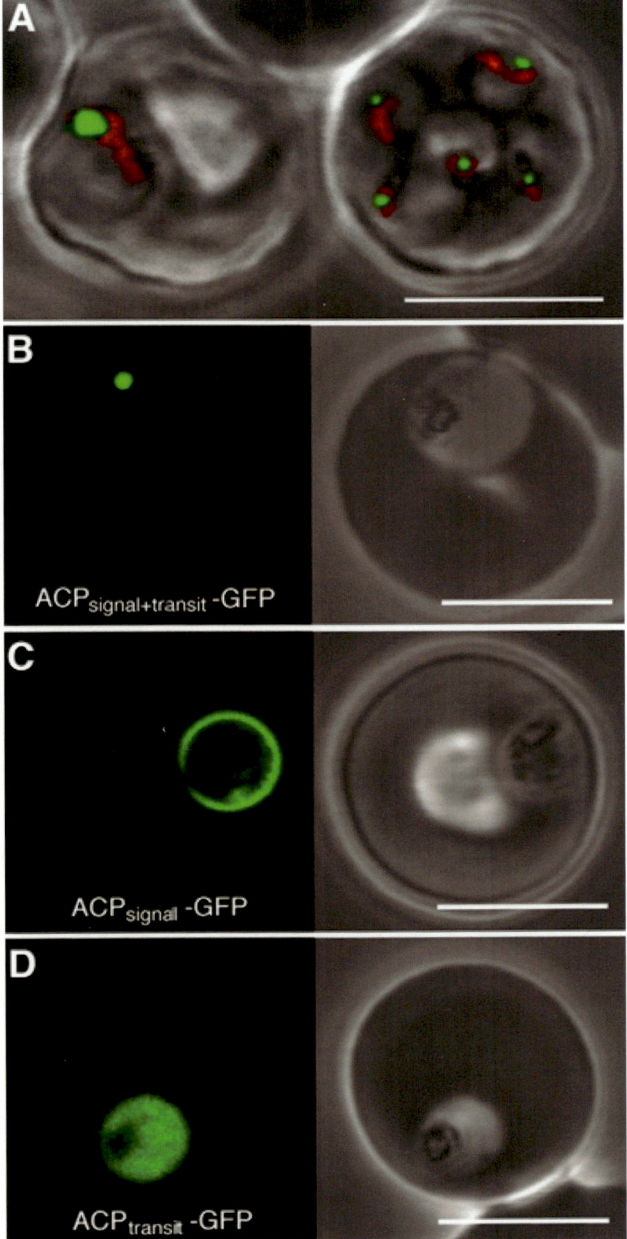

**Chapter 11, Figure 6**. (A) Live intra-erythrocytic *Plasmodium falciparum* parasites expressing the apicoplast-targeted GFP fusion protein $ACP_{presequence}$-GFP and co-stained for mitochondria (red) [right-hand erythrocyte contains multiple (5) infections]. (B) Parasite expressing $ACP_{presequence}$-GFP (presequence containing the signal peptide plus the transit peptide) with GFP accumulating in the apicoplast. (C) Parasite expressing $ACP_{signal}$-GFP with GFP accumulating in the parasitophorous vacuole (some GFP re-enters the parasites into the food vacuole). (D). Parasite expressing $ACP_{signal}$-GFP with GFP accumulating in the parasitophorous vacuole (some GFP re-enters the parasites into the food vacuole). (C) Parasite expressing $ACP_{transit}$-GFP with GFP accumulating in the cytosol. Scale bar = 5µm.

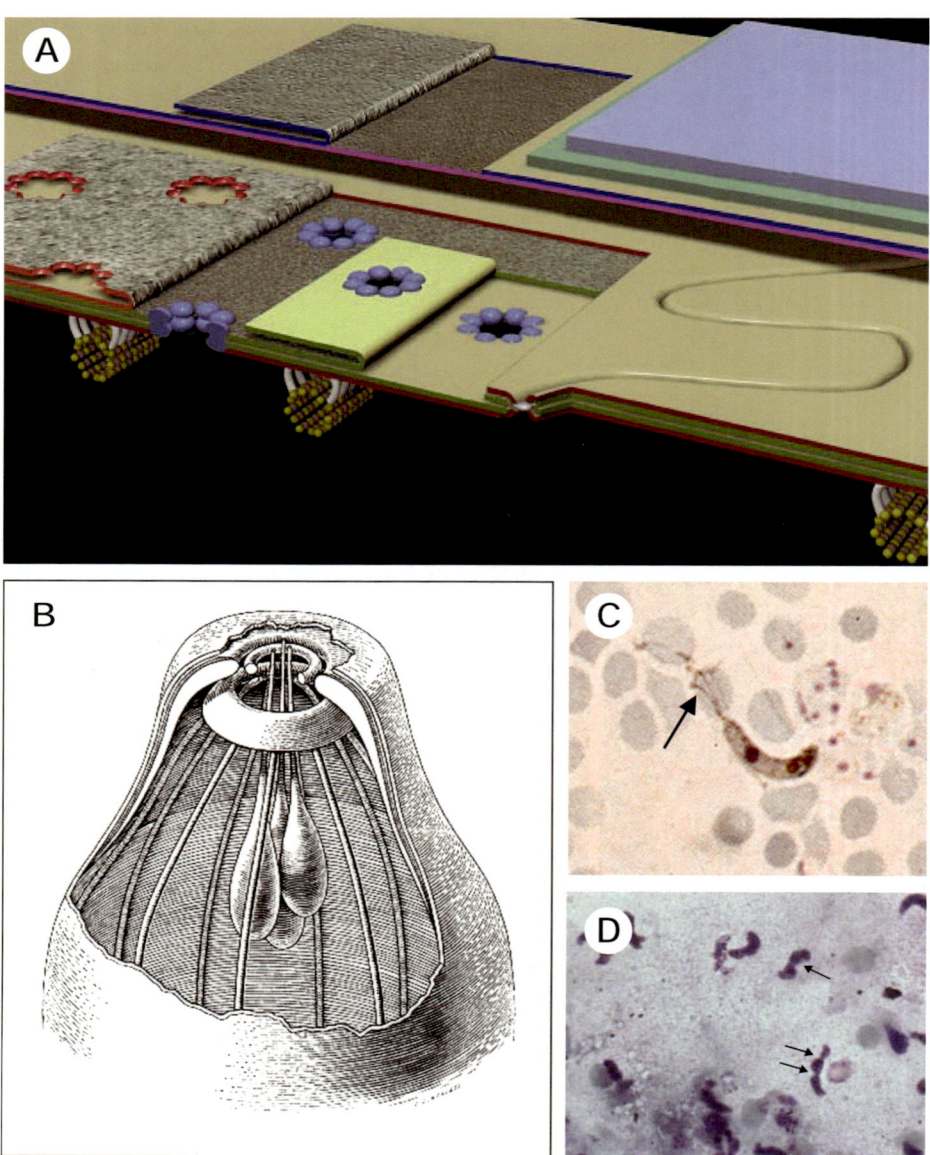

**Chapter 15, Figure 3.** A) Reconstruction from transmission electron micrograph of freeze-fractured ookinete illustrating distribution of large pores in inner pellicle membranes of the apical complex of *P. gallinaceum* (from Raibaud *et al.*, 2001). B) 3D reconstruction from transmission electron micrographs of the ookinete apical complex (from Canning and Sinden, 1973). C) Immunogold silver enhanced micrograph of *P. berghei* ookinetes showing the shedding of Pbs21 by the migrating ookinete (arrow) (from Sinden *et al.*, 1987). D) Giemsa stained preparation of *An. stephensi* peritrophic matrix illustrating the constrictions in *P. berghei* ookinetes (arrows) as they pass through this tissue.